The transfer of spectral line radiation

To Anne, Lisa, John and Jason

C. J. CANNON *

The transfer of spectral line radiation

**Formerly at the Department of Applied Mathematics*
The University of Sydney

The right of the University of Cambridge to print and sell all manner of books was granted by Henry VIII in 1534. The University has printed and published continuously since 1584.

CAMBRIDGE UNIVERSITY PRESS
Cambridge
London New York New Rochelle
Melbourne Sydney

Published by the Press Syndicate of the University of Cambridge
The Pitt Building, Trumpington Street, Cambridge CB2 1RP
32 East 57th Street, New York, NY 10022, USA
10 Stamford Road, Oakleigh, Melbourne 3166, Australia

First published 1985

Printed in Great Britain at the University Press, Cambridge

Library of Congress catalogue card number: 84-5029

British Library cataloguing in publication data

Cannon, C. J.
The transfer of spectral line radiation.

1. Atomic spectra
2. Radiative transitions
I. Title
535.8′4   QC454.A8

ISBN 0 521 25995 9

DJ

CONTENTS

	Preface	*xi*
1	**Basic theory for model 2-level atoms**	1
1.1	The model 2-level atom	1
1.2	The equation of transfer	4
1.3	The evaluation of N_U/N_L	8
1.4	The evaluation of $\phi(\nu, \Omega)$	10
1.5	Complete re-distribution	16
1.6	The complete re-distribution source function	17
1.7	LTE and non-LTE for photons	21
1.8	The qualitative behaviour of the source function	25
1.9	The qualitative behaviour of the emergent intensity	30
	1.9.1. Optical depth	32
	1.9.2. Sample calculation	38
1.10	Simple interesting special cases	38
	1.10.1. Stellar chromospheres	38
	1.10.2. Slab geometry	41
2	**Exact solutions to the transfer equation**	45
2.I	**Singular eigenfunction technique**	45
2.1	Mathematical preliminaries	46
	2.1.1. Plemelj (or Sokhotski) formulae	46
	2.1.2. The homogeneous and inhomogeneous Riemann–Hilbert problems	49
	2.1.3. Green's functions	51
2.2	Reduction of the transfer equation	57
2.3	The general solution	60
	2.3.1. Range I: $\eta \notin [-\gamma, \gamma]$	61
	2.3.2. Range II: $\eta \in [-\gamma, \gamma]$	64
2.4	Full-range completeness	65
2.5	Half-range completeness	74
2.6	Full-range orthogonality	78
	2.6.1. Discrete normalisation	79
	2.6.2. Singular normalisation	80
2.7	Half-range orthogonality	82

2.8 The Green's function solution 90
2.9 Simplified expressions for surface quantities 95
2.10 The thermalisation depth 99
2.10.1. The Doppler profile 100
2.10.2. The Voigt profile 103
2.10.3. Coherent scattering 105
2.II The Fourier transform method 106
2.11 The Green's function 106
2.12 The Fourier transform reduction 109
2.13 The Wiener–Hopf solution 113

3 Numerical methods of solution 122
3.1 Λ iteration 123
3.1.1. Convergence 124
3.1.2. Asymptotic behaviour of $S(\tau)$ 131
3.2 The Feautrier technique 134
3.2.1. Finite differencing 137
3.2.2. Angle and frequency quadrature 140
3.2.3. The recurrence relationship 143
3.2.4. Taylor series expansion at the boundaries 144
3.2.5. Quadrature points and weights 147
3.3 Stability of the Feautrier technique 150
3.3.1. Exact solution with $\Phi(\infty) = B_\nu(\mathrm{T})$ 151
3.3.2. Exact solution with $\tau_{\mathrm{N}_\tau} \not\to \infty$ 151
3.3.3. Scalar finite difference approach 153
3.3.4. Matrix finite difference approach 160
3.4 Variable Eddington factors 161
3.5 The Rybicki re-organisation 165
3.6 Auer's modification 170
3.7 Quadrature perturbations 175
3.7.1. The basic procedure 175
3.7.2. Convergence 179
3.8 Integral equation techniques 184
3.8.1. General method of solution 184
3.8.2. Choice of $w_{j'j}$ 186
3.8.3. The flux divergence method 190
3.9 Linearisation 197
3.9.1. The basic linearisation theory 198
3.9.2. Linearisation of the transfer equation 200
3.9.3. Differential equation formulation 203
3.9.4. Operator linearisation 206
3.10 Newton–Raphson accelerated Λ operators 208

4 Extension to model multi-level atoms 212
4.1 Model 3-level atoms 212
4.2 Qualitative behaviour of 3-level atom source functions 219
4.2.1. Radiative transitions 2–1 and 3–1 219
4.2.2. Radiative transitions 3–1 and 3–2 224

	4.2.3. Radiative transitions 3-2 and 2-1	226
	4.2.4. Corresponding emergent intensities	227
4.3	Transition channelling	228
4.4	The general multi-level atom	230
	4.4.1. Preliminary discussion	230
	4.4.2. The general source function	233
4.5	Solution of the transfer equation	235
5	**Radiation gas dynamics**	241
5.1	The conservation equations	242
	5.1.1. The general Boltzmann equation	242
	5.1.2. The species Boltzmann equation	247
	5.1.3. The general equation of change	250
	5.1.4. The macroscopic conservation equations	253
5.2	The velocity distribution functions	260
	5.2.1. Derivation of $f_i^{(0)}$	261
	5.2.2. Zeroth order conservation equations	265
5.3	First order equations	269
	5.3.1. Derivation of $f_i^{(1)}$	269
	5.3.2. First order conservation equations	276
	5.3.3. Inclusion of inelastic terms	280
5.4	The macroscopic velocity-dependent source function	281
5.5	Numerical solution to the radiative transfer equation	287
	5.5.1. Feautrier's technique	287
	5.5.2. The co-moving frame formulation	290
5.6	Time-dependent solutions	299
	5.6.1. A Eulerian finite difference approach	300
	5.6.2. A Lagrangian finite difference scheme	305
5.7	Velocity-dependent radiative transfer effects	309
	5.7.1. Emergent spectral line profiles	309
	5.7.2. Atom and electron temperature differences	315
5.8	Multi-dimensional radiative transfer	319
	5.8.1. The generalised Feautrier technique for stationary geometry	320
	5.8.2. Radiative channelling in stationary geometry	325
	5.8.3. Multi-dimensional velocity fields	338
6	**Quantum mechanical emission and absorption profiles**	339
6.1	Preliminary discussion (first quantisation)	340
	6.1.1. The Schrödinger wave equation	340
	6.1.2. Simple stationary states and orthogonality	343
	6.1.3. Elements of matrix mechanics	345
	6.1.4. First quantisation	349
6.2	Quantisation of the radiation field	353
	6.2.1. Classical equations for the electromagnetic field	354
	6.2.2. Second quantisation	359
	6.2.3. The general Hamiltonian	363
6.3	Perturbation methods	365
	6.3.1. Time-dependent perturbation theory	365

	6.3.2. Interaction representation	367
	6.3.3. Time-evolution operators	369
6.4	Spectral line profiles (no collisions)	371
	6.4.1. Transition rate for photon emission	374
	6.4.2. Transition rate for photon absorption	379
	6.4.3. Natural broadening – preliminary discussion	380
	6.4.4. Natural broadening – general method	385
	6.4.5. Natural broadening – 2-level atom	392
	6.4.6. Natural broadening profile	394
	6.4.7. Scattering – no collisions	399
6.5	Effect of perturbing collisions	405
	6.5.1. The Schrödinger wave equation	406
	6.5.2. The transition rate $w(1 \to 2 \to 3)$	410
	6.5.3. Time ordering of $\Theta(a, \ldots, d')$	412
	6.5.4. Evaluation of the time-evolution operators	413
	6.5.5. The components of $w(1 \to 2 \to 3)$	417
	6.5.6. Zero collisional perturbations	420
	6.5.7. Collisions – zero width ground state	423
7	**Frequency and angle re-distribution**	427
7.1	The emission profile	428
7.2	Source function simplifications	434
	7.2.1. Complete re-distribution	434
	7.2.2. A linear expression for $S(\mathbf{r}, \nu, \Omega)$	435
	7.2.3. Uncoupling of $\mathbf{v}_L$ and $\mathbf{v}_U$	439
7.3	Several limiting cases of $R(\nu', \Omega'; \nu, \Omega)$	440
	7.3.1. The re-distribution function $R_I(\nu', \Omega'; \nu, \Omega)$	441
	7.3.2. The re-distribution function $R_{II}(\nu', \Omega'; \nu, \Omega)$	445
	7.3.3. The re-distribution function $R_{III}(\nu', \Omega'; \nu, \Omega)$	447
	7.3.4. A simplified re-distribution function	449
7.4	Numerical methods of solution	452
	7.4.1. The linear problem	452
	7.4.2. Inclusion of macroscopic velocities	460
	7.4.3. The re-distribution perturbation technique	463
7.5	Qualitative transfer effects due to partial re-distribution	466
8	**A quantum electrodynamical radiative transfer equation**	473
8.1	The pure state density matrix	473
	8.1.1. Transition rate for photon emission	476
	8.1.2. A Laplace transform description of radiative decay	478
	8.1.3. An alternative approach to radiative decay	482
	8.1.4. Ordering of solutions	484
8.2	The mixed state density matrix	487
	8.2.1. Expectation values	488
	8.2.2. A preliminary approach to the transfer equation	493
8.3	Quantum transfer equations	502
	8.3.1. Specification of the radiation intensity	503
	8.3.2. The interaction Hamiltonian for zero elastic collisions	506

8.3.3. Commutator evaluations 509
8.3.4. Derivation of the transfer equation 512
8.3.5. Population rate equations 518

Appendix A: Half-range orthogonality relationships 523

Appendix B: The $H(z)$ function 529

Appendix C: Expressions for viscosity and heat conductivity 533

Appendix D: Derivation of equation (6.4.41) 534

References 538

PREFACE

The problem we wish to consider involves the interpretation of data from either laboratory plasmas or stellar atmospheres. We receive this data in the form of photons, and if we are to glean any knowledge of the gas from which these observed photons emerged we must know and understand the manner in which photons are created within, and interact with, that gas. In other words, we are necessarily interested in the interaction of radiation (photons) and matter (atoms, ions, molecules maybe, and electrons).

This monograph is an attempt to detail both the physics and mathematics of such interaction from the point of view of the physicist wishing to analyse stellar or laboratory spectral line radiation. It should be particularly stressed that 'classical' approaches to the analysis of spectral line radiation can lead to the complete misinterpretation of observations. We shall see, for example, that relatively hot regions can appear to the observer to be relatively dark; regions moving toward the observer can appear to be red-shifted. A 'classical' analysis would interpret the above relatively hot regions as relatively cold, and those regions moving toward the observer as moving away. There are many such apparently anomalous examples one may cite. Suffice it to say that one can only be sure of the validity of one's interpretation of the data if an adequate understanding of the pertinent interactions giving rise to that data is available.

The presentation here follows a series of lectures given to fourth-year honours undergraduate students at the University of Sydney over the past 12 years. They are intended as an introduction to the subject of gas-dynamic radiative transfer in spectral lines but aim to quickly advance the student to the stage where research in the various related fields may be easily undertaken. It is therefore expected that any student, having completed this fourth-year course, should be immediately able to start Ph.D. thesis work. To achieve this end we structure this monograph by first presenting the basic theory for a simplified case involving model 2-level atoms in a homogeneous stationary atmosphere (along with other simplifying assumptions). An understanding of the qualitative behaviour of the radiation field described by the equation of radiative transfer given in chapter 1 is essential if the student is to appreciate the physics of radiation–matter interaction presented in chapters 4 through 8. Chapter 2 details two distinct methods for the exact solution of the radiative transfer equation, but these solutions only apply to constant property media and therefore only offer the means by which benchmark solutions may be obtained. More physically realistic problems require numerical solution of the appropriate equation of radiative transfer; a variety of such techniques (along with their advantages, disadvantages and numerical stability properties) are discussed in chapter 3. Both chapters 2 and 3 are not required reading for the later chapters (in particular, chapter 2 is intended only for the more mathematically motivated), but knowledge of the material covered in chapter 3 is essential if the student is to undertake adequate research in spectral line formation. Chapter 4 then generalises the model 2-level atom approach of chapter 1 to the model multi-level atom thereby highlighting the effect of the restriction to just two levels has on the qualitative behaviour of the radiative transfer solutions. The need to couple the gas-dynamic conservation equations of mass, linear momentum and energy to the equation of radiative transfer is discussed at some length in chapter 5. There, we present a somewhat simplified approach to the Boltzmann equation (similar to the Chapman–Enskog treatment) in order to illustrate the various terms and physical mechanisms which can be important in a self-consistent analysis of gas dynamics in a radiating plasma. Various interesting velocity-dependent multi-dimensional situations are discussed at the end of chapter 5. Chapter 6 first summarises the theory of the quantisation of the radiation field and this enables the various quantum mechanical absorption, emission and scattering probabilities for individual atoms to be evaluated. These, in turn, are used to detail the corresponding atomic ensemble probabilities presented in chapter 7. In particular, important effects due to physically realistic departures from the complete re-distribution assumption made in chapter 1 are discussed. Chapter 8 then attempts a derivation of the transfer equation using the density matrix formalism of quantum electrodynamics. Such an axiomatic approach is essential if the assumptions and approximations inherent in the phenomenological derivation of the equation of radiative transfer of chapter 1 are to be better understood, but, more particularly, it enables all those higher order interference terms ignored in chapter 1 to be eventually evaluated if so desired.

Discussions with many colleagues over the years have considerably influenced the structuring of the material presented here. In particular, Professors R. N. Thomas and P. R. Wilson have been most helpful indeed both in detailing an overall scientific approach to gas-dynamic radiative transfer and in presenting opportunities in which this approach could be implemented. Finally, I would like to thank Kay Yamamoto for her assistance in preparing this manuscript and to express my appreciation and admiration of her typing skills.

C. J. Cannon

Biographical Note

Christopher John Cannon was born on 13 December 1944 and died of cancer at the age of 39 on 15 August 1984. As a youth he showed promise of international standard as an athlete, principally long-distance running and javelin, but set aside these interests in order to study mathematics at the University of Sydney. He obtained First Class Honours in Applied Mathematics in 1965, and was awarded a PhD in 1970 jointly by the Universities of Sydney and Colorado. It was at the Joint Institute for Laboratory Astrophysics (Boulder) in 1966 that he first met Richard N. Thomas with whom he was later to make very important contributions to modelling of stellar winds and chromospheric heating. He made other significant contributions in a range of problems in theoretical astrophysics, especially the influences of atmospheric inhomogeneities and velocity fields on spectral line radiation transfer. From 1970 until the time of his death he played a key rôle in the Astrophysics Group of the Department of Applied Mathematics in the University of Sydney.

It is a tragedy that he was unable to complete the proof-reading of this book. And it is with great sadness that his family and colleagues, who did complete this task, accept responsibilities for any remaining proof-reading flaws. This book will undoubtedly serve for many years as a memorial to the author's mastery in the field of radiation transfer.

R. W. J.
Department of Applied Mathematics, University of Sydney
February, 1985

Acknowledgement

I wish to thank the following people for their invaluable help in proof-reading my husband's book; his sister Barbara Richards and his colleagues Dr P. W. Buchen, Dr R. G. Crossman, Dr C. J. Durrant, Associate Professor E. D. Fackerell, Dr W. G. Gibson, Dr R. W. James, Dr D. E. Rees, and Professor P. R. Wilson.

Anne Cannon

1

Basic theory for model 2-level atoms

1.1 The model 2-level atom

We shall begin by considering a model stellar atmosphere consisting of atoms of the one species. This is clearly a gross simplification. A real stellar atmosphere presumably consists of atoms (and ions and molecules) of a variety of elements (e.g. hydrogen, calcium, magnesium, iron, etc.) and each atom, in turn, may be thought of as having an infinite number of energy levels. That is to say, the outer-most bound electron may exist in any one of an (essentially) infinite number of quantised states (energy levels) ranging from the ground state of the atom to almost the continuum (at which point the bound electron ceases, in fact, to be bound by the atom). A typical energy level diagram is illustrated in figure 1.1.

A *complete* understanding of the interactions of radiation with matter (i.e. the atom) would therefore require us to include all those levels shown in figure 1.1. This, of course, is an immense problem. We can, however, suggest a simplification by initially considering just two levels, then, with the understanding gained from such an analysis, proceed to three levels (etc.) until the desired 'accuracy of our understanding' is reached.

We thus consider our model atom to have just those two bound energy levels which relate to the radiation of interest. We call these the upper (U) and lower (L) states.

There are five quite distinct physical processes which must now be detailed.

Fig.1.1. Bound quantised electron states and the continuum.

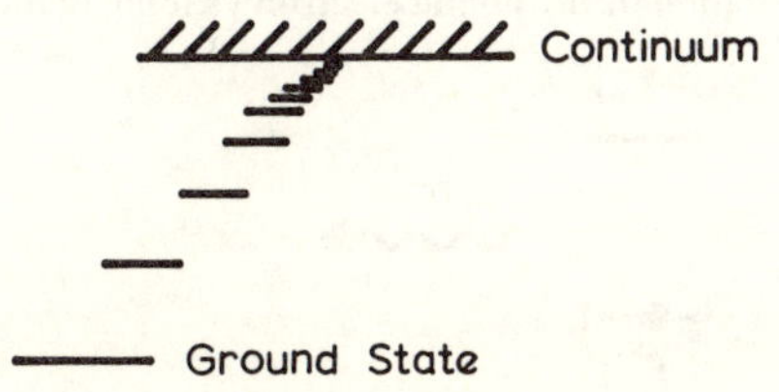

Process 1

Consider a 2-level atom with a bound electron (e) in the lower state. Let the energy difference between the upper and lower states be $h\nu_0$ (where h is the usual Planck constant). It is possible for such an atomic configuration to absorb a photon of frequency ν_0. The photon would then disappear and its energy ($h\nu_0$) would allow the bound electron to move into the upper (or excited) state (see figure 1.2).

This is an absorption process termed radiative excitation.

Process 2

Next consider an atom which already has a bound electron in the upper state. There exists the possibility that this bound electron will move from the upper to the lower state without recourse to any external stimulus (see figure 1.3).

This is known as spontaneous de-excitation. Clearly, the atom in such a de-excitation must lose the energy equivalent to the difference in energies between U and L, i.e. $h\nu_0$. This energy loss generates a photon of frequency ν_0, i.e. the de-excitation of the atom results, in this case, in the emission of radiation of energy $h\nu_0$.

Process 3

As in process 2, the atom initially has a bound electron in the upper state. Here, however, if the atom is bombarded by radiation of frequency ν_0, the bound electron may de-excite, thus generating a further photon of frequency ν_0 (see figure 1.4).

This process is referred to as stimulated or induced de-excitation.

Fig.1.2. Excitation of the bound electron following photon absorption.

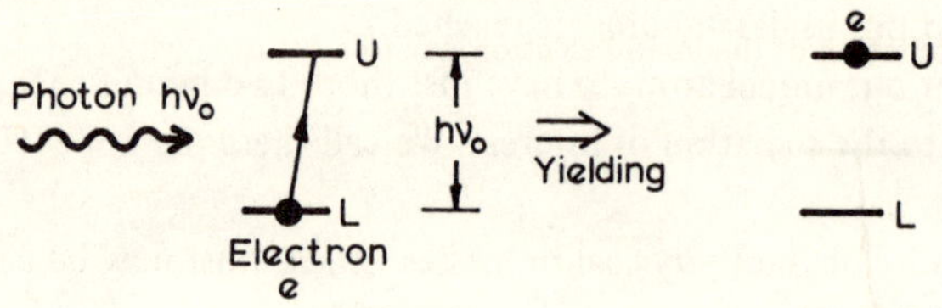

Fig.1.3. Spontaneous de-excitation of the bound electron yielding photon emission.

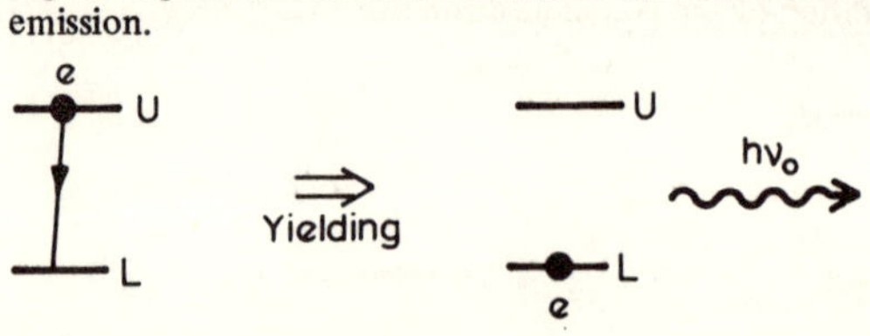

Process 4

An atom with a bound electron in the lower state could undergo a collision with a free electron and this, in turn, could excite the bound electron to the upper state (see figure 1.5).

Here the bound electron would absorb an energy equivalent of $h\nu_0$ from the free electron, i.e. if the energy of the free electron before the collision was E_e, its energy after the collision would be $E_e - h\nu_0$. This is a collisional excitation. (More particularly, it is sometimes referred to as an inelastic collisional excitation.)

Process 5

The effective reverse of process 4 occurs when an atom with the bound electron in the upper state undergoes a collision with a free electron (see figure 1.6).

The resulting collisional de-excitation would lead to the bound electron now existing in the lower or de-excited state with the energy of the free electron having subsequently been increased by $h\nu_0$.

These five processes are fundamental to radiative transfer analyses of spectral line formation. The spectral line is just the radiation of frequency ν_0. The difference between different energy levels leads to different $h\nu_0$ (i.e. different frequencies ν_0) and, thus, different spectral lines are generated. Later, when we

Fig.1.4. Stimulated de-excitation of the bound electron yielding photon emission.

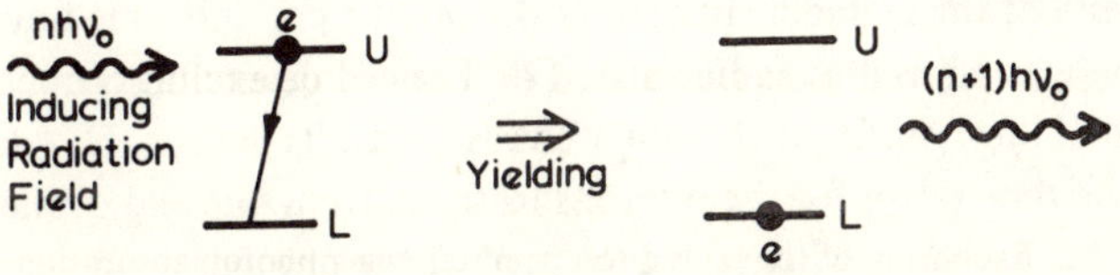

Fig.1.5. Excitation of the bound electron due to a collision with free electrons.

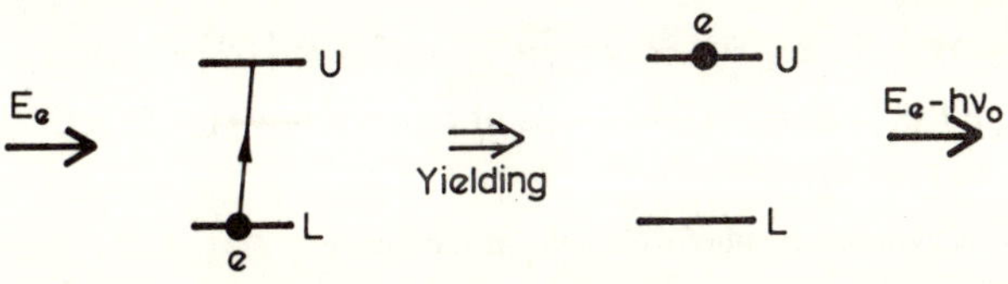

Fig. 1.6. De-excitation of the bound electron due to a collision with free electrons.

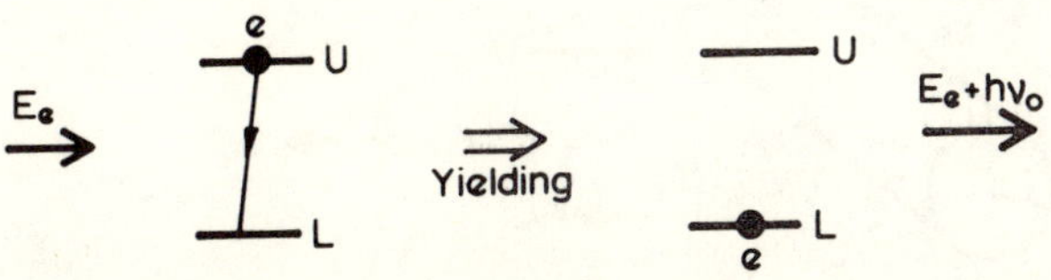

consider other levels, we will find that the spectral line of direct interest can be significantly influenced by processes to and from these other levels. For the moment, however, we shall content ourselves with the above five processes only, the first three of which are, naturally enough, referred to as radiative processes, the last two as collisional processes. In particular, photons may be absorbed (process 1) or emitted (processes 2 and 3) and these therefore constitute the interaction of radiation with matter. The collisional processes have no direct bearing on the radiation but, as we shall see, they are indirectly of the utmost importance. Finally, notice that the energy of the system is conserved in all five processes.

1.2 The equation of transfer

Here we wish to consider spectral line radiation of frequency ν being absorbed and emitted by a gas consisting of simple model 2-level atoms having an energy difference $h\nu_0$ between the upper and lower bound states. We first consider a small cylindrical volume element of length δs and cross-sectional area δA containing the appropriate 2-level atoms (see figure 1.7).

For the moment we shall constrain this model stellar atmosphere to have no mass motion, i.e. the cylindrical volume element is fixed in space and only radiation goes in and out. (This simplification allows us to ignore matter fluxes here - the more complicated problem will be considered in chapter 5.)

We proceed by introducing the quantity $\mathscr{P}$ such that $\mathscr{P}(s, \nu, \Omega, t)\delta A$ is the number of photons of frequency ν travelling in a direction Ω which, at time t, enter one end of the cylinder situated at the point s per unit time interval (note that the area δA is perpendicular to Ω). The corresponding number of photons leaving the cylinder at the other end is then $\mathscr{P}(s + \delta s, \nu, \Omega, t + \delta t)\delta A$, where δt is the time taken for the photons to travel from one end of the cylinder to the other, i.e. $\delta s = c\delta t$, where c is the speed of the photons.

An equation involving the change in the number of photons in the beam from s to δs can be written as

$$[\mathscr{P}(s + \delta s, \nu, \Omega, t + \delta t) - \mathscr{P}(s, \nu, \Omega, t)]\delta A = \{\text{IN}\} - \{\text{OUT}\}, \quad (1.2.1)$$

where {IN} is the total number of photons of frequency ν and direction Ω put

Fig.1.7. A cylindrical volume element in the direction Ω.

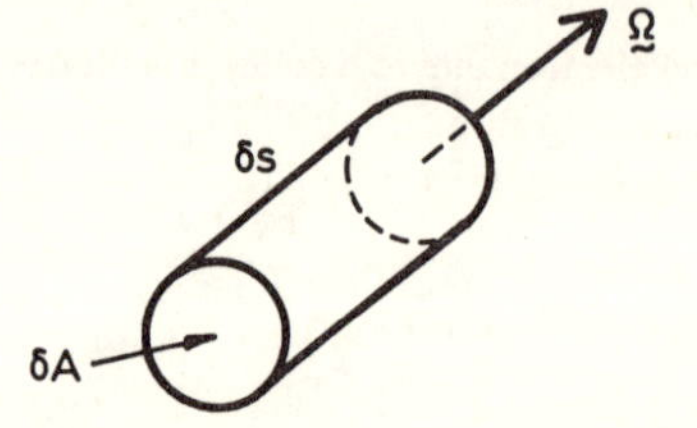

into the beam per unit time and {OUT} is the corresponding number of photons taken out. We know from the preceding section that radiation may be put into the beam by processes 2 and 3, i.e. by spontaneous and stimulated de-excitation of the excited atom. Radiation may be taken out by absorption of photons (process 1). With this information we are therefore in a position to explicitly derive expressions for {IN} and {OUT}.

We first note that the number of photons put into the beam per unit time by spontaneous emissions must be proportional to:

(i) The total number of atoms $N_U \delta s \delta A$ in the cylindrical volume element capable of spontaneous emission, i.e. with an electron in the excited state. We refer to N_U as the number density of the upper level.

(ii) The time rate at which an atom in the excited state will spontaneously de-excite. This rate is usually written as A_{UL} and is known as the Einstein rate coefficient for spontaneous emission. Under the assumption of thermodynamic equilibrium (which we shall discuss later) the quantity A_{UL} is a constant within a spectral line but varies from one spectral line to another. Since A_{UL} is the number of spontaneous de-excitations of an atom per unit time (in fact, it makes more sense to talk about the number of spontaneous de-excitations $N_U A_{UL}$ per unit volume per unit time), A_{UL}^{-1} effectively represents the time, on the average, that a bound electron stays in the upper level if no external influence is placed on it, i.e. A_{UL}^{-1} is the average lifetime of the excited state. Typical values of A_{UL} for strong spectral lines are of order 10^8 sec^{-1}.

(iii) The probability $j(\nu, \Omega)$ of this emission process yielding a photon of frequency ν travelling in the direction Ω. Clearly, a spontaneous de-excitation can lead to the emission of a photon in any direction. However, equation (1.2.1) refers only to photons travelling in the particular direction Ω. Thus, we must introduce the above probability to account for this angle dependence. The same argument applies for the frequency dependence of the emission process although, here, two quite distinct aspects must be considered.

First, if the energy difference between two levels in an atom is $h\nu_0$, then, in the rest frame of the atom, the de-excitation of the atom will lead to the emission of a photon of frequency ν_0. However, the atom is moving relative to the cylindrical volume of gas and, consequently, the frequency of this emitted photon in the rest frame of the cylinder will be Doppler-shifted. All the atoms are moving with their different microscopic velocities, thus contributing differently to the number of photons having a specific frequency ν measured in the rest frame of the cylinder. The evaluation of the corresponding probability $j(\nu, \Omega)$ therefore involves some kind of ensemble average over these microscopic motions.

The second aspect deals with the quantum mechanical specification of the various energy levels. Apart from perhaps the ground state of the atom, the energy states are not infinitesimally sharp (as we have implied) i.e. the excited atom cannot only exist at an energy $h\nu_0$ above the de-excited state, but at a spread of energies $h\nu_0 \pm h\Delta\nu$ where, generally, $\Delta\nu \ll \nu_0$. The various mechanisms (e.g. natural broadening, pressure broadening, etc.) which lead to this spread of energies will be detailed later. For the moment, we simply note that an excited atom can de-excite spontaneously yielding a photon of frequency not just ν_0 in the rest frame of the atom, but $\nu_0 \pm \Delta\nu$. One can then convolve this process with the above-mentioned Doppler shift to yield $j(\nu, \Omega)$. As we shall see, an even more general approach leads to a coupling between ν and Ω in the specification of $j(\nu, \Omega)$.

We now turn to the component of the {IN} term resulting from stimulated or induced emission (process 3). Here, we find the number of photons of frequency ν and direction Ω put into the beam per unit time is proportional to:

(i) $N_U \delta s \delta A$ as before.

(ii) The total number $\mathscr{P}(s, \nu, \Omega, t)$ of photons which cross δA at s per unit time which induce the excited atom to de-excite. Spontaneous emission, as we have said previously, needs no external stimulus for the de-excitation process to occur. The quantity of stimulated emission, however, has been experimentally shown to be directly proportional to the quantity of radiation doing the stimulating and, more particularly, the stimulated radiation is emitted in the same direction and with the same frequency as the stimulating radiation. We restrict ourselves to these experimental facts – indeed, our detailed physical understanding of this precise process is still rather sketchy.

(iii) The probability $\psi(\nu, \Omega)$ of this emission process yielding a photon of frequency ν travelling in the direction Ω.

All the above proportionalities for spontaneous and stimulated emission then yield the result

$$\{\mathrm{IN}\} = \alpha_1 N_U A_{UL} j(\nu, \Omega) \delta s \delta A + \alpha_2 N_U \mathscr{P}(s, \nu, \Omega, t) \psi(\nu, \Omega) \delta s \delta A, \tag{1.2.2}$$

where α_1 and α_2 are proportionality constants.

We now turn to the evaluation of {OUT}. Here, the appropriate proportionalities relate to the absorption of photons (process 1) and are:

(i) $N_L \delta s \delta A$, where N_L is the number density of those atoms capable of absorbing, i.e. those atoms in the lower state.

(ii) The total number $\mathscr{P}(s, \nu, \Omega, t)$ of photons crossing δA at s per unit time capable of being absorbed. This, again, is an experimental result.

(iii) The probability $\phi(\nu, \Omega)$ of a photon of frequency ν travelling in the direction Ω being absorbed. Here, as for $j(\nu, \Omega)$ and $\psi(\nu, \Omega)$, we must take into account the microscopic motions of the individual atoms within the volume element fixed in space and the spread of energy levels within each atom.

We then have

$$\{\text{OUT}\} = \alpha_3 N_{\text{L}} \mathscr{P}(s, \nu, \Omega, t)\phi(\nu, \Omega)\delta s \delta A, \tag{1.2.3}$$

where α_3 is a proportionality constant.

We proceed by defining the intensity I of radiation such that $I(s, \nu, \Omega, t)\delta A$ is the energy flux of photons of frequency ν and direction Ω traversing the area δA (with δA perpendicular to Ω as in figure 1.7) per unit time. Clearly, $I \propto \mathscr{P}$, i.e. we write

$$I = \alpha_4 \mathscr{P}, \tag{1.2.4}$$

so that equation (1.2.1) becomes

$$\begin{aligned} &I(s + \delta s, \nu, \Omega, t + \delta t) - I(s, \nu, \Omega, t) \\ &\quad = \beta_1 N_{\text{U}} A_{\text{UL}} j \delta s + \beta_2 N_{\text{U}} I \psi \delta s - \beta_3 N_{\text{L}} I \phi \delta s. \end{aligned} \tag{1.2.5}$$

The proportionality constant β_1 may be determined by examining the dimensions of equation (1.2.5). The LHS is an energy flux per unit area per unit time whereas $N_{\text{U}} A_{\text{UL}} j \delta s$ is the number of spontaneous de-excitations per unit area per unit time (which result in photon emission (ν, Ω)). However, j is just a probability, i.e. it has no dimensions. Consequently, since these de-excitation processes lead to emission of photons of energy $h\nu$ throughout an angle of 4π radians, one must have $\beta_1 = h\nu/4\pi$.

It is thus customary to write

$$\beta_2 = \frac{h\nu}{4\pi} B_{\text{UL}}, \quad \beta_3 = \frac{h\nu}{4\pi} B_{\text{LU}}, \tag{1.2.6}$$

where B_{UL} and B_{LU} are Einstein's rate coefficients for stimulated emission and absorption. Note that $B_{\text{UL}} I$ and $B_{\text{LU}} I$ have the same dimensions as A_{UL}.

Expanding $I(s + \delta s, \nu, \Omega, t + \delta t)$ in a Taylor series, using $\delta s = c\delta t$, and letting $\delta s \to 0$, enables equation (1.2.5) to be written as

$$\frac{\partial I}{\partial s} + \frac{1}{c}\frac{\partial I}{\partial t} = \frac{h\nu}{4\pi}\,[-N_{\text{L}} B_{\text{LU}} I \phi + N_{\text{U}} B_{\text{UL}} I \psi + N_{\text{U}} A_{\text{UL}} j]. \tag{1.2.7}$$

We refer to equation (1.2.7) as an equation of radiative transfer. Note that $\partial/\partial s$ is a directional derivative and is more commonly written as $(\Omega \cdot \nabla)$.

One can re-write the above equation in the somewhat standard form

$$(\Omega \cdot \nabla) I + \frac{1}{c}\frac{\partial I}{\partial t} = -\kappa(I - S), \tag{1.2.8}$$

where κ is referred to as the opacity and S the source function. Comparison between equations (1.2.7) and (1.2.8) yields

$$\kappa = \frac{h\nu}{4\pi}\,[N_{\text{L}} B_{\text{LU}} \phi - N_{\text{U}} B_{\text{UL}} \psi], \tag{1.2.9}$$

and

$$S = \frac{N_U A_{UL} j}{N_L B_{LU} \phi - N_U B_{UL} \psi}. \tag{1.2.10}$$

Clearly, if we are to solve the equation of transfer (1.2.8), we must have information regarding $N_L, N_U, \phi, \psi, j, B_{LU}, B_{UL}$ and A_{UL}, together with the appropriate boundary conditions. The boundary conditions, however, depend critically on the geometry of the situation being analysed and therefore cannot be considered from a general point of view. The Einstein rate coefficients B_{LU}, B_{UL} and A_{UL} may be determined by appropriate laboratory experiments. The quantities N_U, N_L, ϕ, ψ and j are of uppermost interest and shall be considered in the following sections.

1.3 The evaluation of N_U/N_L

One can re-write the opacity κ and source function S (equations (1.2.9) and (1.2.10)) in the form

$$\kappa = \frac{h\nu}{4\pi} N_L B_{LU} \phi \left[1 - \frac{N_U B_{UL} \psi}{N_L B_{LU} \phi}\right], \tag{1.3.1}$$

and

$$S = \frac{\dfrac{N_U A_{UL} j}{N_L B_{LU} \phi}}{1 - \dfrac{N_U B_{UL} \psi}{N_L B_{LU} \phi}}. \tag{1.3.2}$$

Thus, apart from N_L occurring by itself in equation (1.3.1), the number densities N_U and N_L occur as the ratio N_U/N_L. The thrust behind this section is the evaluation of this ratio.

To do this, we again consider a small cylindrical volume element (see figure 1.7), stationary relative to some fixed reference frame, and assume that there is no flux of atoms into or out of that volume element. (We shall examine non-zero flux situations in chapter 5.)

We then define the term {EXCITATIONS} to be the total number of processes per unit time and volume which enable atoms with bound electrons in the lower state to excite to the upper state. Correspondingly, we define the reverse term {DE-EXCITATIONS} for the total number of de-excitation processes. If there is to be zero flux of atoms in the cylindrical volume element, the only change in the densities N_U and N_L in time will be due to a difference between the number of excitations and de-excitations. For example, if there are more excitations than de-excitations, the number of atoms in the excited state must increase in time. Clearly then, we have the equation stipulating this balance in the form

$$\frac{\Delta N_U}{\Delta t} = \{\text{EXCITATIONS}\} - \{\text{DE-EXCITATIONS}\}. \tag{1.3.3}$$

The total density N of atoms is

$$N = N_U + N_L, \tag{1.3.4}$$

so that, if N is constant (i.e. zero flux of atoms), then

$$\frac{\Delta N_L}{\Delta t} = -\frac{\Delta N_U}{\Delta t}, \tag{1.3.5}$$

as one would expect from equation (1.3.3).

From section 1.1 we know that excitations may result from photon absorption and collisional excitations (processes 1 and 4) whilst de-excitations result from spontaneous and stimulated emissions (processes 2 and 3) and collisional de-excitations (process 5).

Following the arguments detailed in the preceding section, we know that the number of spontaneous de-excitations per unit time and volume which yield a photon of frequency ν (measured in the rest frame of the cylinder) and direction Ω is $N_U A_{UL} j(\nu, \Omega)/4\pi$. However, we are interested not just in that de-excitation process which yields a (ν, Ω) photon, but the total number of spontaneous de-excitations, i.e. we must sum over all angles and frequencies of the emitted photons. Consequently, the total number of spontaneous de-excitations per unit time and volume is

$$\frac{1}{4\pi}\int_{4\pi} \mathrm{d}\Omega \int \mathrm{d}\nu N_U A_{UL} j(\nu, \Omega) = N_U A_{UL}, \tag{1.3.6}$$

where we have used the fact that $j(\nu, \Omega)$ is a probability, viz.

$$\frac{1}{4\pi}\int_{4\pi} \mathrm{d}\Omega \int \mathrm{d}\nu j(\nu, \Omega) = 1. \tag{1.3.7}$$

Equation (1.3.7) holds, of course, because a spontaneous de-excitation of the excited atom must result in the emission of a photon.

Similarly, the total number of stimulated de-excitations per unit time and volume is

$$\frac{N_U B_{UL}}{4\pi}\int_{4\pi} \mathrm{d}\Omega \int \mathrm{d}\nu\, \psi(\nu, \Omega) I(\mathbf{r}, \nu, \Omega, t). \tag{1.3.8}$$

The number of de-excitations per unit time and volume arising from the collision of an atom in the excited state with a free electron is $N_U C_{UL}/4\pi$, where C_{UL} is the time rate of this collisional de-excitation occurring. Indeed, C_{UL} may be directly compared to the time rate A_{UL} for spontaneous de-excitation - in particular, it clearly must be a function of the density and microscopic speed (viz. temperature) of the electrons. However, we must sum over all the directions of the colliding free electrons. Thus, the total number of collisional de-excitations is $N_U C_{UL}$.

Adding all the de-excitation processes then yields

$$\{\text{DE-EXCITATIONS}\} = N_U A_{UL} + \frac{N_U B_{UL}}{4\pi} \int_{4\pi} d\Omega \times \int d\nu \, \psi(\nu, \Omega) I(\mathbf{r}, \nu, \Omega, t) + N_U C_{UL}. \tag{1.3.9}$$

Similarly, we find the total number of excitations to be

$$\{\text{EXCITATIONS}\} = N_L C_{LU} + \frac{N_L B_{LU}}{4\pi} \int_{4\pi} d\Omega \times \int d\nu \, \phi(\nu, \Omega) I(\mathbf{r}, \nu, \Omega, t), \tag{1.3.10}$$

where C_{LU} is the time rate of collisional excitations.

If we define the average intensity $\bar{J}$ such that

$$\bar{J}_{LU} = \frac{1}{4\pi} \int_{4\pi} d\Omega \int d\nu \, \phi(\nu, \Omega) I(\mathbf{r}, \nu, \Omega, t), \tag{1.3.11}$$

and

$$\bar{J}_{UL} = \frac{1}{4\pi} \int_{4\pi} d\Omega \int d\nu \, \psi(\nu, \Omega) I(\mathbf{r}, \nu, \Omega, t), \tag{1.3.12}$$

equation (1.3.3) yields

$$\frac{\Delta N_U}{\Delta t} = N_L(B_{LU} \bar{J}_{LU} + C_{LU}) - N_U(B_{UL} \bar{J}_{UL} + A_{UL} + C_{UL}). \tag{1.3.13}$$

We shall consider the time-independent situation in the remainder of this chapter, i.e. all $\Delta/\Delta t$ terms will be put to zero. Consequently, equation (1.3.13) yields the ratio N_U/N_L required in the specification of the source function S given by equation (1.3.2). Equation (1.3.13) with the LHS equal to zero is often referred to as the equation for a statistically steady state, i.e. the number density of both the upper level and lower level of the model 2-level atom does not change with time and this, in turn, simply states that the number of events which populate a given level must balance the number of events which de-populate it.

1.4 The evaluation of $\phi(\nu, \Omega)$

Having determined the number density ratio N_U/N_L, we may proceed in the specification of the source function by evaluating $\phi(\nu, \Omega)$, $j(\nu, \Omega)$ and

$\psi(\nu, \Omega)$. In section 1.2, we only outlined the concept of these various probabilities. We wish to be more precise here.

The equation of transfer (1.2.8) is an equation specifying the manner in which a beam of radiation interacts with a gas consisting of an ensemble of atoms. In particular, it is an equation for photons, $I(\mathbf{r}, \nu, \Omega, t)$ being the intensity of that beam of photons - it is not an equation for matter. Consequently, when we introduce the quantity ϕ, for example, which attempts to detail the manner in which a photon is absorbed by the gas, we must include not just one atom but the entire ensemble. We therefore define $N_L\phi(\nu, \Omega)$ to be the probability of a photon of frequency ν travelling in the direction Ω being absorbed by an ensemble of N_L atoms with a bound electron in the lower level. We correspondingly define $N_U j(\nu, \Omega)$ and $N_U \psi(\nu, \Omega)$ to be the probabilities of an ensemble of N_U atoms in the excited state emitting a photon of frequency ν and direction Ω as a result of spontaneous and stimulated de-excitations respectively.

We may evaluate $\phi(\nu, \Omega)$ by first considering the absorption process at a microscopic level (i.e. by examining the absorption of a photon by an individual atom), then averaging over the ensemble of atoms. We therefore define $q(\gamma)$ to be the probability of a photon of frequency γ, measured in the rest frame of the atom, being absorbed by an atom in the de-excited state. Since individual atoms have motions relative to one another, it will be necessary to relate all events to some alternative reference frame fixed in space (i.e. the observer's frame of reference).

Let the velocity of the atom be $\mathbf{v}$ measured in this fixed frame (see figure 1.8).

Clearly, if $\mathbf{v}$ is not perpendicular to Ω, the atom will 'see' a Doppler-shifted photon. If the frequency of the photon measured in the rest frame of the observer is ν, then the relativistic Doppler relationship [48]

$$\gamma = \frac{\nu\left(1 - \dfrac{\mathbf{v}\cdot\Omega}{c}\right)}{\left[1 - \dfrac{v^2}{c^2}\right]^{1/2}} \tag{1.4.1}$$

follows.

We now define the new independent variable $\Delta\nu$ such that

$$\Delta\nu = \nu - \nu_0, \tag{1.4.2}$$

Fig.1.8. A photon moving in direction Ω and an atom moving with velocity v.

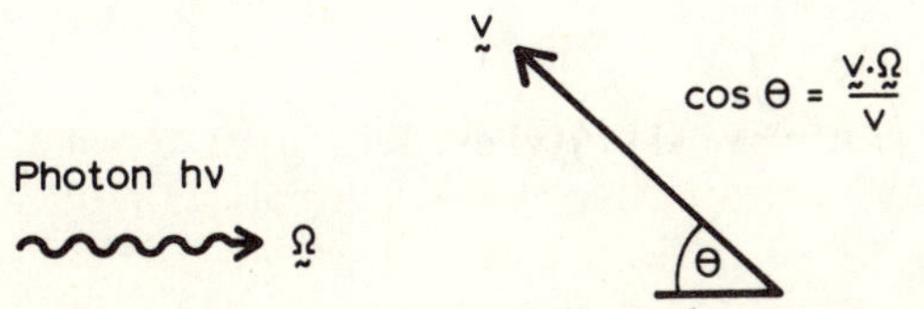

where ν_0 is the frequency of the spectral line in question. For example, if the line arises from an atom having infinitesimally sharp upper and lower levels, then ν_0 is just that frequency corresponding to the energy separation $h\nu_0$ of these levels. Equation (1.4.1) then becomes

$$\Delta\gamma = \Delta\nu - \frac{\nu}{c}\mathbf{v}\cdot\mathbf{\Omega}, \tag{1.4.3}$$

where we have ignored terms of order v^2/c^2 and higher. Further, we are only interested in $|\Delta\nu| \ll \nu$ so that we can also write equation (1.4.3) as $\Delta\gamma = \Delta\nu - \nu_0\mathbf{v}\cdot\mathbf{\Omega}/c$.

Thus, the probability of a photon of frequency $\Delta\nu$ being absorbed by an atom moving with velocity $\mathbf{v}$, where both $\Delta\nu$ and $\mathbf{v}$ are measured in the rest frame of the observer, is $q(\Delta\nu - (\nu/c)\mathbf{v}\cdot\mathbf{\Omega})$.

We must now consider all the other atoms in the ensemble. To do this, we define the velocity distribution function $F_L(\mathbf{v})$ for the de-excited (absorbing) atoms so that $N_L F_L(\mathbf{v})\delta\mathbf{v}$ is the total number of de-excited atoms per unit volume element having a velocity $\mathbf{v}$ in the range $(\mathbf{v}, \mathbf{v}+\delta\mathbf{v})$, i.e. we must have

$$\int_{-\infty}^{\infty} N_L F_L(\mathbf{v})\,d\mathbf{v} = N_L,$$

or

$$\int_{-\infty}^{\infty} F_L(\mathbf{v})\,d\mathbf{v} = 1, \tag{1.4.4}$$

where the above integration is over all velocity space. The function $F_L(\mathbf{v})$ may therefore be considered as the probability of an atom having a velocity $\mathbf{v}$. (In chapter 5, we use the distribution function $f \equiv NF$ where, in general, f is taken as Maxwellian.)

Consequently, the total number (per unit time and volume) of absorptions of photons of frequency $\Delta\nu$ and direction $\mathbf{\Omega}$ by de-excited atoms with velocities in the range $(\mathbf{v}, \mathbf{v}+\delta\mathbf{v})$ is just

$$q\left(\Delta\nu - \frac{\nu}{c}\mathbf{v}\cdot\mathbf{\Omega}\right) N_L F_L(\mathbf{v})\delta\mathbf{v}.$$

The total number of absorptions $N_L\phi(\Delta\nu)$ per unit volume due to all the velocities of the ensemble of moving atoms is obtained by summing over $\mathbf{v}$, i.e. we have

$$N_L\phi(\Delta\nu) = \int_{-\infty}^{\infty} q\left(\Delta\nu - \frac{\nu}{c}\mathbf{v}\cdot\mathbf{\Omega}\right) N_L F_L(\mathbf{v})\,d\mathbf{v},$$

viz.

$$\phi(\Delta\nu) = \int_{-\infty}^{\infty} q\left(\Delta\nu - \frac{\nu}{c}\mathbf{v}\cdot\mathbf{\Omega}\right) F_L(\mathbf{v})\,d\mathbf{v}. \tag{1.4.5}$$

If we integrate the above equation over all $\Delta\nu$ frequency space,

$$\int_{-\infty}^{\infty} \phi(\Delta\nu)\, \mathrm{d}(\Delta\nu) = \int_{-\infty}^{\infty} F_{\mathrm{L}}(\mathbf{v})\, \mathrm{d}\mathbf{v} \int_{-\infty}^{\infty} q\left(\Delta\nu - \frac{\nu}{c}\mathbf{v}\cdot\boldsymbol{\Omega}\right) \mathrm{d}(\Delta\nu), \quad (1.4.6)$$

use the fact that $q(\Delta\gamma)$ is a probability, i.e.

$$\int_{-\infty}^{\infty} q(\Delta\gamma)\, \mathrm{d}(\Delta\gamma) = 1, \quad (1.4.7)$$

and recall equation (1.4.4), we find the important normalisation condition

$$\int_{-\infty}^{\infty} \phi(\Delta\nu)\, \mathrm{d}(\Delta\nu) = 1. \quad (1.4.8)$$

Explicit expressions for $\phi(\Delta\nu)$ may be obtained from the defining equation (1.4.5) once both $q(\Delta\gamma)$ and $F_{\mathrm{L}}(\mathbf{v})$ are known.

It is common practice to assume all the particles constituting the matter in the stellar atmosphere may be represented by a Maxwellian velocity distribution (see chapter 5, equation (5.2.8)), i.e.

$$F_{\mathrm{L}}(\mathbf{v}) = \frac{1}{\bar{V}^3 \pi^{3/2}} \mathrm{e}^{-(\mathbf{v}/\bar{V})^2}, \quad (1.4.9)$$

where

$$\bar{V} = \left[\frac{2kT}{m_{\mathrm{A}}}\right]^{1/2}, \quad (1.4.10)$$

and where T is the temperature of the gas and m_{A} the mass of the atom. The quantity $\bar{V}$ is then the mean speed of the particles. We shall discuss the assumption represented by equation (1.4.9) in chapter 5. For the moment, let us emphasise that this result only applies rigorously for a gas in thermodynamic equilibrium, i.e. for a gas in which no energy is lost from the system. However, as we have seen, the matter (atoms and electrons) is coupled to the radiation field - indeed, there is transfer of energy from matter to radiation and radiation to matter - and, since radiation escapes from the stellar atmosphere (we observe it), the atmosphere cannot be in a precise equilibrium configuration. Equation (1.4.9) must therefore be used with some reservation. (This argument can be extended to show that the definition of temperature, which is necessarily directly associated with the definition of a Maxwellian velocity distribution through equation (1.4.10), is rigorously inconsistent under such non-equilibrium configurations.)

The form $q(\Delta\gamma)$ takes depends upon the atom, and transition within that atom, being examined. For example, if we consider the physically idealistic situation in which both the upper and lower levels of the transition are infinitesimally sharp, i.e. atoms may absorb (and emit) photons of frequency ν_0 in the rest frame of the atom, then we may write

$$q(\Delta\gamma) \equiv \delta(\Delta\gamma), \tag{1.4.11}$$

where the delta function $\delta(\xi)$ has the basic property

$$\int_{-\infty}^{\infty} f(\xi)\delta(\xi - \xi_0)\, d\xi = f(\xi_0), \tag{1.4.12}$$

with, in particular, the normalisation

$$\int_{-\infty}^{\infty} \delta(\xi)\, d\xi = 1. \tag{1.4.13}$$

If we now write $\mathbf{v} = v_1\mathbf{\Omega}_1 + v_2\mathbf{\Omega}_2 + v_3\mathbf{\Omega}_3$ in terms of the orthogonal set of unit vectors $(\mathbf{\Omega}_1, \mathbf{\Omega}_2, \mathbf{\Omega}_3)$ and choose $\mathbf{\Omega}_1$ in the direction of $\mathbf{\Omega}$ such that $\mathbf{v} \cdot \mathbf{\Omega} = v_1$, and note that

$$\int_{-\infty}^{\infty} e^{-v_2^2}\, dv_2 = \sqrt{\pi},$$

for example, equations (1.4.9) and (1.4.11) then yield, after substitution into (1.4.5),

$$\begin{aligned}\phi(\Delta\nu) &= \frac{1}{\bar{V}\sqrt{\pi}} \int_{-\infty}^{\infty} e^{-(v_1/\bar{V})^2}\, \delta\left(\Delta\nu - \frac{\nu}{c} v_1\right) dv_1 \\ &= \frac{c}{\nu\sqrt{\pi}} \int_{-\infty}^{\infty} e^{-(v_1/\bar{V})^2}\, \delta\left(\frac{c\Delta\nu}{\nu\bar{V}} - \frac{v_1}{\bar{V}}\right) d\left(\frac{v_1}{\bar{V}}\right) \\ &= \frac{1}{\Delta\nu_D\sqrt{\pi}}\, e^{-(\Delta\nu/\Delta\nu_D)^2},\end{aligned} \tag{1.4.14}$$

where we have introduced the Doppler width $\Delta\nu_D$ given by

$$\Delta\nu_D = \frac{\nu}{c}\bar{V} = \frac{\nu}{c}\left[\frac{2kT}{m_A}\right]^{1/2}. \tag{1.4.15}$$

The absorption probability given by equation (1.4.14) is known as the Doppler absorption profile and obviously has a Gaussian distribution. Note that it satisfies the normalisation condition (1.4.8).

A more physically realistic situation would allow at least one of the two levels constituting the radiative transition in question to be broadened. For example, the Heisenberg uncertainty principle [83] states that *both* the measured energy of a particle and the time at which that energy was measured cannot be absolutely accurately known due to the influence the measuring process must necessarily exert on the system being measured. Thus, one cannot allocate a precise position for the bound electron in the energy level diagram (figure 1.1).

If, however, we consider a resonance line for which the lower level is the ground state of the atom, then one may still possibly take this lower level to be infinitesimally sharp. The upper level must then be considered broadened as shown in figure 1.9.

The probability of an atom having a bound electron with energy $h\gamma$ above L (or, equivalently, $h\Delta\gamma$ above or below the central energy $h\nu_0$) may be shown to be represented by (see chapter 6, equation (6.4.77))

$$q(\Delta\gamma) = \frac{\delta}{\pi} \cdot \frac{1}{(\Delta\gamma)^2 + \delta^2}, \tag{1.4.16}$$

where

$$\delta = \frac{1}{4\pi} \sum_{i<U} A_{Ui} \tag{1.4.17}$$

is a measure of the width of the broadening. One refers to the above $q(\Delta\gamma)$ as natural broadening and it can be derived using either a classical oscillator approach or the full quantum mechanical analysis (detailed in section 6.5). Note that $q(\Delta\gamma)$ satisfies the normalisation condition (1.4.7), and has a maximum at $\Delta\gamma = 0$, i.e. the maximum of $q(\Delta\gamma)$ occurs at the central frequency ν_0. Thus the most likely position of a bound electron in the excited state is at an energy $h\nu_0$ above L.

Substitution of equation (1.4.16) into (1.4.5), and using (1.4.9), yields

$$\phi(\Delta\nu) = \frac{\delta}{\pi^{3/2}\bar{V}} \int_{-\infty}^{\infty} \frac{e^{-(v_1/V)^2}\, dv_1}{\left(\Delta\nu - \dfrac{\nu}{c} v_1\right)^2 + \delta^2}$$

$$= \frac{\delta}{\pi^{3/2}(\Delta\nu_D)^2} \int_{-\infty}^{\infty} \frac{e^{-(v_1/V)^2}\, d(v_1/V)}{\left(\dfrac{\Delta\nu}{\Delta\nu_D} - \dfrac{v_1}{V}\right)^2 + \left(\dfrac{\delta}{\Delta\nu_D}\right)^2}$$

$$= \frac{1}{\Delta\nu_D\sqrt{\pi}} H\left(a, \frac{\Delta\nu}{\Delta\nu_D}\right), \tag{1.4.18}$$

where we have put

$$a = \frac{\delta}{\Delta\nu_D}. \tag{1.4.19}$$

Fig.1.9. Broadening of the excited state of the atom.

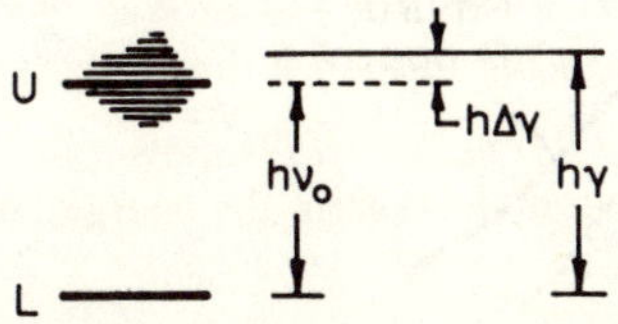

In the foregoing, we have introduced the Voigt function $H(a, \eta)$ defined by [98]

$$H(a, \eta) = \frac{a}{\pi} \int_{-\infty}^{\infty} \frac{e^{-y^2} \, dy}{(y-\eta)^2 + a^2}. \tag{1.4.20}$$

In the limit $\delta \to 0$, i.e. $a \to 0$, the upper level becomes infinitesimally sharp and we then revert to Doppler broadening. It is not difficult to see that the above two expressions for $\phi(\Delta\nu)$ are symmetric about $\nu = \nu_0$, i.e. about $\Delta\nu = 0$, as shown in figure 1.10.

The Doppler profile falls away exponentially for increasing $|\Delta\nu|$, whilst the $\phi(\Delta\nu)$ for combined Doppler and natural broadening exhibits a much slower decrease away from line centre $\Delta\nu = 0$. Indeed, the e^{-y^2} appearing in the integrand of equation (1.4.20) effectively stipulates that the main contribution to the integral occurs for small y. Consequently, when $\eta = \Delta\nu/\Delta\nu_D$ is small, the denominator in the above integral has a behaviour similar to a delta function and $H(a, \eta)$ then has a Gaussian appearance. However, for large $\eta = \Delta\nu/\Delta\nu_D$, $H(a, \eta) \sim 1/\eta^2$. Since $\phi(\Delta\nu)$ is a probability satisfying equation (1.4.8), the areas under both curves in figure 1.10 must be unity. Consequently, the line-centre absorption probability $\phi(0)$ for Doppler broadening must be greater than for combined Doppler and natural broadening.

Finally, there are a variety of other broadening mechanisms (Stark, statistical, etc.) which one can consider. The reader is referred to Mihalas [84] for further details.

1.5 Complete re-distribution [117]

Now that we have determined the general form the absorption probability $\phi(\Delta\nu)$ exhibits (equation (1.4.5)), the final step that must be taken before the equation of radiative transfer may be solved is the determination of the spontaneous and stimulated emission probabilities $j(\Delta\nu)$ and $\psi(\Delta\nu)$. There are several methods, all leading to different results, by which this may be accom-

Fig.1.10. Absorption profiles for both Doppler broadening and Doppler plus natural broadening.

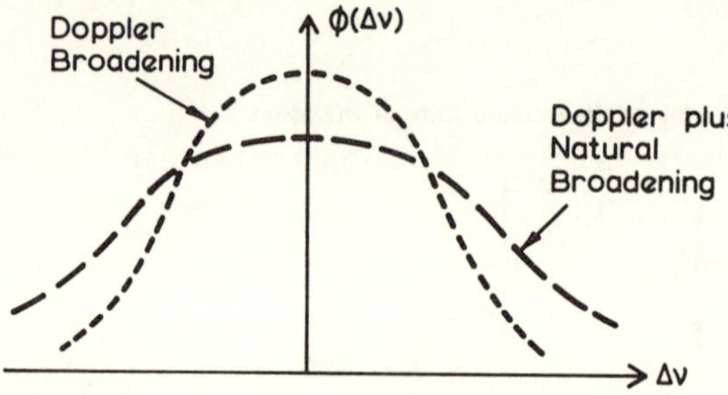

plished. Since, in this chapter, we wish to keep the problem as simple as possible in order to obtain a fundamental understanding of the matter–radiation interaction, we defer the derivation of $j(\Delta\nu)$ and $\psi(\Delta\nu)$ to a later chapter.

For the present, therefore, we choose to make the somewhat physically *ad hoc* assumption

$$\phi(\Delta\nu) \equiv j(\Delta\nu) \equiv \psi(\Delta\nu). \tag{1.5.1}$$

This is known as the complete re-distribution approximation and, as we shall see in chapter 7, holds for the three extreme cases [93]:

(i) Collisions dominate.

(ii) Complete re-distribution in the rest frame of the atom (i.e. the frequency at which a photon is emitted is completely independent of the frequency at which the photon was absorbed). This is also referred to as complete non-coherent scattering.

(iii) Coherence in the rest frame of the atom (i.e. the photon is emitted at a frequency exactly equal to the frequency at which the photon was absorbed) and isotropic frequency-independent radiation.

The third case could apply if, for example, the two levels of the atom were infinitesimally sharp. This is rather a special situation since all transitions between upper and lower levels should involve at least some natural broadening. Thus, the first two physical consequences, i.e. (i) and (ii) above, of equation (1.5.1) are generally the more physically realistic.

1.6 The complete re-distribution source function [118]

The main thrust behind the assumption of complete re-distribution (equation (1.5.1)) is purely mathematical. The opacity and source functions given by equations (1.3.1) and (1.3.2) then become

$$\kappa = \frac{h\nu}{4\pi} N_L B_{LU} \phi \left[1 - \frac{N_U B_{UL}}{N_L B_{LU}}\right], \tag{1.6.1}$$

and

$$S = \frac{\dfrac{N_U A_{UL}}{N_L B_{LU}}}{1 - \dfrac{N_U B_{UL}}{N_L B_{LU}}}. \tag{1.6.2}$$

The source function, in particular, is now independent of frequency within the spectral line (i.e. the frequency variable $\Delta\nu$ does not occur in S) and this simplifies the solution of the equation of radiative transfer enormously.

Further, since we have assumed $\phi(\Delta\nu) \equiv \psi(\Delta\nu)$, equations (1.3.11) and (1.3.12) yield

$$\bar{J}_{LU} = \bar{J}_{UL} = \bar{J} \text{ (say)}. \tag{1.6.3}$$

Thus, if we make the substitution for the ratio N_U/N_L from equation (1.3.13), and consider only the time-independent case so that all $\Delta/\Delta t$ terms go to zero, we find

$$S = \frac{\bar{J} + \dfrac{2h\nu^3}{c^2} \cdot \dfrac{g_L C_{LU}}{g_U A_{UL}}}{1 + \dfrac{C_{UL}}{A_{UL}} \left[1 - \dfrac{g_L C_{LU}}{g_U C_{UL}}\right]}, \tag{1.6.4}$$

where we have used the standard results

$$\frac{A_{UL}}{B_{UL}} = \frac{2h\nu_0^3}{c^2}, \tag{1.6.5}$$

and

$$g_U B_{UL} = g_L B_{LU} \tag{1.6.6}$$

(see chapter 6) where g_U and g_L are the statistical weights of the upper and lower levels respectively.

To proceed further, we require information relating the inelastic collisional excitation and de-excitation rates C_{LU} and C_{UL}. The most commonly accepted method of doing this is to assume these collisional events to be in detailed balance, i.e. we assume the number of collisional de-excitations is exactly equal to the number of collisional excitations for a gas in local thermodynamic equilibrium (LTE). If we take N_L^* and N_U^* to be the respective number densities of the lower and upper levels of our simple model 2-level atom for a gas in LTE (and we shall return to the precise meaning of LTE in the following section), the above statement can be mathematically written as

$$N_L^* C_{LU} = N_U^* C_{UL}. \tag{1.6.7}$$

However, if the gas is in LTE, the number densities N_L^* and N_U^* must satisfy a Boltzmann distribution [29], viz.

$$\frac{N_L^*}{N_U^*} = \frac{g_L}{g_U} \, e^{h\nu_0/kT}, \tag{1.6.8}$$

where T is the temperature of the atoms.

Substitution of equation (1.6.8) into (1.6.7) then leads to the desired ratio between C_{LU} and C_{UL}, viz.

$$\frac{C_{LU}}{C_{UL}} = \frac{g_U}{g_L} \, e^{-h\nu_0/kT}. \tag{1.6.9}$$

As it turns out, the above LTE approximation for the determination of this ratio is quite good for most cases which have, in the past, been of stellar interest. Nevertheless, it is an approximation which should always be viewed with some concern. In particular, for example, equation (1.6.8) is only valid if there is a sufficient number of collisional events per unit volume per unit time which enable the velocity distribution of the atoms to be represented by a Maxwellian. Clearly, as one moves into higher and higher regions of the stellar atmosphere where the particle density becomes smaller and smaller, the number of such collisions must decrease and thus the above assumption becomes increasingly less tenable.

Substitution of equation (1.6.9) into (1.6.4) then yields

$$S = \frac{\bar{J} + \dfrac{2h\nu_0^3}{c^2}\, e^{-h\nu_0/kT}\, \dfrac{C_{UL}}{A_{UL}}}{1 + \dfrac{C_{UL}}{A_{UL}}\,[1 - e^{-h\nu_0/kT}]} . \tag{1.6.10}$$

If we further define ϵ' such that

$$\epsilon' = \frac{C_{UL}}{A_{UL}}\,[1 - e^{-h\nu_0/kT}], \tag{1.6.11}$$

we easily find

$$S = \frac{\bar{J} + \epsilon' B_\nu(T)}{1 + \epsilon'}, \tag{1.6.12}$$

where we have used the well-known relationship for the Planck function [79]

$$B_\nu(T) = \frac{2h\nu_0^3/c^2}{e^{h\nu_0/kT} - 1} . \tag{1.6.13}$$

Equation (1.6.12) is one of two standard forms for the 2-level atom complete re-distribution source function. The alternative form is obtained by putting

$$\epsilon = \frac{\epsilon'}{1 + \epsilon'}, \tag{1.6.14}$$

such that

$$S = (1 - \epsilon)\bar{J} + \epsilon B_\nu(T). \tag{1.6.15}$$

It is of some interest to analyse the physics of each of the terms appearing in equation (1.6.15). To do this, however, some understanding of the actual physical meaning of S is required. The time-independent equation of transfer (see equation (1.2.8)) is just

$$(\Omega \cdot \nabla) I = -\kappa I + \kappa S. \tag{1.6.16}$$

The first term on the RHS of this equation ($-\kappa I$) is effectively the radiation energy taken out of the beam of photons due to absorption. This energy is proportional to the intensity I of radiation so that the opacity κ can be treated as nothing more than a proportionality factor. Note that by writing κ in the form given by equation (1.2.9), i.e. by incorporating the stimulated emission term into κ, we are essentially treating stimulated emissions as negative absorptions.

Similarly, the term κS in equation (1.6.16) is the radiative energy put into the photon beam by just spontaneous emission. Clearly then, if we again treat κ as a proportionality factor, the source function S (having the same dimensions as I) is the corresponding intensity of the radiation field due solely to spontaneous emissions.

We are now in a position to interpret the physical meaning of the terms on the RHS of equation (1.6.15). The first term $\bar{J}$ is simply the average radiation field (averaged over angle and frequency) and effectively represents those processes whereby an atom absorbes a photon then re-emits it. One could call this a source of photons.

The negative sign in the second term $-\epsilon\bar{J}$ enables us to refer to it as a sink of photons. The ratio ϵ may be written as

$$\epsilon = \frac{C_{\mathrm{UL}}\,[1 - \mathrm{e}^{-h\nu_0/kT}]}{A_{\mathrm{UL}} + C_{\mathrm{UL}}\,[1 - \mathrm{e}^{-h\nu_0/kT}]}\,, \tag{1.6.17}$$

and, using the result

$$A_{\mathrm{UL}} + B_{\mathrm{UL}}B_\nu(T) = \frac{A_{\mathrm{UL}}}{1 - \mathrm{e}^{-h\nu_0/kT}}\,, \tag{1.6.18}$$

we find

$$\epsilon = \frac{C_{\mathrm{UL}}}{C_{\mathrm{UL}} + A_{\mathrm{UL}} + B_{\mathrm{UL}}B_\nu(T)}\,. \tag{1.6.19}$$

The quantity ϵ is therefore the ratio of the rate of collisional de-excitation to the *total* (i.e. collisional plus spontaneous plus stimulated) rate of de-excitation. Since a collisional de-excitation does not generate a photon, ϵ is then the effective probability of a de-excitation event not producing a photon. Note that ϵ satisfies $0 < \epsilon < 1$. Thus, with $\bar{J}$ giving the average radiation intensity, $\epsilon\bar{J}$ reflects the number of photons taken out of the radiation field due to photon absorption followed by collisional de-excitation. It is interesting to note that the denominator in equation (1.6.19) includes the term $B_{\mathrm{UL}}B_\nu(T)$. The more appropriate stimulated emission term should be $B_{\mathrm{UL}}\bar{J}$ - the difference may be traced to our use of the LTE approximation (equation (1.6.9)) for the collisional events.

The third term on the RHS of equation (1.6.15) for S, i.e. $\epsilon B_\nu(T)$, is a source of photons and plays the opposite role to the $-\epsilon\bar{J}$ sink term. To see this, we re-write ϵ in the form

$$\epsilon = \frac{C_{\mathrm{LU}}}{C_{\mathrm{LU}} + B_{\mathrm{LU}}B_\nu(T)}\,, \tag{1.6.20}$$

where we have again used equation (1.6.18), together with (1.6.9). Since the numerator in the above expression is the time rate of collisional excitations, and the denominator is the *total* (collisional plus radiative) time rate of excitations, one may also think of ϵ as the probability of an excitation event being due to collisions. When a collisional excitation occurs, energy is taken out of the free electron field and given to the bound electron. Two subsequent events could occur. The thus excited atom could undergo another collision with a free electron and the energy released from the atom due to this de-excitation would be restored to the free electron field - the overall process would involve no radiation. However, if the excited atom undergoes a radiative de-excitation, a photon would result. Thus, the energy taken out of the free electron field due to collisional excitation would be used to create a photon. We know that the energy available in the free electron field for such conversion to photon energy is given by the Planck function $B_\nu(T)$. But, of course, not all the excitation events are collisional, and thus $\epsilon B_\nu(T)$ is that fraction of this available energy $B_\nu(T)$ which is actually used to create photons.

Summary

We can quickly re-capitulate the essentials of the above discussion. The source function S contains three terms. We have a scattering term which details the radiation absorbed thence re-emitted. There is a source $\epsilon B_\nu(T)$ of photons due to collisional excitation, and a sink $-\epsilon\bar{J}$ of photons destroyed by collisional de-excitation.

1.7 LTE and non-LTE for photons [118]

(a) LTE

In the preceding section, we made fleeting reference to the assumption of local thermodynamic equilibrium (LTE). Many present-day analyses incorporate this assumption in a variety of ways and contexts, and it is therefore of considerable benefit to understand its physical meaning and limitations. One can talk about LTE for photons and/or particles. In the present section, however, we shall limit our detailed discussion to LTE for photons only.

Let us first consider a perfectly enclosed and isolated body of gas in thermodynamic equilibrium. By 'perfectly enclosed' we mean that the body does not receive energy, nor does it allow energy to escape, i.e. it is black. It is well known that its associated radiation field is given by the Planck function $B_\nu(T)$, where T is the *one* temperature of the particles within the black body. In other words, if one could view the black body from within, one would observe an intensity of radiation equal to $B_\nu(T)$.

A stellar atmosphere, however, allows energy to escape via non-zero fluxes of radiation and matter, otherwise we would see neither the (photospheric) stars themselves nor their stellar winds. Clearly, therefore, the assumption of thermodynamic equilibrium cannot be directly applied to stellar atmospheres. If, on the other hand, we partition the stellar atmosphere into infinitesimally small volume elements, it might be possible to approximate the state of the gas in each volume element as if each were in thermodynamic equilibrium. If we consider a volume element deep within the stellar atmosphere, so deep, in fact, that very little radiation is able to escape from the 'surface' of the atmosphere, this approximation could be quite satisfactory. Indeed, if we go to depths in the star below the stellar atmosphere, no radiation should escape (by definition of the stellar atmosphere) and the thermodynamic equilibrium approximation should not cause undue concern (unless one is considering, for example, neutrino radiation).

Since, if the above approximation is made at all, it is made locally for each small volume element, it is referred to as local thermodynamic equilibrium. Each volume element is therefore assumed to be a perfectly enclosed and isolated body (which, of course, it can never be in practice), each with its own representative equilibrium temperature. The radiation field in each element is then given by $B_\nu(T)$, where T is that representative temperature. In this way, 'classical astrophysicists' have been able to incorporate a non-zero temperature gradient within their model stellar atmospheres (such a non-zero temperature gradient cannot exist under thermodynamic equilibrium directly) and, at the same time, take advantage of the mathematical simplifications afforded by thermodynamic equilibrium theory.

The question arises, however: how satisfactory is the assumption of LTE, particularly as one considers volume elements higher and higher in the stellar atmosphere where more and more radiation may escape, and where the volume elements become less and less isolated? (Actually, it is tempting to ignore the question entirely and just state that if one observes radiation from a stellar atmosphere it must necessarily have emanated from regions where, by definition of LTE, the assumption of LTE is invalid. It is of some interest, however, to examine the physical consequences of LTE.)

One may attempt an answer to the above question by first considering the equation of a statistically steady state derived in section 1.3 (see equation (1.3.13) with $\Delta N_{\mathrm{U}}/\Delta t$ equal to zero). If we assume complete re-distribution such that $\phi(\Delta\nu) \equiv \psi(\Delta\nu)$, i.e. $\bar{J}_{\mathrm{LU}} \equiv \bar{J}_{\mathrm{UL}} \equiv \bar{J}$ (say), we have

$$N_{\mathrm{L}}(B_{\mathrm{LU}}\bar{J} + C_{\mathrm{LU}}) = N_{\mathrm{U}}(B_{\mathrm{UL}}\bar{J} + A_{\mathrm{UL}} + C_{\mathrm{UL}}). \tag{1.7.1}$$

If we now assume LTE, and in doing this replace N_{L} and N_{U} by their respective LTE values N_{L}^* and N_{U}^*, and replace the radiation field $\bar{J}$ by $B_\nu(T)$ as mentioned above, then

$$N_{\mathrm{L}}^*(B_{\mathrm{LU}}B_\nu(T) + C_{\mathrm{LU}}) = N_{\mathrm{U}}^*(B_{\mathrm{UL}}B_\nu(T) + A_{\mathrm{UL}} + C_{\mathrm{UL}}). \tag{1.7.2}$$

Using thermodynamic equilibrium arguments, the volume elements for which this last equation holds must have atoms satisfying a Maxwellian velocity distribution and, more particularly, the ratio $N_{\mathrm{U}}^*/N_{\mathrm{L}}^*$ must be given by the

Boltzmann formula

$$\frac{N_{\mathrm{U}}^{*}}{N_{\mathrm{L}}^{*}}=\frac{g_{\mathrm{U}}}{g_{\mathrm{L}}}\,\mathrm{e}^{-h\nu_0/kT}. \tag{1.7.3}$$

However, using this equation, together with the standard equations (1.6.5), (1.6.6) and (1.6.13), one finds

$$N_{\mathrm{L}}^{*}B_{\mathrm{LU}}B_{\nu}(T)=N_{\mathrm{U}}^{*}(B_{\mathrm{UL}}B_{\nu}(T)+A_{\mathrm{UL}}), \tag{1.7.4}$$

which, when substituted into equation (1.7.2), immediately yields

$$N_{\mathrm{L}}^{*}C_{\mathrm{LU}}=N_{\mathrm{U}}^{*}C_{\mathrm{UL}}. \tag{1.7.5}$$

Equation (1.7.5) is identical to (1.6.7) used in determining the complete re-distribution source function.

Equations (1.7.4) and (1.7.5) are generally referred to as the equations stipulating detailed balance, i.e. equation (1.7.4) says that the rate of excitation of the lower level by radiative processes (photon absorption) is exactly balanced by the rate of de-excitation of the upper level, also by radiative processes (spontaneous and stimulated photon emission). Likewise, the collisional excitation events in equation (1.7.5) balance collisional de-excitation events. Clearly, these equations are somewhat special cases of the more general equation for a statistically steady state derived in section 1.3 (i.e. equation (1.7.1)) in which *all* excitation processes (not just radiative and collisional excitations *separately*) balance *all* de-excitation processes.

The logic behind this detailed balancing argument is presumably as follows. Deep within the stellar atmosphere collisions are far more important in determining the photon intensity than are spontaneous and stimulated emissions because of the increased particle (electron) density relative to 'surface' regions. Consequently, one can neglect the relatively small terms in equation (1.7.1) and obtain

$$N_{\mathrm{L}}C_{\mathrm{LU}}=N_{\mathrm{U}}C_{\mathrm{UL}}. \tag{1.7.6}$$

If we now regard this as a rigorous mathematical result, the remainder of the terms in equation (1.7.1) then, equally rigorously, yields

$$N_{\mathrm{L}}B_{\mathrm{LU}}\bar{J}=N_{\mathrm{U}}(B_{\mathrm{UL}}\bar{J}+A_{\mathrm{UL}}). \tag{1.7.7}$$

However, if collisions are so dominant that equation (1.7.6) holds, there must be sufficient collisions to maintain Maxwellian velocity distributions for both the upper and lower states, i.e. we must have

$$\frac{N_{\mathrm{U}}}{N_{\mathrm{L}}}=\frac{N_{\mathrm{U}}^{*}}{N_{\mathrm{L}}^{*}}. \tag{1.7.8}$$

Equation (1.7.7), together with equations (1.6.5) and (1.6.6), then yields

$$\bar{J}=\frac{2h\nu_0^3/c^2}{\mathrm{e}^{h\nu_0/kT}-1}=B_{\nu}(T), \tag{1.7.9}$$

i.e. the radiation field is given by the black body equivalent at the appropriate temperature as one would expect directly from LTE arguments. Indeed, we may derive the same result by examining the complete re-distribution source function (equation (1.6.15)) obtained in the preceding section. There we had

$$S = (1-\epsilon)\bar{J} + \epsilon B_\nu(T), \tag{1.7.10}$$

where

$$\begin{aligned}\epsilon &= \frac{C_{\mathrm{UL}}}{C_{\mathrm{UL}} + A_{\mathrm{UL}} + B_{\mathrm{UL}}B_\nu(T)} \\ &= \frac{C_{\mathrm{LU}}}{C_{\mathrm{LU}} + B_{\mathrm{LU}}B_\nu(T)}.\end{aligned} \tag{1.7.11}$$

In the limit as collisional processes completely dominate radiative processes, i.e.

$$\frac{A_{\mathrm{UL}} + B_{\mathrm{UL}}B_\nu(T)}{C_{\mathrm{UL}}} \to 0, \quad \frac{B_{\mathrm{LU}}B_\nu(T)}{C_{\mathrm{LU}}} \to 0, \tag{1.7.12}$$

we find $\epsilon \to 1$, i.e. $S \to B_\nu(T)$. If we now write equation (1.7.10) as

$$\left(\frac{S}{B_\nu(T)} - \epsilon\right) B_\nu(T) = (1-\epsilon)\bar{J}, \tag{1.7.13}$$

and note that $\epsilon \neq 1$ (since $B_{\mathrm{LU}}B_\nu(T) \neq 0$ in equation (1.7.11)), we see that $\bar{J} \to B_\nu(T)$ as $S \to B_\nu(T)$. Again, as expected, the radiation field is just the appropriate Planck function.

However, the concern we must feel for this LTE approximation is immediately apparent. The limit as $\epsilon \to 1$ is purely a mathematical abstraction. In particular, as the density decreases with increasing height in the stellar atmosphere, the number of collisions (or C_{LU} and C_{UL}) must also decrease, whereas A_{UL} (and $B_{\mathrm{UL}}, B_{\mathrm{LU}}$) remain constant. Thus, for physically realistic atmospheres, ϵ decreases with height. The collisional source term $\epsilon B_\nu(T)$ in equation (1.7.10) therefore decreases relative to the corresponding $(1-\epsilon)\bar{J}$ term which, for LTE, is just $(1-\epsilon)B_\nu(T)$, i.e. the radiative processes begin to dominate the collisional processes. This obviously invalidates the concept of detailed balancing whereby the two equations (1.7.4) and (1.7.5) may be obtained separately. Indeed, $\epsilon \sim 10^{-4}$ is not uncommon for spectral line radiation of astrophysical interest (in fact, ϵ can be smaller than 10^{-10} for some very strong lines) and, since LTE is only valid in the limit $\epsilon \to 1$, we immediately see that the assumption of LTE can lead to errors of at least several orders of magnitude.

The point of concern for those astrophysicists who analyse data using LTE radiative transfer is obviously the cut-off or threshold in physical and frequency space where the errors resulting from LTE approximations begin to swamp observational uncertainties.

(b) Non-LTE

Non-LTE, as its name implies, relates to the removal of the assumption of LTE. Such a negative name, however, can cover a multitude of other approximations and assumptions and, for this reason, very little explicit mention is made in this monograph of the concept of non-LTE. It should be stressed, however, that all the analyses presented here incorporate non-LTE, in one way or another, in the determination of the photon distribution.

Nevertheless, one can attach a significant meaning to non-LTE and, in particular, to the word 'non-local'. Photons may be created by collisional excitation (corresponding to the term $\epsilon B_\nu(T)$ in S), thence immediately destroyed (by $-\epsilon\bar{J}$) in the same local region. The energy for this creation would come from the thermal field of free electrons, and the destruction process would restore this energy to the thermal field. However, photons can traverse enormous distances (between emission and absorption) relative to the corresponding mean free path of particle collisions. Thus, energy may be taken out of the thermal field of free electrons at one point (A, say) in the stellar atmosphere by collisional excitation, the resulting photon thence being destroyed by collisional de-excitation at another point B some not inconsiderable distance from A. This is a non-local effect which quickly and efficiently transports thermal energy (via photons) from point A to point B. By contrast, the direct transport of thermal energy, via particle collisions, is purely local.

We can also discuss the concept of LTE and non-LTE for particles whilst still retaining non-LTE for photons. Indeed, equation (1.7.5), for example, used in determining the non-LTE source function S is just an LTE expression for atom–electron processes. The removal of this LTE assumption for particles is only necessary when the individual particles traverse large distances between their collisions - large in the sense that such macroscopic quantities as density and temperature vary appreciably over these mean free collisional path lengths.

1.8 The qualitative behaviour of the source function

Substitution of the results of the preceding sections into the time-independent radiative transfer equation (1.2.8) yields

$$(\Omega \cdot \nabla)I = -N_L B_{LU}\phi(\Delta\nu)\left[1 - \frac{N_U B_{UL}}{N_L B_{LU}}\right][(1-\epsilon)\bar{J} + \epsilon B_\nu(T)]. \quad (1.8.1)$$

If one makes the substitution of the ratio N_U/N_L from equation (1.3.13), the above equation clearly becomes highly non-linear. Even without this somewhat serious complication, equation (1.8.1) poses severe difficulties in its attempted solution because of its integro-differential nature (note that $\bar{J}$ involves integrals over both frequency and angle space).

However, the basic qualitative behaviour of the solution may be obtained by making suitable approximations and simplifications. To do this, we divide the problem into two parts. The first part, consisting of the determination of the qualitative nature of the source function S, constitutes this section. The following section deals with the evaluation of the corresponding emergent intensity.

We therefore begin here by reverting to the original expression for the complete re-distribution source function given by equation (1.6.2), viz.

$$S = \frac{\dfrac{N_U A_{UL}}{N_L B_{LU}}}{1 - \dfrac{N_U B_{UL}}{N_L B_{LU}}}, \tag{1.8.2}$$

which, using equations (1.6.5) and (1.6.6), becomes

$$S = \frac{\dfrac{2h\nu_0^3}{c^2} \cdot \dfrac{g_L N_U}{g_U N_L}}{1 - \dfrac{g_L N_U}{g_U N_L}}. \tag{1.8.3}$$

We now consider a small volume element sufficiently deep within the stellar atmosphere that effectively zero radiation can escape from it to the 'surface', i.e. there is just as much radiation coming into the volume element as there is leaving it. (This is known as radiative equilibrium.) Under these conditions, one may apply the LTE approximation

$$\frac{N_U}{N_L} = \frac{N_U^*}{N_L^*} = \frac{g_U}{g_L}\, e^{-h\nu_0/kT}, \tag{1.8.4}$$

so that equation (1.8.3) becomes

$$S = \frac{2h\nu_0^3/c^2}{e^{h\nu_0/kT} - 1} = B_\nu(T). \tag{1.8.5}$$

Consequently, deep within the stellar atmosphere, $S = B_\nu(T)$. This result will apply to all regions which do not lose radiation to the 'surface'. As we move to higher regions, however, photons may more easily reach the 'surface' and therefore escape. For every such photon that escapes, there is one less photon in the stellar atmosphere capable of being absorbed, and it is this absorption process which results in an atom having a bound electron in the excited or upper state. Thus, as the escape of photons increases, the total number of excited atoms must correspondingly decrease, i.e. N_U decreases. However, if the atoms themselves do not escape from the atmosphere and, more particularly, if there are no mass motions present in the gas (this is the simple model we wish to consider in this chapter), then we must have

$$N_U + N_L = N = N_U^* + N_L^*, \tag{1.8.6}$$

where the total number density N of atoms is a constant at each point in the atmosphere. Thus, if N_U decreases, N_L increases (note that N_U^* and N_L^* remain unchanged since, in LTE, photons do not escape), i.e.

$$N_U < N_U^* \quad \text{and} \quad N_L > N_L^*, \tag{1.8.7}$$

so that

$$\frac{N_U}{N_L} < \frac{N_U^*}{N_L^*}. \tag{1.8.8}$$

We now write

$$\frac{N_U}{N_L} = \gamma \frac{N_U^*}{N_L^*} = \frac{\gamma g_U}{g_L} e^{-h\nu_0/kT}, \tag{1.8.9}$$

where $\gamma < 1$. Equation (1.8.3) then becomes

$$S = \frac{\dfrac{2h\nu_0^3}{c^2}\gamma}{e^{h\nu_0/kT} - \gamma} = \frac{\gamma B_\nu(T)}{1 + \dfrac{1-\gamma}{e^{h\nu_0/kT} - 1}}. \tag{1.8.10}$$

Since $h\nu_0/kT > 0$, i.e. $e^{h\nu_0/kT} > 1$, equation (1.8.10) yields the important result $S < B_\nu(T)$. Clearly, then, as one moves to higher and higher regions in the stellar atmosphere, more and more photons can escape so that N_U decreases, and N_L increases, even further. This results in a smaller and smaller γ which, in turn, leads to a smaller and smaller ratio $S/B_\nu(T)$. (We refer to the decrease in the ratio N_U/N_U^* as a de-population of the excited state (relative to the corresponding LTE value) and the increase in the ratio N_L/N_L^* as an over-population of the lower state.) The qualitative behaviour of S is sketched in figure 1.11 for the simple case in which the temperature, thence Planck function, decreases toward the 'surface'.

Fig.1.11. The Planck function and source function as functions of depth.

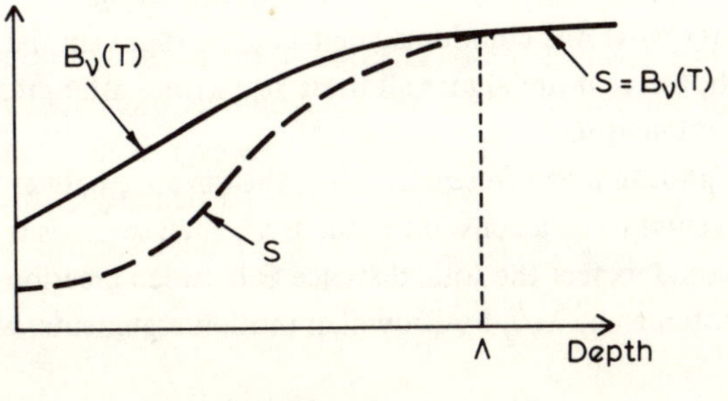

It must now be clear that not only does the distribution of atoms considerably affect the radiation field, but that the radiation field, and its partial loss to the stellar atmosphere, has a correspondingly considerable effect on the atom distribution. This, then, is the basis for radiation–matter interaction. We should also mention that energy is lost to the system once photons escape. Thus, one might expect the time dependence of the problem (which we have neglected) to be of paramount importance. Actually, it is important - stellar evolution takes this into account. Indeed, phenomena changing over time intervals of order of minutes have come under increasing scrutiny because of their obvious importance to our understanding of the structure of both stellar atmospheres and interiors. Here, however, we assume the loss of energy due to this photon escape to be compensated for by energy input from below the atmosphere - we are not concerned here with the nature of this energy input.

Having shown that the source function S has a smaller value than the corresponding Planck function, we now examine the depth Λ at which this departure of S from $B_\nu(T)$ first occurs (note figure 1.11). As we shall see, this is important in determining the qualitative behaviour of the emergent intensity of radiation. However, the above explanation of the behaviour of S is perhaps fractionally oversimplified because it does not take into account the important non-local nature of the problem.

We therefore proceed by recalling that ϵ, given by equation (1.6.19), viz.

$$\epsilon = \frac{C_{\mathrm{UL}}}{C_{\mathrm{UL}} + A_{\mathrm{UL}} + B_{\mathrm{UL}} B_\nu(T)}, \tag{1.8.11}$$

is the probability of a de-excitation event being collisional, in which case the photon is effectively destroyed. For ease in exposition, we take $\epsilon = 10^{-4}$ (a not atypical value). Thus, the probability of a de-excitation being collisional in this case is extremely small. Indeed, for every collisional excitation there would be 9999 radiative (spontaneous and/or stimulated) de-excitations. Therefore, once a photon has been created by a collisional excitation (equivalent to the $\epsilon B_\nu(T)$ term in S), it will be absorbed and emitted ($\bar{J}$ in S), on the average, 9999 times before the atom undergoes a collisional de-excitation with subsequent destruction ($-\epsilon\bar{J}$ in S) of the photon. This emitting and absorbing process (in between collisional excitation and de-excitation) is referred to as scattering.

Figure 1.12 illustrates what one could imagine to be the flight (in the plane of the paper) of the photon as it undergoes all these scatterings after creation at A and before destruction at B.

If we now take the photon mean free path λ (i.e. the distance between successive scattering events) to be a constant (and, as we shall see, this is not true in general), one would expect the total distance Θ between creation and destruction of that photon to be $\lambda\sqrt{n}$ by Brownian motion arguments, where

n is the total number of scatters (9999 in the above example) between creation and destruction. Note that we must have

$$n = \frac{A_{UL} + B_{UL}B_\nu(T)}{C_{UL}}, \tag{1.8.12}$$

$$n = \frac{1}{\epsilon} - 1. \tag{1.8.13}$$

Hence, for $\epsilon \ll 1$, we have

$$\Theta \approx \frac{\lambda}{\sqrt{\epsilon}}. \tag{1.8.14}$$

Let us now consider a point of creation (A_1, say) which is deeper than a distance Θ from the surface (see figure 1.13).

For every photon $(h\nu)_1$ created at A_1 (and subsequently destroyed at B_1), there will be a corresponding destruction of a photon $(h\nu)_2$. This destroyed photon $(h\nu)_2$ would have originally been created at some other point A_2 (say) so that the point A_1 corresponds to B_2. Thus, there will be a balance between photon creation and photon destruction at the one point, i.e. the source term

Fig.1.12. Creation of a photon at A with subsequent scatterings due to photon absorptions and emissions followed eventually by destruction of that photon at B.

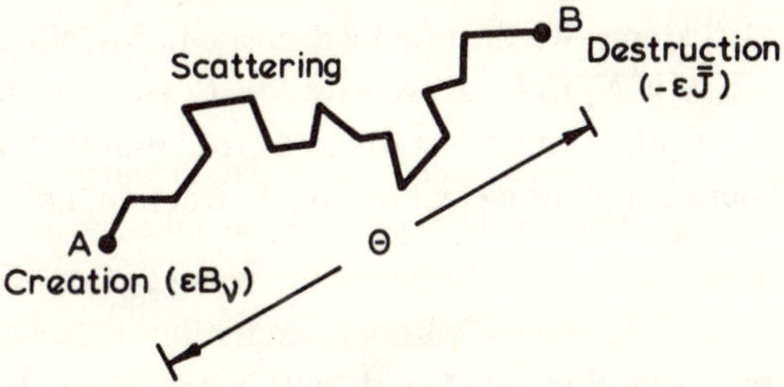

Fig.1.13. Photon creation at A_1 followed by destruction at B_1 depletes the photon energy at A_1. This energy is replaced by photon creation at A_2 followed by destruction at B_2 where B_2 is the same position at A_1.

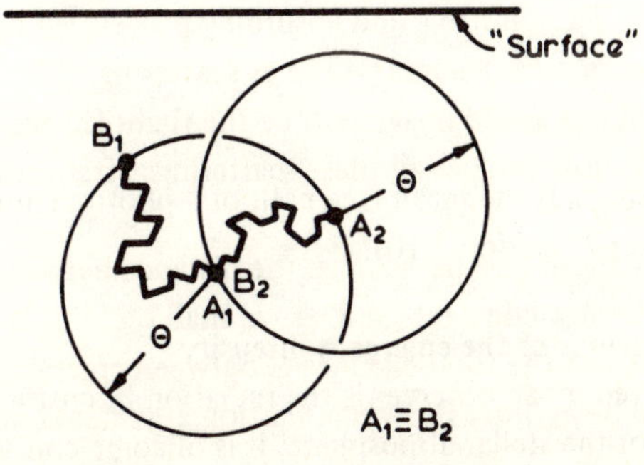

$\epsilon B_\nu(T)$ is exactly balanced by the corresponding sink term $-\epsilon\bar{J}$. This then leads to the LTE result $N_U = N_U^*$ and $N_L = N_L^*$, i.e. $S = \bar{J} = B_\nu(T)$. It should be emphasised, however, that this is still a non-local effect since creation and destruction of a particular photon do not take place at the source point.

If, on the other hand, the point A_2 is located within a distance Θ of the 'surface', then the $(h\nu)_2$ photon created at A_2 may possibly reach the 'surface' of the atmosphere, and thence escape, before being destroyed. If the $(h\nu)_2$ photon does escape from the 'surface' it is clearly incapable of being destroyed at A_1 and, consequently, the balance between photon creation and destruction at A_1 mentioned above does not occur. Thus, the thermal energy used to create the $(h\nu)_1$ photon at A_1 is not replaced by the thermal energy which would otherwise have been deposited by the destruction of the $(h\nu)_2$ photon. However, for every photon creation at A_2 with subsequent destruction at A_1 in a statistically steady state there would, on the average, be a photon creation at A_1 with subsequent destruction at A_2. Thus, any depletion of the thermal energy at A_1 would, through this feedback mechanism, subsequently yield a depletion of the thermal energy at A_2. The free electrons would not suffer this energy loss by themselves but, rather, would share the loss via collisions with the excited and de-excited atoms. Since the total internal and kinetic energy of an atom having a bound electron in the excited state is larger than that in the de-excited state, this energy loss would be accomplished by having a smaller number of excited atoms. We then find a decrease in N_U/N_U^* which, as before, leads to an increase in N_L/N_L^*, i.e. we have $S/B_\nu(T) < 1$, as shown in figure 1.11. Clearly, the depth at which S starts to depart from $B_\nu(T)$ corresponds to the distance Θ between photon creation and destruction, i.e.

$$\Lambda \sim \Theta \sim \frac{\lambda}{\sqrt{\epsilon}}. \tag{1.8.15}$$

We refer to Θ as the thermalisation path length [58] (a destroyed photon is sometimes referred to as being thermalised). However, photons do not travel equal distances between successive emission and absorption processes, i.e. $\lambda \neq$ constant (as we shall see in the following section). In particular, for example, one finds

$$\Lambda \sim \Theta \sim \frac{\lambda_0}{\epsilon} \tag{1.8.16}$$

for Doppler broadening, where λ_0 is the mean free path of a photon emitted at line centre $\Delta\nu = 0$ (see chapter 2, section 2.10).

1.9 The qualitative behaviour of the emergent intensity

Since the data received by an observer is the radiation intensity emergent from the 'surface' of the stellar atmosphere, it is of some considerable benefit to qualitatively examine the behaviour of the emergent intensity corresponding to various source functions. Again, we limit ourselves in this preliminary discussion to the simple model involving 2-level atoms, complete re-distribution and a statistically steady state with no mass motions.

Clearly, the behaviour of the emergent radiation intensity must depend upon the ability of the photons to escape from the 'surface' of the stellar atmosphere and this ability to escape depends, in turn, on the probability $\phi(\Delta\nu)$ of the photons being absorbed. Mathematically, once the source function S is known, one need only solve the transfer equation

$$(\boldsymbol{\Omega}\cdot\nabla)I = -\kappa(I - S), \tag{1.9.1}$$

given the appropriate boundary conditions, where $\phi(\Delta\nu)$ now only appears in κ, viz.

$$\kappa = \frac{h\nu}{4\pi} N_{\mathrm{L}} B_{\mathrm{LU}} \phi(\Delta\nu) \left[1 - \frac{N_{\mathrm{U}} B_{\mathrm{UL}}}{N_{\mathrm{L}} B_{\mathrm{LU}}}\right]. \tag{1.9.2}$$

We have seen in the preceding section that

$$\frac{N_{\mathrm{U}}}{N_{\mathrm{L}}} \leqslant \frac{N_{\mathrm{U}}^*}{N_{\mathrm{L}}^*} = \frac{g_{\mathrm{U}}}{g_{\mathrm{L}}} \, \mathrm{e}^{-h\nu_0/kT}, \tag{1.9.3}$$

i.e.

$$\frac{N_{\mathrm{U}} B_{\mathrm{UL}}}{N_{\mathrm{L}} B_{\mathrm{LU}}} \leqslant \mathrm{e}^{-h\nu_0/kT} \ll 1. \tag{1.9.4}$$

Consequently, for most spectral lines of astrophysical interest, we may neglect the effect of stimulated emissions in the specification of the opacity κ. We therefore re-write equation (1.9.2) as

$$\kappa = \kappa_0 N_{\mathrm{L}} \phi(\Delta\nu), \tag{1.9.5}$$

where $\kappa_0 = h\nu B_{\mathrm{LU}}/4\pi$.

We now take a right-handed orthogonal cartesian coordinate system such that

$$\nabla \equiv \left(\frac{\partial}{\partial x}, \frac{\partial}{\partial y}, \frac{\partial}{\partial z}\right), \tag{1.9.6}$$

and

$$\boldsymbol{\Omega} \equiv (\sin\theta\cos\phi, \sin\theta\sin\phi, \cos\theta), \tag{1.9.7}$$

where θ and ϕ are the usual heliocentric and azimuthal angles respectively, and allow the z direction to be radially outward from the centre of the star. We further simplify equation (1.9.1) by considering only 1-dimensional geometry, as shown in figure 1.14.

Equation (1.9.1) then becomes

$$\mu \frac{\partial I}{\partial z} = -\kappa(I - S), \tag{1.9.8}$$

where it is common notation to write

$$\mu = \cos\theta. \tag{1.9.9}$$

Equation (1.9.8) is a first order differential equation, its solution being uniquely determined given a single boundary condition for all $\mu \in [-1, 1]$ and $\Delta\nu \in (-\infty, \infty)$. If, for example, the medium under consideration is a slab (as in figure 1.14), the boundary condition would stipulate the incident radiation at the top and bottom edges of the slab, i.e. we would initially specify $I(0, \Delta\nu, \mu < 0)$ and $I(z_{\text{SLAB}}, \Delta\nu, \mu > 0)$.

In particular, if the slab has no incident radiation, the solution to equation (1.9.8) for $\mu > 0$ has the form

$$I(z, \Delta\nu, \mu) = \int_{z_{\text{SLAB}}}^{z} \kappa(z', \Delta\nu) S(z') \exp\left[\frac{1}{\mu}\int_{z}^{z'} \kappa(z'', \Delta\nu)\, \mathrm{d}z''\right] \frac{\mathrm{d}z'}{\mu}. \tag{1.9.10}$$

The emergent intensity $I(0, \Delta\nu, \mu > 0)$ can then be obtained simply by putting $z = 0$ in the above equation. However, a major simplification to equation (1.9.10) may be made by a change of independent variable z to an optical path length scale.

1.9.1 Optical depth [84]

We define the optical depth $\tau(z)$ by

$$\tau(z) = -\kappa_0 \int_0^z N_{\text{L}}(z')\, \mathrm{d}z', \tag{1.9.11}$$

i.e.

$$\mathrm{d}\tau = -\kappa_0 N_{\text{L}}\, \mathrm{d}z, \tag{1.9.12}$$

which, using (1.9.5), becomes

$$\mathrm{d}\tau = -\frac{\kappa(z, \Delta\nu)}{\phi(\Delta\nu)}\, \mathrm{d}z. \tag{1.9.13}$$

As we shall see in a later chapter, the concept of optical depth has no unique meaning for model stellar atmospheres which are not 1-dimensional.

Fig.1.14. 1-dimensional slab geometry.

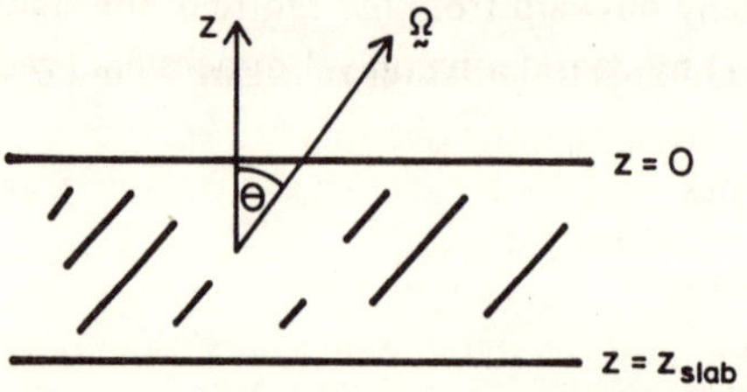

The above change of variable then enables equations (1.9.8) to be written as

$$\mu \frac{\partial I}{\partial \tau} = \phi(\Delta\nu)(I - S), \tag{1.9.14}$$

which, for $\mu > 0$, has the solution

$$I(\tau, \Delta\nu, \mu) = -\int_{\tau_{\mathrm{SLAB}}}^{\tau} S(\tau')\mathrm{e}^{-\phi(\Delta\nu)(\tau'-\tau)/\mu}\,\phi(\Delta\nu)\,\frac{\mathrm{d}\tau'}{\mu}. \tag{1.9.15}$$

If we now extend this slab (having zero radiation incident at both top and bottom edges) so that it has infinite thickness, i.e. $z_{\mathrm{SLAB}} \to -\infty$ or $\tau_{\mathrm{SLAB}} \to \infty$, the emergent radiation intensity may be written as

$$I(0, \Delta\nu, \mu) = \int_0^{\infty} S(\tau')\mathrm{e}^{-\phi(\Delta\nu)\tau'/\mu}\,\phi(\Delta\nu)\,\frac{\mathrm{d}\tau'}{\mu}. \tag{1.9.16}$$

Recall from equations (1.4.14) and (1.4.15), for example, viz.

$$\phi(\Delta\nu) = \frac{1}{\Delta\nu_{\mathrm{D}}\sqrt{\pi}}\,\mathrm{e}^{-(\Delta\nu/\Delta\nu_{\mathrm{D}})^2}, \quad \Delta\nu_{\mathrm{D}} = \frac{\nu}{c}\left[\frac{2kT}{m_{\mathrm{A}}}\right]^{1/2}, \tag{1.9.17}$$

that the absorption probability is depth dependent through its dependence on the temperature T. This depth dependence, however, offers a complication in the discussion of the qualitative nature of $I(0, \Delta\nu, \mu)$. Thus, for ease in exposition, we assume $\phi(\Delta\nu)$ to be depth independent (this approximation, although physically unrealistic, does not change the qualitative validity of the following discussion) and define the new 'frequency-dependent optical depth' $\eta(\tau, \Delta\nu)$ such that

$$\eta(\tau, \Delta\nu) = \phi(\Delta\nu)\tau, \tag{1.9.18}$$

i.e.

$$\mathrm{d}\eta = \phi(\Delta\nu)\,\mathrm{d}\tau. \tag{1.9.19}$$

Equation (1.9.16) then becomes

$$I(0, \Delta\nu, \mu) = \int_0^{\infty} S\left(\frac{\eta}{\phi(\Delta\nu)}\right)\mathrm{e}^{-\eta/\mu}\,\frac{\mathrm{d}\eta}{\mu}. \tag{1.9.20}$$

We are now in a position to qualitatively examine the behaviour of the emergent intensity. We consider two separate cases for μ.

(a) $\mu = 1$

The case $\mu = 1$ (i.e. $\theta = 0$) corresponds to radiation emerging from the atmosphere normal to the surface. Clearly, then, the quantity

$$I(0, \Delta\nu, 1) = \int_0^{\infty} S\left(\frac{\eta}{\phi(\Delta\nu)}\right)\mathrm{e}^{-\eta}\,\mathrm{d}\eta, \tag{1.9.21}$$

is effectively a spatial summation over the source function S weighted by the factor $e^{-\eta}$. Figure 1.15 illustrates the qualitative behaviour of $S(\eta/\phi(\Delta\nu))$ and $e^{-\eta}$ as functions of τ for some fixed value of $\Delta\nu$.

The dotted curve in figure 1.15 represents the qualitative nature of the product $Se^{-\eta}$ appearing as the integrand in equation (1.9.21). This product peaks at the depth $\eta = \eta^*$, say, and it is not unreasonable to suggest that the emergent radiation intensity $I(0, \Delta\nu, 1)$, which is simply the area under the dotted curve in figure 1.15, is controlled more by the source function at η^* than at any other single depth. Certainly, it is clear that the major contribution to $I(0, \Delta\nu, 1)$ comes from depths in the neighbourhood of η^*. In particular, with the exponentially decreasing factor $e^{-\eta}$ ($e^{-1} \sim \frac{1}{3}$ and $e^{-2} \sim \frac{1}{10}$, for example) rapidly curtailing the contributions of S to $I(0, \Delta\nu, 1)$ for $\eta > 1$, we find $\eta^* \lesssim 1$. This then leads to the important result

$$I(0, \Delta\nu, 1) \approx S\left(\frac{\eta^*}{\phi(\Delta\nu)}\right). \tag{1.9.22}$$

If we now define a τ^* and z^* corresponding to η^*, equations (1.9.11) and (1.9.18) yield

$$\tau^* = \frac{\eta^*}{\phi(\Delta\nu)}, \quad z^* = \frac{-\eta^*}{\kappa_0 N_L \phi(\Delta\nu)}, \tag{1.9.23}$$

where, in keeping with the approximation of a depth-independent $\phi(\Delta\nu)$, we have taken N_L in equation (1.9.11) to be depth-independent also. Consequently, if the bulk of the emergent radiation intensity arises from depths $\eta^* \lesssim 1$, then the corresponding physical depths z^* depend on the frequency $\Delta\nu$ through equation (1.9.23), viz.

$$z^* \lesssim \frac{1}{\kappa N_L \phi(\Delta\nu)}. \tag{1.9.24}$$

Since $\phi(\Delta\nu)$ decreases for increasing $|\Delta\nu|$ (see equation (1.9.17), for example), then $|z^*|$ correspondingly increases. We now choose two frequencies $\Delta\nu_1$ and $\Delta\nu_2$, where

$$|\Delta\nu_1| < |\Delta\nu_2|, \tag{1.9.25}$$

Fig.1.15. The integrand of equation (1.9.21) as a function of depth.

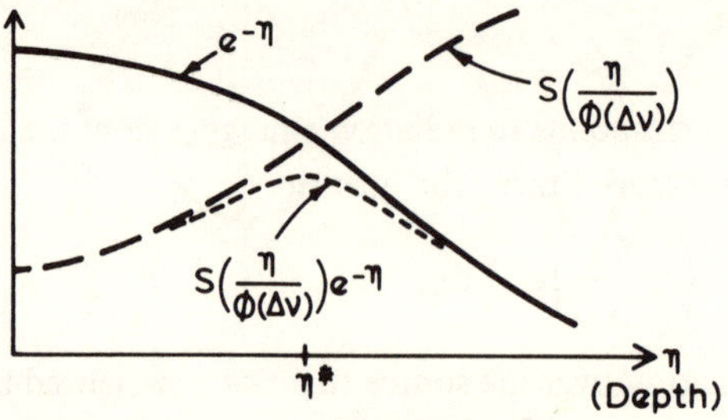

i.e.

$$\phi(\Delta\nu_1) > \phi(\Delta\nu_2), \tag{1.9.26}$$

and write

$$z_i^* = z^*(\Delta\nu_i), \tag{1.9.27}$$

so that

$$|z_1^*| < |z_2^*|. \tag{1.9.28}$$

If we now consider a source function of the form illustrated in figure 1.11, i.e. $S(z)$ increases with increasing physical depth $|z|$ in the stellar atmosphere, then we must have

$$S(z_1^*) < S(z_2^*). \tag{1.9.29}$$

Equation (1.9.22) then yields

$$I(0, \Delta\nu_1, 1) \approx S(z_1^*) < S(z_2^*) \approx I(0, \Delta\nu_2, 1). \tag{1.9.30}$$

Consequently, for the case in which the source function increases with increasing physical depth, the emergent radiation intensity increases for frequencies increasing away from line centre, as shown in figure 1.16.

Clearly, if $\phi(\Delta\nu)$ is symmetric about line centre $\Delta\nu = 0$, i.e.

$$\phi(-\Delta\nu) = \phi(\Delta\nu), \tag{1.9.31}$$

then equation (1.9.17) immediately gives

$$I(0, -\Delta\nu, 1) = I(0, \Delta\nu, 1), \tag{1.9.32}$$

i.e. the emergent spectral line profile will also be symmetric about line centre. As we shall see, however, this does not apply if mass motions occur in the stellar atmosphere.

Before proceeding, one important point should be reiterated. Equation (1.9.22) stipulates that the emergent radiation intensity $I(0, \Delta\nu, 1)$ is approximately equal to $S(z^* \propto \eta^*/\phi(\Delta\nu))$ where $\eta^* \lesssim 1$. Consequently, as we move to larger values of $|\Delta\nu|$, corresponding to smaller values of $\phi(\Delta\nu)$, radiation

Fig.1.16. (*a*) The absorption profile as a function of frequency from line centre; (*b*) The corresponding emergent intensity.

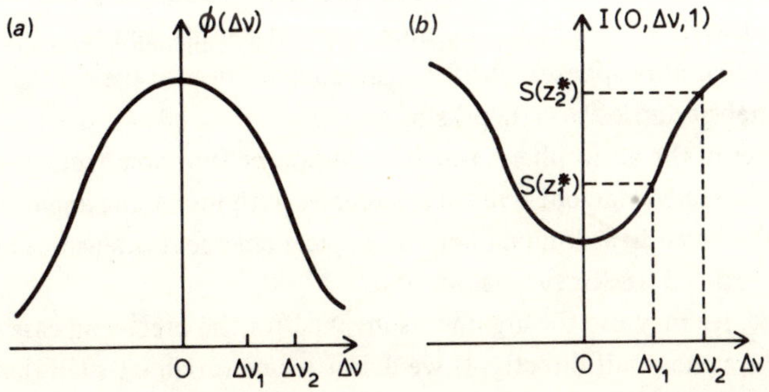

emergent from the 'surface' comes from deeper regions in the stellar atmosphere. This therefore leads to the perhaps unusual concept in which, when we observe at different frequencies within a spectral line, we are, in fact, looking at different depths in the stellar atmosphere.

In this and the preceding section, we have divided the problem of the solution of the equation of radiative transfer into two parts: the first involves the evaluation only of the source function S, the second can then be analysed once S is known. It is, however, illustrative to summarise physically the qualitative nature of the emergent intensity from a composite point of view.

Radiation at frequency $\Delta\nu_1$ experiences an absorption probability $\phi(\Delta\nu_1)$ where, for $|\Delta\nu_1| < |\Delta\nu_2|$, $\phi(\Delta\nu_1) > \phi(\Delta\nu_2)$. Consequently, photons emitted at frequency $\Delta\nu_2$ can travel greater distances before being absorbed. Photons emitted at frequency $\Delta\nu_2$ deep within the stellar atmosphere ($z \sim z_2^*$) therefore have a greater chance of escaping from the 'surface', and thence of being absorbed, than do $\Delta\nu_1$ photons emitted in the same region. The $\Delta\nu_1$ photons, however, may escape from regions ($z \sim z_1^*$) closer to the 'surface' where they do not need to travel such large distances. But as one moves closer and closer to the 'surface', more and more photons of *all* frequencies escape, and this results in a decrease in the radiation field, viz. the number of photons being emitted and absorbed decreases. Therefore, there is a larger pool of photons deeper in the stellar atmosphere and, since $\Delta\nu_2$ photons may escape more easily from this larger and deeper pool, the radiation emerging from the 'surface' is greater at the $\Delta\nu_2$ frequency.

(b) $\mu \neq 1$

Let us now consider radiation emergent from our model 1-dimensional atmosphere at some non-zero angle θ to the normal, i.e. at $\mu \neq 1$. First, consider a photon emitted at frequency $\Delta\nu$. Let the mean free path of this photon be λ so that the main contribution to the emergent radiation intensity at this frequency in the normal direction ($\mu = 1$) comes from a depth $z = z^* = -\lambda$, as shown in figure 1.17.

However, if we now consider radiation emergent at an angle $\theta \neq 0$, the photon still travels a distance λ (assuming constant density and temperature in the model stellar atmosphere), but the main contribution to the emergent radiation intensity arises from depths of order $z = -\lambda \cos\theta = -\mu\lambda$, i.e. from regions higher in the atmosphere. Further, the source function decreases with height and thus the emergent intensity decreases with increasing angle. This is generally referred to as limb-darkening. Typical emergent intensities are sketched in figure 1.18 for two values of μ.

Of course, we may use the arguments invoked for the preceding case $\mu = 1$ to obtain the same result directly. If we define a new variable ζ such that

$$\zeta = \frac{\eta}{\mu} = \frac{\phi(\Delta\nu)\tau}{\mu}, \qquad (1.9.33)$$

equation (1.9.20) becomes

$$I(0, \Delta\nu, \mu) = \int_0^\infty S\left(\frac{\mu\zeta}{\phi(\Delta\nu)}\right) e^{-\zeta} \, d\zeta. \qquad (1.9.34)$$

Clearly, the exponential decrease of the weighting factor $e^{-\zeta}$ indicates that the main contribution to $I(0, \Delta\nu, \mu)$ arises from depths ζ^* where $\zeta^* \lesssim 1$, i.e. we have

$$I(0, \Delta\nu, \mu) \approx S\left(\frac{\mu\zeta^*}{\phi(\Delta\nu)}\right). \qquad (1.9.35)$$

If we now consider the two angles θ_1 and θ_2 with $\theta_1 < \theta_2$, i.e. $\mu_1 > \mu_2$, then, for a fixed frequency $\Delta\nu$ and a source function monotonically increasing with increasing depth,

$$S\left(\frac{\mu_1\zeta^*}{\phi\Delta\nu}\right) > S\left(\frac{\mu_2\zeta^*}{\phi\Delta\nu}\right). \qquad (1.9.36)$$

Fig.1.17. (*a*) The depth from which a photon emerges normal to the surface of the medium; (*b*) The corresponding depth for non-zero angle of emergence.

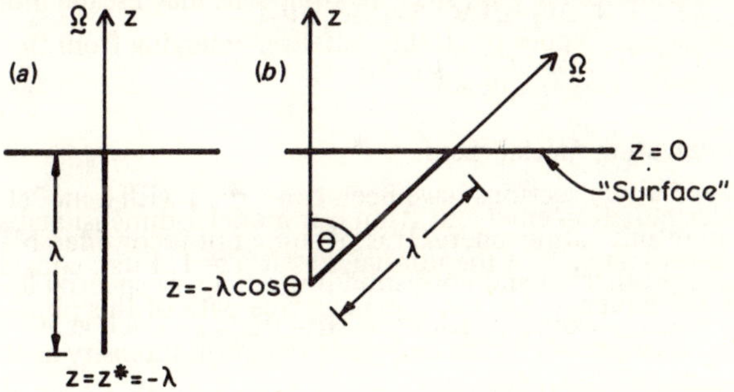

Fig.1.18. The emergent intensity as a function of frequency for two different angles.

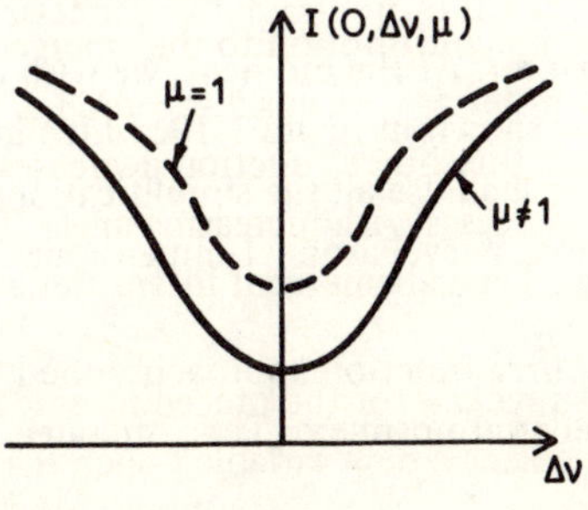

This then leads to the limb-darkening result:

$$I(0, \Delta\nu, \mu_1) > I(0, \Delta\nu, \mu_2), \tag{1.9.37}$$

where $\mu_1 > \mu_2$.

1.9.2 Sample calculation

A quick calculation for a simple source function will convince the reader of the applicability of the foregoing discussion. If we take a monotonically increasing source function of the form

$$S(\tau) = a + b\tau, \tag{1.9.38}$$

where a and b are positive constants, equation (1.9.16) yields

$$I(0, \Delta\nu, \mu) = \int_0^\infty (a + b\tau)\mathrm{e}^{-\phi(\Delta\nu)\tau/\mu}\, \phi(\Delta\nu)\, \frac{\mathrm{d}\tau}{\mu}. \tag{1.9.39}$$

As before, we take $\phi(\Delta\nu)$ to be depth independent so that the integral in equation (1.9.39) may be directly evaluated:

$$I(0, \Delta\nu, \mu) = a + \frac{b\mu}{\phi(\Delta\nu)}. \tag{1.9.40}$$

Clearly, since $a, b > 0$, we immediately have the two basic results already obtained, viz.

$$I(0, \Delta\nu_1, \mu) < I(0, \Delta\nu_2, \mu)\ \forall\ |\Delta\nu_1| < |\Delta\nu_2|, \tag{1.9.41}$$

$$I(0, \Delta\nu, \mu_1) > I(0, \Delta\nu, \mu_2)\ \forall\ \mu_1 > \mu_2. \tag{1.9.42}$$

1.10 Simple interesting special cases

The preceding two sections have been concerned with a model 1-dimensional semi-infinite atmosphere. It is of some not inconsiderable interest to consider the source function and corresponding emergent spectral line intensity for both stellar chromospheric-type structures and slab geometries.

1.10.1 Stellar chromospheres

A stellar chromosphere is defined to be that region of a stellar atmosphere which exhibits an increase in (electron) temperature with increasing height, this temperature increase being a result of non-radiative processes. We shall discuss this more fully in a later chapter. At the moment, we wish only to consider a temperature distribution, as shown in figure 1.19, which includes this temperature increase. In so doing, we shall use all the simplifications made in section 1.8, i.e. complete re-distribution, 2-level atom, 1-dimensional, semi-infinite atmosphere [60].

We know from section 1.8 that the source function approaches the Planck function sufficiently deep within the stellar atmosphere. This, of course, applies no matter what the temperature distribution exhibits. However, as one moves to higher regions in the stellar atmosphere, photons are more able to escape and, following the arguments presented in section 1.8, the source function must now take values smaller than the corresponding Planck function at that physical depth. The point Λ at which this departure of the source function from the Planck function first occurs depends upon the ability of the photons to escape, and this, as we have seen, depends on the opacity. The opacity, in turn, is determined by the number of absorbing atoms N_L and the Einstein rate coefficient $B_{LU}(\propto A_{UL})$. Consequently, the value of Λ depends not only on the atom being considered but on the actual radiative transition within that atom. Thus, one could have two quite distinct situations, as shown in figure 1.20, corresponding to two different spectral transitions.

The Planck function $B_\nu(T)$ shown in figure 1.20 increases with increasing height due to the temperature increase (see figure 1.19). If $N_L^{(1)}$ and $B_{LU}^{(1)}$ for one particular spectral line are such that $\Lambda = \Lambda^{(1)}$ occurs in regions deeper than the temperature minimum $z(T_{\rm min})$, the corresponding source function

Fig.1.19. The qualitative temperature distribution in the neighbourhood of a stellar photospheric–chromospheric temperature minimum.

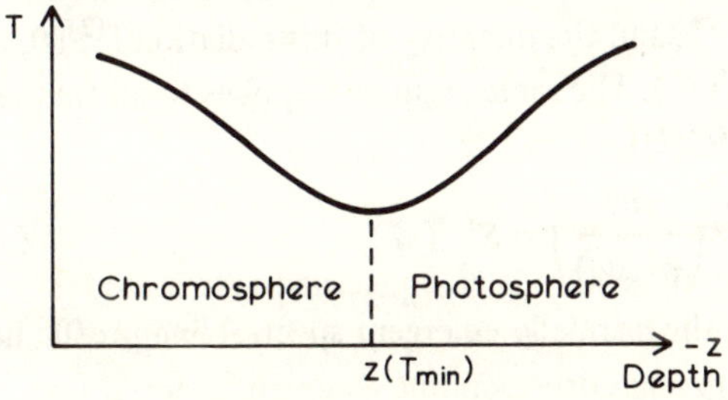

Fig.1.20. Source function distributions as functions of depth for two cases corresponding to photon thermalisation either below the temperature minimum or above.

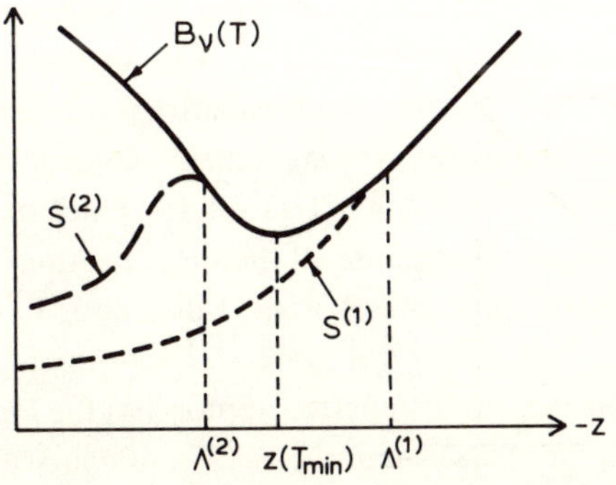

$S^{(1)}$ monotonically decreases towards the 'surface' as shown. If, on the other hand, $B_{\mathrm{LU}}^{(2)}$, for example, is sufficiently large that $\Lambda = \Lambda^{(2)}$ occurs above the temperature minimum, the corresponding source function $S^{(2)}$ would first increase with increasing height in the atmosphere (i.e. $S^{(2)} = B_\nu(T)$ for all depths greater than $\Lambda^{(2)}$) before decreasing towards the 'surface'.

The source function $S^{(1)}$ is qualitatively no different to the source functions considered in the previous sections and one simply observes an emergent radiation intensity having the same structure as shown in figure 1.18. However, $S^{(2)}$ is markedly different, as is the corresponding emergent intensity $I^{(2)}(0, \Delta\nu, \mu)$. To examine the qualitative behaviour of $I^{(2)}(0, \Delta\nu, \mu)$, we consider the four frequencies $\Delta\nu_1$, $\Delta\nu_2$, $\Delta\nu_3$ and $\Delta\nu_4$, where

$$|\Delta\nu_1| < |\Delta\nu_2| < |\Delta\nu_3| < |\Delta\nu_4|, \tag{1.10.1}$$

i.e.

$$\phi(\Delta\nu_1) > \phi(\Delta\nu_2) > \phi(\Delta\nu_3) > \phi(\Delta\nu_4). \tag{1.10.2}$$

Let $z_i^* = z^*(\Delta\nu_i)$ be the point at which the main contribution of $S(\eta/\phi(\Delta\nu))\mathrm{e}^{-\eta}$ to the emergent radiation intensity arises, i.e.

$$|z_1^*| < |z_2^*| < |z_3^*| < |z_4^*|, \tag{1.10.3}$$

as shown in figure 1.21.

Clearly, then, the radiation emerging from the 'surface' at frequency $\Delta\nu_1$ comes from regions near $z = z_1^*$, and the quantity of this radiation $I^{(2)}(0, \Delta\nu_1, 1)$ would be of order $S_1^{(2)} = S^{(2)}(z_1^*)$. The same argument applies to all frequencies, i.e. we have (see equation (1.9.22))

$$I^{(2)}(0, \Delta\nu_i, 1) \approx S^{(2)}\left(\frac{\eta^*}{\phi(\Delta\nu_i)}\right) = S^{(2)}(z_i^*), \tag{1.10.4}$$

where $\eta^* \lesssim 1$. Consequently, the normally emergent spectral line profile has the form shown in figure 1.22.

Fig.1.21. The source function as a function of depth when photon thermalisation occurs above the temperature minimum.

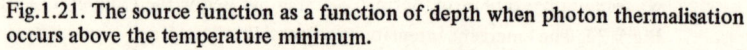

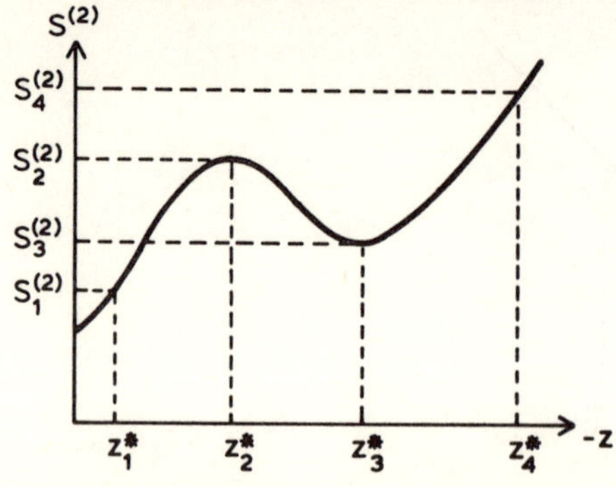

Only half the spectral line profile has been sketched. As already indicated, if the absorption profile $\phi(\Delta\nu)$ is symmetric about $\Delta\nu = 0$, then so is the emergent line profile. It should be noted that, although the depth variation of $\phi(\Delta\nu)$ due to the chromospheric variation in temperature does affect the emergent line intensity considerably, the argument detailed above in illustrating the qualitative structure of the emergent profile remains unaltered.

A comparison of the curves in figures 1.21 and 1.22 shows that the emergent radiation intensity is simply a mapping in frequency (and angle) space of the source function in physical space. In particular, equation (1.9.34), viz.

$$I(0, \Delta\nu, \mu) = \int_0^\infty S\left(\frac{\mu\zeta}{\phi(\Delta\nu)}\right) e^{-\zeta}\, d\zeta, \tag{1.10.5}$$

defines this mapping precisely although the qualitative behaviour of the emergent intensity can always be found from the more approximate mapping (equation (1.9.35))

$$I(0, \Delta\nu, \mu) \approx S\left(\frac{\mu\zeta^*}{\phi(\Delta\nu)}\right), \tag{1.10.6}$$

where $\zeta^* \lesssim 1$, without recourse to the somewhat involved numerical calculations which are otherwise necessary.

(The emergent line profile illustrated in figure 1.22 is qualitatively similar to those observed for the Ca H and K spectral lines and is quite important in the study of stellar atmospheres.)

1.10.2 Slab geometry

Radiative transfer in slab geometry is of some interest because of its applicability to shell-type stars and laboratory plasmas. Here, we shall consider two different slabs, the first being optically thin, the second optically thick.

Fig.1.22. The emergent intensity as a function of frequency for the source function shown in Figure 1.21. Note the local maximum in the wings of the spectral line.

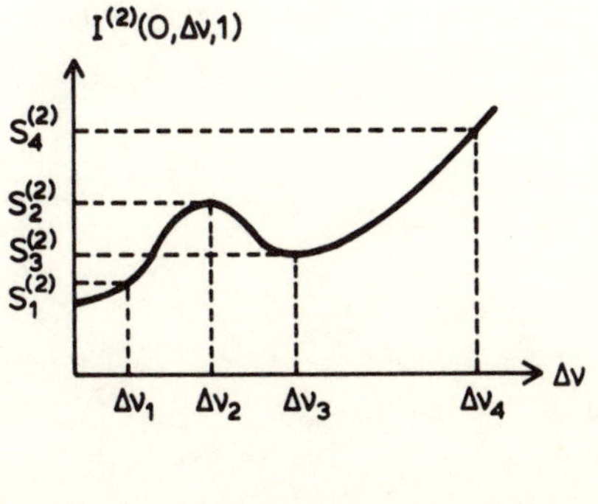

(a) Optically thin slab

An optically thin slab is one in which photons, even those emitted at line centre $\Delta\nu = 0$ where the probability of absorption is greatest, undergo no more than about one or two scatters before escaping. If the slab has zero radiation incident on its upper and lower surfaces, and we take the limit as the thickness of the slab goes to zero so that there is effectively no absorption of photons, then any radiation within the slab arises from a collisional excitation followed by radiative de-excitation, i.e. the source function $S \approx \epsilon B_\nu(T)$. As the thickness of the slab increases, however, the radiation created by these collisional excitations might not all escape but may be absorbed and thence re-emitted. Consequently, the source function will not only effectively include the $\epsilon B_\nu(T)$ term as before, but the $(1-\epsilon)\bar{J}$ term also. However, if the number of such absorptions, after creation, is limited to one (say) before the photon escapes from the slab (i.e. we have an optically thin slab), then $S \approx 2\epsilon B_\nu(T)$. The important point is that radiation can easily escape from *any* position in an optically thin slab, and thus the source function is approximately constant across the slab.

Having thus gained some knowledge of the qualitative behaviour of S, we are now in a position to evaluate the corresponding emergent intensity. Equation (1.9.15), viz.

$$I(0, \Delta\nu, \mu) = \int_0^{\tau_{\mathrm{SLAB}}} S(\tau)\,\mathrm{e}^{-\phi(\Delta\nu)\tau/\mu}\,\phi(\Delta\nu)\,\frac{\mathrm{d}\tau}{\mu}, \qquad (1.10.7)$$

immediately yields

$$I(0, \Delta\nu, \mu) = \bar{S}\left[1 - \mathrm{e}^{-\frac{\phi(\Delta\nu)}{\mu}\tau_{\mathrm{SLAB}}}\right], \qquad (1.10.8)$$

where we have approximated the source function across the slab by $\bar{S}$. The emergent spectral line profile given by equation (1.10.8) is shown in figure 1.23.

Such profiles are referred to as emission profiles. Indeed, if we take the limit $\tau_{\mathrm{SLAB}} \to 0$, equation (1.10.8) becomes

$$I(0, \Delta\nu, \mu) \to \bar{S}\,\frac{\phi(\Delta\nu)}{\mu}\,\tau_{\mathrm{SLAB}}, \qquad (1.10.9)$$

Fig.1.23. The emergent intensity as a function of frequency for an optically thin slab.

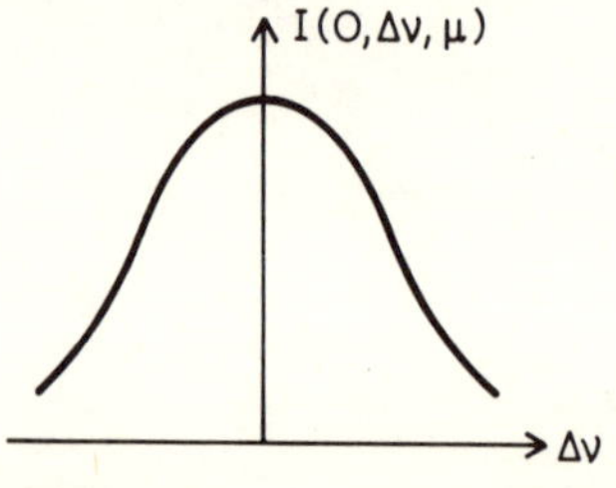

i.e.

$$I(0, \Delta\nu, \mu) \propto \frac{\phi(\Delta\nu)}{\mu} . \tag{1.10.10}$$

The frequency behaviour of the emergent line profile, then, just mimics the frequency dependence of the absorption profile which, we recall, has itself been arbitrarily put equal to the emissions probabilities $j(\Delta\nu)$ and $\psi(\Delta\nu)$ through the assumption of complete re-distribution.

Again, note that we have ignored the fact that the energy loss of the slab due to photon escape should make the problem entirely time dependent. In this sense, the above and following physical models are somewhat idealistic.

(b) Optically thick slab

An optically thick slab is one in which a created photon undergoes more than two absorptions (followed by re-emission) before escaping from the slab. Again, we consider the simplified case in which no radiation is incident on the surfaces of the slab.

If the slab exhibits symmetrical properties (i.e. temperature, densities, etc.) about its centre, the source function should also be symmetric. Further, we know from section 1.8 that the source function increases away from the surface basically because photons experience more difficulty in escaping as one moves further into the slab. It is not surprising, therefore, that the source function takes the qualitative form symmetric about the centre of the slab, as shown in figure 1.24.

We now examine the emergent radiation intensity at the three frequencies $\Delta\nu_1$, $\Delta\nu_2$ and $\Delta\nu_3$ satisfying

$$|\Delta\nu_1| < |\Delta\nu_2| < |\Delta\nu_3|, \tag{1.10.11}$$

i.e.

$$\phi(\Delta\nu_1) > \phi(\Delta\nu_2) > \phi(\Delta\nu_3), \tag{1.10.12}$$

so that

$$|z_1^*| < |z_2^*| < |z_3^*|, \tag{1.10.13}$$

Fig.1.24. The source function in an optically thick slab.

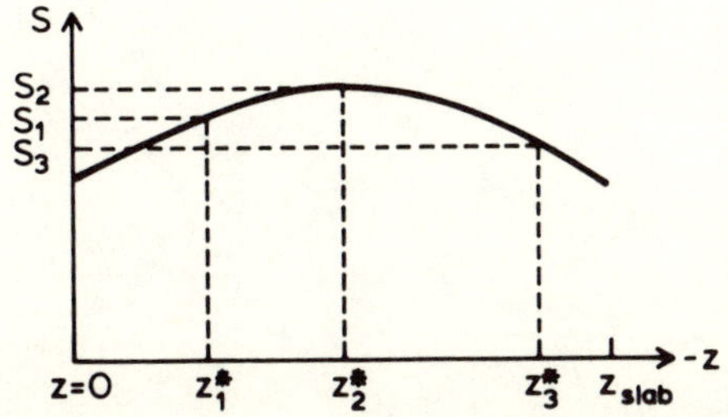

where $z_i^* = z^*(\Delta\nu_i)$ is the physical depth which contributes most to the emergent intensity (for $\mu = 1$) at the $\Delta\nu_i$ frequency. Thus, using equation (1.9.22), viz.

$$I(0, \Delta\nu_i, 1) \approx S\left(\frac{\eta^*}{\phi(\Delta\nu_i)}\right) = S_i, \tag{1.10.14}$$

where $\eta^* \lesssim 1$, we have the mapping shown in figure 1.25 where, again, we only sketch half the spectral line profile.

Clearly, the form this emergent line profile takes is quite different to that for the optically thin case. If, however, the thickness z_{SLAB} of the slab was sufficiently small that $|z_0^*| > z_{\mathrm{SLAB}}/2$, where $z_0^* = z^*(\Delta\nu = 0)$, the above mapping would yield $S_1 > S_2 > S_3$, i.e. there would be no central dip in the emergent intensity. This situation is, of course, nothing more than the optically thin geometry discussed in (a) above. The inequality $z_{\mathrm{SLAB}} \lessgtr 2|z_0^*|$, where z_0^* is effectively the mean free path of a photon at line centre, therefore, distinguishes between the optically thin and optically thick cases.

Fig.1.25. The emergent intensity as a function of frequency for an optically thick slab. Note the central absorption feature in the emission profile.

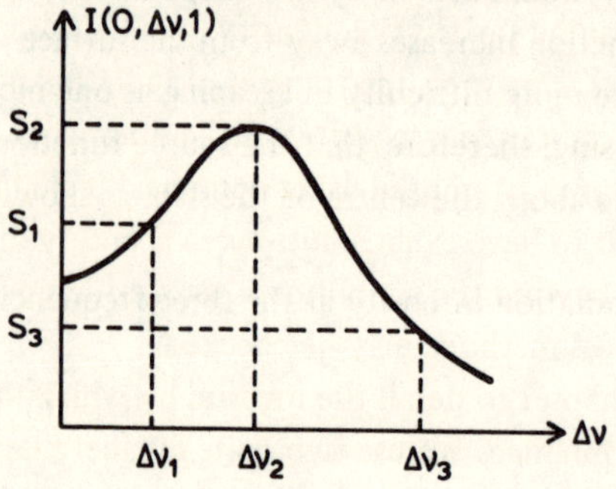

2

Exact solutions to the transfer equation

We are interested here in the solution of integro-differential equations. Although the source function derived in the preceding chapter is a linear functional of the unknown to be determined (i.e. intensity I), the opacity κ exhibits a non-linear dependence on I. Thus, even for the simplest of spectral line transfer problems, an exact closed form analytical solution is not available. Indeed, as we shall see, exact solutions may only be obtained at present for the physically idealistic situation in which all parameters defining the gas (i.e. density, temperature, etc.) are independent of position in the gas. However, the transfer of radiation is a non-local effect and this 'non-localness' can create serious difficulties when the solution to radiative transfer problems is attempted using numerical techniques. It is therefore of considerable benefit to have exact solutions to a variety of simple problems (indeed, simplified beyond the simplifications considered in chapter 1) against which the numerical methods may be checked.

It is the purpose of the present chapter to detail the method by which these exact benchmark solutions may be obtained. We use two quite distinct methods – the singular eigenfunction method [28] and the Fourier transform technique [27] – to solve the one problem of time-independent radiative transfer in a semi-infinite, 1-dimensional medium having constant temperature and density with zero macroscopic velocity fields.

Those not mathematically inclined or motivated might wish to bypass this chapter.

2.1 SINGULAR EIGENFUNCTION TECHNIQUE

Singular eigenfunctions were first introduced into neutron transport theory by Case [26] and, because of the similarity in the defining equations for neutron transport and photon transfer, may be readily generalised to the latter case. However, before we proceed directly to the method of solution, it may be of benefit to discuss some of the more important mathematical results which we shall need.

2.1 Mathematical preliminaries

2.1.1 Plemelj (or Sokhotski) formulae [89]

Consider the function

$$\Phi(z) = \frac{1}{2\pi \mathrm{i}} \int_L \frac{\phi(t)\,\mathrm{d}t}{t-z}, \quad z \notin L, \tag{2.1.1}$$

where L is some contour (closed or open) in the complex z plane, as shown in figure 2.1.

We wish to determine the value of $\Phi(z)$ as $z \to t_0$ from both sides of the contour where $t_0 \in L$.

We define D^+ as the region to the left of the contour (as we look along the direction of the contour) and D^- as the region to the right. For a closed contour, then, D^+ would be the region inside the contour and D^- outside. The above definition of D^+ and D^-, however, loses its meaning when we approach the end-points of an open contour. For the present, we shall only be concerned with the evaluation of $\Phi(z)$ as $z \to t_0$ for t_0 not an endpoint of L.

Take some point $z \in D^+$ or $z \in D^-$ with $z \notin L$. Then define

$$\Phi^{\pm}(t_0) = \lim_{z \to t_0} \Phi(z), \quad z \in D^{\pm}. \tag{2.1.2}$$

We also define the Cauchy principal value integral

$$\Phi(t_0) = \frac{1}{2\pi \mathrm{i}} P \int_L \frac{\phi(t)\,\mathrm{d}t}{t-t_0} \tag{2.1.3}$$

$$= \lim_{\epsilon \to 0} \frac{1}{2\pi \mathrm{i}} \int_{L-L_\epsilon} \frac{\phi(t)\,\mathrm{d}t}{t-t_0}, \tag{2.1.4}$$

where L_ϵ is that part of L contained within a circle of radius ϵ centre t_0, and where $L - L_\epsilon$ is that part of L contained without (see Figure 2.2).

We assume $\phi(t)$ analytic at t_0 and continuous everywhere else. Hence, by analytic continuation, we have $\phi(t)$ analytic in a small neighbourhood of t_0. This analytic continuation, of course, can be extended to the whole complex plane. We now consider a point $z \in D^+$ as shown in figure 2.2. However, we cannot allow $z \to t_0$ directly because $\Phi(z)$ defined by equation (2.1.1) is not valid when $z = t_0 \in L$. We must therefore indent the contour L by removing that

Fig.2.1. The contour L in the complex z plane.

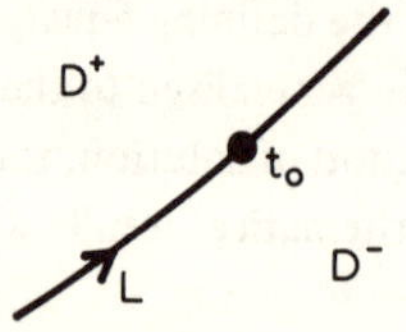

section of L in the neighbourhood of t_0 and connecting the resulting two parts of L by that part of the circle C_1. We take C_1 in D^-, not D^+, to allow $z \to t_0$. We then have

$$\Phi(z) = \frac{1}{2\pi i}\int_{L-L_\epsilon} \frac{\phi(t)\,dt}{t-z} + \frac{1}{2\pi i}\int_{L_\epsilon} \frac{\phi(t)\,dt}{t-z}, \tag{2.1.5}$$

where $z \notin L$. Since $\phi(t)/t-z$ is analytic within the contour bounded by L_ϵ and C_1, Cauchy's theorem [25] yields

$$-\int_{L_\epsilon} \frac{\phi(t)\,dt}{t-z} + \int_{C_1} \frac{\phi(t)\,dt}{t-z} = 0. \tag{2.1.6}$$

Hence

$$\Phi(z) = \frac{1}{2\pi i}\int_{L-L_\epsilon} \frac{\phi(t)\,dt}{t-z} + \frac{1}{2\pi i}\int_{C_1} \frac{\phi(t)\,dt}{t-z}. \tag{2.1.7}$$

We now wish to evaluate $\Phi^+(t_0)$, i.e. we wish to determine $\Phi(z)$ given by equation (2.1.7) as $z \to t_0$. The contour integral over C_1 applies for any ϵ. However, as $\epsilon \to 0$ the contour L_ϵ becomes a straight line, i.e. C_1 becomes a semicircle, and the integral may be easily evaluated. We therefore have, using the definitions given by equations (2.1.3) and (2.1.4),

$$\begin{aligned}\Phi^+(t_0) &= \lim_{\substack{z\to t_0\\ \epsilon\to 0}} \Phi(z), \quad z \in D^+ \\ &= \Phi(t_0) + \tfrac{1}{2}\phi(t_0).\end{aligned} \tag{2.1.8}$$

Similarly, by considering a contour $C_2 \in D^+$ with $z \to t_0$ for $z \in D^-$, as shown in figure 2.3, we find

$$\Phi^-(t_0) = \Phi(t_0) - \tfrac{1}{2}\phi(t_0), \tag{2.1.9}$$

where the negative term in equation (2.1.9) arises from the negative orientation of C_2.

Equations (2.1.8) and (2.1.9) are referred to as either the Plemelj or Sokhotski formulae. They may be equivalently written as

$$\Phi^+(t_0) - \Phi^-(t_0) = \phi(t_0), \tag{2.1.10}$$

Fig.2.2. The contour L with C_1 in D^-.

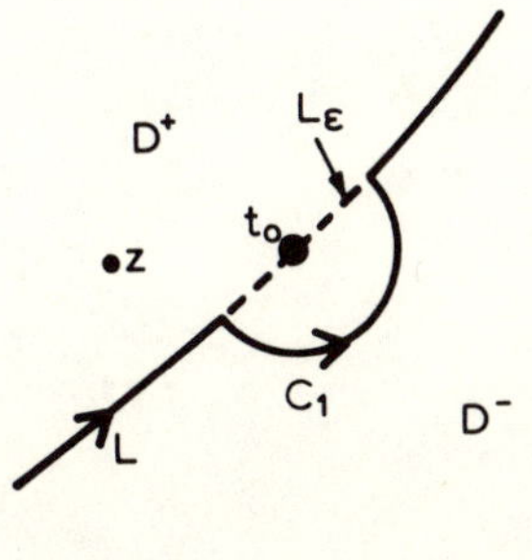

$$\Phi^+(t_0)+\Phi^-(t_0)=\frac{1}{\pi \mathrm{i}}P\int_L \frac{\phi(t)\,\mathrm{d}t}{t-t_0}. \tag{2.1.11}$$

In particular, we see that equation (2.1.10) represents a discontinuity of $\Phi(z)$ at the contour, i.e. the difference between the values of $\Phi(z)$ as we approach t_0 from the left and from the right is non-zero. We refer to the contour L in this case as a cut in the complex z plane. It is not difficult to show that $\Phi^{\pm}(z)$ are analytic functions of z for all $z \in D^{\pm}$.

In the above analysis we assumed $\phi(t)$ to be analytic at t_0. This is, in general, sufficient for problems of interest in radiative transfer theory. It can be shown, however, that the above Plemelj formulae may be obtained under the weaker condition that $\phi(t)$ be continuous on L and satisfy the Lipschitz condition, viz.

$$|\phi(t_2)-\phi(t_1)|<K|t_2-t_1|^{\alpha}, \tag{2.1.12}$$

where $t_1, t_2 \in L$ (t_1 in the neighbourhood of t_2) and where K and α are constants $0<\alpha\leqslant 1$.

The endpoints of the contour L cause considerably more difficulty. A somewhat complicated analysis eventually shows that we require $\phi(t)$ to behave at the endpoints t_A and t_B of L such that, for example,

$$|\phi(t)|\leqslant \frac{K_A^{(1)}}{|t-t_A|^{\alpha_A}}, \quad 0\leqslant\alpha_A<1, \tag{2.1.13}$$

where, again, $K_A^{(1)}$ and α_A are constants. This condition then implies that

$$|\Phi(t)|<\frac{K_A^{(2)}}{|t-t_A|^{\beta_A}}, \tag{2.1.14}$$

where $K_A^{(2)}$ and β_A are constants with $0\leqslant\alpha_A\leqslant\beta_A<1$. The reader is referred to Muskhelishvili [89] for further details. We have simply mentioned the above conditions on $\phi(t)$ for completeness in the statement of the mathematical result. They do not limit the scope of radiative transfer problems we may attempt to solve using singular eigenfunctions.

Fig.2.3. The contour L with C_2 in D^+.

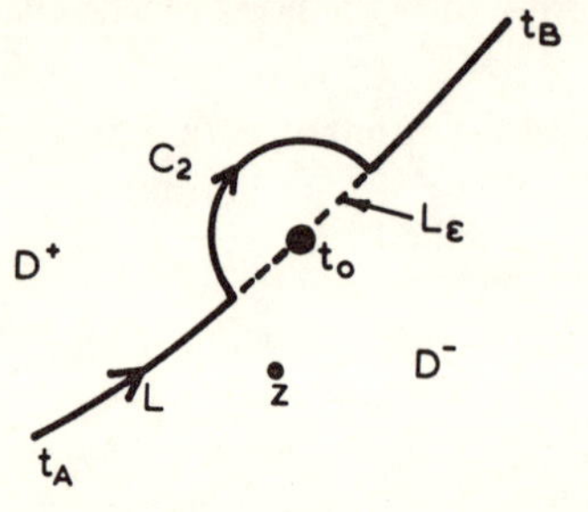

2.1.2 The homogeneous and inhomogeneous Riemann-Hilbert problems[89]

In the following sections, we shall encounter an equation of the form

$$Y^+(z) = g(z)Y^-(z) + f(z), \quad z \in L, \tag{2.1.15}$$

where the unknown $Y(z)$ to be determined is analytic for all $z \notin L$, and where $g(z)$ and $f(z)$ are known functions of the complex variable z. This is known as the inhomogeneous Riemann-Hilbert problem.

The solution of equation (2.1.15) may be effected by first writing

$$X^+(z) = g(z)X^-(z), \quad z \in L, \tag{2.1.16}$$

where $X^+(z)$, $X^-(z)$ and $X(z)$ are all non-zero. Substitution of this expression into equation (2.1.15) yields

$$\frac{Y^+(z)}{X^+(z)} - \frac{Y^-(z)}{X^-(z)} = \frac{f(z)}{X^+(z)}, \quad z \in L, \tag{2.1.17}$$

which, from the Plemelj formulae equation (2.1.10), with

$$\Phi(t_0) \equiv \frac{Y(t_0)}{X(t_0)}, \tag{2.1.18}$$

and

$$\phi(t_0) \equiv \frac{f(t_0)}{X^+(t_0)}, \tag{2.1.19}$$

gives

$$\frac{Y(z)}{X(z)} = \frac{1}{2\pi \mathrm{i}} \int_L \frac{f(t)}{X^+(t)} \cdot \frac{\mathrm{d}t}{t-z} + \mathscr{P}_n(z), \tag{2.1.20}$$

for $z \notin L$, where $\mathscr{P}_n(z)$ is an arbitrary polynomial in z of degree n. Note that any polynomial is continuous across the cut L because of its analyticity, i.e.

$$\mathscr{P}_n^+(z) - \mathscr{P}_n^-(z) \equiv 0. \tag{2.1.21}$$

Thus, equation (2.1.20) does indeed satisfy the first of the Plemelj formulae. The arbitrariness of $\mathscr{P}_n(z)$ is removed usually by examining the behaviour of $Y(z)$ and $X(z)$ at infinity and/or the endpoints of the cut L and thence constructing $\mathscr{P}_n(z)$ such that this prescribed behaviour is satisfied by equation (2.1.20).

Once $X(z)$ is known, the quantity $Y(z)$ to be determined follows immediately from equation (2.1.20). The problem has therefore been reduced to the solution of equation (2.1.16) for $X(z)$ - this is referred to as the homogeneous Riemann-Hilbert problem.

If $g(z) \neq 0$ on L, we may take the natural logarithm of both sides of equation (2.1.16) such that

$$\ln X^+(z) - \ln X^-(z) = \ln g(z), \quad z \in L. \tag{2.1.22}$$

This last equation is again of the form of the first of the Plemelj formulae equation (2.1.11) with

$$\Phi(t_0) = \ln X(t_0), \tag{2.1.23}$$

and

$$\phi(t_0) = \ln g(t_0), \tag{2.1.24}$$

but care must be exercised. Although, for example, $X^+(z)$ is analytic for all $z \in D^+$, the term $\ln X^+(z)$ is not, in general, because of its inherent multi-valued nature. This introduces a complication when L is a closed contour since, then, the value of ln may increase by $2\pi in$, where n is an integer, as one completes a cirtuit of L. In this case, the basic restriction of the Plemelj formulae (i.e. that $\Phi(z)$ be an analytic function of z) would invalidate their use. However, in this chapter, we shall always be interested in cases where L is an open contour so that the above complication does not arise.

We therefore have

$$\ln X(z) = \frac{1}{2\pi i} \int_L \frac{\ln g(t)\, dt}{t - z} + \mathscr{P}_m(z), \quad z \notin L, \tag{2.1.25}$$

here $\mathscr{P}_m(z)$ is another arbitrary polynomial in z of degree m. By writing

$$\Gamma(z) = \frac{1}{2\pi i} \int_L \frac{\ln g(t)\, dt}{t - z}, \tag{2.1.26}$$

the above equation for $X(z)$ yields;

$$X(z) = X_0(z) e^{\Gamma(z)}, \tag{2.1.27}$$

where $X_0(z) = e^{\mathscr{P}_m(z)}$ is analytic for all z in the complex plane. Rather than determine $\mathscr{P}_m(z)$, one evaluates $X_0(z)$ by ensuring that $X(z)$ given by equation (2.1.27) behaves at infinity and the endpoints of L such that the use of the Plemelj formulae is not invalidated.

If the endpoints of the contour L are given by t_A and t_B, as shown in figure 2.3, we may write, for example,

$$\Gamma(z) = \frac{\ln g(z)}{2\pi i} \int_L \frac{dt}{t - z} + \frac{1}{2\pi i} \int_L \frac{\ln g(t) - \ln g(z)}{t - z} \cdot dt$$
$$= \frac{1}{2\pi i} \ln g(z) \ln|z - t_A| + q_A(z), \tag{2.1.28}$$

where $q_A(z)$, incorporating the second integral term on the RHS of equation (2.1.28) and the evaluation of the first term at t_B, is clearly a bounded function of z as $z \to t_A$.

We therefore find that, in the neighbourhood of the endpoint t_A, the behaviour of $X(z)$ is dominated by

$$X(z) \sim X_0(z) |z - t_A|^{-\frac{1}{2\pi i} \ln g(z)}. \tag{2.1.29}$$

However, we have already noted that near the endpoint t_A, we must have

$$|X(t)| < \frac{K_A^{(2)}}{|t - t_A|^{\beta_A}}, \quad 0 \leqslant \beta_A < 1, \tag{2.1.30}$$

for the Plemelj formulae to be valid (see equation (2.1.14)). Thus, $X_0(z)$ must behave in the neighbourhood of t_A as

$$X_0(z) \sim |z - t_A|^{k_A}, \tag{2.1.31}$$

i.e.

$$X(z) \sim |z - t_A|^{k_A - \frac{1}{2\pi i} \ln g(z)}, \tag{2.1.32}$$

so that, if inequality (2.1.30) is to be satisfied, we must have the constant k_A satisfying

$$0 \leqslant \mathcal{R}l\left[\frac{1}{2\pi i} \ln g(z)\right] - k_A < 1, \tag{2.1.33}$$

as $z \to t_A$. In particular, $X_0(z)$ must be analytic across the cut L since it reflects the polynomial $\mathscr{P}_m(z)$, i.e. we must have

$$X_0^+(z) - X_0^-(z) = 0, \quad z \in L, \tag{2.1.34}$$

and thus k_A must be either zero or an integer.

A similar procedure for the upper endpoint t_B yields the final required form for $X_0(z)$, viz.

$$X_0(z) \sim |z - t_A|^{k_A} |z - t_B|^{k_B}, \tag{2.1.35}$$

where k_B is an integer satisfying

$$0 \leqslant -\mathcal{R}l\left[\frac{1}{2\pi i} \ln g(z)\right] - k_B < 1. \tag{2.1.36}$$

Since $g(z)$ is given, k_A and k_B may be determined from inequalities (2.1.33) and (2.1.36), which then enables $X(z)$ to be evaluated. Note that $X(z) \sim e^{\Gamma(z)}$ and is therefore non-vanishing in the complex z plane – thus, the division by $X^+(z)$ which yielded equation (2.1.17) is valid. Knowledge of $X(z)$ then completes the solution for $Y(z)$ given by equation (2.1.20). This solution can be shown to be unique under the conditions already mentioned.

2.1.3 Green's functions [28]

As we shall see in the following section, we will wish to solve the integro-differential equation

$$\xi \frac{\partial}{\partial \tau} \Phi(\tau, \xi) = \Phi(\tau, \xi) - \int_{R_\xi} \psi(\xi') \Phi(\tau, \xi') \, d\xi' - q(\tau, \xi), \tag{2.1.37}$$

where $\psi(\xi)$ and $q(\tau, \xi)$ are given, and where R_ξ specifies the range over which the ξ integration is to be taken.

The above problem is well posed when the appropriate boundary conditions are given. We shall eventually assume zero incident radiation on the surface of the model stellar atmosphere. This is equivalent (as we shall see) to putting

$$\Phi(0, \xi) \equiv 0 \quad \text{for all } \xi < 0. \tag{2.1.38}$$

For clarity in exposition, however, we preliminarily allow non-zero incident radiation on the surface of the atmosphere so that

$$\Phi(0, \xi) = \Phi_{\text{inc}}(\xi), \quad \xi < 0. \tag{2.1.39}$$

Equation (2.1.37) is inhomogeneous for $q(\tau, \xi) \neq 0$; its solution is most readily obtained by considering the corresponding homogeneous equation. To do this, we define the Green's function $G(\tau^*, \xi^* \to \tau, \xi)$ to be a solution of the integro-differential equation

$$\begin{aligned} \xi \frac{\partial G}{\partial \tau}(\tau^*, \xi^* \to \tau, \xi) = G(\tau^*, \xi^* \to \tau, \xi) & \\ - \int_{R_\xi} \psi(\xi') G(\tau^*, \xi^* \to \tau, \xi')\, d\xi' & \\ - \delta(\tau - \tau^*) \delta(\xi - \xi^*), & \end{aligned} \tag{2.1.40}$$

where we have replaced the inhomogeneous term $q(\tau, \xi)$ by the delta functions $\delta(\tau - \tau^*)\delta(\xi - \xi^*)$. The physical picture associated with this mathematical manipulation is quite straightforward. When, for example, $\tau \neq \tau^*$, then $\delta(\tau - \tau^*) = 0$. Similarly, $\delta(\xi - \xi^*) = 0$ for all $\xi \neq \xi^*$. Thus, the inhomogeneous term $\delta(\tau - \tau^*)\delta(\xi - \xi^*)$ in equation (2.1.40) may be thought of as a point source at $\tau = \tau^*$ in τ space and at $\xi = \xi^*$ in ξ space. The Green's function $G(\tau^*, \xi^* \to \tau, \xi)$ then represents the manifestation of this point source at other τ and ξ locations.

Given this interpretation of $G(\tau^*, \xi^* \to \tau, \xi)$, it is not surprising that the manifestation of the point source at $\tau = \tau^*$ must decrease for points $\tau \neq \tau^*$ as the distance $|\tau - \tau^*|$ increases, i.e. we must have

$$G(\tau^*, \xi^* \to \tau, \xi) \to 0, \tag{2.1.41}$$

as $|\tau - \tau^*| \to \infty$.

If we now multiply equation (2.1.40) by $q(\tau^*, \xi^*)$ and integrate over all τ^* and ξ^* space, we easily find

$$\Phi(\tau, \xi) = \int_{R_\tau} d\tau^* \int_{R_\xi} d\xi^*\, q(\tau^*, \xi^*) G(\tau^*, \xi^* \to \tau, \xi). \tag{2.1.42}$$

Thus, $\Phi(\tau, \xi)$ is really only a super-position of all the above-mentioned sources throughout R_τ and R_ξ. It is important to note that the use of Green's functions in the manner discussed above is, in general, only valid for linear systems.

We have not, as yet, taken the boundary conditions into account. There are several methods by which this may be done. One technique is to define volume and surface Green's functions as follows.

Let the *volume Green's function* $G_V(\tau^*, \xi^* \to \tau, \xi)$ be the solution of equation (2.1.37) with $q(\tau, \xi)$ replaced by $\delta(\tau - \tau^*)\delta(\xi - \xi^*)$ and with

$$G_V(\tau^*, \xi^* \to 0, \xi) = 0 \quad \text{for all } \xi < 0. \tag{2.1.43}$$

In contrast, we define the *surface Green's function* $G_S(0, \xi^* \to \tau, \xi)$ to be the solution of equation (2.1.37) with $q(\tau, \xi)$ put identically equal to zero but with

$$G_S(0, \xi^* \to 0, \xi) = \delta(\xi - \xi^*), \quad \xi < 0. \tag{2.1.44}$$

The required solution to equation (2.1.37) then consists of two parts

$$\Phi(\tau, \xi) = \int_0^\infty d\tau^* \int_{R_\xi} d\xi^*\, q(\tau^*, \xi^*) G_V(\tau^*, \xi^* \to \tau, \xi) + \int_{R_\xi(1/2) \equiv \xi < 0} d\xi^*\, \Phi_{inc}(\xi^*) G_S(0, \xi^* \to \tau, \xi), \tag{2.1.45}$$

where $R_\xi(\frac{1}{2}) \equiv \xi < 0$ represents the half-space in ξ^* at the boundary $\tau^* = 0$ over which the ξ^* integration must take place. Furthermore, we shall only be considering semi-infinite model stellar atmospheres so that $R_\tau \equiv [0, \infty)$.

The first term on the RHS of equation (2.1.45) represents the contribution to $\Phi(\tau, \xi)$ due to sources within the atmosphere. The second term represents the corresponding contribution due to the incident radiation – here, we treat the incident radiation as a source of photons placed *at* the surface. Again, since we have a linear system, the above addition of these two terms in equation (2.1.45) is valid. One should check that $\Phi(\tau, \xi)$ given by equation (2.1.45) is indeed the required solution. Direct substitution of $\Phi(\tau, \xi)$ into equation (2.1.37) yields

$$\begin{aligned}
&\int_0^\infty d\tau^* \int_{R_\xi} d\xi^*\, q(\tau^*, \xi^*) \left\{ \xi \frac{\partial G_V}{\partial \tau} (\tau^*, \xi^* \to \tau, \xi) \right. \\
&\quad \left. - G_V(\tau^*, \xi^* \to \tau, \xi) + \int_{R_\xi} \Psi(\xi') G_V(\tau^*, \xi^* \to \tau, \xi')\, d\xi' \right\} \\
&\quad + \int_{R_\xi(1/2)} d\xi^*\, \Phi_{inc}(\xi^*) \left\{ \xi \frac{\partial G_S}{\partial \tau} (0, \xi^* \to \tau, \xi) \right. \\
&\quad \left. - G_S(0, \xi^* \to \tau, \xi) + \int_{R_\xi} \Psi(\xi') G_S(0, \xi^* \to \tau, \xi')\, d\xi' \right\} \\
&\quad + q(\tau, \xi) = 0.
\end{aligned} \tag{2.1.46}$$

By our definitions of G_V and G_S, the first term in braces on the LHS of equation (2.1.46) is just $-\delta(\tau - \tau^*)\delta(\xi - \xi^*)$, whilst the second term is just zero. We therefore have

$$-\int_0^\infty d\tau^* \int_{R_\xi} d\xi^*\, q(\tau^*, \xi^*) \delta(\tau - \tau^*) \delta(\xi - \xi^*) + q(\tau, \xi) = 0, \tag{2.1.47}$$

which, of course, holds true. This therefore proves that $\Phi(\tau, \xi)$ given by equation (2.1.45) is a solution to the integro-differential equation (2.1.37).

Further, we see immediately from equation (2.1.45) for $\Phi(\tau, \xi)$, together with equations (2.1.43) and (2.1.44), that

$$\Phi(0, \xi) = \int_{R_\xi(1/2) \equiv \xi < 0} d\xi^* \, \Phi_{\text{inc}}(\xi^*)\delta(\xi - \xi^*)$$
$$= \Phi_{\text{inc}}(\xi), \tag{2.1.48}$$

for all $\xi < 0$ as required.

Since $\Phi(\tau, \xi)$ satisfies both the defining equation (2.1.37) and the pertinent boundary conditions, it is indeed the required *unique* solution.

The relationship between G_V and G_S

Clearly, the solution $\Phi(\tau, \xi)$ is obtained once $G_V(\tau^*, \xi^* \to \tau, \xi)$ and $G_S(0, \xi^* \to \tau, \xi)$ are determined. However, there exists a somewhat simple relationship between these two Green's functions, which then reduces the problem to the evaluation of just G_V.

We begin by writing equation (2.1.37) satisfied by $\Phi_k(\tau, \xi)$, i.e.

$$\xi \frac{\partial \Phi_k(\tau, \xi)}{\partial \tau} = \Phi_k(\tau, \xi) - \int_{R_\xi} \Psi(\xi')\Phi_k(\tau, \xi') \, d\xi' - q_k(\tau, \xi), \tag{2.1.49}$$

together with a similar expression satisfied by $\Phi_k(\tau, -\xi)$, viz.

$$-\xi \frac{\partial \Phi_k}{\partial \tau}(\tau, -\xi) = \Phi_k(\tau, -\xi) - \int_{R_\xi} \Psi(\xi')\Phi_k(\tau, -\xi') \, d\xi' - q_k(\tau, -\xi), \tag{2.1.50}$$

where we have simply replaced ξ by $-\xi$.

If we now multiply equation (2.1.49) for $k = 1$ by $\Psi(\xi)\Phi_2(\tau, -\xi)$, subtract equation (2.1.50) for $k = 2$ multiplied by $\Psi(\xi)\Phi_1(\tau, \xi)$, and integrate over all ξ and τ space, we find

$$\int_0^\infty d\tau \int_{R_\xi} \xi\Psi(\xi) \, d\xi \left[\Phi_2(\tau, -\xi) \frac{\partial \Phi_1}{\partial \tau}(\tau, \xi) + \Phi_1(\tau, \xi) \frac{\partial \Phi_2}{\partial \tau}(\tau, -\xi)\right]$$
$$= -\int_0^\infty d\tau \int_{R_\xi} \Psi(\xi) \, d\xi [q_1(\tau, \xi)\Phi_2(\tau, -\xi) - q_2(\tau, -\xi)\Phi_1(\tau, \xi)], \tag{2.1.51}$$

where the integral terms appearing in equations (2.1.49) and (2.1.50) clearly cancel. The LHS of equation (2.1.51) may be simplified by direct integration over τ, i.e.

$$\int_{R_\xi} \xi\Psi(\xi) \, d\xi [\Phi_1(\infty, \xi)\Phi_2(\infty, -\xi) - \Phi_1(0, \xi)\Phi_2(0, -\xi)]$$
$$= -\int_0^\infty d\tau \int_{R_\xi} \Psi(\xi) \, d\xi [q_1(\tau, \xi)\Phi_2(\tau, -\xi) - q_2(\tau, -\xi)\Phi_1(\tau, \xi)]. \tag{2.1.52}$$

We shall now consider two different manipulations whereby we replace $\Phi_1(\tau, \xi)$ and $\Phi_2(\tau, -\xi)$ by various combinations of $G_V(\tau^*, \xi^* \to \tau, \xi)$ and $G_S(0, \xi^* \to \tau, \xi)$.

I. Put $\Phi_1(\tau, \xi) = G_V(\tau_1, \xi_1 \to \tau, \xi),$ (2.1.53)

and

$$\Phi_2(\tau, \xi) = G_S(0, \xi_2 \to \tau, \xi). \tag{2.1.54}$$

Our definitions of the two Green's functions then stipulate that

$$q_1(\tau, \xi) = \delta(\tau - \tau_1)\delta(\xi - \xi_1), \tag{2.1.55}$$

whilst

$$q_2(\tau, \xi) \equiv 0. \tag{2.1.56}$$

Furthermore, we must have

$$\Phi_1(0, \xi) \equiv 0 \quad \text{for all } \xi < 0, \tag{2.1.57}$$

and

$$\Phi_2(0, \xi) = \delta(\xi - \xi_2) \quad \text{for all } \xi < 0. \tag{2.1.58}$$

We also know that the Green's functions (both G_V and G_S) must tend to zero as one moves further from the source, i.e. as $|\tau - \tau^*| \to \infty$ (see limit (2.1.41)). Thus, we must also have

$$\Phi_2(\infty, \xi) = 0 \quad \text{for all } \xi. \tag{2.1.59}$$

Substitution of these results into equation (2.1.52), and recognising that

$$\int_{R_\xi} d\xi \equiv \int_{\xi > 0} d\xi + \int_{\xi < 0} d\xi, \tag{2.1.60}$$

yields

$$-\int_{\xi > 0} \xi\Psi(\xi)\, d\xi\, G_V(\tau_1, \xi_1 \to 0, \xi)\delta(-\xi - \xi_2)$$
$$= -\int_0^\infty d\tau \int_{R_\xi} \Psi(\xi)\, d\xi\, \delta(\tau - \tau_1)\delta(\xi - \xi_1) G_S(0, \xi_2 \to \tau, -\xi), \tag{2.1.61}$$

i.e.

$$\xi_2\Psi(-\xi_2)G_V(\tau_1, \xi_1 \to 0, -\xi_2) = -\Psi(\xi_1)G_S(0, \xi_2 \to \tau_1, -\xi_1). \tag{2.1.62}$$

II. Now put

$$\Phi_1(\tau, \xi) = G_V(\tau_1, \xi_1 \to \tau, \xi), \tag{2.1.63}$$
$$\Phi_2(\tau, \xi) = G_V(\tau_2, \xi_2 \to \tau, \xi), \tag{2.1.64}$$

such that

$$q_1(\tau, \xi) = \delta(\tau - \tau_1)\delta(\xi - \xi_1), \tag{2.1.65}$$
$$q_2(\tau, \xi) = \delta(\tau - \tau_2)\delta(\xi - \xi_2), \tag{2.1.66}$$

$$\Phi_1(0, \xi) \equiv 0 \equiv \Phi_2(0, \xi) \quad \text{for all } \xi < 0, \tag{2.1.67}$$

$$\Phi_1(\infty, \xi) \equiv 0 \equiv \Phi_2(\infty, \xi) \quad \text{for all } \xi. \tag{2.1.68}$$

Equation (2.1.52) then gives

$$0 = -\int_0^\infty \mathrm{d}\tau \int_{\xi<0} \Psi(\xi)\, \mathrm{d}\xi\, \delta(\tau - \tau_1)\delta(\xi - \xi_1) G_V(\tau_2, \xi_2 \to \tau, -\xi)$$

$$+ \int_0^\infty \mathrm{d}\tau \int_{\xi>0} \Psi(\xi)\, \mathrm{d}\xi\, \delta(\tau - \tau_2)\delta(-\xi - \xi_2) G_V(\tau_1, \xi_1 \to \tau, \xi),$$

i.e.

$$\Psi(\xi_1) G_V(\tau_2, \xi_2 \to \tau_1, -\xi_1) = \Psi(-\xi_2) G_V(\tau_1, \xi_1 \to \tau_2, -\xi_2). \tag{2.1.69}$$

As we shall see in the following section, the function $\Psi(\xi)$ satisfies the symmetry condition

$$\Psi(-\xi) \equiv \Psi(\xi), \tag{2.1.70}$$

so that equations (2.1.62) and (2.1.69) can be combined to yield

$$G_S(0, \xi_2 \to \tau_1, -\xi_1) = -\xi_2 G_V(0, \xi_2 \to \tau_1, -\xi_1), \tag{2.1.71}$$

which, using the notations appearing in equation (2.1.45) is just

$$G_S(0, \xi^* \to \tau, \xi) = -\xi^* G_V(0, \xi^* \to \tau, \xi). \tag{2.1.72}$$

This is the required relationship between G_S and G_V. Clearly, the solution for $\Phi(\tau, \xi)$ given by equation (2.1.45) then has the form

$$\Phi(\tau, \xi) = \int_0^\infty \mathrm{d}\tau^* \int_{R_\xi} \mathrm{d}\xi^*\, q(\tau^*, \xi^*) G_V(\tau^*, \xi^* \to \tau, \xi)$$

$$- \int_{\xi<0} \xi^*\, \mathrm{d}\xi^*\, \Phi_{\text{inc}}(\xi^*) G_V(0, \xi^* \to \tau, \xi). \tag{2.1.73}$$

The problem has now been reduced to the evaluation of only $G_V(\tau^*, \xi^* \to \tau, \xi)$.

The jump condition

The volume Green's function $G_V(\tau^*, \xi^* \to \tau, \xi)$ is a solution to equation (2.1.40) satisfying the limiting condition (2.1.41) and the boundary condition (2.1.43). The direct solution of equation (2.1.40) is not possible because of the presence of the delta functions $\delta(\tau - \tau^*)\delta(\xi - \xi^*)$. However, these delta functions are zero for all $\tau \neq \tau^*$ and $\xi \neq \xi^*$. Thus, we consider $G_V(\tau^*, \xi^* \to \tau, \xi)$ to be the solution to the homogeneous integro-differential equation

$$\xi \frac{\partial G_V}{\partial \tau}(\tau^*, \xi^* \to \tau, \xi) = G_V(\tau^*, \xi^* \to \tau, \xi)$$

$$- \int_{R_\xi} \Psi(\xi') G_V(\tau^*, \xi^* \to \tau, \xi')\, \mathrm{d}\xi', \tag{2.1.74}$$

when $\tau \neq \tau^*$ and $\xi \neq \xi^*$.

Another equation is required when $\tau = \tau^*$ and $\xi = \xi^*$. To obtain this extra equation, we integrate equation (2.1.40) over τ between $\tau^* - \epsilon$ and $\tau^* + \epsilon$ where we shall let $\epsilon \to 0$, i.e. we have

$$\lim_{\epsilon \to 0} \int_{\tau^*-\epsilon}^{\tau^*+\epsilon} \left\{ \xi \frac{\partial G_V}{\partial \tau} (\tau^*, \xi^* \to \tau, \xi) - G_V(\tau^*, \xi^* \to \tau, \xi) \right.$$

$$\left. + \int_{R_\xi} \Psi(\xi')G_V(\tau^*, \xi^* \to \tau, \xi')\,d\xi' + \delta(\tau - \tau^*)\delta(\xi - \xi^*) \right\} d\tau = 0. \tag{2.1.75}$$

As we shall see, $G_V(\tau^*, \xi^* \to \tau, \xi)$ is a discontinuous function of τ at τ^*, but even so, the integration over this discontinuity yields zero contribution to equation (2.1.75) when we take the limit $\epsilon \to 0$. Equation (2.1.75) then gives

$$\lim_{\epsilon \to 0} \xi[G_V(\tau^*, \xi^* \to \tau^* + \epsilon, \xi) - G_V(\tau^*, \xi^* \to \tau^* - \epsilon, \xi)] = -\delta(\xi - \xi^*). \tag{2.1.76}$$

If we define

$$G_V^{\pm}(\tau^*, \xi^* \to \tau^*, \xi) = \lim_{\epsilon \to 0} G_V(\tau^*, \xi^* \to \tau^* \pm \epsilon, \xi), \tag{2.1.77}$$

we then have the so-called 'jump' condition

$$\xi[G_V^+(\tau^*, \xi^* \to \tau^*, \xi) - G_V^-(\tau^*, \xi^* \to \tau^*, \xi)] = -\delta(\xi - \xi^*). \tag{2.1.78}$$

It is clear from this last equation that $G_V(\tau^*, \xi^* \to \tau, \xi)$ is indeed discontinuous at τ^*.

One then obtains $G_V(\tau^*, \xi^* \to \tau, \xi)$ by solving the homogeneous equation (2.1.74) and forcing the resulting solution to satisfy equation (2.1.78) along with the conditions (2.1.41) and (2.1.43).

Note that at the surface $\tau = \tau^* = 0$, the jump condition (2.1.78) is just

$$\xi G_V^+(0, \xi^* \to 0, \xi) = -\delta(\xi - \xi^*), \tag{2.1.79}$$

so that equation (2.1.73) yields

$$\Phi(0, \xi < 0) = -\int_{\xi<0} \xi^*\,d\xi^*\,\Phi_{inc}(\xi^*)G_V(0, \xi^* \to 0, \xi)$$

$$= \Phi_{inc}(\xi), \tag{2.1.80}$$

as required.

2.2 Reduction of the transfer equation

The 1-dimensional equation of radiative transfer was derived in the preceding chapter – see equation (1.9.1) –

$$\mu \frac{\partial I}{\partial \tau}(\tau, \Delta\nu, \mu) = \phi(\Delta\nu)(I(\tau, \Delta\nu, \mu) - S(\tau)), \tag{2.2.1}$$

where the source function

$$S(\tau) = (1 - \epsilon)\bar{J}(\tau) + \epsilon B_\nu(T), \tag{2.2.2}$$

was obtained for a simple model 2-level atom assuming complete re-distribution. The average radiation intensity $\bar{J}(\tau)$ is then a function only of the depth τ and is given by

$$\bar{J}(\tau) = \tfrac{1}{2} \int_{-\infty}^{\infty} \mathrm{d}(\Delta\nu)\phi(\Delta\nu) \int_{-1}^{1} \mathrm{d}\mu\, I(\tau, \Delta\nu, \mu), \tag{2.2.3}$$

where $\phi(\Delta\nu)$ is the probability of absorption. In general, ϵ and $\phi(\Delta\nu)$ are functions of the depth τ since they both depend, amongst other things, on the local temperature. However, as stated previously, exact analytical solutions are available only for the simple case in which these quantities are independent of τ. The reason for this will become clear once we begin to detail the solution.

We take the boundary condition to be

$$I(0, \Delta\nu, \mu) \equiv 0, \tag{2.2.4}$$

for all $\Delta\nu$ and for all $\mu < 0$, i.e. we choose the atmosphere to have zero radiation flux incident on its surface.

The integro-differential equation (2.2.1) can be simplified [82] by defining the new independent variable ξ where

$$\xi = \frac{\mu}{\phi(\Delta\nu)}, \tag{2.2.5}$$

viz.

$$\xi \frac{\partial I}{\partial \tau} = I - S(\tau). \tag{2.2.6}$$

Clearly, the only independent variables which occur explicitly in equation (2.2.6) are τ and ξ, and we may therefore define the new dependent variable $\Phi(\tau, \xi)$, viz.

$$\Phi(\tau, \xi) \langle \equiv \rangle I(\tau, \Delta\nu, \mu), \tag{2.2.7}$$

so that the transfer equation becomes

$$\xi \frac{\partial \Phi}{\partial \tau}(\tau, \xi) = \Phi(\tau, \xi) - S(\tau). \tag{2.2.8}$$

We now turn our attention to the source function $S(\tau)$ and, in particular, to the average intensity $\bar{J}$. First, in making the change of independent variable given by equation (2.2.5), we choose

$$\mathrm{d}\xi = \frac{\mathrm{d}\mu}{\phi(\Delta\nu)}, \tag{2.2.9}$$

i.e. ξ replaces μ, so that ξ and $\Delta\nu$ remain independent. A complication arises, however, in the definition of ξ since $\phi(\Delta\nu) \to 0$ as $|\Delta\nu| \to \infty$. It is convenient, therefore, to restrict $\Delta\nu$ to $|\Delta\nu| \leqslant \beta$ throughout the analysis so that the important normalisation condition given by equation (1.4.8), viz.

$$\int_{-\infty}^{\infty} \phi(\Delta\nu)\, d(\Delta\nu) = 1, \tag{2.2.10}$$

now becomes

$$\int_{-\beta}^{\beta} \phi(\Delta\nu)\, d(\Delta\nu) = 1, \tag{2.2.11}$$

where, finally, we let $\beta \to \infty$. The range of ξ is then $|\xi| \leqslant 1/\phi(\beta)$, i.e. $\xi \in [-1/\phi(\beta), 1/\phi(\beta)]$ which we write as $\xi \in [-\gamma, \gamma]$. (Note also that we can include a continuum as well as the spectral line, although we have not included this physical effect in the work presented in chapter 1 – the resulting non-zero continuous opacity would eliminate the difficulty discussed above.) We therefore write equation (2.2.3) in the form

$$\bar{J}(\tau) = \tfrac{1}{2}\int_{-\beta}^{\beta} d(\Delta\nu)\phi(\Delta\nu)\int_{-1}^{1} d\mu\, I(\tau, \Delta\nu, \mu). \tag{2.2.12}$$

Further, care must be exercised in specifying the range of integration in equation (2.2.12) when we change the independent variable μ to ξ since $\Delta\nu$, through $\phi(\Delta\nu)$, also appears in the specification of ξ. Consequently, we define

$$\gamma(\Delta\nu) = \frac{1}{\phi(\Delta\nu)}, \tag{2.2.13}$$

so that equation (2.2.12) becomes

$$\bar{J}(\tau) = \tfrac{1}{2}\int_{-\beta}^{\beta} d(\Delta\nu)\phi^2(\Delta\nu)\int_{-\gamma(\Delta\nu)}^{\gamma(\Delta\nu)} d\xi\, \Phi(\tau, \xi). \tag{2.2.14}$$

We wish to change the order of integration on the RHS of equation (2.2.14) where ξ and $\Delta\nu$ are to be considered completely independent of one another. All the functions appearing in the integrand are uniformly continuous and, thus, the proposed interchange is valid – but the range of integrations must be carefully examined. Since $|\mu| \leqslant 1$, we must have $|\xi|\phi(\Delta\nu) \leqslant 1$. However, it is possible that an independent specification of ξ and $\Delta\nu$ would lead to $|\xi|\phi(\Delta\nu) > 1$. (For example, the maximum value of $|\xi|$ is $\gamma = 1/\phi(\beta)$ whereas the maximum value of $\Delta\nu$ is β. Consequently, max $[|\xi|\phi(\Delta\nu)] = \beta/\phi(\beta) \gg 1$.) Thus, when interchanging the order of integration so that we now integrate over $\Delta\nu$ first, we must only include those values of $\Delta\nu$ which satisfy $|\xi|\phi(\Delta\nu) \leqslant 1$. We therefore define the region $R(\xi, \Delta\nu)$ such that $\Delta\nu \in R(\xi, \Delta\nu)$ when $|\xi|\phi(\Delta\nu) \leqslant 1$.

Equation (2.2.14) can then be written as

$$\bar{J}(\tau) = \tfrac{1}{2}\int_{-\gamma}^{\gamma} d\xi\, \Phi(\tau, \xi) \int_{R(\xi,\Delta\nu)} d(\Delta\nu)\phi^2(\Delta\nu). \tag{2.2.15}$$

Note that we integrate over the full-range of $\xi \in [-\gamma, \gamma]$.

We next define $\Psi(\xi)$ by

$$\Psi(\xi)=\frac{1-\epsilon}{2}\int_{R(\xi,\Delta\nu)} \mathrm{d}(\Delta\nu)\phi^2(\Delta\nu), \tag{2.2.16}$$

where, since $\phi(\Delta\nu)$ is an even function of $\Delta\nu$, $\Psi(\xi)$ is an even function of ξ, i.e.

$$\Psi(-\xi)\equiv\Psi(\xi). \tag{2.2.17}$$

Equation (2.2.8), along with (2.2.2), (2.2.15) and (2.2.16), then yields the desired reduced integro-differential equation

$$\xi\frac{\partial\Phi}{\partial\tau}(\tau,\xi)=\Phi(\tau,\xi)-\int_{-\gamma}^{\gamma}\Psi(\xi')\Phi(\tau,\xi')\,\mathrm{d}\xi'-\epsilon B_\nu(T). \tag{2.2.18}$$

The boundary condition given by equation (2.2.4) is simply

$$\Phi(0,\xi)=0, \tag{2.2.19}$$

for all $\xi<0$.

2.3 The general solution

We now wish to determine the general solution to the reduced equation (2.2.18). We already know from section 2.1.3 that this solution may be written as (see equation (2.1.73))

$$\Phi(\tau,\xi)=\int_0^\infty \mathrm{d}\tau^*\int_{-\gamma}^{\gamma}\mathrm{d}\xi^*[\epsilon B_\nu(T)]^*G_\mathrm{V}(\tau^*,\xi^*\to\tau,\xi), \tag{2.3.1}$$

where we have used equation (2.2.19), i.e.

$$\Phi_\mathrm{inc}(\xi)\equiv\Phi(0,\xi)=0,\quad \xi<0. \tag{2.3.2}$$

The volume Green's function $G_\mathrm{V}(\tau^*,\xi^*\to\tau,\xi)$ may be obtained by solving the homogeneous equation

$$\xi\frac{\partial G_\mathrm{V}}{\partial\tau}(\tau^*,\xi^*\to\tau,\xi)=G_\mathrm{V}(\tau^*,\xi^*\to\tau,\xi)-\int_{-\gamma}^{\gamma}\Psi(\xi')G_\mathrm{V}(\tau^*,\xi^*\to\tau,\xi')\,\mathrm{d}\xi', \tag{2.3.3}$$

subject to the jump condition given by equation (2.1.78), viz.

$$\xi[G_\mathrm{V}^+(\tau^*,\xi^*\to\tau^*,\xi)-G_\mathrm{V}^-(\tau^*,\xi^*\to\tau^*,\xi)]=-\delta(\xi-\xi^*), \tag{2.3.4}$$

the limiting condition (equation (2.1.41))

$$G_\mathrm{V}(\tau^*,\xi^*\to\tau,\xi)\to0, \tag{2.3.5}$$

as $|\tau^*-\tau|\to\infty$, and the boundary condition (equation (2.1.43))

$$G_\mathrm{V}(\tau^*,\xi^*\to0,\xi)=0,\quad \xi<0. \tag{2.3.6}$$

We begin the evaluation of G_V by separating the variables τ and ξ such that

$$G_\mathrm{V}(\tau^*,\xi^*\to\tau,\xi)=g(\xi)T(\tau), \tag{2.3.7}$$

where, now, the ξ^* and τ^* variables occur implicitly on the RHS of equation (2.3.7). Substitution into equation (2.3.3) then yields

$$\frac{1}{T(\tau)}\frac{\mathrm{d}T}{\mathrm{d}\tau}(\tau) = \frac{1}{\xi} - \frac{1}{\xi g(\xi)}\int_{-\gamma}^{\gamma} \Psi(\xi')g(\xi')\,\mathrm{d}\xi' \tag{2.3.8}$$

$$= \frac{1}{\eta}, \text{ say,} \tag{2.3.9}$$

where η is an effective separation parameter yet to be determined.

We then have

$$T(\tau) \propto e^{\tau/\eta}. \tag{2.3.10}$$

If we now absorb the proportionality constant in equation (2.3.10) into $g(\xi)$, and re-write equation (2.3.7) as

$$G_V(\tau^*, \xi^* \to \tau, \xi) = g(\eta, \xi)e^{\tau/\eta}, \tag{2.3.11}$$

we find

$$(\eta - \xi)g(\eta, \xi) = \eta \int_{-\gamma}^{\gamma} \Psi(\xi')g(\eta, \xi')\,\mathrm{d}\xi'. \tag{2.3.12}$$

The above equation in $g(\eta, \xi)$ is homogeneous and therefore any of its solutions multiplied by an arbitrary constant is also a solution. We can choose this constant by arbitrarily normalising $g(\eta, \xi)$ such that

$$\int_{-\gamma}^{\gamma} \Psi(\xi)g(\eta, \xi)\,\mathrm{d}\xi = 1, \tag{2.3.13}$$

for all η. Equation (2.3.12) then has the form

$$(\eta - \xi)g(\eta, \xi) = \eta. \tag{2.3.14}$$

(One could, if one wished, replace unity on the RHS of (2.3.13) by a function $\bar{g}(\eta)$, say. This would introduce a factor $\bar{g}(\eta)$ into $g(\eta, \xi)$, but one can readily show that the results presented in the following sections remain unaltered – see equation (2.3.34).)

Since $\xi \in [-\gamma, \gamma]$, the quantity $\eta - \xi$ can be zero if $\eta \in [-\gamma, \gamma]$, and one could not therefore explicitly divide equation (2.3.14) by $\eta - \xi$ to obtain $g(\eta, \xi)$. Consequently, we develop the solution $g(\eta, \xi)$ to equation (2.3.14) separately for the two ranges $\eta \notin [-\gamma, \gamma]$ and $\eta \in [-\gamma, \gamma]$.

2.3.1 ***Range* I:** $\eta \notin [-\gamma, \gamma]$

Here, we have $\eta - \xi \neq 0$ and we immediately obtain

$$g(\eta, \xi) = \frac{\eta}{\eta - \xi}. \tag{2.3.15}$$

The value of η must now be chosen so that equation (2.3.13) is satisfied.

Direct substitution of $g(\eta, \xi)$ into this equation yields

$$\eta \int_{-\gamma}^{\gamma} \frac{\Psi(\xi)}{\eta - \xi} \, d\xi = 1, \tag{2.3.16}$$

which we re-write as

$$\Lambda(\eta) = 1 - \eta \int_{-\gamma}^{\gamma} \frac{\Psi(\xi)}{\eta - \xi} \, d\xi = 0. \tag{2.3.17}$$

Thus, the separation parameter η for $\eta \notin [-\gamma, \gamma]$ must be a zero of the function $\Lambda(\eta)$. To determine the number N of these zeros of $\Lambda(\eta)$, we use the argument principle [25] in the form

$$N = \frac{1}{2\pi} \mid \Delta\{\arg \Lambda(\eta)\} \mid, \tag{2.3.18}$$

where $\Delta\{\arg \Lambda(\eta)\}$ is the change in the argument of $\Lambda(\eta)$ as one traverses the contour surrounding the cut $[-\gamma, \gamma]$ in the complex plane (see figure 2.4).

If we now use the Plemelj formulae given by equations (2.1.8) and (2.1.9), and denote $\Lambda^{\pm}(\eta)$ as the values of $\Lambda(\eta)$ as Im $(\eta) \to 0^{\pm}$, we find

$$\Lambda^{\pm}(\eta) = 1 - \eta P \int_{-\gamma}^{\gamma} \frac{\Psi(\xi)}{\eta - \xi} \, d\xi \pm \pi i \eta \Psi(\eta), \tag{2.3.19}$$

so that

$$\arg \Lambda^{\pm}(\eta) = \tan^{-1} \left[\pm \frac{\pi \eta \Psi(\eta)}{\lambda(\eta)} \right], \tag{2.3.20}$$

where we have defined $\lambda(\eta)$ by

$$\lambda(\eta) = 1 - \eta P \int_{-\gamma}^{\gamma} \frac{\Psi(\xi)}{\eta - \xi} \, d\xi. \tag{2.3.21}$$

If we choose the branch of $\lambda(\eta)$ such that

$$\arg \Lambda^{\pm}(0) = 0, \tag{2.3.22}$$

then we find

$$\arg \Lambda^{+}(\gamma) = \tan^{-1} \left[\frac{\pi \gamma \Psi(\gamma)}{\lambda(\gamma)} \right]. \tag{2.3.23}$$

Fig.2.4. The contour C surrounding the cut from $-\gamma$ to γ in the complex η plane.

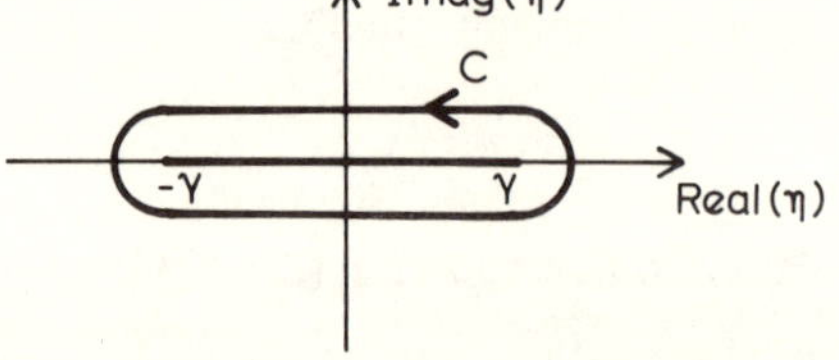

Since $0 < \epsilon < 1$ (from equation (1.6.19), for example) and $\phi(\Delta\nu) > 0$, we must have $\Psi(\eta) > 0$ for all η (see equation (2.2.16)). Equation (2.3.21) shows that $\lambda(\eta) \to -\infty$ as $\eta \to \gamma$, and thus

$$\arg \Lambda^+(\gamma) = \tan^{-1}\left(\frac{\pi\gamma\Psi(\gamma)}{-\infty}\right) = \pi. \tag{2.3.24}$$

Therefore, the change in argument of $\Lambda(\eta)$ going from $\eta = 0$ to $\eta = \gamma$ in the upper-half complex plane is just π. If we now repeat this exercise for the other three quadrants, we find the total change in argument of $\Lambda(\eta)$ as the contour C traverses the cut from $-\gamma$ to γ is 4π. Equation (2.3.18) then stipulates that the number of zeros of $\Lambda(\eta)$, if they exist, is just two.

Since $\Psi(\xi)$ is an even function of ξ, it is not difficult to show that

$$\begin{aligned}\Lambda(\eta) &= 1 - 2\eta^2 \int_0^\gamma \frac{\Psi(\xi)\,\mathrm{d}\xi}{\eta^2 - \xi^2} \\ &= \Lambda(-\eta),\end{aligned} \tag{2.3.25}$$

so that these zeros of $\Lambda(\eta)$ occur symmetrically about the origin. Further, since the number of zeros of $\Lambda(\eta)$ is two, $\Lambda(\eta)$ must be a polynomial in η of degree two. We thence know that the zeros of $\Lambda(\eta)$ must occur as complex conjugate pairs, i.e.

$$\Lambda(\bar{\eta}) = \Lambda(\eta) = 0. \tag{2.3.26}$$

If one of the zeros of $\Lambda(\eta)$ is located on neither the real nor imaginary axes, then equations (2.3.25) and (2.3.26) stipulate that the total number of zeros of $\Lambda(\eta)$ must be four. Thus, if there are to be two zeros of $\Lambda(\eta)$, these zeros must lie either on the real or on the imaginary axes.

Now as $\eta \to \infty$,

$$\Lambda(\eta) \to 1 - \int_{-\gamma}^{\gamma} \Psi(\xi)\,\mathrm{d}\xi, \tag{2.3.27}$$

which, from equation (2.2.16) and interchanging the order of integration, yields

$$\begin{aligned}\Lambda(\eta) &\to 1 - \frac{1-\epsilon}{2}\int_{-\beta}^{\beta} \mathrm{d}(\Delta\nu)\phi^2(\Delta\nu)\int_{-\gamma(\Delta\nu)}^{\gamma(\Delta\nu)} \mathrm{d}\xi \\ &= 1 - (1-\epsilon)\int_{-\beta}^{\beta} \mathrm{d}(\Delta\nu)\phi(\Delta\nu) \\ &= \epsilon,\end{aligned} \tag{2.3.28}$$

i.e. $\Lambda(\eta) \to \epsilon > 0$ as $\eta \to \infty$. However, $\Lambda(\eta) \to -\infty$ as $\eta \to \gamma^+$ so that $\Lambda(\eta)$ changes sign along the real axis between γ and ∞. Consequently, one zero of $\Lambda(\eta)$, which we call η_0, lies on the positive real axis with $\gamma < \eta_0 < \infty$ whilst, from equation (2.3.25), the other zero is just $-\eta_0$.

Hence, the Green's function $G_V(\tau^*, \xi^* \to \tau, \xi)$ in range I ($\eta \notin [-\gamma, \gamma]$) has the two basic forms

$$G_V(\tau^*, \xi^* \to \tau, \xi) = \frac{\pm\eta_0 e^{\pm\tau/\eta_0}}{\pm\eta_0 - \xi}. \tag{2.3.29}$$

We refer to these two solutions as the discrete modes or eigenfunctions.

2.3.2 *Range II:* $\eta \in [-\gamma, \gamma]$

Here, since $\xi \in [-\gamma, \gamma]$, the term $\eta - \xi$ appearing in equation (2.3.14) may be zero and thus division by $\eta - \xi$ must be undertaken with care. It is usual to write the solution to equation (2.3.14) in the form

$$g(\eta, \xi) = P\left(\frac{\eta}{\eta - \xi}\right) + f(\eta)\delta(\eta - \xi), \tag{2.3.30}$$

where the first term on the RHS of equation (2.3.30) represents the solution $g(\eta, \xi)$ when $\eta \neq \xi$. The symbol P has been inserted to emphasise that when an integration of this $g(\eta, \xi)$ over either η or ξ is made, the point $\eta = \xi$ must be excluded from this term. The symbol P will in that sense represent the Cauchy principal value as defined in section 2.1.1 (see equation (2.1.3)). The second term on the RHS of equation (2.3.30) represents the contribution to $g(\eta, \xi)$ when $\eta = \xi$. The function $f(\eta)$ has been included to account for the precise behaviour of $g(\eta, \xi)$ in the small neighbourhood of $\xi = \eta$.

The function $g(\eta, \xi)$ given by equation (2.3.3) is called a singular eigenfunction for obvious reasons: the representation of $g(\eta, \xi)$ by this equation only has a mathematical meaning when it occurs under an integration. Indeed, if we substitute $g(\eta, \xi)$ into the normalisation condition (2.3.13), we find

$$\eta P\int_{-\gamma}^{\gamma} \frac{\Psi(\xi)}{\eta - \xi}\, d\xi + f(\eta)\Psi(\eta) = 1, \tag{2.3.31}$$

which, from the definition of $\lambda(\eta)$ given by equation (2.3.21), yields

$$f(\eta) = \frac{\lambda(\eta)}{\Psi(\eta)}. \tag{2.3.32}$$

The singular eigenfunctions for $\eta \in [-\gamma, \gamma]$ are then of the form

$$g(\eta, \xi) = P\left(\frac{\eta}{\eta - \xi}\right) + \frac{\lambda(\eta)}{\Psi(\eta)}\,\delta(\eta - \xi). \tag{2.3.33}$$

Note that, in contrast to range I, we do not obtain discrete values of η in range II. Consequently, the $g(\lambda, \xi)$ given by equation (2.3.33) are sometimes referred to as continuum modes with $\eta \in [-\gamma, \gamma]$. Note, further, that $\Psi(\eta) > 0$ for all η since $0 < \epsilon < 1$, i.e. $\Psi(\eta) \neq 0$ - see equation (2.2.16).

The basic equation (2.3.3) we are attempting to solve is linear, and thus its general solution will be a linear combination of all the solutions found above in ranges I and II. In particular, we may write

$$G_V(\tau^*, \xi^* \to \tau, \xi) = a^+ g(\eta_0, \xi) e^{\tau/\eta_0} + a^- g(-\eta_0, \xi) e^{-\tau/\eta_0} + \int_{-\gamma}^{\gamma} A(\eta) g(\eta, \xi) e^{\tau/\eta} \, d\eta, \tag{2.3.34}$$

where $g(\pm\eta_0, \xi)$ are given by equation (2.3.13) with $\eta = \pm\eta_0$, and where $g(\eta, \xi)$ will henceforth explicitly refer to the singular eigenfunctions given by equation (2.3.33). The terms a^+, a^- and $A(\eta)$ are expansion coefficients which must be determined from the boundary conditions. However, the Green's function given by equation (2.3.34) above does not satisfy the jump condition (2.3.4) nor the limiting condition (2.3.6), and is therefore not the required Green's function. We shall correct this discrepancy in section 2.8 after we consider a completeness theorem for these eigenfunctions. Equation (2.3.34) has been included here to indicate the fundamental form of the solution to a homogeneous integro-differential equation of the type (2.3.3) using these discrete and continuum modes.

2.4 Full-range completeness

We have shown in the preceding section that the solution to the integro-differential equation under consideration consists of both discrete and continuum modes. We further indicated that the required full solution would be a linear combination of these modes. In the present section, we wish to examine this latter point more carefully. In particular, we wish to show that all the available solutions to the integro-differential equation are contained within the above-mentioned modes and that the general solution so obtained by this linear combination (see equation (2.3.34)) is indeed a complete solution. A non-complete solution is, in general, physically meaningless since it can lead to non-unique results.

We begin by defining the function $r(\xi)$ for all $\xi \in [-\gamma, \gamma]$ to be a linear combination of the discrete and continuum modes such that

$$r(\xi) = a^+ g(\eta_0, \xi) + a^- g(-\eta_0, \xi) + \int_{-\gamma}^{\gamma} A(\eta) g(\eta, \xi) \, d\eta. \tag{2.4.1}$$

In particular, we ask that $r(\xi)$ satisfy the Lipschitz condition

$$|r(\xi_2) - r(\xi_1)| < K |\xi_2 - \xi_1|^{\alpha}, \tag{2.4.2}$$

for all $\xi_1, \xi_2 \neq \gamma, -\gamma$, where K and α are constants with $0 < \alpha \leqslant 1$, whilst at the endpoints satisfying

$$|r(\xi)| < \frac{K_A}{|\xi - \gamma|^{\alpha_A}}, \tag{2.4.3}$$

and

$$|r(\xi)| < \frac{K_B}{|\xi + \gamma|^{\alpha_B}}, \tag{2.4.4}$$

where K_A, K_B, α_A and α_B are constants, and $0 \leqslant \alpha_A < 1$, $0 \leqslant \alpha_B < 1$. These conditions will enable us to use the Plemelj formulae derived in section 2.1.1 and, as we shall see, do not place any restriction on the physical problems which we may wish to solve. The function $r(\xi)$ is otherwise arbitrary. Equation (2.4.1) for $r(\xi)$ may then be compared to an expansion of an arbitrary function in a series of Legendre polynomials or Bessel functions, or an expansion in a Fourier series, etc. We know these expansions are complete, given various restrictions placed on the function being expanded, and are therefore unique. We wish to show here that the expansion given by equation (2.4.1) is likewise complete. In effect, we ask if we can uniquely determine the expansion coefficients a^+, a^- and $A(\eta)$ appearing in equation (2.4.1) such that the resulting expansion is also unique.

We re-write equation (2.4.1) in the form [26]

$$r(\xi) = a^+ g(\eta_0, \xi) + a^- g(-\eta_0, \xi) + \mathscr{F}(\xi), \tag{2.4.5}$$

where

$$\begin{aligned}\mathscr{F}(\xi) &= \int_{-\gamma}^{\gamma} A(\eta) g(\eta, \xi)\, \mathrm{d}\eta \\ &= \frac{\lambda(\xi) A(\xi)}{\Psi(\xi)} + P\int_{-\gamma}^{\gamma} \frac{\eta A(\eta)\, \mathrm{d}\eta}{\eta - \xi},\end{aligned} \tag{2.4.6}$$

where we have used equation (2.3.33) for $g(\eta, \xi)$.

We now introduce a function $Y(z)$ of the complex variable z such that

$$Y(z) = \frac{1}{2\pi \mathrm{i}} \int_{-\gamma}^{\gamma} \frac{\eta A(\eta)\, \mathrm{d}\eta}{\eta - z}, \tag{2.4.7}$$

which satisfies the three conditions:

$C_1(Y)$: $Y(z)$ is analytic in the complex plane cut from $-\gamma$ to γ.

$C_2(Y)$: $Y(z) \sim \frac{1}{z}$ as $|z| \to \infty$.

$C_3(Y)$: $|Y(z)| < \frac{k_1}{|z - \gamma|^{\alpha_1}}$, $|Y(z)| < \frac{k_2}{|z + \gamma|^{\alpha_2}}$,

where k_1, k_2, α_1 and α_2 are constants, and

$0 \leqslant \alpha_1 < 1, \quad 0 \leqslant \alpha_2 < 1.$

These conditions enable us to use the Plemelj formulae (equations (2.1.10) and (2.1.11)) such that

$$Y^+(z) + Y^-(z) = \frac{1}{\pi i} P \int_{-\gamma}^{\gamma} \frac{\eta A(\eta)\, d\eta}{\eta - z}, \tag{2.4.8}$$

and

$$Y^+(z) - Y^-(z) = zA(z). \tag{2.4.9}$$

Clearly, once $Y(z)$ is known, the expansion coefficient $A(z)$ may be determined directly from equation (2.4.9). (As we shall see in section 2.6, however, the use of certain full-range orthogonality relationships simplify enormously the evaluation of $A(\eta)$ - and also a^+ and a^-.) The problem has therefore, in part, been reduced to the determination of a unique $Y(z)$.

Substitution of equations (2.4.8) and (2.4.9) into (2.4.6) yields

$$\mathscr{F}(z) = \frac{\lambda(z)}{z\Psi(z)} [Y^+(z) - Y^-(z)] + \pi i [Y^+(z) + Y^-(z)], \tag{2.4.10}$$

which we re-write as

$$Y^+(z) = g(z) Y^-(z) + \frac{\mathscr{F}(z)}{\dfrac{\lambda(z)}{z\Psi(z)} + \pi i}, \tag{2.4.11}$$

where we have defined

$$g(z) = \frac{\dfrac{\lambda(z)}{z\Psi(z)} - \pi i}{\dfrac{\lambda(z)}{z\Psi(z)} + \pi i}. \tag{2.4.12}$$

Equation (2.4.11) may now be directly compared to the equation to be solved in the inhomogeneous Riemann–Hilbert problem.

We proceed as outlined in section 2.1.2. We introduce the new function $X(z)$ such that

$$X^+(z) = g(z) X^-(z), \tag{2.4.13}$$

where $X(z)$ satisfies the conditions:

$C_1(X)$: $X(z)$ is analytic in the complex plane cut from $-\gamma$ to γ.

$C_2(X)$: $X(z) \neq 0$ in the complex plane.

$C_3(X)$: $|X(z)| < \dfrac{k_3}{|z-\gamma|^{\alpha_3}}$, $|X(z)| < \dfrac{k_4}{|z+\gamma|^{\alpha_4}}$,

where k_3, k_4, α_3 and α_4 are constants, and

$0 \leqslant \alpha_3 < 1$, $0 \leqslant \alpha_4 < 1$.

Substitution of $g(z)$ given by equation (2.4.13) into (2.4.11) yields

$$\frac{Y^+(z)}{X^+(z)} - \frac{Y^-(z)}{X^-(z)} = \frac{\mathscr{F}(z)}{X^+(z)\left(\dfrac{\lambda(z)}{z\Psi(z)} + \pi i\right)}, \tag{2.4.14}$$

which, from the Plemelj formulae, gives

$$Y(z) = \frac{X(z)}{2\pi i} \int_{-\gamma}^{\gamma} \frac{\mathscr{F}(t)\,dt}{X^+(t)\left(\dfrac{\lambda(t)}{t\Psi(t)} + \pi i\right) t - z} + X(z)\,\mathscr{P}_n(z), \tag{2.4.15}$$

where $\mathscr{P}_n(z)$ is an arbitrary polynomial in z of degree n. The explicit evaluation of this polynomial must wait until $X(z)$ has been determined.

To evaluate $X(z)$ (see section 2.1.2), we take the natural logarithm of both sides of equation (2.4.13) to obtain

$$\ln X^+(z) - \ln X^-(z) = \ln g(z), \tag{2.4.16}$$

which, from the Plemelj formulae, yields

$$\ln X(z) = \frac{1}{2\pi i} \int_{-\gamma}^{\gamma} \frac{\ln g(t)\,dt}{t - z} + \mathscr{P}_m(z), \tag{2.4.17}$$

where $\mathscr{P}_m(z)$ is another arbitrary polynomial in z.

Putting

$$\Gamma(z) = \frac{1}{2\pi i} \int_{-\gamma}^{\gamma} \frac{\ln g(t)\,dt}{t - z}, \tag{2.4.18}$$

equation (2.4.17) becomes

$$X(z) = X_0(z)\,e^{\Gamma(z)}, \tag{2.4.19}$$

where

$$\mathscr{P}_m(z) = \ln X_0(z). \tag{2.4.20}$$

The $\Gamma(z)$ given by equation (2.4.18) is clearly analytic for all $z \notin [-\gamma, \gamma]$ and tends to zero as $|z| \to \infty$. Thus, $X(z)$ is similarly analytic for all $z \notin [-\gamma, \gamma]$ and therefore satisfies the first condition $C_1(X)$.

We must now concentrate on the functional dependence of $X_0(z)$ by examining the behaviour of $X(z)$ at the endpoints of the cut $[-\gamma, \gamma]$. Following the procedure outlined in section 2.1.2, we find that for z in the neighbourhood of $-\gamma$, we have

$$\Gamma(z) = \frac{-1}{2\pi i} \ln g(z) \ln |z + \gamma| + q_{-\gamma}, \tag{2.4.21}$$

where $|q_{-\gamma}(z)| < M$ for $|z + \gamma| < \delta$ where M, $\delta > 0$ are constants.

But we know from equation (2.4.12) that

$$\begin{aligned} \ln g(z) &= \ln\left(\frac{\lambda(z)}{z\Psi(z)} - \pi i\right) - \ln\left(\frac{\lambda(z)}{z\Psi(z)} + \pi i\right) \\ &= i \arg\left[\frac{\lambda(z)}{z\Psi(z)} - \pi i\right] - i \arg\left[\frac{\lambda(z)}{z\Psi(z)} + \pi i\right] \\ &= 2i \tan^{-1}\left(\frac{-\pi z\Psi(z)}{\lambda(z)}\right). \end{aligned} \tag{2.4.22}$$

Hence, we have

$$\frac{1}{2\pi i} \ln g(z) = \frac{\theta(z)}{\pi}, \tag{2.4.23}$$

where

$$\theta(z) = \tan^{-1}\left(\frac{-\pi z \Psi(z)}{\lambda(z)}\right), \tag{2.4.24}$$

so that, for z in the neighbourhood of $-\gamma$, $X(z)$ behaves as

$$X(z) = X_0(z) e^{q_{-\gamma}(z)} |z+\gamma|^{-\theta(z)/\pi}. \tag{2.4.25}$$

If $X(z)$ is to obey the second of the conditions $C_3(X)$, then $X_0(z)$ must have the form

$$X_0(z) = |z+\gamma|^{\alpha}, \tag{2.4.26}$$

so that, by direct substitution into equation (2.4.25), we find

$$0 \leqslant \frac{\theta(-\gamma)}{\pi} - \alpha < 1. \tag{2.4.27}$$

Further, if $X(z)$ is to remain analytic for all $z \notin [-\gamma, \gamma]$, then α must be either zero or an integer. (Note that the $e^{q_{-\gamma}(z)}$ term appearing in equation (2.4.25) is well behaved at $z = -\gamma$.)

We know from equation (2.2.16) that $\Psi(z) > 0$ for all z whilst, from equation (2.3.21),

$$\lambda(0) = 1, \tag{2.4.28}$$

and

$$\lambda(z) \to -\infty \quad \text{as } z \to \pm\gamma. \tag{2.4.29}$$

If we therefore take $\theta(0) = 0$, we have

$$\theta(-\gamma) = \pi. \tag{2.4.30}$$

The condition given by the inequality (2.4.27) is then

$$0 \leqslant 1 - \alpha < 1, \tag{2.4.31}$$

which, for integer α, may only be satisfied by $\alpha = 1$. Thus, for $X(z)$ to obey the second of conditions $C_3(X)$, $X_0(z)$ must have the form

$$X_0(z) = z + \gamma. \tag{2.4.32}$$

If we similarly examine the behaviour of $X(z)$ near $z = \gamma$, we find

$$\Gamma(z) = \frac{1}{2\pi i} \ln g(z) \ln |z-\gamma| + q_{\gamma}(z), \tag{2.4.33}$$

where $q_{\gamma}(z)$ is well behaved at $z = \gamma$, so that

$$X(z) = X_0(z) e^{q_{\gamma}(z)} |z-\gamma|^{\theta(z)/\pi}. \tag{2.4.34}$$

Again, if $X(z)$ is to satisfy the first of the conditions $C_3(X)$, then $X_0(z)$ must have the form

$$X_0(z) = (z-\gamma)^{\beta}, \tag{2.4.35}$$

where

$$0 \leqslant -\frac{\theta(\gamma)}{\pi} - \beta < 1, \tag{2.4.36}$$

and β is an integer. But $\theta(\gamma) = -\pi$, and thus we have $\beta = 1$.

We now combine the above two results to find the composite form for $X_0(z)$, viz.

$$X_0(z) = (z+\gamma)(z-\gamma). \tag{2.4.37}$$

Clearly, from this and equation (2.4.19), we have

$$X(z) = (z+\gamma)(z-\gamma)e^{\Gamma(z)}. \tag{2.4.38}$$

Since $\Gamma(z)$ is analytic for all $z \notin [-\gamma, \gamma]$ (see equation (2.4.18)) and tends to zero as $|z| \to \infty$, we must have $X(z)$ given by equation (2.4.38) yielding $X(z) \neq 0$ for all $z \notin [-\gamma, \gamma]$. Thus, condition $C_2(X)$ is also satisfied. This then completes the construction of the required $X(z)$.

We now return our attention to the evaluation of $Y(z)$ given by equation (2.4.15) and, in particular, to the determination of the polynomial $\mathscr{P}_n(z)$.

Since $X(z) \to z^2$ as $|z| \to \infty$, the term $X(z)\,\mathscr{P}_n(z)$ appearing on the RHS of equation (2.4.15) is a polynomial of degree $n+2$. However, we know from condition $C_2(Y)$ on $Y(z)$ that $Y(z) \sim 1/z$ as $|z| \to \infty$ so that, if this condition is to be satisfied, we must have

$$\mathscr{P}_n(z) \equiv 0. \tag{2.4.39}$$

We therefore have

$$Y(z) = \frac{X(z)}{2\pi i} \int_{-\gamma}^{\gamma} \frac{\mathscr{F}(t)\,dt}{X^+(t)\left(\dfrac{\lambda(t)}{t\Psi(t)} + \pi i\right)(t-z)}. \tag{2.4.40}$$

Since $X(z)$ is analytic in the complex plane cut from $-\gamma$ to γ, $Y(z)$ must be likewise analytic, and thus condition $C_1(Y)$ is satisfied by equation (2.4.40). Similarly, because of the behaviour of $X(z)$ in the neighbourhood of the end-points $-\gamma$ and γ of the cut (see condition $C_3(X)$ on $X(z)$), $Y(z)$ also satisfies condition $C_3(Y)$.

We now wish to examine the behaviour of $Y(z)$ as $|z| \to \infty$. To do this, we first simplify the resulting expressions by writing

$$F(t) = \frac{\mathscr{F}(t)}{2\pi i X^+(t)\left(\dfrac{\lambda(t)}{t\Psi(t)} + \pi i\right)}, \tag{2.4.41}$$

so that equation (2.4.40) has the form

$$Y(z) = X(z) \int_{-\gamma}^{\gamma} \frac{F(t)\,dt}{t-z}$$

$$= -\frac{X(z)}{z}\int_{-\gamma}^{\gamma} F(t)\,\mathrm{d}t\left(1+\frac{t}{z}+\frac{t^2}{z^2}+\cdots\right). \tag{2.4.42}$$

However, $X(z) \sim z^2$ as $|z| \to \infty$ so that, as $|z| \to \infty$,

$$Y(z) \sim -\int_{-\gamma}^{\gamma} F(t)\,\mathrm{d}t\left(z+t+\frac{t^2}{z}+\cdots\right). \tag{2.4.43}$$

But condition $C_2(Y)$ on $Y(z)$ stipulates that $Y(z) \sim 1/z$ as $|z| \to \infty$. Consequently, the first two terms in the expansion given by equation (2.4.43) must be identically zero, i.e.

$$\int_{-\gamma}^{\gamma} F(t)\,\mathrm{d}t = 0, \tag{2.4.44}$$

and

$$\int_{-\gamma}^{\gamma} tF(t)\,\mathrm{d}t = 0. \tag{2.4.45}$$

If we substitute $F(t)$ from equation (2.4.41) and $\mathscr{F}(t)$ from equation (2.4.5) into (2.4.44) and (2.4.45), we find

$$\int_{-\gamma}^{\gamma} \frac{r(t) - \dfrac{\eta_0 a^+}{\eta_0 - t} - \dfrac{\eta_0 a^-}{\eta_0 + t}}{X^+(t)\left(\dfrac{\lambda(t)}{t\Psi(t)} + \pi \mathrm{i}\right)}\,\mathrm{d}t = 0, \tag{2.4.46}$$

and

$$\int_{-\gamma}^{\gamma} \frac{r(t) - \dfrac{\eta_0 a^+}{\eta_0 - t} - \dfrac{\eta_0 a^-}{\eta_0 + t}}{X^+(t)\left(\dfrac{\lambda(t)}{t\Psi(t)} + \pi \mathrm{i}\right)}\,t\,\mathrm{d}t = 0, \tag{2.4.47}$$

where we have also substituted for the discrete eigenfunctions, viz.

$$g(\pm\eta_0, \xi) = \frac{\eta_0}{\eta_0 \mp \xi}\,. \tag{2.4.48}$$

It is clear that equations (2.4.46) and (2.4.47) are two equations from which the two as yet unknown expansion coefficients a^+ and a^- may be determined. However, it might be possible that the determinant of this system would be zero, in which case these expansion coefficients could not be uniquely evaluated. We must therefore examine the coefficients of a^+ and a^- appearing in the above two equations more carefully. In particular, we see that if $\eta_0 \to \infty$, the coefficients of a^+ and a^- in both equations would be equal so that the determinant of the system in this case would indeed be zero. This, in fact, is obvious when one examines the explicit form of the discrete eigenfunctions given by equation (2.4.48) – here, as $\eta_0 \to \infty$, we have $g(\eta_0, \xi) \equiv g(-\eta_0, \xi) \equiv 1$ and thus, the $a^- g(-\eta_0, \xi)$ term in the expansion given by equation (2.4.1) may be absorbed into the $a^+ g(\eta_0, \xi)$ term, and this would necessitate the development of another linearly independent solution. However, for all cases of astrophysical interest, $0 < \epsilon < 1$ (see equations (1.6.19) and (1.6.20), for example) so that, from equation (2.3.28) for $\Lambda(\eta)$ as $|\eta| \to \infty$, i.e. $\Lambda(\eta) \to \epsilon$, we cannot have $\eta_0 \to \infty$ as a zero of $\Lambda(\eta)$. Thus, the coefficients of a^+ and a^-, if they are non-zero, are not equal.

It now remains to be shown that the coefficients of a^+ and a^- in equations (2.4.46) and (2.4.47) are non-zero. Let us first consider the coefficient p of a^+ in equation (2.4.46), i.e.

$$p = \eta_0 \int_{-\gamma}^{\gamma} \frac{\mathrm{d}t}{X^+(t)\left(\dfrac{\lambda(t)}{t\Psi(t)} + \pi \mathrm{i}\right)(\eta_0 - t)}. \tag{2.4.49}$$

We know from equation (2.4.38) that

$$X(z) = (z^2 - \gamma^2)\mathrm{e}^{\Gamma(z)}, \tag{2.4.50}$$

where

$$\Gamma(z) = \frac{1}{2\pi \mathrm{i}} \int_{-\gamma}^{\gamma} \frac{\ln g(t)\, \mathrm{d}t}{t - z}.$$

We therefore have

$$X^+(z) = (z^2 - \gamma^2)\mathrm{e}^{\Gamma^+(z)}, \tag{2.4.51}$$

where, using the Plemelj formulae,

$$\Gamma^+(z) = \frac{1}{2\pi \mathrm{i}}\, P \int_{-\gamma}^{\gamma} \frac{\ln g(t)\, \mathrm{d}t}{t - z} + \tfrac{1}{2} \ln g(z), \tag{2.4.52}$$

i.e. using equation (2.4.23),

$$\Gamma^+(z) = \frac{1}{\pi}\, P \int_{-\gamma}^{\gamma} \frac{\theta(t)\, \mathrm{d}t}{t - z} + \tfrac{1}{2} \ln g(z). \tag{2.4.53}$$

Hence,

$$X^+(t) = (t^2 - \gamma^2) \exp\left[\frac{1}{\pi}\, P \int_{-\gamma}^{\gamma} \frac{\theta(t')\, \mathrm{d}t'}{t' - t}\right] \surd[g(t)]$$

$$= (t^2 - \gamma^2) \left[\frac{\dfrac{\lambda(t)}{t\Psi(t)} - \pi \mathrm{i}}{\dfrac{\lambda(t)}{t\Psi(t)} + \pi \mathrm{i}}\right]^{1/2} \exp\left[\frac{1}{\pi}\, P \int_{-\gamma}^{\gamma} \frac{\theta(t')\, \mathrm{d}t}{t' - t}\right], \tag{2.4.54}$$

so that

$$p=\eta_0\int_{-\gamma}^{\gamma}\frac{1}{(t^2-\gamma^2)\left[\left(\frac{\lambda(t)}{t\Psi(t)}\right)^2+\pi^2\right]^{1/2}(\eta_0-t)}$$

$$\times\exp\left[-\frac{1}{\pi}P\int_{-\gamma}^{\gamma}\frac{\theta(t')\,\mathrm{d}t'}{t'-t}\right]\mathrm{d}t. \tag{2.4.55}$$

Since $\pm\eta_0\notin[-\gamma,\gamma]$, i.e. $\eta_0\neq 0$, and since all other terms in the integrand of equation (2.4.55) are positive definite, then $p\neq 0$. One may use a similar argument to show that all the other coefficients of a^+ and a^- in equations (2.4.46) and (2.4.47) are non-zero. Indeed, by expanding the $\eta_0\pm t$ term, it is not difficult to show that these coefficients are unequal and that the determinant of the system is, in general, non-zero. Thus, one can uniquely solve these two equations for a^+ and a^-.

If one therefore wished to expand a function $r(\xi)$ in a series of the discrete and continuum modes in the form given by equation (2.4.1), one could first determine $X(z)$, thence $X^+(z)$, from equation (2.4.38), thence substitute this result into equations (2.4.46) and (2.4.47) to uniquely determine a^+ and a^-. Once these two expansion coefficients are obtained, one may evaluate $F(t)$ from equation (2.4.41) using the $\mathscr{F}(t)$ given by equation (2.4.5) where, of course, $r(t)$ is known. This then enables $Y(z)$ to be determined using equation (2.4.42) and this, in turn, immediately yields $A(z)$ from equation (2.4.9). All these determinations are unique, and thus the expansion of $r(\xi)$ given by equation (2.4.1) is also unique. This effectively completes the proof of completeness. Consequently, one may expand an arbitrary function $r(\xi)$ over the full-range of $\xi\in[-\gamma,\gamma]$, subject to the restrictions already listed, in a series of the discrete and singular eigenfunctions derived in the preceding section knowing that, indeed, all the necessary expansion terms have been found.

Digression:
The evaluation of $X(z)$ could have been simplified by noting from equations (2.3.19) and (2.3.21) that

$$\Lambda^{\pm}(z)=\lambda(z)\pm\pi\mathrm{i}z\Psi(z) \tag{2.4.56}$$

so that

$$\frac{\Lambda^+(z)}{\Lambda^-(z)}=\frac{\dfrac{\lambda(z)}{z\Psi(z)}+\pi\mathrm{i}}{\dfrac{\lambda(z)}{z\Psi(z)}-\pi\mathrm{i}}$$

$$=\frac{1}{g(z)}=\frac{X^-(z)}{X^+(z)}. \tag{2.4.57}$$

If we now put

$$\bar{X}(z) = \frac{1}{X(z)}, \tag{2.4.58}$$

then

$$\bar{X}^{\pm}(z) = \frac{1}{X^{\pm}(z)}, \tag{2.4.59}$$

so that

$$\frac{\Lambda^{+}(z)}{\Lambda^{-}(z)} = \frac{\bar{X}^{-}(z)}{\bar{X}^{+}(z)}. \tag{2.4.60}$$

Since $X(z)$ is analytic in the complex plane cut from $-\gamma$ to γ, $\bar{X}(z)$ must also be analytic for all $z \notin [-\gamma, \gamma]$ which, from equation (2.4.60), implies that

$$\Lambda(z) \equiv \mathscr{P}_n(z)\bar{X}(z) \equiv \mathscr{P}_n(z)/X(z). \tag{2.4.61}$$

Now $X(z) \sim z^2$ as $|z| \to \infty$ whereas $\Lambda(z) \to \epsilon$. Consequently, $n = 2$. We know that $X(z)$ must be analytic for all $z \notin [-\gamma, \gamma]$. Thus, noting that $\Lambda(z)$ has simple poles at $z = \pm\eta_0$, we must have $\mathscr{P}_n(z)$ proportional to $z^2 - \eta_0^2$, i.e.

$$X(z) = \frac{z^2 - \eta_0^2}{\Lambda(z)}. \tag{2.4.62}$$

Further, it is clear from equations (2.4.56) and (2.4.62) that

$$X^{+}(z) = \frac{z^2 - \eta_0^2}{\lambda(z) + \pi i z \Psi(z)}, \tag{2.4.63}$$

so that $Y(z)$ given by equation (2.4.40) has the somewhat simpler form

$$Y(z) = \frac{z^2 - \eta_0^2}{2\pi i \Lambda(z)} \int_{-\gamma}^{\gamma} \frac{t\Psi(t)\mathscr{F}(t)\,dt}{(t^2 - \eta_0^2)(t - z)}, \tag{2.4.64}$$

whilst the two equations (2.4.46) and (2.4.47) in a^{+} and a^{-} may be written as

$$\int_{-\gamma}^{\gamma} \frac{t\Psi(t)}{t^2 - \eta_0^2}\left[r(t) - \frac{\eta_0 a^{+}}{\eta_0 - t} - \frac{\eta_0 a^{-}}{\eta_0 + t}\right] dt = 0 \tag{2.4.65}$$

and

$$\int_{-\gamma}^{\gamma} \frac{t^2\Psi(t)}{t^2 - \eta_0^2}\left[r(t) - \frac{\eta_0 a^{+}}{\eta_0 - t} - \frac{\eta_0 a^{-}}{\eta_0 + t}\right] dt = 0. \tag{2.4.66}$$

It is clear from these last two equations that the coefficients of a^{+} and a^{-} are indeed non-zero ($\Psi(t) > 0$ for all t) and unequal. Indeed, the determinant of the system is non-zero so that one may uniquely determine a^{+} and a^{-} from equations (2.4.65) and (2.4.66).

2.5 Half-range completeness

In the preceding section, we showed that the two discrete modes $g(\pm\eta_0, \xi)$, together with the continuum modes $g(\eta, \xi)$, form a complete set over the full-range $\xi \in [-\gamma, \gamma]$, i.e. one may expand an arbitrary function $r(\xi)$, subject to certain limitations, in a series of these modes for all $\xi \in [-\gamma, \gamma]$. As we shall see in the following sections, it is also necessary to consider half-range completeness.

In particular, we wish to show that one may expand an arbitrary function $r(\xi)$, subject to the conditions given by equations (2.4.2), (2.4.3) and (2.4.4), such that

$$r(\xi) = a^+ g(\eta_0, \xi) + \int_0^\gamma A(\eta) g(\eta, \xi)\, \mathrm{d}\eta, \tag{2.5.1}$$

for all $\xi \in [0, \gamma]$.

We follow the procedure outlined in the preceding section for full-range completeness. We write

$$\begin{aligned} \mathscr{F}(\xi) &= \int_0^\gamma A(\eta) g(\eta, \xi)\, \mathrm{d}\eta \\ &= \frac{\lambda(\xi) A(\eta)}{\Psi(\xi)} + P \int_0^\gamma \frac{\eta A(\eta)\, \mathrm{d}\eta}{\eta - \xi}, \end{aligned} \tag{2.5.2}$$

and define the function $Y(z)$ such that

$$Y(z) = \frac{1}{2\pi \mathrm{i}} \int_0^\gamma \frac{\eta A(\eta)\, \mathrm{d}\eta}{\eta - \xi}, \tag{2.5.3}$$

which satisfies the conditions

$C_1(Y)$: $Y(z)$ is analytic in the complex plane cut from 0 to γ.

$C_2(Y)$: $Y(z) \sim \dfrac{1}{z}$ as $|z| \to \infty$.

$C_3(Y)$: $|Y(z)| < \dfrac{k_1}{|z-\gamma|^{\alpha_1}}$, $Y(z) < \dfrac{k_2}{|z|^{\alpha_2}}$,

where k_1, k_2, α_1 and α_2 are constants, and

$0 \leqslant \alpha_1 < 1, \quad 0 \leqslant \alpha_2 < 1.$

It is not difficult to show that $Y(z)$ has the form (see equation (2.4.15))

$$Y(z) = \frac{X(z)}{2\pi \mathrm{i}} \int_0^\gamma \frac{\mathscr{F}(t)\, \mathrm{d}t}{X^+(t)\left(\dfrac{\lambda(t)}{t\Psi(t)} + \pi \mathrm{i}\right)(t-z)} + X(z)\, \mathscr{P}_m(z), \tag{2.5.4}$$

where

$$X(z) = X_0(z) \mathrm{e}^{\Gamma(z)}, \tag{2.5.5}$$

$$\Gamma(z) = \frac{1}{2\pi \mathrm{i}} \int_0^\gamma \frac{\ln g(t)\, \mathrm{d}t}{t - z}, \tag{2.5.6}$$

with

$$g(z)=\left(\frac{\lambda(z)}{z\Psi(z)}-\pi\mathrm{i}\right)\left(\frac{\lambda(z)}{z\Psi(z)}+\pi\mathrm{i}\right)^{-1} \tag{2.5.7}$$

$$=\frac{X^+(z)}{X^-(z)}. \tag{2.5.8}$$

In particular, $X(z)$ must be analytic for all $z \notin [0, \gamma]$ because of condition $C_1(Y)$ placed on $Y(z)$. This condition is just $C_1(X)$ on $X(z)$. It would be tempting to attempt a relationship between $X(z)$ and $\Lambda(z)$ similar to that given by equation (2.4.62). However, it must be stressed that $\Lambda(z)$ is analytic for all $z \notin [-\gamma, \gamma]$ whereas $1/X(z)$ is analytic for $z \notin [0, \gamma]$. Since these regions of analyticity are different, we cannot, in general, have $X(z)$ proportional to $1/\Lambda(z)$.

The other two conditions on $X(z)$ are

$C_2(X)$: $X(z) \neq 0$ for all $z \notin [0, \gamma]$.

$C_3(X)$: $|X(z)| < \dfrac{k_3}{|z-\gamma|^{\alpha_3}}$, $|X(z)| < \dfrac{k_4}{|z|^{\alpha_4}}$,

where k_3, k_4, α_3 and α_4 are constants, and

$0 \leqslant \alpha_3 < 1$, $0 \leqslant \alpha_4 < 1$.

It is not difficult to repeat the argument leading to equation (2.4.37). In the half-range case, we find

$$X(z)=z-\gamma, \tag{2.5.9}$$

which enables $X(z)$ to satisfy the first of conditions $C_3(X)$. The second of these conditions, however, leads to a different result to that obtained for the full-range completeness. For z in the neighbourhood of zero, we have (see equation (2.4.25))

$$X(z)=X_0(z)\mathrm{e}^{q_0(z)}|z|^{-\theta(z)/\pi}, \tag{2.5.10}$$

where $q_0(z)$ is well behaved in the neighbourhood $z \sim 0$, whilst $\theta(z)$ is given by equation (2.4.24), viz.

$$\theta(z)=\tan^{-1}\left(-\frac{\pi z\Psi(z)}{\lambda(z)}\right). \tag{2.5.11}$$

If $X(z)$ is to obey the second of conditions $C_3(X)$, we must have

$$X_0(z)=z^{\alpha}, \tag{2.5.12}$$

where α is an integer satisfying

$$0 \leqslant \frac{\theta(0)}{\pi}-\alpha<1. \tag{2.5.13}$$

Since we have taken the branch $\theta(0)=0$, this last result implies $\alpha=0$. Thus, the final form for $X(z)$ is just

$$X(z)=(z-\gamma)\mathrm{e}^{\Gamma(z)}. \tag{2.5.14}$$

Since $\Gamma(z) \to 0$ as $|z| \to \infty$ from equation (2.5.6), we must have $X(z) \sim z$ as $|z| \to \infty$. Condition $C_2(Y)$ on $Y(z)$, i.e. $Y(z) \sim 1/z$ and $|z| \to \infty$, would therefore be violated by the $X(z)\ \mathscr{P}_n(z)$ term appearing in equation (2.5.4) unless, of course, we have

$$\mathscr{P}_n(z) \equiv 0. \tag{2.5.15}$$

We therefore have

$$Y(z) = \frac{X(z)}{2\pi \mathrm{i}} \int_0^\gamma \frac{\mathscr{F}(t)\,\mathrm{d}t}{X^+(t)\left(\dfrac{\lambda(t)}{t\Psi(t)} + \pi\mathrm{i}\right)(t-z)}. \tag{2.5.16}$$

With the definition of $F(t)$ given by equation (2.4.41), we find

$$Y(z) \sim -\int_0^\gamma F(t)\,\mathrm{d}t \left\{1 + \frac{t}{z} + \frac{t^2}{z^2} + \cdots\right\} \tag{2.5.17}$$

as $|z| \to \infty$ (note that $X(z) \sim z$ as $|z| \to \infty$) which, if condition $C_2(Y)$ on $Y(z)$ is to be satisfied, implies that

$$\int_0^\gamma F(t)\,\mathrm{d}t = 0,$$

i.e.

$$\int_0^\gamma \frac{r(t) - \dfrac{\eta_0 a^+}{\eta_0 - t}}{X^+(t)\left(\dfrac{\lambda(t)}{t\Psi(t)} + \pi\mathrm{i}\right)}\,\mathrm{d}t = 0. \tag{2.5.18}$$

This is one equation for the one as yet unknown expansion coefficient a^+. As in the preceding section, it may be shown that this equation uniquely defines a^+. Once this a^+ has been determined, $\mathscr{F}(t)$, thence $Y(z)$, thence $A(z)$, may be evaluated uniquely. This then proves that an arbitrary function $r(\xi)$. subject to certain limitations, may be expanded in the half-range $\xi \in [0, \gamma]$ using $g(\eta_0, \xi)$ and $g(\eta, \xi)$ for $\eta \in [0, \gamma]$, i.e. $g(\eta_0, \xi)$ and $g(\eta, \xi)$ form a complete set over the *positive* half-range $\xi \in [0, \gamma]$.

We may similarly prove completeness of the functions $g(-\eta_0, \xi)$ and $g(\eta, \xi)$ for $\xi \in [-\gamma, 0]$ over the *negative* half-range $\xi \in [-\gamma, 0]$. In this case, we expand $r(\xi)$ in the form

$$r(\xi) = a^- g(-\eta_0, \xi) + \int_{-\gamma}^0 A(\eta) g(\eta, \xi)\,\mathrm{d}\eta, \tag{2.5.19}$$

where a^- and $A(\eta)$ may be obtained from the equations

$$Y(z) = \frac{X(z)}{2\pi \mathrm{i}} \int_{-\gamma}^0 \frac{\mathscr{F}(t)\,\mathrm{d}t}{X^+(t)\left(\dfrac{\lambda(t)}{t\Psi(t)} + \pi\mathrm{i}\right)(t-z)}, \tag{2.5.20}$$

$$\mathscr{F}(\xi) = r(\xi) - a^- g(-\eta_0, \xi), \tag{2.5.21}$$

$$X(z) = (z + \gamma)e^{\Gamma_-(z)}, \tag{2.5.22}$$

$$\Gamma_-(z) = \frac{1}{2\pi i}\int_{-\gamma}^{0} \frac{\ln g(t)\, dt}{t - z}, \tag{2.5.23}$$

$$\int_{-\gamma}^{0} \frac{r(t) - \dfrac{\eta_0 a^-}{\eta_0 + t}}{X^+(t)\left(\dfrac{\lambda(t)}{t\Psi(t)} + \pi i\right)}\, dt = 0. \tag{2.5.24}$$

2.6 Full-range orthogonality

In section 2.4 we showed that the functions $g(\pm\eta_0, \xi)$ and $g(\eta, \xi)$ form a complete set for all $\xi \in [-\gamma, \gamma]$, and this then enabled us to expand an arbitrary function $r(\xi)$ as

$$r(\xi) = a^+ g(\eta_0, \xi) + a^- g(-\eta_0, \xi) + \int_{-\gamma}^{\gamma} A(\eta) g(\eta, \xi)\, d\eta. \tag{2.6.1}$$

Our proof of completeness also provided a means by which the above expansion coefficients a^+, a^- and $A(\eta)$, $\eta \in [-\gamma, \gamma]$, may be determined. However, a more direct method of evaluating these quantities may be derived by the use of certain full-range orthogonality conditions.

We begin by re-stating the equation satisfied by $g(\eta, \xi)$, viz.

$$(\eta - \xi) g(\eta, \xi) = \eta \tag{2.6.2}$$

(see equation (2.3.14)). If we re-arrange this equation in the form

$$\left(1 - \frac{\xi}{\eta}\right) g(\eta, \xi) = 1, \tag{2.6.3}$$

and similarly,

$$\left(1 - \frac{\xi}{\eta'}\right) g(\eta', \xi) = 1, \tag{2.6.4}$$

multiply equation (2.6.3) by $\Psi(\xi) g(\eta', \xi)$ and equation (2.6.4) by $\Psi(\xi) g(\eta, \xi)$, subtract then integrate over $\xi \in [-\gamma, \gamma]$, we obtain

$$\left(\frac{1}{\eta'} - \frac{1}{\eta}\right) \int_{-\gamma}^{\gamma} \xi \Psi(\xi) g(\eta, \xi) g(\eta', \xi)\, d\xi = 0, \tag{2.6.5}$$

where we have used the normalisation condition (2.3.13).

Thus, if $\eta \neq \eta'$, we must have

$$\int_{-\gamma}^{\gamma} \xi \Psi(\xi) g(\eta, \xi) g(\eta', \xi)\, d\xi = 0. \tag{2.6.6}$$

This then is part of our full-range orthogonality relationship. Note that equation (2.6.2) holds for all $\eta \in [-\gamma, \gamma]$ and for the two discrete values $\pm\eta_0$. Consequently, the orthogonality relationship given by equation (2.6.6) holds for $g(\eta, \xi)$ and $g(\eta', \xi)$ being either the discrete or singular eigenfunctions, or both, in any combination.

We shall now consider separately the discrete and singular orthogonality condition when $\eta = \eta'$.

2.6.1 *Discrete normalisation*

Let us write

$$N(\pm\eta_0) = \int_{-\gamma}^{\gamma} \xi\Psi(\xi)g^2(\pm\eta_0, \xi)\,\mathrm{d}\xi$$

$$= \eta_0^2 \int_{-\gamma}^{\gamma} \frac{\xi\Psi(\xi)\,\mathrm{d}\xi}{(\pm\eta_0 - \xi)^2}. \tag{2.6.7}$$

From equation (2.3.17) we have

$$\Lambda(\eta) = 1 - \eta \int_{-\gamma}^{\gamma} \frac{\Psi(\xi)\,\mathrm{d}\xi}{\eta - \xi}, \tag{2.6.8}$$

so that

$$\frac{\mathrm{d}\Lambda(\eta)}{\mathrm{d}\eta} = \int_{-\gamma}^{\gamma} \frac{\xi\Psi(\xi)\,\mathrm{d}\xi}{(\eta - \xi)^2}. \tag{2.6.9}$$

We therefore find

$$N(\pm\eta_0) = \eta_0^2 \left[\frac{\mathrm{d}\Lambda(\eta)}{\mathrm{d}\eta}\right]_{\eta = \pm\eta_0}. \tag{2.6.10}$$

Thus, if we have the full-range expansion of $r(\xi)$ in the form

$$r(\xi) = a^+g(\eta_0, \xi) + a^-g(-\eta_0, \xi) + \int_{-\gamma}^{\gamma} A(\eta)g(\eta, \xi)\,\mathrm{d}\eta, \tag{2.6.11}$$

we could immediately determine a^+, for example, by multiplying this equation by $\xi\Psi(\xi)g(\eta_0, \xi)$ and integrating over $\xi \in [-\gamma, \gamma]$. The orthogonality relationship given by equation (2.6.6) then stipulates that the second and third terms in the above expansion disappear leaving

$$\int_{-\gamma}^{\gamma} \xi\Psi(\xi)g(\eta_0, \xi)r(\xi)\,\mathrm{d}\xi = a^+ \int_{-\gamma}^{\gamma} \xi\Psi(\xi)g^2(\eta_0, \xi)\,\mathrm{d}\xi,$$

i.e.

$$a^+ = \frac{1}{N(\eta_0)} \int_{-\gamma}^{\gamma} \xi\Psi(\xi)g(\eta_0, \xi)r(\xi)\,\mathrm{d}\xi. \tag{2.6.12}$$

Similarly, we find

$$a^- = \frac{1}{N(-\eta_0)} \int_{-\gamma}^{\gamma} \xi\Psi(\xi)g(-\eta_0, \xi)r(\xi)\,d\xi. \tag{2.6.13}$$

Clearly, this method of evaluation of a^+ and a^- is somewhat simpler than that presented in section 2.4.

2.6.2 *Singular normalisation*

The determination of $A(\eta)$ is somewhat more difficult than for a^+ and a^-. We begin by multiplying equation (2.6.1) for the expansion of $r(\xi)$ by $\xi\Psi(\xi)g(\eta', \xi)$ where $\eta \in [-\gamma, \gamma]$ and integrating over $\xi \in [-\gamma, \gamma]$. The first two terms on the RHS of equation (2.6.1) then disappear because of equation (2.6.6), i.e. we find

$$\int_{-\gamma}^{\gamma} \xi\Psi(\xi)g(\eta', \xi)r(\xi)\,d\xi = \int_{-\gamma}^{\gamma} \xi\Psi(\xi)g(\eta', \xi)\,d\xi \times \int_{-\gamma}^{\gamma} A(\eta)g(\eta, \xi)\,d\eta. \tag{2.6.14}$$

It is important to note that the functions $g(\eta, \xi)$ and $g(\eta', \xi)$ are singular in nature and are thus not uniformly continuous. It is therefore not possible to change directly the order of integration appearing on the RHS of equation (2.6.14).

Given that

$$g(\eta, \xi) = P\left(\frac{\eta}{\eta - \xi}\right) + \frac{\lambda(\eta)}{\Psi(\eta)}\,\delta(\eta - \xi), \tag{2.6.15}$$

the RHS of equation (2.6.14) has the form

$$\begin{aligned}
\text{RHS (2.6.14)} &= \int_{-\gamma}^{\gamma} \xi\Psi(\xi)\left[P\left(\frac{\eta'}{\eta' - \xi}\right) + \frac{\lambda(\eta')}{\Psi(\eta')}\,\delta(\eta' - \xi)\right] \\
&\quad \times \int_{-\gamma}^{\gamma} A(\eta)\left[P\left(\frac{\eta}{\eta - \xi}\right) + \frac{\lambda(\eta)}{\Psi(\eta)}\,\delta(\eta - \xi)\right] d\eta\,d\xi \\
&= \eta' P\int_{-\gamma}^{\gamma} \frac{\xi\Psi(\xi)\,d\xi}{\eta' - \xi}\, P\int_{-\gamma}^{\gamma} \frac{\eta A(\eta)\,d\eta}{\eta - \xi} \\
&\quad + \eta' P\int_{-\gamma}^{\gamma} \frac{\xi\lambda(\xi)A(\xi)\,d\xi}{\eta' - \xi} + \eta' P\int_{-\gamma}^{\gamma} \frac{\eta\lambda(\eta')A(\eta)\,d\eta}{\eta - \eta'} \\
&\quad + \eta' A(\eta')\frac{\lambda^2(\eta')}{\Psi(\eta')}.
\end{aligned} \tag{2.6.16}$$

One can immediately see that it is explicitly the first term on the RHS of equation (2.6.16) which will now allow the direct interchange of the order of integration.

We may proceed by using the Poincaré-Bertrand formula [89]

$$P\int_{-\gamma}^{\gamma} \frac{\mathrm{d}t}{t-t_0}\, P\int_{-\gamma}^{\gamma} \frac{\phi(t,t_1)\,\mathrm{d}t_1}{t_1-t} = -\pi^2\phi(t_0,t_0)$$

$$+P\int_{-\gamma}^{\gamma} \mathrm{d}t_1 P\int_{-\gamma}^{\gamma} \frac{\phi(t,t_1)\,\mathrm{d}t}{(t-t_0)(t_1-t)}, \tag{2.6.17}$$

where the function $\phi(t, t_1)$ must satisfy a Lipschitz condition.

If we now put $t \equiv \xi$ and $t_1 \equiv \eta$ with

$$\phi(t,t_1) \equiv -\xi\Psi(\xi)\eta A(\eta), \tag{2.6.18}$$

we have, with $t_0 \equiv \eta'$,

$$\text{RHS (2.6.14)} = \pi^2\eta'^3\Psi(\eta')A(\eta')$$

$$+\eta' P\int_{-\gamma}^{\gamma} \eta A(\eta)\,\mathrm{d}\eta\, P\int_{-\gamma}^{\gamma} \frac{\xi\Psi(\xi)\,\mathrm{d}\xi}{(\eta'-\xi)(\eta-\xi)}$$

$$+\eta' P\int_{-\gamma}^{\gamma} [\lambda(\eta')-\lambda(\eta)]\, \frac{\eta A(\eta)\,\mathrm{d}\eta}{\eta-\eta'}$$

$$+\eta'\,\frac{A(\eta')\lambda^2(\eta')}{\Psi(\eta')}. \tag{2.6.19}$$

Now

$$\lambda(\eta) = 1 - \eta P\int_{-\gamma}^{\gamma} \frac{\Psi(\xi)\,\mathrm{d}\xi}{\eta-\xi} \tag{2.6.20}$$

(see equation (2.3.21)), so that

$$\lambda(\eta')-\lambda(\eta) = (\eta'-\eta)P\int_{-\gamma}^{\gamma} \frac{\xi\Psi(\xi)\,\mathrm{d}\xi}{(\eta-\xi)(\eta'-\xi)}. \tag{2.6.21}$$

If we substitute this last result into equation (2.6.19) we see that the second and third terms cancel, and we therefore have

$$\text{RHS (2.6.14)} = \left[\frac{\lambda^2(\eta')}{\Psi^2(\eta')} + \pi^2\eta'^2\right] \eta'\Psi(\eta')A(\eta'). \tag{2.6.22}$$

Thus, if we write

$$N(\eta) = \left(\frac{\lambda^2(\eta)}{\Psi^2(\eta)} + \pi^2\eta^2\right) \eta\Psi(\eta), \tag{2.6.23}$$

equations (2.6.14) and (2.6.22) yield

$$A(\eta) = \frac{1}{N(\eta)}\int_{-\gamma}^{\gamma} \xi\Psi(\xi)g(\eta,\xi)r(\xi)\,\mathrm{d}\xi. \tag{2.6.24}$$

This then completes the determination of the expansion coefficient $A(\eta)$.

One can now summarise the results of this section by writing the general full-range orthogonality result

$$\int_{-\gamma}^{\gamma} \xi\Psi(\xi)g(\eta', \xi)g(\eta, \xi)\, d\xi = N(\eta)\delta(\eta - \eta'), \tag{2.6.25}$$

for η, η' either discrete ($\pm\eta_0$) or continuous ($\eta, \eta' \in [-\gamma, \gamma]$) and where $N(\pm\eta_0)$ and $N(\eta)$, $\eta \in [-\gamma, \gamma]$ are given by equations (2.6.10) and (2.6.23).

2.7 Half-range orthogonality

Let us now consider the evaluation of the expansion coefficients a^+ and $A(\eta)$ for the positive half-range expansion

$$r(\xi) = a^+ g(\eta_0, \xi) + \int_0^{\gamma} A(\eta)g(\eta, \xi)\, d\eta, \tag{2.7.1}$$

for all $\xi \in [0, \gamma]$ using half-range orthogonality. The approach outlined in the preceding section for full-range orthogonality is not applicable here. Instead, we must define a weighting function $W(\xi)$ (to be determined) such that

$$\int_0^{\gamma} W(\xi)g(\eta', \xi)g(\eta, \xi)\, d\xi = 0, \tag{2.7.2}$$

for all $\eta, \eta' \in [0, \gamma]$, or $\eta, \eta' = \eta_0$, or any combination of the two, but with $\eta \neq \eta'$. (Note that, for full-range orthogonality, we had $W(\xi) \equiv \xi\Psi(\xi)$.)

Again, the defining equation for $g(\eta, \xi)$ is

$$\left(1 - \frac{\xi}{\eta}\right)g(\eta, \xi) = 1. \tag{2.7.3}$$

If we now multiply equation (2.7.3) for $g(\eta, \xi)$ by $W(\xi)g(\eta', \xi)/\xi$, multiply equation (2.7.3) for $g(\eta', \xi)$ by $W(\xi)g(\eta, \xi)/\xi$, subtract and integrate over the half-range $\xi \in [0, \gamma]$, we find

$$\left(\frac{1}{\eta'} - \frac{1}{\eta}\right)\int_0^{\gamma} W(\xi)g(\eta, \xi)g(\eta', \xi)\, d\xi = \int_0^{\gamma} \frac{W(\xi)}{\xi}\,[g(\eta', \xi) - g(\eta, \xi)]\, d\xi. \tag{2.7.4}$$

Note that equation (2.7.3) is valid for both discrete and singular eigenfunctions: then η and η' appearing in equation (2.7.4) may be either η_0 or anything in the range $[0, \gamma]$.

If $\eta' \neq \eta$ in equation (2.7.4), we will obtain the required half-range orthogonality condition (2.7.2) only if we also have

$$\int_0^{\gamma} \frac{W(\xi)}{\xi} g(\eta, \xi)\, d\xi = \int_0^{\gamma} \frac{W(\xi)}{\xi} g(\eta', \xi)\, d\xi = \int_0^{\gamma} \frac{W(\xi)}{\xi} g(\eta_0, \xi)\, d\xi, \tag{2.7.5}$$

where $\eta, \eta' \in [0, \gamma]$. Since this last equation holds independent of the value of $\eta \in [0, \gamma]$, we must have the general result

$$\int_0^\gamma \frac{W(\xi)}{\xi} g(\eta, \xi)\, \mathrm{d}\xi = C_1, \tag{2.7.6}$$

where C_1 is a constant independent of η, and where either $\eta \in [0, \gamma]$ or $\eta = \eta_0$.

If we now substitute for the discrete and singular eigenfunctions into equation (2.7.6), we have

$$\eta_0 \int_0^\gamma \frac{W(\xi)\, \mathrm{d}\xi}{\xi(\eta_0 - \xi)} = C_1, \tag{2.7.7}$$

and

$$\frac{\lambda(\eta) W(\eta)}{\eta \Psi(\eta)} + \eta P \int_0^\gamma \frac{W(\xi)\, \mathrm{d}\xi}{\xi(\eta - \xi)} = C_1. \tag{2.7.8}$$

Noting, for example, that

$$\frac{1}{\xi(\eta - \xi)} = \frac{1}{\eta}\left(\frac{1}{\xi} + \frac{1}{\eta - \xi}\right), \tag{2.7.9}$$

the above two equations become

$$\int_0^\gamma \frac{W(\xi)\, \mathrm{d}\xi}{\eta_0 - \xi} = C_2, \tag{2.7.10}$$

and

$$\frac{\lambda(\eta) W(\eta)}{\eta \Psi(\eta)} + P \int_0^\gamma \frac{W(\xi)\, \mathrm{d}\xi}{\eta - \xi} = C_2, \tag{2.7.11}$$

where the new constant C_2 has the form

$$C_2 = C_1 - \int_0^\gamma \frac{W(\xi)\, \mathrm{d}\xi}{\xi}. \tag{2.7.12}$$

We must now solve equations (2.7.10) and (2.7.11) for the required weighting function $W(\xi)$. Clearly, the form of equation (2.7.11) is similar to equation (2.4.6) for $A(\xi)$. However, the differences are sufficiently important to warrant the exposition of the determination of $W(\xi)$.

We proceed as in section 2.4 by defining a function $Y(z)$ such that

$$Y(z) = \frac{1}{2\pi \mathrm{i}} \int_0^\gamma \frac{W(\xi)\, \mathrm{d}\xi}{\xi - z}, \tag{2.7.13}$$

satisfying:

$C_1(Y)$: $Y(z)$ is analytic for all $z \notin [0, \gamma]$.

$C_2(Y)$: $\quad Y(z) \sim \dfrac{1}{z}$ as $|z| \to \infty$.

$C_3(Y)$: $\quad |Y(z)| < \dfrac{k_1}{|z-\gamma|^{\alpha_1}}, \quad |Y(z)| < \dfrac{k_2}{|z|^{\alpha_2}},$

where k_1, k_2, α_1 and α_2 are constants, and

$0 \leqslant \alpha_1 < 1, \quad 0 \leqslant \alpha_2 < 1.$

The Plemelj formulae then give

$$Y^+(z) + Y^-(z) = \frac{1}{\pi i} P \int_0^\gamma \frac{W(\xi)\,d\xi}{\xi - z}, \tag{2.7.14}$$

and

$$Y^+(z) - Y^-(z) = W(z), \tag{2.7.15}$$

which, when substituted into equation (2.7.11), yields

$$\left(\frac{\lambda(z)}{z\Psi(z)} - \pi i\right) Y^+(z) - \left(\frac{\lambda(z)}{z\Psi(z)} + \pi i\right) Y^-(z) = C_2. \tag{2.7.16}$$

We have already defined $X(z)$ for the positive half-range in section 2.5, viz.

$$\frac{X^+(z)}{X^-(z)} = \frac{\dfrac{\lambda(z)}{z\Psi(z)} - \pi i}{\dfrac{\lambda(z)}{z\Psi(z)} + \pi i} \tag{2.7.17}$$

(see equations (2.5.7) and (2.5.8)), where $X(z)$ is analytic for all $z \notin [0, \gamma]$. In particular, it has the explicit form

$$X(z) = (z - \gamma)e^{\Gamma(z)}, \tag{2.7.18}$$

where $\Gamma(z) \to 0$ as $|z| \to \infty$. Equation (2.7.16) then may be written as

$$X^+(z)Y^+(z) - X^-(z)Y^-(z) = \frac{C_2 X^-(z)}{\dfrac{\lambda(z)}{z\Psi(z)} + \pi i}. \tag{2.7.19}$$

Recognising that

$$\Lambda^\pm(z) = \lambda(z) \pm \pi i z\Psi(z), \tag{2.7.20}$$

and

$$\frac{\Lambda^+(z)}{\Lambda^-(z)} = \frac{X^-(z)}{X^+(z)}, \tag{2.7.21}$$

equation (2.7.19) becomes

$$X^+(z)Y^+(z) - X^-(z)Y^-(z) = -\frac{C_2}{2\pi i}(X^+(z) - X^-(z)). \tag{2.7.22}$$

It is at this point that we depart from the procedure outlined in section 2.1.2. One could carry through the analysis as before, but the integral so obtained is not readily evaluated. It is somewhat simpler to define a new function $Q(z)$ such that

$$Q(z) = \frac{X(z)}{z - \eta_0}, \tag{2.7.23}$$

which then has the properties:

(i) $Q(z)$ is analytic in the complex z plane for all $z \notin [0, \gamma]$ except at $z = \eta_0$ at which point it has a pole of order 1.

(ii) $Q(z) \to 1$ as $|z| \to \infty$.

(iii) $Q(z)$ behaves at the endpoints 0 and γ of the cut in precisely the same manner as $X(z)$.

It is not difficult to see that

$$\frac{Q^+(z)}{Q^-(z)} = \frac{X^+(z)}{X^-(z)}, \tag{2.7.24}$$

so that equation (2.7.22) has the form

$$Q^+(z)Y^+(z) - Q^-(z)Y^-(z) = -\frac{C_2}{2\pi \mathrm{i}}(Q^+(z) - Q^-(z)). \tag{2.7.25}$$

The Plemelj formulae then yield

$$Y(z) = \frac{C_2}{4\pi^2 Q(z)} \int_0^\gamma \frac{Q^+(z') - Q^-(z')}{z' - z} \mathrm{d}z' + \frac{\mathscr{P}_n(z)}{Q(z)}, \tag{2.7.26}$$

where $\mathscr{P}_n(z)$ is an arbitrary polynomial of degree n.

Since $Q(z) \to 1$ as $|z| \to \infty$, and condition $C_2(Y)$ on $Y(z)$ stipulates that $Y(z) \sim 1/z$ as $|z| \to \infty$, one would normally put $\mathscr{P}_n(z) \equiv 0$. Note that the first term on the RHS of equation (2.7.26) satisfies condition $C_2(Y)$. However, after reference to the defining equation (2.7.13) for $Y(z)$, and to equation (2.7.10) which must be satisfied by the weighting function $W(\xi)$, we see that $Y(z)$ must also satisfy the extra condition

$$Y(\eta_0) = -\frac{C_2}{2\pi \mathrm{i}}. \tag{2.7.27}$$

Since $1/Q(\eta_0) = 0$ from equation (2.7.23), the first term on the RHS of equation (2.7.26) is always zero for $z = \eta_0$ so that, if the above condition on $Y(\eta_0)$ is to be satisfied, we must have

$$\frac{\mathscr{P}_n(\eta_0)}{Q(\eta_0)} = -\frac{C_2}{2\pi \mathrm{i}},$$

i.e.

$$\mathscr{P}_n(z) = -\frac{C_2 X(\eta_0)}{2\pi i(z-\eta_0)}. \qquad (2.7.28)$$

Note that the condition $C_2(Y)$ on $Y(z)$, i.e. $Y(z) \sim 1/z$ as $|z| \to \infty$. is still satisfied.

The integral occurring in equation (2.7.26) may be readily evaluated using the contours shown in figure 2.5. The contour L_1 surrounds the cut from 0 to γ whilst the contour L_2 is a circle of radius R with the origin as centre.

Clearly, we have

$$\int_0^\gamma \frac{Q^+(z') - Q^-(z')}{z'-z}\,dz' = \int_0^\gamma \frac{Q^+(z')\,dz'}{z'-z} + \int_\gamma^0 \frac{Q^-(z')\,dz'}{z'-z}$$

$$= \int_{L_1} \frac{Q(z')\,dz'}{z'-z}. \qquad (2.7.29)$$

Cauchy's integral theorem stipulates that

$$\int_{L_1} \frac{Q(z')\,dz'}{z'-z} + \lim_{R\to\infty} \int_{L_2} \frac{Q(z')\,dz'}{z'-z} = \int_L \frac{X(z')\,dz'}{(z'-z)(z'-\eta_0)}$$

$$= 2\pi i \sum_j R_j, \qquad (2.7.30)$$

where R_j is the residue of the jth pole contained within the contour $L = L_1 \cup L_2$. Note that we have implicitly assumed that L_1 is sufficiently large to contain the cut $[0, \gamma]$ but sufficiently small that it does not contain any poles which are outside the cut.

We therefore have

$$\int_{L_1} \frac{Q(z')\,dz'}{z'-z} = 2\pi i\left[\frac{X(z)}{z-\eta_0} + \frac{X(\eta_0)}{\eta_0 - z} - 1\right], \qquad (2.7.31)$$

Fig.2.5. The contours L_1 and L_2 in the complex z plane with L_1 surrounding the cut from 0 to γ.

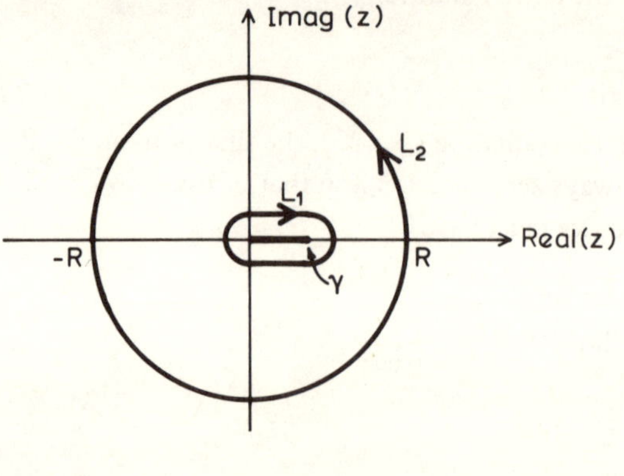

where the last term on the RHS of this last equation is the contribution from the integral over L_2. Note that if we had used equation (2.7.22) directly to obtain $Y(z)$, rather than introduce the new quantity $Q(z)$, the integral over L_2 would not be bounded at infinity.

If we now substitute equations (2.7.28) and (2.7.31) into (2.7.26), we find

$$Y(z) = -\frac{C_2}{2\pi \mathrm{i} Q(z)}\left[\frac{X(z)-X(\eta_0)}{z-\eta_0}-1\right]-\frac{C_2 X(\eta_0)}{2\pi \mathrm{i} Q(z)(z-\eta_0)}$$
$$= \frac{C_2}{2\pi \mathrm{i}}\left(\frac{z-\eta_0}{X(z)}-1\right). \tag{2.7.32}$$

The weighting function $W(z)$ is then immediately determined from equation (2.7.15), viz.

$$W(z) = \frac{C_2(z-\eta_0)}{2\pi \mathrm{i}}\left(\frac{1}{X^+(z)}-\frac{1}{X^-(z)}\right). \tag{2.7.33}$$

The choice of the constant C_2 is arbitrary ($\neq 0$) since its value does not alter the functional dependence of $W(z)$. (Note that, when using the orthogonality relationships, both sides of the equation representing the expansion are multiplied by $W(z)$ and the constant C_2 is effectively cancelled.) We therefore put $C_2 = 1$. This then completes the specification of the required weighting function.

As we shall see in the following section, there are several integrals which need evaluation. The reader is referred to appendix A for details. Here, we simply list the results:

$$\int_0^\gamma W(\xi)g(\eta,\xi)g(-\eta_0,\xi)\,\mathrm{d}\xi = -\frac{2\eta\eta_0^2}{(\eta+\eta_0)X(-\eta_0)}, \tag{2.7.34}$$

$$\int_0^\gamma W(\xi)g(\eta_0,\xi)g(-\eta_0,\xi)\,\mathrm{d}\xi = -\frac{\eta_0^2}{X(-\eta_0)}, \tag{2.7.35}$$

$$\int_0^\gamma W(\xi)g(\eta,\xi)g(-\eta',\xi)\,\mathrm{d}\xi = -\frac{\eta\eta'(\eta'+\eta_0)}{(\eta+\eta')X(-\eta')}, \tag{2.7.36}$$

$$\int_0^\gamma W(\xi)g(\eta_0,\xi)g(-\eta',\xi)\,\mathrm{d}\xi = -\frac{\eta_0\eta'}{X(-\eta')}, \tag{2.7.37}$$

$$\int_0^\gamma W(\xi)g^2(\eta_0,\xi)\,\mathrm{d}\xi = -\frac{\eta_0^2}{X(\eta_0)}, \tag{2.7.38}$$

$$\int_0^\gamma W(\xi)g^2(\eta,\xi)\,\mathrm{d}\xi = \frac{N(\eta)W(\eta)}{\eta\Psi(\eta)}, \tag{2.7.39}$$

where, in the above equations, $\eta, \eta' \in [0, \gamma]$, and

$$\frac{N(\eta)}{\eta\Psi(\eta)} = \frac{\lambda^2(\eta)}{\Psi^2(\eta)} + \pi^2\eta^2 \tag{2.7.40}$$

– see equation (2.6.23).

The weighting function $W(z)$, as with the $X(z)$ function, may be expressed in terms of the $H(z)$ function of Chandrasekhar [29]. This latter quantity is readily evaluated numerically and is of considerable benefit when actually detailing the solution to problems in radiative transfer theory. The conversion from $W(z)$ to $H(z)$ is given in appendix *B*. We find that

$$W(z) = \frac{z\Psi(z)H(z)}{\sqrt{\epsilon}}, \tag{2.7.41}$$

where $H(z)$ satisfies

$$\frac{1}{H(z)} = \sqrt{\epsilon} + \int_0^\gamma \frac{z'\Psi(z')H(z')\,\mathrm{d}z'}{z'+z}. \tag{2.7.42}$$

The convergence of the iterations used to solve numerically the above equation for $H(z)$ is extremely rapid. The relationship between $H(z)$ and $X(z)$ is

$$X(-z) = -\sqrt{\epsilon}\,(z+\eta_0)H(z), \tag{2.7.43}$$

and thus the RHS of the integrals given by equations (2.7.34) through (2.7.39) may be readily obtained.

It is clear from equation (2.7.18) that $X(\pm\eta_0) \neq 0$. We therefore have

$$H(z) \to \infty \text{ as } z \to -\eta_0, \tag{2.7.44}$$

from equation (2.7.43) and thus the direct conversion of $X(\eta_0)$ to $H(-\eta_0)$ in equation (2.7.38) is not justified. In this case, we define $I(\eta, \eta')$ by

$$I(\eta, \eta') = \int_0^\gamma W(\xi)g(\eta, \xi)g(\eta', \xi)\,\mathrm{d}\xi, \tag{2.7.45}$$

where $\eta, \eta' \notin [-\gamma, \gamma]$, i.e.

$$I(\eta, \eta') = \eta\eta' \int_0^\gamma \frac{W(\xi)\,\mathrm{d}\xi}{(\eta-\xi)(\eta'-\xi)}$$

$$= \frac{\eta\eta'}{\sqrt{\epsilon}(\eta'-\eta)} \int_0^\gamma \xi\Psi(\xi)H(\xi)\left(\frac{1}{\eta-\xi} - \frac{1}{\eta'-\xi}\right)\mathrm{d}\xi, \tag{2.7.46}$$

where we have used equation (2.7.41). If we now compare the integrals in this last equation with that appearing in equation (2.7.42), we easily find

$$I(\eta, \eta') = \frac{\eta\eta'}{\sqrt{\epsilon}(\eta'-\eta)}\left[\frac{1}{H(-\eta')} - \frac{1}{H(-\eta)}\right]. \tag{2.7.47}$$

However, equation (B.14) of appendix *B* states that

$$\frac{1}{H(\eta)H(-\eta)} = \Lambda(\eta), \tag{2.7.48}$$

so that, taking the limit as $\eta \to \eta_0$ and $\eta' \to \eta_0$, we have

$$\int_0^\gamma W(\xi) g^2(\eta_0, \xi)\, d\xi = \lim_{\substack{\eta \to \eta_0 \\ \eta' \to \eta_0}} I(\eta, \eta')$$

$$= \frac{\eta_0^2}{\sqrt{\epsilon}} \left[\frac{d}{d\eta} (H(\eta)\Lambda(\eta)) \right]_{\eta=\eta_0}$$

$$= \frac{\eta_0^2}{\sqrt{\epsilon}} H(\eta_0) \left[\frac{d\Lambda(\eta)}{d\eta} \right]_{\eta=\eta_0}, \tag{2.7.49}$$

since $\Lambda(\pm\eta_0) = 0$. The orthogonality relationship then has the simple form

$$\int_0^\gamma \xi \Psi(\xi) H(\xi) g^2(\eta_0, \xi)\, d\xi = \eta_0^2 H(\eta_0) \left[\frac{d\Lambda(\eta)}{d\eta} \right]_{\eta=\eta_0}. \tag{2.7.50}$$

The other relationships given by equations (2.7.34) through (2.7.39) are then

$$\int_0^\gamma \xi \Psi(\xi) H(\xi) g(\eta, \xi) g(-\eta_0, \xi)\, d\xi = \frac{\eta\eta_0}{(\eta + \eta_0) H(\eta_0)}, \tag{2.7.51}$$

$$\int_0^\gamma \xi \Psi(\xi) H(\xi) g(\eta_0, \xi) g(-\eta_0, \xi)\, d\xi = \frac{\eta_0}{2H(\eta_0)}, \tag{2.7.52}$$

$$\int_0^\gamma \xi \Psi(\xi) H(\xi) g(\eta, \xi) g(-\eta', \xi)\, d\xi = \frac{\eta\eta'}{(\eta + \eta') H(\eta')}, \tag{2.7.53}$$

$$\int_0^\gamma \xi \Psi(\xi) H(\xi) g(\eta_0, \xi) g(-\eta', \xi)\, d\xi = \frac{\eta_0\eta'}{(\eta_0 + \eta') H(\eta')}, \tag{2.7.54}$$

$$\int_0^\gamma \xi \Psi(\xi) H(\xi) g^2(\eta, \xi)\, d\xi = H(\eta) N(\eta), \tag{2.7.55}$$

where, in all the above equations, we have $\eta, \eta' \in [0, \gamma]$.

Similar expressions for the negative half-range expansion, viz.

$$r(\xi) = a^- g(-\eta_0, \xi) + \int_{-\gamma}^0 A(\eta) g(\eta, \xi)\, d\eta, \tag{2.7.56}$$

for all $\xi \in [-\gamma, 0]$ may be obtained. The development of the appropriate orthogonality relationships parallels precisely that presented above for the positive half-space.

In particular, we find a weighting function $W(\xi)$, viz.

$$W(\xi) = \frac{\xi \Psi(\xi) H(\xi)}{\sqrt{\epsilon}}, \tag{2.7.57}$$

where here $H(\xi)$ is the negative half-space Chandrasekhar function satisfying

$$\frac{1}{H(z)} = \sqrt{\epsilon} + \int_{-\gamma}^{0} \frac{z'\Psi(z')H(z')\,\mathrm{d}z'}{z'+z}, \tag{2.7.58}$$

such that

$$\int_{-\gamma}^{0} \xi\Psi(\xi)H(\xi)g(\eta,\xi)g(\eta',\xi)\,\mathrm{d}\xi = 0, \tag{2.7.59}$$

for all $\eta, \eta' \in [-\gamma, 0]$, or $\eta, \eta' = -\eta_0$, or any combination of the two, but with $\eta \neq \eta'$. The appropriate negative half-space orthogonality integrals are then

$$\int_{-\gamma}^{0} \xi\Psi(\xi)H(\xi)g(\eta,\xi)g(\eta_0,\xi)\,\mathrm{d}\xi = \frac{\eta\eta_0}{(\eta_0-\eta)H(-\eta_0)}, \tag{2.7.60}$$

$$\int_{-\gamma}^{0} \xi\Psi(\xi)H(\xi)g(-\eta_0,\xi)g(\eta_0,\xi)\,\mathrm{d}\xi = -\frac{\eta_0}{H(-\eta_0)}, \tag{2.7.61}$$

$$\int_{-\gamma}^{0} \xi\Psi(\xi)H(\xi)g(\eta,\xi)g(-\eta',\xi)\,\mathrm{d}\xi = \frac{\eta\eta'}{(\eta+\eta')H(\eta')}, \tag{2.7.62}$$

$$\int_{-\gamma}^{0} \xi\Psi(\xi)H(\xi)g(-\eta_0,\xi)g(-\eta',\xi)\,\mathrm{d}\xi = \frac{\eta_0\eta'}{(\eta_0-\eta')H(\eta')}, \tag{2.7.63}$$

$$\int_{-\gamma}^{0} \xi\Psi(\xi)H(\xi)g^2(\eta,\xi)\,\mathrm{d}\xi = \frac{H(\eta)N(\eta)}{\eta\Psi(\eta)}, \tag{2.7.64}$$

$$\int_{-\gamma}^{0} \xi\Psi(\xi)H(\xi)g^2(-\eta_0,\xi)\,\mathrm{d}\xi = \eta_0^2 H(-\eta_0)\left[\frac{\mathrm{d}\Lambda(\eta)}{\mathrm{d}\eta}\right]_{\eta=-\eta_0}, \tag{2.7.65}$$

where $\eta, \eta' \in [-\gamma, 0]$ in all these integrals, and where $N(\eta)$ is given by equation (2.7.40).

2.8 The Green's function solution

We are now in a position to solve equation (2.3.3), subject to conditions (2.3.4), (2.3.5) and (2.3.6), for the Green's function $G_V(\tau^*, \xi^* \to \tau, \xi)$. We previously found (see equation (2.3.34)) that the general form this solution takes is

$$G_V(\tau^*, \xi^* \to \tau, \xi) = a^+ g(\eta_0, \xi)\mathrm{e}^{\tau/\eta_0} + a^- g(-\eta_0, \xi)\mathrm{e}^{-\tau/\eta_0} + \int_{-\gamma}^{\gamma} A(\eta)g(\eta,\xi)\mathrm{e}^{\tau/\eta}\,\mathrm{d}\eta. \tag{2.8.1}$$

However, this function does not tend to zero as $|\tau - \tau^*| \to \infty$ as required by condition (2.3.5). To rectify this, we re-write $G_V(\tau^*, \xi^* \to \tau, \xi)$ in the form

$$G_V(\tau^*, \xi^* \to \tau, \xi) = a^+ g(\eta_0, \xi) e^{(\tau-\tau^*)/\eta_0} + \int_0^\gamma A(\eta) g(\eta, \xi) e^{(\tau-\tau^*)/\eta} \, d\eta, \tag{2.8.2}$$

for all $\tau < \tau^*$ whilst, for $\tau > \tau^*$, we write

$$G_V(\tau^*, \xi^* \to \tau, \xi) = -a^- g(-\eta_0, \xi) e^{-(\tau-\tau^*)/\eta_0} - \int_{-\gamma}^0 A(\eta) g(\eta, \xi) e^{(\tau-\tau^*)/\eta} \, d\eta. \tag{2.8.3}$$

These two solutions obviously still satisfy the integro-differential equation (2.3.3) since the alterations to the exponents are only multiplicative constants with respect to τ and ξ. Further, it is clear that the above $G_V(\tau^*, \xi^* \to \tau, \xi) \to 0$ as $|\tau - \tau^*| \to \infty$ as required.

If we now substitute the above expressions into the jump condition (2.3.4), viz.

$$\xi[G_V^+(\tau^*, \xi^* \to \tau, \xi) - G_V^-(\tau^*, \xi^* \to \tau, \xi)] = -\delta(\xi - \xi^*), \tag{2.8.4}$$

we find

$$-\frac{1}{\xi}\delta(\xi - \xi^*) = a^+ g(\eta_0, \xi) + a^- g(-\eta_0, \xi) + \int_{-\gamma}^{\gamma} A(\eta) g(\eta, \xi) \, d\eta. \tag{2.8.5}$$

We have already proved that $g(\eta_0, \xi), g(-\eta_0, \xi)$ and $g(\eta, \xi)$ $\eta \in [-\gamma, \gamma]$, form a complete set for $\xi \in [-\gamma, \gamma]$, i.e. the above expansion coefficients a^+, a^- and $A(\eta)$ may be uniquely determined. Use of the full-range orthogonality relationships derived in section 2.6 yields, for example,

$$a^+ N(\eta_0) = -\int_{-\gamma}^{\gamma} \delta(\xi - \xi^*) \Psi(\xi) g(\eta_0, \xi) \, d\xi,$$

i.e.

$$a^\pm = -\frac{g(\pm\eta_0, \xi^*)\Psi(\xi^*)}{N(\pm\eta_0)}, \tag{2.8.6}$$

where $N(\pm\eta_0)$ are given by equation (2.6.10). Similarly, we find

$$A(\eta) = -\frac{g(\eta, \xi^*)\Psi(\xi^*)}{N(\eta)}, \tag{2.8.7}$$

where $N(\eta)$ is given by equation (2.6.23).

The one remaining condition on $G_V(\tau^*, \xi^* \to \tau, \xi)$ is given by equation (2.3.6), viz.

$$G_V(\tau^*, \xi^* \to 0, \xi) = 0, \tag{2.8.8}$$

for all $\xi < 0$. This will not, in general, be satisfied by the above Green's function so that a further modification is necessary. We denote the above Green's function as $G_\infty(\tau^*, \xi^* \to \tau, \xi)$ where the subscript ∞ refers to the fact that, in the absence of any boundaries (i.e. for an infinite medium), the condition given by equation (2.8.8) would not exist and the above solution would suffice as it stands.

We therefore have

$$G_\infty(\tau^*, \xi^* \to \tau, \xi) = -\frac{\Psi(\xi^*)}{N(\eta_0)} g(\eta_0, \xi) g(\eta_0, \xi^*) e^{(\tau-\tau^*)/\eta_0} - \Psi(\xi^*) \int_0^\gamma \frac{g(\eta, \xi) g(\eta, \xi^*)}{N(\eta)} e^{(\tau-\tau^*)/\eta} \, d\eta, \quad (2.8.9)$$

for $\tau < \tau^*$ whilst, for $\tau > \tau^*$, we have

$$G_\infty(\tau^*, \xi^* \to \tau, \xi) = \frac{\Psi(\xi^*)}{N(-\eta_0)} g(-\eta_0, \xi) g(-\eta_0, \xi^*) e^{-(\tau-\tau^*)/\eta_0} + \Psi(\xi^*) \int_{-\gamma}^0 \frac{g(\eta, \xi) g(\eta, \xi^*)}{N(\eta)} e^{(\tau-\tau^*)/\eta} \, d\eta. \quad (2.8.10)$$

To satisfy the boundary condition (2.8.8), we put

$$G_V(\tau^*, \xi^* \to \tau, \xi) = G_\infty(\tau^*, \xi^* \to \tau, \xi) + b^- g(-\eta_0, \xi) e^{-\tau/\eta_0} + \int_{-\gamma}^0 B(\eta) g(\eta, \xi) e^{\tau/\eta} \, d\eta. \quad (2.8.11)$$

This new $G_V(\tau^*, \xi^* \to \tau, \xi)$ certainly satisfies the defining integro-differential equation (2.3.3) and the jump condition (2.3.4) – note that the added terms in equation (2.8.11) involving the expansion coefficients b^- and $B(\eta)$ are continuous at $\tau = \tau^*$. The boundary condition (2.8.8) then takes the form

$$0 = G_\infty(\tau^*, \xi^* \to 0, \xi) + b^- g(-\eta_0, \xi) + \int_{-\gamma}^0 B(\eta) g(\eta, \xi) \, d\eta, \quad (2.8.12)$$

for all $\xi < 0$, where the $G_\infty(\tau^*, \xi^* \to 0, \xi)$ appearing in this last equation must obviously be that given by equation (2.8.9) for $\tau = 0 < \tau^*$.

Again, we know that the expansion given by equation (2.8.12) is complete in the negative half-space. Therefore, if we apply the negative half-space orthogonality relationships (2.7.59) through (2.7.65), i.e. write

$$\int_{-\gamma}^0 \xi \Psi(\xi) H(\xi) g(\eta', \xi) \left\{ b^- g(-\eta_0, \xi) + \int_{-\gamma}^0 B(\eta) g(\eta, \xi) \, d\eta \right\} d\xi$$

$$= \int_{-\gamma}^{0} \xi\Psi(\xi)H(\xi)g(\eta', \xi) \left\{ \frac{\Psi(\xi^*)g(\eta_0, \xi)g(\eta_0, \xi^*)}{N(\eta_0)} \mathrm{e}^{-\tau^*/\eta_0} \right.$$

$$\left. + \int_0^{\gamma} \frac{\Psi(\xi^*)g(\eta, \xi)g(\eta, \xi^*)}{N(\eta)} \mathrm{e}^{-\tau^*/\eta} \, \mathrm{d}\eta \right\} \mathrm{d}\xi, \tag{2.8.13}$$

and, for example, put $\eta' = -\eta_0$, we find

$$b^- N_-(\eta_0) = -\frac{\eta_0}{H(-\eta_0)} \cdot \frac{\Psi(\xi^*)g(\eta_0, \xi^*)}{N(\eta_0)} \mathrm{e}^{-\tau^*/\eta_0}$$

$$-\eta_0 \Psi(\xi^*) \int_0^{\gamma} \frac{\eta g(\eta, \xi^*)}{(\eta_0 + \eta)H(-\eta)N(\eta)} \mathrm{e}^{-\tau^*/\eta} \, \mathrm{d}\eta, \tag{2.8.14}$$

where

$$N_-(\eta_0) = \eta_0^2 H(-\eta_0) \left[\frac{\mathrm{d}\Lambda(\eta)}{\mathrm{d}\eta} \right]_{\eta = -\eta_0}. \tag{2.8.15}$$

Note that the ranges of integration for the double integral on the RHS of equation (2.8.13) do not overlap. Consequently, the integrands are uniformly continuous when $\eta' < 0$ and the order of integration can be reversed. If, now, we put $\eta' \in [-\gamma, 0]$, equation (2.8.13) yields

$$B(\eta') \frac{H(\eta')N(\eta')}{\eta'\Psi(\eta')} = \frac{\eta'\eta_0}{(\eta_0 - \eta')H(-\eta_0)} \frac{\Psi(\xi^*)g(\eta_0, \xi^*)}{N(\eta_0)} \mathrm{e}^{-\tau^*/\eta_0}$$

$$- \int_0^{\gamma} \frac{\eta\eta'}{(\eta' - \eta)H(-\eta)} \frac{\Psi(\xi^*)g(\eta, \xi^*)}{N(\eta)} \mathrm{e}^{-\tau^*/\eta} \, \mathrm{d}\eta. \tag{2.8.16}$$

The above two equations then explicitly define the expansion coefficients b^- and $B(\eta)$, $\eta \in [-\gamma, 0]$. Finally, with these b^- and $B(\eta)$, we see that the (τ, τ^*) dependence of $G_V(\tau^*, \xi^* \to \tau, \xi)$ given by equation (2.8.11) is $\mathrm{e}^{-\tau/\eta_0 - \tau^*/\eta_0}$ and $\mathrm{e}^{+\tau/\eta + \tau^*/\eta'}$ with $\eta, \eta' \in [-\gamma, 0]$. Since, if $|\tau - \tau^*| \to \infty$, we must have at least one of the two conditions $\tau \to \infty$ or $\tau^* \to \infty$ occurring, then the terms involving b^- and $B(\eta)$ in equation (2.8.11) vanish as $|\tau - \tau^*| \to \infty$. Thus, $G_V(\tau^*, \xi^* \to \tau, \xi)$ given by equation (2.8.11) satisfies the defining integro-differential equation (2.3.3) and *all* the appropriate conditions (2.3.4), (2.3.5) and (2.3.6).

The radiation intensity $I(\tau, \Delta\nu, \mu) \equiv \Phi(\tau, \xi)$ may finally be evaluated by substitution of this Green's function into equation (2.3.1). However, if one is to proceed further, one must specify the source term $q(\tau^*, \xi^*)$ appearing in this equation. Here, we illustrate the type of solution one obtains by simply putting

$$q(\tau^*, \xi^*) \equiv 1. \tag{2.8.17}$$

Recognising that

$$\int_{-\gamma}^{\gamma} \Psi(\xi^*) g(\eta, \xi^*)\, \mathrm{d}\xi^* = 1 \tag{2.8.18}$$

(see equation (2.3.13)), we find

$$\begin{aligned}
I(\tau, \Delta\nu, \mu) = & \int_0^{\tau} \left\{ \frac{g(-\eta_0, \xi)}{N(-\eta_0)} \mathrm{e}^{-(\tau-\tau^*)/\eta_0} \right. \\
& \left. + \int_{-\gamma}^{0} \frac{g(\eta, \xi)}{N(\eta)} \mathrm{e}^{(\tau-\tau^*)/\eta}\, \mathrm{d}\eta \right\} \mathrm{d}\tau^* \\
& - \int_{\tau}^{\infty} \left\{ \frac{g(\eta_0, \xi)}{N(\eta_0)} \mathrm{e}^{(\tau-\tau^*)/\eta_0} \right. \\
& \left. + \int_0^{\gamma} \frac{g(\eta, \xi)}{N(\eta)} \mathrm{e}^{(\tau-\tau^*)/\eta}\, \mathrm{d}\eta \right\} \mathrm{d}\tau^* \\
& - \int_0^{\infty} \left\{ \frac{g(-\eta_0, \xi)}{N_{-}(\eta_0)} \mathrm{e}^{-\tau/\eta_0} \left\{ \frac{\eta_0 \mathrm{e}^{-\tau^*/\eta_0}}{H(-\eta_0) N(\eta_0)} \right. \right. \\
& \left. + \eta_0 \int_0^{\gamma} \frac{\eta \mathrm{e}^{-\tau^*/\eta}}{(\eta_0 + \eta) H(-\eta) N(\eta)}\, \mathrm{d}\eta \right\} \\
& - \int_{-\gamma}^{0} \frac{\eta \Psi(\eta) g(\eta, \xi)}{H(\eta) N(\eta)} \mathrm{e}^{\tau/\eta} \left\{ \frac{\eta \eta_0 \mathrm{e}^{-\tau^*/\eta_0}}{(\eta_0 - \eta) H(-\eta_0) N(\eta_0)} \right. \\
& \left. \left. + \int_0^{\gamma} \frac{\eta \eta' \mathrm{e}^{-\tau^*/\eta'}}{(\eta - \eta') H(-\eta') N(\eta')}\, \mathrm{d}\eta' \right\} \mathrm{d}\eta \right\} \mathrm{d}\tau^*,
\end{aligned} \tag{2.8.19}$$

which, after integrating over τ^*, yields

$$\begin{aligned}
I(\tau, \Delta\nu, \mu) = & \frac{\eta_0 g(-\eta_0, \xi)}{N(-\eta_0)} [1 - \mathrm{e}^{-\tau/\eta_0}] - \int_{-\gamma}^{0} \frac{\eta g(\eta, \xi)}{N(\eta)} [1 - \mathrm{e}^{\tau/\eta}]\, \mathrm{d}\eta \\
& - \frac{\eta_0 g(\eta_0, \xi)}{N(\eta_0)} - \int_0^{\gamma} \frac{\eta g(\eta, \xi)}{N(\eta)}\, \mathrm{d}\eta \\
& - \frac{g(-\eta_0, \xi)}{N_{-}(\eta_0)} \mathrm{e}^{-\tau/\eta_0} \left\{ \frac{\eta_0^2}{H(-\eta_0) N(\eta_0)} \right. \\
& \left. + \eta_0 \int_0^{\gamma} \frac{\eta^2}{(\eta_0 + \eta) H(-\eta) N(\eta)}\, \mathrm{d}\eta \right\} \\
& + \int_{-\gamma}^{0} \frac{\eta \Psi(\eta) g(\eta, \xi)}{H(\eta) N(\eta)} \mathrm{e}^{\tau/\eta} \left\{ \frac{\eta \eta_0^2}{(\eta_0 - \eta) H(-\eta_0) N(\eta_0)} \right.
\end{aligned}$$

$$+\int_0^\gamma \frac{\eta\eta'^2}{(\eta-\eta')H(-\eta')N(\eta')}\,d\eta'\Bigg\}\,d\eta, \tag{2.8.20}$$

where, of course, the explicit $\Delta\nu$ and η dependence of $I(\tau, \Delta\nu, \mu)$ arises from

$$\xi = \frac{\mu}{\phi(\Delta\nu)}. \tag{2.8.21}$$

The spectral line source function $S(\tau)$, given by

$$\begin{aligned} S(\tau) &= (1-\epsilon)\bar{J}(\tau) + \epsilon B_\nu(T) \\ &= \frac{1-\epsilon}{2}\int_{-\infty}^{\infty} \phi(\Delta\nu)\,d(\Delta\nu)\int_{-1}^{1} I(\tau, \Delta\nu, \mu)\,d\mu + \epsilon B_\nu(T) \\ &= \int_{-\gamma}^{\gamma} \Psi(\xi)\Phi(\tau, \xi)\,d\xi + \epsilon B_\nu(T), \end{aligned} \tag{2.8.22}$$

may be readily obtained by also using equation (2.8.18). For example, if we again take the source term

$$q(\tau^*, \xi^*) = \epsilon B_\nu(T), \tag{2.8.23}$$

as constant, equations (2.8.21) and (2.8.22), using (2.8.18), yield

$$\begin{aligned} \frac{S(\tau)}{\epsilon B_\nu(T)} &= 1 + \frac{\eta_0}{N(-\eta_0)}\,[1-e^{-\tau/\eta_0}] - \int_{-\gamma}^{0} \frac{\eta}{N(\eta)}\,[1-e^{\tau/\eta}]\,d\eta - \frac{\eta_0}{N(\eta_0)} \\ &\quad - \int_0^\gamma \frac{\eta\,d\eta}{N(\eta)} - \frac{e^{-\tau/\eta_0}}{N_-(\eta_0)}\Bigg\{\frac{\eta_0^2}{H(-\eta_0)N(\eta_0)} \\ &\quad + \eta_0\int_0^\gamma \frac{\eta^2\,d\eta}{(\eta_0+\eta)H(-\eta)N(\eta)}\Bigg\} \\ &\quad + \int_{-\gamma}^{0} \frac{\eta\Psi(\xi)e^{\tau/\eta}}{H(\eta)N(\eta)}\Bigg\{\frac{\eta\eta_0^2}{(\eta_0-\eta)H(-\eta_0)N(\eta_0)} \\ &\quad + \int_0^\gamma \frac{\eta\eta'^2\,d\eta'}{(\eta-\eta')H(-\eta')N(\eta')}\Bigg\}\,d\eta. \end{aligned} \tag{2.8.24}$$

We therefore see that all the Cauchy principal value integrals disappear leaving the relatively simple computer evaluation of the remaining integrals. This, of course, also holds even when $\epsilon B_\nu(T)$ is depth dependent.

2.9 Simplified expressions for surface quantities [82]

The methodology presented in the preceding sections for solution of the equation of radiative transfer is quite general and may therefore be applied regardless of the form of the source term $q(\tau^*, \xi^*)$. As we have seen, even for simple source terms, the resulting expressions for the radiative intensity and source function are somewhat involved. However, certain simplifications are possible when only the surface quantities are required. We illustrate these simplifications for the case of constant $B_\nu(T)$.

The reduced equation to be solved has the form (see equation (2.2.18)),

$$\xi \frac{\partial \Phi}{\partial \tau}(\tau, \xi) = \Phi(\tau, \xi) - \int_{-\gamma}^{\gamma} \Psi(\xi')\Phi(\tau, \xi')\, d\xi' - \epsilon B_\nu(T). \tag{2.9.1}$$

The solution to equation (2.9.1) consists of a homogeneous solution $\Phi_H(\tau, \xi)$ and a particular integral $\Phi_P(\tau, \xi)$ so that

$$\Phi(\tau, \xi) = \Phi_H(\tau, \xi) + \Phi_P(\tau, \xi). \tag{2.9.2}$$

We already know from section 2.3 that the homogeneous solution, satisfying the condition that the radiation field remains finite as $\tau \to \infty$, has the form

$$\Phi_H(\tau, \xi) = c^- g(-\eta_0, \xi) e^{-\tau/\eta_0} + \int_{-\gamma}^{0} C(\eta) g(\eta, \xi) e^{\tau/\eta}\, d\eta, \tag{2.9.3}$$

where the expansion coefficients c^- and $C(\eta)$ are to be determined from the pertinent boundary conditions.

In general, one obtains a particular integral of equation (2.9.1) by the application of Green's functions as presented in the preceding sections. However, when $B_\nu(T)$ is constant, we see that

$$\Phi_P(\tau, \xi) = B_\nu(T) \tag{2.9.4}$$

is a solution to equation (2.9.1) since

$$\int_{-\gamma}^{\gamma} \Psi(\xi)\, d\xi = 1 - \epsilon. \tag{2.9.5}$$

Our general solution is therefore

$$\Phi(\tau, \xi) = B_\nu(T) - c^- g(-\eta_0, \xi) e^{-\tau/\eta_0} + \int_{-\gamma}^{0} C(\eta) g(\eta, \xi) e^{\tau/\eta}\, d\eta. \tag{2.9.6}$$

The boundary condition is just

$$I(0, \Delta\nu, \mu) = 0 \quad \text{for all } \mu < 0,$$

i.e.

$$\Phi(0, \xi) = 0 \quad \text{for all } \xi < 0. \tag{2.9.7}$$

Equation (2.9.6) then yields

$$-B_\nu(T) = c^- g(-\eta_0, \xi) + \int_{-\gamma}^{0} C(\eta) g(\eta, \xi)\, d\eta, \tag{2.9.8}$$

for all $\xi < 0$.

Ordinarily, we would now apply the appropriate half-range orthogonality relationships given by equations (2.7.59) through (2.7.65) to obtain the expansion coefficients c^- and $C(\eta)$. However, the surface quantity $\Phi(0, \xi)$ for all $\xi > 0$, which is simply the emergent radiation intensity, may be obtained without the direct evaluation of c^- and $C(\eta)$.

To do this, we multiply equation (2.9.8) by $\xi\Psi(\xi)H(\xi)g(\xi', \xi)$, where $\xi' > 0$, and integrate over $\xi \in [-\gamma, 0]$. We then find

$$-B_\nu(T)\int_{-\gamma}^{0} \xi\Psi(\xi)H(\xi)g(\xi', \xi)\,\mathrm{d}\xi$$
$$= c^- \int_{-\gamma}^{0} \xi\Psi(\xi)H(\xi)g(\xi', \xi)g(-\eta_0, \xi)\,\mathrm{d}\xi$$
$$+ \int_{-\gamma}^{0} C(\eta)\,\mathrm{d}\eta \int_{-\gamma}^{0} \xi\Psi(\xi)H(\xi)g(\xi', \xi)g(\eta, \xi)\,\mathrm{d}\xi, \tag{2.9.9}$$

for all $\xi' > 0$. Note that in this last equation $g(\xi', \xi)$ is non-singular since $\xi' > 0$ and $\xi < 0$.

The negative half-space orthogonality relations we use here are

$$\int_{-\gamma}^{0} \xi\Psi(\xi)H(\xi)g(\xi', \xi)g(\eta, \xi)\,\mathrm{d}\xi = -\frac{\eta\xi'}{(\eta - \xi')H(-\xi')}, \tag{2.9.10}$$

and

$$\int_{-\gamma}^{0} \xi\Psi(\xi)H(\xi)g(\xi', \xi)g(-\eta_0, \xi)\,\mathrm{d}\xi = -\frac{\eta_0\xi'}{(\eta_0 + \xi')H(-\xi')}, \tag{2.9.11}$$

for all $\xi' > 0$. We also know that if $\xi' > 0$ and $\eta < 0$, then

$$\frac{\eta}{\eta - \xi'} = g(\eta, \xi'), \tag{2.9.12}$$

and

$$\frac{\eta}{\eta_0 + \xi'} = g(-\eta_0, \xi'). \tag{2.9.13}$$

Equation (2.9.9) then becomes

$$B_\nu(T)\int_{-\gamma}^{0} \xi\Psi(\xi)H(\xi)g(\xi', \xi)\,\mathrm{d}\xi$$
$$= \frac{\xi'}{H(-\xi')}\left\{c^- g(-\eta_0, \xi') + \int_{-\gamma}^{0} C(\eta)g(\eta, \xi')\,\mathrm{d}\eta\right\}. \tag{2.9.14}$$

However, the emergent intensity $\Phi(0, \xi)$ for all $\xi > 0$ is given by

$$\Phi(0, \xi) = B_\nu(T) + c^- g(-\eta_0, \xi) + \int_{-\gamma}^{0} C(\eta)g(\eta, \xi)\,\mathrm{d}\eta, \tag{2.9.15}$$

which, by comparison with equation (2.9.14) above, yields

$$\Phi(0, \xi) = B_\nu(T) + B_\nu(T)\frac{H(-\xi)}{\xi}\int_{-\gamma}^{0} \xi'\Psi(\xi')H(\xi')g(\xi, \xi')\,\mathrm{d}\xi'. \quad (2.9.16)$$

One may simplify the integral in equation (2.9.16) by writing

$$\int_{-\gamma}^{0} \xi'\Psi(\xi')H(\xi')g(\xi, \xi')\,\mathrm{d}\xi' = \xi\int_{-\gamma}^{0} \frac{\xi'\Psi(\xi')H(\xi')}{\xi - \xi'}\,\mathrm{d}\xi', \quad (2.9.17)$$

since here $\xi > 0$ and $\xi' < 0$. Equation (2.7.58) defines the negative half-space Chandrasekhar function, viz.

$$\frac{1}{H(\xi)} = \sqrt{\epsilon} + \int_{-\gamma}^{0} \frac{\xi'\Psi(\xi')H(\xi')}{\xi' + \xi}\,\mathrm{d}\xi', \quad (2.9.18)$$

so that equation (2.9.16), together with (2.9.17), yields

$$\Phi(0, \xi) = \sqrt{\epsilon}\, B_\nu(T)H(-\xi),$$

i.e.

$$I(0, \Delta\nu, \mu) = \sqrt{\epsilon}\, B_\nu(T)H\left(-\frac{\mu}{\phi(\Delta\nu)}\right). \quad (2.9.19)$$

This last equation is a particularly simple and convenient expression for the emergent radiation intensity.

The source function $S(0)$ at the surface of the model atmosphere may be obtained by noting that

$$S(\tau) = \int_{-\gamma}^{\gamma} \Psi(\xi)\Phi(\tau, \xi)\,\mathrm{d}\xi + \epsilon B_\nu(T), \quad (2.9.20)$$

which, since $\Phi(0, \xi) = 0$ for all $\xi < 0$, yields

$$S(0) = \int_{0}^{\gamma} \Psi(\xi)\Phi(0, \xi)\,\mathrm{d}\xi + \epsilon B_\nu(T)$$

$$= \sqrt{\epsilon}\, B_\nu(T)\int_{0}^{\gamma} \Psi(\xi)H(-\xi)\,\mathrm{d}\xi + \epsilon B_\nu(T). \quad (2.9.21)$$

If we now change variables from ξ to $-\xi$ in the last integral, and note that $\Psi(-\xi) = \Psi(\xi)$ for all ξ, we have

$$S(0) = \sqrt{\epsilon}\, B_\nu(T)\int_{-\gamma}^{0} \Psi(\xi)H(\xi)\,\mathrm{d}\xi + \epsilon B_\nu(T)$$

$$= \sqrt{\epsilon}\, B_\nu(T)\left(\frac{1}{H(0)} - \sqrt{\epsilon}\right) + \epsilon B_\nu(T),$$

from equation (2.9.18), i.e.

$$S(0) = \frac{\sqrt{\epsilon}\, B_\nu(T)}{H(0)}. \quad (2.9.22)$$

However, from equation (2.7.48), viz.

$$\frac{1}{H(\eta)H(-\eta)} = \Lambda(\eta), \tag{2.9.23}$$

which holds for both positive and negative half-space $H(\eta)$, we find

$$H(0) = 1, \tag{2.9.24}$$

since $\Lambda(0) = 1$. We therefore finally have

$$S(0) = \sqrt{\epsilon}\, B_\nu(T). \tag{2.9.25}$$

It is interesting to note that the surface source function $S(0)$ is independent of the frequency cut-off β and, indeed, independent of the precise form taken by the absorption probability $\phi(\Delta\nu)$.

2.10 The thermalisation depth [58]

The thermalisation depth Θ, as introduced in section 1.8, corresponds to the minimum depth at which $S(\tau) \approx B_\nu(T)$, i.e. we put

$$S(\Theta) \approx B_\nu(T). \tag{2.10.1}$$

The depth dependence of $S(\tau)$ given by equation (2.8.24) exhibits the two forms $e^{-\tau/\eta_0}$ and $e^{-\tau/\eta}$, where $\eta_0 > \gamma$ and $\eta \in [0, \gamma]$. Clearly, since $\eta_0 > \eta$, the asymptotic behaviour of $S(\tau)$ is reflected by the terms containing $e^{-\tau/\eta_0}$. Indeed, the function $S(\tau \gg \eta_0)$ hardly differs from the function $S(\tau \to \infty)$ and, since $S(\tau \to \infty) = B_\nu(T)$, we should have

$$S(\tau \gg \eta_0) \approx B_\nu(T). \tag{2.10.2}$$

Comparison between equations (2.10.1) and (2.10.2) then indicates that

$$\Theta \gg \eta_0. \tag{2.10.3}$$

The purpose of the following analysis is to (approximately) determine the value of η_0. First, we recall that η_0 is a zero of $\Lambda(\eta)$ where

$$\Lambda(\eta) = 1 - \eta \int_{-\gamma}^{\gamma} \frac{\Psi(\xi)}{\eta - \xi}\, d\xi. \tag{2.10.4}$$

For $|\eta| > |\xi|$, we may expand the integrand of equation (2.10.4) in a convergent series of ξ/η, viz.

$$\Lambda(\xi) = 1 - \int_{-\gamma}^{\gamma} \Psi(\xi) \left[1 + \frac{\xi}{\eta} + \left(\frac{\xi}{\eta}\right)^2 + \cdots\right] d\xi. \tag{2.10.5}$$

The function $\Psi(\xi)$ is even, i.e.

$$\Psi(-\xi) = \Psi(\xi), \tag{2.10.6}$$

so that

$$\int_{-\gamma}^{\gamma} \xi\Psi(\xi)\, d\xi = 0. \tag{2.10.7}$$

Consequently, for $|n/\xi| \gg 1$, equation (2.10.5) becomes

$$\Lambda(\eta) \approx 1 - \int_{-\gamma}^{\gamma} \Psi(\xi)\,\mathrm{d}\xi - \frac{1}{\eta^2}\int_{-\gamma}^{\gamma} \xi^2\Psi(\xi)\,\mathrm{d}\xi, \tag{2.10.8}$$

so that, with $\Lambda(\eta_0) = 0$, we have

$$\eta_0 \approx \left[\frac{\displaystyle\int_{-\gamma}^{\gamma} \xi^2\Psi(\xi)\,\mathrm{d}\xi}{1 - \displaystyle\int_{-\gamma}^{\gamma} \Psi(\xi)\,\mathrm{d}\xi}\right]^{1/2}. \tag{2.10.9}$$

The integrals appearing in this last equation may be evaluated by direct substitution of $\Psi(\xi)$ given by equation (2.2.16). We first obtain

$$\begin{aligned}\int_{-\gamma}^{\gamma} \Psi(\xi)\,\mathrm{d}\xi &= \frac{1-\epsilon}{2}\int_{-\gamma}^{\gamma} \mathrm{d}\xi \int_{R(\xi,\Delta\nu)} \mathrm{d}(\Delta\nu)\phi^2(\Delta\nu)\\ &= \frac{1-\epsilon}{2}\int_{-\beta}^{\beta} \phi^2(\Delta\nu)\,\mathrm{d}(\Delta\nu)\int_{-1}^{1}\frac{\mathrm{d}\mu}{\phi(\Delta\nu)}\\ &= 1-\epsilon,\end{aligned} \tag{2.10.10}$$

where, in the above equations, we have interchanged the order of integration and used the normalisation condition (2.2.11), viz.

$$\int_{-\beta}^{\beta} \phi(\Delta\nu)\,\mathrm{d}(\Delta\nu) = 1. \tag{2.10.11}$$

The integral appearing in the numerator of the RHS of equation (2.10.9) is not as readily evaluated. We find

$$\begin{aligned}\int_{-\gamma}^{\gamma} \xi^2\Psi(\xi)\,\mathrm{d}\xi &= \frac{1-\epsilon}{2}\int_{-\beta}^{\beta} \phi^2(\Delta\nu)\,\mathrm{d}(\Delta\nu)\int_{-1}^{1}\left(\frac{\mu}{\phi(\Delta\nu)}\right)^2\frac{\mathrm{d}\mu}{\phi(\Delta\nu)}\\ &= \frac{1-\epsilon}{3}\int_{-\beta}^{\beta}\frac{\mathrm{d}(\Delta\nu)}{\phi(\Delta\nu)}\end{aligned} \tag{2.10.12}$$

To proceed further, we require the explicit form of $\phi(\Delta\nu)$.

2.10.1 The Doppler profile

Equation (1.4.14) gives the form of the probability of absorption $\phi(\Delta\nu)$ for Doppler broadening, i.e.

$$\phi(\Delta\nu) = \frac{1}{\Delta\nu_\mathrm{D}\sqrt{\pi}}\,\mathrm{e}^{-(\Delta\nu/\Delta\nu_\mathrm{D})^2}, \quad \Delta\nu_\mathrm{D} = \frac{\nu}{c}\left[\frac{2kT}{m_\mathrm{A}}\right]^{1/2}. \tag{2.10.13}$$

We therefore have

$$\int_{-\beta}^{\beta} \frac{d(\Delta\nu)}{\phi(\Delta\nu)} = \Delta\nu_D\sqrt{\pi}\int_{-\beta}^{\beta} e^{(\Delta\nu/\Delta\nu_D)^2}\, d(\Delta\nu). \tag{2.10.14}$$

We now put $\sqrt{y} = \Delta\nu/\Delta\nu_D$ so that

$$\begin{aligned}
\int \frac{d(\Delta\nu)}{\phi(\Delta\nu)} &= \frac{\sqrt{\pi}}{2}(\Delta\nu_D)^2 \int \frac{e^y\, dy}{\sqrt{y}} \\
&= \frac{\sqrt{\pi}}{2}(\Delta\nu_D)^2 \left[\frac{e^y}{\sqrt{y}} + \frac{1}{2}\int \frac{e^y\, dy}{y^{3/2}}\right] \\
&= \frac{\sqrt{\pi}}{2}(\Delta\nu_D)^2 \left[1 + \frac{1}{2y} + \frac{3}{4y^2} + \cdots\right]\frac{e^y}{\sqrt{y}} \\
&= \frac{\sqrt{\pi}}{2}(\Delta\nu_D)^3 \left[1 + \frac{1}{2}\left(\frac{\Delta\nu_D}{\Delta\nu}\right)^2 + \frac{3}{4}\left(\frac{\Delta\nu_D}{\Delta\nu}\right)^4 \right. \\
&\quad \left. + \cdots\right]\frac{e^{(\Delta\nu/\Delta\nu_D)^2}}{\Delta\nu}.
\end{aligned} \tag{2.10.15}$$

Hence we have

$$\int_{-\beta}^{\beta} \frac{d(\Delta\nu)}{\phi(\Delta\nu)} \approx \sqrt{\pi}\,(\Delta\nu_D)^3\, \frac{e^{(\beta/\Delta\nu_D)^2}}{\beta} = \frac{(\Delta\nu_D)^2}{\beta\phi(\beta)}, \tag{2.10.16}$$

where, in obtaining equation (2.10.16), we have neglected the higher order terms on the RHS of equation (2.10.15) since we always have $\beta > \Delta\nu_D$. Indeed, as we shall see in the following chapter when discussing numerical methods of solution, we must have $\beta > 3\Delta\nu_D$ for Doppler absorption profiles. The question arises, however – what value do we take for the frequency cut-off? In particular, we note that if we explicitly consider a Doppler profile, then

$$\int_{-\infty}^{\infty} \frac{1}{\Delta\nu_D\sqrt{\pi}}\, e^{-(\Delta\nu/\Delta\nu_D)^2}\, d(\Delta\nu) = 1, \tag{2.10.17}$$

as required by the normalisation (1.4.8) of the absorption profile. Clearly, therefore, we must have

$$\int_{-\beta}^{\beta} \frac{1}{\Delta\nu_D\sqrt{\pi}}\, e^{-(\Delta\nu/\Delta\nu_D)^2}\, d(\Delta\nu) \neq 1, \tag{2.10.18}$$

in violation of the normalisation condition. Consequently, we must choose the frequency cut-off β sufficiently large so that the error due to finite β is sufficiently small. To develop a criterion for the choice of β, we first note that the model 2-level atom complete re-distribution source function with this frequency cut-off has the form

$$S(\tau) = \frac{1-\epsilon}{2} \int_{-\beta}^{\beta} \phi(\Delta\nu)\, d(\Delta\nu) \int_{-1}^{1} I(\tau, \Delta\nu, \mu)\, d\mu + \epsilon B_\nu(T), \quad (2.10.19)$$

where, for example, the $\epsilon B_\nu(T)$ represents the source of photons due to creation by collisional excitation (followed by spontaneous emission). Consequently, any error in the integral appearing in equation (2.10.19) arising from the use of finite β could be thought of as an extra source or sink of photons (source if the error is positive, sink if negative), but this extra source would be purely numerical with no physical meaning. Thus, since the error in the integral appearing in equation (2.10.18) is just

$$2 \int_{\beta}^{\infty} \frac{1}{\Delta\nu_D \sqrt{\pi}}\, e^{-(\Delta\nu/\Delta\nu_D)^2}\, d(\Delta\nu),$$

and the maximum value of $I(\tau, \Delta\nu, \mu)$ is $B_\nu(T)$, we would expect β to be sufficiently large so that, from equation (2.10.19), we have

$$(1-\epsilon) \int_{\beta}^{\infty} \frac{1}{\Delta\nu_D \sqrt{\pi}}\, e^{-(\Delta\nu/\Delta\nu_D)^2}\, d(\Delta\nu) \ll \epsilon, \quad (2.10.20)$$

if our solution is to be physically meaningful. If we evaluate this last integral by putting $\sqrt{y} = \Delta\nu/\Delta\nu_D$ again, we find

$$\frac{1-\epsilon}{2\sqrt{\pi}}\, \frac{\Delta\nu_D e^{-(\beta/\Delta\nu_D)^2}}{\beta} \ll \epsilon,$$

i.e.

$$(\Delta\nu_D)^2\, \frac{\phi(\beta)}{2\beta} \ll \frac{\epsilon}{1-\epsilon}. \quad (2.10.21)$$

Since $\phi(\beta)$ varies far more dramatically than β, we eliminate $\phi(\beta)$ from equations (2.10.16) and (2.10.21) to obtain

$$\int_{-\beta}^{\beta} \frac{d(\Delta\nu)}{\phi(\Delta\nu)} \gg \frac{(\Delta\nu_D)^4(1-\epsilon)}{2\epsilon\beta^2},$$

i.e.

$$\int_{-\gamma}^{\gamma} \xi^2\, \Psi(\xi)\, d\xi \gg \frac{(\Delta\nu_D)^4(1-\epsilon)^2}{6\epsilon\beta^2}. \quad (2.10.22)$$

Substitution of this last result into equation (2.10.9), with equation (2.10.10), yields

$$\eta_0 \gg \frac{(\Delta\nu_D)^2(1-\epsilon)}{\sqrt{6\epsilon}\beta}. \quad (2.10.23)$$

If we note that $\epsilon \ll 1$ and $\beta > 3\Delta\nu_D$, the thermalisation depth Θ_D for

Doppler broadening is then

$$\Theta_D \gg \frac{\Delta\nu_D}{\epsilon}. \tag{2.10.24}$$

The probability of absorption $\phi(0)$ at line centre is $1/\Delta\nu_D\sqrt{\pi}$ so that the mean distance λ_0 travelled by a photon at line centre is of order $\Delta\nu_D$. We therefore find

$$\Theta_D \gg \frac{\lambda_0}{\epsilon}, \tag{2 10.25}$$

in keeping with the discussion presented in section 1.8 - see equation (1.8.16).

2.10.2 *The Voigt profile*

We now turn our attention to the profile

$$\phi(\Delta\nu) = \frac{1}{\Delta\nu_D\sqrt{\pi}} H\left(a, \frac{\Delta\nu}{\Delta\nu_D}\right), \tag{2.10.26}$$

where

$$H(a, \eta) = \frac{a}{\pi}\int_{-\infty}^{\infty} \frac{e^{-y^2}\,dy}{(y-\eta)^2 + a^2} \tag{2.10.27}$$

(see equations (1.4.18) and (1.4.20)). It is not possible to perform the same direct type of analysis as in section 2.10.1 above. We therefore proceed by considering two limiting formulations of equations (2.10.26) and (2.10.27).

First, if we consider small $\Delta\nu$, and recognise that $a \ll 1$, the main contribution to the integral appearing in equation (2.10.27) comes from $y \approx \Delta\nu/\Delta\nu_D$. We then re-write $\phi(\Delta\nu)$ such that

$$\phi(\Delta\nu) = \frac{a}{\Delta\nu_D \pi^{3/2}}\, e^{-(\Delta\nu/\Delta\nu_D)^2} \int_{-\infty}^{\infty} \frac{dy}{\left(y - \dfrac{\Delta\nu}{\Delta\nu_D}\right)^2 + a^2}$$

$$+ \frac{a}{\Delta\nu_D \pi^{3/2}} \int_{-\infty}^{\infty} \frac{e^{-y^2} - e^{-(\Delta\nu/\Delta\nu_D)^2}}{\left(y - \dfrac{\Delta\nu}{\Delta\nu_D}\right)^2 + a^2}\, dy,$$

i.e.

$$\phi(\Delta\nu) = \frac{a}{\Delta\nu_D \pi^{3/2}}\, e^{-(\Delta\nu/\Delta\nu_D)^2} \left[\frac{1}{a}\tan^{-1}\left(\frac{y - \dfrac{\Delta\nu}{\Delta\nu_D}}{a}\right)\right]_{-\infty}^{\infty}$$

$$+ \text{ higher order terms}$$

$$= \frac{1}{\Delta\nu_D\sqrt{\pi}}\, e^{-(\Delta\nu/\Delta\nu_D)^2} + \text{higher order terms.} \tag{2.10.28}$$

Thus, the Voigt profile behaves as a Gaussian (i.e. a Doppler profile) for small $\Delta\nu$ - we have already examined the thermalisation depth for just such profiles in the preceding sub-section.

If we now consider large $\Delta\nu$, and note that the appearance of e^{-y^2} stipulates the consideration of only small y, the Voigt profile becomes

$$\phi(\Delta\nu) = \frac{a}{\Delta\nu_D \pi^{3/2}} \int_{-\infty}^{\infty} \frac{e^{-y^2}\, dy}{\left(y - \dfrac{\Delta\nu}{\Delta\nu_D}\right)} \left[1 - \frac{a^2}{\left(y - \dfrac{\Delta\nu}{\Delta\nu_D}\right)^2} + \cdots\right]$$

$$= \frac{a\Delta\nu_D}{\pi^{3/2}(\Delta\nu)^2} \int_{-\infty}^{\infty} e^{-y^2}\, dy \left[1 + \frac{2\Delta\nu_D y}{\Delta\nu} + \cdots\right] [1 + O(a^2)]$$

$$= \frac{a\Delta\nu_D}{\pi(\Delta\nu)^2} + O\left[a^3, \left(\frac{\Delta\nu_D}{\Delta\nu}\right)^3\right]. \tag{2.10.29}$$

We now wish to use this form of $\phi(\Delta\nu)$ in equation (2.10.12), viz.

$$\int_{-\gamma}^{\gamma} \xi^2 \Psi(\xi)\, d\xi = \frac{1-\epsilon}{3} \int_{-\beta}^{\beta} \frac{d(\Delta\nu)}{\phi(\Delta\nu)}$$

$$\approx \frac{(1-\epsilon)\pi}{3a\Delta\nu_D} \int_{-\beta}^{\beta} (\Delta\nu)^2\, d(\Delta\nu)$$

$$= \frac{2(1-\epsilon)\pi\beta^3}{9a\Delta\nu_D}. \tag{2.10.30}$$

From the above discussion of errors due to the frequency cut-off β (see equation (2.10.20)), we have

$$(1-\epsilon) \int_{\beta}^{\infty} \frac{a\Delta\nu_D}{\pi(\Delta\nu)^2}\, d(\Delta\nu) \ll \epsilon. \tag{2.10.31}$$

Note that we are integrating over $\Delta\nu \in [\beta, \infty)$ and it is therefore valid to use the asymptotic form for $\phi(\Delta\nu)$ given by the leading term on the RHS of equation (2.10.29). The above inequality is then

$$\beta \gg \frac{a(1-\epsilon)\Delta\nu_D}{\pi\epsilon}. \tag{2.10.32}$$

Hence, equation (2.10.30) yields

$$\int_{-\gamma}^{\gamma} \xi^2 \Psi(\xi)\, d\xi \gg \frac{2a^2(\Delta\nu_D)^2(1-\epsilon)^4}{9\pi^2\epsilon^3}. \tag{2.10.33}$$

Substitution of this result and equation (2.10.10) into equation (2.10.9) then leads to

$$\eta_0 \gg \frac{\sqrt{2}a\Delta\nu_D(1-\epsilon)^2}{3\pi\epsilon^2}. \tag{2.10.34}$$

However, this result only applies for large $|\Delta\nu|$; we must also take into account the smaller frequencies for which the Gaussian profile is applicable. We therefore have

$$\Theta_V \gg \max\left(\frac{\Delta\nu_D}{\epsilon}, \frac{a\Delta\nu_D}{\epsilon^2}\right), \tag{2.10.35}$$

where Θ_V is the thermalisation depth for the Voigt profile. Again, we have assumed $\epsilon \ll 1$. If we measure Θ_V in units of optical path length λ_0 at line centre, we find

$$\Theta_V \gg \max\left(\frac{\lambda_0}{\epsilon}, \frac{a\lambda_0}{\epsilon^2}\right), \tag{2.10.36}$$

see inequality (2.10.25).

2.10.3 Coherent scattering

It is of some interest to consider the case of completely coherent scattering in which the photon is absorbed and emitted at the one frequency measured in the rest frame of the observer. This is a somewhat physically idealistic case because it does now allow for the Doppler change in frequency due to the individual motions of the atoms. However, it does yield a result for the thermalisation depth Θ_C for coherent scattering which can be related to those derived above.

If we let $\Delta\nu_0$ be the one frequency at which photons are scattered, then

$$\phi(\Delta\nu) \equiv \delta(\Delta\nu - \Delta\nu_0). \tag{2.10.37}$$

Substitution of this $\phi(\Delta\nu)$ into equation (2.10.12) clearly presents difficulties. It is therefore more pertinent to evaluate $\Psi(\xi)$ explicitly. Equation (2.2.16), viz.

$$\Psi(\xi) = \frac{1-\epsilon}{2} \int_{R(\xi,\Delta\nu)} \phi^2(\Delta\nu)\, \mathrm{d}(\Delta\nu),$$

yields

$$\Psi(\xi) = \frac{1-\epsilon}{2}, \tag{2.10.38}$$

so that equation (2.10.4) becomes

$$\Lambda(\eta) = 1 - \frac{\eta(1-\epsilon)}{2} \int_{-1}^{1} \frac{\mathrm{d}\mu}{\eta - \mu}, \tag{2.10.39}$$

where we have put

$$\xi = \frac{\mu}{\phi(\Delta\nu)} \equiv \frac{\mu}{\phi(\Delta\nu_0)} = \mu. \tag{2.10.40}$$

In this case $\eta_0 > 1 \geqslant \mu$, so that

$$\Lambda(\eta) = 1 - \frac{1-\epsilon}{2}\int_{-1}^{1}\left(1 + \frac{\mu}{\eta_0} + \frac{\mu^2}{\eta_0^2} + \cdots\right) d\mu$$

$$= 1 - (1-\epsilon)\left[1 + \frac{1}{3\eta_0^2} + \cdots\right], \tag{2.10.41}$$

which, with

$$\Lambda(\eta_0) = 0, \tag{2.10.42}$$

yields

$$\eta_0 \approx \left(\frac{1-\epsilon}{3\epsilon}\right)^{1/2}. \tag{2.10.43}$$

Again taking $\epsilon \ll 1$, we have

$$\Theta_C \gg \frac{\lambda_0}{\sqrt{\epsilon}}, \tag{2.10.44}$$

where, of course, we put $1 = \lambda_0 \propto 1/\phi(\Delta\nu_0)$. This last result (2.10.44) was indicated in the discussion leading to equation (1.8.15).

2.II THE FOURIER TRANSFORM METHOD [27]

We now present an alternative method of finding exact closed form analytical solutions to the equation of radiative transfer. We consider Green's functions similar to those derived in the preceding sections, but satisfying partial differential equations as opposed to partial integro-differential equations. In this case, one need not work with the singular eigenfunctions $g(\eta, \xi)$.

Again, we consider a semi-infinite constant ϵ atmosphere with zero incident radiation on the surface.

2.11 The Green's function

We take equation (2.2.18) as our starting point, viz.

$$\xi\,\frac{\partial\Phi}{\partial\tau}(\tau, \xi) = \Phi(\tau, \xi) - \int_{-\gamma}^{\gamma} \Psi(\xi')\Phi(\tau, \xi')\, d\xi' - \epsilon B_\nu(T), \tag{2.11.1}$$

where

$$\Phi(\tau, \xi) \equiv I(\tau, \Delta\nu, \mu).$$

We now define the average intensity (multiplied by $1-\epsilon$) $\bar{\Phi}(\tau)$ as

$$\bar{\Phi}(\tau) = \int_{-\gamma}^{\gamma} \Psi(\xi)\Phi(\tau, \xi)\, d\xi, \tag{2.11.2}$$

i.e.

$$\bar{\Phi}(\tau) = (1-\epsilon)\bar{J}(\tau) = \frac{1-\epsilon}{2}\int_{-\infty}^{\infty} \phi(\Delta\nu)\, \mathrm{d}(\Delta\nu) \int_{-1}^{1} \mathrm{d}\mu\, I(\tau, \Delta\nu, \mu). \tag{2.11.3}$$

Equation (2.11.1) then has the form

$$\xi \frac{\partial \Phi}{\partial \tau}(\tau, \xi) = \Phi(\tau, \xi) - \bar{\Phi}(\tau) - q(\tau), \tag{2.11.4}$$

where

$$q(\tau) = \epsilon B_\nu(T).$$

If we now define the average Green's function $\bar{G}(\tau^* \to \tau, \xi)$ such that

$$\xi \frac{\partial \bar{G}}{\partial \tau}(\tau^* \to \tau, \xi) = \bar{G}(\tau^* \to \tau, \xi) - \delta(\tau - \tau^*), \tag{2.11.5}$$

multiply this last equation by $\bar{\Phi}(\tau^*) + q(\tau^*)$ and integrate over all τ^* space, then we immediately see that a solution to equation (2.11.4) is given by

$$\Phi(\tau, \xi) = \int_0^{\infty} \mathrm{d}\tau^* \bar{G}(\tau^* \to \tau, \xi)\, [\bar{\Phi}(\tau^*) + q(\tau^*)]. \tag{2.11.6}$$

However, this solution does not, in general, satisfy the required boundary condition given by

$$\Phi(0, \xi) = \Phi_{\mathrm{inc}}(\xi), \tag{2.11.7}$$

for all $\xi < 0$.

Again, as in section 2.1, we define volume and surface Green's functions $\bar{G}_{\mathrm{V}}(\tau^* \to \tau, \xi)$ and $\bar{G}_{\mathrm{S}}(0 \to \tau, \xi)$ which satisfy the differential equation

$$\xi \frac{\partial \bar{G}}{\partial \tau}(\tau^* \to \tau, \xi) = \bar{G}(\tau^* \to \tau, \xi) - \beta(\tau, \tau^*), \tag{2.11.8}$$

where

$$\beta_{\mathrm{V}}(\tau, \tau^*) = \delta(\tau - \tau^*) \tag{2.11.9}$$

$$\beta_{\mathrm{S}}(\tau, \tau^*) = 0, \tag{2.11.10}$$

and where

$$\bar{G}_{\mathrm{V}}(\tau^* \to 0, \xi) = 0 \quad \text{for all } \xi < 0, \tag{2.11.11}$$

and

$$\bar{G}_{\mathrm{S}}(0 \to 0, \xi) = 1 \quad \text{for all } \xi < 0. \tag{2.11.12}$$

Note that we must also have

$$\bar{G}(\tau^* \to \tau, \xi) \to 0 \quad \text{as } |\tau - \tau^*| \to \infty. \tag{2.11.13}$$

The solution for $\Phi(\tau, \xi)$ then has the form (see equation (2.1.45))

$$\Phi(\tau, \xi) = \int_0^\infty d\tau^* \bar{G}_V(\tau^* \to \tau, \xi)\, [\bar{\Phi}(\tau^*) + q(\tau^*)] + \bar{G}_S(0 \to \tau, \xi)\Phi_{inc}(\xi). \tag{2.11.14}$$

The first term on the RHS of equation (2.11.14) is just a particular solution of equation (2.11.4) whilst the second term is a solution to the corresponding homogeneous equation. Consequently, the above $\Phi(\tau, \xi)$ satisfies the equation (2.11.4) to be solved and the required boundary condition (2.11.7). It is therefore the complete solution.

Again, as in section 2.1, we can derive a relationship between $\bar{G}_V(\tau^* \to \tau, \xi)$ and $\bar{G}_S(0 \to \tau, \xi)$. Let $\chi_1(\tau, \xi)$ and $\chi_2(\tau, -\xi)$ satisfy the equations

$$\xi \frac{\partial \chi_1}{\partial \tau}(\tau, \xi) = \chi_1(\tau, \xi) - \beta_1(\tau), \tag{2.11.15}$$

$$-\xi \frac{\partial \chi_2}{\partial \tau}(\tau, -\xi) = \chi_2(\tau, -\xi) - \beta_2(\tau). \tag{2.11.16}$$

If we now multiply equation (2.11.15) by $\chi_2(\tau, -\xi)$ and equation (2.11.16) by $\chi_1(\tau, \xi)$, subtract then integrate over all τ space, we find

$$\int_0^\infty d\tau \left[\xi\chi_2(\tau, -\xi)\frac{\partial \chi_1}{\partial \tau}(\tau, \xi) + \xi\chi_1(\tau, \xi)\frac{\partial \chi_2}{\partial \tau}(\tau, -\xi)\right] = -\int_0^\infty d\tau\, [\beta_1(\tau)\chi_2(\tau, -\xi) - \beta_2(\tau)\chi_1(\tau, \xi)],$$

i.e.

$$\xi[\chi_1(\infty, \xi)\chi_2(\infty, -\xi) - \chi_1(0, \xi)\chi_2(0, -\xi)] = -\int_0^\infty d\tau\, [\beta_1(\tau)\chi_2(\tau, -\xi) - \beta_2(\tau)\chi_1(\tau, \xi)]. \tag{2.11.17}$$

As in section 2.1, we consider two distinct manipulations.

I. Put $\chi_1(\tau, \xi) = \bar{G}_V(\tau_1 \to \tau, \xi)$, (2.11.18)

and

$$\chi_2(\tau, \xi) = \bar{G}_S(0 \to \tau, \xi), \tag{2.11.19}$$

which, because of our definitions of the volume and surface Green's functions, yields

$$\beta_1(\tau) = \delta(\tau - \tau_1), \tag{2.11.20}$$

$$\beta_2(\tau) = 0, \tag{2.11.21}$$

with

$$\chi_1(0, \xi) = 0 \quad \text{for all } \xi < 0, \tag{2.11.22}$$

and

$$\chi_2(0, \xi) = 1 \quad \text{for all } \xi < 0. \tag{2.11.23}$$

Substitution of these quantities into equation (2.11.17), and noting condition (2.11.13), results in

$$\xi \bar{G}_V(\tau_1 \to 0, \xi) = \bar{G}_S(0 \to \tau_1, -\xi), \tag{2.11.24}$$

for all $\xi > 0$.

II. We next put

$$\chi_1(\tau, \xi) = \bar{G}_V(\tau_1 \to \tau, \xi), \tag{2.11.25}$$

and

$$\chi_2(\tau, \xi) = \bar{G}_V(\tau_2 \to \tau, \xi), \tag{2.11.26}$$

so that

$$\beta_1(\tau) = \delta(\tau - \tau_1), \tag{2.11.27}$$

$$\beta_2(\tau) = \delta(\tau - \tau_2), \tag{2.11.28}$$

with

$$\chi_1(0, \xi) = 0 = \chi_2(0, \xi) \quad \text{for all } \xi < 0, \tag{2.11.29}$$

whilst

$$\chi_1(\infty, \xi) = 0 = \chi_2(\infty, \xi). \tag{2.11.30}$$

Substitution into equation (2.11.17) then yields

$$0 = -\bar{G}_V(\tau_2 \to \tau_1, -\xi) + \bar{G}_V(\tau_1 \to \tau_2, \xi). \tag{2.11.31}$$

If we now combine equations (2.11.24) and (2.11.31), we find

$$\bar{G}_S(0 \to \tau_1, -\xi) = \xi \bar{G}_V(0 \to \tau_1, -\xi), \tag{2.11.32}$$

so that, replacing τ_1 by τ and ξ by $-\xi$, we have

$$\bar{G}_S(0 \to \tau, \xi) = -\xi \bar{G}_V(0 \to \tau, \xi). \tag{2.11.33}$$

Thus, the solution $\Phi(\tau, \xi)$, given by equation (2.11.14), may be re-written as

$$\Phi(\tau, \xi) = \int_0^\infty d\tau^* \bar{G}_V(\tau^* \to \tau, \xi) \, [\Phi(\tau^*) + q(\tau^*)] \\ - \xi \bar{G}_V(0 \to \tau, \xi) \Phi_{\text{inc}}(\xi). \tag{2.11.34}$$

Clearly, then, the solution requires the evaluation of the single Green's function $\bar{G}_V(\tau^* \to \tau, \xi)$.

2.12 The Fourier transform reduction

To proceed further, we define the infinite space Fourier transform pair [25]

$$\tilde{\Phi}_\infty(\kappa) = \mathscr{F}_\infty \{\bar{\Phi}(\tau)\} = \int_{-\infty}^{\infty} d\tau \, e^{i\kappa\tau} \bar{\Phi}(\kappa), \tag{2.12.1}$$

and

$$\bar{\Phi}(\tau) = \mathscr{F}_{\infty}^{-1}\{\tilde{\bar{\Phi}}_{\infty}(\kappa)\} = \frac{1}{2\pi}\int_{-\infty + i\alpha}^{\infty + i\alpha} d\kappa\, e^{-i\kappa\tau}\tilde{\bar{\Phi}}_{\infty}(\kappa), \tag{2.12.2}$$

where α appearing in the limits of integration in equation (2.12.2) defines the contour of integration in the usual strip of analyticity of the function $\tilde{\bar{\Phi}}_{\infty}(\kappa)$, i.e. $\alpha \in (a, b)$, where

$$|\bar{\Phi}(\tau)| \leqslant K_1 e^{a\tau}, \quad \tau \in (-\infty, 0), \tag{2.12.3}$$

and

$$|\bar{\Phi}(\kappa)| \leqslant K_2 e^{-b\tau}, \quad \tau \in (0, \infty), \tag{2.12.4}$$

where K_1 and K_2 are constants.

We next define the positive half-space Fourier transform pair

$$\tilde{\bar{\Phi}}^{+}(\kappa) = \mathscr{F}_{+}\{\bar{\Phi}(\tau)\} = \int_0^{\infty} d\tau\, e^{i\kappa\tau}\bar{\Phi}(\tau), \tag{2.12.5}$$

and

$$\bar{\Phi}(\tau) = \mathscr{F}_{+}^{-1}\{\tilde{\bar{\Phi}}^{+}(\kappa)\} = \frac{1}{2\pi}\int_{-\infty + i\alpha}^{\infty + i\alpha} d\kappa\, e^{-i\kappa\tau}\tilde{\bar{\Phi}}^{+}(\kappa). \tag{2.12.6}$$

Note that the integration is over the complete range of κ for both equations (2.12.2) and (2.12.6) even though the integration over κ is different in each Fourier transform pair, i.e. we have, for some general function $f(x)$ where

$$F(\kappa) = \int_a^b dx\, f_1(x)\, e^{i\kappa x}, \tag{2.12.7}$$

and

$$f_2(x) = \frac{1}{2\pi}\int_{-\infty + i\alpha}^{\infty + i\alpha} d\kappa\, F(\kappa)\, e^{-i\kappa x}, \tag{2.12.8}$$

that

$$f_1(x) = f_2(x), \tag{2.12.9}$$

only for all $x \in [a, b]$.

If we now apply the infinite space Fourier transform to equation (2.11.5), we find

$$\xi[\bar{G}_V(\tau^* \to \tau, \xi)e^{i\kappa\tau}]_{-\infty}^{\infty} - \xi\kappa i\,\mathscr{F}_{\infty}\{\bar{G}_V(\tau^* \to \tau, \xi)\}$$
$$= \mathscr{F}_{\infty}\{\bar{G}_V(\tau^* \to \tau, \xi)\} - e^{i\kappa\tau^*}. \tag{2.12.10}$$

However, we know that

$$\bar{G}_V(\tau^* \to \tau, \xi) \to 0 \quad \text{as } |\tau - \tau^*| \to \infty, \tag{2.12.11}$$

so that the above equation becomes

$$(1 + i\xi\kappa)\,\mathscr{F}_{\infty}\{\bar{G}_V(\tau^* \to \tau, \xi)\} = e^{i\kappa\tau^*}. \tag{2.12.12}$$

Application of the inverse infinite space Fourier transform given by equation (2.12.2) then yields

$$\bar{G}_V(\tau^* \to \tau, \xi) = \frac{1}{2\pi} \int d\kappa \, \frac{e^{-i\kappa(\tau - \tau^*)}}{1 + i\xi\kappa}. \tag{2.12.13}$$

Thus, the required Green's function is easily determined. Substitution of this result into equation (2.11.34) gives

$$\Phi(\tau, \xi) = \frac{1}{2\pi} \int_0^\infty d\tau^* \int \frac{d\kappa' e^{-\kappa'(\tau - \tau^*)}}{1 + i\xi\kappa'} \, [\bar{\Phi}(\tau^*) + q(\tau^*)]$$
$$- \frac{\xi}{2\pi} \int \frac{d\kappa' e^{-i\kappa'\tau}}{1 + i\xi\kappa'} \, \Phi_{\text{inc}}(\xi). \tag{2.12.14}$$

We now apply the infinite space Fourier transform to equation (2.12.14) to obtain

$$\mathscr{F}_\infty\{\Phi(\tau, \xi)\} = \frac{1}{2\pi} \int_{-\infty}^{\infty} d\tau \, e^{i\kappa\tau} \int_0^\infty d\tau^*$$
$$\times \int \frac{d\kappa' e^{-\kappa'(\tau - \tau^*)}}{1 + i\xi\kappa'} \, [\bar{\Phi}(\tau^*) + q(\tau^*)]$$
$$- \frac{\xi}{2\pi} \int_{-\infty}^{\infty} d\tau \, e^{i\kappa\tau} \int \frac{d\kappa' e^{-i\kappa'\tau}}{1 + i\xi\kappa'} \, \Phi_{\text{inc}}(\xi), \tag{2.12.15}$$

where we must carefully distinguish between the use of κ and κ'.

The functions appearing in the above integrands are uniformly continuous, and thus the integrals may be re-ordered to yield

$$\mathscr{F}_\infty\{\Phi(\tau, \xi)\} = \frac{1}{2\pi} \int_0^\infty d\tau^* \, [\bar{\Phi}(\tau^*) + q(\tau^*)]$$
$$\times \int_{-\infty}^{\infty} d\tau \, e^{i\kappa\tau} \int d\kappa' e^{-i\kappa'\tau} \, \frac{e^{i\kappa'\tau^*}}{1 + i\kappa'\xi}$$
$$- \frac{\xi}{2\pi} \Phi_{\text{inc}}(\xi) \int_{-\infty}^{\infty} d\tau \, e^{i\kappa\tau} \int d\kappa' e^{-i\kappa'\tau} \, \frac{1}{1 + i\kappa'\xi}$$
$$= \int_0^\infty d\tau^* \, [\bar{\Phi}(\tau^*) + q(\tau^*)] \int_{-\infty}^{\infty} d\tau \, e^{i\kappa\tau} \mathscr{F}_\infty^{-1} \left\{ \frac{e^{i\kappa'\tau^*}}{1 + i\kappa'\xi} \right\}$$
$$- \xi \Phi_{\text{inc}}(\xi) \int_{-\infty}^{\infty} d\tau \, e^{i\kappa\tau} \mathscr{F}_\infty^{-1} \left\{ \frac{1}{1 + i\kappa'\xi} \right\}. \tag{2.12.16}$$

Since

$$\mathscr{F}\{\mathscr{F}^{-1}\{F(\kappa)\}\} = F(\kappa), \tag{2.12.17}$$

the above equation yields

$$\mathscr{F}_{\infty}\{\Phi(\tau,\xi)\} = \int_0^{\infty} \mathrm{d}\tau^* [\bar{\Phi}(\tau^*) + q(\tau^*)] \frac{\mathrm{e}^{\mathrm{i}\kappa\tau^*}}{1+\mathrm{i}\kappa\xi} - \frac{\xi\Phi_{\mathrm{inc}}(\xi)}{1+\mathrm{i}\kappa\xi}$$

$$= \frac{1}{1+\mathrm{i}\kappa\xi}\,\mathscr{F}_{+}\{\bar{\Phi}(\tau)\} + \frac{1}{1+\mathrm{i}\kappa\xi}\,\mathscr{F}_{+}\{q(\tau)\} - \frac{\xi\Phi_{\mathrm{inc}}(\xi)}{1+\mathrm{i}\kappa\xi}. \tag{2.12.18}$$

We now multiply equation (2.12.18) by $\Psi(\xi)$ and integrate over $\xi \in [-\gamma, \gamma]$ to obtain

$$\mathscr{F}_{\infty}\{\bar{\Phi}(\tau)\} = \{\mathscr{F}_{+}\{\bar{\Phi}(\tau)\} + \mathscr{F}_{+}\{q(\tau)\}(1-\epsilon)\} \int_{-\gamma}^{\gamma} \frac{\Psi(\xi)\,\mathrm{d}\xi}{1+\mathrm{i}\kappa\xi} - \int_{-\gamma}^{\gamma} \frac{\xi\Psi(\xi)\Phi_{\mathrm{inc}}(\xi)\,\mathrm{d}\xi}{1+\mathrm{i}\kappa\xi}, \tag{2.12.19}$$

which we re-write as

$$\tilde{\Phi}_{\infty}(\kappa) = [1 - \tilde{\Lambda}(\kappa)]\tilde{\Phi}^{+}(\kappa) + \tilde{P}(\kappa), \tag{2.12.20}$$

where

$$\tilde{\Lambda}(\kappa) = 1 - \int_{-\gamma}^{\gamma} \frac{\Psi(\xi)\,\mathrm{d}\xi}{1+\mathrm{i}\kappa\xi}, \tag{2.12.21}$$

and

$$\tilde{P}(\kappa) = (1-\epsilon)[1-\tilde{\Lambda}(\kappa)]\,\mathscr{F}_{+}\{q(\tau)\} - \int_{-\gamma}^{\gamma} \frac{\xi\Psi(\xi)\Phi_{\mathrm{inc}}(\xi)\,\mathrm{d}\xi}{1+\mathrm{i}\kappa\xi}. \tag{2.12.22}$$

The newly defined function $\tilde{\Lambda}(\kappa)$ is known, as is the function $\tilde{P}(\kappa)$ once the boundary condition $\Phi_{\mathrm{inc}}(\xi)$ and the internal sources $q(\tau)$ are given. It now remains to solve equation (2.12.20) for the required unknown $\tilde{\Phi}^{+}(\tau)$, then apply the half-space inverse Fourier transform to obtain $\bar{\Phi}(\tau)$.

Before proceeding in the development of the solution, we examine the properties of $\tilde{\Lambda}(\kappa)$ more closely. Recall that, when using the discrete and singular eigenfunction technique as detailed in section 2.3, we defined the function $\Lambda(\eta)$ by

$$\Lambda(\eta) = 1 - \eta \int_{-\gamma}^{\gamma} \frac{\Psi(\xi)\,\mathrm{d}\xi}{\eta - \xi} \tag{2.12.23}$$

(see equation (2.3.17)). Clearly, then, we immediately have the simple relationship

$$\tilde{\Lambda}\left(\kappa = \frac{\mathrm{i}}{\eta}\right) = \Lambda(\eta). \tag{2.12.24}$$

Thus, all the theory developed for $\Lambda(\eta)$ may be used in studying $\tilde{\Lambda}(\kappa)$. In particular, since $\Lambda(\eta)$ is analytic for all $\eta \notin [-\gamma, \gamma]$, we see that $\tilde{\Lambda}(\kappa)$ is analytic in the complex κ plane cut from $i\gamma^{-1}$ to $i\infty$ and from $-i\gamma^{-1}$ to $-i\infty$ as shown in figure 2.6, i.e. $\pm i\gamma^{-1}$ are branch points.

We denote the upper and lower half-spaces, excluding the cuts, as D^+ and D^- respectively. Thus, $\tilde{\Lambda}(\kappa)$ is analytic for all $\kappa \in D^+ \cup D^-$.

Note that $\Lambda(\eta)$ has zeros at $\pm\eta_0$ so that $\tilde{\Lambda}(\kappa)$ has zeros at $\pm i/\eta_0$ which we write as $\pm i\kappa_0$. Since $\eta_0 \notin [-\gamma, \gamma]$ is real, i.e. $\eta_0 > \gamma$, then $\kappa_0 < \gamma^{-1}$ so that the zeros $\pm i\kappa_0$ of $\tilde{\Lambda}(\kappa)$ lie on the imaginary axis as shown in figure 2.6.

2.13 The Wiener-Hopf solution

We now wish to solve equation (2.12.20) for $\tilde{\Phi}^+(\kappa)$, where $\tilde{\Lambda}(\kappa)$ and $\tilde{P}(\kappa)$ are known.

We know from the theory of Fourier transforms that $\mathscr{F}_+\{\bar{\Phi}(\tau)\}$ is analytic in the upper half plane, i.e. the superscript ($^+$) in $\tilde{\Phi}^+(\kappa)$ now denotes upper-half plane analyticity. Similarly, if we write

$$\tilde{\Phi}^-(\kappa) = \int_{-\infty}^{0} d\tau\, e^{i\kappa\tau}\, \bar{\Phi}(\tau), \tag{2.13.1}$$

then $\tilde{\Phi}^-(\kappa)$ is analytic for all κ in the lower-half plane.

Thus, since we must have

$$\tilde{\Phi}_\infty(\kappa) = \tilde{\Phi}^+(\kappa) + \tilde{\Phi}^-(\kappa), \tag{2.13.2}$$

from equation (2.12.1), equation (2.12.20) may be written as

$$\tilde{\Lambda}(\kappa)\tilde{\Phi}^+(\kappa) = -\tilde{\Phi}^-(\kappa) + \tilde{P}(\kappa). \tag{2.13.3}$$

This last equation is of the same form solved in section 2.1 for the inhomogeneous Riemann-Hilbert problem. However, the region of analyticity of $\tilde{\Lambda}(\kappa)$ is relatively complicated (when compared with that for $\Lambda(\eta)$) and, although we could map this region onto another domain consisting of the whole complex

Fig.2.6. The region of analyticity of $\tilde{\Lambda}$.

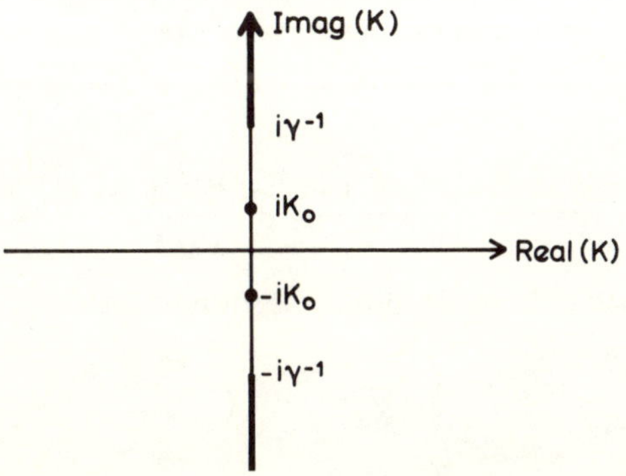

plane cut simply from $-\gamma$ to γ (which is the domain of analyticity of $\Lambda(\eta)$), thereby enabling the theory developed in section 2.1 to be easily applied, we find it more straightforward to use the standard Wiener-Hopf method of solution [25].

To do this, we must first find a Wiener-Hopf factorisation of the form

$$\tilde{\Lambda}(\kappa) = \frac{\tilde{\Lambda}^+(\kappa)}{\tilde{\Lambda}^-(\kappa)}, \tag{2.13.4}$$

where $\tilde{\Lambda}^\pm(\kappa)$ are analytic for all $\mathrm{Im}(\kappa) \gtrless 0$.

We shall return to the explicit evaluation of $\tilde{\Lambda}^+(\kappa)$ and $\tilde{\Lambda}^-(\kappa)$ later in this section. For the moment, we see that the substitution of equation (2.13.4) into (2.13.3) yields

$$\tilde{\Lambda}^+(\kappa)\tilde{\Phi}^+(\kappa) + \tilde{\Lambda}^-(\kappa)\tilde{\Phi}^-(\kappa) = \tilde{P}(\kappa)\tilde{\Lambda}^-(\kappa). \tag{2.13.5}$$

We next define a function $S(\kappa)$ such that

$$S(\kappa) = \tilde{\Lambda}^+(\kappa)\tilde{\Phi}^+(\kappa), \tag{2.13.6}$$

for all $\mathrm{Im}(\kappa) > 0$, and

$$S(\kappa) = -\tilde{\Lambda}^-(\kappa)\tilde{\Phi}^-(\kappa), \tag{2.13.7}$$

for all $\mathrm{Im}(\kappa) < 0$. Hence, $S(\kappa)$ is analytic for all κ in the complex plane, except for $\mathrm{Im}(\kappa) = 0$ where we have, from equation (2.13.5),

$$S^+(\kappa) - S^-(\kappa) = \tilde{P}(\kappa)\tilde{\Lambda}^-(\kappa). \tag{2.13.8}$$

The Plemelj formulae derived in section 2.1 then yield

$$S(\kappa) = \frac{1}{2\pi \mathrm{i}} \int_{-\infty}^{\infty} \frac{\tilde{P}(\kappa')\tilde{\Lambda}^-(\kappa')\,\mathrm{d}\kappa'}{\kappa' - \kappa} + \mathscr{P}_n(\kappa), \tag{2.13.9}$$

where $\mathscr{P}_n(\kappa)$ is a polynomial of degree n. The actual form of $\mathscr{P}_n(\kappa)$ will depend not only on the form of $\tilde{\Lambda}^\pm(\kappa)$, but also on the form of $\tilde{P}(\kappa)$ and this, in turn, depends upon the source term $q(\tau)$ and the boundary quantity $\Phi_{\mathrm{inc}}(\xi)$.

It should be stressed that the result given by equation (2.13.9) obtained from (2.13.8) is only valid if the composite function $\tilde{P}(\kappa)\tilde{\Lambda}^-(\kappa)$ satisfies a Lipshitz condition on the real axis. Since we have not, as yet, detailed the form $\tilde{\Lambda}^-(\kappa)$ takes, we must return to this consideration later.

The required function $\tilde{\Phi}^+(\kappa)$ may be immediately obtained from equations (2.13.6) and (2.13.9), viz.

$$\tilde{\Phi}^+(\kappa) = \frac{1}{2\pi \mathrm{i}\tilde{\Lambda}^+(\kappa)} \int_{-\infty}^{\infty} \frac{\tilde{P}(\kappa')\tilde{\Lambda}^-(\kappa')\,\mathrm{d}\kappa'}{\kappa' - \kappa} + \frac{\mathscr{P}_n(\kappa)}{\tilde{\Lambda}^+(\kappa)}. \tag{2.13.10}$$

The solution $\bar{\Phi}(\tau)$ is then obtained by direct substitution into

$$\bar{\Phi}(\tau) = \frac{1}{2\pi} \int_{-\infty + \mathrm{i}\alpha}^{\infty + \mathrm{i}\alpha} \mathrm{d}\kappa\, \mathrm{e}^{-\mathrm{i}\kappa\tau}\, \tilde{\Phi}^+(\kappa). \tag{2.13.11}$$

We now return to the explicit evaluation of $\tilde{\Lambda}^+(\kappa)$ and $\tilde{\Lambda}^-(\kappa)$. To do this, we first note that, since $\Lambda(\eta)$ has simple zeros at $\pm\eta_0$, then $\tilde{\Lambda}(\kappa)$ has simple zeros at $\pm i\eta_0^{-1}$, which we write as $\pm i\kappa_0$ - see figure 2.6. We also note that $\tilde{\Lambda}(\kappa)$ has branch points at $\kappa = i\gamma^{-1}$, which we write as $\pm i\kappa_B$. Thus, the function $T(\kappa)$ given by

$$T(\kappa) = \ln\left[\frac{\tilde{\Lambda}(\kappa)(\kappa^2 + \kappa_B^2)}{\kappa^2 + \kappa_0^2}\right], \tag{2.13.12}$$

is analytic for all $\kappa \in D^+ \cup D^-$ and, since $\tilde{\Lambda}(\kappa) \to 1$ as $|\kappa| \to \infty$, is bounded at infinity. Indeed, with an appropriate choice of branch of ln, we have

$$T(\kappa) \to 0 \quad \text{as } |\kappa| \to \infty. \tag{2.13.13}$$

If we now decompose this ln term, again with the same appropriate choice of branch of ln, we have

$$T(\kappa) = \ln\left[\frac{\tilde{\Lambda}^+(\kappa)}{\tilde{\Lambda}^-(\kappa)} \cdot \frac{\kappa^2 + \kappa_B^2}{\kappa^2 + \kappa_0^2}\right], \tag{2.13.14}$$

which we write as

$$T(\kappa) = T^+(\kappa) - T^-(\kappa), \tag{2.13.15}$$

where

$$T^+(\kappa) = \ln\left[\tilde{\Lambda}^+(\kappa) \cdot \frac{\kappa + i\kappa_B}{\kappa + i\kappa_0}\right], \tag{2.13.16}$$

and

$$T^-(\kappa) = \ln\left[\tilde{\Lambda}^-(\kappa) \cdot \frac{\kappa - i\kappa_0}{\kappa - i\kappa_B}\right]. \tag{2.13.17}$$

Note that $T^{\pm}(\kappa)$ are analytic for all $\text{Im}(\kappa) \gtrless 0$. We must now solve for $T^+(\kappa)$ and $T^-(\kappa)$.

Let us first consider a function $f(\kappa)$ analytic within the contour L, as shown in figure 2.7, where $L = L_+ \cup L_- \cup L_1 \cup L_2$.

Fig.2.7. The contour L in the complex κ plane.

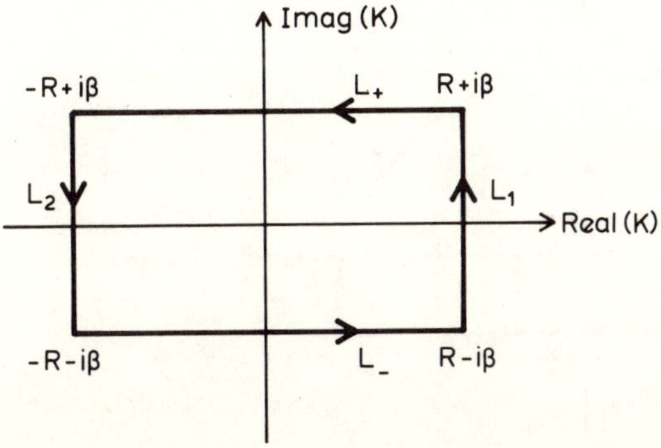

We then have from Cauchy's integral theorem that

$$f(z)=\frac{1}{2\pi \mathrm{i}}\int_L \frac{f(\kappa)\,\mathrm{d}\kappa}{\kappa-z}. \tag{2.13.18}$$

If we now allow $R\to\infty$ and $f(z)$ is such that $f(z)\to 0$ as $|z|\to\infty$, we have

$$\lim_{R\to\infty}\int_{L_1}\frac{f(\kappa)\,\mathrm{d}\kappa}{\kappa-z}=0=\lim_{R\to\infty}\int_{L_2}\frac{f(\kappa)\,\mathrm{d}\kappa}{\kappa-z}. \tag{2.13.19}$$

We now consider the contour $L=L_-\cup L_3$ as shown in figure 2.8 where L_3 is that part of a circle of radius R with centre at $(0,-\mathrm{i}\beta)$ that joins L_-.

Further, if we consider the analytic continuation of $f(\kappa)$, which we call $f^+(\kappa)$, for all $\mathrm{Im}(\kappa)>\beta$ where $f^+(\kappa)=f(\kappa)$ for all $\mathrm{Im}(\kappa)\in[-\beta,\beta]$, then Cauchy's integral theorem gives

$$f^+(z)=\frac{1}{2\pi \mathrm{i}}\lim_{R\to\infty}\int_{L_-\cup L_3}\frac{f(\kappa)\,\mathrm{d}\kappa}{\kappa-z}=\frac{1}{2\pi \mathrm{i}}\lim_{R\to\infty}\int_{L_-}\frac{f(\kappa)\,\mathrm{d}\kappa}{\kappa-z}, \tag{2.13.20}$$

where we have used the fact that the integral over L_3 vanishes as $R\to\infty$ since $f(z)\to 0$ as $|z|\to\infty$.

Thus, the integral over L_- as $R\to\infty$ represents a function analytic for all $\mathrm{Im}(\kappa)>-\beta$. Similarly, one may show that the integral over L_+ as $R\to\infty$ represents a function $f^-(\kappa)$ analytic for all $\mathrm{Im}(\kappa)$.

Equation (2.13.18), together with (2.13.19), then yields

$$f(z)=\frac{1}{2\pi \mathrm{i}}\int_{L_-\cup L_+}\frac{f(\kappa)\,\mathrm{d}\kappa}{\kappa-z}=f^+(\kappa)+f^-(\kappa). \tag{2.13.21}$$

We now take the contour $L=L_+\cup L_-\cup L_1\cup L_2$ such that L_+ and L_- enclose the branch cuts as shown in figure 2.9, whilst the contours L_1 and L_2

Fig.2.8. The contours L_- and L_3 in the complex κ plane.

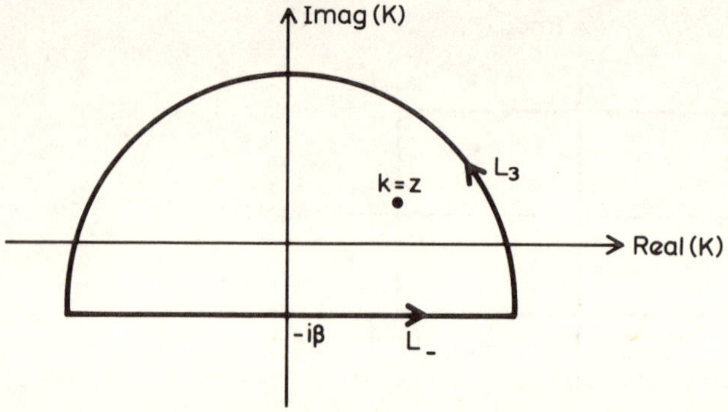

are now those parts of a circle of radius R with the centre as origin which join L_+ and L_-.

Since $T(\kappa) \to 0$ as $|\kappa| \to \infty$, we have

$$\lim_{R\to\infty} \int_{L_1} \frac{T(\kappa)\,\mathrm{d}\kappa}{\kappa - z} = 0 = \lim_{R\to\infty} \int_{L_2} \frac{T(\kappa)\,\mathrm{d}\kappa}{\kappa - z}, \tag{2.13.22}$$

so that

$$T(z) = \frac{1}{2\pi\mathrm{i}} \int_L \frac{T(\kappa)\,\mathrm{d}\kappa}{\kappa - z} = \frac{1}{2\pi\mathrm{i}} \int_{L_-\cup L_+} \frac{T(\kappa)\,\mathrm{d}\kappa}{\kappa - z}. \tag{2.13.23}$$

However, $T(\kappa)$ given by equation (2.13.15) consists of the sum of two parts $T^{\pm}(\kappa)$ analytic for all $\mathrm{Im}(\kappa) \gtrless 0$. Hence, if we compare equation (2.13.15) with (2.13.21), note the similarity of equation (2.13.20), for example, and the integral over L_- appearing in equation (2.13.23), and recall that both $T(z)$ and $f(z) \to 0$ as $|z| \to \infty$, then we immediately have

$$T^{\pm}(z) = \pm \frac{1}{2\pi\mathrm{i}} \int_{L_{\mp}} \frac{T(\kappa)\,\mathrm{d}\kappa}{\kappa - z}. \tag{2.13.24}$$

We may now substitute for $T(\kappa)$ from equation (2.13.12) to obtain

$$T^{+}(z) = \frac{1}{2\pi\mathrm{i}} \int_{L_-} \ln\left[\tilde{\Lambda}(\kappa)\cdot\frac{\kappa^2 + \kappa_B^2}{\kappa^2 + \kappa_0^2}\right] \frac{\mathrm{d}\kappa}{\kappa - z}, \tag{2.13.25}$$

and

$$T^{-}(z) = \frac{-1}{2\pi\mathrm{i}} \int_{L_+} \ln\left[\tilde{\Lambda}(\kappa)\cdot\frac{\kappa^2 + \kappa_B^2}{\kappa^2 + \kappa_0^2}\right] \frac{\mathrm{d}\kappa}{\kappa - z}. \tag{2.13.26}$$

Fig.2.9. The contour L in the complex κ plane indented at the cuts from $\mathrm{i}\kappa_B$ to $\mathrm{i}\infty$ and from $-\mathrm{i}\kappa_B$ to $-\mathrm{i}\infty$.

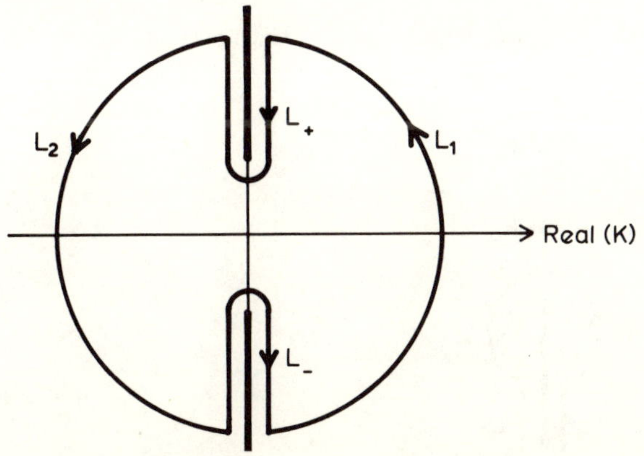

The above expression for $T^{+}(z)$ may be simplified by using the contour L_{-} as shown in figure 2.10, where $L_{-} = L_4 \cup L_5 \cup L_6$.

The contour L_6 is that part of the circle of radius with centre $(0, -\mathrm{i}\kappa_{\mathrm{B}})$ which joins L_4 and L_5.

We then define a parameter t such that $\kappa = t\mathrm{e}^{3\pi\mathrm{i}/2}$ on L_4 for $t \in (-\mathrm{i}\kappa_B, -\mathrm{i}\infty)$ as $\delta \to 0$ whilst on L_5, $\kappa = t\mathrm{e}^{-\pi\mathrm{i}/2}$. We then have

$$\lim_{\delta \to 0} \int_{L_4 \cup L_5} \ln\left(\frac{\kappa^2 + \kappa_{\mathrm{B}}^2}{\kappa^2 + \kappa_0^2}\right) \frac{\mathrm{d}\kappa}{\kappa - z}$$

$$= \int_{-\mathrm{i}\infty}^{-\mathrm{i}\kappa_{\mathrm{B}}} \ln\left(\frac{t^2 \mathrm{e}^{3\pi\mathrm{i}} + \kappa_{\mathrm{B}}^2}{t^2 \mathrm{e}^{3\pi\mathrm{i}} + \kappa_0^2}\right) \frac{\mathrm{e}^{3\pi\mathrm{i}/2}\,\mathrm{d}t}{t\mathrm{e}^{3\pi\mathrm{i}/2} - z}$$

$$+ \int_{-\mathrm{i}\kappa_{\mathrm{B}}}^{-\mathrm{i}\infty} \ln\left(\frac{t^2 \mathrm{e}^{-\pi\mathrm{i}} + \kappa_{\mathrm{B}}^2}{t^2 \mathrm{e}^{-\pi\mathrm{i}} + \kappa_0^2}\right) \frac{\mathrm{e}^{-\pi\mathrm{i}/2}\,\mathrm{d}t}{t\mathrm{e}^{-\pi\mathrm{i}/2} - z}$$

$$= 0, \tag{2.13.27}$$

where $z \notin R^{-}$ as shown in figure 2.10.

Further, the function $\ln\,[(\kappa - \mathrm{i}\kappa_{\mathrm{B}})/(\kappa^2 + \kappa_0^2)]/\kappa - z$ is analytic within L_6, i.e. for all $z \notin R^{-}$ so that, from Cauchy's integral theorem, we have

$$\int_{L_6} \ln\left(\frac{\kappa - \mathrm{i}\kappa_{\mathrm{B}}}{\kappa^2 + \kappa_0^2}\right) \frac{\mathrm{d}\kappa}{\kappa - z} = 0. \tag{2.13.28}$$

On L_6 we have $\kappa = -\mathrm{i}\kappa_{\mathrm{B}} + \delta \mathrm{e}^{\mathrm{i}\theta}$ so that

$$\lim_{\delta \to 0} \int_{L_6} \ln(\kappa + \mathrm{i}\kappa_{\mathrm{B}}) \frac{\mathrm{d}\kappa}{\kappa - z} = \lim_{\delta \to 0} \int_0^{2\pi} \frac{\ln(\delta \mathrm{e}^{\mathrm{i}\theta})\,\delta \mathrm{i}\mathrm{e}^{\mathrm{i}\theta}\,\mathrm{d}\theta}{-\mathrm{i}\kappa_{\mathrm{B}} + \delta \mathrm{e}^{\mathrm{i}\theta} - z},$$

$$= 0, \tag{2.13.29}$$

since $\delta \ln \delta \to 0$ as $\delta \to 0$.

Fig.2.10. The contour L_6 surrounding the branch point at $-\mathrm{i}\kappa_B$.

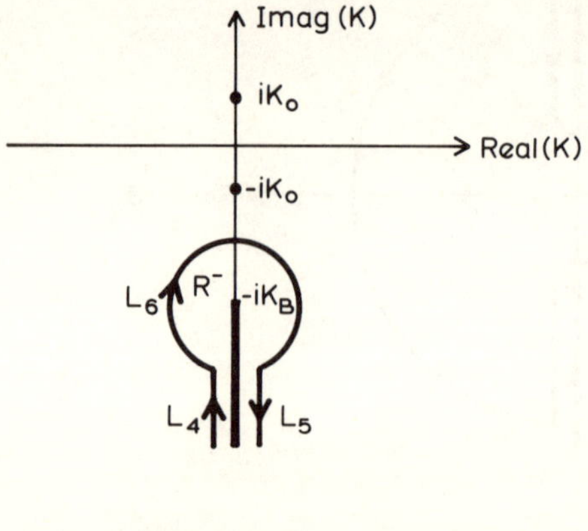

Combining all these results we have

$$\int_{L_-=L_4\cup L_5\cup L_6} \ln\left(\frac{\kappa^2+\kappa_B^2}{\kappa^2+\kappa_0^2}\right)\frac{d\kappa}{\kappa-z}=0 \qquad (2.13.30)$$

for all $z \notin R^-$.

If we now substitute this result into equation (2.13.25), we find

$$T^+(z)=\frac{1}{2\pi i}\int_{L_-}\ln\tilde{\Lambda}(\kappa)\cdot\frac{d\kappa}{\kappa-z}. \qquad (2.13.31)$$

Similarly, we find

$$T^-(z)=\frac{-1}{2\pi i}\int_{L_+}\ln\tilde{\Lambda}(\kappa)\cdot\frac{d\kappa}{\kappa-z}. \qquad (2.13.32)$$

This effectively completes the specification of $\tilde{\Lambda}^+(\kappa)$ and $\tilde{\Lambda}^-(\kappa)$. In particular, we see from equations (2.13.16) and (2.13.17) that

$$\tilde{\Lambda}^+(\kappa)=\frac{\kappa+i\kappa_0}{\kappa+i\kappa_B}\,e^{T^+(\kappa)}, \qquad (2.13.33)$$

and

$$\tilde{\Lambda}^-(\kappa)=\frac{\kappa-i\kappa_B}{\kappa-i\kappa_0}\,e^{T^-(\kappa)}. \qquad (2.13.34)$$

These $\tilde{\Lambda}^{\pm}(\kappa)$ are clearly analytic for all $\text{Im}(\kappa) \gtrless 0$ as required for the Wiener-Hopf factorisation and, in particular, tend to unity as $|\kappa|\to\infty$ since $T^{\pm}(\kappa)\to 0$ as $|\kappa|\to\infty$.

We may obtain $T^+(\kappa)$ and $T^-(\kappa)$ in terms of functions already evaluated using the singular eigenfunction technique. If we now map from the complex κ plane to the complex η plane using $\kappa = i/\eta$, we see from equation (2.12.24) that

$$\tilde{\Lambda}(\kappa)=\Lambda(\eta), \qquad (2.13.35)$$

so that the contour L_- in the complex κ plane maps onto the contour L_- as shown in figure 2.11.

Fig.2.11. The contour L_- surrounding the cut from $-\gamma$ to 0.

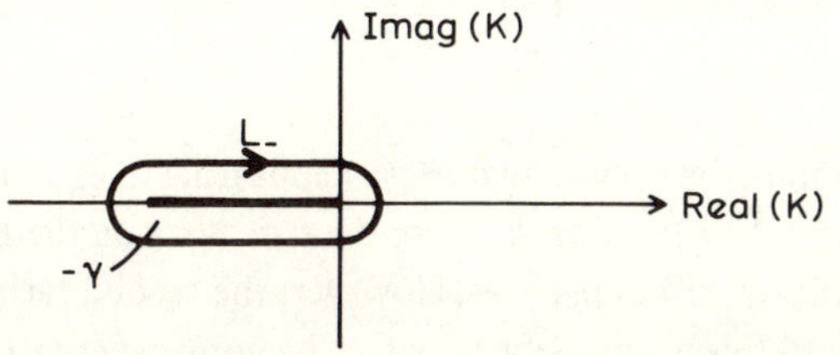

In particular, we have

$$\frac{1}{2\pi i}\int_{L_-} \ln \Lambda(\kappa)\,\frac{d\kappa}{\kappa - z} \to \frac{-1}{2\pi i}\int_{L_-} \ln \frac{\Lambda(\eta)\,d\eta}{\eta^2\left(\dfrac{1}{\eta}-\dfrac{1}{z}\right)}$$

$$= \frac{1}{2\pi i}\int_{L_-} \ln \Lambda(\eta)\,\frac{z\,d\eta}{\eta(\eta - z)}$$

$$= \frac{1}{2\pi i}\int_{L_-} \ln \Lambda(\eta)\left(\frac{1}{\eta - z} - \frac{1}{\eta}\right)d\eta$$

$$= \frac{1}{2\pi i}\int_{-\gamma}^{0} [\ln \Lambda^+(\eta) - \ln \Lambda^-(\eta)]\left(\frac{1}{\eta - z} - \frac{1}{\eta}\right)d\eta. \tag{2.13.36}$$

We know from section 2.4 that

$$\frac{\Lambda^+(\eta)}{\Lambda^-(\eta)} = \frac{1}{g(\eta)} \tag{2.13.37}$$

(see equation (2.4.57)) so that

$$\frac{1}{2\pi i}\int_{L_-} \ln \tilde{\Lambda}(\kappa)\,\frac{d\kappa}{\kappa - z} \to \frac{-1}{2\pi i}\int_{-\gamma}^{0} \ln g(\eta)\left(\frac{1}{\eta - z} - \frac{1}{\eta}\right)d\eta$$

$$= \Gamma_-(0) - \Gamma_-(z), \tag{2.13.38}$$

where we have used equation (2.5.23), viz.

$$\Gamma_-(z) = \frac{1}{2\pi i}\int_{-\gamma}^{0} \frac{\ln g(t)\,dt}{t - z} = \tilde{\Gamma}_-\left(\frac{i}{z}\right). \tag{2.13.39}$$

The above equations can then be combined to yield

$$\tilde{\Lambda}^+(\kappa) = \frac{\kappa + i\kappa_0}{\kappa + i\kappa_B}\, e^{\tilde{\Gamma}_-(i\infty) - \tilde{\Gamma}_-(i/\kappa)}. \tag{2.13.40}$$

A similar result may be obtained for $\tilde{\Lambda}^-(\kappa)$, viz.

$$\tilde{\Lambda}^-(\kappa) = \frac{\kappa - i\kappa_B}{\kappa - i\kappa_0}\, e^{\tilde{\Gamma}_+(i/\kappa) - \tilde{\Gamma}_+(i\infty)}, \tag{2.13.41}$$

where

$$\Gamma_+(z) = \frac{1}{2\pi i}\int_{0}^{\gamma} \frac{\ln g(t)\,dt}{t - z} = \tilde{\Gamma}_+\left(\frac{i}{z}\right) \tag{2.13.42}$$

- see equation (2.5.6).

We need now only examine the polynomial $\mathscr{P}_n(\kappa)$ appearing in equation (2.13.10). We see that, since $\tilde{\Lambda}^+(\kappa) \to 1$ as $|\kappa| \to \infty$, the first term on the RHS of equation (2.13.10) tends to zero as $|\kappa| \to \infty$. However, the second term behaves as κ^n as $|\kappa| \to \infty$ and this is unsatisfactory for the convergence of the integral appearing in equation (2.12.6) constituting the inverse Fourier transform. Thus, we must have

$$\mathscr{P}_n(\kappa) \equiv 0. \tag{2.13.43}$$

Further, we note that $\tilde{\Lambda}^-(\kappa)$ satisfies a Lipschitz condition on the real axis so that, for physically realistic $q(\tau)$ and $\Phi_{\mathrm{inc}}(\xi)$, the composite function $P(\kappa)\tilde{\Lambda}^-(\kappa)$ also satisfies this Lipschitz condition. Consequently, the use of the Plemelj formulae leading to equation (2.13.9) is valid.

The final solution is therefore

$$\bar{\Phi}(\tau) = \frac{1}{(2\pi)^2 \mathrm{i}} \int_{-\infty+\mathrm{i}\alpha}^{\infty+\mathrm{i}\alpha} \frac{\mathrm{d}\kappa\, \mathrm{e}^{-\mathrm{i}\kappa\tau}}{\tilde{\Lambda}^+(\kappa)} \int_{-\infty}^{\infty} \frac{\tilde{P}(\kappa')\tilde{\Lambda}^-(\kappa')\, \mathrm{d}\kappa'}{\kappa' - \kappa}, \tag{2.13.44}$$

where $\tilde{\Lambda}^{\pm}(\kappa)$ are given by equations (2.13.40) and (2.13.41).

This is a somewhat more compact closed form analytical solution than that given by equation (2.8.24), for example, using the singular eigenfunction expansion technique, but it would appear that in practice the Fourier transform method might be slightly more cumbersome when performing the final numerical evaluation of the integrals appearing in equation (2.13.44).

3

Numerical methods of solution

The simple equation of radiative transfer derived in chapter 1 is non-linear partial integro-differential and cannot be solved in closed analytical form. Thus there is a need for numerical methods of solution even for the simplest cases of astrophysical interest. In the present chapter, we consider only the 1-dimensional equation solved exactly in the preceding chapter, i.e.

$$\mu \frac{\partial I}{\partial \tau}(\tau, \Delta\nu, \mu) = \phi(\Delta\nu)[I(\tau, \Delta\nu, \mu) - S(\tau)], \tag{3.I}$$

where

$$S(\tau) = \frac{1-\epsilon}{2} \int_{-\infty}^{\infty} \phi(\Delta\nu)\, \mathrm{d}(\Delta\nu) \int_{-1}^{1} I(\tau, \Delta\nu, \mu)\, \mathrm{d}\mu + \epsilon B_\nu(T). \tag{3.II}$$

These two equations have been obtained by assuming complete re-distribution, i.e. we have equated the emission and absorption probabilities $\psi(\Delta\nu)$, $j(\Delta\nu)$ and $\phi(\Delta\nu)$. We further note therefore that

$$\mathrm{d}\tau = -\frac{h\nu_0}{4\pi}(N_L B_{LU} - N_U B_{UL})\, \mathrm{d}z \tag{3.III}$$

- see equations (1.9.2) and (1.9.12). We have seen, however, that $N_U B_{UL} \ll N_L B_{LU}$ in general. Indeed, it is the $N_U B_{UL}$ term in equation (3.III) which introduces the non-linearity in the unknown $I(\tau, \Delta\nu, \mu)$ through equation (1.3.13) via the ratio N_U/N_L. Thus, if we neglect the stimulated emission component in equation (3.III), we have

$$\mathrm{d}\tau = -\frac{h\nu_0}{4\pi} N_L B_{LU}\, \mathrm{d}z, \tag{3.IV}$$

where, since N_L is a function of depth only, the optical depth τ and physical depth $-z$ are simply related.

Equations (3.I) and (3.II) were solved exactly in chapter 2 only for constant ϵ and depth-independent $\phi(\Delta\nu)$. However, this ϵ involves collisions between atoms and free electrons and must therefore be a function of position (depth). Further, reference to section 1.4, particularly to equations (1.4.14) and (1.4.15), shows that $\phi(\Delta\nu)$ contains the Doppler width $\Delta\nu_D$ which, itself, depends upon position via the temperature T. In the following, we therefore present a variety of numerical methods which not only enable us to solve equations (3.I) and (3.II) for depth-dependent ϵ and $\phi(\Delta\nu)$ but, more importantly, can be generalised to solve still more complicated problems involving mult-dimensional geometry, partial re-distribution, multi-level atoms, velocity-dependent source functions, etc. It should be stressed, however, that the accuracy of each method will probably depend upon the particular problem under consideration, i.e. some methods could perform well under certain physical situations but fare rather badly under others. It is important therefore to check each numerical solution either by comparing results with those obtained by an alternative method or by testing the method on a 'reduced' problem for which the solution is known exactly. Of course, basic computer limitations will not enable us to eliminate all errors, but at least the confidence we place in our solution will be increased.

3.1 Λ iteration [68]

The most straightforward numerical method of solution of the equation of radiative transfer is Λ iteration.

Equation (3.I) may be immediately written in the form

$$I(\tau, \Delta\nu, \mu) = \int_{\tau}^{\infty} \phi(\Delta\nu) S(\tau') \mathrm{e}^{-\phi(\Delta\nu)(\tau'-\tau)/\mu} \frac{\mathrm{d}\tau'}{\mu}, \qquad (3.1.1)$$

for all $\mu > 0$ whilst, for $\mu < 0$, we have

$$I(\tau, \Delta\nu, \mu) = -\int_{0}^{\tau} \phi(\Delta\nu) S(\tau') \mathrm{e}^{-\phi(\Delta\nu)(\tau'-\tau)/\mu} \frac{\mathrm{d}\tau'}{\mu}, \qquad (3.1.2)$$

where we have taken zero incident radiation flux at the surface of the atmosphere.

If we now substitute these expressions into the source function $S(\tau)$ given by equation (3.II), we find

$$S(\tau) = \frac{1-\epsilon}{2} \int_{-\infty}^{\infty} \phi(\Delta\nu)\, \mathrm{d}(\Delta\nu) \left\{ \int_{0}^{1} \frac{\mathrm{d}\mu}{\mu} \int_{\tau}^{\infty} \mathrm{d}\tau' \right.$$

$$\left. - \int_{-1}^{0} \frac{\mathrm{d}\mu}{\mu} \int_{0}^{\tau} \mathrm{d}\tau' \right\} S(\tau') \phi(\Delta\nu) \mathrm{e}^{-\phi(\Delta\nu)(\tau'-\tau)/\mu} + \epsilon B_\nu(T). \quad (3.1.3)$$

Replacing μ by $-\mu$ in the second term in braces appearing in equation (3.1.3) then yields

$$S(\tau) = \frac{1-\epsilon}{2}\int_{-\infty}^{\infty} \phi(\Delta\nu)\,\mathrm{d}(\Delta\nu)\int_0^1 \frac{\mathrm{d}\mu}{\mu}\int_0^{\infty} \mathrm{d}\tau' S(\tau')$$
$$\times\, \phi(\Delta\nu)\mathrm{e}^{-\phi(\Delta\nu)|\tau'-\tau|/\mu} + \epsilon B_\nu(T), \tag{3.1.4}$$

which we re-write as

$$S(\tau) = (1-\epsilon)\Lambda S + \epsilon B_\nu(T), \tag{3.1.5}$$

where the operator Λ obviously has the form

$$\Lambda \equiv \tfrac{1}{2}\int_{-\infty}^{\infty} \phi^2(\Delta\nu)\,\mathrm{d}(\Delta\nu)\int_0^1 \frac{\mathrm{d}\mu}{\mu}\int_0^{\infty} \mathrm{e}^{-\phi(\Delta\nu)|\tau'-\tau|/\mu}\,\mathrm{d}\tau'. \tag{3.1.6}$$

The Λ iteration procedures then involve an initial guess at $S(\tau)$, substitution of this guess into the RHS of equation (3.1.5) to obtain a new $S(\tau)$, then re-substitution, etc. until convergence is attained. We therefore have the nth iteration term

$$S_n(\tau) = (1-\epsilon)\Lambda S_{n-1} + \epsilon B_\nu(T). \tag{3.1.7}$$

Note that in the above expression for Λ, and in equations (3.1.1) and (3.1.2), we have taken $\phi(\Delta\nu)$ to be depth independent. If $\phi(\Delta\nu)$ depends upon τ, we easily find

$$\Lambda \equiv \tfrac{1}{2}\int_{-\infty}^{\infty} \phi(\Delta\nu,\tau)\,\mathrm{d}(\Delta\nu)\int_0^1 \frac{\mathrm{d}\mu}{\mu}\int_0^{\infty} \mathrm{d}\tau'$$
$$\times\, \phi(\Delta\nu,\tau')\exp\left[-\frac{1}{\mu}\left|\int_{\tau'}^{\tau} \phi(\Delta\nu,\tau'')\,\mathrm{d}\tau''\right|\right]. \tag{3.1.8}$$

3.1.1 Convergence

The rate of convergence [9] of the Λ iterative scheme can be simply understood physically by first re-writing equation (3.1.5) in the form

$$S(\tau) = [1-(1-\epsilon)\Lambda]^{-1}[\epsilon B_\nu(T)], \tag{3.1.9}$$

where, it must be remarked, Λ is an operator dependent upon τ and τ' and does not, in general, commute. Equation (3.1.9) can then be formally written as

$$S(\tau) = \{1 + (1-\epsilon)\Lambda + (1-\epsilon)\Lambda(1-\epsilon)\Lambda + \cdots$$
$$+ [(1-\epsilon)\Lambda]^n + \cdots\}\,[\epsilon B_\nu(T)]. \tag{3.1.10}$$

If we choose as our zeroth order solution

$$S_0(\tau) = \epsilon B_\nu(T), \tag{3.1.11}$$

we see that

$$S_1(\tau) = (1-\epsilon)\Lambda S_0(\tau) + \epsilon B_\nu(T)$$
$$= \{1 + (1-\epsilon)\Lambda\}\,[\epsilon B_\nu(T)], \tag{3.1.12}$$

$$S_2(\tau) = \{1 + (1-\epsilon)\Lambda + (1-\epsilon)\Lambda(1-\epsilon)\Lambda\}\,[\epsilon B_\nu(T)], \tag{3.1.13}$$

etc., so that

$$S_n(\tau) = \{1 + (1-\epsilon)\Lambda + (1-\epsilon)\Lambda(1-\epsilon)\Lambda + \cdots + \cdots + [(1-\epsilon)\Lambda]^n\}\,[\epsilon B_\nu(T)]. \tag{3.1.14}$$

Thus, we see that each individual term in the series given by equation (3.1.10) essentially represents one iteration on the sum of the preceding terms. Clearly, if $\epsilon = 1$, this series would converge in one iteration to yield $S(\tau) = B_\nu(T)$. Thus, the larger the value of ϵ (note that $\epsilon < 1$) the faster the rate of convergence.

However, there exists a large class of astrophysical problems, particularly those involving strong spectral lines, for which $\epsilon \ll 1$. It is therefore of some interest to delve more deeply into the convergence properties of equation (3.1.14).

First, from a physical point of view, we see that the first term on the RHS of equation (3.1.10), i.e. $\epsilon B_\nu(T)$, represents the contribution to the radiation field due to collisional excitation of a de-excited atom followed by one emission process. This has been discussed in some detail in section 1.6. The second term on the RHS of equation (3.1.10), i.e.

$$(1-\epsilon)\Lambda[\epsilon B_\nu(T)] = \Lambda[\epsilon B_\nu(T)] - \epsilon\Lambda[\epsilon B_\nu(T)], \tag{3.1.15}$$

then consists of two parts: the first corresponds to the absorption of the photon just created followed by re-emission, whilst the second part corresponds to the destruction of that created photon. Thus, the second term on the RHS of equation (3.1.10) represents the total number of newly created photons which manage to survive the first scattering event, i.e. the first absorption followed by re-emission. Continuing this argument, we see that the $(n + 1)$th term in equation (3.1.10) represents the number of those photons which experience n scatters after their creation. We know from section 1.8, however, that photons undergo approximately $1/\epsilon$ scatters between creation and destruction for Doppler $\phi(\Delta\nu)$, for example (and even more for Voigt $\phi(\Delta\nu)$), so that, if we are to obtain an adequate representation for the radiation field from equation (3.1.10), we must include all those terms which reflect the real physics of the situation. Adequate convergence of the mathematical series (3.1.10) is merely equivalent to this adequate physical representation. Thus, convergence is only attained after at least $1/\epsilon$ terms and this, of course, corresponds to $1/\epsilon$ iterations. Since it is not unusual for ϵ to be of order 10^{-4}, the number of iterations required using Λ iteration can be quite prohibitive.

The situation can be somewhat different for finite slab (as opposed to semi-infinite) geometry. If the slab has an optical thickness of order N, then photons can scatter approximately N times after they are created before they escape – here we take $N < 1/\epsilon$. On the average, the number of scatters between creation and escape will be less than N since photons created in the middle of the slab, for example, will only require approximately $N/2$ scatters before escape. Note that since $N < 1/\epsilon$ the photon is more likely to escape than be destroyed. The radiation field within the slab would therefore be adequately represented by the sum of all those photons undergoing less than or equal to about N scatters between creation and escape (or destruction). Thus, we should have adequate convergence of the series (3.1.10) after approximately N iterations.

Since we have equated each term in the Λ iterative scheme with a precise physical process, it is not unreasonable to suggest that the Λ iteration method of solution always converges (albeit slowly at times).

(a) Mathematical considerations

The Λ operator given by equation (3.1.6) (and (3.1.8)) has been derived for a semi-infinite medium. If, however, we consider the atmosphere to be represented by a slab of finite optical width W, we have

$$\Lambda \equiv \tfrac{1}{2}\int_{-\infty}^{\infty} \phi^2(\Delta\nu)\,\mathrm{d}(\Delta\nu)\int_0^1 \frac{\mathrm{d}\mu}{\mu}\int_0^W \mathrm{d}\tau'\mathrm{e}^{-\phi(\Delta\nu)|\tau'-\tau|/\mu}, \tag{3.1.16}$$

which we write as

$$\Lambda \equiv \int_0^W K_1(|\tau'-\tau|)\,\mathrm{d}\tau', \tag{3.1.17}$$

where

$$K_1(\tau) = \tfrac{1}{2}\int_{-\infty}^{\infty} \phi^2(\Delta\nu)E_1(\tau\phi(\Delta\nu))\,\mathrm{d}(\Delta\nu). \tag{3.1.18}$$

Again, we have taken $\phi(\Delta\nu)$ to be depth independent.

The first exponential integral $E_1(x)$ is given by

$$E_1(x) = \int_0^1 \mathrm{e}^{-x/\mu}\,\frac{\mathrm{d}\mu}{\mu}, \tag{3.1.19}$$

which, replacing μ by $1/t$, may be re-written as

$$E_1(x) = \int_1^{\infty} \mathrm{e}^{-xt}\,\frac{\mathrm{d}t}{t}. \tag{3.1.20}$$

For ease in exposition, we take ϵ and $B_\nu(T)$ to be also depth independent so that equation (3.1.10) becomes

$$S(\tau) = \epsilon B_\nu(\tau)\sum_{n=0}^{\infty}(1-\epsilon)^n\Lambda^n. \tag{3.1.21}$$

We then define $k_n(\tau)$ such that

$$k_0(\tau) = 1, \tag{3.1.22}$$

and

$$\begin{aligned} k_n(\tau) &= \int_0^W K_1(|\tau'-\tau|)k_{n-1}(\tau')\,\mathrm{d}\tau' \\ &\equiv \Lambda k_{n-1}(\tau), \end{aligned} \tag{3.1.23}$$

for all $n \geqslant 1$. We then see that

$$\Lambda^0 \equiv k_0(\tau), \tag{3.1.24}$$

whilst, from equation (3.1.18),

$$\Lambda \equiv k_1(\tau). \tag{3.1.25}$$

Indeed, we have the general result

$$\Lambda^n \equiv k_n(\tau) \quad \text{for all } n. \tag{3.1.26}$$

Equation (3.1.21) may then be re-written as

$$S(\tau) = \epsilon B_\nu(T) \sum_{n=0}^{\infty} (1-\epsilon)^n k_n(\tau). \tag{3.1.27}$$

It is convenient to define the function $K_2(\tau)$ in a similar manner to $K_1(\tau)$ such that

$$K_2(\tau) = \int_{-\infty}^{\infty} \phi(\Delta\nu) E_2(\tau\phi(\Delta\nu))\, \mathrm{d}(\Delta\nu), \tag{3.1.28}$$

where the second exponential integral $E_2(x)$ has the form

$$E_2(x) = \int_1^{\infty} \mathrm{e}^{-xt} \frac{\mathrm{d}t}{t^2}\,. \tag{3.1.29}$$

It is not difficult to see that

$$\begin{aligned} \frac{\mathrm{d}K_2}{\mathrm{d}\tau}(\tau) &= -\int_{-\infty}^{\infty} \phi^2(\Delta\nu)\, \mathrm{d}(\Delta\nu) \int_1^{\infty} \mathrm{e}^{-\tau\phi(\Delta\nu)t} \frac{\mathrm{d}t}{t} \\ &= -\int_{-\infty}^{\infty} \phi^2(\Delta\nu) E_1(\tau\phi(\Delta\nu)\, \mathrm{d}(\Delta\nu) \\ &= -2K_1(\tau), \end{aligned} \tag{3.1.30}$$

so that

$$\begin{aligned} k_1(\tau) &= \int_0^W K_1(|\tau'-\tau|)\, \mathrm{d}\tau' \\ &= \int_0^\tau K_1(\tau-\tau')\, \mathrm{d}\tau' + \int_\tau^W K_1(\tau'-\tau)\, \mathrm{d}\tau' \\ &= \int_0^\tau K_1(y)\, \mathrm{d}y + \int_0^{W-\tau} K_1(y)\, \mathrm{d}y \\ &= -\tfrac{1}{2}[K_2(\tau)-K_2(0)] - \tfrac{1}{2}[K_2(W-\tau)-K_2(0)] \\ &= 1 - \tfrac{1}{2}[K_2(\tau)+K_2(W-\tau)], \end{aligned} \tag{3.1.31}$$

where we have used the result

$$\begin{aligned} K_2(0) &= \int_{-\infty}^{\infty} \phi(\Delta\nu)\, \mathrm{d}(\Delta\nu) \int_1^{\infty} \frac{\mathrm{d}t}{t^2} = \int_{-\infty}^{\infty} \phi(\Delta\nu)\, \mathrm{d}(\Delta\nu) \\ &= 1 \end{aligned} \tag{3.1.32}$$

from our normalisation condition on the absorption profile $\phi(\Delta\nu)$. Furhter, we see that

$$K_2(\infty) = 0, \tag{3.1.33}$$

so that $K_2(\tau)$ monotonically decreases as shown in figure 3.1.

It is clear from the above behaviour of $K_2(\tau)$ that

$$K_2(\tau) + K_2(W - \tau) \leqslant K_2(0) + K_2(W), \tag{3.1.34}$$

and

$$2K_2\left(\frac{W}{2}\right) \leqslant K_2(\tau) + K_2(W - \tau). \tag{3.1.35}$$

We then have the inequalities

$$0 < \tfrac{1}{2}[1 - K_2(W)] \leqslant k_1(\tau) \leqslant 1 - K_2\left(\frac{W}{2}\right) < 1. \tag{3.1.36}$$

If we now perform a Λ operation on these inequalities, and note that Λ is a positive definite operator, then we have

$$0 < \tfrac{1}{2}[1 - K_2(W)]\Lambda(1) < \Lambda k_1(\tau) < \left[1 - K_2\left(\frac{W}{2}\right)\right]\Lambda(1) < \Lambda(1),$$

i.e. using equation (3.1.23),

$$0 < \tfrac{1}{2}[1 - K_2(W)]k_1(\tau) < k_2(\tau) < \left[1 - K_2\left(\frac{W}{2}\right)\right]k_1(\tau) < k_1(\tau). \tag{3.1.37}$$

Note that the equality sign in (3.1.36) does not appear in (3.1.37) since, when performing the Λ operation, we are effectively summing over all τ space. (The left equality in (3.1.36) only applies at the one point $\tau = 0$ whereas the right equality applies only at $\tau = W/2$.)

Fig.3.1. The monotonic decrease of K_2 as a function of depth.

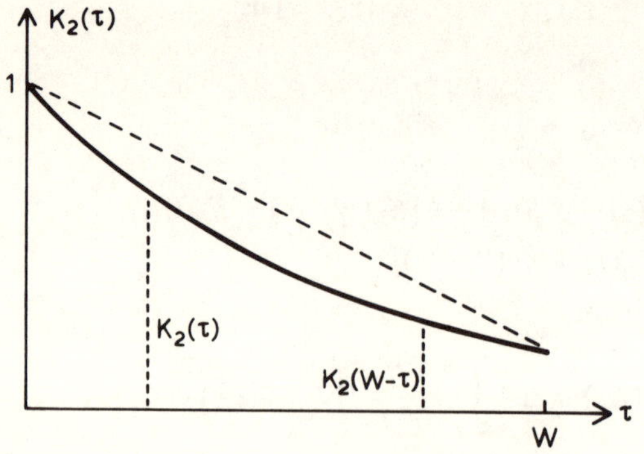

If we repeat the above process, we have

$$0<\tfrac{1}{2}[1-K_2(W)]k_2(\tau)<k_3(\tau)<\left[1-K_2\left(\frac{W}{2}\right)\right]k_2(\tau)<k_2(\tau). \tag{3.1.38}$$

However, we know from (3.1.37) that

$$k_2(\tau)>\tfrac{1}{2}[1-K_2(W)]k_1(\tau),$$

for example, so that (3.1.38) yields

$$k_3(\tau)>\{\tfrac{1}{2}[1-K_2(W)]\}^2k_1(\tau). \tag{3.1.39}$$

Similarly, we find

$$k_3(\tau)<\left\{1-K_2\left(\frac{W}{2}\right)\right\}^2k_1(\tau)<1. \tag{3.1.40}$$

Continuing this process, we have, for all $n\geqslant 2$,

$$0<\{\tfrac{1}{2}[1-K_2(W)]\}^{n-1}k_1(\tau)<k_n(\tau)<\left\{1-K_2\left(\frac{W}{2}\right)\right\}^{n-1}k_1(\tau)<1. \tag{3.1.41}$$

If we now substitute these inequalities into the series given by equation (3.1.27), and note that all terms are positive so that one may make the appropriate summation, we have

$$\begin{aligned}S(\tau)&=\epsilon B_\nu(T)\left[1+(1-\epsilon)k_1(\tau)+\sum_{n=2}^{\infty}(1-\epsilon)^nk_n(\tau)\right]\\&>\epsilon B_\nu(T)\left[1+(1-\epsilon)k_1(\tau)+\sum_{n=2}^{\infty}(1-\epsilon)^n\right.\\&\quad\left.\times\{\tfrac{1}{2}[1-K_2(W)]\}^{n-1}k_1(\tau)\right],\end{aligned}$$

i.e.

$$S(\tau)>\epsilon B_\nu(T)\left[1+(1-\epsilon)k_1(\tau)\sum_{n=0}^{\infty}(1-\epsilon)^n\{\tfrac{1}{2}[1-K_2(W)]\}^n\right]. \tag{3.1.42}$$

Similarly, we find

$$S(\tau)<\epsilon B_\nu(T)\left\{1+(1-\epsilon)k_1(\tau)\sum_{n=0}^{\infty}(1-\epsilon)^n\left[1-K_2\left(\frac{W}{2}\right)\right]^n\right\}, \tag{3.1.43}$$

The inequalities given by (3.1.42) and (3.1.43) then represent the bounding series for the Λ iteration. One can immediately see that if $\epsilon\ll 1$ and W is sufficiently large that $K_2(W/2) \ll 1$ (recall that $K_2(\infty) = 0$), then the convergence given by inequality (3.1.43) is extremely slow. Nevertheless, since $0 < \epsilon < 1$ and $0 \leqslant K_2(\tau) \leqslant 1$, convergence of the Λ iteration is assured. If, however, W is small so that $K_2(W/2) \lesssim 1$, then convergence of (3.1.43) is relatively rapid. Likewise, convergence is rapid for large ϵ. These two cases have already been mentioned from a physical point of view in part (a) above. However, the convergence of the series (3.1.42) is always quite rapid for all ϵ and W because of the factor $1/2^n$.

We may re-write the above inequalities by recognising that

$$\frac{1}{1-\lambda} = \sum_{n=0}^{\infty} \lambda^n,$$

so that

$$S(\tau) > \epsilon B_\nu(T) \left\{1 + \frac{(1-\epsilon)k_1(\tau)}{1 - \dfrac{1-\epsilon}{2}\,[1 - K_2(W)]}\right\}, \tag{3.1.44}$$

and

$$S(\tau) < \epsilon B_\nu(T) \left\{1 + \frac{(1-\epsilon)k_1(\tau)}{1 - (1-\epsilon)\left[1 - K_2\left(\dfrac{W}{2}\right)\right]}\right\}. \tag{3.1.45}$$

After some slight re-arrangement using equation (3.1.31), we finally have the required inequalities

$$\epsilon B_\nu(T) \left\{1 + \frac{1 - \frac{1}{2}[K_2(\tau) + K_2(W-\tau)]}{\dfrac{\epsilon}{1-\epsilon} + \frac{1}{2}[1 + K_2(W)]}\right\}$$

$$< S(\tau) < \epsilon B_\nu(T) \left\{1 + \frac{1 - \frac{1}{2}[K_2(\tau) + K_2(W-\tau)]}{\dfrac{\epsilon}{1-\epsilon} + K_2\left(\dfrac{W}{2}\right)}\right\}. \tag{3.1.46}$$

In particular, we see that as $W \to \infty$, $K_2(W) \to 0$, and this implies

$$S(\infty) < B_\nu(T), \tag{3.1.47}$$

whilst

$$S(0) > \frac{2\epsilon B_\nu(T)}{1+\epsilon}. \tag{3.1.48}$$

The inequality (3.1.47) is quite accurate for semi-infinite media and this therefore indicates that the boundary series (3.1.43) is also accurate. Thus, the convergence properties of (3.1.43) give some insight into the convergence of Λ iteration. As we have seen in section 2.9, $S(0) = \sqrt{\epsilon} B_\nu(T)$ for constant

ϵ and T in a semi-infinite atmosphere so that, for small ϵ, inequality (3.1.48) is quite inaccurate. In contrast to (3.1.43), therefore, the convergence of (3.1.42), whilst being quite rapid, does not accurately represent the convergence of the Λ iterative scheme for the semi-infinite situation. The reverse holds true for an optically thin slab, i.e. for small W.

3.1.2 *Asymptotic behaviour of* S(τ)

We have already considered the behaviour of the source function for large τ using the exact solutions derived in chapter 2. Here, we wish to repeat the analysis using the form of $S(\tau)$ given by equation (3.1.5), i.e.

$$S(\tau) = (1-\epsilon)\int_0^W K_1(|\tau'-\tau|)S(\tau')\,\mathrm{d}\tau' + \epsilon B_\nu(T), \tag{3.1.49}$$

where

$$K_1(\tau) = \tfrac{1}{2}\int_{-\infty}^{\infty} \phi^2(\Delta\nu)\,\mathrm{d}(\Delta\nu)\int_0^{\infty} \mathrm{e}^{-\tau\phi(\Delta\nu)t}\,\frac{\mathrm{d}t}{t}. \tag{3.1.50}$$

Since all terms appearing in the integrand of equation (3.1.50) are positive, the maximum value of $K_1(\tau)$ occurs at $\tau = 0$. Consequently, in considering the asymptotic behaviour of $S(\tau)$, we are then able to approximate the RHS of equation (3.1.49) by taking $S(\tau)$ outside the integral at $\tau' = \tau$ such that

$$S(\tau) \approx (1-\epsilon)S(\tau)\int_0^W K_1(|\tau'-\tau|)\,\mathrm{d}\tau' + \epsilon B_\nu(T). \tag{3.1.51}$$

In this way, the $K_1(|\tau'-\tau|)$ term acts as a type of delta function. Using equation (3.1.31) we find

$$S(\tau) \approx \frac{\epsilon B_\nu(T)}{1-(1-\epsilon)\{1-\frac{1}{2}[K_2(\tau)+K_2(W-\tau)]\}}. \tag{3.1.52}$$

For a semi-infinite atmosphere, i.e. for $W \to \infty$, we have $K_2(W) \to 0$ so that

$$S(\tau) \approx \frac{B_\nu(T)}{1-\frac{1}{2}K_2(\tau)+\dfrac{1}{2\epsilon}K_2(\tau)}. \tag{3.1.53}$$

We now need to determine the asymptotic behaviour of $K_2(\tau)$ and this, of course, depends upon the precise form of the absorption profile $\phi(\Delta\nu)$. We therefore consider the following two cases.

(a) Doppler profile

Here we have

$$\phi(\Delta\nu) = \frac{1}{\Delta\nu_\mathrm{D}\sqrt{\pi}}\mathrm{e}^{-(\Delta\nu/\Delta\nu_\mathrm{D})^2}, \tag{3.1.54}$$

so that, from equation (3.1.28) and noting that $\phi(\Delta\nu)$ is an even function of $\Delta\nu$, we have

$$K_2(\tau) = 2\int_0^\infty \frac{e^{-(\Delta\nu/\Delta\nu_D)^2}}{\Delta\nu_D\sqrt{\pi}}\, E_2\left[\frac{\tau e^{-(\Delta\nu/\Delta\nu_D)^2}}{\Delta\nu_D\sqrt{\pi}}\right] d(\Delta\nu). \tag{3.1.55}$$

We now define

$$u = \frac{\tau}{\Delta\nu_D\sqrt{\pi}}\, e^{-(\Delta\nu/\Delta\nu_D)^2} \quad \text{and} \quad \lambda = \frac{\tau}{\Delta\nu_D\sqrt{\pi}},$$

so that

$$\Delta\nu = \Delta\nu_D \sqrt{\left[\ln\left(\frac{\lambda}{u}\right)\right]} \tag{3.1.56}$$

Substitution of this $\Delta\nu$ into equation (3.1.55) then yields

$$K_2(\tau) = \frac{\Delta\nu_D}{\tau}\int_0^\lambda \frac{E_2(u)\, du}{\sqrt{[\ln(\lambda/u)]}}$$

$$= \frac{\Delta\nu_D}{\tau\sqrt{(\ln\lambda)}}\int_0^\lambda E_2(u)\left[1 + \frac{\ln u}{2\ln\lambda} + \cdots\right] du. \tag{3.1.57}$$

Note that we are interested here only in the asymptotic form of $K_2(\tau)$. Consequently, when taking large τ in the integrand of equation (3.1.57), the $(\ln u/2\ln\lambda)^n$ terms decrease for increasing n, i.e.

$$K_2(\tau) \sim \frac{\Delta\nu_D}{\tau\sqrt{(\ln\lambda)}}\int_0^\lambda E_2(u)\, du. \tag{3.1.58}$$

Now

$$E_2(u) = \int_1^\infty e^{-ut}\frac{dt}{t^2},$$

which, integrating by parts, yields

$$E_2(u) = e^{-u} - u\int_1^\infty e^{-ut}\frac{dt}{t}. \tag{3.1.59}$$

Substitution of this result into equation (3.1.58) then gives

$$K_2(\tau) \sim \frac{\Delta\nu_D}{\tau\sqrt{(\ln\lambda)}}\left\{1 - e^{-\lambda} + \int_1^\infty \frac{dt}{t}\left[\frac{\lambda e^{-\lambda t}}{t} + \frac{e^{-\lambda t} - 1}{t^2}\right]\right\},$$

which, for large τ (i.e. large λ) becomes

$$K_2(\tau) \sim \frac{\Delta\nu_D}{2\tau\sqrt{(\ln\lambda)}} = \frac{\Delta\nu_D}{2\tau\sqrt{\left[\ln\left(\frac{\tau}{\Delta\nu_D\sqrt{\pi}}\right)\right]}}. \tag{3.1.60}$$

Equation (3.1.53) then yields

$$S(\tau) \sim \frac{B_\nu(T)}{1 - \dfrac{\Delta\nu_D}{4\tau\sqrt{\left[\ln\left(\dfrac{\tau}{\Delta\nu_D\sqrt{\pi}}\right)\right]}}\left(1 - \dfrac{1}{\epsilon}\right)}. \tag{3.1.61}$$

Equation (3.1.61) now represents the asymptotic behaviour of $S(\tau)$. In particular, we see that the approach of $S(\tau)$ to $B_\nu(T)$ is quite slow. More particularly still, we see that $S(\tau) \approx B_\nu(T)$ for

$$\left|\frac{\Delta\nu_D}{4\tau\sqrt{\left[\ln\left(\dfrac{\tau}{\Delta\nu_D\sqrt{\pi}}\right)\right]}}\left(1 - \frac{1}{\epsilon}\right)\right| \ll 1, \tag{3.1.62}$$

which, ignoring the extremely small dependence of τ within the $\sqrt{(\ln)}$ term, yields

$$\tau \gg \frac{\Delta\nu_D(1-\epsilon)}{4\epsilon}. \tag{3.1.63}$$

Consequently, the thermalisation depth Θ_D for Doppler broadening satisfies

$$\Theta_D \gg \frac{\Delta\nu_D}{\epsilon}, \tag{3.1.64}$$

where we have taken $\epsilon \ll 1$. This result is essentially the same as that given by equation (2.10.24).

(b) Voigt profile

Here, we have

$$\begin{aligned}\phi(\Delta\nu) &= \frac{1}{\Delta\nu_D\sqrt{\pi}} H\left(a, \frac{\Delta\nu}{\Delta\nu_D}\right)\\ &= \frac{a}{\Delta\nu_D\pi^{3/2}}\int_{-\infty}^{\infty}\frac{e^{-y^2}\,dy}{\left(y - \dfrac{\Delta\nu}{\Delta\nu_D}\right)^2 + a^2}.\end{aligned} \tag{3.1.65}$$

Equation (2.10.28) stipulates the behaviour of $\phi(\Delta\nu)$ for small $|\Delta\nu|$, viz.

$$\phi(\Delta\nu) \approx \frac{1}{\Delta\nu_D\sqrt{\pi}}\, e^{-(\Delta\nu/\Delta\nu_D)^2}. \tag{3.1.66}$$

We have already examined the asymptotic form of $S(\tau)$ for just such profiles in part (a) above.

However, for large $|\Delta\nu|$, we have, from equation (2.10.29),

$$\phi(\Delta\nu) \approx \frac{a\Delta\nu_D}{\pi(\Delta\nu)^2}. \tag{3.1.67}$$

We now substitute this last equation into (3.1.28) to obtain

$$K_2(\tau) = \frac{2a\Delta\nu_D}{\pi}\int_0^\infty E_2\left(\frac{a\Delta\nu_D\tau}{\pi(\Delta\nu)^2}\right)\frac{d(\Delta\nu)}{(\Delta\nu)^2}$$

$$= \frac{2a\Delta\nu_D}{\pi}\int_1^\infty \frac{dt}{t^2}\int_0^\infty \exp\left[-\frac{\tau a\Delta\nu_D t}{\pi(\Delta\nu)^2}\right]\frac{d(\Delta\nu)}{(\Delta\nu)^2}. \qquad (3.1.68)$$

We then make the change of variable

$$y = \sqrt{\left(\frac{a\Delta\nu_D}{\pi}\right)}\frac{1}{\Delta\nu},$$

so that equation (3.1.68) becomes

$$K_2(\tau) = 2\sqrt{\left(\frac{a\Delta\nu_D}{\pi}\right)}\int_1^\infty e^{-\tau t y^2}\,dy$$

$$= \sqrt{\left(\frac{a\Delta\nu_D}{\pi}\right)}\int_1^\infty \frac{dt}{t^2}\sqrt{\left(\frac{\pi}{\tau t}\right)}$$

$$= \tfrac{2}{3}\sqrt{\left(\frac{a\Delta\nu_D}{\tau}\right)}. \qquad (3.1.69)$$

In this case, the asymptotic form for $S(\tau)$ becomes

$$S(\tau) \sim \frac{B_\nu(T)}{1-\frac{1}{3}\left(\sqrt{\left(\frac{a\Delta\nu_D}{\tau}\right)}\right)\left(1-\frac{1}{\epsilon}\right)}. \qquad (3.1.70)$$

so that $S(\tau) \approx B_\nu(T)$ for

$$\tau \gg \frac{a\Delta\nu_D(1-\epsilon)^2}{9\epsilon^2}. \qquad (3.1.71)$$

This result differs by only a constant factor to that given by equation (2.10.34) using the exact solution for $S(\tau)$. As in section 2.10, we must also take into account the smaller frequencies when discussing the asymptotic behaviour of $S(\tau)$. Thus, the thermalisation depth Θ_V for Voigt broadening satisfies

$$\Theta_V \gg \max\left(\frac{\Delta\nu_D}{\epsilon}, \frac{a\Delta\nu_D}{\epsilon^2}\right), \qquad (3.1.72)$$

where we have again taken $\epsilon \ll 1$.

3.2 The Feautrier technique [39]

The equation of radiative transfer is a first order integro-differential equation and, as can be readily seen from the form of the exact solution derived in chapter 2, its solution exhibits an exponentially increasing component with increasing depth. It is not unreasonable, therefore, to imagine a corresponding exponential increase with depth of any small error one might generate due to, for example, depth discretisation in a numerical solution. Such errors, of course, cannot be completely eliminated if one is to attempt a numerical solution and thus, regardless of the care one might exercise in minimising these errors, we can expect their amplification as one proceeds to greater depths. This amplification, as it turns out, usually swamps the solution, thus rendering it meaningless.

The Feautrier technique overcomes this difficulty by transforming the transfer equation into second order integro-differential form which must therefore have two boundary conditions for its complete solution. We have one boundary condition at the surface of the model stellar atmosphere (for example, zero incident radiation) with another deep within the atmosphere (for example, $S \to B_\nu(T)$ as $\tau \to \infty$). In this case, the above-mentioned amplification of any numerical errors cannot continue unchecked as one proceeds into the atmosphere because the second boundary condition places a constraint on the solution deep within the atmosphere and this, in turn, implicitly places a constraint on the errors. The Feautrier technique therefore offers an extremely powerful tool for the solution of the equation of radiative transfer.

To illustrate the method, we first define the radiation intensity in the positive and negative directions such that

$$I^{\pm}(\tau, \Delta\nu, \mu) = I(\tau, \Delta\nu, \pm\mu) \;\forall\; \mu > 0. \tag{3.2.1}$$

These intensities must therefore satisfy the equations

$$\frac{\mu}{\phi(\Delta\nu)} \frac{\partial I^+}{\partial \tau}(\tau, \Delta\nu, \mu) = I^+(\tau, \Delta\nu, \mu) - S(\tau), \tag{3.2.2}$$

$$-\frac{\mu}{\phi(\Delta\nu)} \frac{\partial I^-}{\partial \tau}(\tau, \Delta\nu, \mu) = I^-(\tau, \Delta\nu, \mu) - S(\tau) \;\forall\; \mu > 0. \tag{3.2.3}$$

It is clear that, since the source function (assuming complete re-distribution and zero macroscopic velocity fields) is independent of angle, we must have $S^+(\tau) \equiv S^-(\tau) \equiv S(\tau)$. Indeed, from equation (3.II), we have

$$\begin{aligned} S(\tau) &= \frac{1-\epsilon}{2} \int_{-\infty}^{\infty} \phi(\Delta\nu)\, \mathrm{d}(\Delta\nu) \\ &\quad \times \left\{ \int_0^1 I(\tau, \Delta\nu, \mu) + \int_{-1}^0 I(\tau, \Delta\nu, \mu) \right\} \mathrm{d}\mu + \epsilon B_\nu(T) \\ &= \frac{1-\epsilon}{2} \int_{-\infty}^{\infty} \phi(\Delta\nu)\, \mathrm{d}(\Delta\nu) \int_0^1 [I^+(\tau, \Delta\nu, \mu) + I^-(\tau, \Delta\nu, \mu)]\, \mathrm{d}\mu \\ &\quad + \epsilon B_\nu(T). \end{aligned} \tag{3.2.4}$$

We now define the two new dependent variables Φ and Ψ satisfying

$$\Phi(\tau, \Delta\nu, \mu) = \tfrac{1}{2}[I^+(\tau, \Delta\nu, \mu) + I^-(\tau, \Delta\nu, \mu)], \tag{3.2.5}$$

$$\Psi(\tau, \Delta\nu, \mu) = \tfrac{1}{2}[I^+(\tau, \Delta\nu, \mu) - I^-(\tau, \Delta\nu, \mu)], \tag{3.2.6}$$

for all $\mu > 0$.

Equation (3.2.4) then becomes

$$S(\tau) = (1-\epsilon)\int_{-\infty}^{\infty} \phi(\Delta\nu)\,\mathrm{d}(\Delta\nu)\int_0^1 \Phi(\tau, \Delta\nu, \mu) + \epsilon B_\nu(T), \tag{3.2.7}$$

whilst, adding and subtracting equations (3.2.2) and (3.2.3), we obtain

$$\frac{\mu}{\phi(\Delta\nu)}\frac{\partial\Psi}{\partial\tau}(\tau, \Delta\nu, \mu) = \Phi(\tau, \Delta\nu, \mu) - S(\tau), \tag{3.2.8}$$

$$\frac{\mu}{\phi(\Delta\nu)}\frac{\partial\Phi}{\partial\tau}(\tau, \Delta\nu, \mu) = \Psi(\tau, \Delta\nu, \mu). \tag{3.2.9}$$

If we now substitute Ψ given by equation (3.2.9) into (3.2.8) we find

$$\left(\frac{\mu}{\phi(\Delta\nu)}\frac{\partial}{\partial\tau}\right)\left(\frac{\mu}{\phi(\Delta\nu)}\frac{\partial}{\partial\tau}\right)\Phi(\tau, \Delta\nu, \mu) = \Phi(\tau, \Delta\nu, \mu) - S(\tau)$$

$$= \Phi(\tau, \Delta\nu, \mu) - (1-\epsilon)\int_{-\infty}^{\infty} \phi(\Delta\nu)\,\mathrm{d}(\Delta\nu)$$

$$\times \int_0^1 \Phi(\tau, \Delta\nu, \mu')\,\mathrm{d}\mu' - \epsilon B_\nu(T). \tag{3.2.10}$$

Equation (3.2.10) is the second order integro-differential equation we now wish to solve for the new unknown intensity sum $\Phi(\tau, \Delta\nu, \mu)$.

The two required boundary conditions are effectively given by equation (3.2.9), viz.

$$\frac{\mu}{\phi(\Delta\nu)}\frac{\partial\Phi}{\partial\tau}(\tau, \Delta\nu, \mu) = \Phi(\tau, \Delta\nu, \mu) - I^-(\tau, \Delta\nu, \mu) \tag{3.2.11}$$

$$= I^+(\tau, \Delta\nu, \mu) - \Phi(\tau, \Delta\nu, \mu). \tag{3.2.12}$$

We readily obtain from equation (3.2.11)

$$\frac{\mu}{\phi(\Delta\nu)}\left[\frac{\partial\Phi}{\partial\tau}(\tau, \Delta\nu, \mu)\right]_{\tau=0} = \Phi(0, \Delta\nu, \mu) - I^-(0, \Delta\nu, \mu), \tag{3.2.13}$$

where $I^-(0, \Delta\nu, \mu) = I_{\text{inc}}(\Delta\nu, \mu)$ is the incident ($\mu < 0$) radiation intensity at the surface of the atmosphere.

If we take as our boundary condition deep within the atmosphere $I^+(\tau, \Delta\nu, \mu) \to B_\nu(T)$ as $\tau \to \infty$, i.e. we approach an LTE configuration as discussed in section 1.8, we have

$$\frac{\mu}{\phi(\Delta\nu)}\left[\frac{\partial\Phi}{\partial\tau}(\tau,\Delta\nu,\mu)\right]_{\tau\to\infty}=B_\nu(T)-\Phi(\infty,\Delta\nu,\mu), \tag{3.2.14}$$

from equation (3.2.12).

This then completes the specification of the problem for the Feautrier technique.

3.2.1 *Finite differencing*

We must now consider the details of the numerical method. We begin by constructing an optical depth grid $\{\tau_j\}$ for $j = 1, N_\tau$ where N_τ is the number of τ points to be included in this grid. In particular, we have $\tau_1 = 0$ and τ_{N_τ} corresponding to some point deep within the model atmosphere. We therefore replace $\Phi(\tau, \Delta\nu, \mu)$ by $\Phi_j(\Delta\nu, \mu)$.

Next, we initially replace derivatives by the simple expedient

$$\left[\frac{\partial\Phi}{\partial\tau}(\tau,\Delta\nu,\mu)\right]_{\tau=\tau_j}\approx\frac{\Phi_{j+1}(\Delta\nu,\mu)-\Phi_j(\Delta\nu,\mu)}{\tau_{j+1}-\tau_j}. \tag{3.2.15}$$

The error in such depth discretisation is obvious when one writes the equally valid backward representation

$$\left[\frac{\partial\Phi}{\partial\tau}(\tau,\Delta\nu,\mu)\right]_{\tau=\tau_j}\approx\frac{\Phi_{j}(\Delta\nu,\mu)-\Phi_{j-1}(\Delta\nu,\mu)}{\tau_{j}-\tau_{j-1}}. \tag{3.2.16}$$

Clearly, the above two formulae must yield different results, certainly within the neighbourhood $\tau = \tau_j$. We shall detail a method in the following section which dramatically reduces the errors generated by this discrepancy. For the moment, we use equation (3.2.15) to represent the derivative in the surface boundary condition (3.2.13) (note that τ_0, which would otherwise occur in equation (3.2.16) when $j = 1$, is not defined in our $\{\tau_j\}$ grid), and equation (3.2.16) to represent the derivative in (3.2.14).

The second order derivative appearing in equation (3.2.10) does not present such severe difficulties. Since we have

$$\frac{\partial^2 f}{\partial x^2}=\lim_{\Delta x\to 0}\frac{\left[\dfrac{\partial f}{\partial x}\right]_{x+\Delta x/2}-\left[\dfrac{\partial f}{\partial x}\right]_{x-\Delta x/2}}{\Delta x},$$

and $\phi(\Delta\nu)$ is taken here to be independent of τ (note that $\phi(\Delta\nu)$ is, in general, a function of position since it is dependent upon the temperature appearing in the Doppler width term $\Delta\nu_D$ - see section 1.4), we readily have

$$\left[\left(\frac{\mu}{\phi(\Delta\nu)}\frac{\partial}{\partial\tau}\right)^2\Phi\right]_{\tau=\tau_j} = \left(\frac{\mu}{\phi(\Delta\nu)}\right)^2\left[\frac{\partial^2\Phi}{\partial\tau^2}\right]_{\tau=\tau_j}$$

$$\approx \left(\frac{\mu}{\phi(\Delta\nu)}\right)^2\left\{\left[\frac{\partial\Phi}{\partial\tau}\right]_{\tau=\tau_j+\frac{\tau_{j+1}-\tau_j}{2}} - \left[\frac{\partial\Phi}{\partial\tau}\right]_{\tau=\tau_j-\frac{\tau_j-\tau_{j-1}}{2}}\right\}\frac{1}{\dfrac{\tau_{j+1}-\tau_j}{2}+\dfrac{\tau_j-\tau_{j-1}}{2}}$$

$$\approx \left(\frac{\mu}{\phi(\Delta\nu)}\right)^2\left\{\frac{\Phi_{j+1}-\Phi_j}{\tau_{j+1}-\tau_j} - \frac{\Phi_j-\Phi_{j-1}}{\tau_j-\tau_{j-1}}\right\} \times \frac{2}{\tau_{j+1}-\tau_{j-1}}. \tag{3.2.17}$$

If we now put

$$\Delta_j = \tau_{j+1}-\tau_j, \quad \nabla_j = \tau_j - \tau_{j-1}, \tag{3.2.18}$$

equation (3.2.17) yields

$$\left[\left(\frac{\mu}{\phi(\Delta\nu)}\frac{\partial}{\partial\tau}\right)^2\Phi(\tau,\Delta\nu,\mu)\right]_{\tau=\tau_j}$$
$$\approx \mu^2\,[a_j(\Delta\nu)\Phi_{j-1}(\Delta\nu,\mu) + b_j(\Delta\nu)\Phi_j(\Delta\nu,\mu) + c_j(\Delta\nu)\Phi_{j+1}(\Delta\nu,\mu)], \tag{3.2.19}$$

where

$$a_j(\Delta\nu) = \frac{2}{\nabla_j(\Delta_j+\nabla_j)\phi^2(\Delta\nu)}, \tag{3.2.20}$$

$$c_j(\Delta\nu) = \frac{2}{\Delta_j(\Delta_j+\nabla_j)\phi^2(\Delta\nu)}, \tag{3.2.21}$$

and

$$b_j(\Delta\nu) = -a_j(\Delta\nu) - c_j(\Delta\nu). \tag{3.2.22}$$

Before we proceed, it is important to stress that we have taken $\phi(\Delta\nu)$ to be depth independent. If, however, $\phi(\Delta\nu)$ is a function of τ, we can re-write

$$\frac{1}{\phi(\Delta\nu)}\frac{\partial\Phi}{\partial\tau}(\tau,\Delta\nu,\mu) = -\frac{1}{\kappa}\frac{\partial\Phi}{\partial z}(z,\Delta\nu,\mu), \tag{3.2.23}$$

and

$$\left(\frac{1}{\phi(\Delta\nu)}\frac{\partial}{\partial\tau}\right)^2\Phi(\tau,\Delta\nu,\mu) = \left(\frac{1}{\kappa}\frac{\partial}{\partial z}\right)^2\Phi(z,\Delta\nu,\mu), \tag{3.2.24}$$

where we have used the definition of optical depth (1.9.14)

$$d\tau = -\frac{\kappa\,dz}{\phi(\Delta\nu)}. \tag{3.2.25}$$

We may then write

$$\left[\left(\frac{\mu}{\phi(\Delta\nu)}\frac{\partial}{\partial\tau}\right)^2\Phi\right]_{\tau=\tau_j}$$

$$\approx \mu^2\left\{\left[\frac{1}{\kappa}\frac{\partial\Phi}{\partial\tau}\right]_{z=z_j+\frac{z_{j+1}-z_j}{2}}-\left[\frac{1}{\kappa}\frac{\partial\Phi}{\partial z}\right]_{z=z_j-\frac{z_j-z_{j-1}}{2}}\right\}$$

$$\times\ \{\tfrac{1}{4}(\kappa_{j+1}+\kappa_j)(z_{j+1}-z_j)+\tfrac{1}{4}(\kappa_j+\kappa_{j-1})(z_j-z_{j-1})\}^{-1}$$

$$\approx\mu^2\frac{\left\{\dfrac{\Phi_{j+1}-\Phi_j}{\frac{1}{2}(\kappa_{j+1}+\kappa_j)(z_{j+1}-z_j)}-\dfrac{\Phi_j-\Phi_{j-1}}{\frac{1}{2}(\kappa_j+\kappa_{j-1})(z_j-z_{j-1})}\right\}}{\frac{1}{4}(\kappa_{j+1}+\kappa_j)(z_{j+1}-z_j)+\frac{1}{4}(\kappa_j+\kappa_{j-1})(z_j-z_{j-1})}. \tag{3.2.26}$$

If we now re-define Δ_j and ∇_j such that

$$\Delta_j=\tfrac{1}{2}(\kappa_{j+1}+\kappa_j)(z_{j+1}-z_j), \tag{3.2.27}$$

$$\nabla_j=\tfrac{1}{2}(\kappa_j+\kappa_{j-1})(z_j-z_{j-1}), \tag{3.2.28}$$

noting that Δ_j and ∇_j are now functions of $\phi(\Delta\nu)$, we have equation (3.2.19) again, but with

$$a_j(\Delta\nu)=\frac{2}{\nabla_j(\Delta_j+\nabla_j)}, \tag{3.2.29}$$

$$c_j(\Delta\nu)=\frac{2}{\Delta_j(\Delta_j+\nabla_j)}, \tag{3.2.30}$$

and

$$b_j(\Delta\nu)=-a_j(\Delta\nu)-c_j(\Delta\nu). \tag{3.2.31}$$

An alternative method of specifying the second order derivative is

$$\left[\left(\frac{1}{\phi(\Delta\nu)}\frac{\partial}{\partial\tau}\right)^2\Phi\right]_{\tau_j}=\left[\left(\frac{1}{\kappa}\frac{\partial}{\partial z}\right)\left(\frac{1}{\kappa}\frac{\partial}{\partial z}\right)\Phi\right]_{z_j}$$

$$=\left[\frac{1}{\kappa}\left(\frac{1}{\kappa}\frac{\partial^2\Phi}{\partial z}-\frac{1}{\kappa^2}\frac{\partial\kappa}{\partial z}\frac{\partial\Phi}{\partial z}\right)\right]_{z_j}. \tag{3.2.32}$$

If, for example, we write

$$\left[\frac{\partial\kappa}{\partial z}\right]_{z_j}=\tfrac{1}{2}\left\{\left[\frac{\partial\kappa}{\partial z}\right]_{z_j+\frac{z_{j+1}-z_j}{2}}+\left[\frac{\partial\kappa}{\partial z}\right]_{z_j+\frac{z_j-z_{j-1}}{2}}\right\}$$

$$=\tfrac{1}{2}\left[\frac{\Delta\kappa_j}{\Delta z_j}+\frac{\nabla\kappa_j}{\nabla z_j}\right], \tag{3.2.33}$$

where we have defined, for example,

$$\Delta\kappa_j=\kappa_{j+1}-\kappa_j,\quad \nabla\kappa_j=\kappa_j-\kappa_{j-1}, \tag{3.2.34}$$

equation (3.2.32) yields

$$\left[\left(\frac{1}{\phi(\Delta\nu)}\frac{\partial}{\partial\tau}\right)^2\Phi\right]_{\tau_j}=\frac{2}{\kappa_j^2(\Delta z_j+\nabla z_j)}\left[\frac{\Phi_{j+1}-\Phi_j}{\Delta z_j}-\frac{\Phi_j-\Phi_{j-1}}{\nabla z_j}\right]$$
$$-\frac{1}{4\kappa_j^3}\left[\frac{\Delta\kappa_j}{\Delta z_j}+\frac{\nabla\kappa_j}{\nabla z_j}\right]\left[\frac{\Phi_{j+1}-\Phi_j}{\Delta z_j}+\frac{\Phi_j-\Phi_{j-1}}{\nabla z_j}\right]. \tag{3.2.35}$$

This again leads to an equation of the form (3.2.19) but with

$$a_j(\Delta\nu)=\frac{2}{\kappa_j^2(\Delta z_j+\nabla z_j)\nabla z_j}+\frac{1}{4\kappa_j^3\nabla z_j}\left[\frac{\Delta\kappa_j}{\Delta z_j}+\frac{\nabla\kappa_j}{\nabla z_j}\right],$$

$$c_j(\Delta\nu)=\frac{2}{\kappa_j^2(\Delta z_j+\nabla z_j)\Delta z_j}-\frac{1}{4\kappa_j^3\Delta z_j}\left[\frac{\Delta\kappa_j}{\Delta z_j}+\frac{\nabla\kappa_j}{\nabla z_j}\right],$$

and

$$b_j(\Delta\nu)=-a_j(\Delta\nu)-c_j(\Delta\nu).$$

3.2.2 *Angle and frequency quadrature*

We now construct a grid in both frequency $\{\Delta\nu_i\}$ and angle $\{\mu_k\}$ space such that $i=1, N_\nu$ and $k=1, N_\mu$. The dependent variable $\Phi(\tau, \Delta\nu, \mu)$ is then replaced by the discretised quantity $\Phi_{jik}=\Phi_j(\Delta\nu_i, \mu_k)=\Phi(\tau_j, \Delta\nu_i, \mu_k)$.

We next consider the integral term appearing in equation (3.2.10). In particular, we specify the quadrature formulae for some arbitrary function $f(\tau, \Delta\nu, \mu)$ of the form

$$\int_{-\infty}^{\infty}\phi(\Delta\nu)f(\tau, \Delta\nu, \mu)\,\mathrm{d}(\Delta\nu)\approx\sum_{i=1}^{N_\nu}w_i^{(1)}f(\tau, \Delta\nu_i, \mu), \tag{3.2.36}$$

$$\int_0^1 f(\tau, \Delta\nu, \mu)\,\mathrm{d}\mu\approx\sum_{k=1}^{N_\mu}w_k^{(2)}f(\tau, \Delta\nu, \mu_k), \tag{3.2.37}$$

where $\{w_i^{(1)}\}$ and $\{w_k^{(2)}\}$ are the quadrature weights associated with quadrature points $\{\Delta\nu_i\}$ and $\{\mu_k\}$ respectively. These frequency and angle grids are usually constructed independent of one another, but are generally chosen with the above quadrature formulae in mind. We shall discuss appropriate quadratures more explicitly later.

We then make the further approximation

$$\int_{-\infty}^{\infty}\phi(\Delta\nu)\,\mathrm{d}(\Delta\nu)\int_0^1 f(\tau, \Delta\nu, \mu)\,\mathrm{d}\mu\approx\sum_{i=1}^{N_\nu}\sum_{k=1}^{N_\mu}\alpha_i\beta_k f(\tau, \Delta\nu_i, \mu_k), \tag{3.2.38}$$

where, clearly, α_i and β_k will involve the quadrature weights $w_i^{(1)}$ and $w_k^{(2)}$.

Substitution of this result, together with equation (3.2.19), into (3.2.10) then yields the discretised equation of radiative transfer

$$\mu_k^2[a_{ji}\Phi_{j-1,ik} + b_{ji}\Phi_{jik} + c_{ji}\Phi_{j+1,ik}]$$
$$= \Phi_{jik} - (1-\epsilon_j)\sum_{i'=1}^{N_\nu}\sum_{k'=1}^{N_\mu} \alpha_{i'}\beta_{k'}\Phi_{ji'k'} - [\epsilon B_\nu(T)]_j. \tag{3.2.39}$$

We therefore have a linear non-homogeneous system of $N_\tau N_\nu N_\mu$ equations to solve for the $N_\tau N_\nu N_\mu$ unknowns Φ_{jik}. However, these equations are coupled in frequency and angle via the summation term on the RHS of equation (3.2.39).

The system can be simplified by defining the vector Φ_j, having $N_\nu N_\mu$ elements, of the form

$$\Phi_j = \begin{pmatrix} \Phi_j(\Delta\nu_1, \mu_1) \\ \vdots \\ \Phi_j(\Delta\nu_{N_\nu}, \mu_1) \\ \Phi_j(\Delta\nu_1, \mu_2) \\ \vdots \\ \Phi_j(\Delta\nu_{N_\nu}, \mu_{N_\mu}) \end{pmatrix}. \tag{3.2.40}$$

We next define the diagonal matrices $\mathbf{A}_j$ and $\mathbf{C}_j$ (of size $N_\nu N_\mu$) satisfying

$$\mathbf{A}_j = \begin{pmatrix} a_{j1}\mu_1^2 & & & & & 0 \\ & \ddots & & & & \\ & & a_{jN_\nu}\mu_1^2 & & & \\ & & & a_{j1}\mu_2^2 & & \\ & 0 & & & \ddots & \\ & & & & & a_{jN_\nu}\mu_{N_\mu}^2 \end{pmatrix}. \tag{3.2.41}$$

and

$$\mathbf{C}_j = \begin{pmatrix} c_{j1}\mu_1^2 & & & & & 0 \\ & \ddots & & & & \\ & & c_{jN_\nu}\mu_1^2 & & & \\ & & & c_{j1}\mu_2^2 & & \\ & 0 & & & \ddots & \\ & & & & & c_{jN_\nu}\mu_{N_\mu}^2 \end{pmatrix}. \tag{3.2.42}$$

together with the full matrix $\mathbf{B}_j$ given by

$$\mathbf{B}_j = \mathbb{1} - (1-\epsilon_j)\mathbf{W}_j + \mathbf{A}_j + \mathbf{C}_j, \tag{3.2.43}$$

where $\mathbb{1}$ is the identity matrix and

$$\mathbf{W}_j = \begin{pmatrix} \alpha_1\beta_1, & \alpha_2\beta_1, & \cdots & \alpha_{N_\nu}\beta_{N_\mu} \\ \vdots & & & \\ \alpha_1\beta_1, & \alpha_2\beta_1, & & \alpha_{N_\nu}\beta_{N_\mu} \end{pmatrix}. \tag{3.2.44}$$

If we now define the vector $\mathbf{L}_j$ such that

$$\mathbf{L}_j = [\epsilon B_\nu(T)]_j \, \mathbb{1}, \tag{3.2.45}$$

where Π is the unit column vector, then the system of equations (3.2.39) can be written in the general matrix form

$$-\mathbf{A}_j\mathbf{\Phi}_{j-1} + \mathbf{B}_j\mathbf{\Phi}_j - \mathbf{C}_j\mathbf{\Phi}_{j+1} = \mathbf{L}_j. \tag{3.2.46}$$

We must now consider the appropriate boundary conditions before we proceed in the development of the numerical solution of equation (3.2.46). Indeed, it is not difficult to see that if we use equation (3.2.15) to evaluate the first order derivative appearing in equation (3.2.13), we obtain

$$\frac{\mu_k}{\Delta_{1i}} (\Phi_{2ik} - \Phi_{1ik}) = \Phi_{1ik} - I_{\text{inc}}(\Delta\nu_i, \mu_k), \tag{3.2.47}$$

where

$$\Delta_{1i} \equiv \phi(\Delta\nu_i)(\tau_2 - \tau_1).$$

If we now put

$$\mathbf{C}_1 = \begin{pmatrix} \mu_1/\Delta_{11} & & 0 \\ & \ddots & \\ 0 & & \mu_{N_\nu}/\Delta_{1N_\nu} \end{pmatrix}, \tag{3.2.48}$$

and

$$\mathbf{B}_1 = \mathbb{1} + \mathbf{C}_1, \tag{3.2.49}$$

with

$$\mathbf{L}_1 = \begin{pmatrix} I_{\text{inc}}(\Delta\nu_1, \mu_1) \\ \vdots \\ I_{\text{inc}}(\Delta\nu_{N_\nu}, \mu_{N_\mu}) \end{pmatrix}, \tag{3.2.50}$$

then the system of equations (3.2.47) has the matrix form

$$\mathbf{B}_1\mathbf{\Phi}_1 - \mathbf{C}_1\mathbf{\Phi}_2 = \mathbf{L}_1, \tag{3.2.51}$$

which is identical in structure to equation (3.2.46) but with

$$\mathbf{A}_1 \equiv \mathbf{0}. \tag{3.2.52}$$

Similarly, we obtain a matrix equation for the boundary condition at $\tau = \tau_{N_\tau}$ given by equation (3.2.14) using the first order derivative approximation (3.2.16), viz.

$$-\mathbf{A}_{N_\tau}\mathbf{\Phi}_{N_\tau-1} + \mathbf{B}_{N_\tau}\mathbf{\Phi}_{N_\tau} = \mathbf{L}_{N_\tau}, \tag{3.2.53}$$

where

$$\mathbf{A}_{N_\tau} = -\begin{pmatrix} \mu_1/\nabla_{N_\tau 1} & & 0 \\ & \ddots & \\ 0 & & \mu_{N_\mu}/\nabla_{N_\tau N_\nu} \end{pmatrix}, \tag{3.2.54}$$

$$\mathbf{B}_{N_\nu} = \mathbb{1} + \mathbf{A}_{N_\tau}, \tag{3.2.55}$$

with

$$\mathbf{L}_{N_\tau} = [B_\nu(T)]_{N_\tau}\mathbb{1}. \tag{3.2.56}$$

Again, equation (3.2.53) is similar in form to (3.2.46) but with

$$\mathbf{C}_{N_\tau} \equiv \mathbf{0}. \tag{3.2.57}$$

We are now in a position to solve equation (3.2.46), together with the boundary conditions (3.2.51) and (3.2.53), using Gaussian elimination.

3.2.3 The recurrence relationship [44]

We now wish to solve the matrix equation

$$-\mathbf{A}_j\mathbf{\Phi}_{j-1} + \mathbf{B}_j\mathbf{\Phi}_j - \mathbf{C}_j\mathbf{\Phi}_{j+1} = \mathbf{L}_j, \tag{3.2.58}$$

for all $j = 1, N_\tau$ where

$$\mathbf{A} \equiv \mathbf{0} \equiv \mathbf{C}_{N_\tau}. \tag{3.2.59}$$

Using equation (3.2.59), we have

$$\mathbf{B}_1\mathbf{\Phi}_1 - \mathbf{C}_1\mathbf{\Phi}_2 = \mathbf{L}_1,$$

i.e.

$$\mathbf{\Phi}_1 = \mathbf{B}_1^{-1}[\mathbf{L}_1 + \mathbf{C}_1\mathbf{\Phi}_2]. \tag{3.2.60}$$

This last equation has the basic recurrence form

$$\mathbf{\Phi}_j = \mathbf{U}_j + \mathbf{V}_j\mathbf{\Phi}_{j+1}, \tag{3.2.61}$$

where

$$\mathbf{U}_1 = \mathbf{B}_1^{-1}\mathbf{L}_1, \tag{3.2.62}$$

and

$$\mathbf{V}_1 = \mathbf{B}_1^{-1}\mathbf{C}_1. \tag{3.2.63}$$

If we substitute the general recurrence relationship (3.2.61) into (3.2.58), we find

$$-\mathbf{A}_j\mathbf{U}_{j-1} + (\mathbf{B}_j - \mathbf{A}_j\mathbf{V}_{j-1})\mathbf{\Phi}_j - \mathbf{C}_j\mathbf{\Phi}_{j+1} = \mathbf{L}_j,$$

i.e.

$$\mathbf{\Phi}_j = (\mathbf{B}_j - \mathbf{A}_j\mathbf{V}_{j-1})^{-1}[\mathbf{L}_j + \mathbf{A}_j\mathbf{U}_{j-1} + \mathbf{C}_j\mathbf{\Phi}_{j+1}], \tag{3.2.64}$$

which, itself, has the general form (3.2.61) with

$$\mathbf{U}_j = (\mathbf{B}_j - \mathbf{A}_j\mathbf{V}_{j-1})^{-1}(\mathbf{L}_j + \mathbf{A}_j\mathbf{U}_{j-1}), \tag{3.2.65}$$

and

$$\mathbf{V}_j = (\mathbf{B}_j - \mathbf{A}_j\mathbf{V}_{j-1})^{-1}\mathbf{C}_j. \tag{3.2.66}$$

Equation (3.2.61) therefore represents the general recurrence relationship at each jth point in the model atmosphere where $\mathbf{U}_j$ and $\mathbf{V}_j$ given by equations (3.2.65) and (3.2.66) are known functions. In particular, equation (3.2.59) stipulates that these two equations reduce to equations (3.2.62) and (3.2.63) for $j = 1$ whilst, for $j = N_\tau$, we have

$$\mathbf{V}_{N_\tau} \equiv \mathbf{0}. \tag{3.2.67}$$

This last result then yields

$$\mathbf{\Phi}_{N_\tau} = \mathbf{U}_{N_\tau}. \tag{3.2.68}$$

The computational procedure therefore requires the evaluation of $\mathbf{U}_1$ and $\mathbf{V}_1$ first, followed by $\mathbf{U}_2$ and $\mathbf{V}_2$ etc., using equations (3.2.65) and (3.2.66), until $\mathbf{U}_{N_\tau}$ is obtained. Once this latter quantity has been determined, equation (3.2.68) then immediately yields $\mathbf{\Phi}_{N_\tau}$. This $\mathbf{\Phi}_{N_\tau}$ can now be substituted into equation (3.2.61) to obtain $\mathbf{\Phi}_{N_\tau - 1}$, viz.

$$\mathbf{\Phi}_{N_\tau - 1} = \mathbf{U}_{N_\tau - 1} + \mathbf{V}_{N_\tau - 1}\mathbf{\Phi}_{N_\tau}. \tag{3.2.69}$$

The resulting $\mathbf{\Phi}_{N_\tau - 1}$ can then be used to obtain $\mathbf{\Phi}_{N_\tau - 2}$ etc. (by back substitution) until all the $\mathbf{\Phi}_j$ have been evaluated. This then completes the solution. The above process is referred to as Gaussian elimination.

3.2.4 *Taylor series expansion at the boundaries [3]*

The system of equations used in the Feautrier technique involves second order derivatives at all points *not* on the boundaries. The differencing of these second order equations presents no difficulties. However, the defining equations at the boundaries are first order (see equations (3.2.13) and (3.2.14)) and, as we have seen from equations (3.2.15) and (3.2.16), there is considerable inaccuracy in the specification of the appropriate finite difference approximation. This may be even more readily seen by expanding the function $\Phi(\tau, \Delta\nu, \mu)$ at the jth point in a Taylor series, viz.

$$\Phi(\tau_j + \Delta\tau_j, \Delta\nu, \mu) = \Phi(\tau_j, \Delta\nu, \mu) + \Delta\tau_j \left[\frac{\partial \Phi}{\partial \tau}(\tau, \Delta\nu, \mu)\right]_{\tau_j} + \cdots, \tag{3.2.70}$$

i.e. if we neglect terms $O(\delta\tau_j)^2$, we have

$$\left[\frac{\partial \Phi}{\partial \tau}\right]_{\tau_j} \approx \frac{\Phi_{j+1} - \Phi_j}{\Delta\tau_j}, \tag{3.2.71}$$

identical to equation (3.2.15).

However, there exists a simple method which, while considerably improving the accuracy of the numerical solution, effectively entails zero extra computing.

We write

$$\Phi_2(\Delta\nu, \mu) = \Phi(\tau_2, \Delta\nu, \mu) = \Phi(\tau_1 + \Delta\tau_1, \Delta\nu, \mu)$$

$$= \Phi(\tau_1, \Delta\nu, \mu) + \Delta\tau_1 \left[\frac{\partial \Phi}{\partial \tau}\right]_{\tau_1} + \frac{(\Delta\tau_1)^2}{2!}\left[\frac{\partial^2 \Phi}{\partial \tau^2}\right]_{\tau_1} + O(\Delta\tau_1)^3. \quad (3.2.72)$$

The derivatives appearing on the RHS of equation (3.2.72) may be eliminated using the boundary condition (3.2.13) and the defining equation (3.2.10), viz.

$$\left[\frac{\mu}{\phi(\Delta\nu)}\frac{\partial \Phi}{\partial \tau}\right]_{\tau_1} = \Phi(\tau_1, \Delta\nu, \mu) - I_{\text{inc}}(\Delta\nu, \mu), \quad (3.2.73)$$

$$\left[\left(\frac{\mu}{\phi(\Delta\nu)}\right)^2 \frac{\partial^2 \Phi}{\partial \tau^2}\right]_{\tau_1} = \Phi(\tau_1, \Delta\nu, \mu) - S(\tau_1), \quad (3.2.74)$$

where, for ease in exposition, we have taken $\phi(\Delta\nu)$ to be depth independent in this last equation. We will consider the more general result using depth-dependent $\phi(\Delta\nu)$ at the end of this section.

Equation (3.2.72) then becomes

$$\Phi_2(\Delta\nu, \mu) = \Phi_1(\Delta\nu, \mu) + \frac{\phi(\Delta\nu)\Delta\tau_1}{\mu}\left[\Phi(\Delta\nu, \mu) - I_{\text{inc}}(\Delta\nu, \mu)\right] + \tfrac{1}{2}\left(\frac{\phi(\Delta\nu)\Delta\tau_1}{\mu}\right)^2 [\Phi_1(\Delta\nu, \mu) - S(\tau_1)], \quad (3.2.75)$$

where we have neglected terms $O(\Delta\tau_1)^3$.

Using the quadrature formula (3.2.38), we find

$$S(\tau_1) \approx (1-\epsilon_1)\sum_{i=1}^{N_\nu}\sum_{k=1}^{N_\mu} \alpha_i\beta_k\Phi_{1ik} + [\epsilon B_\nu(T)]_1, \quad (3.2.76)$$

so that equation (3.2.75) may be re-arranged to yield

$$\left[1 + \frac{2\mu_k}{\phi_i\Delta\tau_1} + 2\left(\frac{\mu_k}{\phi_i\Delta\tau_1}\right)^2\right]\Phi_{1ik} - 2\left(\frac{\mu_k}{\phi_i\Delta\tau_1}\right)^2\Phi_{2ik} - (1-\epsilon_1)\sum_{i'=1}^{N_\nu}\sum_{k'=1}^{N_\mu}\alpha_{i'}\beta_{k'}\Phi_{1i'k'} = [\epsilon B_\nu(T)]_1 + \frac{2\mu_k}{\phi_i\Delta\tau_1} I_{\text{inc}}(\Delta\nu_i, \mu_k). \quad (3.2.77)$$

If we now put

$$\mathbf{C}_1 = \frac{2}{(\Delta\tau_1)^2}\begin{pmatrix}(\mu_1/\phi_1)^2 & & 0 \\ & \ddots & \\ 0 & & (\mu_{N_\mu}/\phi_{N_\nu})^2\end{pmatrix}, \quad (3.2.78)$$

and

$$\mathbf{B}_1 = \mathbb{1} + \mathbf{C}_1 + \mathbf{D}_1 - (1-\epsilon_1)\mathbf{W}_1, \quad (3.2.79)$$

where $\mathbf{W}_1$ is defined by equation (3.2.44) and

$$\mathbf{D}_1 = \frac{2}{\Delta\tau_1}\begin{pmatrix} \mu_1/\phi_1 & & 0 \\ & \ddots & \\ 0 & & \mu_{N_\mu}/\phi_{N_\mu} \end{pmatrix},$$

$$= \sqrt{2}\,\mathbf{C}_1^{1/2}, \tag{3.2.80}$$

then the system of equations (3.2.77) may be written in matrix form

$$\mathbf{B}_1\mathbf{\Phi}_1 - \mathbf{C}_1\mathbf{\Phi}_2 = \mathbf{L}_1, \tag{3.2.81}$$

where

$$\mathbf{L}_1 = [\epsilon B_\nu(T)]_1\mathbb{1} + \mathbf{D}_1\begin{pmatrix} I_{\text{inc}}(\Delta\nu_1, \mu_1) \\ \vdots \\ I_{\text{inc}}(\Delta\nu_{N_\nu}, \mu_{N_\mu}) \end{pmatrix}. \tag{3.2.82}$$

Equation (3.2.81) has exactly the same explicit form as (3.2.51) and, therefore, may be used in precisely the same manner in developing the recurrence relationship. The only effective difference between equations (3.2.51) and (3.2.81) lies in the structure of the $\mathbf{B}_1$ matrix which is now full not diagonal. Consequently, we must now invert N_τ, rather than $N_{\tau-1}$ matrices. The extra computing this involves is not significant since N_τ is usually at least 25.

When $\phi(\Delta\nu)$ is depth dependent, we write

$$\mathrm{d}\tau = -\frac{\kappa(\Delta\nu)}{\phi(\Delta\nu)}\,\mathrm{d}z, \tag{3.2.83}$$

so that

$$\left[\frac{\mu}{\kappa}\frac{\partial\Phi}{\partial z}\right]_{z_1} = I_{\text{inc}}(\Delta\nu, \mu) - \Phi(z_1, \Delta\nu, \mu), \tag{3.2.84}$$

and

$$\left[\left(\frac{\mu}{\kappa}\frac{\partial}{\partial z}\right)^2\Phi\right]_{z_1} = \Phi(z_1, \Delta\nu, \mu) - S(z_1). \tag{3.2.85}$$

The appropriate Taylor series expansion is then

$$\Phi(z_2, \Delta\nu, \mu) = \Phi(z_1, \Delta\nu, \mu) + \kappa_1(\Delta\nu)\Delta z_1\left[\frac{1}{\kappa}\frac{\partial\Phi}{\partial z}\right]_{z_1} + \frac{(\kappa_1(\Delta\nu)\Delta z_1)^2}{2}\left[\left(\frac{1}{\kappa}\frac{\partial}{\partial z}\right)^2\Phi\right]_{z_1} + O(\kappa_1\Delta z_1)^3. \tag{3.2.86}$$

The above three equations can then be combined as before to yield an equation of the same form as (3.2.81). However, here, the $\mathbf{B}_1$ and $\mathbf{C}_1$ matrices, and the vector $\mathbf{L}_1$, are obtained simply by replacing $\phi(\Delta\nu_i)\Delta\tau_1$ by $-\kappa_1(\Delta\nu_i)\Delta z_1$. Note that $\phi(\Delta\nu_i)\Delta\tau_1$ is not, in general, equal to $-\kappa_1(\Delta\nu_i)\Delta z_1$ even for depth-independent $\phi(\Delta\nu)$ since

$$\phi(\Delta\nu_i)\Delta\tau_1 = \phi(\Delta\nu_i)(\tau_2 - \tau_1) = -\phi(\Delta\nu_i)\int_{z_1}^{z_2} \frac{\kappa(z, \Delta\nu_i)}{\phi(\Delta\nu_i)}\,dz$$

$$= -\frac{h\nu}{4\pi} B_{LU}\phi(\Delta\nu_i)\int_{z_1}^{z_2} N_L(z)\,dz$$

$$\neq -\kappa(z_1, \Delta\nu_i)(z_2 - z_1),$$

unless $N_L(z)$ is independent of depth (the reader is referred to equations (1.9.5) and (1.9.12) for the definitions used in the above manipulations).

A similar analysis may be used to improve the accuracy of the differencing approximation at the other boundary deep within the model atmosphere although this is not necessary in general because all quantities vary only slowly at such depths.

3.2.5 *Quadrature points and weights [55]*

We now briefly consider an important computational detail relating to equation (3.2.38), viz.

$$\int_{-\infty}^{\infty} \phi(\Delta\nu)\,d(\Delta\nu)\int_0^1 f(\tau, \Delta\nu, \mu)\,d\mu \cong \sum_{i=1}^{N_\nu}\sum_{k=1}^{N_\mu} \alpha_i\beta_k f(\tau, \Delta\nu_i, \mu_k), \tag{3.2.38}$$

where we have constructed the quadrature points $\{\Delta\nu_i\}$ and $\{\mu_k\}$ for the frequency and angle integrations.

We could, of course, combine equations (3.2.36) and (3.2.37) to yield

$$\int_{-\infty}^{\infty} \phi(\Delta\nu)\,d(\Delta\nu)\int_0^1 f(\tau, \Delta\nu, \mu)\,d\mu$$

$$\approx \sum_{i=1}^{N_\nu}\sum_{k=1}^{N_\mu} w_i^{(1)} w_k^{(2)} f(\tau, \Delta\nu_i, \mu_k), \tag{3.2.87}$$

where $w_i^{(1)}$ and $w_k^{(2)}$ are the weights corresponding to individual quadratures over frequency and angle respectively. We usually take $\{\mu_k\}$ and $\{w_k^{(2)}\}$ corresponding to Gaussian integration [1], while the $\{\Delta\nu_i\}$ and $\{w_k^{(1)}\}$ are those pertaining to trapezoidal quadrature. The $\{\Delta\nu_i\}$ for Doppler broadening are evenly spaced over the frequency bandwidth $[-\beta, \beta]$ with

$$w_i^{(1)} = h\phi(\Delta\nu_i) \quad \text{for } i \neq 1, i \neq N_\nu,$$

$$w_1^{(1)} = \frac{h}{2}\,\phi(\Delta\nu_1) \quad w_{N_\nu}^{(1)} = \frac{h}{2}\,\phi(\Delta\nu_{N_\nu}), \tag{3.2.88}$$

where

$$h = \Delta\nu_i - \Delta\nu_{i-1} = \frac{2\beta}{N_\nu - 1},$$

i.e.

$$\begin{aligned}\int_{-\infty}^{\infty} \phi(\Delta\nu) f(\tau, \Delta\nu, \mu)\, \mathrm{d}(\Delta\nu) &\approx \int_{-\beta}^{\beta} \phi(\Delta\nu) f(\tau, \Delta\nu, \mu)\, \mathrm{d}(\Delta\nu)\\ &\approx \sum_{i=1}^{N_\nu} w_i^{(1)} f(\tau, \Delta\nu_i, \mu)\\ &= \frac{h}{2} [\phi(\Delta\nu_1) f(\tau, \Delta\nu_1, \mu)\\ &\quad + 2\phi(\Delta\nu_2) f(\tau, \Delta\nu_2, \mu)\\ &\quad + \cdots + 2\phi(\Delta\nu_{N_\nu - 1}) f(\tau\ \Delta\nu_{N_\nu - 1}, \mu)\\ &\quad + \phi(\Delta\nu_{N_\nu}) f(\tau, \Delta\nu_{N_\nu}, \mu)]. \end{aligned} \tag{3.2.89}$$

Further, experimental computations show that we should have

$$h < \tfrac{1}{2} \min (\Delta\nu_\mathrm{D}), \tag{3.2.90}$$

where min $(\Delta\nu_\mathrm{D})$ is the minimum value of the Doppler width in the model atmosphere and, since $\phi(\Delta\nu)$ rapidly decreases with increasing $|\Delta\nu|$ for Doppler broadening, we need only take $\beta \gtrsim 3 \max (\Delta\nu_\mathrm{D})$. More points are required when using the Voigt profile because the much slower decrease of $\phi(\Delta\nu)$ in the wings of the line relative to Doppler broadening (see figure 1.10 for example) necessitates a much larger value of the bandwidth factor β. Here, one usually takes h satisfying equation (3.2.90) for $|\Delta\nu_i| \lesssim 3 \max (\Delta\nu_\mathrm{D})$ but, for $\Delta\nu$ outside this region, take $\{\Delta\nu_i\}$ satisfying

$$\Delta\nu_i = 2(\Delta\nu_{i-1} - \Delta\nu_{i-2}). \tag{3.2.91}$$

It should be stressed that one can readily check the validity of the angle and frequency quadrature simply by increasing the number of these quadrature points (by both decreasing h and increasing β) until the changes in the solution so obtained are less than the desired accuracy.

However, a most important computational detail remains. We have already discussed in section 2.10 the need to accurately evaluate the integrals over $\Delta\nu$ and μ since any errors in this evaluation are physically equivalent to spurious sources and/or sinks of photons and this, of course, will lead to spurious emergent line radiation.

We know from equation (1.4.8) that

$$\int_{-\infty}^{\infty} \phi(\Delta\nu)\, \mathrm{d}(\Delta\nu) = 1, \tag{3.2.92}$$

so that, if we take $f(\tau, \Delta\nu, \mu) \equiv 1$ for example, equation (3.2.87) yields

$$\int_{-\infty}^{\infty} \phi(\Delta\nu)\, d(\Delta\nu) \int_0^1 f(\tau, \Delta\nu, \mu)\, d\mu = 1,$$

i.e. we require

$$\sum_{i=1}^{N_\nu} \sum_{k=1}^{N_\mu} w_i^{(1)} w_k^{(2)} = 1. \tag{3.2.93}$$

Although, for Gaussian quadrature,

$$\sum_{k=1}^{N_\mu} w_k^{(2)} = 1, \tag{3.2.94}$$

the above frequency weights $w_i^{(1)}$ yield

$$\sum_{i=1}^{N_\nu} w_i^{(1)} \neq 1 \tag{3.2.95}$$

- see equation (3.2.88).

Consequently, equation (3.2.87) is clearly unsatisfactory. However, if we now take the weights α_i and β_k such that

$$\alpha_i = c w_i^{(1)} \quad \text{and} \quad \beta_k = w_k^{(2)}, \tag{3.2.96}$$

where c is a constant, then equation (3.2.38) with $f(\tau, \Delta\nu, \mu) \equiv 1$ yields

$$c \sum_{i=1}^{N_\nu} \sum_{k=1}^{N_\mu} w_i^{(1)} w_k^{(2)} = 1. \tag{3.2.97}$$

Therefore, the weighting factor $\alpha_i \beta_k$ appearing in equation (3.2.38) should be

$$\alpha_i \beta_k = \frac{w_i^{(1)} w_k^{(2)}}{\sum_{i'=1}^{N_\nu} \sum_{k'=1}^{N_\mu} w_{i'}^{(1)} w_{k'}^{(2)}}. \tag{3.2.98}$$

This clearly ensures, therefore, that the integration given by equation (3.2.38) of a function $f(\tau, \Delta\nu, \mu)$ independent of $\Delta\nu$ and μ is absolutely accurate. Although it does not imply the corresponding accuracy when $f(\tau, \Delta\nu, \mu)$ is a function of $\Delta\nu$ and μ (and this is the case of interest) it does suggest that any errors which do arise are minimised.

Finally, if there are no velocity or magnetic fields, the radiative intensities are symmetric about $\Delta\nu = 0$ as, of course, is $\phi(\Delta\nu)$. Thus, writing $\phi(-\Delta\nu) = \phi(\Delta\nu)$ and $f(\tau, -\Delta\nu, \mu) = f(\tau, \Delta\nu, \mu)$, we have

$$\int_{-\infty}^{\infty} \phi(\Delta\nu)\, d(\Delta\nu) \int_0^1 f(\tau, \Delta\nu, \mu)\, d\mu$$

$$= 2 \int_0^{\infty} \phi(\Delta\nu)\, d(\Delta\nu) \int_0^1 f(\tau, \Delta\nu, \mu)\, d\mu. \tag{3.2.99}$$

Consequently, if we take $\Delta\nu_i = -\Delta\nu_{N_\nu - i+1}$ with $w_i^{(1)} = w_{N_\nu - i+1}^{(1)}$, only half the frequency bandwidth $[-\beta, \beta]$ need be considered in the computations. The number of frequency quadrature points N_ν required is then reduced by a factor of two, as is the size of the matrices to be inverted - see equations (3.2.65) and (3.2.66), and this results in significant savings in both computer time and storage. We stress, however, that the above simplification only applies for stationary, non-magnetic media.

3.3 Stability of the Feautrier technique

One may discuss the stability of the second order differential system, given in the preceding sections, from several different points of view.

First, we consider closed form analytical solutions to the simplified case in which we take $N_\nu = 2$ (with $\Delta\nu_2 = -\Delta\nu_1$) and $N_\mu = 1$. Equations (3.2.38) and (3.2.99) then become

$$\int_{-\infty}^{\infty} \phi(\Delta\nu)\, \mathrm{d}(\Delta\nu) \int_0^1 f(\tau, \Delta\nu, \mu)\, \mathrm{d}\mu \approx \alpha_1 \beta_1 f(\tau, \Delta\nu_1, \mu_1), \tag{3.3.1}$$

where our normalisation condition

$$\int_{-\infty}^{\infty} \phi(\Delta\nu)\, \mathrm{d}(\Delta\nu) \int_0^1 \mathrm{d}\mu = 1,$$

discussed in the preceding sub-section, stipulates that the weights α_1 and β_1 must satisfy

$$\alpha_1 \beta_1 = 1. \tag{3.3.2}$$

If we now drop the subscript 1 on $\Delta\nu_1$ and μ_1, the equation of radiative transfer (3.2.10) for this 1-point quadrature approximation becomes

$$\begin{aligned} \frac{\mu^2}{\phi^2(\Delta\nu)} \frac{\partial^2 \Phi}{\partial \tau^2} &= \Phi - (1-\epsilon)\Phi - \epsilon B_\nu(T) \\ &= \epsilon[\Phi - B_\nu(T)], \end{aligned} \tag{3.3.3}$$

where $\Phi = \Phi(\tau, \Delta\nu, \mu)$. One-point Gaussian quadrature over $\mu \in [0, 1]$ is just $\mu = \mu_1 = 1/\sqrt{3}$. The equation to be solved then has the form

$$\frac{\partial^2 \Phi}{\partial \tau^2} = 3\epsilon\phi^2(\Delta\nu)[\Phi - B_\nu(T)]. \tag{3.3.4}$$

We shall consider two different closed form analytical solutions to this last equation corresponding to two different boundary conditions deep within the model atmosphere. The stability of the system from a finite difference approach is then examined.

3.3.1 *Exact solution with $\Phi(\infty) = B_\nu(T)$*

The solution to equation (3.3.4) with $\epsilon_1 B_\nu(T)$ and $\phi = \phi(\Delta\nu)$ constant is just

$$\Phi(\tau) = A e^{\sqrt{(3\epsilon)}\phi\tau} + C e^{-\sqrt{(3\epsilon)}\phi\tau} + B_\nu(T), \quad (3.3.5)$$

where A and C are the two constants to be determined from the two boundary conditions. It is important to note here that the solution has both exponentially increasing and decreasing components and that such components could possibly allow for an exponential amplification of any numerical error due to round-off or quadrature or finite differencing.

If we consider the above exact solution to equation (3.3.4) with the lower boundary condition $\Phi(\tau) \to B_\nu(T)$ as $\tau \to \infty$ we must have $A = 0$.

Consequently, the exponentially increasing factor disappears and the amplification of errors would perhaps not arise. However, it would be unreasonable to assume that under a numerical scheme our finite differencing quadrature and round-off would allow $A = 0$ exactly.

The boundary condition at the surface with zero incident radiation (see equation (3.2.13)) is just

$$\frac{1}{\sqrt{3}\phi}\frac{\partial\Phi}{\partial\tau} = \Phi \quad \text{at } \tau = 0, \quad (3.3.6)$$

and this yields

$$C = -\frac{B_\nu(T)}{1 + \sqrt{\epsilon}}. \quad (3.3.7)$$

The complete solution is then

$$\Phi(\tau) = B_\nu(T)\left[1 - \frac{e^{-\sqrt{(3\epsilon)}\phi\tau}}{1 + \sqrt{\epsilon}}\right], \quad (3.3.8)$$

so that

$$\begin{aligned} S(\tau) &= (1 - \epsilon)\Phi + \epsilon B_\nu(T) \\ &= B_\nu(T)\left[1 - \frac{1 - \epsilon}{1 + \sqrt{\epsilon}} e^{-\sqrt{(3\epsilon)}\phi\tau}\right]. \end{aligned} \quad (3.3.9)$$

It is interesting to note two points. First, $S(0) = \sqrt{\epsilon} B_\nu(T)$, a result which was obtained from the exact solutions derived in chapter 2 (see equation (2.9.25)). Second, $S(\tau) \approx B_\nu(T)$ for all τ satisfying $\sqrt{(3\epsilon)}\phi\tau \gg 1$, i.e. $\phi\tau \gg 1/\sqrt{\epsilon}$. This, not surprisingly, is the same result as that obtained in section 2.10.3 using the exact solutions for coherent scattering (see equation (2.10.44)).

3.3.2 *Exact solution with $\tau_{N_\tau} \not\to \infty$*

Clearly, computer limitations will now allow us to specify infinity as the deepest τ point in our depth grid. We therefore consider the situation in which the boundary condition $\Phi(\tau) \to B_\nu(T)$ as $\tau \to \infty$ is replaced by the Feautrier condition (3.2.14), viz.

$$\frac{1}{\sqrt{3\phi}} \frac{\partial \Phi}{\partial \tau} = B_\nu(T) - \Phi \quad \text{at } \tau = \tau_\infty, \tag{3.3.10}$$

where $\tau_\infty = \tau_{N_\tau}$ is now the deepest point in our $\{\tau_j\}$ grid. This, together with the boundary condition (3.3.6), then yields the two equations in A and C:

$$\sqrt{\epsilon}(A - C) = A + C + B_\nu(T),$$

and

$$\sqrt{\epsilon}[A e^{\sqrt{(3\epsilon)}\phi\tau_\infty} - C e^{-\sqrt{(3\epsilon)}\phi\tau_\infty}] = -A e^{\sqrt{(3\epsilon)}\phi\tau_\infty} - C e^{-\sqrt{(3\epsilon)}\phi\tau_\infty},$$

i.e.

$$A = \frac{(1-\sqrt{\epsilon})B_\nu(T)}{(1-\sqrt{\epsilon})^2 + (1+\sqrt{\epsilon})^2 e^{2\sqrt{(3\epsilon)}\phi\tau_\infty}}, \tag{3.3.11}$$

and

$$C = \frac{-(1+\sqrt{\epsilon})B_\nu(T) e^{2\sqrt{(3\epsilon)}\phi\tau_\infty}}{(1-\sqrt{\epsilon})^2 + (1+\sqrt{\epsilon})^2 e^{2\sqrt{(3\epsilon)}\phi\tau_\infty}}. \tag{3.3.12}$$

The important point to now notice is that $A \neq 0$ and that any errors due to quadrature and differencing, or round-off, may be exponentially amplified.

It is of some interest to test the accuracy of the solution at $\tau = 0$. We eventually find from the above equations that

$$\frac{S(0)}{B_\nu(T)} = 1 + \frac{(1-\epsilon)[1 - \sqrt{\epsilon} - (1+\sqrt{\epsilon}) e^{2\sqrt{(3\epsilon)}\phi\tau_\infty}]}{(1-\sqrt{\epsilon})^2 + (1+\sqrt{\epsilon})^2 e^{2\sqrt{(3\epsilon)}\phi\tau_\infty}}, \tag{3.3.13}$$

so that, for $\sqrt{(3\epsilon)}\phi\tau_\infty$ sufficiently large

$$\frac{S(0)}{B_\nu(T)} \approx 1 - \frac{1-\epsilon}{1+\sqrt{\epsilon}} = \sqrt{\epsilon}, \tag{3.3.14}$$

which, as one would expect, corresponds to the exact result. If, however, $\sqrt{(3\epsilon)}\phi\tau_\infty$ was not sufficiently large, i.e. $e^{2\sqrt{(3\epsilon)}\phi\tau_\infty} \not\gg$. Then the result given

Table 3.1. *Values of* S*(0) for various* τ_∞

$S(0)/B_\nu(T)$		τ_∞
$\epsilon = 10^{-4}$	$\epsilon = 10^{-6}$	
0.819	0.982	10
0.0684	0.828	10^2
0.0100	0.0615	10^3
0.0100	0.001 00	10^4
0.01	0.001	∞

by this last equation would not be obtained. Thus, it is essential to construct the depth $\{\tau_j\}$ grid sufficiently deep within the model atmosphere to obtain accurate solutions not only at large depths but at the surface also. To emphasise this point quantitatively, we compute $S(0)/B_\nu(T)$ for the not atypical values $\epsilon = 10^{-4}$ and 10^{-6} for several values of τ_∞. Note that our 1-point frequency quadrature effectively states that $\phi(\Delta\nu) \equiv \delta(\Delta\nu - \Delta\nu_0)$ which, in the present context, implies $\phi = 1$. The results are shown in table 3.1 where the exact value $S(0)/B_\nu(T) = \sqrt{\epsilon}$ is listed for $\tau_\infty = \infty$.

Clearly, there exist large errors in $S(0)$ if τ_∞ is not sufficiently large.

3.3.3 *Scalar finite difference approach [35]*

If we approximate the second order derivative appearing in equation (3.3.4) by finite differences, and this is the situation of most interest in this section, we have

$$\frac{\dfrac{\Phi_{j+1} - \Phi_j}{\Delta_j} - \dfrac{\Phi_j - \Phi_{j-1}}{\nabla_j}}{\frac{1}{2}(\Delta_j + \nabla_j)} = 3\phi^2 \epsilon_j (\Phi_j - B_j), \tag{3.3.15}$$

where

$$\Delta_j = \tau_{j+1} - \tau_j, \quad \nabla_j = \tau_j - \tau_{j-1}, \tag{3.3.16}$$

and

$$B_j = [B_\nu(T)]_{\tau_j}. \tag{3.3.17}$$

If we now put

$$a_j = \frac{2}{\nabla_j(\Delta_j + \nabla_j)}, \quad c_j = \frac{2}{\Delta_j(\Delta_j + \nabla_j)}, \tag{3.3.18}$$

with

$$b_j = -a_j - c_j, \tag{3.3.19}$$

$$l_j = 3\phi^2 \epsilon_j B_j, \tag{3.3.20}$$

and

$$\beta_j = 3\phi^2 \epsilon_j - b_j, \tag{3.3.21}$$

equation (3.3.15) may be written in the form

$$-a_j \Phi_{j-1} + \beta_j \Phi_j - c_j \Phi_{j+1} = l_j. \tag{3.3.22}$$

This last equation is merely the scalar representation of the matrix equation (3.2.46). We may therefore solve it using the same method as that presented in section 3.2.3, i.e. we write

$$\Phi_j = u_j + v_j \Phi_{j+1}, \tag{3.3.23}$$

which, when substituted into equation (3.3.22), yields

$$-a_j u_{j-1} - a_j v_{j-1} \Phi_j + \beta_j \Phi_j - c_j \Phi_{j+1} = l_j,$$

i.e.

$$\Phi_j = \frac{l_j + a_j u_{j-1} + c_j \Phi_{j+1}}{\beta_j - a_j v_{j-1}}, \tag{3.3.24}$$

which, by comparison with equation (3.3.23), suggests the recurrence formulae:

$$u_j = \frac{l_j + a_j u_{j-1}}{\beta_j - a_j v_{j-1}}, \tag{3.3.25}$$

and

$$v_j = \frac{c_j}{\beta_j - a_j v_{j-1}}. \tag{3.3.26}$$

Our boundary conditions give

$$a_1 = 0 = C_N, \tag{3.3.27}$$

where N refers to the last depth grid point. The method of solution then requires the u_j and v_j quantities to be determined from u_{j-1} and v_{j-1} for all j where, in particular $u_1 = l_1/\beta_1$ and $v_1 = c_1/\beta_1$. Once all these u_j and v_j are known, one effectively solves for Φ_j by back substitution using equation (3.3.23). In particular, since $C_N = 0$, then $v_N = 0$ so that $\Phi_N = u_N$ which has been determined. One then solves for Φ_{N-1}, etc.

We examine the stability of this system by considering the manifestation of an error Δv_j, say, in v_j as applied to the recurrence relationship (3.3.26) due to numerical round-off or finite differencing, i.e. we put

$$v_j = \bar{v}_j + \Delta v_j, \tag{3.3.28}$$

where $\bar{v}_j$ is the exact quantity one would obtain if the equations could be solved to infinite accuracy, and v_j is the corresponding (approximate) numerical value. We therefore have

$$\bar{v}_j = \frac{c_j}{\beta_j - a_j \bar{v}_{j-1}}. \tag{3.3.29}$$

Substitution of equation (3.3.28) into (3.3.26) then yields

$$\bar{v}_j + \Delta v_j = \frac{c_j}{\beta_j - a_j \bar{v}_{j-1} - a_j \Delta v_{j-1}},$$

i.e.

$$\Delta v_j \approx \frac{a_j c_j \Delta v_{j-1}}{(\beta_j - a_j \bar{v}_{j-1})^2}, \tag{3.3.30}$$

where, at each jth step, we assume that the error generated at that step is sufficiently small such that

$$\left| \frac{a_j \Delta v_{j-1}}{\beta_j - a_j \bar{v}_{j-1}} \right| \ll 1. \tag{3.3.31}$$

Of course, if this last inequality did not hold, our errors at each depth would be swamping our solution for v at that depth and we would know, without further analysis, that our system is unstable.

Since the v_j terms are determined successively for increasing values of j, the system will be stable if

$$\left|\frac{\Delta v_j}{\Delta v_{j-1}}\right| < 1, \tag{3.3.32}$$

i.e. the errors decrease with increasing j. From equation (3.3.30), we have the equivalent condition:

$$\left|\frac{a_j c_j}{(\beta_j - a_j \bar{v}_{j-1})^2}\right| < 1. \tag{3.3.33}$$

If this last condition was violated, the system would amplify any numerical errors and instability would then occur. Therefore, we would need to either reject the technique or administer some correcting procedure. If condition (3.3.33) was satisfied, however, instabilities might yet occur due to the amplification of errors in the recurrence relationship defining the quantities u_j. Consequently, we again consider the error u_j such that

$$u_j = \bar{u}_j + \Delta u_j. \tag{3.3.34}$$

Equation (3.3.25) then yields

$$\frac{\Delta u_j}{\Delta u_{j-1}} = \frac{a_j}{\beta_j - a_j \bar{v}_{j-1}}, \tag{3.3.35}$$

where we have assumed that no errors are propagated in the v_j recurrence formula, i.e. we have replaced v_{j-1} by $\bar{v}_{j-1}$. The condition for stability in the u_j recurrence system is then

$$\left|\frac{a_j}{\beta_j - a_j \bar{v}_{j-1}}\right| < 1. \tag{3.3.36}$$

Let us now consider the 'backward' recurrence relationship (3.3.23), i.e.

$$\Phi_j = u_j + v_j \Phi_{j+1}. \tag{3.3.23}$$

Assuming the u_j and v_j recurrence relationships to be stable so that we may replace u_j and v_j by $\bar{u}_j$ and $\bar{v}_j$ respectively, we put

$$\Phi_j = \bar{\Phi}_j + \Delta\Phi_j, \tag{3.3.37}$$

and equation (3.3.23) then becomes

$$\bar{\Phi}_j + \Delta\Phi_j = \bar{u}_j + \bar{v}_j(\bar{\Phi}_{j+1} + \Delta\Phi_{j+1}),$$

i.e.

$$\frac{\Delta\Phi_j}{\Delta\Phi_{j+1}} = \bar{v}_j = \frac{c_j}{\beta_j - a_j \bar{v}_{j-1}}. \tag{3.3.38}$$

Since we obtain Φ_j from Φ_{j+1}, the recurrence system determining Φ_j is stable if

$$\left|\frac{\Delta\Phi_j}{\Delta\Phi_{j+1}}\right| < 1,$$

i.e.

$$\left|\frac{c_j}{\beta_j - a_j\bar{v}_{j-1}}\right| < 1. \tag{3.3.39}$$

We now have the three conditions (3.3.33), (3.3.36) and (3.3.39) which must be satisfied if the Feautrier technique is to be stable. Note that if conditions (3.3.36) and (3.3.39) are both satisfied, then so is condition (3.3.33).

Recall that

$$\beta_j = 3\epsilon_j - b_j = 3\epsilon_j + a_j + c_j, \tag{3.3.40}$$

where, as explained earlier, $\phi = 1$. We then have

$$\bar{v}_j = \frac{c_j}{\beta_j - a_j\bar{v}_{j-1}} = \frac{c_j}{3\epsilon_j + a_j(1-\bar{v}_{j-1}) + c_j}. \tag{3.3.41}$$

Now

$$\bar{v}_1 = \frac{c_1}{\beta_1} = \frac{c_1}{3\epsilon_1 + c_1} \tag{3.3.42}$$

(note that $a_1 = 0$) so that, since $\epsilon_j > 0$ and $c_j \geqslant 0$ for all j, $0 < \bar{v}_1 < 1$. Equation (3.3.41) then stipulates that $0 < \bar{v}_2 < 1$ (note that $a_j > 0$ for all $j > 1$). One may continue this argument to show that $0 < v_j < 1$ for all $j < N$.

We next consider condition (3.3.36). We first construct a $\{\tau_j\}$ grid satisfying

$$\frac{\tau_{j+1} - \tau_j}{\tau_j - \tau_{j-1}} = f, \tag{3.3.43}$$

where $f \geqslant 1$ is a constant factor. (Putting $f = 2$ yields adequate results in practice.) Equation (3.3.16) gives

$$f = \frac{\Delta_j}{\nabla_j}, \tag{3.3.44}$$

so that, from equation (3.3.18),

$$a_j = fc_j. \tag{3.3.45}$$

We then have

$$\left|\frac{a_j}{\beta_j - a_j\bar{v}_{j-1}}\right| = \left|\frac{a_j}{3\epsilon_j + a_j(1-\bar{v}_{j-1}) + c_j}\right|$$
$$= \left|\frac{fc_j}{3\epsilon_j + fc_j(1-\bar{v}_{j-1}) + c_j}\right|. \tag{3.3.46}$$

Since $f \geqslant 1$ with $\epsilon_j > 0$, $c_j > 0$ and $0 < v_j < 1$ for all $j < N$, then

$$\left|\frac{a_j}{\beta_j - a_j\bar{v}_{j-1}}\right| < \frac{f}{f(1-\bar{v}_{j-1})+1}. \tag{3.3.47}$$

Thus, if $f\bar{v}_{j-1} < 1$, then

$$\left|\frac{a_j}{\beta_j - a_j\bar{v}_{j-1}}\right| < 1. \tag{3.3.48}$$

This last inequality is, of course, one of the three required stability conditions. Since $f \geqslant 1$, we must have $a_j \geqslant c_j$ from equation (3.3.45). Thus, if (3.3.48) is satisfied, then so must (3.3.39) and these two together will ensure that condition (3.3.33) is satisfied. We therefore focus our attention on the product $f\bar{v}_{j-1}$. Indeed, we require $f\bar{v}_{j-1} < 1$. Note that

$$f\bar{v}_{j-1} = \frac{fc_j}{3\epsilon_j + fc_j(1-\bar{v}_{j-2}) + c_j},$$

i.e.

$$f\bar{v}_{j-1} < \frac{f}{f(1-\bar{v}_{j-2})+1}, \tag{3.3.49}$$

so that if $f\bar{v}_{j-2} < 1$ then $f\bar{v}_{j-1} < 1$. This leads to $f\bar{v}_j < 1$ for all j.

However, we do not always have $f\bar{v}_{j-2} < 1$. For example,

$$\bar{v}_1 = \frac{c_1}{3\epsilon_1 + c_1}, \tag{3.3.50}$$

where, using the boundary condition (3.3.6),

$$c_1 = \frac{1}{\sqrt{3}\Delta_1}. \tag{3.3.51}$$

This finite differencing of the first order derivative appearing in equation (3.3.6) can, of course, be more accurately approximated the smaller we take the first depth step. In this case, we would have $\Delta_1 \ll 1$, i.e. $c_1 \gg 1$ which, with $\epsilon < 1$ (generally, we are interested in $\epsilon \ll 1$), yields $\bar{v}_1 \lesssim 1$ from equation (3.3.50). We then have $f\bar{v}_1 \lesssim f$ so that, for the not unreasonable value $f = 2$, $f\bar{v}_1 \nless 1$ and therefore condition (3.3.48) would be violated. Consequently, any small error in u_j would then be amplified (at least for $j \gtrsim 1$) and it would appear that the system would be unstable.

However, $\bar{v}_j$ decreases for increasing j. We can show this by noting that

$$\frac{c_{j+1}}{c_j} = \frac{\Delta_j(\Delta_j + \nabla_j)}{\Delta_{j+1}(\Delta_{j+1} + \nabla_{j+1})}, \tag{3.3.52}$$

where, from equation (3.3.44), we have

$$\Delta_{j+1} = f\nabla_{j+1} = f\Delta_j, \tag{3.3.53}$$

whilst

$$\nabla_{j+1} = \Delta_j = f\nabla_j. \tag{3.3.54}$$

Substitution of these results into equation (3.3.52) yields

$$\frac{c_{j+1}}{c_j} = \frac{1}{f^2}, \tag{3.3.55}$$

so that

$$\frac{\bar{v}_{j+1}}{\bar{v}_j} = \frac{c_{j+1}}{c_j} \cdot \frac{3\epsilon_j + fc_j(1-\bar{v}_{j-1}) + c_j}{3\epsilon_{j+1} + fc_{j+1}(1-\bar{v}_j) + c_{j+1}}$$
$$= \frac{3\epsilon_j + fc_j(1-\bar{v}_{j-1}) + c_j}{3\epsilon_{j+1}f^2 + fc_j(1-\bar{v}_j) + c_j}. \tag{3.3.56}$$

Usually ϵ_j increases with j since the number of collisions between atoms and free electrons increases with increasing τ due to the increase in electron density. Consequently, with $f \geqslant 1$, we should have $\epsilon_{j+1}f^2 \geqslant \epsilon_j$. Equation (3.3.56) then says that, if $\bar{v}_j < \bar{v}_{j-1}$, then

$$\bar{v}_{j+1} < \bar{v}_j. \tag{3.3.57}$$

We have already seen that

$$\bar{v}_1 = \frac{c_1}{3\epsilon_1 + c_1} \lesssim 1, \tag{3.3.58}$$

whilst

$$\bar{v}_2 = \frac{c_2}{3\epsilon_2 + fc_2(1-\bar{v}_1) + c_2} = \frac{c_1}{3\epsilon_2 f^2 + fc_1(1-\bar{v}_1) + c_1}, \tag{3.3.59}$$

so that, since $\bar{v}_1 > 0$ and assuming $\epsilon_2 f^2 > \epsilon_1$ as above, then $\bar{v}_2 < \bar{v}_1$. Repeating this process yields the inequality (3.3.57).

It is instructive to follow the trend of $\bar{v}_j$ further. In particular, we know from section 3.3.2 that our depth grid must cover a sufficiently large range in τ. Indeed, the results from that section indicate the magnitudes of the errors we might expect in $S(\tau)$ if we do not have $\tau_\infty \gg \Theta$ where Θ is the usual thermalisation distance. For Doppler broadening, for example, we must have $\tau_\infty \gg 1/\epsilon$. The construction of the $\{\tau_j\}$ grid given by equation (3.3.43) then indicates that

$$\Delta_N \sim \frac{1}{\epsilon}, \tag{3.3.60}$$

so that c_{N-1} given by

$$c_{N-1} = \frac{2}{\Delta_{N-1}(\Delta_{N-1} + \nabla_{N-1})}, \tag{3.3.61}$$

behaves as $c_{N-1} \sim \epsilon^2$. Consequently,

$$\bar{v}_{N-1} = \frac{c_{N-1}}{3\epsilon_{N-1} + fc_{N-1}(1-\bar{v}_{N-2}) + c_{N-1}}, \tag{3.3.62}$$

yields

$$\bar{v}_{N-1} \sim \epsilon (\ll 1). \tag{3.3.63}$$

Thus, for large j (which corresponds to large τ), we have $\bar{v}_j \sim \epsilon$ so that $f\bar{v}_j \sim f\epsilon < 1$ for all $f < 1/\epsilon$. Clearly, the value $f = 2$, for example, satisfies this last inequality for all $\epsilon < 0.5$ and, as stated previously, we are basically interested in the case $\epsilon \ll 1$. Of course, a smaller value of f requires a larger number of $\{\tau_j\}$ grid points to cover the same depth. However, the larger the value of ϵ, the smaller the thermalisation distance Θ and this leads to a smaller τ_∞. Consequently, the condition $f\epsilon < 1$ is hardly restrictive.

It is clear, therefore, that somewhere between $j = 1$ and large j the inequality $f\bar{v}_j > 1$ changes to $f\bar{v}_j < 1$. It is this latter inequality which ensures condition (3.3.36) and this, in turn, ensures stability of the Feautrier technique. We therefore have the situation in which the errors in u_j first amplify, then, after a certain distance, damp out entirely. We may obtain an approximate value of the distance over which these errors amplify by noting that if $f\bar{v}_{j-1} = 1$, say, then

$$f\bar{v}_j = \frac{fc_j}{3\epsilon_j + fc_j}, \tag{3.3.64}$$

and if $f\bar{v}_j \approx 1$ (one should expect this since, at the previous jth point, $f\bar{v}_{j-1} = 1$), then $fc_j \gg 3\epsilon_j$. However, $c_j \sim 1/\Delta_j^2$ so that

$$\Delta_j \ll \sqrt{\left(\frac{f}{3\epsilon_j}\right)}. \tag{3.3.65}$$

This effectively gives an upper bound on the distance over which u_j is amplified. Note that if $\tau_\infty \gg 1/\epsilon$, then this distance $\Delta_j \ll \sqrt{\epsilon\tau_\infty}$ so that for $\epsilon = 10^{-4}$, for example, the distance of error amplification is less than 1/100th of the overall grid and is therefore hardly significant. Indeed, this amplification distance corresponds to approximately 100 optical path lengths so that, if we choose $\Delta\tau_1 = 10^{-3}$ (a typical value) with $f = 2$, the number n of depth steps required to cover the depth range 0–100 is given by equation (3.3.43), viz. $10^{-3}2^n = 10^2$ which has the solution $n \sim 17$. This is the maximum value of n for that particular $\{\tau_j\}$ grid since the depth steps become larger for larger j. For example, the propagation of an error generated at $j = 10$, say, where $\Delta_j = 10^{-3}2^{10} \sim 1$ would require a value of n given by $\Delta_{10}2^n = 10^2$, i.e. a number of depth steps $n \sim 7$ would be required to cover an amplification distance ranging from $\tau \sim 1$ to $\tau \sim 100$. The maximum error due to round-off would then result in the loss of no more than $\log_{10} 2^n$ significant figures which, on the usual 14-significant-figure machine, corresponds to a reduction of the accuracy of the numerical evaluation (for $n \sim 17$) to approximately nine significant figures. Again, we stress that this is the worst possible situation. It does, however, suggest that one should not take too small a value for the initial depth step $\Delta\tau_1$. For example, if we took $\Delta\tau_1 = 10^{-10}$, then $n \sim 40$, which would correspond to a loss of 12 significant figures due to round-off. Clearly, the error situation becomes worse for smaller values of ϵ since the amplification distance then increases. However, once this problem has been recognised, an appropriate choice of $\Delta\tau_1$ given ϵ can be made which will always ensure stability *and* yield sufficient figure significance.

Finally, not only must we have $\Delta\tau_1$ sufficiently small for the accurate representation of the boundary derivative by finite differences (and $\Delta\tau_1 \sim 10^{-3}$ is not an unreasonable value for these purposes), but we must construct a $\{\tau_j\}$ grid satisfying $\tau_\infty \gg \Theta$ (see section 3.3.2). We have shown stability of the Feautrier system for $f \geqslant 1$. Therefore, if we take, for example, $f = 2$ with $\Delta\tau_1 = 10^{-3}$ and $N = 30$ depth steps, we find $\tau_\infty \sim \Delta\tau_1 2^{N-1} \sim 5 \times 10^5$. Consequently, for $\epsilon = 10^{-4}$, we have $\tau_\infty \gg 1/\epsilon$ as required for Doppler broadening. If, however, we now take $\epsilon = 10^{-8}$, we would require $N \gtrsim 41$ with $\Delta\tau = 10^{-3}$ for $\tau_\infty \gg 1/\epsilon$. Thus, although the depth grid for $\epsilon = 10^{-8}$ must extend to depths a factor of 10^4 greater than that required for $\epsilon = 10^{-4}$, the number of $\{\tau_j\}$ grid points required only increases from $N \sim 30$ to $N \sim 40$. This does not constitute an overwhelming increase in computer requirements. The importance of this particular advantage of the Feautrier technique cannot be too highly stressed, particularly as we consider smaller and smaller ϵ.

3.3.4 *Matrix finite difference approach*

The preceding sub-section illustrated the stability inherent in the Feautrier method for scalar systems. However, problems of astro-physical interest will invariably involve matrix difference equations and, unfortunately, the preceding analysis is not readily adaptable to such systems except by way of analogy.

Naturally, if all the off-diagonal terms in $\mathbf{B}_j$ for all j are zero, then the system of matrix equations (3.2.46) simply reduces to a larger system of scalar equations for which stability has been proven. In this case, the radiation fields at different angles and frequencies would be completely de-coupled from one another and 'local' considerations would dominate. It is the off-diagonal elements which couple the radiation field at different μ and $\Delta\nu$, and which present the difficulty in analysing the stability of the matrix system.

Indeed, the least we can do is state that the more dominant the diagonal elements in $\mathbf{B}_j$, the higher the probability of stability. We can examine this point further. Equation (3.2.43) gives

$$\text{diag}\,\{\mathbf{B}_j\} = 1 - (1 - \epsilon_j)\alpha_i\beta_k + a_j + c_j, \tag{3.3.66}$$

where

$$a_j + b_j = \frac{1}{\Delta_j + \nabla_j}\left(\frac{1}{\Delta_j} + \frac{1}{\nabla_j}\right), \tag{3.3.67}$$

and $\alpha_i\beta_k$ are weights for the frequency and angle quadrature. Since our normalisation condition on the absorption profile gives

$$\sum_{i=1}^{N_\nu} \sum_{k=1}^{N_\mu} \alpha_i \beta_k = 1 \tag{3.3.68}$$

(see equation (3.2.98)), where $\alpha_i > 0$ and $\beta_k > 0$ for all i and k, then we must have $\alpha_i \beta_k < 1$. We therefore have

$$|\text{off-diag}\,\{\mathbf{B}_j\}| = (1 - \epsilon_j)\alpha_i \beta_k < 1. \tag{3.3.69}$$

We have seen in the scalar situation studied in the preceding subsection that any numerical error can be amplified for small j. However, $\Delta_j \ll 1$ and $\nabla_j \gg 1$ for small j so that $(a_j + b_j) \gg 1$, i.e. diag $\{\mathbf{B}_j\} \gg 1$ for small j. Clearly, therefore,

$$\text{diag}\,\{\mathbf{B}_j\} \gg \text{off-diag}\,\{\mathbf{B}_j\}, \tag{3.3.70}$$

for small j. Consequently, in the region where our scalar analysis indicates instability, the diagonal elements of $\mathbf{B}_j$ are strongly dominant, and thus one should not be concerned with instabilities arising from off-diagonal elements. On the other hand, for large j where one might expect strong stability from the scalar analysis, $\Delta_j \gg 1$ and $\nabla_j \gg 1$ so that

$$\text{diag}\,\{\mathbf{B}_j\} \approx \text{off-diag}\,\{\mathbf{B}_j\}. \tag{3.3.71}$$

Thus, the expected strong stability in this region could be lessened by the presence of these non-zero off-diagonal terms.

Although one cannot be more definite than this from a theoretical consideration, particularly since it is the matrix $\mathbf{B}_j - \mathbf{A}_j\mathbf{V}_{j-1}$, not $\mathbf{B}_j$ solely, which requires inversion (see equation (3.2.64)) and we cannot detail the explicit effect of the $\mathbf{A}_j\mathbf{V}_{j-1}$ elements on the structure of the composite matrix, practical experience shows the system still to be highly stable even when the off-diagonal elements become dominant. In this case, however, there is a limit on the size of the matrices one can invert.

3.4 Variable Eddington factors [6]

The basic disadvantages of the Feautrier method of solution of the matrix system of equations (3.2.46) is the need to invert matrices of size $N_\nu N_\mu$. This disadvantage is not important when N_ν and N_μ are sufficiently small that $n = N_\nu N_\mu \lesssim 30$, for example. However, there exists a large range of problems of astrophysical interest for which $n > 30$; indeed, it is not unreasonable to have $N_\mu = 3$ and, to cover the extended wings of Voigt profiles, $N_\nu \sim 100$. The inaccuracy of the matrix inversion due to the accumulation of round-off errors could be quite prohibitive for large matrices, particularly if the diagonal elements are not dominant, and thus any technique which overcomes this difficulty can be of considerable numerical importance.

There are several methods which do, indeed, eliminate (or at least reduce) the problem. The method we introduce in the present section involves an iterative procedure and is only directly applicable when the spectral line source function and opacity are independent of angle. Further, it does not attempt to reduce the size of the matrices due to the frequency quadrature; it is usually the frequency, not angle, grid which accounts for the large size. Nevertheless, it has sufficient merit to be included here.

We write the second order differential equation of radiative transfer derived in section 3.2 in the form

$$\mu^2 \left(\frac{1}{\kappa}\frac{\partial}{\partial z}\right)^2 \Phi(z, \Delta\nu, \mu) = \Phi(z, \Delta\nu, \mu) - S(z), \tag{3.4.1}$$

subject to the boundary conditions

$$\frac{\mu}{\kappa}\frac{\partial \Phi}{\partial z}(z, \Delta\nu, \mu) = -\Phi(z, \Delta\nu, \mu) + I_{\text{inc}}(\Delta\nu, \mu), \tag{3.4.2}$$

at $z = 0$ whilst, as $z \to -\infty$ (i.e. $\tau \to \infty$),

$$\frac{\mu}{\kappa}\frac{\partial \Phi}{\partial z}(z, \Delta\nu, \mu) = \Phi(z, \Delta\nu, \mu) - B_\nu(T). \tag{3.4.3}$$

We now define

$$J(z, \Delta\nu) = \tfrac{1}{2}\int_{-1}^{1} I(z, \Delta\nu, \mu)\, \mathrm{d}\mu = \int_0^1 \Phi(z, \Delta\nu, \mu)\, \mathrm{d}\mu, \tag{3.4.4}$$

$$K(z, \Delta\nu) = \tfrac{1}{2}\int_{-1}^{1} \mu^2 I(z, \Delta\nu, \mu)\, \mathrm{d}\mu = \int_0^1 \mu^2 \Phi(z, \Delta\nu, \mu)\, \mathrm{d}\mu, \tag{3.4.5}$$

with

$$H(z, \Delta\nu) = \int_0^1 \mu\Phi(z, \Delta\nu, \mu)\, \mathrm{d}\mu. \tag{3.4.6}$$

If we now suitably integrate equations (3.4.1), (3.4.2) and (3.4.3) over $\mu \in [0, 1]$, and take κ to be independent of μ, we find

$$\left(\frac{1}{\kappa}\frac{\partial}{\partial z}\right)^2 K(z, \Delta\nu) = J(z, \Delta\nu) - S(z), \tag{3.4.7}$$

$$\frac{1}{\kappa}\frac{\partial K}{\partial z}(z, \Delta\nu) = -H(z, \Delta\nu) + \int_0^1 \mu I_{\text{inc}}(\Delta\nu, \mu)\, \mathrm{d}\mu, \tag{3.4.8}$$

at $z = 0$ whilst, as $z \to -\infty$,

$$\frac{1}{\kappa}\frac{\partial K}{\partial z}(z, \Delta\nu) = H(z, \Delta\nu) - B_\nu(T). \tag{3.4.9}$$

This system of equations may be closed by introducing the two factors $f_K(z, \Delta\nu)$ and $f_H(z, \Delta\nu)$ such that

$$f_K(z, \Delta\nu) = \frac{K(z, \Delta\nu)}{J(z, \Delta\nu)}, \tag{3.4.10}$$

and

$$f_H(z, \Delta\nu) = \frac{H(z, \Delta\nu)}{J(z, \Delta\nu)} \,. \tag{3.4.11}$$

Equations (3.4.7), (3.4.8) and (3.4.9) then yield

$$\left(\frac{1}{\kappa}\frac{\partial}{\partial z}\right)^2 (f_K J) = J - S, \tag{3.4.12}$$

$$\left(\frac{1}{\kappa}\frac{\partial}{\partial z}\right)(f_K J) = -f_H J + \int_0^1 \mu I_{\text{inc}}(\Delta\nu, \mu)\, \mathrm{d}\mu, \tag{3.4.13}$$

at $z = 0$ whilst, as $z \to -\infty$,

$$\left(\frac{1}{\kappa}\frac{\partial}{\partial z}\right)(f_K J) \doteq f_H J - B_\nu(T). \tag{3.4.14}$$

The advantage of the method is realised when we re-formulate the source function, viz.

$$S(z) = (1-\epsilon) \sum_{-\infty}^{\infty} \phi(\Delta\nu)\, \mathrm{d}(\Delta\nu) \int_0^1 \Phi(z, \Delta\nu, \mu)\, \mathrm{d}\mu + \epsilon B_\nu(T)$$

$$= (1-\epsilon) \sum_{-\infty}^{\infty} \phi(\Delta\nu) J(z, \Delta\nu)\, \mathrm{d}(\Delta\nu) + \epsilon B_\nu(T). \tag{3.4.15}$$

Equations (3.4.12) through (3.4.14), together with (3.4.15), constitute a system effectively identical to that obtained using the Feautrier technique except that the angle μ has been eliminated. Thus, when one approximates the derivatives appearing in the above equations by finite differences, and replaces the frequency integral by the appropriate quadrature, one obtains precisely the same form of matrix equations (3.2.46) as before which, in turn, may be solved by the same recurrence method of Gaussian elimination, i.e. we have

$$-\mathbf{A}_j \mathbf{J}_{j-1} + \mathbf{B}_j \mathbf{J}_j - \mathbf{C}_j \mathbf{J}_{j+1} = \mathbf{L}_j, \tag{3.4.16}$$

where

$$\mathbf{A}_1 \equiv \mathbf{0} \equiv \mathbf{C}_N, \tag{3.4.17}$$

so that

$$\mathbf{J}_j = \mathbf{U}_j + \mathbf{V}_j \mathbf{J}_{j+1}, \tag{3.4.18}$$

with

$$\mathbf{U}_j = (\mathbf{B}_j - \mathbf{A}_j \mathbf{V}_{j-1})^{-1} (\mathbf{L}_j + \mathbf{A}_j \mathbf{U}_{j-1}), \tag{3.4.19}$$

and

$$\mathbf{V}_j = (\mathbf{B}_j - \mathbf{A}_j \mathbf{V}_{j-1})^{-1} \mathbf{C}_j. \tag{3.4.20}$$

In particular, we see that

$$S(z) \approx (1-\epsilon) \sum_{i=1}^{N_\nu} \alpha_i J(z, \Delta\nu_i) + \epsilon B_\nu(T), \tag{3.4.21}$$

so that the size of the matrices to be inverted in equation (3.4.20) has been reduced to N_ν. Herein lies the power of the technique.

The computing procedure involves an initial choice of $f_K^{(1)}$ and $f_H^{(1)}$ where the superscript (1) refers to the first iteration. This then enables $J^{(1)}(z_j, \Delta\nu_i)$ to be obtained which, when substituted into equation (3.4.21), yields $S^{(1)}(z_j)$. One can then determine the intensities $I^{(1)}(z_j, \Delta\nu_i, \mu_k)$ from the original equation of transfer, viz.

$$\frac{\mu}{\kappa}\frac{\partial I^{(1)}}{\partial z}(z, \Delta\nu, \mu) = -I^{(1)}(z, \Delta\nu, \mu) + S^{(1)}(z), \tag{3.4.22}$$

and these, in turn, can be used to evaluate new values of f_K and f_H, i.e. we obtain

$$f_K^{(2)}(z_j, \Delta\nu_i) = \frac{\sum_{k=1}^{N_\mu} \beta_k \mu_k^2 \Phi^{(1)}(z_j, \Delta\nu_i, \mu_k)}{\sum_{k=1}^{N_\mu} \beta_k \Phi^{(1)}(z_j, \Delta\nu_i, \mu_k)}, \tag{3.4.23}$$

$$f_H^{(2)}(z_j, \Delta\nu_i) = \frac{\sum_{k=1}^{N_\mu} \beta_k \mu_k \Phi^{(1)}(z_j, \Delta\nu_i, \mu_k)}{\sum_{k=1}^{N_\mu} \beta_k \Phi^{(1)}(z_j, \Delta\nu_i, \mu_k)}, \tag{3.4.24}$$

where, of course, we have

$$\Phi(z_j, \Delta\nu_i, \mu_k) = \tfrac{1}{2}[I(z_j, \Delta\nu_i, \mu_k) + I(z_j, \Delta\nu_i, -\mu_k)]. \tag{3.4.25}$$

These new values of f_K and f_H are substituted into the second order system to obtain the second iterative $J^{(2)}(z_j, \Delta\nu_i)$ and the entire process is repeated until convergence is attained.

Since the time taken to invert a matrix of size $(n \times n)$ is of order n^3, the time saved by reducing the size of the matrices to be inverted can be quite substantial. For example, if we take $N_\mu = 3$, then the saving in computer time would be of order 3^3, particularly since almost all the time taken in the Feautrier technique is in matrix inversion. However, one must balance this saving in computer time (and storage) with the extra time required to solve equation (3.4.22) once $S^{(1)}(z)$ is known. The solution to this equation is formally given by

$$I^{(1)}(z, \Delta\nu, \mu \gtrless 0) = \int_{\substack{\infty\\0}}^{z} \kappa(z', \Delta\nu) S^{(1)}(z') \times \exp\left[-\frac{1}{\mu}\left|\int_z^{z'} \kappa(z'', \Delta\nu)\, dz''\right|\right]\frac{dz'}{\mu}, \tag{3.4.26}$$

which, since $S^{(1)}(z)$ is given, may be solved by straightforward quadrature. The computing time required to do this is insignificant compared with the time taken to invert the matrices (albeit reduced in size).

Let us now examine the initial choice of f_K and f_H. If we choose 1-point quadrature over angle, then equations (3.4.4), (3.4.5) and (3.4.6) become

$$J(z, \Delta\nu) = \beta_1 \Phi(z, \Delta\nu, \mu_1), \tag{3.4.27}$$

$$K(z, \Delta\nu) = \beta_1 \mu_1^2 \Phi(z, \Delta\nu, \mu_1), \tag{3.4.28}$$

and

$$H(z, \Delta\nu) = \beta_1 \mu_1 \Phi(z, \Delta\nu, \mu_1), \tag{3.4.29}$$

where μ_1 is the one quadrature point and β_1 the corresponding quadrature weight. For Gaussian quadrature over $\mu \in [0, 1]$ we have $\mu_1 = 1/\sqrt{3}$ (and $\beta_1 = 1$) so that the above equations yield

$$f_K^{(1)} = \frac{1}{3}; \quad f_H^{(1)} = \frac{1}{\sqrt{3}} \, . \tag{3.4.30}$$

These are the ordinary Eddington factors. The iterative procedure described above simply allows these factors to vary with frequency and position.

An analysis of the iterative scheme's convergence properties is not available. In practice, good convergence is attained with solutions usually converging to within several significant figures after only several iterations. It has been argued that the Eddington factors f_K and f_H are determined by the ratios of summations over the radiation field (see equations (3.4.23) and (3.4.24)) so that any errors in the determination of $\Phi(z_j, \Delta\nu_i, \mu_k)$ effectively cancel. This, together with the fact that the basic matrix equation (3.4.16) is stable under a Gaussian elimination scheme, in a sense ensures stability of the entire procedure.

3.5 The Rybicki re-organisation [102]

The method presented in the preceding section was devised to overcome (partially) the problem associated with the computation of matrix inverses in the Feautrier technique when the size of the matrices are large. The use of variable Eddington factors f_K and f_H, as presented in that section, suffers from the severe limitation that the opacity and source function must be independent of angle. As we shall see, this limitation renders the method useless for a large range of problems of present-day astrophysical interest in radiative transfer theory. Further, the size of the matrices can still be prohibitively large since it is only the angle dependence of the matrices which has been eliminated.

An alternative method, which does not suffer from the above disadvantages, uses precisely the same components in the matrix system as in the ordinary Feautrier technique, but re-organises the manner in which the solution is obtained. As we shall see, it only becomes applicable when the size $(n \times n)$ of the matrices, where $n = N_\nu N_\mu$, is such that $n > N$ where N is the number of $\{\tau_j\}$ grid points.

We begin the exposition of the Rybicki re-organisation by first re-vamping the ordinary Feautrier system and writing the second order differential equation (after finite differencing and numerical quadrature – see equation (3.2.39)) in the form

$$\mu_k^2[a_{ji}\Phi_{j-1,ik} + b_{ji}\Phi_{jik} + c_{ji}\Phi_{j+1,ik}]$$
$$= \Phi_{jik} - (1-\epsilon_j)\sum_{i'=1}^{N_\nu}\sum_{k'=1}^{N_\mu}\alpha_{i'}\beta_{k'}\Phi_{ji'k'} - [\epsilon B_\nu(T)]_j, \tag{3.5.1}$$

for all $i = 1, N_\nu$ and $k = 1, N_\mu$ and for all $j \neq 1, j \neq N$.

We now define the new quantities A_{jm}, B_{jm}, C_{jm} and W_{jm}, where

$$m = (k-1)N_\nu + i, \tag{3.5.2}$$

for all $i = 1, N_\nu$ and $k = 1, N_\mu$ (i.e. $1 \leqslant m \leqslant n$) such that

$$\left.\begin{aligned} A_{jm} &= \mu_k^2 a_{ji}, & C_{jm} &= \mu_k^2 c_{ji}, \\ B_{jm} &= 1 + A_{jm} + C_{jm}, \quad \text{and} & W_{jm} &= \alpha_i \beta_k. \end{aligned}\right\} \tag{3.5.3}$$

Equation (3.5.1) then has the form

$$-A_{jm}\Phi_{j-1,m} + B_{jm}\Phi_{jm} - C_{jm}\Phi_{j+1,m}$$
$$= [\epsilon B_\nu(T)]_j + (1-\epsilon_j)\sum_{m'=1}^{n} W_{jm'}\Phi_{jm'}, \tag{3.5.4}$$

where

$$\Phi_{jm} \equiv \Phi_{jik}. \tag{3.5.5}$$

The ordinary Feautrier system uses the vector Φ_j, where

$$\Phi_j = \begin{pmatrix} \Phi_{j1} \\ \vdots \\ \Phi_{jn} \end{pmatrix}, \tag{3.5.6}$$

such that, writing

$$\mathbf{W}_j = (W_{j1}, \ldots, W_{jn}), \tag{3.5.7}$$

equation (3.5.4) becomes

$$-A_{jm}\Phi_{j-1,m} + B_{jm}\Phi_{jm} - C_{jm}\Phi_{j+1,m} - (1-\epsilon_j)\mathbf{W}_j\Phi_j = [\epsilon B_\nu(T)]_j. \tag{3.5.8}$$

The above system for all $m = 1, n$ is then put into matrix form by writing

$$\mathbf{A}_j = \begin{pmatrix} A_{j1} & & 0 \\ & \ddots & \\ 0 & & A_{jn} \end{pmatrix}, \tag{3.5.9}$$

$$\mathbf{C}_j = \begin{pmatrix} C_{j1} & & 0 \\ & \ddots & \\ 0 & & C_{jn} \end{pmatrix}, \tag{3.5.10}$$

$$\mathbf{B}_j = \begin{pmatrix} B_{j1} & & 0 \\ & \ddots & \\ 0 & & B_{jn} \end{pmatrix} - (1-\epsilon_j)\begin{pmatrix} W_{j1}, & \ldots, & W_{jn} \\ \vdots & & \\ W_{j1}, & & W_{jn} \end{pmatrix}, \tag{3.5.11}$$

and

$$\mathbf{L}_j = [\epsilon B_\nu(T)]_j \mathbb{1}. \tag{3.5.12}$$

We then obtain

$$-\mathbf{A}_j\mathbf{\Phi}_{j-1} + \mathbf{B}_j\mathbf{\Phi}_j - \mathbf{C}_j\mathbf{\Phi}_{j+1} = \mathbf{L}_j, \tag{3.5.13}$$

where the matrices to be inverted are clearly of size ($n \times n$).

The Rybicki re-organisation approaches the problem differently. Here, we construct the $\mathbf{\Phi}^{(m)}$ vector (different from $\mathbf{\Phi}_j$) such that

$$\mathbf{\Phi}^{(m)} = \begin{pmatrix} \Phi_{1m} \\ \vdots \\ \Phi_{Nm} \end{pmatrix}. \tag{3.5.14}$$

Further, we must explicitly include the equations for $j = 1$ and $j = N$. The boundary conditions (3.2.13) and (3.2.14) are

$$\left(\frac{\mu}{\kappa}\frac{\partial\Phi}{\partial z}\right)_1 = I_{\text{inc}}(\Delta\nu, \mu) - \Phi_1, \tag{3.5.15}$$

and

$$\left(\frac{\mu}{\kappa}\frac{\partial\Phi}{\partial z}\right)_N = \Phi_N - [B_\nu(T)]_N. \tag{3.5.16}$$

If we now take a Taylor series expansion at these boundaries as discussed in section 3.2.4, the above two equations can be written as

$$B_{1m}\Phi_{1m} - C_{1m}\Phi_{2m} - (1-\epsilon)\mathbf{W}_1\mathbf{\Phi}_1 = [\epsilon B_\nu(T)]_1 + L_1 \tag{3.5.17}$$

(see equation (3.2.77) for example) and

$$-A_{Nm}\Phi_{N-1,m} + B_{Nm}\Phi_{Nm} - (1-\epsilon_N)\mathbf{W}_N\mathbf{\Phi}_N = [\epsilon B_\nu(T)]_N + L_N, \tag{3.5.18}$$

where the B_{1m}, C_{1m}, A_{Nm} and B_{Nm} have different functional forms to those A_{jm}, B_{jm} and C_{jm} given by equations (3.5.3). In particular, the L_1 term will involve I_{inc}, and L_N will be a function of $[B_\nu(T)]_N$. The reader is referred to section 3.2.4 for further details regarding the explicit form of these quantities.

Next, we note that the source function S_j at z_j may be written as

$$\begin{aligned} S_j &= (1-\epsilon_j)\sum_{m=1}^{n} W_{jm}\Phi_{jm} + [\epsilon B_\nu(T)]_j \\ &= (1-\epsilon_j)\mathbf{W}_j\mathbf{\Phi}_j + [\epsilon B_\nu(T)]_j. \end{aligned} \tag{3.5.19}$$

We then obtain the set of equations

$$B_{1m}\Phi_{1m} - C_{1m}\Phi_{2m} = S_1 + L_1', \tag{3.5.20}$$

$$-A_{jm}\Phi_{j-1,m} + B_{jm}\Phi_{jm} - C_{jm}\Phi_{j+1,m} = S_j, \tag{3.5.21}$$

for all $j \neq 1, j \neq N$ (from equation (3.5.4)) and

$$-A_{Nm}\Phi_{N-1,m} + B_{Nm}\Phi_{Nm} = S_N + L'_N. \tag{3.5.22}$$

These equations may be put into matrix form by defining the tri-diagonal matrix $\mathbf{P}^{(m)}$ such that

$$\mathbf{P}^{(m)} = \begin{pmatrix} B_{1m}, & -C_{1m} & & & & \\ -A_{2m}, & B_{2m}, & -C_{2m} & & 0 & \\ & \ddots & \ddots & \ddots & & \\ & 0 & & -A_{N-1,m}, & B_{N-1,m}, & -C_{N-1,m} \\ & & & & -A_{Nm}, & B_{Nm} \end{pmatrix}. \tag{3.5.23}$$

Equations (3.5.20) through (3.5.22) then yield

$$\mathbf{P}^{(m)}\mathbf{\Phi}^{(m)} = \mathbf{S} + \mathbf{L}_\mathrm{R}, \tag{3.5.24}$$

where

$$\mathbf{S} = \begin{pmatrix} S_1 \\ \vdots \\ \vdots \\ \vdots \\ S_N \end{pmatrix}, \quad \mathbf{L}_\mathrm{R} = \begin{pmatrix} L'_1 \\ 0 \\ \vdots \\ 0 \\ L'_N \end{pmatrix}. \tag{3.5.25}$$

However, equation (3.5.24) is but one equation for the two unknown vectors $\mathbf{\Phi}^{(m)}$ and $\mathbf{S}$. We therefore require another equation relating these two quantities. This can be done simply by noting that

$$S_j = (1 - \epsilon_j) \sum_{m=1}^{n} W_{jm}\Phi_{jm} + [\epsilon B_\nu(T)]_j,$$

so that

$$S = \begin{pmatrix} (1-\epsilon_1)[W_{11}\Phi_{11} + \cdots + W_{1n}\Phi_{1n}] \\ \vdots \\ (1-\epsilon_N)[W_{N1}\Phi_{N1} + \cdots + W_{Nn}\Phi_{Nn}] \end{pmatrix} + \begin{pmatrix} [\epsilon B_\nu(T)]_1 \\ \vdots \\ [\epsilon B_\nu(T)]_N \end{pmatrix}$$

$$= \mathbf{W}^{(1)}\mathbf{\Phi}^{(1)} + \cdots + \mathbf{W}^{(N)}\mathbf{\Phi}^{(N)} + \mathscr{L}, \tag{3.5.26}$$

where

$$\mathbf{W}^{(m)} = \begin{pmatrix} (1-\epsilon_1)W_{1m} & & 0 \\ & \ddots & \\ 0 & & (1-\epsilon_N)W_{Nm} \end{pmatrix}, \tag{3.5.27}$$

and

$$\mathscr{L} = \begin{pmatrix} [\epsilon B_\nu(T)]_1 \\ \\ [\epsilon B_\nu(T)]_N \end{pmatrix}. \tag{3.5.28}$$

We therefore have the required second relationship

$$\mathbf{S} = \sum_{m=1}^{n} \mathbf{W}^{(m)}\boldsymbol{\Phi}^{(m)} + \mathscr{L}, \tag{3.5.29}$$

which, using equation (3.5.24), yields

$$\mathbf{S} = \sum_{m=1}^{n} \mathbf{W}^{(m)} [\mathbf{P}^{(m)}]^{-1}(\mathbf{S} + \mathbf{L}_\mathrm{R}) + \mathscr{L},$$

i.e.

$$\mathbf{S} = \left\{ \mathbb{1} - \sum_{m=1}^{n} \mathbf{W}^{(m)} [\mathbf{P}^{(m)}]^{-1} \right\}^{-1} \left\{ \mathscr{L} + \sum_{m=1}^{n} \mathbf{W}^{(m)} [\mathbf{P}^{(m)}]^{-1} \mathbf{L}_\mathrm{R} \right\}. \tag{3.5.30}$$

This completes the derivation of the Rybicki re-organisation. Note that $n + 1$ matrices of size $(N \times N)$ must now be inverted whereas, under the ordinary Feautrier scheme, one must invert N matrices of size $(n \times n)$. Clearly, since the bulk of the computing time required to solve the transfer equation using these two techniques involves matrix inversion, the present method offers a saving in computer time (and storage) for $N > n$. As we have seen in section 3.3.3, $N \lesssim 40$ for most cases of astrophysical interest whilst $n = N_\nu N_\mu$ can (but not always) be of order several hundred. In this case, the saving in computer requirements is immense. Indeed, the matrix inversion of $\mathbf{P}^{(m)}$ is considerably simpler than ordinary inversion of full matrices - note that $\mathbf{P}^{(m)}$ is only tri-diagonal. In practice, one determines the inverse $\mathbf{X}^{(m)}$ of $\mathbf{P}^{(m)}$, i.e.

$$\mathbf{P}^{(m)}\mathbf{X}^{(m)} = \mathbb{1}, \tag{3.5.31}$$

by writing

$$\mathbf{X}^{(m)} = (\mathbf{X}_1^{(m)}, \ldots, \mathbf{X}_N^{(m)}), \tag{3.5.32}$$

such that

$$\mathbf{X}_j^{(m)} = \begin{pmatrix} X_{1j}^{(m)} \\ \vdots \\ X_{Nj}^{(m)} \end{pmatrix}. \tag{3.5.33}$$

One then obtains each $X_{jj'}^{(m)}$ element of $\mathbf{X}^{(m)}$ by solving the system of equations

$$B_{1m}X_{1j'}^{(m)} - C_{1m}X_{2j'}^{(m)} = \delta_{ij'}, \tag{3.5.34}$$

$$-A_{jm}X^{(m)}_{j-1,j'} + B_{jm}X^{(m)}_{jj'} - C_{jm}X^{(m)}_{j+1,j'} = \delta_{jj'}, \tag{3.5.35}$$

for $j \neq 1, j \neq N$, and

$$-A_{Nm}X^{(m)}_{N-1,j'} + B_{Nm}X^{(m)}_{Nj'} = \delta_{Nj'}, \tag{3.5.36}$$

where $\delta_{jj'}$ is the Kronecker delta function.

Gaussian elimination then yields

$$X^{(m)}_{jj'} = u^{(m)}_{jj'} + v^{(m)}_{jj'}X^{(m)}_{j+1,j'}, \tag{3.5.37}$$

where

$$u^{(m)}_{jj'} = \frac{\delta_{jj'} + A_{jm}u^{(m)}_{j-1,j'}}{B_{jm} - A_{jm}v^{(m)}_{j-1,j'}}, \tag{3.5.38}$$

and

$$v^{(m)}_{jj'} = \frac{C_{jm}}{B_{jm} - A_{jm}v^{(m)}_{j-1,j'}}. \tag{3.5.39}$$

Note that

$$A_{1m} = 0 = C_{Nm}. \tag{3.5.40}$$

The above system is solved for all $j = 1, N$ and $m = 1, n$. The recurrence relationships (3.5.37) through (3.5.39) are scalar in nature, their stability being shown in section 3.3.3. Thus, the determination of the $[\mathbf{P}^{(m)}]^{-1}$ appearing in equation (3.5.30) is both straightforward and stable.

One cannot *a priori* say a great deal about the inverse of the matrix in braces in equation (3.5.30). In practice, however, the matrix is well conditioned, and the computation of its inverse presents no difficulty. Note that there is only one such matrix inversion needed.

Finally, once the source function vector **S** has been determined from equation (3.5.30), the individual $\boldsymbol{\Phi}^{(m)}$ terms may be immediately obtained from equation (3.5.24) where, of course, the $[\mathbf{P}^{(m)}]^{-1}$ have already been evaluated.

3.6 Auer's modification [4]

The accuracy of the numerical solution using Feautrier's method may be improved by a further modification. The ordinary Feautrier technique uses the differencing formula

$$\left[\frac{\partial^2 \Phi}{\partial \tau^2}\right]_j = a_j\Phi_{j-1} + b_j\Phi_j + c_j\Phi_{j+1}, \tag{3.6.1}$$

for the second order derivatives where

$$\left.\begin{aligned} a_j &= \frac{2}{\nabla_j(\Delta_j + \nabla_j)}, \quad c_j = \frac{2}{\Delta_j(\Delta_j + \nabla_j)} \\ b_j &= -a_j - c_j, \end{aligned}\right\} \tag{3.6.2}$$

with

$$\Delta_j = \tau_{j+1} - \tau_j, \quad \nabla_j = \tau_j - \tau_{j-1}. \tag{3.6.3}$$

One best understands the thrust behind the Auer modification by first re-deriving the formula (3.6.1) using a Taylor series expansion, i.e. we have

$$\Phi_{j+1} = \Phi_j + \Delta_j \left[\frac{\partial \Phi}{\partial \tau}\right]_j + \frac{\Delta_j^2}{2!}\left[\frac{\partial^2 \Phi}{\partial \tau^2}\right]_j + \cdots, \tag{3.6.4}$$

and

$$\Phi_{j-1} = \Phi_j - \nabla_j \left[\frac{\partial \Phi}{\partial \tau}\right]_j + \frac{\nabla_j^2}{2!}\left[\frac{\partial^2 \Phi}{\partial \tau^2}\right]_j + \cdots. \tag{3.6.5}$$

If we now ignore all terms higher than Δ_j^2 and ∇_j^2, and eliminate $[\partial\Phi/\partial\tau]_j$ from these last two equations, we find

$$\nabla_j \Phi_{j+1} + \Delta_j \Phi_{j-1} - (\Delta_j + \nabla_j)\Phi_j = \tfrac{1}{2}\Delta_j \nabla_j (\Delta_j + \nabla_j)\left[\frac{\partial^2 \Phi}{\partial \tau^2}\right]_j, \tag{3.6.6}$$

which then reduces to equation (3.6.1).

Let us now consider the possibility of retaining the higher order terms Δ_j^3, ∇_j^3, Δ_j^4 and ∇_j^4 in equations (3.6.4) and (3.6 5). Writing

$$\Phi_j^{(n)} \equiv \left[\frac{\partial^n \Phi}{\partial \tau^n}\right]_j, \tag{3.6.7}$$

these equations become

$$\Phi_{j+1} = \Phi_j + \Delta_j \Phi_j^{(1)} + \frac{\Delta_j^2}{2!}\Phi_j^{(2)} + \frac{\Delta_j^3}{3!}\Phi_j^{(3)} + \frac{\Delta_j^4}{4!}\Phi_j^{(4)} + \cdots, \tag{3.6.8}$$

and

$$\Phi_{j-1} = \Phi_j - \nabla_j \Phi_j^{(1)} + \frac{\nabla_j^2}{2!}\Phi_j^{(2)} - \frac{\nabla_j^3}{3!}\Phi_j^{(3)} + \frac{\nabla_j^4}{4!}\Phi_j^{(4)} - \cdots. \tag{3.6.9}$$

We may evaluate $\Phi_j^{(3)}$ and $\Phi_j^{(4)}$ by expanding $\Phi_{j+1}^{(2)}$ and $\Phi_{j-1}^{(2)}$ in a Taylor series such that

$$\Phi_{j+1}^{(2)} = \Phi_j^{(2)} + \Delta_j \Phi_j^{(3)} + \frac{\Delta_j^2}{2!}\Phi_j^{(4)} + \cdots, \tag{3.6.10}$$

$$\Phi_{j-1}^{(2)} = \Phi_j^{(2)} - \nabla_j \Phi_j^{(3)} + \frac{\nabla_j^2}{2!}\Phi_j^{(4)} - \cdots. \tag{3.6.11}$$

If we ignore all terms $\Phi_j^{(n)}$ with $n > 4$, equations (3.6.8) through (3.6.11) are four equations from which we may eliminate the three quantities $\Phi_j^{(1)}$, $\Phi_j^{(3)}$ and $\Phi_j^{(4)}$. In particular, if we eliminate $\Phi_j^{(1)}$ from equations (3.6.8) and (3.6.9), we find

$$\nabla_j \Phi_{j+1} + \Delta_j \Phi_{j-1} = (\Delta_j + \nabla_j)\Phi_j + \frac{\Delta_j \nabla_j}{2}(\Delta_j + \nabla_j)\Phi_j^{(2)}$$
$$+ \frac{\Delta_j \nabla_j}{3!}(\Delta_j^2 - \nabla_j^2)\Phi_j^{(3)} + \frac{\Delta_j \nabla_j}{4!}(\Delta_j^3 + \nabla_j^3)\Phi_j^{(4)}, \tag{3.6.12}$$

which, using the definitions of a_j, b_j and c_j above, yields

$$a_j\Phi_{j-1} + b_j\Phi_j + c_j\Phi_{j+1} - \Phi_j^{(2)}$$
$$= \tfrac{1}{3}(\Delta_j - \nabla_j)\Phi_j^{(3)} + \frac{\Delta_j^3 + \nabla_j^3}{12(\Delta_j + \nabla_j)}\Phi_j^{(4)}. \qquad (3.6.13)$$

We now eliminate $\Phi_j^{(3)}$ and $\Phi_j^{(4)}$ from equations (3.6.10) and (3.6.11) to obtain

$$\Phi_j^{(4)} = a_j\Phi_{j-1}^{(2)} + b_j\Phi_j^{(2)} + c_j\Phi_{j+1}^{(2)}, \qquad (3.6.14)$$

and

$$\Phi_j^{(3)} = -\frac{\Delta_j}{2}a_j\Phi_{j-1}^{(2)} + \tfrac{1}{2}(\Delta_j a_j - \nabla_j c_j)\Phi_j^{(2)} + \frac{\nabla_j}{2}c_j\Phi_{j+1}^{(2)}. \qquad (3.6.15)$$

Substitution of these $\Phi_j^{(3)}$ and $\Phi_j^{(4)}$ into equation (3.6.13) eventually yields

$$a_j\Phi_{j-1} + b_j\Phi_j + c_j\Phi_{j+1} + p_j\Phi_{j-1}^{(2)} + (q_j - 1)\Phi_j^{(2)} + r_j\Phi_{j+1}^{(2)} = 0, \qquad (3.6.16)$$

where

$$p_j = \frac{a_j\Delta_j^2}{12} - \frac{1}{6}, \qquad (3.6.17)$$

$$r_j = \frac{c_j\nabla_j^2}{12} - \frac{1}{6}, \qquad (3.6.18)$$

and

$$q_j = -p_j - r_j. \qquad (3.6.19)$$

Equation (3.6.16), derived using a Taylor series expansion up to Δ_j^4 and ∇_j^4, is accurate to order Δ_j^4 and ∇_j^4, whilst equation (3.6.1) is only accurate to order Δ_j^2 and ∇_j^2. Clearly, if one puts $p_j = 0 = r_j$, equation (3.6.16) reduces to (3.6.1).

Before proceeding in the development of the method and, in particular, the discussion of the changes resulting from the extra terms $\Phi_{j-1}^{(2)}$ and $\Phi_{j+1}^{(2)}$, we again briefly outline the procedure for the Feautrier technique.

We have the second order equation of radiative transfer in the form

$$\frac{\mu^2}{\phi^2(\Delta\nu)}\frac{\partial^2\Phi}{\partial\tau^2} = \Phi - S,$$

i.e.

$$\frac{\mu^2}{\phi^2(\Delta\nu)}\Phi_j^{(2)} = \Phi_j - S_j, \qquad (3.6.20)$$

where, for ease in exposition, we have again taken $\phi(\Delta\nu)$ to be depth independent.

One then substitutes this $\Phi_j^{(2)}$ into equation (3.6.1) to obtain

$$a_j\Phi_{j-1} + b_j\Phi_j + c_j\Phi_{j+1} - \frac{\phi^2(\Delta\nu)}{\mu}(\Phi_j - S_j) = 0, \qquad (3.6.21)$$

which has the form

$$-A_{jm}\Phi_{j-1,m} + B_{jm}\Phi_{jm} - C_{jm}\Phi_{j+1,m} - (1-\epsilon_j) \times \sum_{m'=1}^{n} W_{jm'}\Phi_{jm'} = [\epsilon B_\nu(T)]_j, \tag{3.6.22}$$

where we have used the definitions of A_{jm}, B_{jm}, C_{jm}, W_{jm} and Φ_{jm} given by equations (3.5.3) and (3.5.5).

If we perform the same manipulation, but with equation (3.6.16), we find

$$a_j\Phi_{j-1} + b_j\Phi_j + c_j\Phi_{j+1} + \frac{p_j\phi^2(\Delta\nu)}{\mu^2}(\Phi_{j-1} - S_{j-1}) + (q_j - 1)\frac{\phi^2(\Delta\nu)}{\mu^2}(\Phi_j - S_j) + \frac{r_j\phi^2(\Delta\nu)}{\mu^2}(\Phi_{j+1} - S_{j+1}) = 0,$$

i.e.

$$\begin{aligned} -(A_{jm} + P_{jm})\Phi_{j-1,m} &+ (B_{jm} + Q_{jm})\Phi_{jm} \\ -(C_{jm} + R_{jm})\Phi_{j+1,m} &+ \sum_{m'=1}^{n} \{(1-\epsilon_{j-1})P_{jm'}W_{j-1,m'}\Phi_{j-1,m'} \\ -(1-\epsilon_j)(1-Q_{jm})W_{jm'}\Phi_{jm'} &+ (1-\epsilon_{j+1})R_{jm}W_{j+1,m'}\Phi_{j+1,m'}\} \\ = -P_{jm}[\epsilon B_\nu(T)]_{j-1} &+ (1-Q_{jm})[\epsilon B_\nu(T)]_j - R_{jm}[\epsilon B_\nu(T)]_{j+1}, \end{aligned} \tag{3.6.23}$$

where, for example, we have defined

$$P_{jm} = p_j, \tag{3.6.24}$$

for $m = (k-1)N_\nu + i$. Although, here, p_j, q_j and r_j are independent of μ and $\Delta\nu$, we have introduced the generalisation represented by equation (3.6.24) to account for depth-dependent $\phi(\Delta\nu)$. In this case, the p_j, q_j and r_j would be functions of $\Delta\nu$ and the need for the specification of a P_{jm}, Q_{jm} and R_{jm} as functions of i (thence m) would then arise.

By defining the vector

$$\boldsymbol{\Phi}_j = \begin{pmatrix} \Phi_{j1} \\ \vdots \\ \Phi_{jm} \end{pmatrix}, \tag{3.6.25}$$

we know that equation (3.6.22) may be written in the matrix form:

$$-\mathbf{A}_j\boldsymbol{\Phi}_{j-1} + \mathbf{B}_j\boldsymbol{\Phi}_j - \mathbf{C}_j\boldsymbol{\Phi}_{j+1} = \mathbf{L}_j, \tag{3.6.26}$$

where $\mathbf{A}_j$, $\mathbf{B}_j$, $\mathbf{C}_j$ and $\mathbf{L}_j$ are given by equations (3.5.9) through (3.5.12). Similarly, we may write equation (3.6.23) in matrix form:

$$-\mathbf{A}_j^{(A)}\boldsymbol{\Phi}_{j-1} + \mathbf{B}_j^{(A)}\boldsymbol{\Phi}_j - \mathbf{C}_j^{(A)}\boldsymbol{\Phi}_{j+1} = \mathbf{L}_j^{(A)}, \tag{3.6.27}$$

where, defining

$$\mathbf{P}_j = \begin{pmatrix} P_{j1} & & 0 \\ & \cdot & \\ & & \cdot \\ 0 & & P_{jn} \end{pmatrix}, \quad \mathbf{Q}_j = \begin{pmatrix} Q_{j1} & & 0 \\ & \cdot & \\ & & \cdot \\ 0 & & Q_{jn} \end{pmatrix}, \quad \mathbf{R}_j = \begin{pmatrix} R_{j1} & & 0 \\ & \cdot & \\ & & \cdot \\ 0 & & R_{jn} \end{pmatrix}, \tag{3.6.28}$$

we have

$$\mathbf{A}_j^{(A)} = \mathbf{A}_j + [\mathbb{1} - (1 - \epsilon_{j-1})\mathbf{W}_{j-1}]\mathbf{P}_j, \tag{3.6.29}$$

$$\mathbf{B}_j^{(A)} = \mathbf{B}_j + [\mathbb{1} - (1 - \epsilon_j)\mathbf{W}_j]\mathbf{Q}_j, \tag{3.6.30}$$

$$\mathbf{C}_j^{(A)} = \mathbf{C}_j + [\mathbb{1} - (1 - \epsilon_{j+1})\mathbf{W}_{j+1}]\mathbf{R}_j, \tag{3.6.31}$$

and

$$\mathbf{L}_j^{(A)} = \mathbf{L}_j - [\epsilon B_\nu(T)]_{j-1}\mathbf{P}_j\Pi - [\epsilon B_\nu(T)]_j\mathbf{Q}_j\Pi - [\epsilon B_\nu(T)]_{j+1}\mathbf{R}_j\Pi. \tag{3.6.32}$$

The $\mathbf{W}_j$ matrix is defined by equation (3.2.44).

It is important to notice that equations (3.6.27) and (3.6.26) have identical structures and may therefore be solved by the same recurrence technique using Gaussian elimination. The main difference between the two matrix equations is that $\mathbf{A}_j^{(A)}$ and $\mathbf{C}_j^{(A)}$ are now dense matrices whereas $\mathbf{A}_j$ and $\mathbf{C}_j$ are pure diagonal. However, the matrices being inverted are of the form $\mathbf{B}_j^{(A)} - \mathbf{A}_j^{(A)}\mathbf{V}_{j-1}^{(A)}$ where $\mathbf{B}_j^{(A)}$ is a full matrix anyway. Thus, this difference is hardly significant as a computational aspect. Further, since the extra terms in these matrices effectively result from higher order terms in the Taylor expansion, one would not expect the Auer modification to greatly affect the stability of the Feautrier technique.

We conclude this section by detailing the modified difference formulae for the boundary conditions. The Taylor series expansion, accurate to $\Phi_1^{(3)}$, of Φ_2 and $\Phi_2^{(2)}$ are of the form

$$\Phi_2 = \Phi_1 + \Delta_1\Phi_1^{(1)} + \frac{\Delta_1^2}{2!}\Phi_1^{(2)} + \frac{\Delta_1^3}{3!}\Phi_1^{(3)}, \tag{3.6.33}$$

and

$$\Phi_2^{(2)} = \Phi_1^{(2)} + \Delta_1\Phi_1^{(3)}. \tag{3.6.34}$$

Eliminating $\Phi_1^{(3)}$ yields

$$\Phi_2 = \Phi_1 + \Delta_1\Phi_1^{(1)} + \frac{\Delta_1^2}{2}\Phi_1^{(2)} + \frac{\Delta_1^2}{6}(\Phi_2^{(2)} - \Phi_1^{(2)}). \tag{3.6.25}$$

The boundary condition at $\tau = 0$, for example, is given by equation (3.2.13), i.e.

$$\left[\frac{\mu}{\phi(\Delta\nu)}\frac{\partial\Phi}{\partial\tau}\right]_1 = \Phi_1 - I_{\text{inc}}(\Delta\nu, \mu), \tag{3.6.36}$$

so that substitution into equation (3.6.35), together with (3.6.20), yields

$$\Phi_2 = \Phi_1 + \frac{\Delta_1\phi(\Delta\nu)}{\mu}(\Phi_1 - I_{\text{inc}}) + \frac{\Delta_1^2\phi^2(\Delta\nu)}{2\mu^2}(\Phi_1 - S_1) + \frac{\Delta_1^2\phi^2(\Delta\nu)}{6\mu^2}(\Phi_2 - S_2 - \Phi_1 + S_1). \quad (3.6.37)$$

This last equation may also be written in matrix form (3.6.27), but with

$$\mathbf{A}_1^{(A)} = \mathbf{0}. \quad (3.6.38)$$

Of course, the same procedure may be applied to the boundary condition (3.2.14) at infinity.

Finally, depth-dependent $\phi(\Delta\nu)$ offers no difficulty. Here, one simply uses the finite differencing formulae discussed at the end of section 3.2.1 - see equation (3.2.26) for example. In particular, equation (3.2.86) illustrates the appropriate Taylor series expansion but only to second order.

3.7 Quadrature perturbations [16, 17]

As we have already mentioned, the main disadvantage of the Feautrier technique is the need to invert large matrices if the numbers of angle and frequency quadrature points are large. The use of variable Eddington factors does reduce the problem slightly although, since it is the number of required frequency points which presents the difficulty, this improvement is only minimal. The Rybicki re-organisation is a major improvement, however, when the number of depth grid points N is smaller than $n = N_\nu N_\mu$ where N_ν and N_μ are the number of frequency and angle quadrature points respectively (see section 3.5). The size of the matrices to be inverted in this latter case is $(N \times N)$.

A further substantial improvement in computing time and accuracy may be obtained using quadrature perturbations. We present the method within the context of the differential equation of radiative transfer but stress that the procedure also applies to the corresponding integral equation given, for example, in section 3.1 for Λ iteration. We shall return to this latter formulation later in this section.

3.7.1 The basic procedure

We take, as our defining equation, the second order differential transfer equation

$$\frac{\mu^2}{\phi^2(\Delta\nu)}\frac{\partial^2\Phi}{\partial\tau^2} = \Phi - S, \quad (3.7.1)$$

where, again for ease in exposition, we take $\phi(\Delta\nu)$ to be depth independent. If we replace integrals by numerical quadrature, we have

$$S(\tau) \equiv (1-\epsilon)\sum_{i=1}^{N_\nu}\sum_{k=1}^{N_\mu} \alpha_i\beta_k\Phi(\tau, \Delta\nu_i, \mu_k) + \epsilon B_\nu(T), \quad (3.7.2)$$

where α_i and β_k are the frequency and angle quadrature weights (see section 3.2.5).

We now define the operator L such that

$$L = \sum_{i=1}^{N_\nu} \sum_{k=1}^{N_\mu} \alpha_i \beta_k, \tag{3.7.3}$$

which then yields

$$S(\tau) = (1-\epsilon) L\Phi + \epsilon B_\nu(T). \tag{3.7.4}$$

In actually performing the computations, one chooses N_ν and N_μ sufficiently large to obtain a solution of sufficient accuracy. However, let us now consider the new values N_ν^* and N_μ^* where $N_\nu^* < N_\nu; N_\mu^* < N_\mu$.

Correspondingly, we define the new operator L^* such that

$$L^* = \sum_{i=1}^{N_\nu^*} \sum_{k=1}^{N_\mu^*} \alpha_i^* \beta_k^*,$$

i.e.

$$L^*\Phi = \sum_{i=1}^{N_\nu} \sum_{k=1}^{N_\mu} \alpha_i^* \beta_k^* \Phi(\tau, \Delta\nu_i^*, \mu_k^*), \tag{3.7.5}$$

where $\{\Delta\nu_i^*\}$ and $\{\mu_k^*\}$ are the new frequency and angle grids, and α_i^* and β_k^* the new frequency and angle weights, associated with this lower order N_ν^* and N_μ^* numerical quadrature.

Equations (3.7.1) and (3.7.4) may then be written in the form

$$\begin{aligned} \frac{\mu^2}{\phi^2(\Delta\nu)} \frac{\partial^2 \Phi}{\partial \tau^2} &= \Phi - (1-\epsilon)L\Phi - \epsilon B_\nu(T) \\ &= \Phi - (1-\epsilon)L^*\Phi - \epsilon B_\nu(T) - (1-\epsilon)(L - L^*)\Phi. \end{aligned} \tag{3.7.6}$$

We now construct the perturbation series in Φ of the form

$$\Phi = \Phi^{(0)} + \lambda\Phi^{(1)} + \lambda^2\Phi^{(2)} + \cdots, \tag{3.7.7}$$

where λ is a parameter (later put equal to unity) which enables us to isolate terms of approximately equal magnitude. We assume, for the moment, that the series (3.7.7) is convergent. If we further assume that the composite operator $L - L^* \sim O(\lambda)$, substitution of equation (3.7.7) into (3.7.6), and equating equal powers of λ, yields

$$\frac{\mu^2}{\phi^2(\Delta\nu)} \frac{\partial^2 \Phi^{(0)}}{\partial \tau^2} = \Phi^{(0)} - (1-\epsilon)L^*\Phi^{(0)} - \epsilon B_\nu(T), \tag{3.7.8}$$

and

$$\frac{\mu^2}{\phi^2(\Delta\nu)} \frac{\partial^2 \Phi^{(l)}}{\partial \tau^2} = \Phi^{(l)} - (1-\epsilon)L^*\Phi^{(l)} - E^{(l)}, \tag{3.7.9}$$

for all $l \geqslant 1$, where

$$E^{(l)} = (1-\epsilon)(L-L^*)\Phi^{(l-1)}. \tag{3.7.10}$$

The computational procedure then involves the initial solution of equation (3.7.8) for $\Phi^{(0)}$, thence the determination of $E^{(1)}$ from equation (3.7.10). This $E^{(1)}$ may then be used in equation (3.7.9) to obtain $\Phi^{(1)}$. The process is then repeated until a suitably accurate solution using series (3.7.7) is attained. Equations (3.7.8) and (3.7.9) are identical in structure except that the non-homogeneous 'source' term differs - consequently, the Green's function is the same for each equation. Thus, for example, if one were to use the Feautrier technique (or any of the modifications of this technique as presented in sections 3.4 through 3.6), one would generate the matrix difference equations of the form

$$-\mathbf{A}_j^*\Phi_{j-1}^{(l)} + \mathbf{B}_j^*\Phi_j^{(l)} - \mathbf{C}_j^*\Phi_{j+1}^{(l)} = \mathscr{L}_j^{(l)}, \tag{3.7.11}$$

where $\mathbf{A}_j^*$, $\mathbf{B}_j^*$ and $\mathbf{C}_j^*$ have the same form as $\mathbf{A}_j$, $\mathbf{B}_j$ and $\mathbf{C}_j$ given by equations (3.2.41) through (3.2.43), except that the size of these matrices is $(n^* \times n^*)$ where $n^* = N_\nu^* N_\mu^*$.

We also have

$$\mathscr{L}_j^{(0)} = [\epsilon B_\nu(T)]_j \,\mathbb{1}, \tag{3.7.12}$$

where Π is the unit column vector of size n^* whereas, as for $l \geqslant 1$

$$\mathscr{L}_j^{(l)} = (1-\epsilon_j)(L-L^*)\Phi_j^{(l-1)}. \tag{3.7.13}$$

Recalling that equation (3.7.11) may be solved using Gaussian elimination, we have

$$\Phi_j^{(l)} = \mathbf{U}_j^{(l)} + \mathbf{V}_j\Phi_{j+1}^{(l)}, \tag{3.7.14}$$

where

$$\mathbf{U}_j^{(l)} = (\mathbf{B}_j^* - \mathbf{A}_j^*\mathbf{V}_{j-1})^{-1}|\mathscr{L}_j^{(l)} + \mathbf{A}_j^*\mathbf{U}_{j-1}^{(l)}|, \tag{3.7.15}$$

and

$$\mathbf{V}_j = (\mathbf{B}_j^* - \mathbf{A}_j^*\mathbf{V}_{j-1})^{-1}\mathbf{C}_j^*. \tag{3.7.16}$$

It is clear from the above recurrence relationships that the matrix inverses $(\mathbf{B}_j - \mathbf{A}_j\mathbf{V}_{j-1})^{-1}$ need only be computed for the solution of the equation corresponding to $l = 0$, i.e. (3.7.8) and thence stored for subsequent use when $l \geqslant 1$, i.e. (3.7.9). Consequently, only one set of matrix inversions is required. The $\mathbf{U}_j^{(l)}$ differ for different l since the terms $\mathbf{E}_j^{(l)}$ differ and therefore these $\mathbf{U}_j^{(l)}$ must be evaluated separately for each l.

The main advantage of the method, however, is that the matrices to be inverted are of size $(n^* \times n^*)$ not $(n \times n)$ and, since the time taken to invert a matrix of size n is proportional to n^3, the decrease in computer time (particularly since the computation of matrix inverses effectively constitutes all the computer time used) is of order $(n/n^*)^3$. Test problems have been run with $n \sim 30$ and $n^* \sim 3$, for example, with good convergence; the decrease in computer time in this case is of order 10^3. Indeed, good convergence has been attained for the same test problems, but with $n^* = 1$, i.e. with only one angle and one frequency quadrature point, although, here, an inappropriate choice of $\Delta\nu_1^*$ can lead to divergence.

Before proceeding to a discussion of convergence, however, let us first consider the evaluation of $(L - L^*)\Phi^{(l-1)}$ appearing in $E^{(l)}$ given by equation (3.7.10). The computation of $L^*\Phi^{(l-1)}$ is straightforward since $\Phi^{(l-1)}$ has already been obtained at the quadrature points $\{\Delta\nu_i^*\}$ and $\{\mu_k^*\}$ at which the L^* operator is defined, i.e. we have

$$L^*\Phi^{(l-1)} = \sum_{i=1}^{N_\nu^*} \sum_{k=1}^{N_\mu^*} \alpha_i^* \beta_k^* \Phi^{(l-1)}(\tau, \Delta\nu_i^*, \mu_k^*). \tag{3.7.17}$$

The corresponding evaluation of $L\Phi^{(l-1)}$ is not as straightforward since

$$L\Phi^{(l-1)} = \sum_{i=1}^{N_\nu} \sum_{k=1}^{N_\mu} \alpha_i \beta_k \Phi^{(l-1)}(\tau, \Delta\nu_i, \mu_k), \tag{3.7.18}$$

i.e. one must know $\Phi^{(l-1)}$ at the quadrature points $\{\Delta\nu_i\}$ and $\{\mu_k\}$ whereas it has only been determined at $\{\Delta\nu_i^*\}$ and $\{\mu_k^*\}$. One could possibly obtain the $\Phi^{(l-1)}(\tau, \Delta\nu_i, \mu_k)$ by interpolating (perhaps extrapolating) on the $\{\Delta\nu_i^*\}$ and $\{\mu_k^*\}$ grids from $\Phi^{(l-1)}(\tau, \Delta\nu_i^*, \mu_k^*)$. However, the thrust behind the perturbation method involves the use of values of N_ν^* and N_μ^* as small as possible and thus interpolation could be quite inaccurate. An alternative procedure is to first define the lth source function $S^{(l)}$ where

$$S^{(l)} = (1 - \epsilon)L^*\Phi^{(l)} + E^{(l)}, \tag{3.7.19}$$

for all $l \geqslant 0$ (note that $E^{(0)} \equiv \epsilon B_\nu(T)$). Equations (3.7.8) and (3.7.9) are then simply of the form

$$\frac{\mu^2}{\phi^2(\Delta\nu)} \frac{\partial^2 \Phi^{(l)}}{\partial\tau^2} = \Phi^{(l)} - S^{(l)}, \tag{3.7.20}$$

which, of course, is equivalent to the first order equation of transfer

$$\frac{\mu}{\phi(\Delta\nu)} \frac{\partial I^{(l)}}{\partial\tau} = I^{(l)} - S^{(l)}, \tag{3.7.21}$$

for all $l \geqslant 0$.

One may then obtain $L\Phi^{(l-1)}$ appearing in equation (3.7.10) by evaluating $S^{(l-1)}$ using equation (3.7.19) – note that $L^*\Phi^{(l-1)}$ has already been computed – thence solving equation (3.7.21) explicitly for $I^{(l-1)}(\tau, \Delta\nu_i, \mu_k)$ at the higher order quadrature points $\{\Delta\nu_i\}$ and $\{\mu_k\}$. Of course, since $S^{(l-1)}$ has already been determined, equation (3.7.21) may be considered as a first order ordinary differential equation in $I^{(l-1)}$ which may therefore be easily solved. The quantity $L\Phi^{(l-1)}$ is then finally obtained by direct substitution of $I^{(l-1)}(\tau, \Delta\nu_i, \pm\mu_k)$ or, equivalently, $\Phi^{(l-1)}(\tau, \Delta\nu_i, \mu_k)$, into equation (3.7.18).

3.7.2 *Convergence*

Equation (3.7.6) is linear in Φ so that the series (3.7.7), if it converges, converges to the correct solution. This can be seen simply by adding equation (3.7.8) to (3.7.9) for all $l \geqslant 1$, i.e. we obtain

$$\begin{aligned}\frac{\partial^2}{\phi^2(\Delta\nu)}\frac{\partial^2}{\partial\tau^2}(\Phi^{(0)}+\Phi^{(1)}+\cdots)\\ &= \Phi^{(0)}+\Phi^{(1)}+\cdots-(1-\epsilon)L^*(\Phi^{(0)}+\Phi^{(1)}+\cdots)-\epsilon B_\nu(T)\\ &\quad -(1-\epsilon)(L-L^*)(\Phi^{(0)}+\Phi^{(1)}+\cdots)\\ &= \Phi^{(0)}+\Phi^{(1)}+\cdots-(1-\epsilon)L(\Phi^{(0)}+\Phi^{(1)}+\cdots)-\epsilon B_\nu(T),\end{aligned} \tag{3.7.22}$$

i.e.

$$\frac{\mu^2}{\phi^2(\Delta\nu)}\frac{\partial^2\Phi}{\partial\tau^2} = \Phi-(1-\epsilon)L\Phi-\epsilon B_\nu(T), \tag{3.7.23}$$

where

$$\Phi = \sum_{l=0}^{\infty}\Phi^{(l)}. \tag{3.7.24}$$

Indeed, the above perturbation series is rigorously equivalent to an iterative procedure for, if we define

$$\Phi_l = \sum_{l'=0}^{l}\Phi^{(l')}, \tag{3.7.24a}$$

equation (3.7.22) may be written as

$$\frac{\mu^2}{\phi^2(\Delta\nu)}\frac{\partial^2\Phi_l}{\partial\tau^2} = \Phi_l-(1-\epsilon)L^*\Phi_l-\epsilon B_\nu(T)-(1-\epsilon)(L-L^*)\Phi_{l-1}, \tag{3.7.25}$$

for all $l \geqslant 0$ where we take $\Phi_{-1}\equiv 0$. This then has the form

$$\frac{\mu^2}{\phi^2(\Delta\nu)}\frac{\partial^2\Phi_l}{\partial\tau^2} = \Phi_l-(1-\epsilon)L^*\Phi_l-\mathcal{E}_l, \tag{3.7.26}$$

where

$$\mathcal{E}_0 = \epsilon B_\nu(T), \tag{3.7.27}$$

and, for all $l \geqslant 1$,

$$\mathcal{E}_l = (1-\epsilon)(L-L^*)\Phi_{l-1}. \tag{3.7.28}$$

The comments following on from equation (3.7.10) also apply to the above system. Clearly, when convergence has been attained, i.e. $\Phi_l \to \Phi_{l-1}$ or $\Phi^{(l)} \to 0$ as $l \to \infty$, equation (3.7.26) reduces again to

$$\frac{\mu^2}{\phi^2(\Delta\nu)}\frac{\partial^2\Phi_\infty}{\partial\tau^2} = \Phi_\infty-(1-\epsilon)L\Phi_\infty-\epsilon B_\nu(T). \tag{3.7.29}$$

However, the procedure (either equation (3.7.9) or (3.7.26)) might not converge if n^* is too small. To see this, we re-write the equation of radiative transfer in integral form – see equation (3.1.5) – viz,

$$S = (1-\epsilon)\Lambda S + \epsilon B_\nu(T), \tag{3.7.30}$$

where Λ is the operator defined by equation (3.1.6) (or (3.1.8) for depth-dependent $\phi(\Delta\nu)$), i.e.

$$\Lambda = \tfrac{1}{2}\int_{-\infty}^{\infty} \phi^2(\Delta\nu)\,\mathrm{d}(\Delta\nu)\int_0^1 \frac{\mathrm{d}\mu}{\mu}\int_0^\infty \mathrm{e}^{-\phi(\Delta\nu)|\tau'-\tau|/\mu}\,\mathrm{d}\tau'. \tag{3.7.31}$$

In particular, we note that this Λ operator involves an integration over angle and frequency and, thus, may be approximated by numerical quadrature. We therefore define Λ to be the operator resulting from numerical quadrature using N_ν and N_μ frequency and angle points respectively, and Λ^* the corresponding operator using N_ν^* and N_μ^* points.

We thence write equation (3.7.30) as

$$\begin{aligned}
S &= [1-(1-\epsilon)\Lambda]^{-1}\epsilon B_\nu(T)\\
&= \{1-(1-\epsilon)\Lambda^*-(1-\epsilon)(\Lambda-\Lambda^*)\}^{-1}\epsilon B_\nu(T)\\
&= \{[1-(1-\epsilon)\Lambda^*]\{1-[1-(1-\epsilon)\Lambda^*]^{-1}(1-\epsilon)(\Lambda-\Lambda^*)\}\}^{-1}\\
&\quad\times \epsilon B_\nu(T)\\
&= \{1-[1-(1-\epsilon)\Lambda^*]^{-1}(1-\epsilon)(\Lambda-\Lambda^*)\}^{-1}[1-(1-\epsilon)\Lambda^*]^{-1}\\
&\quad\times \epsilon B_\nu(T)\\
&= \left\{1+\sum_{l=1}^{\infty}\{[1-(1-\epsilon)\Lambda^*]^{-1}(1-\epsilon)(\Lambda-\Lambda^*)\}^l\right\}\\
&\quad\times [1-(1-\epsilon)\Lambda^*]^{-1}\epsilon B_\nu(T) \qquad (3.7.31)\\
&= \sum_{l=0}^{\infty} S^{(l)}, \qquad (3.7.32)
\end{aligned}$$

where we put

$$S^{(0)} = [1-(1-\epsilon)\Lambda^*]^{-1}\epsilon B_\nu(T), \tag{3.7.33}$$

and

$$S^{(l)} = [1-(1-\epsilon)\Lambda^*]^{-1}(1-\epsilon)(\Lambda-\Lambda^*)S^{(l-1)}, \tag{3.7.34}$$

for all $l \geqslant 1$. We may, of course, re-write these equations in the form

$$S^{(0)} = (1-\epsilon)\Lambda^* S^{(0)} + \epsilon B_\nu(T), \tag{3.7.35}$$

and

$$S^{(l)} = (1-\epsilon)\Lambda^* S^{(l)} + E^{(l)}, \tag{3.7.36}$$

where

$$E^{(l)} = (1-\epsilon)(\Lambda-\Lambda^*)S^{(l-1)}, \tag{3.7.37}$$

for all $l \geqslant 1$. Again, one notices that equations (3.7.35) and (3.7.36) are identical except for the non-homogeneous 'source' terms $\epsilon B_\nu(T)$ and $E^{(l)}$, and thus the Green's function used to solve equation (3.7.35) may be used to solve (3.7.36). Note that one could have obtained the same result by constructing a perturbation series in S analogous to the series (3.7.7) in Φ. Indeed, the above equations are themselves analogous to equations (3.7.8) through (3.7.10) derived using the differential equation approach. For example, one sees that equation (3.7.35) is simply the integral equation form of (3.7.8). Thus, $S^{(0)}$ defined by equation (3.7.35) is the same $S^{(0)}$ as that defined by (3.7.19). This then implies that E_1 defined by equation (3.7.37) is identical to that given by (3.7.10). The above argument may be repeated for $S^{(l)}$ and $E^{(l)}$ for all $l \geqslant 2$.

Again, if we re-formulate the series (3.7.32) as an iterative procedure by writing

$$S_l = \sum_{l'=0}^{l} S^{(l')}, \tag{3.7.38}$$

we find

$$S_l = (1-\epsilon)\Lambda^* S_l + \epsilon B_\nu(T) + (1-\epsilon)(\Lambda - \Lambda^*)S_{l-1}, \tag{3.7.39}$$

where, in obtaining our zeroth order solution S_0, we put $S_{-1} = 0$.

By comparison, Λ iteration takes the form (see equation (3.1.7))

$$S_l^{(\Lambda)} = (1-\epsilon)S_{l-1}^{(\Lambda)} + \epsilon B_\nu(T), \tag{3.7.40}$$

where the superscript (Λ) has been included to distinguish the source function obtained using Λ iteration from that using the perturbation series. If we wish to obtain a series for the Λ iteration analogous to equation (3.7.38), we write

$$S_l^{(\Lambda)} = \sum_{l'=0}^{l} S_\Lambda^{(l')}, \tag{3.7.41}$$

where, from equation (3.1.10), for example,

$$S_\Lambda^{(l)} = (1-\epsilon)\Lambda S_\Lambda^{(l-1)}, \tag{3.7.42}$$

for all $l \geqslant 1$ with

$$S_\Lambda^{(0)} = \epsilon B_\nu(T). \tag{3.7.43}$$

If we now consider an infinite medium, and this constitutes the worst possible situation for convergence of the two series (3.7.38) and (3.7.41) - by contrast, convergence in an optically thin slab is rapid - we have

$$\Lambda \equiv \int_{-\infty}^{\infty} K_1(|\tau' - \tau|)\, \mathrm{d}\tau' \tag{3.7.44}$$

(see equation (3.1.17)), where

$$K_1(\tau) = \tfrac{1}{2}\int_{-\infty}^{\infty} \phi^2(\Delta\nu) E_1(\tau\phi(\Delta\nu))\, \mathrm{d}(\Delta\nu) \tag{3.7.45}$$

is to be represented by numerical quadrature using N_ν and N_μ frequency and angle quadrature points respectively. Correspondingly, we define $K_1^*(\tau)$ to be that obtained from equation (3.7.45) using N_ν^* and N_μ^* quadrature points, i.e.

$$\Lambda^* \equiv \int_{-\infty}^{\infty} K_1^*(|\tau' - \tau|)\, d\tau'. \tag{3.7.46}$$

We now take the infinite space 1-dimensional Fourier transform of equation (3.7.36) coupled to (3.7.37), viz.

$$S^{(l)} = (1-\epsilon)\Lambda^* S^{(l)} + (1-\epsilon)(\Lambda - \Lambda^*)S^{(l-1)}, \tag{3.7.47}$$

to obtain, using the Fourier transform convolution theorem,

$$\tilde{S}^{(l)} = (1-\epsilon)\tilde{K}_1^* \tilde{S}^{(l)} + (1-\epsilon)(\tilde{K}_1 - \tilde{K}_1^*)\tilde{S}^{(l-1)}, \tag{3.7.48}$$

where the tilde denotes the Fourier transform, and where we have assumed ϵ to be independent of τ.

The rate of convergence using the perturbation series can then be seen to be

$$\begin{aligned} \frac{\tilde{S}^{(l)}}{\tilde{S}^{(l-1)}} &= \frac{(1-\epsilon)(\tilde{K}_1 - \tilde{K}_1^*)}{1-(1-\epsilon)\tilde{K}_1^*} \\ &= 1 - \frac{1-(1-\epsilon)\tilde{K}_1}{1-(1-\epsilon)\tilde{K}_1^*}. \end{aligned} \tag{3.7.49}$$

Clearly, if the procedure is to converge, we must have

$$\left|\frac{\tilde{S}^{(l)}}{\tilde{S}^{(l-1)}}\right| < 1,$$

i.e.

$$-1 < 1 - \frac{1-(1-\epsilon)\tilde{K}_1}{1-(1-\epsilon)\tilde{K}_1^*} < 1. \tag{3.7.50}$$

The right-hand inequality (3.7.50) yields

$$(1-\epsilon)\tilde{K}_1 < 1 \tag{3.7.51}$$

(use the first of equations (3.7.49)) whereas the left-hand inequality is just

$$1-(1-\epsilon)\tilde{K}_1^* > \tfrac{1}{2}[1-(1-\epsilon)\tilde{K}_1]. \tag{3.7.52}$$

If we now apply the same Fourier transform to equations (3.7.42), we obtain

$$\tilde{S}_\Lambda^{(l)} = (1-\epsilon)\tilde{K}_1 \tilde{S}_\Lambda^{(l-1)}, \tag{3.7.53}$$

which, for convergence, must satisfy

$$\left|\frac{\tilde{S}_\Lambda^{(l)}}{\tilde{S}_\Lambda^{(l-1)}}\right| < 1,$$

i.e.

$$(1-\epsilon)\tilde{K}_1 < 1. \tag{3.7.54}$$

This last result is identical to inequality (3.7.51) which the perturbation series procedure must satisfy. We have shown that Λ iteration always converges and, thus, the right-hand inequality (3.7.51) is always satisfied.

However, inequality (3.7.52) may be violated by an unfortunate choice of K_1^* (leading to $\tilde{K}_1^*$). For example, consider the extreme case when $n^* = 1$, i.e. we take 1-point angle ($\mu^* = 1/\sqrt{3}$) and frequency ($\Delta\nu^*$) quadrature. We then have

$$K_1^*(\tau) = \frac{\sqrt{3}}{2}\,\phi(\Delta\nu^*)\mathrm{e}^{-\sqrt{3}|\tau|\phi(\Delta\nu^*)}, \tag{3.7.55}$$

since, under 1-point quadrature,

$$\int_0^1 f(\mu)\,\mathrm{d}\mu \approx f(\mu^*),$$

and

$$\int_{-\infty}^{\infty} \phi(\Delta\nu) f(\Delta\nu)\,\mathrm{d}(\Delta\nu) \approx f(\Delta\nu^*)$$

- note that

$$\int_{-\infty}^{\infty} \phi(\Delta\nu)\,\mathrm{d}(\Delta\nu) = 1.$$

We then have

$$\begin{aligned}\tilde{K}_1^*(\kappa) &= \int_{-\infty}^{\infty} \mathrm{e}^{\mathrm{i}\kappa\tau} K_1^*(\tau)\,\mathrm{d}\tau \\ &= \frac{\sqrt{3}}{2}\,\phi(\Delta\nu^*)\left\{\int_0^{\infty} \mathrm{e}^{\mathrm{i}\kappa\tau - \sqrt{3}\tau\phi(\Delta\nu^*)}\,\mathrm{d}\tau \right. \\ &\quad \left. + \int_{-\infty}^{0} \mathrm{e}^{\mathrm{i}\kappa\tau + \sqrt{3}\tau\phi(\Delta\nu^*)}\,\mathrm{d}\tau\right\} \\ &= \frac{3\phi^2(\Delta\nu^*)}{3\phi^2(\Delta\nu^*) + \kappa^2}.\end{aligned} \tag{3.7.56}$$

Consequently, for κ sufficiently small and $\Delta\nu^*$ sufficiently small, then $3\phi^2(\Delta\nu^*) \gg \kappa^2$. This leads to $\tilde{K}^*(\kappa) \cong 1$ for small κ, which might violate inequality (3.7.52), particularly if ϵ is extremely small. Thus, in this case, there is a critical frequency $\Delta\nu_\mathrm{c}$ below which 1-point quadrature in frequency will yield a diverging perturbation series. (Note, however, that in the limit as $\epsilon \to 1$, inequality (3.7.52) must be satisfied for all $\tilde{K}_1^*$.)

If, on the other hand, we choose a 1-point quadrature $N_\nu^* = 1$ with $\Delta\nu^*$ sufficiently large (i.e. $\phi(\Delta\nu^*)$ sufficiently small), such that $\tilde{K}_1^* \ll \tilde{K}_1$, then we see from equation (3.7.49) that

$$\frac{\tilde{S}^{(l)}}{\tilde{S}^{(l-1)}} \approx (1-\epsilon)\tilde{K}, \tag{3.7.57}$$

i.e. we must have

$$(1-\epsilon)\tilde{K}_1 < 1, \tag{3.7.58}$$

for convergence. This, however, is precisely the condition for convergence of Λ iteration (inequality (3.7.54)). Thus, although the perturbation approach will converge for sufficiently large $\Delta\nu^*$ using 1-point frequency quadrature, the convergence rate could mimic that for Λ iteration and this, of course, is unsatisfactory for semi-infinite atmospheres in which $\epsilon \ll 1$.

Clearly, then, there exists a $\Delta\nu^* \gtrsim \Delta\nu_c$ at which good convergence is attained in semi-infinite atmosphere computations using 1-point frequency quadrature. However, it is generally more satisfactory to use $N_\nu^* \sim 5$ which, with $N_\mu^* = 1$, still only requires the inversion of (5×5) matrices whilst overcoming the above-mentioned convergence difficulties.

In slab geometry, one may choose a $\Delta\nu^*$ using 1-point quadrature such that $\tilde{K}_1^* \approx \tilde{K}_1$, i.e. $\Lambda^* \approx \Lambda$ and, thus, convergence is extremely rapid. This effectively reflects the dependence of the physics on the local conditions in such an atmosphere, as opposed to the non-local considerations which come into effect in a semi-infinite medium.

3.8 Integral equation techniques

Section 3.2 developed the numerical procedure for solution of the equation of radiative transfer using the differential equation approach. In that case, the unknown to be determined was the radiation field $\Phi(\tau, \Delta\nu, \mu)$ at depth τ, frequency $\Delta\nu$ and angle $\cos^{-1}\mu$. The integral equation approach, however, solves for the source function S which, under the assumption of complete re-distribution, is a function only of τ. Thus, one might suspect, the reduction in the number of independent variables in the dependent unknown could lead to a corresponding reduction in the difficulty of numerical solution. Indeed, one could compare this supposed reduction in degree of difficulty with the two approaches used to solve the transfer equation exactly in chapter 2 (compare the singular eigenfunction technique using the differential equation with the Fourier transform method using the integral equation). More complicated (and more physically realistic) situations arise when we allow departures from complete re-distribution (see chapter 7) since, in this case, we have $S(\tau, \Delta\nu, \mu)$ dependent also on frequency and angle.

3.8.1 General method of solution [55]

We may outline the general method of solution by first writing the integral form of the equation of radiative transfer for a semi-infinite medium as

$$S(\tau) = (1 - \epsilon)\Lambda S(\tau) + \epsilon B_\nu(T) \tag{3.8.1}$$

(see equation (3.1.5)) where the operator Λ has the form (taking $\phi(\Delta\nu)$ to be depth independent)

$$\Lambda \equiv \int_0^\infty K_1(|\tau' - \tau|)\, \mathrm{d}\tau', \tag{3.8.2}$$

where

$$K_1(\tau) = \tfrac{1}{2} \int_{-\infty}^{\infty} \phi^2(\Delta\nu)\, \mathrm{d}(\Delta\nu) \int_0^1 \mathrm{e}^{-\tau\phi(\Delta\nu)/\mu} \frac{\mathrm{d}\mu}{\mu}. \tag{3.8.3}$$

The reader is referred to section 3.1 for a derivation of the above equations.

We know from chapter 1 that

$$S(\tau) \approx B_\nu(T), \tag{3.8.4}$$

for all $\tau \gg \Theta$ where Θ is the thermalisation path length for the radiative transition in question. Thus, if we put $\tau_N \gg \Theta$, where τ_N is the Nth (and last) point in a depth discretisation grid $\tau_1, \tau_2, \ldots, \tau_N$, then we may write

$$\Lambda S = \int_0^{\tau_N} K_1(|\tau' - \tau|) S(\tau')\, \mathrm{d}\tau' + \int_{\tau_N}^{\infty} K_1(|\tau' - \tau|) B_\nu(T)\, \mathrm{d}\tau'. \tag{3.8.5}$$

We now approximate the first integral on the RHS of equation (3.8.5) by a quadrature sum such that

$$\int_0^{\tau_N} K_1(|\tau' - \tau|) S(\tau')\, \mathrm{d}\tau' \approx \sum_{j'=1}^{N} w_{j'}(\tau) S_{j'}, \tag{3.8.6}$$

where $w_{j'}(\tau)$ are the quadrature weights at the depth τ. Thus, writing $w_{j'j} \equiv w_{j'}(\tau_j)$, equation (3.8.6) becomes

$$\int_0^{\tau_N} K_1(|\tau' - \tau_j|) S(\tau') \approx \sum_{j'=1}^{N} w_{j'j} S_{j'}, \tag{3.8.7}$$

which, when substituted into equation (3.8.1), yields

$$S_j = (1 - \epsilon_j) \left\{ \sum_{j'=1}^{N} w_{j'j} S_{j'} + \int_{\tau_N}^{\infty} K_1(|\tau' - \tau|) B_\nu(T)\, \mathrm{d}\tau' \right\} + [\epsilon B_\nu(T)]_j. \tag{3.8.8}$$

We may now write the above system of equations (for all $j = 1, N$) in matrix form by putting

$$\mathbf{S} = \begin{pmatrix} S_1 \\ \vdots \\ S_N \end{pmatrix}, \tag{3.8.9}$$

$$\mathbf{W} = \begin{pmatrix} W_{11}, & \ldots, & W_{N1} \\ \vdots & & \\ W_{1N} & & W_{NN} \end{pmatrix}, \tag{3.8.10}$$

and

$$\mathbf{L} = \begin{pmatrix} [\epsilon B_\nu(T)]_1 + \mathscr{L}_1 \\ \vdots \\ [\epsilon B_\nu(T)]_N + \mathscr{L}_N \end{pmatrix}, \tag{3.8.11}$$

where

$$W_{j'j} = (1-\epsilon)_j w_{j'j}, \tag{3.8.12}$$

and

$$\mathscr{L}_j = \int_{\tau_N}^{\infty} K_1(|\tau' - \tau|) B_\nu(T)\, \mathrm{d}\tau'. \tag{3.8.13}$$

Equations (3.4.9) for all $k = 1, N$ then yield

$$\mathbf{S} = \mathbf{WS} + \mathbf{L}, \tag{3.8.14}$$

which has the solution

$$\mathbf{S} = (\mathbb{1} - \mathbf{W})^{-1}\mathbf{L}. \tag{3.8.15}$$

Clearly, the computational procedure involves the inversion of the one $(N \times N)$ matrix $(\mathbb{1} - \mathbf{W})$. This is effectively equivalent to the Rybicki reorganisation presented in section 3.5. The problem has now been reduced to the evaluation of the weighting terms $w_{j'j}$.

3.8.2 Choice of $w_{j'j}$

There exists a variety of methods, each yielding different results, by which one may obtain the weighting functions $w_{j'j}$. Here, for purposes of illustration, we present just three procedures. Further schemes may be readily obtained simply by increasing the order of the polynomial used to describe the integrand appearing in equation (3.8.6). A fourth scheme is given in section 3.8.3 for a slightly differently formulated integral equation specification. The reader is also referred to section 3.10.

(a) Constant segments

The most straightforward manner of evaluating a set of $w_{j'j}$ is simply to take the source function to be constant in each of the $N-1$ intervals as shown in figure 3.2.

We thence put $S(\tau) = S(\bar{\tau}_j)$ for all $\tau \in [\tau_j, \tau_{j+1}]$ so that

$$\int_0^{\tau_N} K_1(|\tau' - \tau|) S(\tau')\, \mathrm{d}\tau' = \sum_{j'=1}^{N-1} \int_{\tau_{j'}}^{\tau_{j'+1}} K_1(|\tau' - \tau|) S(\tau')\, \mathrm{d}\tau'$$

$$\approx \sum_{j'=1}^{N-1} S(\bar{\tau}_{j'}) \int_{\tau_{j'}}^{\tau_{j'+1}} K_1(|\tau' - \tau|)\, d\tau', \tag{3.8.16}$$

which, from equation (3.8.7), immediately yields

$$w_{j'j} = \int_{\tau_{j'}}^{\tau_{j'+1}} K_1(|\tau' - \tau_j|)\, d\tau', \tag{3.8.17}$$

with

$$S_{j'} = S(\bar{\tau}_{j'}).$$

Note that this case is slightly different to the general situation discussed in section 3.8.1 in that only $N-1$, not N, values of S_j are obtained.

The value of $\bar{\tau}_j$ is somewhat arbitrary except that $\bar{\tau}_j \in (\tau_j, \tau_{j+1})$ of course. Effectively, therefore, one obtains the $N-1$ source functions $S(\bar{\tau}_j)$ at the $N-1$ values of $\bar{\tau}_j$ where $\bar{\tau}_j$ can be written as

$$\bar{\tau}_j = \lambda\tau_j + (1-\lambda)\tau_{j+1}, \tag{3.8.18}$$

where $\lambda \in (0, 1)$. The most obvious choice of λ is $\lambda = \frac{1}{2}$. Note that one need not choose $\bar{\tau}_j$ to evaluate the weighting functions.

One can evaluate the integral appearing in equation (3.8.17) analytically in part. Using the definition (3.8.3) of $K_1(\tau)$, we find

$$w_{j'j} = \tfrac{1}{2} \int_{-\infty}^{\infty} \phi^2(\Delta\nu)\, d(\Delta\nu) \int_0^1 \frac{d\mu}{\mu} \int_{\tau_{j'}}^{\tau_{j'+1}} e^{-|\tau' - \tau_j|\phi(\Delta\nu)/\mu}\, d\tau', \tag{3.8.19}$$

which, for $\tau_j < \tau_{j'}$, for example, yields

$$w_{j'j} = \tfrac{1}{2} \int_{-\infty}^{\infty} \phi(\Delta\nu)\, d(\Delta\nu) \int_0^1 d\mu \left[e^{-(\tau_{j'} - \tau_j)\phi(\Delta\nu)/\mu} - e^{-(\tau_{j'+1} - \tau_j)\phi(\Delta\nu)/\mu}\right]. \tag{3.8.20}$$

Fig.3.2. The source function is represented by a constant between the depths τ_j and τ_{j+1}.

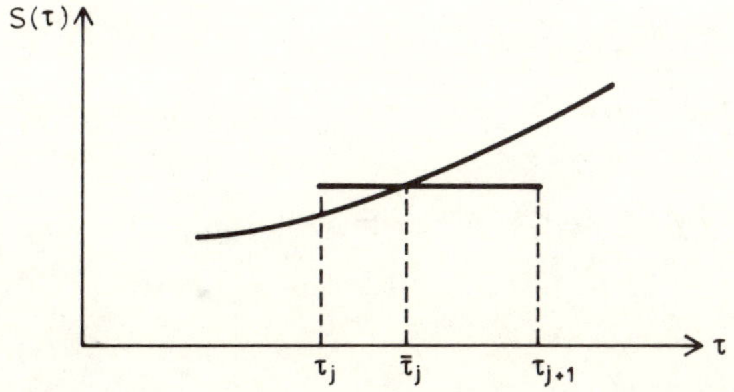

Similar expressions can be found for $\tau_j \in (\tau_{j'}, \tau_{j'+1})$ and $\tau_j > \tau_{j'+1}$. One cannot, in general, proceed with the analytical evaluation of the double integral over angle and frequency appearing in this last equation. It is necessary, therefore, to utilise numerical quadrature over these independent variables as in the differential equation formulation. It is important to note, however, that one may take as many angle and frequency quadrature points as is necessary for the desired accuracy without changing the size ($N \times N$) of the matrix requiring inversion. This is one of the main advantages of the integral equation formulation.

(b) Linear segments

Here we write $S(\tau)$ for all $\tau \in [\tau_j, \tau_{j+1}]$ in the form

$$S(\tau) = \frac{\tau_{j+1} - \tau}{\Delta_j} \cdot S_j + \frac{\tau - \tau_j}{\Delta_j} \cdot S_{j+1}, \tag{3.8.21}$$

where

$$\Delta_j = \tau_{j+1} - \tau_j.$$

It is clear from equation (3.8.21) that $S(\tau)$ has been taken to be a linear function of τ between τ_j and τ_{j+1} with $S(\tau_j) = S_j$ and $S(\tau_{j+1}) = S_{j+1}$.

In determining the weighting functions appearing in equation (3.8.7), we again use the first of equations (3.8.16):

$$\int_0^{\tau_N} K_1(|\tau' - \tau_j|) S(\tau')\,\mathrm{d}\tau'$$

$$\approx \sum_{j'=1}^{N-1} \left\{ \frac{S_{j'}}{\Delta_{j'}} \int_{\tau_{j'}}^{\tau_{j'+1}} K_1(|\tau' - \tau_j|)(\tau_{j'+1} - \tau')\,\mathrm{d}\tau' + \frac{S_{j'+1}}{\Delta_{j'}} \int_{\tau_{j'}}^{\tau_{j'+1}} K_1(|\tau' - \tau_j|)(\tau' - \tau_{j'})\,\mathrm{d}\tau' \right\}$$

$$= \sum_{N}^{j'=1} w_{j'j} S_{j'}, \tag{3.8.22}$$

where

$$w_{1j} = \frac{1}{\Delta_1} \int_{\tau_1}^{\tau_2} K_1(|\tau' - \tau_j|)(\tau_2 - \tau')\,\mathrm{d}\tau', \tag{3.8.23}$$

$$w_{Nj} = \frac{1}{\nabla_N} \int_{\tau_{N-1}}^{\tau_N} K_1(|\tau' - \tau_j|)(\tau' - \tau_N)\,\mathrm{d}\tau', \tag{3.8.24}$$

and for all $2 \leqslant j' \leqslant N-1$,

$$w_{j'j} = \frac{1}{\Delta_{j'}} \int_{\tau_{j'}}^{\tau_{j'+1}} K_1(|\tau' - \tau_j|)(\tau_{j'+1} - \tau')\,\mathrm{d}\tau'$$

$$+ \frac{1}{\nabla_{j'}} \int_{\tau_{j'}}^{\tau_{j'+1}} K_1(|\tau' - \tau_j|)(\tau' - \tau_{j'-1})\,\mathrm{d}\tau'. \tag{3.8.25}$$

Some of the integrals appearing in equations (3.8.23) through (3.8.25) may be analytically evaluated as before. However, the integrals over angle and frequency will require numerical quadrature. Again, this may be done with as many angle and frequency quadrature points as desired without altering the size of the final matrix to be inverted.

(c) Quadratic segments [64]

A more accurate representation for the source function $S(\tau)$ for all $\tau \in [\tau_j, \tau_{j+1}]$ is the quadratic form:

$$S(\tau) = S_j + A_j(\tau - \tau_j) + B_j(\tau - \tau_j)(\tau - \tau_{j+1}), \tag{3.8.26}$$

where A_j and B_j are constants yet to be evaluated.

We may obtain A_j and B_j by placing constraints on the points through which $S(\tau)$ passes. For example, if we consider a so-called leading parabola so that $S(\tau)$ passes through S_j, S_{j+1} and S_{j+2}, we find

$$S(\tau_j) = S_j,$$

$$S(\tau_{j+1}) = S_j + A_j(\tau_{j+1} - \tau_j),$$

and

$$S(\tau_{j+2}) = S_j + A_j(\tau_{j+2} - \tau_j) + B_j(\tau_{j+2} - \tau_j)(\tau_{j+2} - \tau_{j+1}).$$

The first of these three equations is satisfied identically. The last two can be solved for the two unknowns A_j and B_j, viz.

$$\left.\begin{aligned} A_j &= \frac{S_{j+1} - S_j}{\Delta_j}, \\ &\text{and} \\ B_j &= \alpha_j S_j + \beta_j S_{j+1} + \gamma_j S_{j+2}, \end{aligned}\right\} \tag{3.8.27}$$

where

$$\alpha_j = \frac{1}{\Delta_j(\Delta_j + \Delta_{j+1})}, \quad \beta_j = \frac{-1}{\Delta_j \Delta_{j+1}}, \quad \gamma_j = \frac{1}{\Delta_{j+1}(\Delta_j + \Delta_{j+1})}. \tag{3.8.28}$$

Substitution of these A_j and B_j into equation (3.8.26) finally yields

$$S(\tau) = \bar{\alpha}_j S_j + \bar{\beta}_j S_{j+1} + \bar{\gamma}_j S_{j+2}, \tag{3.8.29}$$

where

$$\bar{\alpha}_j = 1 - \frac{\tau - \tau_j}{\Delta_j} + (\tau - \tau_j)(\tau - \tau_{j+1})\alpha_j,$$

$$\bar{\beta}_j = \frac{\tau - \tau_j}{\Delta_j} + (\tau - \tau_j)(\tau - \tau_{j+1})\beta_j,$$

and

$$\bar{\gamma}_j = (\tau - \tau_j)(\tau - \tau_{j+1})\gamma_j. \tag{3.8.30}$$

As in the preceding two sub-sections, one then substitutes equation (3.8.29) into the first of equations (3.8.16) to obtain $w_{j'j}$.

If, on the other hand, we choose to derive A_j and B_j by forcing $S(\tau)$ to pass through the points S_{j-1}, S_j and S_{j+1}, we again find

$$\left.\begin{aligned} A_j &= \frac{S_{j+1} - S_j}{\Delta_j}, \\ &\text{but with} \\ B_j &= \alpha_{j-1}S_{j-1} + \beta_{j-1}S_j + \gamma_{j-1}S_{j+1}, \end{aligned}\right\} \tag{3.8.31}$$

where the α_j, β_j and γ_j terms are defined by equations (3.8.28). One then proceeds in the evaluation of the corresponding $w_{j'j}$ as discussed above.

3.8.3 *The flux divergence method [2]*

A slightly different approach to the integral equation formulation of the transfer equation involves the flux term $H(\tau, \Delta\nu)$ defined by

$$H(\tau, \Delta\nu) = \tfrac{1}{2}\int_{-1}^{1} \mu I(\tau, \Delta\nu, \mu)\,\mathrm{d}\mu. \tag{3.8.32}$$

Further, if we define the angle-averaged intensity $J(\tau, \Delta\nu)$ by

$$J(\tau, \Delta\nu) = \tfrac{1}{2}\int_{-1}^{1} I(\tau, \Delta\nu, \mu)\,\mathrm{d}\mu, \tag{3.8.33}$$

and integrate the equation of radiative transfer

$$\mu\frac{\partial I}{\partial\tau}(\tau, \Delta\nu, \mu) = \phi(\Delta\nu)[I(\tau, \Delta\nu, \mu) - S(\tau)], \tag{3.8.34}$$

over angle (note that, for ease in exposition, we take $S(\tau)$ to be independent of angle - the method, however, is not restricted to angle-independent source functions), then we find

$$\frac{\partial H}{\partial\tau}(\tau, \Delta\nu) = \phi(\Delta\nu)[J(\tau, \Delta\nu) - S(\tau)]. \tag{3.8.35}$$

If we now integrate over frequency, noting that

$$\int_{-\infty}^{\infty} \phi(\Delta\nu)\, \mathrm{d}(\Delta\nu) = 1,$$

$$\bar{J}(\tau) = \int_{-\infty}^{\infty} \phi(\Delta\nu) J(\tau, \Delta\nu)\, \mathrm{d}(\Delta\nu),$$

and

$$S(\tau) = (1-\epsilon)\bar{J}(\tau) + \epsilon B_\nu(T)$$

we find

$$\frac{\epsilon}{1-\epsilon}\,[S(\tau) - B_\nu(T)] = \int_{-\infty}^{\infty} \frac{\partial H}{\partial \tau}(\tau, \Delta\nu)\, \mathrm{d}(\Delta\nu). \tag{3.8.36}$$

We must now obtain an expression for the flux divergence (derivative) as a function of the source function. We may do this in the same manner as used to obtain the Λ operator in section 3.1.

Equations (3.1.1) and (3.1.2) are

$$I(\tau, \Delta\nu, \mu) = \int_{\tau}^{\infty} \phi(\Delta\nu) S(\tau') \mathrm{e}^{-\phi(\Delta\nu)(\tau'-\tau)/\mu} \frac{\mathrm{d}\tau'}{\mu}, \tag{3.1.1}$$

for all $\mu > 0$ and, for all $\mu < 0$,

$$I(\tau, \Delta\nu, \mu) = -\int_{0}^{\tau} \phi(\Delta\nu) S(\tau') \mathrm{e}^{-\phi(\Delta\nu)(\tau'-\tau)/\mu} \frac{\mathrm{d}\tau'}{\mu}, \tag{3.1.2}$$

where we have again taken zero incident radiation on the boundary of the model atmosphere with $\phi(\Delta\nu)$ depth independent.

Equation (3.8.32) then yields

$$\begin{aligned} H(\tau, \Delta\nu) &= \tfrac{1}{2}\int_0^1 \mathrm{d}\mu \int_{\tau}^{\infty} \phi(\Delta\nu) S(\tau') \mathrm{e}^{-\phi(\Delta\nu)(\tau'-\tau)/\mu}\, \mathrm{d}\tau' \\ &\quad - \tfrac{1}{2}\int_{-1}^{0} \mathrm{d}\mu \int_{0}^{\tau} \phi(\Delta\nu) S(\tau') \mathrm{e}^{-\phi(\Delta\nu)(\tau'-\tau)/\mu}\, \mathrm{d}\tau' \\ &= \tfrac{1}{2}\int_{\tau}^{\infty} \phi(\Delta\nu) E_2[(\tau'-\tau)\phi(\Delta\nu)] S(\tau')\, \mathrm{d}\tau' \\ &\quad - \tfrac{1}{2}\int_{0}^{\tau} \phi(\Delta\nu) E_2[(\tau-\tau')\phi(\Delta\nu)] S(\tau')\, \mathrm{d}\tau', \end{aligned} \tag{3.8.37}$$

where $E_2(x)$ is the second exponential integral, viz.

$$E_2(x) = \int_0^1 \mathrm{e}^{-x/\mu}\, \mathrm{d}\mu.$$

One can then obtain $\partial H(\tau, \Delta\nu)/\partial\tau$ by direct differentiation of equation (3.8.37), viz.

$$\begin{aligned}
\frac{\partial H}{\partial \tau}(\tau, \Delta\nu) &= -\phi(\Delta\nu)S(\tau) \\
&\quad + \tfrac{1}{2}\int_{\tau}^{\infty} \phi(\Delta\nu)\,\frac{\partial E_2}{\partial \tau}\,[(\tau'-\tau)\phi(\Delta\nu)]S(\tau')\,\mathrm{d}\tau' \\
&\quad - \tfrac{1}{2}\int_{0}^{\tau} \phi(\Delta\nu)\,\frac{\partial E_2}{\partial \tau}\,[(\tau-\tau')\phi(\Delta\nu)]S(\tau')\,\mathrm{d}\tau' \\
&= -\phi(\Delta\nu)S(\tau) \\
&\quad + \tfrac{1}{2}\int_{\tau}^{\infty} \phi^2(\Delta\nu)S(\tau')\,\mathrm{d}\tau' \int_0^1 \frac{\mathrm{d}\mu}{\mu}\,\mathrm{e}^{-(\tau'-\tau)\phi(\Delta\nu)/\mu} \\
&\quad + \tfrac{1}{2}\int_{0}^{\tau} \phi^2(\Delta\nu)S(\tau')\,\mathrm{d}\tau' \int_0^1 \frac{\mathrm{d}\mu}{\mu}\,\mathrm{e}^{-(\tau-\tau')\phi(\Delta\nu)/\mu} \\
&= -\phi(\Delta\nu)S(\tau) + \tfrac{1}{2}\int_0^{\infty} \phi^2(\Delta\nu)S(\tau')E_1(|\tau'-\tau|\phi(\Delta\nu))\,\mathrm{d}\tau'.
\end{aligned} \tag{3.8.38}$$

Substitution of this result into equation (3.8.36) yields

$$\begin{aligned}
\frac{\epsilon}{1-\epsilon}[S(\tau) - B_\nu(T)] &= -\int_{-\infty}^{\infty} \phi(\Delta\nu)S(\tau)\,\mathrm{d}(\Delta\nu) \\
&\quad + \tfrac{1}{2}\int_{-\infty}^{\infty} \phi^2(\Delta\nu)\,\mathrm{d}(\Delta\nu) \\
&\quad \times \int_0^{\infty} S(\tau')E_1(|\tau'-\tau|\phi(\Delta\nu))\,\mathrm{d}\tau' \\
&= -S(\tau) + \int_0^{\infty} K_1(|\tau'-\tau|)S(\tau')\,\mathrm{d}\tau',
\end{aligned} \tag{3.8.39}$$

which, of course, has the form

$$S(\tau) = (1-\epsilon)\Lambda S + \epsilon B_\nu(T), \tag{3.8.40}$$

where the operator Λ has been defined, for example, by equation (3.8.2). Consequently, there is no rigorous mathematical difference between the flux divergence approach and the ordinary integral equation formulation of the problem.

However, a somewhat differently structured equation results when we change variables from τ' to y in equation (3.8.37) where

$$\begin{aligned} y &= \tau' - \tau \quad \text{for } \tau' > \tau, \\ &= \tau - \tau' \quad \text{for } \tau' < \tau. \end{aligned} \tag{3.8.41}$$

We then have

$$\begin{aligned} H(\tau, \Delta\nu) = \tfrac{1}{2} \int_0^\infty \phi(\Delta\nu) E_2[y\phi(\Delta\nu)] S(y+\tau)\, \mathrm{d}y \\ - \tfrac{1}{2} \int_0^\tau \phi(\Delta\nu) E_2[y\phi(\Delta\nu)] S(\tau - y)\, \mathrm{d}y, \end{aligned} \tag{3.8.42}$$

such that

$$\begin{aligned} \frac{\partial H}{\partial \tau}(\tau, \Delta\nu) = \tfrac{1}{2} \int_0^\infty \phi(\Delta\nu) E_2[y\phi(\Delta\nu)] \frac{\partial S}{\partial \tau}(\tau + y)\, \mathrm{d}y \\ - \tfrac{1}{2} \int_0^\tau \phi(\Delta\nu) E_2[y\phi(\Delta\nu)] \frac{\partial S}{\partial \tau}(\tau - y)\, \mathrm{d}y \\ + \tfrac{1}{2}\phi(\Delta\nu) E_2[\tau\phi(\Delta\nu)] S(0). \end{aligned} \tag{3.8.43}$$

Substitution of this last result into equation (3.8.36) yields

$$\begin{aligned} \frac{\epsilon}{1-\epsilon}[S(\tau) - B_\nu(T)] = \tfrac{1}{2} \int_0^\infty K_2(y) \frac{\partial S}{\partial \tau}(\tau + y)\, \mathrm{d}y \\ - \tfrac{1}{2} \int_0^\tau K_2(y) \frac{\partial S}{\partial \tau}(\tau - y)\, \mathrm{d}y \\ + \tfrac{1}{2} K_2(\tau) S(0), \end{aligned} \tag{3.8.44}$$

where $K_2(\tau)$ is defined by

$$K_2(\tau) = \int_{-\infty}^\infty \phi(\Delta\nu) E_2[\tau\phi(\Delta\nu)]\, \mathrm{d}(\Delta\nu) \tag{3.8.45}$$

- see equation (3.1.28).

Equation (3.8.44) then represents the new integral (in fact, integro-differential) equation to be solved for $S(\tau)$. The main advantage in using this equation revolves around the function $K_2(\tau)$ which does not have the logarithmic singularity at $\tau = 0$ exhibited by $K_1(\tau)$, i.e.

$$K_1(0) = \tfrac{1}{2} \int_{-\infty}^{\infty} \phi^2(\Delta\nu)\, \mathrm{d}(\Delta\nu) \int_0^1 \frac{\mathrm{d}\mu}{\mu}$$

$$\sim - \lim_{\mu \to 0^+} \ln \mu, \tag{3.8.46}$$

whereas

$$K_2(0) = 1. \tag{3.8.47}$$

Consequently, one would hope that the weighting functions $w_{j'j}$ used to solve equation (3.8.44) could be more accurately obtained than those for the ordinary integral equation approach involving $K_1(\tau)$. Although the weights $w_{j'j}$ given by equation (3.8.20), for example, using constant segments have had the logarithmic singularity removed by direct integration over τ', the problem still arises when this direct integration is not possible and this occurs, of course, for the important case of depth-dependent $\phi(\Delta\nu)$. Thus, the integral equation technique previously discussed in section 3.8.1 has severe limitations.

Evaluation of $w_{j'j}$

We have already presented three methods in section 3.8.2 for obtaining the weighting functions $w_{j'j}$. A further alternative [7, 55] is provided by writing

$$S(\tau) = \sum_{j=1}^{N} c_j f_j(\tau), \tag{3.8.48}$$

where the $f_j(\tau)$ functions are explicitly given by

$$\begin{aligned} f_j(\tau) &= (1-\lambda)\left(1 - \frac{\tau}{\tau_j}\right) + \lambda\left(1 - \frac{\tau}{\tau_j}\right)^2, \quad \tau < \tau_j, \\ &= 0, \quad \tau \geqslant \tau_j, \end{aligned} \tag{3.8.49}$$

for all $j = 2, N-1$ with

$$f_1(\tau) = 1, \quad f_N(\tau) = \tau.$$

The parameter λ is an arbitrary quantity satisfying $\lambda \in [0, 1]$. One chooses λ to obtain maximum stability in the resulting equations. Usually, one has $\lambda = 1$.

We now detail a method for determining the coefficients c_j. We first write the matrix $\mathbf{F}$ such that

$$\mathbf{F} = \begin{pmatrix} f_{11}, & \ldots, & f_{N1} \\ \vdots & & \\ f_{1N} & & f_{NN} \end{pmatrix}, \tag{3.8.50}$$

where $f_{j'j} = f_{j'}(\tau_j)$, and define the inverse $\mathbf{G} = \mathbf{F}^{-1}$ of $\mathbf{F}$ by

$$\mathbf{G}=\begin{pmatrix} g_{11}, & \dots, & g_{N1} \\ \vdots & & \\ g_{1N} & & g_{NN} \end{pmatrix}. \tag{3.8.51}$$

If we now put

$$\mathbf{S}=\begin{pmatrix} S_1 \\ \vdots \\ S_N \end{pmatrix} \quad \text{and} \quad \mathbf{c}=\begin{pmatrix} c_1 \\ \vdots \\ c_N \end{pmatrix}, \tag{3.8.52}$$

then

$$\mathbf{S}=\mathbf{Fc}, \tag{3.8.53}$$

since

$$S_j=\sum_{j'=1}^{N} f_{j'j}c_{j'}. \tag{3.8.54}$$

Consequently,

$$\mathbf{c}=\mathbf{F}^{-1}\mathbf{S}=\mathbf{GS},$$

i.e.

$$c_j=\sum_{j'=1}^{N} g_{j'j}S_{j'}. \tag{3.8.55}$$

A combination of equations (3.8.48) and (3.8.55) yields

$$S(\tau)=\sum_{j=1}^{N}\sum_{j'=1}^{N} g_{j'j}S_{j'}f_j(\tau), \tag{3.8.56}$$

which, writing

$$h_{j'}(\tau)=\sum_{j=1}^{N} g_{j'j}f_j(\tau), \tag{3.8.57}$$

becomes

$$S(\tau)=\sum_{j'=1}^{N} h_{j'}(\tau)S_{j'}. \tag{3.8.58}$$

This last equation is merely a generalisation of equation (3.8.26) where we represented $S(\tau)$ by quadratic segments.

Once the $h_{j'}$ are determined, one substitutes the above $S(\tau)$ into the integral equation (3.8.44) to obtain an equation of the form (3.8.14). Indeed, if we write

$$w_{j'j}=\tfrac{1}{2}\int_0^\infty K_2(y)\left\{\left[\frac{\partial h_{j'}}{\partial\tau}(\tau+y)\right]_{\tau=\tau_j} - \left[\frac{\partial h_{j'}}{\partial\tau}(\tau-y)\right]_{\tau=\tau_j}\right\}\mathrm{d}y+\tfrac{1}{2}K_2(\tau_j)h_{j'}(0), \tag{3.8.59}$$

equations (3.8.44) and (3.8.58) yield

$$S(\tau_j) = [B_\nu(T)]_{\tau_j} + \frac{1-\epsilon_j}{\epsilon_j} \sum_{j'=1}^{N} w_{j'j} S_{j'}. \tag{3.8.60}$$

Thus, writing

$$\mathbf{W} = \begin{pmatrix} W_{11}, & \ldots, & W_{N1} \\ \vdots & & \\ W_{1N} & & W_{NN} \end{pmatrix}, \tag{3.8.61}$$

where

$$W_{j'j} = \frac{1-\epsilon_j}{\epsilon_j} w_{j'j}, \tag{3.8.62}$$

and

$$\mathbf{L} = \begin{pmatrix} [B_\nu(T)]_1 \\ \vdots \\ [B_\nu(T)]_N \end{pmatrix}, \tag{3.8.63}$$

equation (3.8.60) becomes

$$\mathbf{S} = \mathbf{WS} + \mathbf{L}, \tag{3.8.64}$$

as before.

Finally, we note that the matrix **F** has a large number of zeros and this enables its inverse **G** to be found analytically. For example, if we take $\lambda = 1$, we find

$$\mathbf{F} = \begin{pmatrix} 1 & f_{21} & f_{31} & \ldots & f_{N-1,1} & \tau_1 \\ 1 & 0 & f_{32} & & f_{N-1,2} & \tau_2 \\ 1 & 0 & 0 & f_{42} & & \\ \vdots & & \vdots & & & \\ 1 & & & & 0 & \tau_{N-1} \\ 1 & 0 & 0 & & 0 & \tau_N \end{pmatrix}, \tag{3.8.65}$$

where

$$f_{j'j} = \left(1 - \frac{\tau_j}{\tau_{j'}}\right)^2. \tag{3.8.66}$$

One may easily begin the analytical evaluation of $\mathbf{F}^{-1}$ by noting that the last two rows have only two non-zero elements. We immediately have, for example,

$$g_{N1} + \tau_{N-1} g_{NN} = 0,$$

and

$$g_{N1} + \tau_N g_{NN} = 1,$$

i.e.

$$g_{NN} = \frac{1}{\nabla_N}, \quad g_{N1} = -\frac{\tau_{N-1}}{\nabla_N}.$$

The third last row then gives

$$g_{N1} + f_{N-1,N-2}g_{N,N-1} + \tau_{N-2}g_{NN} = 0,$$

which then enables $g_{N,N-1}$ to be immediately obtained. The process is then continued.

3.9 Linearisation

The equation of radiative transfer for a model 2-level atom assuming complete re-distribution is given in chapter 1, i.e.

$$\mu \frac{\partial I}{\partial z}(z, \Delta\nu, \mu) = \frac{h\nu_0}{4\pi}[-N_L B_{UL} I(z, \Delta\nu, \mu)$$

$$+ N_U B_{UL} I(z, \Delta\nu, \mu) + N_U A_{UL}]\phi(\Delta\nu), \tag{3.9.1}$$

where the number densities N_U and N_L may be obtained from the equation of statistical equilibrium

$$N_L(B_{LU}\bar{J}(z) + C_{LU}) = N_U(B_{UL}\bar{J}(z) + C_{UL} + A_{UL}). \tag{3.9.2}$$

In all the preceding work on the solution of the transfer equation, the ratio N_L/N_U was eliminated from the above two equations yielding

$$\mu \frac{\partial I}{\partial z}(z, \Delta\nu, \mu) = -\kappa(z)\phi(\Delta\nu)[I(z, \Delta\nu, \mu) - (1-\epsilon)\bar{J}(z) - \epsilon B_\nu(T)], \tag{3.9.3}$$

where

$$\kappa(z) = \frac{h\nu_0}{4\pi} N_L B_{LU}\left[1 - \frac{N_U B_{UL}}{N_L B_{LU}}\right]\phi(\Delta\nu). \tag{3.9.4}$$

Thus, if $\kappa(z)$ is given, equation (3.9.3) is a linear equation in $I(z, \Delta\nu, \mu)$. However, it is clear from equation (3.9.2) that the populations N_U and N_L are functions of the radiation field $\bar{J}(z)$ and, thus, equation (3.9.3) is not rigorously linear in the unknown $I(z, \Delta\nu, \mu)$ because of the occurrence of N_U and N_L in the opacity $\kappa(z)$ term.

Indeed, equations (3.9.1) and (3.9.2) are but two equations to be solved for the three unknowns $I(z, \Delta\nu, \mu)$, N_L and N_U, and thus we require another equation if the problem is to be well posed. The obvious choice for this extra equation is

$$N_L + N_U = N_T, \tag{3.9.5}$$

where N_T is the total number density presumably given, along with the temperature, as part of the model atmosphere. (One should, of course, couple the solution of the equation of radiative transfer with the solutions of the equations specifying conservation of mass, energy and linear momentum since it is these latter three equations which detail the density, temperature and velocity distributions throughout the model atmosphere - see chapter 5.)

Equations (3.9.2) and (3.9.5) may then be combined to give

$$N_L = \frac{N_T}{1 + \dfrac{N_U}{N_L}} = \frac{N_T(B_{UL}\bar{J} + A_{UL} + C_{UL})}{(B_{UL} + B_{LU})\bar{J} + A_{UL} + C_{UL} + C_{LU}} \,. \tag{3.9.6}$$

Substitution of this result into the opacity term $\kappa(z)$ given by equation (3.9.4) explicitly shows the non-linearity in $I(z, \Delta\nu, \mu)$ exhibited by equation (3.9.3).

Consequently, the solution to the full equation of radiative transfer may only be obtained by an iterative procedure. As we shall see in later chapters, other non-linearities enter the problem when we consider multi-level atom situations and departures from complete re-distribution. The purpose of the present section, therefore, is to detail a procedure by which the solution to the non-linear equation of radiative transfer may be obtained.

3.9.1 The basic linearisation theory

The presentation of the method involves somewhat algebraically messy equations, particularly for the differential form of the transfer equation, and this tends to obscure the thrust behind the technique. We therefore begin the exposition of the iterative procedure by considering the solution to an equation of the form

$$f(x) = 0. \tag{3.9.7}$$

The technique we shall use is equivalent to the Newton–Raphson [51] method where we guess an initial solution x_0 and derive a correction δx_0. Our first iteration then yields

$$x_1 = x_0 + \delta x_0, \tag{3.9.8}$$

where x_1, hopefully, is a more accurate solution to $f(x) = 0$ than is x_0. The process is then repeated until the solution converges to the desired accuracy. The correction δx_0 is obtained by noting that

$$f(x_1) \approx 0,$$

i.e.

$$f(x_0) + \delta x_0 f'(x_0) + \cdots \approx 0. \tag{3.9.9}$$

Neglecting the higher order terms $O(\delta x_0)^2$, we find

$$x_1 = x_0 - \frac{f(x_0)}{f'(x_0)} \,. \tag{3.9.10}$$

It can be shown that a necessary and sufficient condition for convergence of the above iterative scheme is

$$\left|\left[\frac{\mathrm{d}}{\mathrm{d}x}\left(\frac{f(x)}{f'(x)}\right)\right]_{x=\xi}\right| < 1, \tag{3.9.11}$$

where ξ is in the neighbourhood of the solution. Note that the procedure does not always converge. However, when convergence is attained, it is quadratic, i.e. $\delta x_n \sim (\delta x_{n-1})^2$ and convergence is therefore extremely rapid.

We may use the same approach to solve the equation

$$Ly(x) = f(y(x), x), \tag{3.9.12}$$

where L is an operator (integral, differential or integro-differential) operating on the unknown $y(x)$, and where $f(y, x)$ is a known function of y and x.

Again, we guess a solution $y_0(x)$ for all x in the domain of interest and determine a correction $\delta y_0(x)$ by writing

$$\begin{aligned} L[y_0(x) + \delta y_0(x)] &= f(y_0(x) + \delta y_0(x), x) \\ &= f(y_0(x), x) + \delta y_0(x)\frac{\partial f}{\partial y_0}(y_0(x), x) \\ &\quad + O(\delta y_0(x)). \end{aligned} \tag{3.9.13}$$

Consequently, a correction $\delta y_0(x)$ may be obtained from the now *linear* equation

$$L\delta y_0(x) = \delta y_0(x)\frac{\partial f}{\partial y_0}(y_0(x), x) - Ly_0(x) + f(y_0(x), x). \tag{3.9.14}$$

This δy_0, once determined, is then used to construct the first iteration solution $y_1(x)$ and the process is continued. The procedure outlined above always produces a linear equation and, for this reason, is referred to as a linearisation.

If we have two equations of the form (3.9.12) in the two unknowns $y^{(1)}(x)$ and $y^{(2)}(x)$, i.e.

$$L_i y^{(i)}(x) = f_i(y^{(1)}(x), y^{(2)}(x), x), \tag{3.9.15}$$

for $i = 1, 2$, we may write

$$y_1^{(i)}(x) = y_0^{(i)}(x) + \delta y_0^{(i)}(x), \tag{3.9.16}$$

where the corrections $\delta y_0^{(i)}(x)$ to the initial guesses $y_0^{(i)}(x)$ may be obtained from the two equations

$$\begin{aligned} L_i\delta y_0^{(i)}(x) &= \delta y_0^{(1)}(x)\left[\frac{\partial f_i}{\partial y^{(1)}}(y^{(1)}, y^{(2)}, x)\right]_{(y_0^{(1)}, y_0^{(2)})} \\ &\quad + \delta y_0^{(2)}(x)\left[\frac{\partial f_i}{\partial y^{(2)}}(y^{(1)}, y^{(2)}, x)\right]_{(y_0^{(1)}, y_0^{(2)})} \\ &\quad - L_i y_0^{(i)}(x) + f_i(y_0^{(1)}, y_0^{(2)}, x), \end{aligned} \tag{3.9.17}$$

for $i = 1, 2$ where we have again used a Taylor series expansion about $y_0^{(1)}(x)$ and $y_0^{(2)}(x)$. Of course, the resulting equations (3.9.17) are linear in the unknowns to be determined.

Finally, consider the case when the operator L_i is a function of $y^{(i')}(x)$ where $i = 1, 2; i' = 2, 1$. Equations (3.9.15) then have the form

$$L_1(y^{(2)})y^{(1)} = f_1(y^{(1)}, y^{(2)}, x), \tag{3.9.18}$$

and

$$L_2(y^{(1)})y^{(2)} = f_2(y^{(1)}, y^{(2)}, x), \tag{3.9.19}$$

where, in equation (3.9.18), for example, the operator L_1, which is a function of $y^{(2)}$, operates on $y^{(1)}$. If we just consider equation (3.9.18), we have

$$\begin{aligned} L_1(y_0^{(2)} + \delta y_0^{(2)})(y_0^{(1)} + \delta y_0^{(1)}) \equiv{}& L_1(y_0^{(2)})y_0^{(1)} + L_1(y_0^{(2)})\delta y_0^{(1)} \\ &+ \delta y_0^{(2)} \left[\frac{\partial L_1 y}{\partial y^{(2)}}^{(2)}\right]_{y_0^{(2)}} y_0^{(1)} \\ &+ O(\delta y_0^{(1)}\delta y_0^{(2)}, (\delta y_0^{(1)})^2, (\delta y_0^{(2)})^2). \end{aligned} \tag{3.9.20}$$

Thus, if we define the new operator L_3 such that

$$L_3(y^{(2)}) \equiv \frac{\partial L_1}{\partial y^{(2)}} (y^{(2)}), \tag{3.9.21}$$

the linearisation of equation (3.9.18) eventually yields

$$\begin{aligned} L_1(y_0^{(2)})\delta y_0^{(1)} = {}& \delta y_0^{(1)} \left[\frac{\partial f_1}{\partial y^{(1)}} (y^{(1)}, y^{(2)}, x)\right]_{(y_0^{(1)}, y_0^{(2)})} \\ &+ \partial y_0^{(2)} \left[\frac{\partial f_1}{\partial y^{(2)}} (y^{(1)}, y^{(2)}, x)\right]_{(y_0^{(1)}, y_0^{(2)})} \\ &- \delta y_0^{(2)} L_3(y_0^{(2)})y_0^{(1)} - L_1(y_0^{(2)})y_0^{(1)} + f_1(y_0^{(1)}, y_0^{(2)}, x). \end{aligned} \tag{3.9.22}$$

A similar equation may be obtained from (3.9.19).

This completes the basic theory of the linearisation process.

3.9.2 Linearisation of the transfer equation [5, 84]

One could substitute N_L given by equation (3.9.6) into the opacity $\kappa(z)$ (equation (3.9.4)) so that the transfer equation would then involve only the one unknown $I(z, \Delta\nu, \mu)$. However, the problems we shall consider in later chapters involve somewhat more complicated terms both in the transfer equation and the rate equation (3.9.2), and thus the algebraic manipulations required to eliminate all the unknowns except $I(z, \Delta\nu, \mu)$ from the transfer equation become prohibitive. Here, therefore, we consider the separate linearisation of the three basic equations defining the three coupled unknowns $I(z, \Delta\nu, \mu)$, N_L and N_U, viz. equations (3.9.1), (3.9.2) and (3.9.5). Equation (3.9.5) is particularly simple and could be used to immediately eliminate either N_U or N_L from the other two equations. For purposes of exposition, however, we retain (3.9.5) as a separate equation.

Following the linearisation procedure outlined in the preceding sub-section, we guess the initial values I_0, $N_L^{(0)}$ and $N_U^{(0)}$ and then proceed in the determination of their corrections δI_0, $\delta N_L^{(0)}$ and $\delta N_U^{(0)}$. The three equations (3.9.1), (3.9.2) and (3.9.5) then yield

$$\mu\frac{\partial\delta I_0}{\partial z}=\frac{h\nu_0}{4\pi}\phi(\Delta\nu)\{-N_L^{(0)}B_{LU}\delta I_0+N_U^{(0)}B_{UL}\delta I_0$$
$$-B_{LU}I_0\delta N_L^{(0)}+B_{UL}I_0\delta N_U^{(0)}+A_{UL}\delta N_U^{(0)}\}$$
$$-\mu\frac{\partial I_0}{\partial z}+\frac{h\nu_0}{4\pi}\phi(\Delta\nu)\{-N_L^{(0)}B_{LU}I_0+N_U^{(0)}B_{UL}I_0+N_U^{(0)}A_{UL}\},\tag{3.9.23}$$

$$N_L^{(0)}B_{LU}\delta\bar{J}_0+(B_{LU}\bar{J}_0+C_{LU})\delta N_L^{(0)}+N_L^{(0)}(B_{LU}\bar{J}_0+C_{LU})$$
$$=N_U^{(0)}B_{UL}\delta\bar{J}_0+(B_{UL}\bar{J}_0+C_{UL}+A_{UL})\delta N_U^{(0)}$$
$$+N_U^{(0)}(B_{UL}\bar{J}_0+C_{UL}+A_{UL}),\tag{3.9.24}$$

and

$$N_L^{(0)}+N_U^{(0)}+\delta N_L^{(0)}+\delta N_U^{(0)}=N_T,\tag{3.9.25}$$

where

$$\delta\bar{J}_0=\tfrac{1}{2}\int_{-\infty}^{\infty}\phi(\Delta\nu)\,d(\Delta\nu)\int_{-1}^{1}d\mu\delta I_0(z,\Delta\nu,\mu).\tag{3.9.26}$$

However, since N_T is known, we may readily take our initial guesses $N_L^{(0)}$ and $N_U^{(0)}$ such that

$$N_L^{(0)}+N_U^{(0)}=N_T,$$

i.e.

$$\delta N_L^{(0)}+\delta N_U^{(0)}=0.\tag{3.9.27}$$

Equations (3.9.23), (3.9.24) and (3.9.27) may now be solved for the corrections δI_0, $\delta N_L^{(0)}$ and $\delta N_U^{(0)}$.

In general, certain algebraic simplifications are possible. Equations (3.9.24) and (3.9.27) may be used to eliminate $\delta N_U^{(0)}$, for example, viz.

$$\delta N_L^{(0)}=\frac{(N_U^{(0)}B_{UL}-N_L^{(0)}B_{LU})\delta\bar{J}_0}{(B_{LU}+B_{UL})\bar{J}_0+C_{LU}+C_{UL}+A_{UL}}+q_1,\tag{3.9.28}$$

where

$$q_1=\frac{N_U^{(0)}(B_{UL}\bar{J}_0+C_{UL}+A_{UL})-N_L^{(0)}(B_{LU}\bar{J}_0+C_{LU})}{(B_{LU}+B_{UL})\bar{J}_0+C_{LU}+C_{UL}+A_{UL}}.\tag{3.9.29}$$

The q_1 term has been isolated since it contains only the zeroth order quantities $\bar{J}_0$, $N_L^{(0)}$ and $N_U^{(0)}$ and none of the zeroth order corrections.

Next, we write

$$\kappa_0(z) = \frac{h\nu}{4\pi} N_L^{(0)} B_{LU} \left[1 - \frac{N_U^{(0)} B_{UL}}{N_L^{(0)} B_{LU}}\right] \phi(\Delta\nu), \tag{3.9.30}$$

so that equation (3.9.28) becomes

$$\mu \frac{\partial \delta I_0}{\partial z} = -\kappa_0(z)[\delta I_0 - \delta \mathscr{S}_0], \tag{3.9.31}$$

where the 'source function correction' has the form

$$\delta \mathscr{S}_0 = \frac{(B_{UL} I_0 + A_{UL})\delta N_U^{(0)} - B_{LU} I_0 \delta N_L^{(0)}}{N_L^{(0)} B_{LU} - N_U^{(0)} B_{UL}} + q_2, \tag{3.9.32}$$

with

$$q_2 = \frac{N_U^{(0)}(B_{UL} I_0 + A_{UL}) - N_L^{(0)} B_{LU} I_0 - \dfrac{4\pi\mu}{h\nu_0 \phi(\Delta\nu)} \dfrac{\partial I_0}{\partial z}}{N_L^{(0)} B_{LU} - N_U^{(0)} B_{UL}}. \tag{3.9.33}$$

Substitution of $\delta N_L^{(0)}$ from equation (3.9.28) into $\delta \mathscr{S}_0$ gives

$$\delta \mathscr{S}_0 = f(I_0, N_L^{(0)}, N_U^{(0)}) \delta \bar{J}_0 + q_3, \tag{3.9.34}$$

where

$$f(I_0, N_L^{(0)}, N_U^{(0)}) = \frac{(B_{LU} + B_{UL}) I_0 + A_{UL}}{(B_{LU} + B_{UL}) \bar{J}_0 + C_{LU} + C_{UL} + A_{UL}}, \tag{3.9.35}$$

and

$$q_3 = q_2 - \frac{(B_{LU} + B_{UL}) I_0 + A_{UL}}{N_L^{(0)} B_{LU} - N_U^{(0)} B_{UL}} \cdot q_1 \tag{3.9.36}$$

The iterative procedure is then as follows:

(i) guess values of I_0, $N_L^{(0)}$ and $N_U^{(0)}$ - one could use, for example, the corresponding LTE values $B_\nu(T)$, N_L^* and N_U^*;

(ii) solve the linearised transfer equation (3.9.31) for δI_0 with $\delta \mathscr{S}_0$ given by (3.9.34);

(iii) obtain $\delta \bar{J}_0$ using (3.9.26);

(iv) evaluate $\delta N_L^{(0)}$ from (3.9.28) thence $\delta N_U^{(0)}$ from (3.9.27);

(v) obtain

$$\left.\begin{aligned} I_1 &= I_0 + \delta I_0, \\ N_L^{(1)} &= N_L^{(0)} + \delta N_L^{(0)}, \\ &\text{and} \\ N_U^{(1)} &= N_U^{(0)} + \delta N_U^{(0)}, \end{aligned}\right\} \tag{3.9.37}$$

(vi) re-evaluate $\kappa_0(z), f, q_1, q_2$ and q_3, and repeat the process until convergence is found.

Note that, as the solution converges, q_1 and q_2 (thence q_3) tend to zero since their numerators are merely the terms appearing in the equation of statistical equilibrium and the radiative transfer equation respectively, and their denominators are clearly non-zero. Consequently, the corrections δI, δN_L and δN_U likewise tend to zero as expected. One can compare this with equation (3.9.10) where, as the solution converges, $f(x) \to 0$, thence $x_{n+1} \to x_n$, i.e. $\delta x \to 0$.

In the following section, we detail the formulation of the linearised radiative transfer equation (3.9.31) for the differential specification of the numerical method of solution.

3.9.3 *Differential equation formulation*

We now propose to solve equation (3.9.31) using the Feautrier technique. We begin by defining δI^+ and δI^- such that

$$\delta I^{\pm} = \delta I(z, \Delta\nu, \mu \gtrless 0), \tag{3.9.38}$$

together with $I^{\pm} = I(z, \Delta\nu, \mu \gtrless 0)$. Equations (3.9.31) and (3.9.34) then yield

$$\frac{\mu}{\kappa}\frac{\partial \delta I^+}{\partial z} = -\delta I^+ + f^+ \delta \bar{J} + q_3^+, \tag{3.9.39}$$

and

$$-\frac{\mu}{\kappa}\frac{\partial \delta I^-}{\partial z} = -\delta I^- + f^- \delta \bar{J} + q_3^-, \tag{3.9.40}$$

where, since we have assumed complete re-distribution, $\delta \bar{J}$ is independent of μ (and $\Delta\nu$). We have also introduced the quantities

$$f^{\pm} = f(I^{\pm}, N_L, N_U), \tag{3.9.41}$$

and

$$q_3^{\pm} = q_3(I^{\pm}, N_L, N_U). \tag{3.9.42}$$

If we now define

$$\delta\Phi = \tfrac{1}{2}(\delta I^+ + \delta I^-), \tag{3.9.43}$$

and

$$\delta\Psi = \tfrac{1}{2}(\delta I^+ - \delta I^-), \tag{3.9.44}$$

equations (3.9.39) and (3.9.40) yield

$$\left(\frac{\mu}{\kappa}\frac{\partial}{\partial z}\right)\delta\Phi = -\delta\Psi + \tfrac{1}{2}(f^+ - f^-)\delta\bar{J} + \tfrac{1}{2}(q_3^+ - q_3^-), \tag{3.9.45}$$

and

$$\left(\frac{\mu}{\kappa}\frac{\partial}{\partial z}\right)\delta\Psi = -\delta\Phi + \tfrac{1}{2}(f^+ + f^-)\delta\bar{J} + \tfrac{1}{2}(q_3^+ + q_3^-) \tag{3.9.46}$$

If, for ease in exposition, we write

$$F^{\pm} = \tfrac{1}{2}(f^+ \pm f^-), \tag{3.9.47}$$

and

$$Q_3^{\pm} = \tfrac{1}{2}(q_3^+ \pm q_3^-), \tag{3.9.48}$$

the above equations yield the required second order differential equation for $\delta\Phi$, viz.

$$\left(\frac{\mu}{\kappa}\frac{\partial}{\partial z}\right)^2 \delta\Phi = \delta\Phi - F^+\delta\bar{\Phi} - Q^+ + \frac{\mu}{\kappa}\frac{\partial}{\partial z}(F^-\delta\bar{\Phi} + Q^-), \tag{3.9.49}$$

where

$$\delta\bar{\Phi}(z) = \delta\bar{J}(z) = \int_{-\infty}^{\infty} \phi(\Delta\nu)\,\mathrm{d}(\Delta\nu) \int_0^1 \mathrm{d}\mu\,\delta\Phi(z, \Delta\nu, \mu). \tag{3.9.50}$$

Equation (3.9.49) is of the same form as that derived in section 3.2 except for the fourth term on the RHS involving F^- and Q^-. This fourth term does not present any difficulty – equation (3.9.49) can still be put into the standard matrix form

$$-\mathbf{A}_j\delta\mathbf{\Phi}_{j-1} + \mathbf{B}_j\delta\mathbf{\Phi}_j - \mathbf{C}_j\delta\mathbf{\Phi}_{j+1} = \mathbf{L}_j, \tag{3.9.51}$$

by appropriate differencing (see section 3.2.2). The matrices $\mathbf{A}_j$ and $\mathbf{C}_j$, which are purely diagonal in the ordinary Feautrier application, now become dense because of the non-zero F^- (and Q^-) terms.

The two required boundary conditions may be obtained from equation (3.9.45), i.e.

$$\left(\frac{\mu}{\kappa}\frac{\partial}{\partial z}\right) \delta\Phi = -\delta\Phi + F^-\delta\bar{\Phi} + Q^-, \tag{3.9.52}$$

at $z = 0$ whilst, as $z \to -\infty$,

$$\left(\frac{\mu}{\kappa}\frac{\partial}{\partial z}\right) \delta\Phi = \delta\Phi + F^-\delta\bar{\Phi} + Q^-. \tag{3.9.53}$$

In deriving the above two equations, we take $\delta I^- \equiv 0$ at $z = 0$ and $\delta I^+ \equiv 0$ as $z \to -\infty$ since no correction should need to be applied to any quantity, such as a boundary condition, which is accurately known. Indeed, the initial guesses $I(0, \Delta\nu, \mu < 0)$ and $I(-\infty, \Delta\nu, \mu > 0)$ are specified *a priori.*

The matrix equations at the boundaries have the form

$$\mathbf{B}_1\delta\mathbf{\Phi}_1 - \mathbf{C}_1\delta\mathbf{\Phi}_2 = \mathbf{L}_1, \tag{3.9.54}$$

and

$$-\mathbf{A}_N\delta\mathbf{\Phi}_{N-1} + \mathbf{B}_N\delta\mathbf{\Phi}_N = \mathbf{L}_N, \tag{3.9.55}$$

where one may use a Taylor series expansion to improve the accuracy of the differencing (see section 3.2.4).

Again, the recurrence solution is of the form

$$\delta\mathbf{\Phi}_j = \mathbf{U}_j + \mathbf{V}_j\delta\mathbf{\Phi}_{j+1}, \tag{3.9.56}$$

where

$$\mathbf{U}_j = (\mathbf{B}_j - \mathbf{A}_j\mathbf{V}_{j-1})^{-1}(\mathbf{L}_j + \mathbf{A}_j\mathbf{U}_{j-1}), \tag{3.9.57}$$

and

$$\mathbf{V}_j = (\mathbf{B}_j - \mathbf{A}_j\mathbf{V}_{j-1})^{-1}\mathbf{C}_j. \tag{3.9.58}$$

Clearly, the computer time and programming effort required to solve the above system for δI at *each* iteration in the linearisation procedure is effectively the same as that required for the ordinary Feautrier technique presented in section 3.2. Note, however, that the matrices $\mathbf{A}_j$, $\mathbf{B}_j$ and $\mathbf{C}_j$ change at each iteration because $F^{\pm}$ and $Q^{\pm}$ (which are dependent upon I, N_{L} and N_{U}) change and, thus, a separate set of matrix inversion is required at *each* iteration.

We conclude this section by briefly mentioning the stability of the above scheme. We have seen in section 3.3 that stability is attained under various conditions satisfied by the $\mathbf{U}_j$ and $\mathbf{V}_j$ quantities. These, in turn, led to other conditions which A_j, B_j and C_j must satisfy and, since the matrices $\mathbf{A}_j$ and $\mathbf{C}_j$ are now dense, it is even more difficult to carry the stability analysis over from the scalar to the matrix case. However, implicit in the stability analysis presented in section 3.3 was the fact that $0<\epsilon<1$, i.e. $0<1-\epsilon<1$. In the present context, this is equivalent to the requirement $F^+\in(0, 1)$. From equations (3.9.35) and (3.9.47) we have

$$F^+ = \frac{(B_{\mathrm{LU}}+B_{\mathrm{UL}})(I^+ + I^-)/2 + A_{\mathrm{UL}}}{(B_{\mathrm{LU}}+B_{\mathrm{UL}})\bar{J} + C_{\mathrm{LU}} + C_{\mathrm{UL}} + A_{\mathrm{UL}}}. \tag{3.9.59}$$

Clearly, $F^+>0$ whilst a sufficient condition for $F^+<1$ is

$$\tfrac{1}{2}(I^+ + I^-)<\bar{J}. \tag{3.9.60}$$

This last condition does not always hold since the LHS (3.9.60) is merely the average of the radiation field at the particular angles $\pm\mu$ at a particular frequency. For example, if we consider I^+ and I^- in the wings of the line at the surface of a model 1-dimensional semi-infinite atmosphere, we have $I^+\sim B_\nu(T)$ and $I^-\sim 0$, i.e. we find $\max(I^+ + I^-)/2\sim B_\nu(T)/2$. However, we have seen that the source function at the surface for constant property media is $\sqrt{\epsilon}B_\nu(T)$. Thus, at the surface $\min \bar{J}\sim\sqrt{\epsilon}B_\nu(T)$.

We therefore see that inequality (3.9.60) is violated for wing photons at the surface for $\epsilon<\frac{1}{4}$.

A necessary condition for $F^+<1$ is just

$$\frac{I^+ + I^-}{2}<\bar{J}+\frac{C_{\mathrm{LU}}+C_{\mathrm{UL}}}{B_{\mathrm{LU}}+B_{\mathrm{UL}}}, \tag{3.9.61}$$

where

$$\frac{C_{LU}+C_{UL}}{B_{LU}+B_{UL}}=\frac{2h\nu^3}{c^2}\cdot\frac{C_{UL}}{A_{UL}}\cdot\frac{g_L+g_U e^{-h\nu/kT}}{g_L+g_U},$$

$$=\frac{\epsilon B_\nu(T)}{1-\epsilon}\,e^{h\nu/kT}\cdot\frac{g_L+g_U e^{-h\nu/kT}}{g_L+g_U}, \tag{3.9.62}$$

where, in obtaining this last equation, we have used (1.6.17), viz.

$$\epsilon=\frac{C_{UL}(1-e^{-h\nu/kT})}{A_{UL}+C_{UL}(1-e^{-h\nu/kT})}. \tag{1.6.17}$$

Since $e^{-h\nu/kT}\ll 1$ for cases of astrophysical interest generally and the statistical weights g_L and g_U are of order unity, inequality (3.9.61) is approximately

$$\frac{I^+ + I^-}{2}<\bar{J}+\frac{\epsilon B_\nu(T)}{1-\epsilon}\,e^{h\nu/kT}. \tag{3.9.63}$$

Clearly, then, the larger the value of ϵ, the greater the chance of stability. Further, since $I^+, I^- \leqslant B_\nu(T)$ at all frequencies, angles and positions for model 2-level atom situations, a sufficient condition for the last inequality to be satisfied is

$$\frac{\epsilon}{1-\epsilon}>e^{-h\nu/kT}, \tag{3.9.64}$$

which, using $e^{-h\nu/kT}\ll 1$, yields

$$\epsilon>e^{-h\nu/kT}. \tag{3.9.65}$$

This last condition is, however, somewhat more stringent than is necessary for stability. In practice, stability can still be achieved even if (3.9.65) is violated.

3.9.4 *Operator linearisation*

Section 3.9.2 developed the linearisation of the equation of radiative transfer, the statistical equilibrium equation and the total density equation. Section 3.9.3 then presented the re-structuring of these linearised equations in the form enabling numerical solution by differential equation techniques.

An alternative method is to first re-structure the unlinearised equations in their integral and/or differential form, then perform the linearisation. As an example of the procedure that can be followed, we consider the second order differential equation using the Feautrier technique, viz.

$$\left(\frac{\mu}{\kappa}\frac{\partial}{\partial z}\right)^2\Phi=\Phi-S, \tag{3.9.66}$$

given the two boundary conditions

$$\left(\frac{\mu}{\kappa}\frac{\partial}{\partial z}\right)\Phi = -\Phi + I_{\mathrm{inc}}, \tag{3.9.67}$$

at $z = 0$ whilst, as $z \to -\infty$,

$$\left(\frac{\mu}{\kappa}\frac{\partial}{\partial z}\right)\Phi = \Phi - B_\nu(T). \tag{3.9.68}$$

One could now perform the linearisation as discussed in section 3.9.1 (see, for example, equation (3.9.22)).

However, it has been more common in the literature to proceed one step further before linearising these equations. We first write them in difference form. We illustrate the procedure on equation (3.9.67) and, for ease in exposition, do not introduce the Taylor series expansion beyond $\Delta_1 = z_2 - z_1$ (see section 3.2.4). Equation (3.9.67) then has the difference form

$$\frac{2\mu(\Phi_2 - \Phi_1)}{(\kappa_1 + \kappa_2)\Delta_1} = -\Phi_1 + I_{\mathrm{inc}}, \tag{3.9.69}$$

where

$$\kappa_j = \frac{h\nu_0}{4\pi}[N_{\mathrm{L}j}B_{\mathrm{LU}} - N_{\mathrm{U}j}B_{\mathrm{UL}}]\phi(\Delta\nu) \tag{3.9.70}$$

- again, we take $\phi(\Delta\nu)$ to be depth independent.

We now linearise equation (3.9.69) using the theory developed in section 3.9.1 (leading up to equation (3.9.22)). However, we note that equation (3.9.69) contains the six quantities Φ_1, Φ_2, $N_{\mathrm{L}1}$, $N_{\mathrm{L}2}$, $N_{\mathrm{U}1}$ and $N_{\mathrm{U}2}$ to be linearised, i.e. we have

$$\begin{aligned}\frac{2\mu(\Phi_2^{(0)} - \Phi_1^{(0)})}{(\kappa_1^{(0)} + \kappa_2^{(0)})\Delta_1} &+ \frac{2\mu(\delta\Phi_2^{(0)} - \delta\Phi_1^{(0)})}{(\kappa_1^{(0)} + \kappa_2^{(0)})\Delta_1} \\ &+ \frac{2\mu(\Phi_2^{(0)} - \Phi_1^{(0)})}{\Delta_1}\sum_{j=1}^{2}\left\{\frac{\partial}{\partial N_{\mathrm{L}j}}\left(\frac{1}{\kappa_1+\kappa_2}\right)\delta N_{\mathrm{L}j}^{(0)}\right. \\ &\left.+ \frac{\partial}{\partial N_{\mathrm{U}j}}\left(\frac{1}{\kappa_1+\kappa_2}\right)\delta N_{\mathrm{U}j}^{(0)}\right\}_{(N_{\mathrm{L}j}^{(0)},\, N_{\mathrm{U}j}^{(0)})} \\ &= -\Phi_1^{(0)} - \delta\Phi_1^{(0)} + I_{\mathrm{inc}}.\end{aligned} \tag{3.9.71}$$

Equation (3.9.70) yields

$$\frac{\partial}{\partial N_{\mathrm{L}j}}\left(\frac{1}{\kappa_1+\kappa_2}\right) = \frac{-1}{(\kappa_1+\kappa_2)^2}\frac{h\nu_0}{4\pi}B_{\mathrm{LU}}\phi(\Delta\nu), \tag{3.9.72}$$

and

$$\frac{\partial}{\partial N_{\mathrm{U}j}}\left(\frac{1}{\kappa_1+\kappa_2}\right) = \frac{1}{(\kappa_1+\kappa_2)^2}\frac{h\nu_0}{4\pi}B_{\mathrm{UL}}\phi(\Delta\nu). \tag{3.9.73}$$

Substitution of these last two results into equation (3.9.71) then completes the linearisation of (3.9.69).

One can immediately see that the process becomes somewhat lengthy (albeit simple) when one attempts the linearisation of the difference form of the second order equation (3.9.66) because:

(i) there are now nine quantities Φ, N_L and N_U at $j-1, j$ and $j+1$ to be linearised, and

(ii) the denominator in (3.9.66) contains terms involving the product of κ_j with κ_{j-1}, κ_j and κ_{j+1}.

3.10 Newton–Raphson accelerated Λ operators

We complete this chapter by applying the linearisation techniques of the previous section 3.9 to the Λ operator detailed in section 3.1. In particular, we develop a method which only requires a trivially small change to any computer code written for the extremely slow Λ iteration technique, but which exhibits a rapid rate of convergence.

We start by re-writing equation (3.1.5), viz.

$$S(\tau) = (1-\epsilon)\Lambda S + \epsilon B_\nu(T), \tag{3.10.1}$$

where the integral operator Λ is given by equation (3.1.8). If we now construct a depth grid $\{\tau_j; j=1, N\}$, then equation (3.10.1) becomes

$$S(\tau_j) = (1-\epsilon_j) \sum_{j'=1}^{N} w_{j'j} S(\tau_{j'}) + [\epsilon B_\nu(T)]_j, \tag{3.10.2}$$

as discussed in section 3.8. We next write $S_j = S(\tau_j)$ and define the function $f(S_j)$ by

$$f(S_j) = S_j - (1-\epsilon_j) \sum_{j'=1}^{N} w_{j'j} S_{j'} - [\epsilon B_\nu(T)]_j. \tag{3.10.3}$$

Clearly, we wish to solve the equation

$$f(S_j) = 0. \tag{3.10.4}$$

Thus, using the Newton–Raphson procedure outlined by equation (3.9.10), we have the $(l+1)$th iterative solution

$$\begin{aligned} S_j^{(l+1)} &= S_j^{(l)} - \frac{f(S_j^{(l)})}{\partial f(S_j^{(l)})/\partial S_j^{(l)}} \\ &= S_j^{(l)} - \frac{S_j^{(l)} - (1-\epsilon_j) \sum_{j'=1}^{N} w_{j'j} S_{j'} - [\epsilon B_\nu(T)]_j}{1-(1-\epsilon_j) w_{jj}} \\ &= \frac{(1-\epsilon_j) \sum_{j' \neq j}^{N} w_{j'j} S_{j'}^{(l)} + [\epsilon B_\nu(T)]_j}{1-(1-\epsilon_j) w_{jj}}. \end{aligned} \tag{3.10.5}$$

One can use quadratic segments as in section 3.8.2 (c) to evaluate the weights $w_{j'j}$ – note that this step is required for Λ iteration, the integral equation technique of section 3.8.1 and, effectively, the perturbation approaches detailed in section 3.7 – then, using an initial guess of $\epsilon B_\nu(T)$, for example, iterate directly using the above scheme (3.10.5). Note, in particular, that one may include as many angle and frequency quadrature points in the evaluation of the $w_{j'j}$ as deemed necessary for the required accuracy. The resulting weights in this case are, of course, dependent only on the depth coordinate. The changes in any computer code written for the corresponding Λ iteration technique, viz.

$$S_j^{(l+1,\Lambda)} = (1-\epsilon_j)\sum_{j'=1}^{N} w_{j'j} S_{j'}^{(l,\Lambda)} + [\epsilon B_\nu(T)]_j, \tag{3.10.6}$$

are therefore clearly trivial and, since Λ iteration is by far the simplest method for solving the integro-differential radiative transfer equation, the above scheme (3.10.5) certainly has the important advantage of ease in development.

Trial calculations for semi-infinite geometries (convergence is even faster in slab geometries) display extremely rapid convergence deep within the atmosphere with a somewhat slower rate near the surface. We can examine the behaviour of this convergence more closely by considering a constant property medium and, for the purposes of this discussion, evaluate an approximate set of $w_{j'j}$ using the constant segments described in section 3.8.2 (a). However, we approach the problem a little differently here by constructing a second depth grid $\{\tau_{j+(1/2)}; j=1, N-1\}$ where $\tau_{j+(1/2)} \in (\tau_j, \tau_{j+1})$. We then have for $\tau_{j'} < \tau_j$

$$\begin{aligned} w_{j'j} &= \tfrac{1}{2}\int_{-\infty}^{\infty} \phi(\Delta\nu)\,\mathrm{d}(\Delta\nu) \int_0^1 \mathrm{d}\mu \int_{\tau_{j'-(1/2)}}^{\tau_{j'+(1/2)}} \mathrm{e}^{-(\tau_j-\tau')\phi(\Delta\nu)/\mu} \\ &\quad \times \frac{\phi(\Delta\nu)}{\mu}\,\mathrm{d}\tau' \\ &= \tfrac{1}{2}\int_{-\infty}^{\infty} \phi(\Delta\nu)\,\mathrm{d}(\Delta\nu) \int_0^1 \mathrm{d}\mu\, \mathrm{e}^{-(\tau_j-\tau_{j'+(1/2)})\phi(\Delta\nu)/\mu} \\ &\quad \times [1-\mathrm{e}^{-(\tau_{j'+(1/2)}-\tau_{j'-(1/2)})\phi(\Delta\nu)/\mu}], \end{aligned} \tag{3.10.7}$$

whereas, for $\tau_{j'} > \tau_j$, we similarly find

$$\begin{aligned} w_{j'j} &= \tfrac{1}{2}\int_{-\infty}^{\infty} \phi(\Delta\nu)\,\mathrm{d}(\Delta\nu) \int_0^1 \mathrm{d}\mu\, \mathrm{e}^{-(\tau_{j'-(1/2)}-\tau_j)\phi(\Delta\nu)/\mu} \\ &\quad \times [1-\mathrm{e}^{-(\tau_{j'+(1/2)}-\tau_{j'-(1/2)})\phi(\Delta\nu)/\mu}]. \end{aligned} \tag{3.10.8}$$

Finally, w_{jj} satisfies

$$w_{jj} = \tfrac{1}{2}\int_{-\infty}^{\infty} \phi(\Delta\nu)\,\mathrm{d}(\Delta\nu) \int_0^1 \mathrm{d}\mu \left[\int_{\tau_j}^{\tau_{j+(1/2)}} \mathrm{e}^{-(\tau'-\tau_j)\phi(\Delta\nu)/\mu}\right.$$

$$\times \frac{\phi(\Delta\nu)}{\mu}\,\mathrm{d}\tau' + \int_{\tau_{j-(1/2)}}^{\tau_j} \mathrm{e}^{-(\tau_j-\tau')\phi(\Delta\nu)/\mu}\,\frac{\phi(\Delta\nu)}{\mu}\,\mathrm{d}\tau'\Bigg]$$

$$= 1 - \tfrac{1}{2}\int_{-\infty}^{\infty} \phi(\Delta\nu)\,\mathrm{d}(\Delta\nu) \int_0^1 \mathrm{d}\mu\,[\mathrm{e}^{-(\tau_{j+(1/2)}-\tau_j)\phi(\Delta\nu)/\mu}$$

$$+ \mathrm{e}^{-(\tau_j-\tau_{j-(1/2)})\phi(\Delta\nu)/\mu}]. \qquad (3.10.9)$$

It is worth noting, before proceeding, that even for physically realistic computations involving non-constant property media, one can structure the atmosphere into slabs of thickness $\tau_{j+(1/2)} - \tau_{j-(1/2)}$ in which the macroscopic properties such as density and temperature of each slab can be taken as constant and given by the appropriate value at τ_j. This is, in effect, the procedure one adopts anyway when discretising each coordinate space. Consequently, allowing the source function $S(\tau)$ to be represented as a quadratic function of τ for $\tau \in (\tau_{j-(1/2)}, \tau_{j+(1/2)})$ and using the values S_{j-1}, S_j and S_{j+1} to obtain this quadratic, the weights $w_{j'j}$ which one should, in practice, use have a form similar to the simplified $w_{j'j}$ given by equations (3.10.7) through (3.10.9). Simple integration by parts then determines all the τ integrals; however, numerical quadrature is still required for the integrations over $\Delta\nu$ and μ. The details are left to the reader.

It is not unreasonable to construct a semi-infinite medium τ grid with $\tau_1 = 0$, $\tau_2 = 0.01$ and, for $j > 2$, $\tau_{j+1} = 2\tau_j$. The subsequent large differences $\tau_{j'+(1/2)} - \tau_j$ and $\tau_j - \tau_{j'-(1/2)}$, for large τ_j, appearing in equations (3.10.7) through (3.10.9), then stipulate that $w_{jj} \approx 1$ and $w_{j'j} \approx 0$ for $j' \neq j$. Substitution of these values into the iterative scheme (3.10.5) immediately yields $S_j \approx B_\nu(T)$ for large τ_j thereby illustrating the exhibited rapid convergence deep within the medium. At the surface $\tau_{j+(1/2)} - \tau_j$ and $\tau_j - \tau_{j-(1/2)}$ are somewhat smaller so that $w_{jj} \ll 1$. Thus, the denominator in the iterative scheme (3.10.5) remains at approximately unity here. However, the rapid convergence of $S(\tau)$ to $B_\nu(T)$ for large τ rapidly affects the summation term in the numerator of equation (3.10.5) and, although convergence is not as rapid at the surface, it can certainly be considered satisfactory. For example, computations involving model semi-infinite geometries with $\epsilon = 10^{-4}$ (thus necessitating more than $\epsilon^{-1} = 10^4$ Λ iterations using scheme (3.10.6) - see section 3.1.1) exhibit convergence to 0.01% after two iterations at $\tau = 5 \times 10^5$, even when using an initial guess of $\epsilon B_\nu(T)$ for all τ, whereas convergence to 1% is attained after four iterations at $\tau = 10^4$, 12 iterations at $\tau = 10^3$, 20 iterations at $\tau = 10^2$, 26 iterations at $\tau = 10$ and 29 iterations at the surface. Notice that convergence is quadratic for the Newton–Raphson process. Convergence to three significant figures at the surface takes less than 40 iterations. Consequently, solutions of this accuracy may be obtained more quickly using the above iterative scheme than when using the integral equation formulation requiring matrix inversion discussed in section 3.8.1. Indeed, these Newton–Raphson accelerated Λ operators may be used in conjunction with any of methods discussed in the following chapters relating to more complicated physical models.

4

Extension to model multi-level atoms

We considered model atoms in the last chapter consisting of just two levels in order to facilitate the basic understanding of the interaction of radiation with matter. Here, we wish to remove this simplification by first including a third level in the model atom where, now, three quite distinct physical situations can be discussed. We then derive the general multi-level atom source function. The chapter is completed with a brief discussion of the Feautrier technique applied to these models and the development of a 'benchmark' exact solution for a particular case.

4.1 Model 3-level atoms

Figure 4.1 shows our new model atom with the three levels separated by energies $h\nu_{21}, h\nu_{31}$ and $h\nu_{32}$ where we take $\nu_{21} \neq \nu_{32} \neq \nu_{31}$.

If we wish to determine the specific radiation intensity I_{21} in the 2–1 transition, for example, we must derive the appropriate 2–1 radiative transfer equation. One does this in precisely the same manner as discussed in section 1.2 for the 2-level atom. We thence find

$$\frac{1}{c}\frac{\partial I_{21}}{\partial t} + (\Omega \cdot \nabla) I_{21} = -\kappa_{21}(I_{21} - S_{21}), \qquad (4.1.1)$$

Fig.4.1. A model atom having three bound electron states.

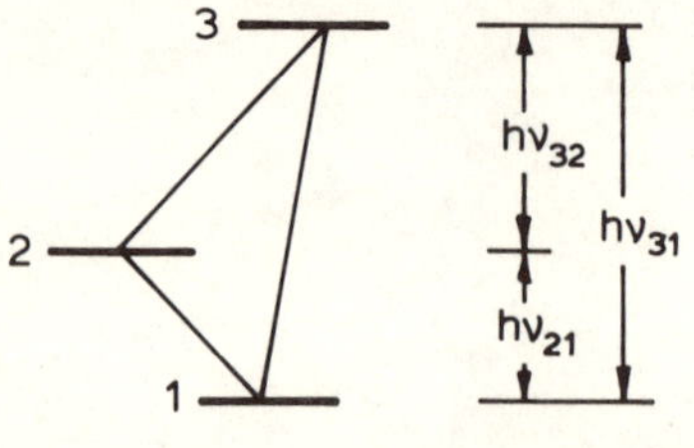

where

$$\kappa_{21} = \frac{h\nu_{21}}{4\pi} [N_1 B_{12} \phi_{12}(\Delta\nu) - N_2 B_{21} \psi_{12}(\Delta\nu)] \tag{4.1.2}$$

and

$$S_{21} = \frac{N_2 A_{21} j_{12}(\Delta\nu)}{N_1 B_{12} \phi_{12}(\Delta\nu) - N_2 B_{21} \psi_{12}(\Delta\nu)}. \tag{4.1.3}$$

Note that, in detailing the above equations, we have only considered the explicit radiative processes (absorption, stimulated emission, spontaneous emission) experienced by photons of energy $h\nu_{21}$ (and direction Ω). Level 3 does not explicitly enter the analysis here.

We thence proceed by evaluating the source function S_{21}. Again, we assume complete re-distribution such that

$$\phi_{12}(\Delta\nu) \equiv \psi_{12}(\Delta\nu) \equiv j_{12}(\Delta\nu). \tag{4.1.4}$$

The determination of S_{21} is therefore reduced to the determination of the density ratio N_2/N_1 since

$$S_{21} = \frac{A_{21} N_2/N_1}{B_{12} - B_{21} N_2/N_1}. \tag{4.1.5}$$

Again, we shall consider a statistically steady state so that the number densities N_1 and N_2 do not change with time. Thus, we may balance all those processes which populate a given level with those that act to de-populate it. If we now define R_{ij} as the total time rate at which an atom with a bound electron in level i is excited to level j, we must have, for example,

$$R_{12} = B_{12} \bar{J}_{12} + C_{12}, \tag{4.1.6}$$

where $\bar{J}_{12}$ is the average radiation intensity in the 2–1 transition, viz.

$$\bar{J}_{12} = \frac{1}{4\pi} \int_{-\infty}^{\infty} \mathrm{d}(\Delta\nu) \int_{4\pi} \mathrm{d}\Omega\, \phi_{12} I(\mathbf{r}, \Delta\nu, \Omega, t). \tag{4.1.7}$$

Note that complete re-distribution stipulates $\bar{J}_{12} = \bar{J}_{21}$.

Similarly, we have the general results

$$R_{21} = A_{21} + B_{21} \bar{J}_{21} + C_{21}, \tag{4.1.8}$$

$$R_{13} = B_{13} \bar{J}_{31} + C_{13}, \tag{4.1.9}$$

$$R_{31} = A_{31} + B_{31} \bar{J}_{31} + C_{31}, \tag{4.1.10}$$

$$R_{23} = B_{23} \bar{J}_{32} + C_{23}, \tag{4.1.11}$$

and

$$R_{32} = A_{32} + B_{32} \bar{J}_{32} + C_{32}, \tag{4.1.12}$$

where we have also assumed complete re-distribution in the 3–1 and 3–2 transitions (separately).

Obviously, the values of the various spontaneous emission rate coefficients A_{ji}, and thence the corresponding stimulated emission and absorption rate coefficients B_{ji} and B_{ij}, depend upon the atom under consideration, but still satisfy the general relationships

$$\frac{A_{ji}}{B_{ji}} = \frac{2h\nu_{ji}^3}{c^2}, \quad g_i B_{ij} = g_j B_{ji}. \tag{4.1.13}$$

For model 3-level atoms as shown in figure 4.1, quantum selection rules usually result in at least one of the three values A_{21}, A_{31} or A_{32} being effectively zero. In this case, the corresponding radiation field would likewise not exist and this would then necessitate the elimination of the appropriate radiative terms in the above R_{ij} and R_{ji}. This is a trivial point. For the moment, we shall retain the use of the general R_{ij} and R_{ji} without recourse to its more explicit form.

The equation of a statistically steady state for level 1 may then be written as

$$N_1(R_{12} + R_{13}) = N_2 R_{21} + N_3 R_{31}. \tag{4.1.14}$$

The LHS (4.1.14) is the total rate of de-population of level 1 (due to excitations to levels 2 and 3) whilst the RHS (4.1.14) is the total rate of population of level 1 (due to de-excitations from levels 2 and 3).

Similarly, the corresponding equation for level 2 has the form

$$N_2(R_{21} + R_{23}) = N_1 R_{12} + N_3 R_{32}. \tag{4.1.15}$$

Equations (4.1.14) and (4.1.15) are two equations in the three unknowns N_1, N_2 and N_3. We could include a third equation by equating the rates of population and de-population of level 3, viz.

$$N_3(R_{31} + R_{32}) = N_1 R_{13} + N_2 R_{23}, \tag{4.1.16}$$

but this last result can be shown to be nothing more than a super-position of the previous two equations. Clearly, therefore, we may only determine ratios of the desired three unknowns. In particular, we find

$$\frac{N_1}{N_2} = \frac{R_{21} + R_{23} \cdot \dfrac{R_{31}}{R_{31} + R_{32}}}{R_{12} + R_{13} \cdot \dfrac{R_{32}}{R_{31} + R_{32}}}. \tag{4.1.17}$$

The result presented in this last equation is not surprising. It we re-write it as

$$N_1\left[R_{12} + R_{13} \cdot \frac{R_{32}}{R_{31} + R_{32}}\right] = N_2\left[R_{21} + R_{23} \cdot \frac{R_{31}}{R_{31} + R_{32}}\right], \tag{4.1.18}$$

we see that the first term on the LHS (i.e. $N_1 R_{12}$) is the total rate of *direct* excitation from level 1 to level 2, whilst the second term is the total rate of excitation from level 1 to level 2 *via level* 3. Note that the ratio $R_{32}/(R_{31} + R_{32})$ is the probability, given a bound electron in level 3, of a de-excitation event yielding a bound electron in level 2. Thus, $R_{13}R_{32}/(R_{31} + R_{32})$ is the composite rate of excitation from level 1 to level 3 thence from level 3 to level 2. Clearly, then, the LHS (4.1.18) is the total rate of excitation (direct and indirect) from level 1 to level 2 whilst the RHS (4.1.18) represents the reverse process.

To simplify the following algebra, and to take into account the composite rates discussed above, we define

$$R_{132} = R_{13}\,\frac{R_{32}}{R_{31}+R_{32}}\,, \qquad R_{231} = R_{23}\,\frac{R_{31}}{R_{31}+R_{32}} \tag{4.1.19}$$

so that

$$\frac{N_2}{N_1} = \frac{R_{12}+R_{132}}{R_{21}+R_{231}}\,. \tag{4.1.20}$$

Substitution of these results into equation (4.1.5) for S_{21} yields;

$$S_{21} = \frac{\bar{J}_{21} + \dfrac{g_1}{g_2 B_{21}}(C_{12}+R_{132})}{1 + \dfrac{C_{21}}{A_{21}}\left(1-\dfrac{g_1 C_{12}}{g_2 C_{21}}\right) + \dfrac{R_{231}}{A_{21}}\left(1-\dfrac{g_1 R_{132}}{g_2 R_{231}}\right)}\,. \tag{4.1.21}$$

Clearly, if there was no level 3, the composite rates R_{132} and R_{231} would be zero, and the form for S_{21} would then be identical to that obtained in section 1.6 (equation (1.6.4)) for the 2-level atom source function as one would expect. Further, it is interesting to note the parallel occurrence of R_{132} and C_{12} (and, similarly, R_{231} and C_{21}) in the above source function. Indeed, if we put

$$\mathscr{C}_{12} = C_{12} + R_{132}\,, \quad \mathscr{C}_{21} = C_{21} + R_{231}\,, \tag{4.1.22}$$

equation (4.1.21) becomes

$$S_{21} = \frac{\bar{J}_{21} + \dfrac{g_1 \mathscr{C}_{12}}{g_2 B_{21}}}{1 + \dfrac{\mathscr{C}_{21}}{A_{21}}\left(1-\dfrac{g_1 \mathscr{C}_{12}}{g_2 \mathscr{C}_{21}}\right)}\,, \tag{4.1.23}$$

and this has precisely the same structure as the 2-level atom source function (1.6.4). As we shall see, this parallel occurrence has a quite straightforward physical interpretation.

We next recall the assumption of LTE collisional excitation and de-excitation events used in section 1.6, viz.

$$N_1^* C_{12} = N_2^* C_{21}\,, \tag{4.1.24}$$

where N_1^* and N_2^* are the respective LTE number densities for levels 1 and 2. One could possibly argue that this latter expression would not be appropriate for a model 3-level atom since we should have an equation which takes into account the third level in much the same manner as given by our statistically steady state equations (4.1.14–16), i.e.

$$N_1^*(C_{12} + C_{13}) = N_2^* C_{21} + N_3^* C_{31}\,, \tag{4.1.25}$$

However, as stated in section 1.7, LTE configurations may be mathematically described by detailed balance (see equations (1.7.4) and (1.7.5)). Thus, not only would we have equation (4.1.24) above, but also

$$N_1^* C_{13} = N_3^* C_{31}. \tag{4.1.26}$$

It is clear that equation (4.1.25) is still satisfied by these 'more detailed' LTE equations.

Thus, if we assume equation (4.1.24) and, as in section 1.6 for the 2-level atom source function, define

$$\epsilon'_{21} = \frac{C_{21}}{A_{21}}\left(1 - \frac{g_1 C_{12}}{g_2 C_{21}}\right), \tag{4.1.27}$$

we have

$$\epsilon'_{21} = \frac{C_{21}}{A_{21}}\,(1 - \mathrm{e}^{-h\nu_{21}/kT}) = \frac{C_{21}}{A_{21} + B_{21} B_{\nu_{21}}(T)}, \tag{4.1.28}$$

where

$$B_{\nu_{21}}(T) = \frac{2h\nu_{21}^3/c^2}{\mathrm{e}^{h\nu_{21}/kT} - 1}. \tag{4.1.29}$$

Further, if we put

$$\epsilon'_{231} = \frac{R_{231}}{A_{21}}\left(1 - \frac{g_1 R_{132}}{g_2 R_{231}}\right), \tag{4.1.30}$$

and

$$[\epsilon B_\nu(T)]_{231} = \frac{g_1 R_{132}}{g_2 B_{21}}, \tag{4.1.31}$$

then we find

$$S_{21} = \frac{\bar{J}_{21} + \epsilon'_{21} B_{\nu_{21}}(T) + [\epsilon B_\nu]_{231}}{1 + \epsilon'_{21} + \epsilon'_{231}}. \tag{4.1.32}$$

Let us quickly recall the physical interpretation of the terms appearing in the 2-level atom source function. There we had

$$S_{21}\ (\text{2-level}) = \frac{\bar{J}_{21} + \epsilon'_{21} B_{\nu_{21}}(T)}{1 + \epsilon'_{21}}, \tag{4.1.33}$$

which, using equation (4.1.28) and

$$\epsilon'_{21} = \frac{C_{12}}{B_{21} B_{\nu_{21}}(T)}, \tag{4.1.34}$$

becomes

$$S_{21}\ (\text{2-level}) = \frac{A_{21} + B_{21}B_{\nu_{21}}(T)}{A_{21} + B_{21}B_{\nu_{21}}(T) + C_{21}} \cdot \bar{J}_{21} + \frac{C_{12}}{B_{21}B_{\nu_{21}}(T) + C_{12}} \cdot B_{\nu_{21}}(T). \tag{4.1.35}$$

The first term on the RHS (4.1.35) represents the fraction of the absorbed radiation field $\bar{J}_{21}$ which is re-emitted (as opposed to being destroyed by collisional de-excitation). The second term is simply the source of photons due to collisional excitation. We write these terms as $(1 - \epsilon_{21})\bar{J}_{21}$, or $\bar{J}_{21}/(1 + \epsilon'_{21})$, and $\epsilon_{21}B_{\nu_{21}}(T)$, or $\epsilon'_{21}B_{\nu_{21}}(T)/(1 + \epsilon'_{21})$.

If we now return to the 3-level atom source function given by equation (4.1.32), we can immediately see that the $[\epsilon B_\nu]_{231}$ term appearing in the numerator represents an adjustment to the effective source term $\epsilon'_{21}B_{\nu_{21}}(T)$. This adjustment is necessary since we must take into account the fact that the bound electron can find itself in level 2 not only as a result of photon absorption or collisional excitation from level 1, but also of radiative and collisional de-excitation from level 3. Similarly, the ϵ'_{231} term appearing in the denominator of equation (4.1.32) represents an adjustment to ϵ'_{21} which takes into account de-excitation from level 2 to level 1, not by collisional de-excitation directly, but by indirect collisional *and* radiative processes involving level 3.

Reference to equations (4.1.9–12) and (4.1.19) shows that R_{132} and R_{231} involve the average radiation intensities $\bar{J}_{31}$ and $\bar{J}_{32}$ (if they exist non-zero). Thus, any further simplification of the 3-level atom source function S_{21} given by equation (4.1.32) would require first a parallel derivation of S_{31} and $\bar{S}_{32}$ with subsequent solution of the respective transfer equations for I_{31} and I_{32}. But these, in turn, involve S_{21}, etc., and thus some form of iteration is, in general, required. Simplification of the above S_{21} is therefore not available without recourse to further approximation.

We should stress that, in practice, *no* further approximation is generally made in the determination of S_{21}. However, the *physical* interpretation of the terms appearing in S_{21} may be made more mathematically attractive by *assuming* levels 1 and 3 to balance one another independent of level 2, and levels 2 and 3 to balance one another independent of level 1. We then have

$$N_1^*R_{13} = N_3^*R_{31}, \tag{4.1.36}$$

and

$$N_2^*R_{23} = N_3^*R_{32}, \tag{4.1.37}$$

so that

$$\frac{g_1R_{132}}{g_2R_{231}} = \frac{g_1R_{13}R_{32}}{g_2R_{23}R_{31}} = \frac{g_1N_2^*}{g_2N_1^*} = \mathrm{e}^{-h\nu_{21}/kT}. \tag{4.1.38}$$

Thus

$$\epsilon'_{231} = \frac{R_{231}}{A_{21} + B_{21} B_{\nu_{21}}(T)}, \tag{4.1.39}$$

and

$$[\epsilon B_\nu]_{231} = \frac{R_{132}}{B_{12}}. \tag{4.1.40}$$

We then find

$$S_{21} = \frac{A_{21} + B_{21} B_{\nu_{21}}(T)}{A_{21} + B_{21} B_{\nu_{21}}(T) + C_{21} + R_{231}} \cdot \bar{J}_{21} + \frac{C_{12} + R_{132}}{B_{12} B_{\nu_{21}}(T) + C_{12} + R_{132}} \cdot B_{\nu_{21}}(T). \tag{4.1.41}$$

One may now compare this form with the corresponding 2-level atom source function given by equation (4.1.35). One sees that C_{21} is now replaced by $C_{21} + R_{231}$ and C_{12} replaced by $C_{12} + R_{132}$ so that the general term $C_{ij} + R_{i3j}$ represents all processes from level i to level j except those involving *direct radiative* transitions between i and j. Consequently, the interpretation of equation (4.1.35) may now be easily modified for the 3-level atom source function (4.1.41).

Finally, it is not difficult to show that if radiative transitions exist between levels 1 and 3, and levels 2 and 3, the corresponding source functions have the form

$$S_{31} = \frac{\bar{J}_{31} + \epsilon'_{31} B_{\nu_{31}}(T) + [\epsilon B_\nu]_{321}}{1 + \epsilon'_{31} + \epsilon'_{321}}, \tag{4.1.42}$$

$$S_{32} = \frac{\bar{J}_{32} + \epsilon'_{32} B_{\nu_{32}}(T) + [\epsilon B_\nu]_{312}}{1 + \epsilon'_{32} + \epsilon'_{312}} \tag{4.1.43}$$

where, for example,

$$\epsilon'_{31} = \frac{C_{31}}{A_{31}} \left(1 - \frac{g_1 C_{13}}{g_3 C_{31}}\right), \tag{4.1.44}$$

$$\epsilon'_{321} = \frac{R_{321}}{A_{31}} \left(1 - \frac{g_1 R_{123}}{g_3 R_{321}}\right), \tag{4.1.45}$$

and

$$[\epsilon B_\nu]_{321} = \frac{g_1 R_{123}}{g_3 B_{31}}, \tag{4.1.46}$$

for

$$R_{321} = R_{32} \cdot \frac{R_{21}}{R_{21} + R_{23}}, \quad R_{123} = R_{12} \cdot \frac{R_{23}}{R_{21} + R_{23}}. \tag{4.1.47}$$

Indeed, the derivation of these source functions can be facilitated by appropriate cyclical re-arrangements of the level subscripts.

4.2 Qualitative behaviour of 3-level atom source functions [8,42]

Section 1.8 of chapter 1 was devoted solely to the determination of the qualitative behaviour of the spectral line source function for model 2-level atoms. Here, we wish to perform the same approximate analysis for the 3-level atom source functions derived in the preceding section. There are three different models which we may discuss; the three resulting source functions, and their corresponding emergent line intensities, are quite different and are therefore considered separately. In particular, we examine those source functions arising only from 1-dimensional semi-infinite stellar atmospheres (as opposed to slab geometry).

4.2.1 Radiative transitions 2-1 and 3-1

Figure 4.2 shows the model 3-level atom in which radiative transitions occur between levels 2 and 1 and between levels 3 and 1 (shown as solid lines) whilst collisions (dotted lines) occur between all levels.

Here, the source function S_{21} is given by equation (4.1.32) but with $R_{23} = C_{23}$ and $R_{32} = C_{32}$. The 3-1 source function S_{31} is given by equation (4.1.42). However, in discussing the qualitative behaviour of S_{21}, the formula given by equation (4.1.32) has little immediate value. Indeed, it is somewhat easier to revert to its original form (4.1.5) and, as in section 1.8 when discussing the 2-level atom source function, ignore stimulated emissions. We then have

$$S_{21} \propto \frac{N_2}{N_1}, \qquad (4.2.1)$$

and, similarly,

$$S_{31} \propto \frac{N_3}{N_1}, \qquad (4.2.2)$$

where, as before, we have assumed complete re-distribution.

Since we have two source functions to consider, we shall be interested in their relative behaviour and this, in turn, must depend (amongst other things) on the relative opacity κ_{21} and κ_{31} for each line. Clearly, we have

$$\kappa_{21} \propto \nu_{21} N_1 B_{12} \phi_{12}(\Delta\nu), \qquad (4.2.3)$$

Fig.4.2. A model 3-level atom with radiative transitions only between levels 1 and 2 and between levels 1 and 3.

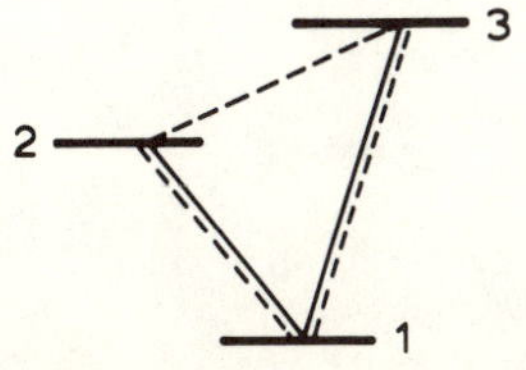

and

$$\kappa_{31} \propto \nu_{31} N_1 B_{13} \phi_{13}(\Delta\nu), \tag{4.2.4}$$

i.e.

$$\frac{\kappa_{21}}{\kappa_{31}} = \frac{\nu_{21}}{\nu_{31}} \cdot \frac{A_{21}}{A_{31}} \cdot \frac{g_2}{g_3} \cdot \frac{\nu_{31}^3}{\nu_{21}^3} \cdot \frac{\phi_{12}(\Delta\nu)}{\phi_{13}(\Delta\nu)}, \tag{4.2.5}$$

where we have used the relationships between the Einstein rate coefficients given by equation (4.1.13).

One usually finds $A_{21} > A_{31}$ (basically because there is a larger energy difference between levels 3–1 than between levels 2–1; thus the atom finds it easier to de-excite spontaneously in the 2–1 transition). Consequently, with $\nu_{31} > \nu_{21}$, it would appear that we have $\kappa_{21} > \kappa_{31}$. However, the relative behaviour of the absorption probabilities $\phi_{12}(\Delta\nu)$ and $\phi_{13}(\Delta\nu)$ introduces a complicating factor. For example, if we only consider Doppler broadening, we have

$$\phi_{12}(\Delta\nu) = \frac{1}{\sqrt{\pi}(\Delta\nu_D)_{12}} \exp\left\{-\left[\frac{\Delta\nu}{(\Delta\nu_D)_{12}}\right]^2\right\}, \tag{4.2.6}$$

where the Doppler width $(\Delta\nu_D)_{12}$ has the form

$$(\Delta\nu_D)_{12} = \frac{\nu_{21}}{c}\left[\frac{2kT}{m_A}\right]^{1/2} \tag{4.2.7}$$

– see equation (1.4.15). Clearly, with a similar formula for $\phi_{13}(\Delta\nu)$, the appearance of the central frequencies ν_{21} and ν_{31} within the exponentials could have quite a profound effect. Indeed, we find

$$\phi_{12}(\Delta\nu) \gtrless \phi_{13}(\Delta\nu) \;\forall\; |\Delta\nu| \lessgtr |\Delta\nu_c|, \tag{4.2.8}$$

where $\Delta\nu_c$ is some 'change-over' frequency as shown in figure 4.3. Note that in figure 4.3, the area under each of $\phi_{12}(\Delta\nu)$ and $\phi_{13}(\Delta\nu)$ must be unity because of the normalisation condition

$$\int_{-\infty}^{\infty} \phi(\Delta\nu)\, d(\Delta\nu) = 1. \tag{4.2.9}$$

Fig.4.3. The absorption profiles for the 1–2 and 1–3 transitions.

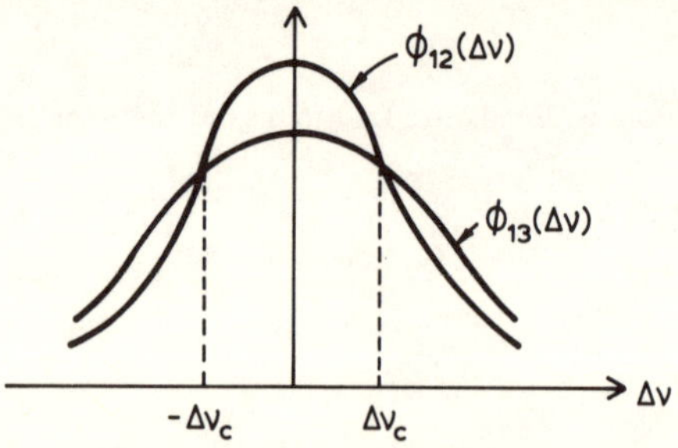

This complicating feature of the relative behaviour of the opacities is, of course, quite important in detailing the solution to the radiative transfer equation for physically realistic situations. Here, however, we are interested only in the qualitative behaviour of S_{21} and S_{31} and therefore we simply take $\kappa_{21} > \kappa_{31}$ for all frequencies $\Delta\nu$ away from the respective line centres. As we shall see, any modifying variation due to the relative behaviour of $\phi_{12}(\Delta\nu)$ and $\phi_{13}(\Delta\nu)$ is not important to the following physical arguments.

We shall begin by considering regions sufficiently deep within the stellar atmosphere from which effectively no radiation may escape to the surface. We know that, in this situation, the approximation of LTE for photons is reasonably valid so that we immediately have

$$\bar{J}_{21} \approx S_{21} \approx B_{\nu_{21}}(T), \tag{4.2.10}$$

and

$$\bar{J}_{31} \approx S_{31} \approx B_{\nu_{31}}(T). \tag{4.2.11}$$

As we move from these deeper regions toward the stellar surface, radiation has an ever increasing probability of escape. Since we are (somewhat arbitrarily) assuming the opacity in the 2-1 transition to be greater than that in the 3-1 transition (i.e. photons can travel further between emission and absorption in the 3-1 transition), radiation will first escape via the 3-1 mode. Figure 4.4 shows the region over which the $h\nu_{21}$ and $h\nu_{31}$ photons may range between their creation, at point A say, and their destruction (represented by the circles). The radii of the circles shown in figure 4.4 are the respective thermalisation path lengths Θ_{21} and Θ_{31} for those photons (see section 1.8).

Thus, a new dependence has been introduced into the discussion. These thermalisation path lengths themselves not only depend upon the opacities κ_{21} and κ_{31}, but on the number of photon scatters (i.e. photon absorptions followed by photon emissions) between photon creation and photon destruction. For

Fig.4.4. Photons created at A travel a distance Θ_{21} in the 1–2 transition before being destroyed by collisional de-excitation but scatter a distance Θ_{31} if absorbed and emitted in the 1–3 transition.

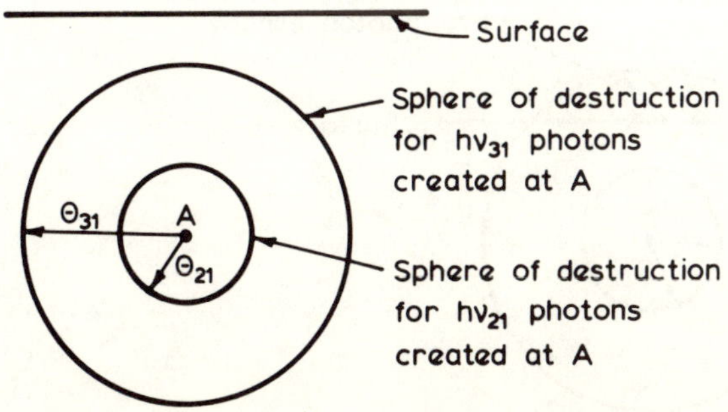

the model 2-level atom source function, creation followed collisional excitation and destruction followed collisional de-excitation, and we had (see equation (1.8.16))

$$\Theta_{21} \sim \frac{\lambda_{21}}{\epsilon_{21}}, \tag{4.2.12}$$

for Doppler broadening, for example, where λ_{21} is the mean free path of a photon emitted at line centre in the 2–1 transition.

For transitions between levels 2 and 1 for a model 3-level atom, however, we may also have creation of a photon due to excitation (collisional or radiative from level 1) to level 3 followed by collisional de-excitation to level 2 (i.e. R_{132}). These processes supplement the creation due to direct collisional excitation from level 1 to level 2. Similarly, photons may be destroyed by collisional excitation from level 2 to level 3, thence de-excitation (radiative or collisional) to level 1 (i.e. R_{231}). Again, these processes supplement the destruction of photons due to direct collisional de-excitation from level 2 to 1.

Thus, Θ_{21} for the above model 3-level atom would not take the explicit form given by equation (4.2.12), but could be implicitly written as

$$\Theta_{21} = \frac{\lambda_{21}}{f(\epsilon_{21}, R_{231}, R_{132})}. \tag{4.2.13}$$

It is clear, however, that the composite rates R_{231} and R_{132} given by equations (4.1.19) and (4.1.20) involve the radiation field $\bar{J}_{31}$; thus any clear algebraic expression for Θ_{21} is not forthcoming. Similarly, Θ_{31} depends upon $\bar{J}_{21}$. It is not unreasonable to suggest, however, that if $\kappa_{21} > \kappa_{31}$, i.e. $\lambda_{21} < \lambda_{31}$, and $\epsilon_{21} \sim \epsilon_{31}$ with $R_{231} \sim R_{321}$, then $\Theta_{21} < \Theta_{31}$ as shown in figure 4.4.

Keeping all the above discussion in mind, as we move from regions deep within the stellar atmosphere towards the surface, the source function in the 3–1 transition first becomes affected by the influence of the surface, and this occurs at a depth of order Θ_{31} as illustrated in figure 4.5.

Fig.4.5. Photons scattering in the 1–3 transition escape from the medium more easily than those in the 1–2 transition.

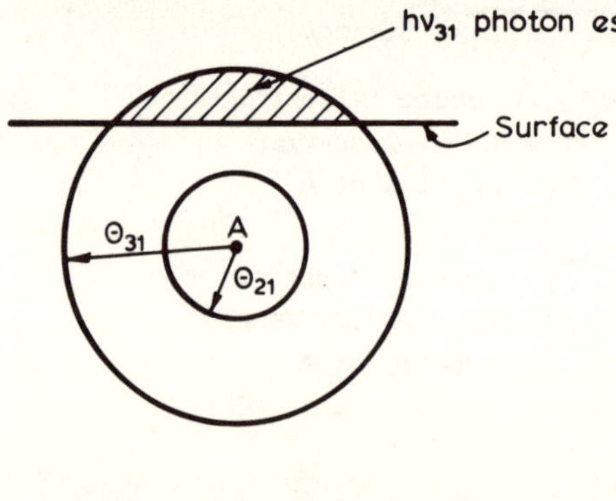

As discussed in section 1.8, this escape of photons will then lead to a depopulation of the upper level of the transition, i.e. $N_3 \rightarrow N_3^- < N_3^*$ and, since the total number of atoms does not change (i.e. $N_1 + N_2 + N_3 =$ constant in a statistically steady state), an overpopulation of the lower level results so that $N_1 \rightarrow N_1^+ > N_1^*$. Thus, we have

$$S_{31} \propto \frac{N_3}{N_1} \rightarrow \frac{N_3^-}{N_1^+} < \frac{N_3^*}{N_1^*} \propto B_{\nu_{31}}(T), \tag{4.2.14}$$

and this then implies that the source function S_{31} continues to decrease away from the corresponding Planck function with increasing height in the stellar atmosphere.

One must now take into account the effect this has on the radiative transfer in the 2–1 transition. Two immediate effects may be discerned. First, if $N_3 \rightarrow N_3^-$ whilst N_2 remains unchanged, the populations N_3 and N_2 would be out of balance relative to their respective LTE values N_3^* and N_2^*. There is no *a priori* reason why these levels should not be out of such balance - however, collisional processes between levels 2 and 3 would attempt to restore the balance. Indeed, it is implicit in the derivation of S_{31} and S_{32} that

$$N_2^* C_{23} = N_3^* C_{32}. \tag{4.2.15}$$

Thus, the loss of atoms with the bound electron in level 3 due to escape of $h\nu_{31}$ photons discussed above will be partly compensated by a gain due to collisional excitation from level 2. We then have $N_2 \rightarrow N_2^- < N_2^*$.

The second effect has already been mentioned implicitly. We had $N_1 \rightarrow N_1^+$. However, since photons emitted in the 2-1 transition cannot reach the surface from depths of order Θ_{31} (recall $\Theta_{21} < \Theta_{31}$), the radiation field I_{21} would attempt to maintain an LTE distribution whereby $S_{21} \approx B_{\nu_{21}}(T)$. Thus, some of the N_1^+ gain would be lost to N_2 via collisional *and* radiative processes. This N_2 gain would offset, to some extent, the loss of level 2 atoms due to collisional excitation to level 3. Nevertheless, N_2 decreases to below its corresponding LTE value N_2^* so that

$$S_{21} \propto \frac{N_2}{N_1} \rightarrow \frac{N_2^-}{N_1^+} < \frac{N_2^*}{N_1^*} \propto B_{\nu_{21}}(T). \tag{4.2.16}$$

Consequently, both S_{31} *and* S_{21} decrease below their corresponding Planck functions once $h\nu_{31}$ photons begin to escape from the atmosphere as shown in figure 4.6, and this occurs at the physical depth z_{31} corresponding to the thermalisation optical path length Θ_{31}.

Naturally, as we move to still higher regions in the stellar atmosphere, more and more photons in the 3-1 transition escape and thus S_{31}, thence S_{21}, decrease still further. This decrease eventually becomes enhanced by the loss of $h\nu_{21}$ photons, this loss occurring within a distance of order Θ_{21} from the surface.

4.2.2 *Radiative transitions 3-1 and 3-2*

The model 3-level atom in this particular case is shown in figure 4.7. Again, the solid lines denote radiative transitions whereas the dotted lines represent collisional processes. Thus, we have $R_{21} = C_{21}$ and $R_{12} = C_{12}$.

Most of the preliminary physical discussion for this model atom has already been included in section 4.2.1 above. Let us therefore begin here by arbitrarily setting $\kappa_{31} < \kappa_{32}$ with $\Theta_{31} > \Theta_{32}$. The appropriate source function proportionalities are $S_{31} \propto N_3/N_1$ and $S_{32} \propto N_3/N_2$. Following the preceding discussion, we see that deep within the stellar atmosphere $S_{31} \to B_{\nu_{31}}(T)$ and $S_{32} \to B_{\nu_{32}}(T)$. As one moves to higher regions in the stellar atmosphere, photons in the most transparent (i.e. least opaque) transition begin to escape from the surface. Thus, at a depth of order Θ_{31}, we find $N_3 \to N_3^- < N_3^*$ with $N_1 \to N_1^+ > N_1^*$. However, collisions between levels 1 and 2 would enable level 2 to share some of this population increase given to level 1. Whilst this is occurring, radiative transitions between levels 2 and 3 would attempt to restore the LTE balance between N_2

Fig.4.6. (*a*) The qualitative behaviour of the Planck and source functions in the 1–3 transition; (*b*) The corresponding Planck and source functions in the 1–2 transition.

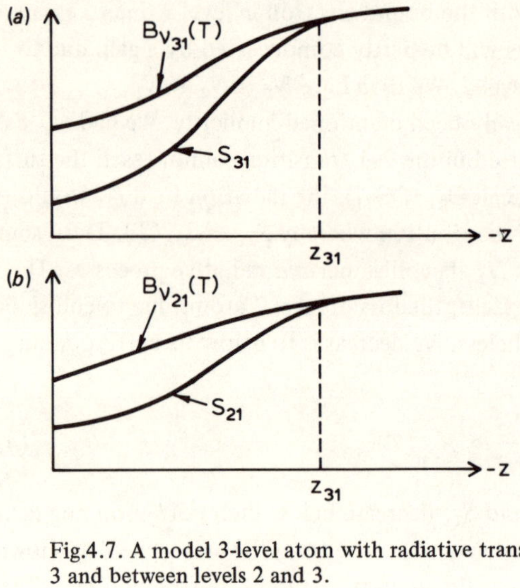

Fig.4.7. A model 3-level atom with radiative transitions only between levels 1 and 3 and between levels 2 and 3.

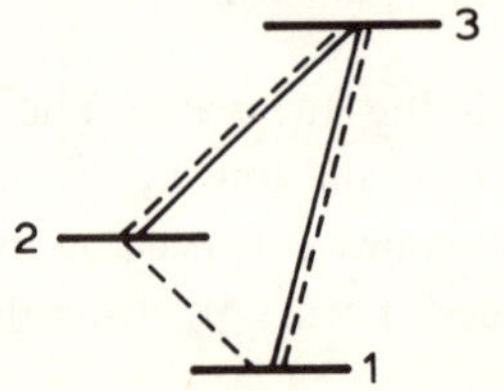

and N_3^- thus resulting in part of the N_3^- loss being compensated for by a corresponding gain from N_2. (Of course, this balancing process is attempted by levels 2 and 3 since radiation in the 3-2 transition cannot escape to the surface at depth of order Θ_{31} (recall that $\Theta_{32} < \Theta_{31}$ in our model) and thus levels 2 and 3 are effectively in LTE at depths between Θ_{32} and Θ_{31}.)

As in section 4.2.1 above, this is a somewhat cyclical phenomenon resulting in $N_3 \to N_3^- < N_3^*$, $N_1 \to N_1^+ > N_1^*$ and $N_2 \to N_2^\pm \approx N_2^*$. We therefore find

$$S_{31} \propto \frac{N_3}{N_1} \to \frac{N_3^-}{N_1^+} < \frac{N_3^*}{N_1^*} \propto B_{\nu_{31}}(T), \tag{4.2.17}$$

and

$$S_{32} \propto \frac{N_3}{N_2} \to \frac{N_3^-}{N_2^\pm} < \frac{N_3^*}{N_2^*} \propto B_{\nu_{32}}(T), \tag{4.2.18}$$

so that S_{31} and S_{32} begin to decrease below their respective LTE values as shown in figure 4.8. Again, we notice that S_{32} beging to depart from $B_{\nu_{32}}(T)$ at depths of order Θ_{31}, not Θ_{32}.

As we move to still higher regions in the stellar atmosphere, radiation in the 3-2 transition, which was hitherto unable to reach the surface, is now capable of escaping. This, of course, occurs at physical depths of order z_{32} corresponding to the thermalisation depth Θ_{32}. Now, however, not only do we have a depletion of level 3 atoms due to the escape of $h\nu_{31}$ photons, but also a de-population due to this 3-2 radiation escape. The consequent double depletion increases the rate

Fig.4.8. (*a*) The qualitative behaviour of the Planck and source functions in the 1-3 transition; (*b*) The corresponding Planck and source functions in the 2-3 transition.

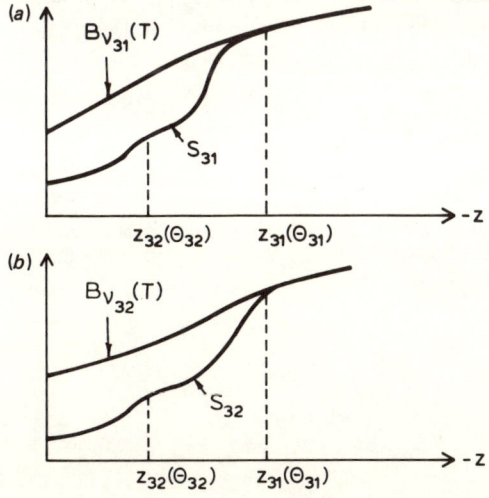

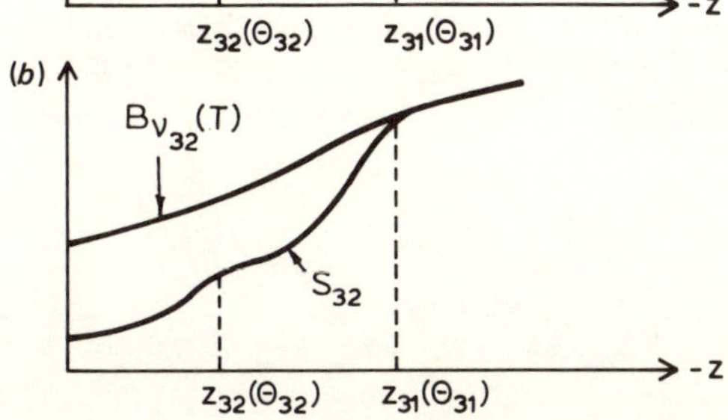

of de-population of level 3 thus yielding $N_2^{\pm} \to N_2^{+} > N_2^{*}$. This overpopulation of N_2 would then, in turn, be shared with level 1 atoms due to collisional processes between levels 2 and 1. Thus, corresponding to the increased rate of depletion of N_3, we have an increased rate of overpopulation of N_1, and this results in the change in the gradients of S_{31} and S_{32} at depths of order z_{32} (corresponding to Θ_{32}), as shown in figure 4.8. Clearly, if $\Theta_{32} \sim \Theta_{31}$, the second dip shown in that diagram would not be apparent.

4.2.3 *Radiative transitions 3–2 and 2–1*

An interesting situation arises when we allow radiative transitions only between levels 3 and 2 and between levels 2 and 1 as shown in figure 4.9.

Again, deep within the atmosphere we have $S_{32} \to B_{\nu_{32}}(T)$ and $S_{21} \to B_{\nu_{21}}(T)$. If we take $\kappa_{32} < \kappa_{21}$ with $\Theta_{32} > \Theta_{21}$, the escape of radiation in the 3–2 transition, as we move toward the surface, will force a de-population of level 3 and an overpopulation of level 2, i.e. $N_3 \to N_3^{-} < N_3^{*}$ and $N_2 \to N_2^{+} > N_2^{*}$. Some of this level 2 overpopulation will be shared at depths z between z_{32} and z_{21} by level 1 atoms since the 2–1 transition will attempt to remain in LTE. Nevertheless, we find

$$S_{32} \propto \frac{N_3}{N_2} \to \frac{N_3^{-}}{N_2^{+}} < \frac{N_3^{*}}{N_2^{*}} \propto B_{\nu_{32}}(T), \qquad (4.2.19)$$

whilst

$$S_{21} \propto \frac{N_2}{N_1} \to \frac{N_2^{+}}{N_1} > \frac{N_2^{*}}{N_1^{*}} \propto B_{\nu_{21}}(T). \qquad (4.2.20)$$

Thus, the decrease in S_{32} at depths of order z_{32} is accompanied by a corresponding increase in S_{21} as shown in figure 4.10.

This increase in S_{21} will continue with increasing height in the stellar atmosphere until regions of order Θ_{21} from the surface are reached. Here, photons in the 2–1 transition begin to escape to the surface so that N_2 tends to decrease whilst N_1 increases, i.e. $N_2^{+} \to N_2^{-} < N_2^{*}$ with $N_1 \to N_1^{+} > N_1^{*}$. We then have S_{21} decreasing at z_{21} as shown in figure 4.10.

Fig.4.9. A model 3-level atom with radiative transitions only between levels 1 and 2 and between levels 2 and 3.

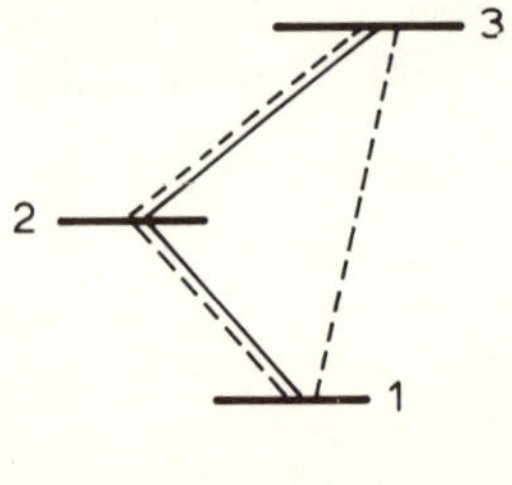

We have not included a second dip in S_{32} at depths z_{32} (nor, for that matter, second dips in S_{21} and S_{31} of section 4.2.1) similar to those shown in figure 4.8. In section 4.2.2, we had level 3 as a common upper level to the two available radiative transitions 3–1 and 3–2. At depths of order z_{32} in that case we had a *direct* double depletion of N_3 due to both $h\nu_{31}$ and $h\nu_{21}$ photon escape to the surface. In the cases considered in section 4.2.1 and here, no such common upper level exists and so no *direct* double depletion can occur. *Indirect* double depletion can arise via another level, but this other level would always tend to smooth out any such dips which might occur simply because one of the two depletion processes would involve collisions and these, of course, are independent of Θ. One could argue that there is double overpopulation of level 1 for the case considered in section 4.2.1 at z_{21} (Θ_{21}) where photons in the 2–1 transition begin to escape. However, the population of the lower state is significantly larger than the population of the upper states so that, whilst a depletion of an upper level could be a significant fraction of that upper level population, the corresponding increase in the lower level population would be a much smaller fraction. Thus, a direct double depletion of N_3 is sufficient to cause a noticeable second dip in the source function (if $\Theta_{31} \not\approx \Theta_{32}$), whilst a corresponding double over-population of level 1 is generally insufficient.

4.2.4 *Corresponding emergent intensities*

The preceding discussion highlighted the importance of the third level in controlling the structure of the source function. Although the source functions

Fig.4.10. (*a*) The qualitative behaviour of the Planck and source functions in the 2–3 transition; (*b*) The corresponding Planck and source functions in the 1–2 transition.

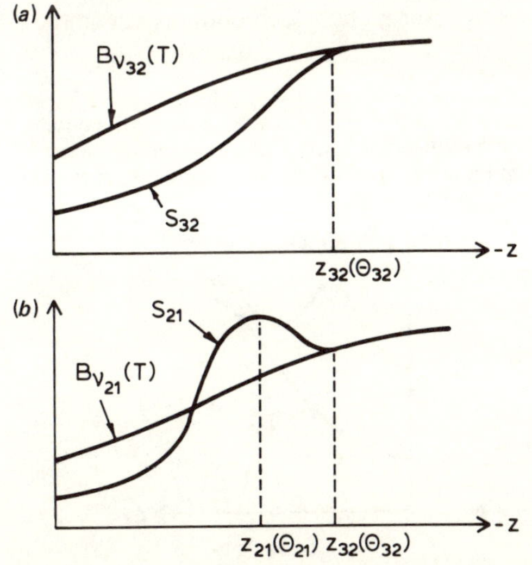

of section 4.2.1, which we henceforth refer to as case (a), resemble those for a model 2-level atom, the source functions of section 4.2.2 (case (b)) and 4.2.3 (case (c)) are markedly different. Since the emergent intensity is essentially a mapping of the source function from physical to frequency space (see section 1.9), it is not difficult to see that the emergent intensities corresponding to the source functions of case (b) and case (c) exhibit the qualitative behaviour illustrated in figure 4.11.

It is interesting to note that the effective emission peaks in $I_{21}(0, \Delta\nu, \mu)$ in case (c) are similar to those obtained for chromospheric-type stellar situations shown in figure 1.22. Clearly, if one observed emergent intensities of the form given by $I_{21}(0, \Delta\nu, \mu)$ above, and attempted to derive the temperature structure of the stellar atmosphere using only a model 2-level atom when, in fact, the model 3-level atom is more appropriate, one would misinterpret entirely the form such a temperature distribution should take. Indeed, in this case, one would obtain a false chromospheric-type temperature rise rather than the true photospheric-type temperature decrease with increasing height. Perhaps this is a bad example since most of the observed stellar spectral line intensities which have these central emission peaks arise from model atoms of the form of case (a) or case (b), but not case (c), and may generally be explained only by an outward temperature increase. (Actually, as we shall see in chapter 7, an alternative explanation for these emergent intensity emission peaks is available using departures from complete re-distribution.) Nevertheless, it is clear that care is needed when analysing spectral line data.

4.3 Transition channelling

Implicit in the preceding discussion of the qualitative behaviour of the 3-level atom source functions is the process by which photons in one transition may be channelled into another.

Fig.4.11. (*a*) The qualitative behaviour of the emergent intensities obtained from the source functions shown in Figure 4.8; (*b*) The corresponding emergent intensities obtained from Figure 4.10. Note the local maximum in the emergent intensity for the 1–2 transition.

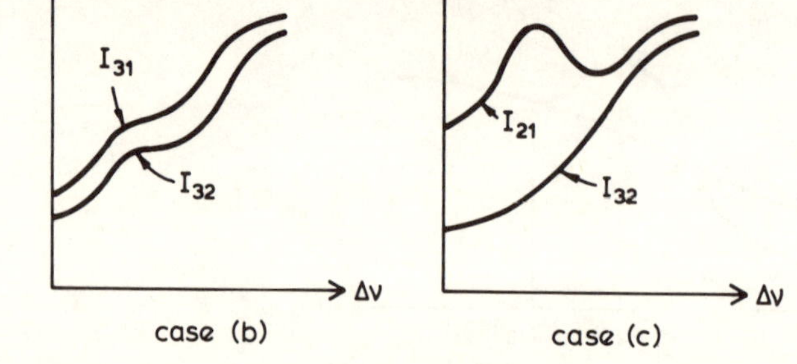

In section 4.2.1 (case (a)), for example, we found that for $\kappa_{31} < \kappa_{21}$ with $\Theta_{31} > \Theta_{21}$, $h\nu_{31}$ photons could escape from the atmosphere at depths Θ_{31} whereas $h\nu_{21}$ photons could not. This meant that $N_3 \to N_3^-$ and $N_1 \to N_1^+$. However, levels 2 and 3 would attempt to maintain an LTE ratio in their population densities through collisional excitation and de-excitation, as would levels 2 and 1 through both radiative and collisional processes, particularly since $h\nu_{21}$ photons cannot reach the surface from Θ_{31}. Thus, not only the 3–1 source function but also S_{21} decrease below their corresponding Planck functions. However, if we consider a 2–1 transition for only a model 2-level atom, such a decrease in S_{21} below $B_{\nu_{21}}(T)$ could only be interpreted as a result of $h\nu_{21}$ photon escape. We know, however, that in the preceding 3-level atom case (a) at depths of order Θ_{31}, these $h\nu_{21}$ photons cannot reach the surface directly. Thus, we must interpret the decrease of S_{21} below $B_{\nu_{21}}(T)$ at depth Θ_{31} as a channelling of $h\nu_{21}$ photons into the 3–1 transition, from which they may thence escape.

Of course, the physical process is quite straightforward. Photons in the 2–1 transition may be absorbed, thus exciting the atom to level 2. Collisions between this atom and free electrons could then further excite the bound electron to level 3. Spontaneous or stimulated de-excitation from level 3 to level 1 would then generate a $h\nu_{31}$ photon (which could escape from depths of order Θ_{31}) thus completing the conversion of $h\nu_{21}$ to $h\nu_{31}$.

This channelling is important from a diagnostic point of view in attempting to interpret the temperature structure, for example, of the stellar atmosphere giving rise to the spectral line data at hand. However, it is also important in the determination of the energy structure and, more importantly, the energy loss, of the stellar atmosphere. We shall study the energy transfer problem in some detail in chapter 5. For the moment, we should note that the computed energy loss from the stellar atmosphere due to photon escape in the 2–1 and 3–1 transitions is greater when one takes these transitions to be coupled in a model 3-level atom than when one assumes the transitions to arise from two quite independent model 2-level atoms. For example, if we considered the $h\nu_{21}$ photons interacting with only model 2-level atoms, such photons could not escape from the stellar atmosphere below depths of order Θ_{21}. However, if we now allow this 2–1 transition to be coupled to the 3–1 transition by the use of a model 3-level atom, the $h\nu_{21}$ photons may channel into the 3–1 transition and thus escape below depths Θ_{21}. This deeper escape mechanism for $h\nu_{21}$, which is not available for the equivalent 2-level atom situation, must necessarily increase the energy loss from the atmosphere.

Thus, these channelling processes further enhance the non-LTE effects mentioned in section 1.7. Energy may be transferred through a gas by direct elastic collisions where the internal state of the atom remains unchanged; the distance over which this transfer occurs is of the order of the mean free path for such collisions. We refer to this as a local phenomenon - it enables us to attach some physical meaning to the LTE assumption. An alternative form of energy transfer involves the creation of a photon followed by a series of scatterings thence destruction. Energy would be taken out of the thermal field by the creation event and replaced by the destruction. Since the distance ($\sim\Theta$) between photon creation and destruction is generally far greater than the above-mentioned collisional mean free path, this process is relatively non-local. Further, if photons are created in the 2–1 transition, thence channelled into the 3–1 transition where $\kappa_{31} < \kappa_{21}$, the distance between photon creation and destruction is increased and this, in turn, enhances the non-local energy transfer.

4.4 The general multi-level atom

The evaluation of the ratio N_j/N_i required in the determination of the j–i transition source function S_{ji} given by

$$S_{ji} = \frac{A_{ji}N_j/N_i}{B_{ij} - B_{ji}N_j/N_i}, \tag{4.4.1}$$

becomes progressively more complicated as one includes more and more levels in the model atom. In practice, however, it might be necessary to include a large number of levels in the analysis and thus some alternative, and certainly easier, derivation of the multi-level atom source function than that suggested in section 4.1 is warranted.

4.4.1 Preliminary discussion

To illustrate an alternative formulation, we first return to the model 3-level atom discussed in section 4.1. We again have the statistically steady state rate equations

$$N_1(R_{12} + R_{13}) = N_2R_{21} + N_3R_{31}, \tag{4.4.2}$$

and

$$N_2(R_{21} + R_{23}) = N_1R_{12} + N_3R_{32}. \tag{4.4.3}$$

We can evaluate the ratio N_2/N_1 either from equation (4.4.2) or (4.4.2). Indeed, if we use equation (4.4.2), we may write either

$$\frac{N_2}{N_1} = \frac{R_{12} + R_{13}}{R_{21}} - \frac{N_3R_{31}}{N_1R_{21}}, \tag{4.4.4}$$

or

$$\frac{N_2}{N_1} = \left[\frac{R_{21}}{R_{12} + R_{13}} + \frac{N_3R_{31}}{N_2(R_{12} + R_{13})}\right]^{-1}. \tag{4.4.5}$$

The use of these two ratios yields the two numerically equivalent, but structurally different, forms of the 2–1 source function:

$$S_{21}^{(1)} = \frac{\bar{J}_{21} + \epsilon'_{21} B_{\nu_{21}}(T) + \dfrac{1}{B_{12}}\left(R_{13} - \dfrac{N_3 R_{31}}{N_1}\right)}{1 + \epsilon'_{21} + \dfrac{g_1}{g_2 A_{21}}\left(\dfrac{N_3 R_{31}}{N_1} - R_{13}\right)}, \tag{4.4.6}$$

$$S_{21}^{(2)} = \frac{\bar{J}_{21} + \epsilon'_{21} B_{\nu_{21}}(T) + \dfrac{R_{13}}{B_{12}}}{1 + \epsilon'_{21} + \dfrac{1}{A_{21}}\left(\dfrac{N_3 R_{31}}{N_2} - \dfrac{g_1}{g_2} R_{13}\right)}. \tag{4.4.7}$$

Another two forms of S_{21} may be obtained by noting the two values of N_2/N_1 derived from equation (4.4.3), viz.

$$\frac{N_2}{N_1} = \frac{R_{12}}{R_{21} + R_{23}} + \frac{N_3 R_{32}}{N_1(R_{21} + R_{23})}, \tag{4.4.8}$$

or

$$\frac{N_2}{N_1} = \left[\frac{R_{21} + R_{23}}{R_{12}} - \frac{N_3 R_{32}}{N_2 R_{12}}\right]^{-1}. \tag{4.4.9}$$

We then have

$$S_{21}^{(3)} = \frac{\bar{J}_{21} + \epsilon'_{21} B_{\nu_{21}}(T) + \dfrac{N_3 R_{32}}{N_1 B_{12}}}{1 + \epsilon'_{21} + \dfrac{1}{A_{21}}\left(R_{23} - \dfrac{g_1 N_3 R_{32}}{g_2 N_1}\right)}, \tag{4.4.10}$$

$$S_{21}^{(4)} = \frac{\bar{J}_{21} + \epsilon'_{21} B_{\nu_{21}}(T)}{1 + \epsilon'_{21} + \dfrac{1}{A_{21}}\left(R_{23} - \dfrac{N_3 R_{32}}{N_2}\right)}. \tag{4.4.11}$$

In all four cases, the 2–1 source function may be generally written [119] as

$$S_{21} = \frac{\bar{J}_{21} + \epsilon'_{21} B_{\nu_{21}}(T) + (\eta B^*)_{21}}{1 + \epsilon'_{21} + \eta_{21}}, \tag{4.4.12}$$

although $(\eta B^*)_{21} \equiv 0$ for $S_{21}^{(4)}$. In particular, we see that both $(\eta B^*)_{21}$ and η_{21} do not contain any terms *directly* involving the 2–1 transition but only those terms which affect the 2–1 transition *indirectly* via level 3. It is not difficult to also see that if we now reduce our model 3-level atom to just two levels, we have $(\eta B^*)_{21} \to 0$ and $\eta_{21} \to 0$, thus yielding the normal 2-level atom result as expected.

We can choose any of the above four forms $S_{21}^{(m)}$ $(m = 1,4)$ in numerically evaluating the source function for the 2–1 transition. If we choose to use $S_{21}^{(1)}$, for example, then we require $\bar{J}_{31}$ since it appears both in the rates R_{13} and R_{31}, and the ratio N_3/N_1, viz.

$$\left.\begin{aligned} R_{13} &= B_{13}\bar{J}_{31} + C_{13}, \\ R_{31} &= B_{31}\bar{J}_{31} + A_{31} + C_{31}, \end{aligned}\right\} \tag{4.4.13}$$

and

$$\frac{N_3}{N_1} = \frac{g_3}{g_1}\left[1 + \frac{2h\nu_{31}^3}{c^2 S_{31}}\right]^{-1}, \tag{4.4.14}$$

where

$$S_{31} = S_{31}(\bar{J}_{31}).$$

Clearly, therefore, we find it necessary, in general, to iterate between the various radiative transitions allowable within the model atom. The iterative procedure might not converge for all four source functions and this, of course, narrows our choice of $S_{21}^{(m)}$. Convergence properties of any iterative procedure designed to solve the equation of radiative transfer for model multi-level atoms depend upon a variety of factors not least of which is the actual structure of the model atom being used. For example, if we consider the model 3-level atom above in which only the two radiative transitions 2–1 and 3–1 are possible, then $S_{21}^{(1)}$ or $S_{21}^{(3)}$ might be a more appropriate choice for S_{21} than $S_{21}^{(2)}$ or $S_{21}^{(4)}$ since the ratio N_3/N_2 occurs in these latter two source functions, and this ratio is not specifically related to either S_{21} or S_{31}. We could write, however,

$$\begin{aligned} \frac{N_3}{N_2} &= \frac{N_3}{N_1}\cdot\frac{N_1}{N_2} \\ &= \frac{g_3}{g_2}\left[1 + \frac{2h\nu_{31}^3}{c^2 S_{31}}\right]^{-1}\left[1 + \frac{2h\nu_{21}^3}{c^2 S_{21}}\right], \end{aligned} \tag{4.4.15}$$

but this would introduce S_{21} into $(\eta B^*)_{21}$ and η_{21} resulting, in turn, in the introduction of an essential non-linearity in the equation of transfer for the 2–1 radiation field. One could include the S_{21} appearing in equation (4.4.15) in the iterative scheme mentioned above, but it is not at all clear how this non-linearity will affect the stability of the numerical procedure used.

We further illustrate this point by recalling the sensitivity of the numerical methods presented in chapter 2 for the 2-level atom problem to the value of ϵ' appearing in the denominator of the source function

$$S\,(\text{2-level}) = \frac{\bar{J} + \epsilon' B_\nu(T)}{1 + \epsilon'} = (1 - \epsilon)\bar{J} + \epsilon B_\nu(T).$$

The smaller the value of ϵ, the more severe are the demands made on the numerical methods of solution. Thus, it would appear more feasible to choose $S_{21}^{(m)}$ such that $\epsilon_{21}' + \eta_{21}$ is maximised. Indeed, it was also found necessary for ϵ to satisfy $0 < \epsilon < 1$ which, with $\epsilon = \epsilon'/(1 + \epsilon')$, implies $\epsilon' > 0$. Relating this to the 3-level atom problem then yields the inequality necessary for numerical stability:

$$\epsilon_{21}' + \eta_{21} > 0. \qquad (4.4.16)$$

We always have $\epsilon_{21}' > 0$ since, from equation (4.1.34) for example, $\epsilon_{21}' \equiv C_{12}/B_{12}B(T)$ where all terms are positive definite. However, reference to equations (4.4.6,7) and (4.4.10,11) indicates that η_{21} is not necessarily positive. Thus, condition (4.4.16) should also be kept in mind when choosing S_{21} from the four $S_{21}^{(m)}$ detailed above.

4.4.2 *The general source function*

Let us now consider a model atom having an infinite number of levels. Let U and L denote the respective upper and lower levels of the transition in question as illustrated in figure 4.12.

The statistically steady state rate equation for level L then has the form

$$N_{\mathrm{L}}\left[R_{\mathrm{LU}} + \sum_{i \neq \mathrm{U}} R_{\mathrm{L}i}\right] = N_{\mathrm{U}}R_{\mathrm{UL}} + \sum_{i \neq \mathrm{U}} N_i R_{i\mathrm{L}}. \qquad (4.4.17)$$

Similarly, for the upper U level, we have

$$N_{\mathrm{U}}\left[R_{\mathrm{UL}} + \sum_{i \neq L} R_{\mathrm{U}i}\right] = N_{\mathrm{L}}R_{\mathrm{LU}} + \sum_{i \neq \mathrm{L}} N_i R_{i\mathrm{U}}. \qquad (4.4.18)$$

The rates R_{ij} appearing in these last two equations must satisfy the general expressions

$$R_{ij} = B_{ji}\bar{J}_{ji} + A_{ji} + C_{ji} \quad \forall \ j > i, \qquad (4.4.19)$$

$$R_{ij} = B_{ij}\bar{J}_{ji} + C_{ij} \quad \forall \ i > j. \qquad (4.4.20)$$

If we now use equation (4.4.17) to evaluate the ratio $N_{\mathrm{U}}/N_{\mathrm{L}}$ appearing in the U-L source function

$$S_{\mathrm{UL}} = \frac{A_{\mathrm{UL}}N_{\mathrm{U}}/N_{\mathrm{L}}}{B_{\mathrm{LU}} - B_{\mathrm{UL}}N_{\mathrm{U}}/N_{\mathrm{L}}}, \qquad (4.4.21)$$

Fig. 4.12. A model multi-level atom with the upper level U and lower level *L* isolated.

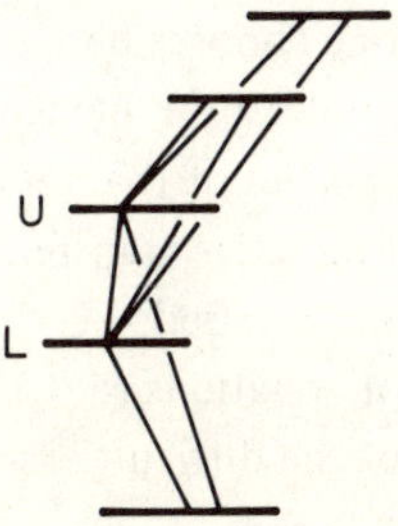

which we formally write as

$$S_{\mathrm{UL}} = \frac{\bar{J}_{\mathrm{UL}} + \epsilon'_{\mathrm{UL}} B_{\nu_{\mathrm{UL}}}(T) + (\eta B^*)_{\mathrm{UL}}}{1 + \epsilon'_{\mathrm{UL}} + \eta_{\mathrm{UL}}}, \tag{4.4.22}$$

where

$$\epsilon_{\mathrm{UL}} = \frac{C_{\mathrm{UL}}}{A_{\mathrm{UL}}}\left(1 - \frac{g_{\mathrm{L}} C_{\mathrm{LU}}}{g_{\mathrm{U}} C_{\mathrm{UL}}}\right), \tag{4.4.23}$$

then we have the two possibilities

$$(\eta B^*)^{(1)}_{\mathrm{UL}} = \frac{1}{B_{\mathrm{LU}}} \sum_{i \neq \mathrm{U}} \left(R_{\mathrm{L}i} - \frac{N_i R_{i\mathrm{L}}}{N_{\mathrm{L}}}\right), \tag{4.4.24}$$

$$\eta^{(1)}_{\mathrm{UL}} = \frac{g_{\mathrm{L}}}{g_{\mathrm{U}} A_{\mathrm{UL}}} \sum_{i \neq \mathrm{U}} \left(\frac{N_i R_{i\mathrm{L}}}{N_{\mathrm{L}}} - R_{\mathrm{L}i}\right), \tag{4.4.25}$$

or

$$(\eta B^*)^{(2)}_{\mathrm{UL}} = \frac{1}{B_{\mathrm{LU}}} \sum_{i \neq \mathrm{U}} R_{\mathrm{L}i}, \tag{4.4.26}$$

$$\eta^{(2)}_{\mathrm{UL}} = \frac{1}{A_{\mathrm{UL}}} \sum_{i \neq \mathrm{U}} \left(\frac{N_i R_{i\mathrm{L}}}{N_{\mathrm{U}}} - \frac{g_{\mathrm{L}} R_{\mathrm{L}i}}{g_{\mathrm{U}}}\right), \tag{4.4.27}$$

analogous to $S_{21}^{(1)}$ and $S_{21}^{(2)}$ given by equations (4.4.6) and (4.4.7).

If, on the other hand, we use equation (4.4.18) to evaluate the ratio $N_{\mathrm{L}}/N_{\mathrm{U}}$, we find

$$(\eta B^*)^{(3)}_{\mathrm{UL}} = \frac{1}{B_{\mathrm{LU}}} \sum_{i \neq \mathrm{L}} \frac{N_i R_{i\mathrm{U}}}{N_{\mathrm{L}}}, \tag{4.4.28}$$

$$\eta^{(3)}_{\mathrm{UL}} = \frac{1}{A_{\mathrm{UL}}} \sum_{i \neq \mathrm{L}} \left(R_{\mathrm{U}i} - \frac{g_{\mathrm{L}} N_i R_{i\mathrm{U}}}{g_{\mathrm{U}} N_{\mathrm{L}}}\right), \tag{4.4.29}$$

or

$$(\eta B^*)^{(4)}_{\mathrm{UL}} = 0, \tag{4.4.30}$$

$$\eta^{(4)}_{\mathrm{UL}} = \frac{1}{A_{\mathrm{UL}}} \sum_{i \neq \mathrm{L}} \left(R_{\mathrm{U}i} - \frac{N_i R_{i\mathrm{U}}}{N_{\mathrm{U}}}\right). \tag{4.4.31}$$

In all four cases, the $(\eta B^*)^{(m)}_{\mathrm{LU}}$ and $\eta^{(m)}_{\mathrm{UL}}$ terms do not directly involve the radiative transition U–L under consideration. Again, one chooses between the above four forms of S_{UL} by taking into account the model atom being used in the analysis and, coupled to this, the convergence properties of the iterative procedure designed to solve the pertinent equation of radiative transfer. In particular, S_{UL} must be so chosen that inequality (4.4.16) is satisfied.

Finally, we should mention that the linearisation of equations (4.4.17) and (4.4.18), as discussed in chapter 3, can resolve some of the difficulties associated with making the above-mentioned choice.

4.5 Solution of the transfer equation

We have seen in the preceding section that it is generally necessary to solve the equation of radiative transfer for multi-level atom analyses using iterative procedures. Probably the most satisfactory numerical method available incorporates the Feautrier technique (see section 3.2) whereby the equation of radiative transfer for the 2–1 transition, for example, may be written as

$$\left(\frac{\mu}{\kappa_{21}}\frac{\partial}{\partial z}\right)^2 \Phi_{21}^{(l)} = \Phi_{21}^{(l)} - S_{21}^{(l)}, \tag{4.5.1}$$

where

$$\Phi_{21}^{(l)}(z, \Delta\nu, \mu) = \tfrac{1}{2}[I_{21}^{(l)}(z, \Delta\nu, \mu > 0) + I_{21}^{(l)}(z, \Delta\nu, \mu < 0)], \tag{4.5.2}$$

and

$$S_{21}^{(l)} = \frac{\bar{J}_{21}^{(l)} + \epsilon'_{21} B_{\nu_{21}}(T) + (\eta B^*)_{21}^{(l-1)}}{1 + \epsilon'_{21} + \eta_{21}^{(l-1)}}. \tag{4.5.3}$$

The superscript l in these equations now refers to the iterative step. Clearly, the ηB^* and η terms contain all the allowable radiative transitions other than that between levels 2 and 1, and thus take the values computed at the previous iteration. One could incorporate a relaxation technique [81, 120] (see equation (5.6.34), for example, applied to conservation of linear momentum) if convergence difficulties are otherwise insurmountable.

Numerical accuracy can be a problem and it is of some value therefore to be able to test a computer code by comparing results (albeit for a simpler reduced problem) with those known to be exact. It is possible to obtain a closed form analytical solution to the equation of radiative transfer when the two allowable radiative transitions in a model 3-level atom, for example, have a common lower level (e.g. NaD lines) and stimulated emission can be ignored. Here, using the notation of section 4.2.1, we have

$$\left.\begin{aligned} R_{12} &= B_{12}\bar{J}_{21} + C_{12}, & R_{21} &= A_{21} + C_{21},\\ R_{13} &= B_{13}\bar{J}_{31} + C_{13}, & R_{31} &= A_{31} + C_{31},\\ R_{23} &= C_{23}, & R_{32} &= C_{32}, \end{aligned}\right\} \tag{4.5.4}$$

and, neglecting stimulated emission in S_{21} and S_{31},

$$S_{21} = \frac{N_2 A_{21}}{N_1 B_{12}}, \quad S_{21} = \frac{N_3 A_{31}}{N_1 B_{13}}. \tag{4.5.5}$$

Consequently, using equation (4.1.17) for the density ratio N_2/N_1, we have the *linear* form

$$S_{21} = \frac{\bar{J}_{21} + \dfrac{C_{12}}{B_{12}} + \dfrac{C_{32}(B_{13}\bar{J}_{31} + C_{13})}{B_{12}(A_{31} + C_{31} + C_{32})}}{1 + \dfrac{C_{21}}{A_{21}} + \dfrac{C_{23}(A_{31} + C_{31})}{A_{21}(A_{31} + C_{31} + C_{32})}}, \tag{4.5.6}$$

with a similar expression for S_{31} involving a *linear* combination of $\bar{J}_{31}$ and $\bar{J}_{21}$.

The 1-dimensional time-dependent equations of radiative transfer for the 2–1 and 3–1 transitions then have the general form

$$\mu \frac{\partial I_{21}}{\partial \tau} = \sigma\phi(\Delta\nu)I_{21} - \phi(\Delta\nu)\int_{-\infty}^{\infty} d(\Delta\nu')\phi(\Delta\nu')\int_{-1}^{1} d\mu' \times [w_{11}I_{21}(\tau, \Delta\nu', \mu') + w_{12}I_{31}(\tau, \Delta\nu', \mu')] - \phi(\Delta\nu)q_{21}, \tag{4.5.7}$$

$$\mu \frac{\partial I_{31}}{\partial \tau} = \phi(\Delta\nu)I_{31} - \phi(\Delta\nu)\int_{-\infty}^{\infty} d(\Delta\nu')\phi(\Delta\nu')\int_{-1}^{1} d\mu' \times [w_{21}I_{21}(\tau, \Delta\nu', \mu') + w_{22}I_{31}(\tau, \Delta\nu', \mu')] - \phi(\Delta\nu)q_{31}, \tag{4.5.8}$$

where we have taken

$$\phi_{21}(\Delta\nu) \equiv \phi_{31}(\Delta\nu) \equiv \phi(\Delta\nu), \tag{4.5.9}$$

and written

$$d\tau = \frac{h\nu_{31}}{4\pi} N_1 B_{13}\, dz, \tag{4.5.10}$$

with $\sigma = \kappa_{21}/\kappa_{31}$. We arbitrarily take $\sigma > 1$. Clearly, the precise form of the w_{ij} and q_{ij} may be ascertained from equation (4.5.6) for S_{21} and its corresponding expression for S_{31}.

We illustrate the solution to the above coupled integro-differential equations for a semi-infinite constant property (i.e. σ, w_{ij} and q_{ij} are all constant) atmosphere. Following the procedure detailed in chapter 2, we define $\xi = \mu/\phi(\Delta\nu)$ such that, with $\gamma = \max(\xi)$ and $I(\tau, \Delta\nu, \mu) \Longleftrightarrow \phi(\tau, \xi)$, equations (4.5.7) and (4.5.8) in vector notation then become

$$\xi \frac{\partial \mathbf{\Phi}}{\partial \tau} = \sigma\mathbf{\Phi} - \mathbf{P}\int_{-\gamma}^{\gamma} \Psi(\xi')\mathbf{\Phi}(\tau, \xi')d\xi' - \mathbf{Q}, \tag{4.5.11}$$

where

$$\Psi(\xi) = \int_{R(\xi, \Delta\nu)} \phi^2(\Delta\nu)\,d\Delta\nu, \tag{4.5.12}$$

with

$$\mathbf{\Phi} = \begin{pmatrix} \Phi_{21} \\ \Phi_{31} \end{pmatrix}, \quad \sigma = \begin{pmatrix} \sigma & 0 \\ 0 & 1 \end{pmatrix}, \tag{4.5.13}$$

$$\mathbf{P} = \begin{pmatrix} w_{11} & w_{12} \\ w_{21} & w_{22} \end{pmatrix}, \quad \mathbf{Q} = \begin{pmatrix} q_{21} \\ q_{31} \end{pmatrix}. \tag{4.5.14}$$

The domain $R(\xi, \Delta\nu)$ is defined such that $\Delta\nu \in R(\xi, \Delta\nu)$ if and only if $|\xi|\phi(\Delta\nu) \leqslant 1$.

We now define a vector Green's function $\mathbf{G}(\tau^*, \xi^* \to \tau, \xi)$ satisfying

$$\xi \frac{\partial \mathbf{G}}{\partial \tau} = \sigma\mathbf{G} - \mathbf{P}\int_{-\gamma}^{\gamma} \Psi(\xi')\mathbf{G}(\tau^*, \xi^* \to \tau, \xi')d\xi' - \delta(\tau - \tau^*)\delta(\xi - \xi^*)\begin{pmatrix} 1 \\ 1 \end{pmatrix} \tag{4.5.15}$$

(see equation (2.1.40)) so that, writing

$$\mathbf{G}(\tau^*, \xi^* \to \tau, \xi) = \begin{pmatrix} G_1(\tau^*, \xi^* \to \tau, \xi) & 0 \\ 0 & G_2(\tau^*, \xi^* \to \tau, \xi) \end{pmatrix}, \tag{4.5.16}$$

where

$$\mathbf{G}(\tau^*, \xi^* \to \tau, \xi) = \begin{pmatrix} G_1(\tau^*, \xi^* \to \tau, \xi) \\ G_2(\tau^*, \xi^* \to \tau, \xi) \end{pmatrix}, \tag{4.5.17}$$

the required vector unknown $\mathbf{\Phi}(\tau, \xi)$ for a semi-infinite atmosphere may be obtained from

$$\mathbf{\Phi}(\tau, \xi) = \int_0^\infty d\tau^* \int_{-\gamma}^{\gamma} d\xi^* \, \mathbf{G}(\tau^*, \xi^* \to \tau, \xi)\mathbf{Q}(\tau^*, \xi^*)$$
$$- \int_{-\gamma}^{0} \xi^* \, d\xi^* \, \mathbf{G}(0, \xi^* \to \tau, \xi)\mathbf{\Phi}_{\text{inc}}(\xi^*) \tag{4.5.18}$$

(see equation (2.1.73)), where $\mathbf{\Phi}_{\text{inc}}(\xi^*)$ is the radiation field incident at $\tau^* = 0$.

Clearly, integrating equation (4.5.15) with respect to τ 'at the point' τ^* yields the 'jump' condition on $\mathbf{G}$:

$$\xi[G^+(\tau^*, \xi^* \to \tau^*, \xi) - G^-(\tau^*, \xi^* \to \tau^*, \xi)] = -\delta(\xi - \xi^*)\begin{pmatrix} 1 \\ 1 \end{pmatrix} \tag{4.5.19}$$

(see equation (2.1.78)) where $\mathbf{G}^+$ and $\mathbf{G}^-$ refer to values of $\mathbf{G}$ as $\tau \to \tau^*$ from above and below respectively.

Finally, $\mathbf{G}$ must satisfy the two further conditions

$$\mathbf{G}(\tau^*, \xi^* \to \tau, \xi) \to 0 \quad \text{as } |\tau - \tau^*| \to \infty \tag{4.5.20}$$

(see condition (2.1.41)), and

$$\mathbf{G}(\tau^*, \xi \to 0, \xi) = 0 \;\; \forall \;\; \xi < 0 \tag{4.5.21}$$

(see equation (2.1.43)).

We are now in a position to use the normal mode expansion technique discussed in some detail in chapter 2. In particular, we require a solution vector $\mathbf{G}$ to equation (4.5.15) with $\tau \neq \tau^*$ (i.e. the homogeneous equation) satisfying equation (4.5.19). To do this, we propose a solution of the form [110]

$$\mathbf{G}(\tau^*, \xi^* \to \tau, \xi) = \mathbf{g}(\eta, \xi)e^{\tau/\eta}. \tag{4.5.22}$$

Substitution into equation (4.5.15) with $\tau \neq \tau^*$ yields

$$\begin{pmatrix} \sigma\eta - \xi & 0 \\ 0 & \eta - \xi \end{pmatrix} \mathbf{g}(\eta, \xi) = \eta\mathbf{PM}(\eta), \tag{4.5.23}$$

where

$$\mathbf{M}(\eta) = \int_{-\gamma}^{\gamma} \Psi(\xi)g(\eta, \xi)\,d\xi. \tag{4.5.24}$$

We must now consider three distinct regions over which η may vary.

Range I:

$$\eta \notin [-\gamma, \gamma].$$

Clearly, for η in this range (recall $\sigma > 1$), equation (4.5.23) gives

$$\mathbf{g}^{(1)}(\eta, \xi) = \begin{pmatrix} \dfrac{\eta}{\sigma\eta - \xi} & 0 \\ 0 & \dfrac{\eta}{\eta - \xi} \end{pmatrix} \mathbf{PM}(\eta), \tag{4.5.25}$$

so that, defining $\mathbf{\Lambda}(\eta)$ by

$$\mathbf{\Lambda}(\eta) = \mathbb{1} - \int_{-\gamma}^{\gamma} \Psi(\xi) \begin{pmatrix} \dfrac{\eta}{\sigma\eta - \xi} & 0 \\ 0 & \dfrac{\eta}{\eta - \xi} \end{pmatrix} \mathbf{P}\, d\xi, \tag{4.5.26}$$

where $\mathbb{1}$ is the identity matrix, equation (4.5.24) yields

$$\mathbf{\Lambda}(\eta)\mathbf{M}(\eta) = 0. \tag{4.5.27}$$

We therefore have

$$\text{determinant } [\mathbf{\Lambda}(\eta)] = 0, \tag{4.5.28}$$

which has an even number $2N_r$ of roots (usually two or four depending upon σ and the w_{ij}). It is not difficult to show that these roots are real and symmetric about $\eta = 0$, and we therefore refer to them as $\pm\eta_{0k}$ for $k = 1, N_r$.

Range II:

$$\eta \in \left[-\frac{\gamma}{\sigma}, \frac{\gamma}{\sigma}\right]$$

In this region, both $\eta - \xi$ and $\sigma\eta - \xi$ can have zero value (again recall that we have taken $\sigma > 1$). This corresponding difficulty of inverting equation (4.5.23) may be overcome in the same manner as proposed in chapter 2. The resulting equation is

$$\mathbf{g}^{(2)}(\eta, \xi) = \mathbf{T}^{(2)}(\eta, \xi)\mathbf{PM}(\eta), \tag{4.5.29}$$

where

$$\mathbf{T}^{(2)}(\eta, \xi) = \begin{pmatrix} T_1^{(2)}(\eta, \xi) & 0 \\ 0 & T_2^{(2)}(\eta, \xi) \end{pmatrix}, \tag{4.5.30}$$

with

$$T_1^{(2)}(\eta, \xi) = P\left(\frac{\eta}{\sigma\eta - \xi}\right) + \lambda_1^{(2)}(\eta)\delta(\sigma\eta - \xi), \tag{4.5.31}$$

and

$$T_2^{(2)}(\eta, \xi) = P\left(\frac{\eta}{\eta - \xi}\right) + \lambda_2^{(2)}(\eta)\delta(\eta - \xi). \tag{4.5.32}$$

The symbol P indicates that any ensuing integration is to be taken in the Cauchy principal value sense.

As in range I, we use equation (4.5.24) to obtain the condition

$$\det\left[\mathbb{1} - \int_{-\gamma}^{\gamma} \Psi(\xi) T^{(2)}(\eta, \xi) \mathbf{P} \mathrm{d}\xi\right] = 0. \tag{4.5.33}$$

This equation is sufficient to determine the two distribution functions $\lambda_1^{(2)}(\eta)$ and $\lambda_2^{(2)}(\eta)$.

Range III:

$$\eta \in \left[-\gamma, -\frac{\gamma}{\sigma}\right], \quad \eta \in \left[\frac{\gamma}{\sigma}, \gamma\right]$$

The inversion of equation (4.5.23) in this domain gives

$$\mathbf{g}^{(3)}(\eta, \xi) = \mathbf{T}^{(3)}(\eta, \xi)\mathbf{P}\mathbf{M}(\eta), \tag{4.5.34}$$

where

$$\mathbf{T}^{(3)}(\eta, \xi) = \begin{pmatrix} \dfrac{\eta}{\sigma\eta - \xi} & 0 \\ 0 & P\left(\dfrac{\eta}{\eta - \xi}\right) + \lambda^{(3)}(\eta)\delta(\eta - \xi) \end{pmatrix}. \tag{4.5.35}$$

Equation (4.5.24), in turn, yields

$$\det\left[\mathbb{1} - \int_{-\lambda}^{\lambda} \Psi(\xi) \mathbf{T}^{(3)}(\eta, \xi) \mathbf{P} \mathrm{d}\xi\right] = 0 \tag{4.5.36}$$

which then gives two values of $\lambda^{(3)}(\eta)$ denoted by $\lambda_1^{(3)}(\eta)$ and $\lambda_2^{(3)}(\eta)$. This two-fold degeneracy therefore produces two linearly independent functional forms for $\mathbf{g}^{(3)}$ which we refer to as $g_1^{(3)}$ and $g_2^{(3)}$.

Hence, the solution for $\mathbf{G}(\tau^*, \xi^* \to \tau, \xi)$ satisfying $\mathbf{G} \to 0$ as $|\tau - \tau^*| \to \infty$ is given by

$$\begin{aligned} \mathbf{G}(\tau^*, \xi^* \to \tau, \xi) = &\sum_{k=1}^{N_r} \overset{+}{a}_k \mathbf{g}^{(1)}(\eta_{0k}, \xi)\, \mathrm{e}^{(\tau - \tau^*)/\eta_{0k}} \\ &+ \int_0^{\gamma} A^{(2)}(\eta) \mathbf{g}^{(2)}(\eta, \xi)\, \mathrm{e}^{(\tau - \tau^*)/\eta} \mathrm{d}\eta \\ &+ \int_{\gamma/\sigma}^{\gamma} [A_1^{(3)}(\eta) \mathbf{g}_1^{(3)}(\eta, \xi) + A_2^{(3)}(\eta, \xi)]\, \mathrm{e}^{(\tau - \tau^*)/\eta} \mathrm{d}\eta, \end{aligned} \tag{4.5.37}$$

for $\tau < \tau^*$ whilst, for $\tau > \tau^*$, we have

$$\mathbf{G}(\tau^*, \xi^* \to \tau, \xi) = -\sum_{k=1}^{N_r} a_k^- \mathbf{g}^{(1)}(-\eta_{0k}, \xi)\, e^{-(\tau-\tau^*)/\eta_{0k}} - \int_{-\gamma}^{0} A^{(2)}(\eta)\mathbf{g}^{(2)}(\eta, \xi)\, e^{(\tau-\tau^*)/\eta}\, d\eta - \int_{-\gamma}^{-\gamma/\sigma} [A_1^{(3)}(\eta)\mathbf{g}_1^{(3)}(\eta, \xi) + A_2^{(3)}(\eta)\mathbf{g}_2^{(3)}(\eta, \xi)]\, e^{(\tau-\tau^*)/\eta}\, d\eta. \tag{4.5.38}$$

The expansion coefficients $a_k^+, a_k^-, A^{(2)}(\eta), A_1^{(3)}(\eta)$ and $A_2^{(3)}(\eta)$ may be determined as in section 2.8 by substituting these last two equations into the 'jump' condition (4.5.19) and then applying certain full-range orthogonality relationships (not derived here). This completes the specification of the infinite space vector Green's function $\mathbf{G}_\infty(\tau^*, \xi^* \to \tau, \xi)$.

However, $\mathbf{G}_\infty$ does not satisfy the condition (4.5.21) of zero incident radiation flux at the surface of the atmosphere. We therefore define the required half-space vector Green's function $\mathbf{G}$ such that

$$\mathbf{G}(\tau^*, \xi^* \to \tau, \xi) = \mathbf{G}_\infty(\tau^*, \xi^* \to \tau, \xi) + \sum_{k=1}^{N_r} b_k^- \mathbf{g}^{(1)}(-\eta_{0k}, \xi)\, e^{-\tau/\eta_{0k}} + \int_{-\gamma}^{0} B^{(2)}(\eta)\mathbf{g}^{(2)}(\eta, \xi)\, e^{\tau/\eta}\, d\eta + \int_{-\gamma}^{-\gamma/\sigma} [B_1^{(3)}(\eta)\mathbf{g}_1^{(3)}(\eta, \xi) + B_2^{(3)}(\eta)\mathbf{g}_2^{(3)}(\eta, \xi)]\, e^{\tau/\eta}\, d\eta. \tag{4.5.39}$$

Clearly, substitution of equation (4.5.39) into (4.5.21) and the application of the appropriate half-range orthogonality relationships (again, not derived here) will then enable the new expansion coefficients b_k^-, $B^{(2)}(\eta)$, $B_1^{(3)}(\eta)$ and $B_2^{(3)}(\eta)$ to be immediately determined.

The above solution can be generalised further for a multi-level atom but still only when all the allowable radiative transitions have the one common lower level.

5

Radiation gas dynamics

In all the preceding work we have, of necessity, allowed the atoms interacting with the photons and electrons to each exhibit their individual non-zero motions, but have assumed the average of these motions to be zero. In this sense, the gas could be considered stationary. Although this simplification enabled us to gain some understanding of the basic transfer processes, it neglects a series of effects which have a profound influence on our interpretation of the emergent line spectra. Indeed, there exists a large variety of velocity fields in stellar atmospheres ranging from 'quiet' oscillations with velocity amplitudes of approximately 1 km sec^{-1} through shock waves where the velocities are of the order of the local speed of sound, to mass motions exceeding 2000 km sec^{-1}. Since these macroscopic velocities are important to our understanding of the structure of stellar atmospheres (both in the determination of the temperature and density distributions of the gas giving rise to the observed line spectra, and in the specification of the physical processes that control that structure), a detailed knowledge of the manner in which mass motions alter the qualitative and quantitative nature of the radiation field is essential.

In the following sections, therefore, we first discuss the derivation of the pertinent equations stipulating conservation of mass, linear momentum and energy (these are required for a self-consistent specification of densities, temperatures and macroscopic velocities), and correspondingly isolate those assumptions and approximations regarding Maxwellian microscopic velocity distributions, macroscopic velocity gradients, LTE for particles, etc. which are inherent in such equations. We are then able to derive the simple 2-level atom complete re-distribution source function. Following this, we illustrate several methods of solution to the consequent equation of radiative transfer (together with the appropriate gas-dynamical conservation equations) and, finally, discuss the influence of different macroscopic velocity distributions on the shape and magnitude of the emergent line intensity.

5.1 The conservation equations

We wish to introduce a macroscopic velocity **V** into the equation of radiative transfer. This, together with the other macroscopic parameters density (for example, N_L, N_U, N_e) and temperature (T_L, T_U, T_e), cannot be arbitrarily specified independent of one another, but must be so chosen that the appropriate equations specifying conservation of mass, linear momentum and energy are not violated. This stipulation is nothing more than a desire for Newton's law of motion to be satisfied. Clearly, therefore, a discussion of the derivation (and consequent assumptions and approximations) of these equations is necessary for our understanding of the manner in which radiation interacts not only with the microscopic particles, but with the ensemble of moving gas as a whole. However, our main interest in this monograph lies in the structure of the radiative transfer equation and the form of its solutions; the matter particle equations specifically are of a secondary interest and the following discussion of their derivation is therefore only of a somewhat cursory phenomenological nature. Indeed, the subjects of gas kinetics and (quantum) statistical mechanics are so immense that, out of necessity, our discussion is limited only to those points which bear directly on the incorporation of such theories into our model of the interaction between radiation and matter.

5.1.1 The general Boltzmann equation [30, 40, 52]

We begin by defining the distribution function $f_i(\mathbf{r}, \mathbf{v}_i, t)$ of the ith species such that $f_i(\mathbf{r}, \mathbf{v}_i, t)\delta\mathbf{r}\delta\mathbf{v}_i$ represents the number of the ith species particles moving with velocities (measured in the rest frame of the observer) in the range $\delta\mathbf{v}_i$ about $\mathbf{v}_i$ located in the volume element $\delta\mathbf{r}$ about $\mathbf{r}$ at time t. It is clear that $\mathbf{r}$ and t are independent quantities. We next assume that the velocity of each individual particle is independent of its previous history. This history, of course, involves all the interactions each particle has with all the other particles in the ensemble. In this sense, therefore, we assume that there exists zero correlation between particle velocities at previous times $t' < t$, or at least any correlations that do exist decrease sufficiently rapidly that they can be ignored for times of interest (i.e. for times which are large compared with the time duration of collisions). Consequently, we assume $\mathbf{v}_i$ to be independent of both $\mathbf{r}$ and t.

Thence, the ith species number density $N_i(\mathbf{r}, t)$ is obtained by simply summing over all the individual velocities, viz.

$$N_i(\mathbf{r}, t) = \int f_i(\mathbf{r}, \mathbf{v}_i, t)\, \mathrm{d}\mathbf{v}_i. \tag{5.1.1}$$

Note that this velocity distribution f_i differs from that used in chapter 1 in that the former function F_i was treated as a probability, i.e.

$$F_i = \frac{f_i}{N_i} \tag{5.1.2}$$

- see equation (1.4.4).

The momentum of the ith species particles occupying a unit volume element about $\mathbf{r}$ is written as $N_i m_i \bar{\mathbf{v}}_i$ where m_i is the mass of the ith species and $\bar{\mathbf{v}}_i$ is the average velocity (at $\mathbf{r}$ and t). This momentum can be obtained by summing over all the individual momenta, viz.

$$N_i m_i \bar{\mathbf{v}}_i = \int m_i \mathbf{v}_i f_i(\mathbf{r}, \mathbf{v}_i, t)\, d\mathbf{v}_i,$$

i.e.

$$\bar{\mathbf{v}}_i = \frac{1}{N_i} \int \mathbf{v}_i f_i\, d\mathbf{v}_i. \tag{5.1.3}$$

It is usually the momentum, not velocity, which is measured; it is therefore standard procedure to use equation (5.1.3) to define the average velocity (relative to the observer). Indeed, all averaged quantities $\bar{\phi}_i(\mathbf{r}, t)$ of an integrable function $\phi_i(\mathbf{r}, \mathbf{v}_i, t)$ may be defined by

$$\bar{\phi}_i(\mathbf{r}, t) = \frac{1}{N_i} \int \phi_i f_i\, d\mathbf{v}_i. \tag{5.1.4}$$

The total number density N is the sum of all the individual number densities N_i, i.e.

$$N(\mathbf{r}, t) = \sum_{i \neq I} N_i(\mathbf{r}, t), \tag{5.1.5}$$

where the summation in this last equation extends over all the different species except the photons (I). The total particle momentum is the sum of all the individual species momenta $\rho_i \bar{\mathbf{v}}_i$ where $\rho_i = N_i m_i$ is the mass density of the ith species, i.e.

$$\rho \mathbf{V}(\mathbf{r}, t) = \sum_{i \neq I} \rho_i \bar{\mathbf{v}}_i$$

$$= \sum_{i \neq I} m_i \int \mathbf{v}_i f_i\, d\mathbf{v}_i, \tag{5.1.6}$$

where

$$\rho(\mathbf{r}, t) = \sum_{i \neq I} \rho_i = \sum_{i \neq I} N_i m_i$$

$$= \sum_{i \neq I} m_i \int f_i\, d\mathbf{v}_i. \tag{5.1.7}$$

As we shall see, it is the macroscopic velocity $\mathbf{V}(\mathbf{r}, t)$ which appears in the spectral line source function.

Finally, we define the peculiar velocity $\mathbf{V}_i$ of the individual particles belonging to the ith species by

$$\mathbf{V}_i = \mathbf{v}_i - \mathbf{V}, \tag{5.1.8}$$

so that, using equation (5.1.4), we have the diffusion velocity $\mathbf{V}_i$ satisfying

$$\bar{\mathbf{V}}_i = \bar{\mathbf{v}}_i - \mathbf{V} \tag{5.1.9}$$

(note that $\mathbf{V}$ is independent of $\mathbf{v}_i$). Thus, summing over all the individual species diffusion velocities,

$$\begin{aligned}\sum_{i \neq I} \rho_i \bar{\mathbf{V}}_i &= \sum_{i \neq I} \rho_i (\bar{\mathbf{v}}_i - \mathbf{V}) = \rho \mathbf{V} - \mathbf{V} \sum_{i \neq I} \rho_i \\ &= 0,\end{aligned} \tag{5.1.10}$$

as expected.

We now examine the derivation of the defining equation for the distribution function f_i. First, consider the time evolution of a volume element $\delta\mathbf{r}$ about $\mathbf{r}$ of gas from time t to $t + \delta t$. If an ith species particle in this ensemble does not interact with any other particle in the time interval δt, then its position vector will change from $\mathbf{r}$ to $\mathbf{r} + \mathbf{v}_i \delta t$ (note that $\mathbf{r}$, $\mathbf{v}_i$ and t are all independent variables). Further, if there exists a resultant external force $\mathbf{F}_i$ acting on all the individual particles in the ith species, and $\mathbf{F}_i$ is considered constant over the time interval δt, then the velocity of an ith species particle will change from $\mathbf{v}_i$ to $\mathbf{v}_i + \mathbf{a}_i \delta t$ where $\mathbf{a}_i = \mathbf{F}_i / m_i$ is the acceleration due to $\mathbf{F}_i$. (The case of a magnetic field $\mathbf{B}$ for which $\mathbf{F}_i = \mathbf{v}_i \times \mathbf{B}$ dependent upon $\mathbf{v}_i$ can be easily treated [30]. The inclusion of non-zero $\mathbf{B}$ introduces an extra term in the Boltzmann equation.) Finally, the volume element will change from $\delta\mathbf{r}$ to $\delta\mathbf{r} + \mathbf{v}_i \delta t$. Thus, the number of ith species particles at time t represented by $f_i(\mathbf{r}, \mathbf{v}_i, t)\delta\mathbf{r}\delta\mathbf{v}_i$ will, at time $t + \delta t$ in the absence of collisions or interactions of any kind, be represented by $f_i(\mathbf{r} + \mathbf{v}_i \delta t, \mathbf{v}_i + \mathbf{a}_i \delta t, t + \delta t) \times (\delta\mathbf{r} + \mathbf{v}_i, t)\,(\delta\mathbf{v}_i + \mathbf{a}_i \delta t)$. Statistically, these two numbers should be the same. However, if encounters do occur in the time interval δt, and this will result not only in particles altering their velocities but, in the case of interactions with photons, for example, altering their state (e.g. from the de-excited L state to the excited U state), then the number of particles at time $t + \delta t$ will, in general, differ from that at time t. An inherent assumption is immediately made here; although we now allow encounters in the time interval δt, we still represent the new position vector by $\mathbf{r} + \mathbf{v}_i \delta t$ and the new velocity by $\mathbf{v}_i + \mathbf{a}_i \delta t$. We then write

$$\begin{aligned}&f_i(\mathbf{r} + \mathbf{v}_i \delta t, \mathbf{v}_i + \mathbf{a}_i \delta t, t + \delta t)(\delta\mathbf{r} + \mathbf{v}_i + \mathbf{a}_i \delta t) \\ &\quad - f_i(\mathbf{r}, \mathbf{v}_i, t)\delta\mathbf{r}\,\delta\mathbf{v}_i = \sum_j \Phi_{ij} \delta\mathbf{r}\,\delta\mathbf{v}_i \delta t,\end{aligned} \tag{5.1.11}$$

where the quantity $\Phi_{ij}\delta\mathbf{r}\delta\mathbf{v}_i\delta t$ on the RHS (5.1.11) represents the effect of changes due to the above-mentioned interactions with other (jth) species and with itself ($j = i$). Note that we include photons as interacting ‘particles’.

We can expand the LHS (5.1.11) in a Taylor series, divide through by $\delta\mathbf{r}\delta\mathbf{v}_i\delta t$, and let $\delta t \to 0$ to obtain

$$\frac{\partial f_i}{\partial t} + \mathbf{v}_i \cdot \nabla f_i + \mathbf{a}_i \cdot \nabla_{\mathbf{v}_i} f_i = \sum_j \Phi_{ij}, \tag{5.1.12}$$

where we have written

$$\nabla_{\mathbf{v}_i} \equiv \frac{\partial}{\partial \mathbf{v}_i}.$$

The equation of radiative transfer already derived in chapter 1 can be thought of as a special case of equation (5.1.12). Here, with $\mathbf{v}_i \equiv \mathbf{v}_I = c\mathbf{\Omega}$ (where c is the speed of light and $\mathbf{\Omega}$ the direction of propagation), $\mathbf{a}_i \equiv \mathbf{a}_I \equiv 0$ (since $\mathbf{a}_I = \dot{\mathbf{v}}_I \equiv 0$ where we have ignored all relativistic effects) and $f_i \equiv f_I = I/h\nu c$ (recall that I represents the energy in a beam of photons each with energy $h\nu$ - we also divide I by c to obtain f_I since f_I should represent the number of photons per unit volume), we obtain

$$\text{LHS (5.1.12)} = \frac{1}{h\nu c}\frac{\partial I}{\partial t} + \frac{1}{h\nu}(\mathbf{\Omega} \cdot \nabla)I, \tag{5.1.13}$$

while, by comparison with equation (1.2.7), we must have

$$\sum_j \Phi_{Ij} = -N_{\mathrm{L}} I \bar{\sigma}_{I\mathrm{L}} + N_{\mathrm{U}} I \bar{\sigma}_{I\mathrm{U}} + N_{\mathrm{U}} \bar{\sigma}_{0\mathrm{U}}, \tag{5.1.14}$$

where the 'encounter cross-sections' $\bar{\sigma}_{ij}$ are

$$\bar{\sigma}_{I\mathrm{L}} = B_{\mathrm{LU}}\phi/4\pi, \quad \bar{\sigma}_{I\mathrm{U}} = B_{\mathrm{UL}}\psi/4\pi, \quad \bar{\sigma}_{0\mathrm{U}} = A_{\mathrm{UL}}j/4\pi. \tag{5.1.15}$$

Recalling equation (5.1.1), we see that

$$N_{\mathrm{U}} = \int f_{\mathrm{U}}(\mathbf{r}, \mathbf{v}_{\mathrm{U}}, t)\, \mathrm{d}\mathbf{v}_{\mathrm{U}}, \tag{5.1.16}$$

and

$$N_{\mathrm{L}} = \int f_{\mathrm{L}}(\mathbf{r}, \mathbf{v}_{\mathrm{L}}, t)\, \mathrm{d}\mathbf{v}_{\mathrm{L}}, \tag{5.1.17}$$

so that equation (5.1.14), with $I = h\nu c f_I$, is a special case of

$$\sum_j \Phi_{ij} = \sum_j \int (\sigma'_{ij} f'_i f'_j\, \mathrm{d}\mathbf{v}'_j - \sigma_{ij} f_i f_j\, \mathrm{d}\mathbf{v}_j), \tag{5.1.18}$$

where the $-\sigma_{ij}f_if_j$ term represents sinks of $f_i(\mathbf{r}, \mathbf{v}_i, t)$ and the dashed quantity $\sigma'_{ij}f'_if'_j$ represents corresponding sources.

The $\Sigma\Phi_{ij}$ encounter term given by equation (5.1.18) also applies to particles and is discussed in great detail in all the standard texts on the general Boltzmann equation. Here, we only wish to examine those points pertinent to the analysis of the interaction between radiation and matter.

The first and most important assumption in the above encounter term relates to the binary $(f_i f_j)$ nature of the interactions. To see this, we first describe the sink $-\sigma_{ij} f_i f_j \, d\mathbf{v}_j$. It contains the factor $f_j(\mathbf{r}, \mathbf{v}_j, t)\delta\mathbf{v}_j$ representing the number of jth species particles per unit volume in the velocity range $\delta\mathbf{v}_j$ about $\mathbf{v}_j$ which are capable of interacting with the ith species (f_i), the time rate of these encounters being σ_{ij}. These encounters will deflect the ith species particle *out of* $\delta\mathbf{v}_i$ about $\mathbf{v}_i$, thus reducing the number of these particles with velocities in this range. Consequently, summing over all the jth species velocities, $\delta\mathbf{v}_i \delta\mathbf{r} \int \sigma_{ij} f_i f_j \, d\mathbf{v}_j$ represents the time rate of decrease of ith species particles in $\delta\mathbf{r}$ about $\mathbf{r}$ with velocities in $\delta\mathbf{v}_i$ about $\mathbf{v}_i$ due to interactions with the jth species. Note that we sum over all the jth species interactions in equation (5.1.18).

Clearly, we have assumed the gas to be sufficiently dilute so that three (or more) particle encounters are rare and may therefore be neglected.

Correspondingly, ith species particles outside the velocity range $\delta\mathbf{v}_i$ about $\mathbf{v}_i$ will interact with jth species particles with the possible deflection *into* $\delta\mathbf{v}_i$ about $\mathbf{v}_i$. The resulting source may be obtained in much the same way as that leading to the sinks described above. However, here, the velocities $\delta\mathbf{v}_i$ about $\mathbf{v}_i$ clearly arise only after the collision. Thus, to distinguish between velocities before and after encounters, we introduce the dashed notation in equation (5.1.18) relating to sources only where f_i', for example, has the function dependence $f_i(\mathbf{r}, \mathbf{v}_i', t)$ with $\mathbf{v}_i' \equiv \mathbf{v}_i$. As we shall see, this notation is convenient in deriving the required conservation equations. Note, however, that we still retain the assumption of only binary collisions.

If the colliding particles are perfectly elastic and symmetric (which they are *not*) and one can ignore the effect of the forces $\mathbf{F}_j$ on the particles during a collision (or encounter), then one can invoke conservation of energy and linear momentum arguments for individual particles to show that [30]

$$\sigma_{ij}' \, d\mathbf{v}_j' \equiv \sigma_{ij} \, d\mathbf{v}_j. \qquad (5.1.19)$$

This statement effectively insists that for every encounter (in which two particles with velocities $\mathbf{v}_i$ and $\mathbf{v}_j$ before the encounter have velocities $\mathbf{v}_i'$ and $\mathbf{v}_j'$ after) there is a corresponding inverse encounter (whereby the velocities before and after are $\mathbf{v}_i'$ and $\mathbf{v}_j'$, and $\mathbf{v}_i$ and $\mathbf{v}_j$ respectively), and this, in turn, assumes there exists some type of equilibrium in the gas. Clearly, as one approaches the boundary of the medium, any equilibrium condition *must* be violated. We have already discussed, at some length in chapter 1, for example, the non-equilibrium situation in which photons escaping from the surface of a radiating gas destroy any equilibrium configuration which might have otherwise existed. Further, the assumption (5.1.19) will not hold, for example, for spontaneous de-excitation; the inverse encounter would be spontaneous excitation and this, of course, is not physically realistic (i.e. $A_{\mathrm{UL}} \neq 0$ but $A_{\mathrm{LU}} = 0$

or, $\sigma_{UL} \neq 0$, $\sigma_{LU} = 0$). Indeed, since $B_{UL} \neq B_{LU}$ in general and $\phi \not\equiv \psi$, we must have $\bar{\sigma}_{IL} \neq \bar{\sigma}_{IU}$.

The above discussion is somewhat phenomenological. One could perhaps be more precise by considering Louiville's equation and the consequent BBGKY hierarchy [80]. However, the generalisations offered by such an analysis are of little practical importance for our needs. A still more axiomatic approach to the detailed interactions and all their effects is contained in the quantum statistical treatment of gas kinetics, and this will be discussed in chapter 8.

The above equations (5.1.12), (5.1.18) and (5.1.19) then yield the general Boltzmann equation

$$\frac{\partial f_i}{\partial t} + \mathbf{v}_i \cdot \nabla f_i + \mathbf{a}_i \cdot \nabla_{\mathbf{v}_i} f_i$$
$$= \sum_{j \neq I}^{(1)} \int \sigma_{ij}(f_i' f_j' - f_i f_j)\, \mathrm{d}\mathbf{v}_j + \sum_{j}^{(2)} \int (\sigma_{ij}' f_j'\, \mathrm{d}\mathbf{v}_j' - \sigma_{ij} f_i f_j\, \mathrm{d}\mathbf{v}_j), \quad (5.1.20)$$

where the superscripted (1) summation refers to all those 'elastic' encounters for which equation (5.1.19) is reasonably appropriate whilst the superscripted (2) summation contains all other (inelastic) terms describing a change of state.

5.1.2 *The species Boltzmann equation*

We begin by again examining the equation of radiative transfer. For example, if $q(\gamma_L)$ is the probability of a photon of frequency $\gamma_L = \Delta\nu - \nu \mathbf{v}_L \cdot \boldsymbol{\Omega}/c$ (γ_L measured in the rest frame of the atom) being absorbed, then (assuming the normalisation factor $h\nu c B_{LU}/4\pi$) with $f_i \equiv f_I = I/h\nu c$, we have

$$\sigma_{IL} = \frac{h\nu c B_{LU}}{4\pi} q(\gamma_L). \quad (5.1.21)$$

Consequently,

$$\sigma_{IL} f_I f_L \, \mathrm{d}\mathbf{v}_L = \frac{B_{LU} I}{4\pi} \int q(\gamma_L) f_L \, \mathrm{d}\mathbf{v}_L = \frac{B_{LU} I N_L \phi}{4\pi}, \quad (5.1.22)$$

where we have defined the ensemble absorption probability

$$\phi = \frac{1}{N_L} \int q(\gamma_L) f_L \, \mathrm{d}\mathbf{v}_L. \quad (5.1.23)$$

This last result is consistent with equation (1.4.5) with $f_L = N_L F_L$ (see equation (5.1.2)).

Similarly, we have

$$\sigma_{IU}' f_I' f_U' \, \mathrm{d}\mathbf{v}_U' = \frac{B_{UL} I N_U \psi}{4\pi}, \quad (5.1.24)$$

$$\psi = \frac{1}{N_U} \int e_1(\gamma_U) f_U \, d\mathbf{v}_U, \tag{5.1.25}$$

where $e_1(\gamma_U)$ is an emission probability analogous to $q(\gamma_L)$. We shall examine $e_1(\gamma_U)$ in some detail in chapters 6 and 7.

Finally, putting $f'_0 \equiv 1$ since spontaneous emission is a singular (not binary) encounter, we have

$$\sigma'_{0U} f'_0 f'_U \, d\mathbf{v}'_U = \frac{A_{UL} N_U j}{4\pi}, \tag{5.1.26}$$

where

$$j = \frac{1}{N_U} \int e_2(\gamma_U) f_U \, d\mathbf{v}_U. \tag{5.1.27}$$

(As we shall see, $e_1(\gamma_U) \equiv e_2(\gamma_U)$.) Note that the emission probabilities $e_1(\gamma)$ and $e_2(\gamma)$ must be normalised to unity, i.e.

$$\int e_1(\gamma) \, d\gamma = 1 = \int e_2(\gamma) \, d\gamma. \tag{5.1.28}$$

Putting $f_i = f_I = I/h\nu c$ and $\mathbf{v}_i = \mathbf{v}_I = c\mathbf{\Omega}$, the substitution of the above results into equation (5.1.20), with the superscripted (1) summation yielding zero contribution (we are not considering such processes as electron scattering whereby photons 'bounce off' particles), then yields

$$\frac{1}{c}\frac{\partial I}{\partial t} + (\mathbf{\Omega} \cdot \nabla) I = \frac{h\nu}{4\pi} [-N_L B_{LU} I \phi + N_U B_{UL} I \psi + N_U A_{UL} j], \tag{5.1.29}$$

as obtained in chapter 1. The above analysis again emphasises the phenomenological nature of the classical equation of radiative transfer. Indeed, we have effectively used the same arguments here as those given in section 1.2 of chapter 1 to arrive at equation (1.2.7).

We next examine the Boltzmann equation for the de-excited (L) atoms. The inelastic term $\Sigma(2)$ now includes interactions of the form L-I and L-e which change the state L to the excited (U) state (e refers to free electrons), and U-I, U-e and U-0 which produce state L from state U (the last interaction mentioned above refers to spontaneous emission). We therefore have

$$\sum_j^{(2)} \int (\sigma'_{ij} f'_i f'_j \, d\mathbf{v}'_j - \sigma_{ij} f_i f_j \, d\mathbf{v}_j) \equiv \int [\sigma_{IU} f_I f_U \, d\mathbf{v}_I + \sigma_{eU} f_e f_U \, d\mathbf{v}_e$$
$$+ \sigma_{0U} f_0 f_U \, d\mathbf{v}_0 - \sigma_{IL} f_I f_L \, d\mathbf{v}_I - \sigma_{eL} f_e f_L \, d\mathbf{v}_e], \tag{5.1.30}$$

where we have assumed the velocity of the atom (both excited and de-excited) remains unchanged after an inelastic collision with a free electron. This is not unreasonable since $m_e \ll m_A$. Note that the σ_{0U} cross-section refers to spontaneous emission so that $f_0 \equiv 1$ and $\int f_0 \, d\mathbf{v}_0 = 1$. Further, we have been somewhat abstract in detailing the integration over $d\mathbf{v}_I$. Reference to chapter 1 emphasises that this integration is over both angle ($\mathbf{v}_I = c\boldsymbol{\Omega}$ explicitly) and frequency $\Delta\nu$ – this, in fact, is implicit when we write $d\mathbf{v}_j$ in general because we must include encounters with *all* photons and these encounters will be both velocity and energy ($h\nu$) dependent. Encounters between ordinary matter particles are only velocity dependent in our model.

Again, with $\sigma_{IL} = h\nu c B_{LU} q(\gamma_L)/4\pi$ and $f_I = I/h\nu c$, the fourth term on RHS (5.1.30) becomes

$$-\int \sigma_{IL} f_I f_L \, d\mathbf{v}_I = -\frac{B_{LU}}{4\pi} f_L \iint q(\gamma_L) I \, d(\Delta\nu) \, d\Omega, \tag{5.1.31}$$

whilst the collisional excitation term is

$$-\int \sigma_{eL} f_e f_L \, d\mathbf{v}_e = -f_L C_{LU}, \tag{5.1.32}$$

where

$$C_{LU} = \int \sigma_{eL} f_e \, d\mathbf{v}_e, \tag{5.1.33}$$

which we assume to be independent of $\mathbf{v}_L$.

The other three terms on RHS (5.1.30), corresponding to de-excitation processes, are also quite straightforward. We therefore have the Boltzmann equation for the de-excited L state:

$$\frac{\partial f_L}{\partial t} + \mathbf{v}_L \cdot \nabla f_L + \mathbf{a}_L \cdot \nabla_{\mathbf{v}_L} f_L = \sum_{j \neq I}^{(1)} \int \sigma_{Lj} (f_L' f_j' - f_L f_j) \, d\mathbf{v}_j + \sum_j^{(2)} \Phi_{Lj}, \tag{5.1.34}$$

where

$$\sum_j^{(2)} \Phi_{Lj} = \frac{B_{UL}}{4\pi} f_U \iint e_1(\gamma_U) I \, d(\Delta\nu) \, d\Omega + A_{UL} f_U + C_{UL} f_U$$
$$- \frac{B_{LU}}{4\pi} f_L \iint q(\gamma_L) I \, d(\Delta\nu) \, d\Omega - C_{LU} f_L, \tag{5.1.35}$$

where we have used equation (5.1.28).

The corresponding equation for the excited U state may be similarly obtained:

$$\frac{\partial f_U}{\partial t} + \mathbf{v}_U \cdot \nabla f_U + \mathbf{a}_U \cdot \nabla_{\mathbf{v}_U} f_U$$
$$= \sum_{j \neq I}^{(1)} \int \sigma_{Uj} (f_U' f_j' - f_U f_j) \, d\mathbf{v}_j + \sum_j^{(2)} \Phi_{Uj}, \tag{5.1.36}$$

where we find

$$\sum_{j}^{(2)} \Phi_{Uj} = -\sum_{j}^{(2)} \Phi_{Lj}. \tag{5.1.37}$$

We finally examine the Boltzmann equation for the free electrons. The inelastic terms $\Sigma(2)$ used in deriving the Boltzmann equation for f_U and f_L above incorporated the change in state $L \to U$ and $U \to L$ due to interactions with photons and colliding free electrons. All other encounters were included in the elastic term $\Sigma(1)$. Free electrons, however, do not change state in our model, they simply change their velocities (and energies), even when they interact inelastically with an atom. Clearly, if we include continuum radiation in our model, then radiative and collisional ionisation, plus re-combination processes, would need insertion into the inelastic $\Sigma(2)$ term for free electrons. This would also necessitate the inclusion of extra terms in equation (5.1.35). A complete analysis would, of course, incorporate continuum radiation (indeed, most of the energy of the radiation field in stellar atmospheres is generally carried in the continuum) but, as we have stressed in chapter 1, our desire here is to simplify the model (even beyond that required by physical realism, if necessary) in order to isolate and clarify those fundamentals describing the interaction between radiation and matter. Consequently, the inelastic $\Sigma(2)$ term does not apply here to free electrons, and we simply have

$$\frac{\partial f_e}{\partial t} + \mathbf{v}_e \cdot \nabla f_e + \mathbf{a}_e \cdot \nabla_{\mathbf{v}_e} f_e = \sum_{j \neq I}^{(1)} \int \sigma_{ej}(f_e' f_j' - f_e f_j)\, d\mathbf{v}_j. \tag{5.1.38}$$

One should now include a similar equation for the ions. Clearly, if we assume the gaseous ensemble to be electrically neutral, the total charge of the ions must balance that of the electrons. In reality, the ions and electrons move with different microscopic velocities (note that $m_e \ll$ mass of atoms and ions) and this can, in turn lead to different macroscopic velocities of these two species. If ions and electrons do diffuse differently through physical space, charge separation occurs and this must introduce a further microscopic force (of attraction) into the analysis. For ease in exposition, therefore, we ignore charge separation and assume the velocity distribution of the ions obeys an equation similar to (5.1.38). Note, however, that any inclusion of continuum radiation necessitates a detailed specification of the Boltzmann equation for these ions since inelastic terms involving radiative and collisional ionisation and re-combination processes would then become important [130].

5.1.3 The general equation of change

We now proceed in the derivation of the pertinent conservation equations by multiplying the general equation (5.1.20) by a function $\psi_i(\mathbf{r}, \mathbf{v}_i, t)$, to be specified later, and integrating over $\mathbf{v}_i$, i.e.

$$\int \psi_i \left[\frac{\partial f_i}{\partial t} + (\mathbf{v}_i \cdot \nabla) f_i + (\mathbf{a}_i \cdot \nabla_{\mathbf{v}_i}) f_i \right] \mathrm{d}\mathbf{v}_i$$

$$= \sum_{j \neq I}^{(1)} \iint \psi_i \sigma_{ij} (f_i' f_j' - f_i f_j) \, \mathrm{d}\mathbf{v}_i \, \mathrm{d}\mathbf{v}_j + \sum_{j}^{(2)} \int \psi_i \Phi_{ij} \, \mathrm{d}\mathbf{v}_i. \qquad (5.1.39)$$

Next, we consider each of the terms on the LHS (5.1.39) in turn. First,

$$\int \psi_i \frac{\partial f_i}{\partial t} \mathrm{d}\mathbf{v}_i = \frac{\partial}{\partial t} \int \psi_i f_i \, \mathrm{d}\mathbf{v}_i - \int f_i \frac{\partial \psi_i}{\partial t} \mathrm{d}\mathbf{v}_i$$

$$= \frac{\partial}{\partial t} (N_i \bar{\psi}_i) - N_i \overline{\frac{\partial \psi_i}{\partial t}}, \qquad (5.1.40)$$

where we have used equation (5.1.4).

If we examine just the x component of $\mathbf{v}_i \cdot \nabla$ in the second term on LHS (5.1.39) we find

$$\int \psi_i v_{ix} \frac{\partial f_i}{\partial x} \mathrm{d}\mathbf{v}_i = \frac{\partial}{\partial x} \int \psi_i v_{ix} f_i \, \mathrm{d}\mathbf{v}_i - \int f_i v_{ix} \frac{\partial \psi_i}{\partial x} \mathrm{d}\mathbf{v}_i$$

$$= \frac{\partial}{\partial x} (N_i \overline{\psi_i v_{ix}}) - N_i \overline{v_{ix} \frac{\partial \psi_i}{\partial x}}, \qquad (5.1.41)$$

where, in obtaining the first of equations (5.1.41), we recall that $\mathbf{r}$, $\mathbf{v}_i$ and t are independent variables. Similar expressions may be obtained for the y and z components of $\mathbf{v}_i \cdot \nabla$; the net result is

$$\int \psi_i (\mathbf{v}_i \cdot \nabla) f_i \, \mathrm{d}\mathbf{v}_i = \nabla \cdot (N_i \overline{\psi_i \mathbf{v}_i}) - N_i \overline{(\mathbf{v}_i \cdot \nabla) \psi_i}. \qquad (5.1.42)$$

A similar procedure may be used to reduce the last term on LHS (5.1.39), viz.

$$\left[\int \psi_i (\mathbf{a}_i \cdot \nabla_{\mathbf{v}_i}) f_i \, \mathrm{d}\mathbf{v}_i \right]_x = \left[\mathbf{a}_i \cdot \int \psi_i \nabla_{\mathbf{v}_i} f_i \, \mathrm{d}\mathbf{v}_i \right]_x$$

$$= a_{ix} \iiint \psi_i \frac{\partial f_i}{\partial v_{ix}} \mathrm{d}v_{ix} \, \mathrm{d}v_{iy} \, \mathrm{d}v_{iz}$$

$$= a_{ix} \iint [\psi_i f_i]_{v_{ix} \to -\infty}^{v_{ix} \to \infty} \mathrm{d}v_{iy} \, \mathrm{d}v_{iz}$$

$$- a_{ix} \iiint f_i \frac{\partial \psi_i}{\partial v_{ix}} \mathrm{d}v_{ix} \, \mathrm{d}v_{iy} \, \mathrm{d}v_{iz}$$

$$= -\left[\mathbf{a}_i \cdot N_i \overline{\frac{\partial \psi_i}{\partial \mathbf{v}_i}} \right]_x, \qquad (5.1.43)$$

where we have assumed $f_i \to 0$ as $|v_{ix}| \to \infty$. This last condition is intuitively obvious. Adding the corresponding y and z components then yields

$$\int \psi_i (\mathbf{a}_i \cdot \nabla_{\mathbf{v}_i}) f_i \, d\mathbf{v}_i = -N_i a_i \cdot \overline{\nabla_{\mathbf{v}_i} \psi_i}. \tag{5.1.44}$$

The above results, when put together in equation (5.1.39), then yield Enskog's general equation of change [30], viz.

$$\frac{\partial}{\partial t}(N_i \bar{\psi}_i) + \nabla \cdot (N_i \overline{\psi_i \mathbf{v}_i}) - N_i \left\{ \overline{\frac{\partial \psi_i}{\partial t}} + \overline{(\mathbf{v}_i \cdot \nabla)\psi_i} + \frac{\mathbf{F}_i}{m_i} \cdot \overline{\nabla_{\mathbf{v}_i} \psi_i} \right\}$$
$$= \sum_{j \neq I}^{(1)} \iint \psi_i \sigma_{ij} (f_i' f_j' - f_i f_j) \, d\mathbf{v}_i \, d\mathbf{v}_j + \sum_j^{(2)} \int \psi_i \Phi_{ij} \, d\mathbf{v}_i. \tag{5.1.45}$$

The first term on RHS (5.1.45) can also be manipulated. If we write

$$\Phi_1(ij) = \iint \psi_i \sigma_{ij} (f_i' f_j' - f_i f_j) \, d\mathbf{v}_i \, d\mathbf{v}_j, \tag{5.1.46}$$

and assume that every encounter has an exact inverse (we certainly cannot make this assumption for the superscripted (2) summation terms), i.e.

$$\Phi_2(ij) = \iint \psi_i' \sigma_{ij}' (f_i f_j - f_i' f_j') \, d\mathbf{v}_i' \, d\mathbf{v}_j', \tag{5.1.47}$$

then

$$\Phi_1(ij) = \Phi_2(ij). \tag{5.1.48}$$

However, use of equation (5.1.19) shows that if

$$\Phi_3(ij) = -\iint \psi_i' \sigma_{ij} (f_i' f_j' - f_i f_j) \, d\mathbf{v}_i \, d\mathbf{v}_j, \tag{5.1.49}$$

then

$$\Phi_2(ij) = \Phi_3(ij). \tag{5.1.50}$$

Thus, with $2\Phi_1 = \Phi_1 + \Phi_3$, we have

$$\Phi_1(ij) = \tfrac{1}{2} \iint (\psi_i - \psi_i') \sigma_{ij} (f_i' f_j' - f_i f_j) \, d\mathbf{v}_i \, d\mathbf{v}_j. \tag{5.1.51}$$

This then enables equation (5.1.45) to be written as

$$\frac{\partial}{\partial t}(N_i \bar{\psi}_i) + \nabla \cdot (N_i \overline{\psi_i \mathbf{v}_i}) - N_i \left\{ \overline{\frac{\partial \psi_i}{\partial t}} + \overline{(\mathbf{v}_i \cdot \nabla)\psi_i} + \frac{\mathbf{F}_i}{m_i} \cdot \overline{\nabla_{\mathbf{v}_i} \psi_i} \right\}$$
$$= \tfrac{1}{2} \sum_{j \neq I}^{(1)} \iint (\psi_i - \psi_i') \sigma_{ij} (f_i' f_j' - f_i f_j) \, d\mathbf{v}_i \, d\mathbf{v}_j + \sum_j^{(2)} \int \psi_i \Phi_{ij} \, d\mathbf{v}_i. \tag{5.1.52}$$

We further define

$$\Phi_4 = \sum_{ij \neq I}^{(1)} \Phi_1(ij), \quad \Phi_5 = \sum_{ij \neq I}^{(1)} \Phi_1(ji).$$

Since the ij and ji in $\Phi_1(ij)$ are dummy subscripts when appearing in summations over both i and j, we must have

$$\Phi_4 = \Phi_5 = \tfrac{1}{2}(\Phi_4 + \Phi_5),$$

i.e.

$$\sum_{ij \neq I}^{(1)} \iint \psi_i \sigma_{ij}(f_i' f_j' - f_i f_j)\, \mathrm{d}\mathbf{v}_i\, \mathrm{d}\mathbf{v}_j$$

$$= \tfrac{1}{4} \sum_{ij \neq I}^{(1)} \iint (\psi_i + \psi_j - \psi_i' - \psi_j')\sigma_{ij}(f_i' f_j' - f_i f_j)\, \mathrm{d}\mathbf{v}_i\, \mathrm{d}\mathbf{v}_j. \qquad (5.1.53)$$

We therefore have the second important result

$$\sum_{i \neq I} \left\{ \frac{\partial}{\partial t}(N_i \bar{\psi}_i) + \nabla \cdot (N_i \overline{\mathbf{v}_i \psi_i}) - N_i \left[\overline{\frac{\partial \psi_i}{\partial t}} + \overline{(\mathbf{v}_i \cdot \nabla)\psi_i} + \frac{\mathbf{F}_i}{m_i} \cdot \overline{\nabla_{\mathbf{v}_i} \psi_i} \right] \right\}$$

$$= \tfrac{1}{4} \sum_{ij \neq I}^{(1)} \iint (\psi_i + \psi_j - \psi_i' - \psi_j')\sigma_{ij}(f_i' f_j' - f_i f_j)\, \mathrm{d}\mathbf{v}_i\, \mathrm{d}\mathbf{v}_j$$

$$+ \sum_{\substack{i \neq I \\ j}}^{(2)} \int \psi_i \Phi_{ij}\, \mathrm{d}\mathbf{v}_i. \qquad (5.1.54)$$

This last result, together with equation (5.1.52), forms the basis of the following sub-section.

5.1.4 The macroscopic conservation equations

The $\Sigma(1)$ term in equation (5.1.52) vanishes when ψ_i is a constant during a collision, i.e. when

$$\psi_i = \psi_i'. \qquad (5.1.55)$$

Clearly, neglecting relativistic effects, equation (5.1.55) is satisfied when ψ_i is the mass m_i of an ith species particle (conservation of mass). Similarly, the $\Sigma(1)$ term in equation (5.1.54) vanishes when ψ_i is a summational invariant during a collision, i.e. when

$$\psi_i + \psi_j = \psi_i' + \psi_j'. \qquad (5.1.56)$$

This last equation simply states that the sum of ψ_i and ψ_j (relating to two separate particles) before and after an encounter is the same. As already mentioned, it is standard practice to ignore the effect of the external force $\mathbf{F}_i$ on the ith species particle *during* an *elastic* collision. This then enables equation (5.1.56) to also be satisfied by $\psi_i = m_i \mathbf{v}_i$ (conservation of linear momentum) and $\psi_i = \frac{1}{2} m_i \mathbf{v}_i^2$ (conservation of energy).

We shall detail the effect of these ψ_i in turn.

(a) Conservation of mass

We first consider the de-excited state with $i = L$, i.e. $\psi_i = m_L$. Equations (5.1.52) and (5.1.55), together with equation (5.1.35), yield

$$\frac{\partial N_L}{\partial t} + \nabla \cdot (N_L \bar{\mathbf{v}}_L) = \frac{B_{UL}}{4\pi} \int I \, d(\Delta\nu) \, d\Omega \int f_U e_1(\gamma_U) \, d\mathbf{v}_L$$

$$+ N_U A_{UL} + N_U C_{UL} - \frac{B_{LU}}{4\pi} \int I \, d(\Delta\nu) \, d\Omega$$

$$\times \int f_L q(\gamma_L) \, d\mathbf{v}_L - N_L C_{LU}. \tag{5.1.57}$$

Photons will have a linear momentum magnitude $|h\nu/c| \sim 10^{-21}$ cgs units for spectral lines of interest whereas $|m_L \mathbf{v}_L| \sim 10^{-18}$. Consequently, we can assume the velocity of the atom remains unchanged during an encounter with a photon (either photon emission or absorption) so that $\mathbf{v}_L = \mathbf{v}_U$. (This has, in fact, already been used to obtain the $N_U A_{UL}$ and $N_U C_{UL}$ terms appearing in equation (5.1.57).) Thus, recalling that

$$\gamma_U = \gamma(\mathbf{v}_U) = \Delta\nu - \frac{\nu}{c} \mathbf{v}_U \cdot \mathbf{\Omega}, \tag{5.1.58}$$

and

$$\gamma_L = \gamma(\mathbf{v}_L) = \Delta\nu - \frac{\nu}{c} \mathbf{v}_L \cdot \mathbf{\Omega}, \tag{5.1.59}$$

equation (5.1.57) can be written as

$$\frac{\partial N_L}{\partial t} + \nabla \cdot (N_L \bar{\mathbf{v}}_L) = N_U B_{UL} \bar{J}_{UL} + N_U A_{UL} + N_U C_{UL}$$
$$- N_L B_{LU} \bar{J}_{LU} - N_L C_{LU}, \tag{5.1.60}$$

where

$$\bar{J}_{UL} = \frac{1}{4\pi} \int \psi I \, d\nu \, d\Omega, \tag{5.1.61}$$

and

$$\bar{J}_{LU} = \frac{1}{4\pi} \int \phi I \, d\nu \, d\Omega. \tag{5.1.62}$$

The probabilities ψ and ϕ have already been defined by equations (5.1.25) and (5.1.23) respectively.

Equation (5.1.60) stipulates conservation of mass for the de-excited state. A similar analysis for the excited state yields

$$\frac{\partial N_U}{\partial t} + \nabla \cdot (N_U \bar{\mathbf{v}}_U) = N_L B_{LU} \bar{J}_{LU} + N_L C_{LU} - N_U B_{UL} \bar{J}_{UL} - N_U A_{UL} - N_U C_{UL}, \tag{5.1.63}$$

whereas, for the free electrons, we have

$$\frac{\partial N_e}{\partial t} + \nabla \cdot (N_e \bar{\mathbf{v}}_e) = 0, \tag{5.1.64}$$

with a similar equation for the ions.

We stress again that this last equation only applies when we can neglect ionisation and re-combination processes.

Equation (5.1.7) stipulates

$$\rho = N_L m_L + N_U m_U + N_e m_e + N_+ m_+, \tag{5.1.65}$$

where the subscript + refers to ions, whereas, from equation (5.1.10),

$$\rho_L \bar{\mathbf{V}}_L + \rho_U \bar{\mathbf{V}}_U + \rho_e \bar{\mathbf{V}}_e + \rho_+ \bar{\mathbf{V}}_+ = 0. \tag{5.1.66}$$

Thus, using $\bar{\mathbf{v}}_i = \bar{\mathbf{V}}_i + \mathbf{V}$, the above results yield the *total* conservation equation

$$\frac{\partial \rho}{\partial t} + \nabla \cdot (\rho \mathbf{V}) = 0, \tag{5.1.67}$$

as expected.

(b) Conservation of linear momentum

Here we start from equation (5.1.54) with $\psi_i = m_i \mathbf{v}_i$ and the principle of linear momentum, viz. the linear momentum of a system is conserved if there is a zero resultant external force acting on the system. Thus, neglecting the effect of $\mathbf{F}_i$ during an elastic encounter, we again have the first term on the RHS (5.1.54) vanishing for all elastic encounters leaving the second term to account for the inelastic processes. However, use of equation (5.1.37), and the assumption $\mathbf{v}_U = \mathbf{v}_L$ discussed in the previous sub-section, yields zero contribution for the $\Sigma(2)$ component also. Thus, we have

$$\sum_{i \neq I} \left\{ \frac{\partial}{\partial t} (N_i m_i \bar{\mathbf{v}}_i) + \nabla \cdot (N_i m_i \overline{\mathbf{v}_i \mathbf{v}_i}) - N_i m_i \left[\overline{\frac{\partial \mathbf{v}_i}{\partial t}} + \overline{(\mathbf{v}_i \cdot \nabla)\mathbf{v}_i} + \frac{\mathbf{F}_i}{m_i} \cdot \overline{\nabla_{\mathbf{v}_i} \mathbf{v}_i} \right] \right\} = 0, \tag{5.1.68}$$

which, recalling the fact that $\mathbf{r}$, t and $\mathbf{v}_i$ are independent variables, yields

$$\sum_{i \neq I} \left\{ \frac{\partial}{\partial t} (\rho_i \bar{\mathbf{v}}_i) + \nabla \cdot [\rho_i \overline{(\mathbf{V}_i + \mathbf{V})(\mathbf{V}_i + \mathbf{V})}] - N_i \mathbf{F}_i \right\} = 0. \tag{5.1.69}$$

Thus, defining the pressure tensor **P** by

$$\mathbf{P} = \sum_{i \neq I} \rho_i \overline{\mathbf{V}_i \mathbf{V}_i} = \sum_{i=I} m_i \int \mathbf{V}_i \mathbf{V}_i f_i \, d\mathbf{v}_i, \tag{5.1.70}$$

and using equations (5.1.6) and (5.1.66), equation (5.1.69) is just

$$\frac{\partial}{\partial t}(\rho \mathbf{V}) + \nabla \cdot (\rho \mathbf{V}\mathbf{V}) + \nabla \cdot \mathbf{P} = \sum_{i \neq I} N_i \mathbf{F}_i. \tag{5.1.71}$$

This last result is one of the two standard gas-dynamic forms of the equation stipulating conservation of total particle linear momentum. The alternative form may be obtained from equation (5.1.71) using (5.1.67), i.e.

$$\rho \frac{\partial \mathbf{V}}{\partial t} + \rho(\mathbf{V} \cdot \nabla)\mathbf{V} + \nabla \cdot \mathbf{P} = \sum_{i \neq I} N_i \mathbf{F}_i. \tag{5.1.72}$$

We shall be more explicit in detailing the pressure tensor **P** once the microscopic velocity distributions f_i are known. It is important to note here, however, that some authors tend to discuss a radiation pressure term by putting $\mathbf{V}_i \equiv \mathbf{V}_I = c\boldsymbol{\Omega}$ into equation (5.1.70) where $\boldsymbol{\Omega}$ refers to the direction of propagation of the photon. This is not valid because the summation in this equation is over $i \neq I$. Radiation terms appearing in the above conservation equation for *total* particle linear momentum arise only in the pressure tensor **P** via the velocity distribution as we shall see.

It should be noted, however, that there is a radiation contribution on the RHS (5.1.71) due to the $\Sigma(2)$ term when we remove the assumption $\mathbf{v}_L = \mathbf{v}_U$, and this can be an important factor for some geometries. (In this case, we would put $\mathbf{v}_L \neq \mathbf{v}_U$ both in the conservation of linear momentum equation above *and* the conservation of mass equations (5.1.60) and (5.1.63), as well as the conservation of energy equations following.) Of course, it is tempting to introduce an explicit separate radiation pressure term. The momentum of a photon of energy $h\nu$ is $h\nu\boldsymbol{\Omega}/c$. Clearly, if an atom absorbs this photon, the momentum of that atom must be increased by this amount. Similarly, the emission of a photon will generate a 'recoil' effect on the emitting atom. If the radiation field being absorbed is completely isotropic *and* photons are emitted isotropically, then the net change in momentum of a unit volume element of the gas will be statistically zero. This will occur deep within the stellar atmosphere where equilibrium considerations are reasonably appropriate. (There will also be a zero net change in linear momentum of the system if the photon is emitted at the same frequency and in the same direction it had before being absorbed.) However, as one considers regions closer to the 'surface' of the gas where the photons can escape more readily, the radiation field becomes more anisotropic, and thus significant linear momentum can be transferred from the radiation field to the gas ensemble. This is particularly clear for a shell surrounding a core star. The radiation from the core can be absorbed by the shell thence re-emitted. Statistically, half the re-emitted radiation will propagate away from the core and the other half will propagate inwards. Consequently, the shell gas will have gained linear momentum equivalent to half that momentum associated with the radiation emitted from the core.

But it is important to recognise that this momentum exchange occurs specifically at the microscopic level where atoms absorb and emit photons (i.e. $\mathbf{v}_L \neq \mathbf{v}_U$). Only when the ensemble average (5.1.4) is performed does one have the gas as a whole moving macroscopically. Consequently, radiation pressure should not be considered as a separate explicit macroscopic force but, rather, as an accumulation of linear momentum transfer processes associated with the absorption and emission of individual photons by individual atoms [104, 111]. One could perhaps model this total momentum transfer to the shell described above using an empirical radiation force acting on the boundary of the shell, but this should not be confused with the more exact macroscopic conservation equations derived from the microscopic internal description afforded by the Boltzmann equations.

(c) Conservation of energy

For the case considered here, we put $\psi_L = \frac{1}{2}m_L\mathbf{v}_L^2$, $\psi_e = \frac{1}{2}m_e\mathbf{v}_e^2$ and $\psi_U = \frac{1}{2}m_U\mathbf{v}_U^2 + h\nu$. Note that in detailing ψ_U we include the energy $h\nu$ of the excited state since this energy has the potential to produce a photon of frequency ν. Indeed, we make use of the principle of energy, viz. the kinetic energy plus potential energy of a system is constant if the work done by the resultant external force on the system is zero, although, when considering elastic collisions only, the above potential energy term $h\nu$ need not be included in the specification of ψ_i. Thus, recalling those assumptions already discussed, the elastic collision term vanishes so that equations (5.1.54) and (5.1.34-37) yield

$$\sum_{i\neq I}\left\{\frac{\partial}{\partial t}\,[\tfrac{1}{2}N_i m_i\overline{(\mathbf{V}_i+\mathbf{V})\cdot(\mathbf{V}_i+\mathbf{V})}]\right.$$
$$+\nabla\cdot\left[\frac{N_i m_i}{2}\,\overline{(\mathbf{V}_i+\mathbf{V})\cdot(\mathbf{V}_i+\mathbf{V})(\mathbf{V}_i+\mathbf{V})}\right]$$
$$\left.-\frac{N_i m_i}{2}\left[\overline{\frac{\partial v_i^2}{\partial t}}+\overline{(\mathbf{v}_i\cdot\nabla)\mathbf{v}_i}+\frac{\mathbf{F}_i}{m_i}\cdot\overline{\nabla_{\mathbf{v}_i}v_i^2}\right]\right\}$$
$$+\frac{\partial}{\partial t}(N_U h\nu)+\nabla\cdot[N_U h\nu(\bar{\mathbf{V}}_U+\mathbf{V})]$$
$$=h\nu[N_L B_{LU}\bar{J}_{LU}+N_L C_{LU}-N_U B_{UL}\bar{J}_{UL}-N_U A_{UL}-N_U C_{UL}]. \tag{5.1.73}$$

If we now define the internal energy $\rho\epsilon$ by

$$\rho\epsilon = \sum_{i=I} \tfrac{1}{2}\rho_i \overline{V_i^2} + N_{\mathrm{U}}h\nu$$

$$= \sum_{i \neq I} \tfrac{1}{2}m_i \int V_i^2 f_i \, \mathrm{d}\mathbf{v}_i + N_{\mathrm{U}}h\nu, \tag{5.1.74}$$

where the first term corresponds to internal kinetic energy and the second to internal potential energy, then equation (5.1.73), together with equation (5.1.66), yields

$$\frac{\partial}{\partial t}(\tfrac{1}{2}\rho V^2 + \rho\epsilon) + \sum_{i \neq I} \left\{ \nabla \cdot \left[\frac{\rho_i}{2} (\overline{V_i^2 \mathbf{V}_i} + V^2 \mathbf{V} + \overline{V_i^2}\mathbf{V} + 2\overline{(\mathbf{V} \cdot \mathbf{V}_i)\mathbf{V}_i}) \right] - N_i \mathbf{F}_i(\bar{\mathbf{V}}_i + \mathbf{V}) \right\} + \nabla \cdot [N_{\mathrm{U}}h\nu(\bar{\mathbf{V}}_{\mathrm{U}} + \mathbf{V})]$$
$$= h\nu\,[N_{\mathrm{L}}B_{\mathrm{LU}}\bar{J}_{\mathrm{LU}} + N_{\mathrm{L}}C_{\mathrm{LU}} - N_{\mathrm{U}}B_{\mathrm{UL}}\bar{J}_{\mathrm{UL}} - N_{\mathrm{U}}A_{\mathrm{UL}} - N_{\mathrm{U}}C_{\mathrm{UL}}]. \tag{5.1.75}$$

If we further define the energy flux vector $\mathbf{q}$ by

$$\mathbf{q} = \sum_{i \neq I} \tfrac{1}{2}\rho_i \overline{V_i^2 \mathbf{V}_i} = \sum_{i \neq I} \tfrac{1}{2}m_i \int V_i^2 \mathbf{V}_i f_i \, \mathrm{d}\mathbf{v}_i, \tag{5.1.76}$$

then equation (5.1.75) has the form

$$\frac{\partial}{\partial t}(\tfrac{1}{2}\rho V^2 + \rho\epsilon) + \nabla \cdot [\mathbf{q} + \tfrac{1}{2}\rho V^2 \mathbf{V} + \rho\epsilon\mathbf{V} + \mathbf{P} \cdot \mathbf{V} + N_{\mathrm{U}}h\nu\bar{\mathbf{V}}_{\mathrm{U}}]$$
$$- \sum_{i \neq I} N_i \mathbf{F}_i \cdot (\bar{\mathbf{V}}_i + \mathbf{V})$$
$$= h\nu[N_{\mathrm{L}}B_{\mathrm{LU}}\bar{J}_{\mathrm{LU}} + N_{\mathrm{L}}C_{\mathrm{LU}} - N_{\mathrm{U}}B_{\mathrm{UL}}\bar{J}_{\mathrm{UL}} - N_{\mathrm{U}}A_{\mathrm{UL}} - N_{\mathrm{U}}C_{\mathrm{UL}}]. \tag{5.1.77}$$

This last result can be written in a more standard form by noting that the integration of the equation of radiative transfer (or the Boltzmann equation with $f_I = I/h\nu c$; see equation (5.1.29)) over both angle and frequency within the spectral line yields

$$\frac{1}{c}\frac{\partial \bar{I}}{\partial t} + \nabla \cdot F_{\mathrm{RAD}} = -\frac{h\nu}{4\pi}[N_{\mathrm{L}}B_{\mathrm{LU}}\bar{J}_{\mathrm{LU}} - N_{\mathrm{U}}B_{\mathrm{UL}}\bar{J}_{\mathrm{UL}} - N_{\mathrm{U}}A_{\mathrm{UL}}], \tag{5.1.78}$$

where

$$F_{\mathrm{RAD}} = \frac{1}{4\pi}\iint \mathbf{\Omega} I \, \mathrm{d}(\Delta\nu) \, \mathrm{d}\Omega, \tag{5.1.79}$$

is the flux of radiative energy and

$$\bar{I} = \frac{1}{4\pi} \iint I \, \mathrm{d}(\Delta\nu) \, \mathrm{d}\Omega, \tag{5.1.80}$$

is an average intensity term not to be confused with $\bar{J}_{\mathrm{LU}}$ and $\bar{J}_{\mathrm{UL}}$.

We are then able to write the conservation of energy equation (5.1.77) in the required form:

$$\frac{\partial}{\partial t}(\tfrac{1}{2}\rho V^2 + \rho\epsilon) + \nabla \cdot [\mathbf{q} + \tfrac{1}{2}\rho V^2 \mathbf{V} + \rho\epsilon\mathbf{V} + \mathbf{P}\cdot\mathbf{V} + N_{\mathrm{U}} h\nu \bar{\mathbf{V}}_{\mathrm{U}}]$$

$$- \sum_{i \neq I} N_i \mathbf{F}_i \cdot (\bar{\mathbf{V}}_i + \mathbf{V}) = -4\pi\nabla \cdot F_{\mathrm{RAD}} - \frac{4\pi}{c}\frac{\partial \bar{I}}{\partial t}$$

$$+ h\nu[N_{\mathrm{L}} C_{\mathrm{LU}} - N_{\mathrm{U}} C_{\mathrm{UL}}]. \tag{5.1.81}$$

It is common in the literature for the RHS (5.1.81) to be replaced by the lone term $-4\pi\nabla \cdot F_{\mathrm{RAD}}$. Clearly, this approximation ignores the above $\partial\bar{I}/\partial t$ term (it has little practical importance in general) and the two inelastic collisional terms. The collisional terms balance one another, and can therefore be neglected, only if the assumption of LTE for photons is valid (see section 1.7).

An alternative form of equation (5.1.81) may be obtained by using our results for conservation of mass (5.1.67) and conservation of linear momentum (5.1.72), viz.

$$\rho\frac{\partial\epsilon}{\partial t} + \rho(\mathbf{V}\cdot\nabla)\epsilon + \nabla\cdot(\mathbf{q} + N_{\mathrm{U}} h\nu\bar{\mathbf{V}}_{\mathrm{U}}) + \mathbf{P}\cdot\nabla\mathbf{V} - \sum_{i\neq I} N_i \mathbf{F}_i \cdot \bar{\mathbf{V}}_i$$

$$= -4\pi\nabla\cdot F_{\mathrm{RAD}} - \frac{4\pi}{c}\frac{\partial\bar{I}}{\partial t} + h\nu[N_{\mathrm{L}} C_{\mathrm{LU}} - N_{\mathrm{U}} C_{\mathrm{UL}}]. \tag{5.1.82}$$

We will find it necessary to use the energy conservation equation not only for the gaseous ensemble as a whole described above, but for each individual species taken separately. The appropriate results may be obtained by starting from equation (5.1.52) where now the elastic $\Sigma(1)$ term does not vanish (note that the energy of each individual particle is not conserved during a collision). However, if we take $j = i$ in this summation, we clearly have

$$\left[\iint (\psi_i - \psi_i')\sigma_{ij}(f_i' f_j' - f_i f_j) \, \mathrm{d}\mathbf{v}_i \, \mathrm{d}\mathbf{v}_j\right]_{j=i} = 0. \tag{5.1.83}$$

Thus, substitution of

$$\psi_i = \tfrac{1}{2} m_i v_i^2 + h\nu\delta_{i\mathrm{U}}, \tag{5.1.84}$$

where $\delta_{i\mathrm{U}}$ is the Kronecker delta function, into equation (5.1.52) yields

$$\frac{\partial}{\partial t}[\overline{\tfrac{1}{2}\rho_i(\mathbf{V}_i+\mathbf{V})\cdot(\mathbf{V}_i+\mathbf{V})}+N_i h\nu\delta_{i\mathrm{U}}]$$
$$+\nabla\cdot[\overline{\tfrac{1}{2}\rho_i(\mathbf{V}_i+\mathbf{V})\cdot(\mathbf{V}_i+\mathbf{V})(\mathbf{V}_i+\mathbf{V})}+N_i h\nu(\bar{\mathbf{V}}_i+\mathbf{V})\delta_{i\mathrm{U}}]$$
$$-N_i\mathbf{F}_i\cdot(\bar{\mathbf{V}}_i+\mathbf{V})=\frac{m_i}{4}\sum_{j\neq i,I}^{(1)}\iint(v_i^2-v_i'^2)\sigma_{ij}(f_i'f_j'-f_if_j)\,\mathrm{d}\mathbf{v}_i\,\mathrm{d}\mathbf{v}_j$$
$$+\frac{m_i}{2}(1-\delta_{i\mathrm{e}})\sum_j^{(2)}\int v_i^2\Phi_{ij}\,\mathrm{d}\mathbf{v}_i. \qquad (5.1.85)$$

Note that the $\Sigma(2)$ term in this last equation vanishes when $i=\mathrm{e}$ (or +); see equation (5.1.38).

We now define

$$\rho_i\epsilon_i=\tfrac{1}{2}\rho_i\overline{V_i^2}+N_{\mathrm{U}}h\nu\delta_{i\mathrm{U}}, \qquad (5.1.86)$$
$$\mathbf{q}_i=\tfrac{1}{2}\rho_i\overline{V_i^2\mathbf{V}_i}, \qquad (5.1.87)$$

and

$$\mathbf{P}_i=\rho_i\overline{\mathbf{V}_i\mathbf{V}_i}, \qquad (5.1.88)$$

so that equation (5.1.85) may be written in the form

$$\frac{\partial}{\partial t}(\tfrac{1}{2}\rho_iV^2+\rho_i\epsilon_i+\rho_i\bar{\mathbf{V}}_i\cdot\mathbf{V})+\nabla\cdot[\mathbf{q}_i+\tfrac{1}{2}\rho_iV^2\mathbf{V}+\rho_i\epsilon_i\mathbf{V}+\mathbf{P}_i\cdot\mathbf{V}$$
$$+N_{\mathrm{U}}h\nu\bar{\mathbf{V}}_{\mathrm{U}}\delta_{i\mathrm{U}}+\tfrac{1}{2}\rho_iV^2\bar{\mathbf{V}}_i+\rho_i(\mathbf{V}\cdot\bar{\mathbf{V}}_i)\mathbf{V}]-N_i\mathbf{F}_i\cdot(\bar{\mathbf{V}}_i+\mathbf{V})$$
$$=\frac{m_i}{4}\sum_{j\neq i,I}^{(1)}\iint(v_i^2-v_i'^2)\sigma_{ij}(f_i'f_j'-f_if_j)\,\mathrm{d}\mathbf{v}_i\,\mathrm{d}\mathbf{v}_j$$
$$-\frac{m_i}{2}(1-\delta_{i\mathrm{e}})\sum_j^{(2)}\int v_i^2\Phi_{ij}\,\mathrm{d}\mathbf{v}_j. \qquad (5.1.89)$$

We shall return to this last equation, and to equations (5.1.67), (5.1.72) and (5.1.82), after the velocity distributions f_i have been examined in some detail.

5.2 The velocity distribution functions

An exact solution to the Boltzmann equation (5.1.20) for f_i is not available. However, it is possible to generate a perturbation series solution for f_i for which the zeroth order term corresponds to a complete equilibrium configuration. To illustrate this, we gather the interaction terms into $F^{(1)}(f_i,f_j)$ and $F^{(2)}(f_i,f_j)$ and write the general equation (5.1.20) in the form

$$\frac{\partial f_i}{\partial t}+\mathbf{v}_i\cdot\nabla f_i+\mathbf{a}_i\cdot\nabla_{\mathbf{v}_i}f_i=\sum_{j\neq I}^{(1)}F^{(1)}(f_i,f_j)+\sum_j^{(2)}F^{(2)}(f_i,f_j). \qquad (5.2.1)$$

To emphasise the importance of elastic collisions, we consider a modified Boltzmann equation

$$\frac{\partial f_i}{\partial t} + \mathbf{v}_i \cdot \nabla f_i + \mathbf{a}_i \cdot \nabla_{\mathbf{v}_i} f_i = \frac{1}{\lambda} \sum_{j \neq I}^{(1)} F^{(1)}(f_i, f_j) + \sum_{j}^{(2)} F^{(2)}(f_i, f_j), \quad (5.2.2)$$

where the smaller the value of λ, the more important are elastic collisions (i.e. $1/\lambda$ is a measure of the number of elastic collisions). Corresponding to this parameter λ, we construct the perturbation series

$$f_i = f_i^{(0)} + \lambda f_i^{(1)} + \lambda^2 f_i^{(2)} + \cdots. \quad (5.2.3)$$

Substitution of this last equation into (5.2.2), and equating coefficients of λ^l to zero, yields

$$\sum_{j \neq I}^{(1)} F^{(1)}(f_i^{(0)}, f_j^{(0)}) = 0, \quad (5.2.4)$$

$$\sum_{j \neq I}^{(1)} [F^{(1)}(f_i^{(0)}, f_j^{(1)}) + F^{(1)}(f_i^{(1)}, f_j^{(0)})]$$
$$= \frac{\partial f_i^{(0)}}{\partial t} + (\mathbf{v}_i \cdot \nabla) f_i^{(0)} + \mathbf{a}_i \cdot \nabla_{\mathbf{v}_i} f_i^{(0)} - \sum_{j}^{(2)} F^{(2)}(f_i^{(0)}, f_j^{(0)}), \quad (5.2.5)$$

etc.

One can solve equation (5.2.4) for $f_i^{(0)}$, thence equation (5.2.5) for $f_i^{(1)}$. The higher order terms $f_i^{(l)}$ for $l \geqslant 2$ may be similarly evaluated (although each successive term, as we shall see, involves a great deal more calculation than its predecessor). All the $f_i^{(l)}$ may then be substituted into equation (5.2.3) with $\lambda = 1$ to finally obtain f_i to the desired accuracy. Clearly, as elastic collisions become more important, the more dominant $f_i^{(0)}$ becomes, and one would thence expect correspondingly faster convergence of series (5.2.3). However, this collisional dominance must be limited. We cannot have elastic collisions of such magnitude in number (per unit time and volume) that our basic assumption, in which we regard the gas as sufficiently dilute that only binary (not many-body) collisions may be retained in the analysis, is invalidated. On the other hand, we must have a sufficient number of these binary elastic collisions to enable us to state that the $\Sigma(1)$ term dominates the $\Sigma(2)$. The dominance of $\Sigma(2)$ over $\Sigma(1)$ presents extreme difficulties in the determination of the pertinent velocity distribution functions – indeed, the coupling of these two terms represents the interaction between radiation and matter.

5.2.1 *Derivation of $f_i^{(0)}$ [30]*

Notwithstanding the above fundamental questions, we attempt a solution of equation (5.2.4), viz.

$$\sum_{j \neq I}^{(1)} \int \sigma_{ij} (f_i'^{(0)} f_j'^{(0)} - f_i^{(0)} f_i^{(0)}) \, d\mathbf{v}_j = 0, \quad (5.2.6)$$

by writing

$$f_i'^{(0)} f_j'^{(0)} = f_i^{(0)} f_j^{(0)}$$

i.e.

$$\ln f_i'^{(0)} + \ln f_j'^{(0)} = \ln f_i^{(0)} + \ln f_j^{(0)}. \tag{5.2.7}$$

This last result states that $\ln f_i^{(0)}$ is a summational invariant and must therefore consist of a linear super-position of the known summational invariants $m_i, m_i\mathbf{v}_i$ and $\frac{1}{2}m_i v_i^2 + h\nu\delta_{i\mathrm{U}}$ satisfying equation (5.1.56). We thence write

$$\ln f_i^{(0)} = m_i\alpha_i + m_i\mathbf{v}_i \cdot \boldsymbol{\beta}_i + (\tfrac{1}{2}m_i v_i^2 + h\nu\delta_{i\mathrm{U}})\gamma_i, \tag{5.2.8}$$

where α_i, $\boldsymbol{\beta}_i$ and γ_i are the expansion coefficients. Clearly, equation (5.2.6) is nothing more than the Boltzmann equation with all the streaming terms $\partial f_i/\partial t$, $(\mathbf{v}_i \cdot \nabla) f_i$ and $(\mathbf{F}_i \cdot \nabla_{\mathbf{v}_i}) f_i$, plus the superscripted (2) summation, put equal to zero. This then corresponds to the total equilibrium configuration in which f_i is independent of both $\mathbf{r}$ and t (i.e. uniform and time independent), external forces $\mathbf{F}_i$ are all zero (or at least they do *no* work), and no inelastic encounters exist. Consequently, α_i, $\boldsymbol{\beta}_i$ and γ_i must also be independent of time and position.

If we write

$$A_i = \exp\left(m_i\alpha_i + \frac{m_i\beta_i^2}{2\gamma_i} + h\nu\delta_{i\mathrm{U}}\gamma_i\right), \tag{5.2.9}$$

equation (5.2.8) becomes

$$f_i^{(0)} = A_i \exp(\tfrac{1}{2}m_i C_i^2 \gamma_i), \tag{5.2.10}$$

where

$$\mathbf{C}_i = \mathbf{v}_i + \frac{\boldsymbol{\beta}_i}{\gamma_i}. \tag{5.2.11}$$

The three expansion coefficients A_i, $\boldsymbol{\beta}_i$ and γ_i may be evaluated by choosing three constraints. We could, for example, use the three equations

$$N_i = \int f_i \, \mathrm{d}\mathbf{v}_i, \tag{5.2.12}$$

$$\rho\mathbf{V} = \sum_{i \neq I} m_i \int \mathbf{v}_i f_i \, \mathrm{d}\mathbf{v}_i, \tag{5.2.13}$$

and

$$\rho\epsilon = \sum_{i \neq I} \tfrac{1}{2}m_i \int V_i^2 f_i \, \mathrm{d}\mathbf{v}_i + N_{\mathrm{U}} h\nu \tag{5.2.14}$$

– see equations (5.1.1), (5.1.6) and (5.1.74). Before proceeding further, it is appropriate to define the temperature scalar T such that

$$\tfrac{3}{2}NkT = \sum_{i \neq I} \tfrac{1}{2}m_i \int V_i^2 f_i \, \mathrm{d}\mathbf{v}_i, \tag{5.2.15}$$

where k is the usual Boltzmann constant. Clearly, T is simply a measure of the mean microscopic (as opposed to macroscopic) kinetic energy of the particles.

The above system of equations thence defines the required macroscopic variables N_i, $\mathbf{V}$ and T which, as we have seen, appear in the equation of radiative transfer. It is particularly convenient to choose the above-mentioned required three constraints by replacing f_i in the above three integrals by $f_i^{(0)}$, i.e.

$$N_i = \int f_i^{(0)}\, \mathrm{d}\mathbf{v}_i, \tag{5.2.16}$$

$$\rho\mathbf{V} = \sum_{i \neq I} m_i \int \mathbf{v}_i f_i^{(0)}\, \mathrm{d}\mathbf{v}_i, \tag{5.2.17}$$

and

$$\tfrac{3}{2}NkT = \sum_{i \neq I} \tfrac{1}{2}m_i \int V_i^2 f_i^{(0)}\, \mathrm{d}\mathbf{v}_i. \tag{5.2.18}$$

In this manner, the values of these N_i, $\mathbf{V}$ and T are not directly altered by higher order terms in the series (5.2.3), but they still retain some consistency in the physical interpretation of their macroscopic properties. However, the quantities $\mathbf{P}$, $\mathbf{q}$ and $\bar{\mathbf{V}}_i$ defined by equations (5.1.70), (5.1.76) and (5.1.9) will be directly affected by these higher order $f_i^{(l)}$ and this will lead to extra (differential) terms in the equations stipulating conservation of linear momentum (5.1.72) and energy (5.1.82). (As we shall see, the zeroth order term in $\mathbf{P}$ will simply yield the hydrostatic pressure p whilst the first order term will contain a shear component in the pressure tensor related to the effects of viscosity.) The number of these extra differential terms in the conservation equations is directly related to the number of terms taken in the series (5.2.3) and this, of course, depends upon the desired accuracy of the required solution. We stress, however, that all the conservation equations still involve *only* the macroscopic quantities N_i, $\mathbf{V}$ and T defined by equations (5.2.16–18).

Equations (5.2.10) and (5.2.16) give

$$N_i = A_i \int \mathrm{e}^{1/2 m_i C_i^2 \gamma_i}\, \mathrm{d}\mathbf{v}_i, \tag{5.2.19}$$

where $\mathrm{d}\mathbf{v}_i = \mathrm{d}\mathbf{C}_i$. It is immediately clear that the defining integral is divergent unless $\gamma_i < 0$. If we transform $\mathbf{C}_i$ from cartesian to spherical polar coordinates, i.e. $\mathrm{d}\mathbf{C}_i \equiv C_i^2 \sin\theta\, \mathrm{d}\theta\, \mathrm{d}\phi\, \mathrm{d}C_i$, then

$$\begin{aligned} N_i &= A_i \int_0^\infty C_i^2\, \mathrm{e}^{1/2 m_i C_i^2 \gamma_i}\, \mathrm{d}C_i \int_0^\pi \sin\theta\, \mathrm{d}\theta \int_0^{2\pi} \mathrm{d}\phi \\ &= 2\pi A_i \Gamma(\tfrac{3}{2}) \left(\frac{2}{-m_i\gamma_i}\right)^{3/2} \\ &= A_i \left(\frac{2\pi}{-m_i\gamma_i}\right)^{3/2}, \end{aligned} \tag{5.2.20}$$

where we have used the result

$$\int_0^\infty x^n \mathrm{e}^{-ax}\,\mathrm{d}x = \frac{\Gamma(n+1)}{a^{n+1}}.$$

Equation (5.2.17), together with (5.2.10–11) yields

$$\rho\mathbf{V} = \sum_{i\neq I} m_i \left[\int \mathbf{C}_i f_i^{(0)}\,\mathrm{d}\mathbf{C}_i - \frac{\beta_i}{\gamma_i}\int f_i^{(0)}\,\mathrm{d}\mathbf{v}_i\right]$$

$$= \sum_{i\neq I} m_i A_i \int \mathbf{C}_i\,\mathrm{e}^{1/2 m_i C_i^2 \gamma_i}\,\mathrm{d}\mathbf{C}_i - \sum_{i\neq I} m_i N_i \frac{\beta_i}{\gamma_i}. \qquad (5.2.21)$$

The first term on the RHS (5.2.21) is zero since the integrand is an odd function. Next, if we assume β_i/γ_i is independent of i, we thence have

$$\rho\mathbf{V} = -\frac{\beta}{\gamma}\sum_{i\neq I}\rho_i,$$

i.e.

$$\frac{\beta}{\gamma} = -\mathbf{V}, \qquad (5.2.22)$$

which, recalling equation (5.2.11), stipulates

$$\mathbf{C}_i = \mathbf{v}_i - \mathbf{V}. \qquad (5.2.23)$$

Comparison with equation (5.1.8) then shows that

$$\mathbf{C}_i = \mathbf{V}_i. \qquad (5.2.24)$$

Finally, equation (5.2.18) yields

$$\tfrac{3}{2}NkT = \sum_{i\neq I} \tfrac{1}{2}m_i \int C_i^2 f_i^{(0)}\,\mathrm{d}\mathbf{v}_i$$

$$= \sum_{i\neq I} \tfrac{1}{2}m_i A_i \int_0^\infty C_i^4\,\mathrm{e}^{1/2 m_i C_i^2\gamma_i}\,\mathrm{d}C_i \int_0^\pi \sin\theta\,\mathrm{d}\theta \int_0^{2\pi}\mathrm{d}\phi$$

$$= \sum_{i\neq I} m_i A_i \pi\Gamma(\tfrac{5}{2})\left(\frac{2}{-m_i\gamma_i}\right)^{5/2}$$

$$= -\tfrac{3}{2}\sum_{i\neq I}\frac{N_i}{\gamma_i}, \qquad (5.2.25)$$

where we have used equation (5.2.20) to eliminate A_i. We now write

$$NT = \sum_{i\neq I} N_i T_i$$

$$= N_\mathrm{L}T_\mathrm{L} + N_\mathrm{U}T_\mathrm{U} + N_\mathrm{e}T_\mathrm{e} + N_+T_+, \qquad (5.2.26)$$

so that

$$\gamma_i = -\frac{1}{kT_i}. \qquad (5.2.27)$$

This last result, together with equations (5.2.20) and (5.2.23), yields the Maxwellian velocity distributions

$$f_i^{(0)} = N_i \left(\frac{m_i}{2\pi k T_i}\right)^{3/2} \mathrm{e}^{-\frac{m_i}{2kT_i}(\mathbf{v}_i - \mathbf{V})^2}. \tag{5.2.28}$$

The macroscopic quantities N_i, $\mathbf{V}$ and T have already been defined as functions of this zeroth order velocity distribution - see equations (5.2.16–18).

5.2.2 *Zeroth order conservation equations*

We may now evaluate the zeroth order pressure tensor $\mathbf{P}^{(0)}$, energy flux vector $\mathbf{q}^{(0)}$ and diffusion velocity $\bar{\mathbf{V}}_i^{(0)}$. From equation (5.1.70), we see that $\mathbf{P}^{(0)}$ is a (3×3) matrix of the form

$$\mathbf{P}^{(0)} = \begin{pmatrix} p_{xx} & p_{xy} & p_{xz} \\ p_{yx} & p_{yy} & p_{yz} \\ p_{zx} & p_{zy} & p_{zz} \end{pmatrix}, \tag{5.2.29}$$

where

$$p_{\alpha\beta} = \sum_{i \neq I} m_i \int V_{i\alpha} V_{i\beta} f_i^{(0)} \, \mathrm{d}\mathbf{v}_i. \tag{5.2.30}$$

Each individual $p_{\alpha\beta}$ term therefore corresponds to the force per unit area in the α direction on a plane surface perpendicular to the β direction. Clearly, the diagonal elements in the above matrix are normal forces whereas the off-diagonal elements relate to shearing forces. Further,

$$\begin{aligned} p_{xx} + p_{yy} + p_{zz} &= \sum_{i \neq I} m_i N_i \overline{V_i^2} \\ &= 3p, \end{aligned} \tag{5.2.31}$$

where this last equation defines the average pressure p (sometimes called the hydrostatic pressure) exerted by the individual particle velocities within the gas, i.e.

$$p = \tfrac{1}{3} \sum_{i \neq I} \rho_i \overline{V_i^2} = \sum_{i \neq I} p_i, \tag{5.2.32}$$

where p_i are the partial pressures corresponding to each individual species.

However, we have already defined

$$\tfrac{3}{2} kNT = \sum_{i \neq I} \tfrac{1}{2} m_i \int V_i^2 f_i^{(0)} \, \mathrm{d}\mathbf{v}_i = \sum_{i \neq I} \tfrac{1}{2} \rho_i \overline{V_i^2}, \tag{5.2.33}$$

so that, using equation (5.2.32), we have

$$p = NkT. \tag{5.2.34}$$

This last result is commonly referred to as the perfect gas law or the equation of state. It is not difficult to see that, using equation (5.2.26),

$$p_i = N_i k T_i. \tag{5.2.35}$$

Recall that in deriving $f_i^{(0)}$ we first obtained equation (5.2.4) and, as was discussed previously, this result only applies rigorously when the gas is uniform, time independent and in complete equilibrium. Consequently, when in this equilibrium configuration, all shearing forces must be zero, i.e. $p_{\alpha\beta}^{(0)} = 0$ for all $\alpha \neq \beta$, and

$$p_{xx}^{(0)} = p_{yy}^{(0)} = p_{zz}^{(0)} = p; \tag{5.2.36}$$

otherwise all volume elements in the gas would experience uneven internal forces acting to displace it from equilibrium. This may be seen mathematically by noting the following results: For any integrable scalar $g(V_i)$ where $V_i = |\mathbf{V}_i| = (V_{ix}^2 + V_{iy}^2 + V_{iz}^2)^{1/2}$,

$$\begin{aligned}\int g(V_i)\mathbf{V}_i \, \mathrm{d}\mathbf{v}_i &= \int g(V_i)\mathbf{V}_i \, \mathrm{d}\mathbf{V}_i \\ &= \iiint g(V_i)(V_{ix}, V_{iy}, V_{iz}) \, \mathrm{d}V_{ix} \, \mathrm{d}V_{iy} \, \mathrm{d}V_{iz} \\ &= 0,\end{aligned} \tag{5.2.37}$$

since the integrand is odd in each component of $\mathbf{V}_i$. Similarly, for example,

$$\begin{aligned}\int g(V_i) V_{ix} V_{iy} \, \mathrm{d}\mathbf{v}_i &= \iiint g(V_i) V_{ix} V_{iy} \, \mathrm{d}V_{ix} \, \mathrm{d}V_{iy} \, \mathrm{d}V_{iz} \\ &= 0.\end{aligned} \tag{5.2.38}$$

Reference to equation (5.2.28), along with equation (5.1.8), shows that $f_i^{(0)} \equiv g(V_i)$ so that $p_{\alpha\beta}^{(0)} = 0$ for all $\alpha \neq \beta$ as discussed above. Equation (5.2.29) then yields

$$\mathbf{P}^{(0)} = p\mathbb{1}, \tag{5.2.39}$$

where $\mathbb{1}$ is the identity matrix having unit diagonal and zero off-diagonal elements.

We have seen that the vector $\mathbf{q}$ represents a flux of energy across the surface of the volume element. Consequently, in equilibrium where no non-zero fluxes of energy are possible, we must have $\mathbf{q}^{(0)} = 0$. This can also be shown mathematically using equation (5.2.37) with

$$\mathbf{q}^{(0)} = \sum_{i \neq I} \tfrac{1}{2} m_i \int V_i^2 \mathbf{V}_i f_i^{(0)} \, \mathrm{d}\mathbf{V}_i \tag{5.2.40}$$

- see equation (5.1.76).

Similarly, we find

$$\bar{\mathbf{V}}_i^{(0)} = \frac{1}{N_i} \int \mathbf{V}_i f_i^{(0)} \, \mathrm{d}\mathbf{V}_i = 0. \tag{5.2.41}$$

These special cases of $\mathbf{P}$, $\mathbf{q}$ and $\bar{\mathbf{V}}_i$ only relate to the equilibrium configuration for which $f_i \equiv f_i^{(0)}$. As we shall see, the inclusion of higher order terms in f_i using series (5.2.3), corresponding to departures from complete thermodynamic equilibrium, leads to significant mathematical and physical modification of the above results.

The zeroth order conservation equations [95] may then be derived from equations (5.1.60, 63, 64), (5.1.72) and (5.1.82), viz.

$$\frac{\partial N_L}{\partial t} + \nabla \cdot (N_L \mathbf{V}) = N_U B_{UL} \bar{J}_{UL} + N_U A_{UL} + N_U C_{UL} - N_L B_{LU} \bar{J}_{LU} - N_L C_{LU}, \quad (5.2.42)$$

$$\frac{\partial N_U}{\partial t} + \nabla \cdot (N_U \mathbf{V}) = N_L B_{LU} \bar{J}_{LU} + N_L C_{LU} - N_U B_{UL} \bar{J}_{UL} - N_U A_{UL} - N_U C_{UL}, \quad (5.2.43)$$

$$\frac{\partial N_e}{\partial t} + \nabla \cdot (N_e \mathbf{V}) = 0, \quad (5.2.44)$$

$$\rho \frac{\partial \mathbf{V}}{\partial t} + \rho(\mathbf{V} \cdot \nabla)\mathbf{V} + \nabla p = \sum_{i \neq I} N_i \mathbf{F}_i, \quad (5.2.45)$$

and

$$\rho \frac{\partial \epsilon}{\partial t} + \rho(\mathbf{V} \cdot \nabla)\epsilon + p\nabla \cdot \mathbf{V} = -4\pi \nabla \cdot F_{RAD} - \frac{4\pi}{c}\frac{\partial \bar{I}}{\partial t} + h\nu[N_L C_{LU} - N_U C_{UL}], \quad (5.2.46)$$

where

$$p = kNT, \quad (5.2.47)$$

$$\rho\epsilon = \tfrac{3}{2}kNT + N_U h\nu. \quad (5.2.48)$$

The energy conservation equation (5.1.89) for each species then has the form

$$\frac{\partial}{\partial t}(\tfrac{1}{2}\rho_i V^2 + \rho_i \epsilon_i) + \nabla \cdot [\tfrac{1}{2}\rho_i V^2 \mathbf{V} + \rho_i \epsilon_i \mathbf{V} + p_i \mathbf{V}] - N_i \mathbf{F}_i \cdot \mathbf{V}$$

$$= \frac{m_i}{4} \sum_{j \neq i, I}^{(1)} \iint (v_i^2 - v_i'^2)\sigma_{ij}(f_i' f_j' - f_i f_j)\, \mathrm{d}\mathbf{v}_i\, \mathrm{d}\mathbf{v}_j + \frac{m_i}{2}(1 - \delta_{ie}) \sum_j^{(2)} \int v_i^2 \Phi_{ij}\, \mathrm{d}\mathbf{v}_i, \quad (5.2.49)$$

where

$$p_i = kN_i T_i, \quad (5.2.50)$$

and

$$p_i \epsilon_i = \tfrac{3}{2}kN_i T_i + N_U h\nu \delta_{iU}. \quad (5.2.51)$$

A difficulty now arises. The above derivation of the Maxwellian velocity distribution $f_i^{(0)}$ assumed a uniform, time-independent equilibrium configuration for the gas, and this implied that the expansion coefficients $\alpha_i(A_i)$, β_i and γ_i are not only independent of $\mathbf{v}_i$ but of $\mathbf{r}$ and t. Equations (5.2.20), (5.2.22) and (5.2.27) then stipulate that the macroscopic quantities N_i, $\mathbf{V}$ and T should likewise be independent of $\mathbf{r}$ and t. However, this would violate (or, at least, make trivial) the above zeroth order conservation equations where a dependence of N_i, $\mathbf{V}$ and T on $\mathbf{r}$ and t is implicit; consequently, some contradiction is apparent. We overcome this difficulty by introducing the concept of local thermodynamic equilibrium (LTE) for particles. Here, we simply state that each individual volume element is in equilibrium and that the properties of the gas in that volume element are described by the *local* values of N_i, $\mathbf{V}$ and T. Clearly, if the element is in equilibrium locally, it must not only be uniform and time independent, but also isolated in the sense that the work done by external forces on the element must be zero. (Of course, we have already used this detail both in ignoring the effect of $\mathbf{F}_i$ during collisions and in neglecting the term $\mathbf{F}_i \cdot \nabla_{\mathbf{v}_i} f_i^{(0)}$ in obtaining equation (5.2.8).) It is important to bear these assumptions in mind when using conservation equations (of any order!) in ionised plasmas, for example, since they may be readily violated, particularly in the presence of large gradients in N_i, $\mathbf{V}$ and T accompanying shock waves.

If the particle LTE assumption is violated, and one should clearly distinguish between LTE for particles and LTE for photons (see section 1.7), one must use the appropriate non-LTE equations, be they Boltzmann (we have already seen that the Boltzmann equation yields the non-LTE radiative transfer equation for photons), Louiville, BBGKY hierarchy, quantum statistical, etc. Indeed, as one moves to higher regions in stellar atmospheres, the gas becomes increasingly dilute and it is not inconceivable that there would not exist a sufficient number of elastic collisions per unit time and volume to maintain the assumption leading to equation (5.2.4), i.e. the elastic collision terms on the RHS of the Boltzmann equation might not dominate the streaming terms on the LHS nor the inelastic $F^{(2)}(f_i, f_j)$ terms. Since this equation (5.2.4) was the starting point in the analysis leading to the Maxwellian velocity distribution (5.2.28), this lack of an adequate number of elastic collisions would also require the rejection of the particle LTE conservation equations. Clearly, somewhere between the stellar photospheric regions and the interstellar matter, the particle LTE assumption must be invalidated and care must therefore be exercised in the use of the conservation equations. We already have a similar invalidation of LTE for photons, but this occurs somewhere between the deep photosphere and the temperature minimum in stellar atmospheres (depending upon the opacity of the spectral line in question).

We complete this section by briefly discussing the opposite case in which the inelastic $F^{(2)}(f_i, f_j)$ term dominates the $F^{(1)}(f_i, f_j)$ quantity corresponding to elastic collisions. Following the same procedure leading to equation (5.2.4), we find

$$\sum_{j}^{(2)} F^{(2)}(f_i^{(0)}, f_j^{(0)}) = 0, \qquad (5.2.52)$$

which, using equations (5.1.34) and (5.3.35), yields

$$\frac{B_{\mathrm{UL}}}{4\pi} f_{\mathrm{U}}^{(0)} \iint e_1(\gamma_{\mathrm{U}}) I \,\mathrm{d}(\Delta\nu)\,\mathrm{d}\Omega + (A_{\mathrm{UL}} + C_{\mathrm{UL}}) f_{\mathrm{U}}^{(0)}$$

$$= \frac{B_{\mathrm{LU}}}{4\pi} f_{\mathrm{L}}^{(0)} \iint q(\gamma_{\mathrm{L}}) I \,\mathrm{d}(\Delta\nu)\,\mathrm{d}\Omega + C_{\mathrm{LU}} f_{\mathrm{L}}^{(0)}. \qquad (5.2.53)$$

Clearly, even if one ($f_{\mathrm{L}}^{(0)}$, say) of the two velocity distributions were Maxwellian, this last equation stipulates that the other distribution ($f_{\mathrm{U}}^{(0)}$) must be non-Maxwellian. This therefore necessitates a complete re-examination of the microscopic distributions for very dilute radiating media. Indeed, it is necessary, in such situations, to solve each individual species Boltzmann equation *coupled* both to all the other particle Boltzmann equations *and* to the Boltzmann equation of radiative transfer. This is beyond the scope of this monograph – but at least the reader has been made aware of the difficulties.

5.3 First order equations [40]

Keeping the discussion at the end of the preceding section in mind, we now turn our attention to the evaluation of the next term $f_i^{(1)}$ in the series (5.2.3). The inclusion of the radiative terms in the analysis somewhat complicates the exposition. Consequently, we first examine the development of $f_i^{(1)}$ when there are no inelastic encounters, then, having generated some understanding of the effect of such an $f_i^{(1)}$ on the first order conservation equations for mass, energy and linear momentum, we briefly outline the manner in which the radiative terms influence the results.

5.3.1 Derivation of $f_i^{(1)}$

We begin by writing

$$f_i^{(1)}(\mathbf{r}, \mathbf{v}_i, t) = f_i^{(0)}(\mathbf{r}, \mathbf{v}_i, t)\phi_i(\mathbf{r}, \mathbf{v}_i, t), \qquad (5.3.1)$$

so that equation (5.2.5), with the inelastic term $F^{(2)}(f_i^{(0)}, f_j^{(0)})$ put equal to zero, becomes

$$\sum_{j \neq I}^{(1)} \int \sigma_{ij} f_i^{(0)} f_j^{(0)} [\phi_i + \phi_j - \phi_i - \phi_j] \,\mathrm{d}\mathbf{v}_j$$

$$= \frac{\partial f_i^{(0)}}{\partial t} + (\mathbf{v}_i \cdot \nabla) f_i^{(0)} + (\mathbf{a}_i \cdot \nabla_{\mathbf{v}_i}) f_i^{(0)}. \qquad (5.3.2)$$

We must now substitute $f_i^{(0)}$ given by equation (5.2.28) into this last result. It is most convenient to write

$$\ln f_i^{(0)} = \ln N_i - \tfrac{3}{2}\ln T_i - \frac{m_i}{2kT_i}\mathbf{V}_i\cdot\mathbf{V}_i + K, \tag{5.3.3}$$

where K is a constant. Since $\mathbf{V}_i = \mathbf{v}_i - \mathbf{V}$, we have

$$\frac{1}{f_i^{(0)}}\,\text{LHS (5.3.2)} = \left(\frac{\mathrm{d}}{\mathrm{d}t}\right)\ln f_i^{(0)} + \mathbf{V}_i\cdot\nabla\ln f_i^{(0)} + \mathbf{a}_i\cdot\nabla_{\mathbf{v}_i}\ln f_i^{(0)}, \tag{5.3.4}$$

where we have defined the Lagrangian time derivative $\mathrm{d}/\mathrm{d}t$ by

$$\frac{\mathrm{d}}{\mathrm{d}t} \equiv \frac{\partial}{\partial t} + \mathbf{V}\cdot\nabla. \tag{5.3.5}$$

Thus, using equation (5.3.3),

$$\frac{\mathrm{d}}{\mathrm{d}t}\ln f_i^{(0)} = \frac{\mathrm{d}}{\mathrm{d}t}\ln N_i + \left(\frac{m_iV_i^2}{2kT_i} - \tfrac{3}{2}\right)\frac{1}{T_i}\frac{\mathrm{d}T_i}{\mathrm{d}t} - \frac{m_i}{kT_i}\mathbf{V}_i\cdot\frac{\mathrm{d}\mathbf{V}_i}{\mathrm{d}t}, \tag{5.3.6}$$

$$\nabla\ln f_i^{(0)} = \nabla\ln N_i + \left(\frac{m_iV_i^2}{2kT_i} - \tfrac{3}{2}\right)\nabla\ln T_i - \frac{m_i}{kT_i}(\mathbf{V}_i\cdot\nabla)\mathbf{V}_i, \tag{5.3.7}$$

and

$$\nabla_{\mathbf{v}_i}\ln f_i^{(0)} = -\frac{m_i}{kT_i}(\mathbf{V}_i\cdot\nabla_{\mathbf{v}_i})\mathbf{V}_i = -\frac{m_i\mathbf{V}_i}{kT_i}, \tag{5.3.8}$$

where, in obtaining this last equation, we recall that $\mathbf{V}(\mathbf{r}, t)$ is independent of $\mathbf{v}_i$.

The macroscopic conservation equations (5.2.42, 5.2.43, 5.2.44), (5.2.45) and (5.2.49), neglecting inelastic terms, may be written in the form

$$\frac{\mathrm{d}N_i}{\mathrm{d}t} = -N_i\cdot\nabla\mathbf{V}, \tag{5.3.9}$$

$$\rho\frac{\mathrm{d}\mathbf{V}}{\mathrm{d}t} = -\nabla p + \sum_{i\neq I} N_i\mathbf{F}_i, \tag{5.3.10}$$

and, using $\rho_i\epsilon_i = \tfrac{3}{2}kN_iT_i$ and $p_i = kN_iT_i$,

$$\frac{\mathrm{d}T_i}{\mathrm{d}t} = -\tfrac{3}{2}T_i\cdot\nabla\mathbf{V}. \tag{5.3.11}$$

In obtaining this last result, we have also put the first integral term appearing on the RHS (5.2.49) to zero. This only applies if $T_i = T_j$ for all i, j, i.e. if the temperature is the same for all species and this, in turn, can only occur in a complete equilibrium configuration. However, this is not inconsistent with our description of the zeroth order macroscopic equations of conservation.

Since $\mathbf{r}$, $\mathbf{v}_i$ and t are independent variables,

$$\frac{\mathrm{d}\mathbf{V}_i}{\mathrm{d}t} = -\frac{\mathrm{d}\mathbf{V}}{\mathrm{d}t} \quad\text{and}\quad (\mathbf{V}_i\cdot\nabla)\mathbf{V}_i = -(\mathbf{V}_i\cdot\nabla)\mathbf{V}. \tag{5.3.12}$$

Thus, using

$$\nabla \ln N_i = \nabla \ln \left(\frac{N_i}{N}\right) + \nabla \ln N$$

$$= \nabla \ln \left(\frac{N_i}{N}\right) + \nabla \ln p - \nabla \ln T, \quad (5.3.13)$$

where we put $T_i = T$ for all i, and writing

$$G_{ij}(\phi_i, \phi_j) = \int \sigma_{ij} f_i^{(0)} f_j^{(0)} [\phi_i' + \phi_j' - \phi_i - \phi_j] \, \mathrm{d}\mathbf{v}_j, \quad (5.3.14)$$

all the above equations can be combined to yield the desired result:

$$\sum_{j \neq I}^{(1)} G_{ij}(\phi_i, \phi_j) = f_i^{(0)} \left\{ \frac{N}{N_i} \mathbf{V}_i \cdot \mathbf{d}_i + \frac{m_i}{kT} (\mathbf{V}_i \mathbf{V}_i - \tfrac{1}{3} V_i^2 \Pi) : \nabla \mathbf{V} \right.$$

$$\left. + \left(\frac{m_i V_i^2}{2kT} - \tfrac{5}{2} \right) \mathbf{V}_i \cdot \nabla \ln T \right\}, \quad (5.3.15)$$

where

$$\mathbf{d}_i = \nabla \left(\frac{N_i}{N}\right) + \left(\frac{N_i}{N} - \frac{N_i m_i}{\rho}\right) \nabla \ln p - \frac{N_i m_i}{p\rho} \left(\rho \mathbf{a}_i - \sum_{i \neq I} N_i \mathbf{F}_i\right), \quad (5.3.16)$$

and where we have re-written $\mathbf{V}_i \cdot (\mathbf{V}_i \cdot \nabla)\mathbf{V}$ as $\mathbf{V}_i \mathbf{V}_i : \nabla \mathbf{V}$, and $V_i^2 \nabla \cdot \mathbf{V}$ as $V_i^2 \Pi : \nabla \mathbf{V}$. Note that the double dot product of two tensors $\mathbf{w}^{(1)}$ and $\mathbf{w}^{(2)}$ is

$$\mathbf{w}^{(1)} : \mathbf{w}^{(2)} = \sum_{\alpha\beta} w_{\alpha\beta}^{(1)} w_{\beta\alpha}^{(2)}. \quad (5.3.17)$$

We now wish to solve equation (5.3.15) for the unknown factor ϕ_i. Since ϕ_i only appears in the integral in this expression, it is clear that if any solution for ϕ_i is supplemented by a summational invariant satisfying

$$\psi_i' + \psi_j' = \psi_i + \psi_j, \quad (5.3.18)$$

then the resulting function will also be a solution to equation (5.3.15). This introduces a non-uniqueness into the evaluation of ϕ_i. However, the difficulty can be overcome by recalling the *two* definitions of the macroscopic quantities N_i, $\rho\mathbf{V}$ and $\rho\epsilon$ given by equations (5.2.12–14) and (5.2.16–18). If we substitute $f_i = f_i^{(0)} + f_i^{(0)} \phi_i$ into the former set of equations, the latter set stipulates that

$$\int f_i^{(0)} \phi_i \, \mathrm{d}\mathbf{v}_i = 0, \quad (5.3.19)$$

$$\sum_{i \neq I} m_i \int \mathbf{v}_i f_i^{(0)} \phi_i \, \mathrm{d}\mathbf{v}_i = 0, \quad (5.3.20)$$

$$\sum_{i \neq I} \tfrac{1}{2} m_i \int V_i^2 f_i^{(0)} \phi_i \, \mathrm{d}\mathbf{v}_i = 0. \quad (5.3.21)$$

These three equations then offer additional constraints on ϕ_i – indeed, it can be shown [30, 40] that they induce uniqueness in ϕ_i.

We have already used the summational invariants m_i, $m_i\mathbf{v}_i$ and $\frac{1}{2}m_i\mathbf{v}_i^2$ in detailing the Maxwellian velocity distribution and the corresponding zeroth order macroscopic conservation equations. It can be readily shown using the summational invariance of m_i and $\mathbf{v}_i$ that $\frac{1}{2}m_iV_i^2$ is also summationally invariant – we prefer its use here to that of $\frac{1}{2}m_i\mathbf{v}_i^2$.

Further, it is clear that the three quantities $\mathbf{d}_i$, $\nabla\mathbf{V}$ and $\nabla \ln T$ appearing on the RHS (5.3.15) are independent of $\mathbf{v}_i$ and are therefore of a macroscopic nature. Consequently, the linear super-position

$$\phi_i = \mathbf{A}_i \cdot \nabla \ln T + \mathbf{B}_i : \nabla\mathbf{V} + \sum_j \mathbf{D}_{ij} \cdot \mathbf{d}_j + \eta_i m_i + \boldsymbol{\xi}_i \cdot m_i\mathbf{v}_i + \zeta_i \tfrac{1}{2}m_iV_i^2, \tag{5.3.22}$$

is a solution to equation (5.3.15). All the expansion coefficients $\mathbf{A}_i$, $\mathbf{B}_i, \ldots,$ ζ_i must be explicit functions of the macroscopic variables N_i, $\mathbf{V}$ and T, and only implicit functions of $\mathbf{r}$ and t.

It is not difficult to see that the $\mathbf{d}_j$ are not linearly independent since, from equation (5.3.16),

$$\sum_{j\neq I} \mathbf{d}_j = 0. \tag{5.3.23}$$

Consequently, one of the $\mathbf{d}_j$ terms for each ϕ_i may be absorbed into all the other $\mathbf{d}_j$, and we reflect this absorption by choosing to put

$$\mathbf{D}_{ii} = 0 \quad \text{for all } i. \tag{5.3.24}$$

The $\nabla \ln T$, $\nabla\mathbf{V}$ and $\mathbf{d}_j$ terms in ϕ_i are now linearly independent so that the substitution of this ϕ_i into equation (5.3.15), and equating coefficients of $\nabla \ln T$, $\nabla\mathbf{V}$ and $\mathbf{d}_j$ to zero, yields

$$\sum_{j\neq I} G_{ij}(\mathbf{A}_i, \mathbf{A}_j) = f_i^{(0)} \left(\frac{m_iV_i^2}{2kT} - \tfrac{5}{2}\right) \mathbf{V}_i, \tag{5.3.25}$$

$$\sum_{j\neq I} G_{ij}(\mathbf{B}_i, \mathbf{B}_j) = f_i^{(0)} \frac{m_i}{kT} (\mathbf{V}_i\mathbf{V}_i - \tfrac{1}{3}V_i^2 \mathbb{1}), \tag{5.3.26}$$

and

$$\sum_{j\neq I} G_{ij}\left(\sum_k \mathbf{D}_{ik} \cdot \mathbf{d}_k, \sum_k \mathbf{D}_{jk} \cdot \mathbf{d}_k\right) = f_i^{(0)} \frac{N}{N_i} \mathbf{V}_i \cdot \mathbf{d}_i,$$

i.e.

$$\sum_{j\neq I} G_{ij}(0, \mathbf{D}_{ji}) = f_i^{(0)} \frac{N}{N_i} \mathbf{V}_i; \tag{5.3.27}$$

note that the summationally invariant terms in ϕ_i disappear when substituted into equation (5.3.15).

The RHS (5.3.25) and (5.3.27) are vectors in the direction $\mathbf{V}_i$. Consequently, the LHS of these two equations must also be in this direction. Since the $G_{ij}(\phi_i, \phi_j)$ term is linear in ϕ_i and ϕ_j, the $\mathbf{A}_i$ and $\mathbf{D}_{ij}$ vectors must all be parallel to $\mathbf{V}_i$. Thus, we write

$$\mathbf{A}_i = A(V_i)\mathbf{V}_i \quad \text{and} \quad \mathbf{D}_{ij} = D_j(V_i)\mathbf{V}_i, \tag{5.3.28}$$

where the proportionality factors $A(V_i)$ and $D_j(V_i)$ are scalars yet to be determined.

If we consider the $\alpha\beta$ element in the matrix equation (5.3.26), we find

$$\sum_{j\neq I} G_{ij}(B_{i\alpha\beta}, B_{j\alpha\beta}) = f_i^{(0)} \frac{m_i}{kT}\left(V_{i\alpha}V_{i\beta} - \frac{V_i^2\delta_{\alpha\beta}}{3}\right), \tag{5.3.29}$$

where now

$$V_i^2 = V_{ixx}^2 + V_{iyy}^2 + V_{izz}^2, \tag{5.3.30}$$

is the trace of the matrix $\mathbf{V}_i\mathbf{V}_i$.

Adding equations (5.3.29) for just the diagonal elements of $\mathbf{V}_i\mathbf{V}_i$ therefore yields

$$\sum_{j\neq I} G_{ij}(B_{ixx} + B_{iyy} + B_{izz}, B_{jxx} + B_{jyy} + B_{jzz}) \tag{5.3.31}$$

for all i, which can only be satisfied by

$$B_{ixx} + B_{iyy} + B_{izz} = 0. \tag{5.3.32}$$

Further, a straightforward manipulation gives

$$\sum_{j\neq I} G_{ij}(B_{i\alpha\beta} - B_{i\beta\alpha}, B_{j\alpha\beta} - B_{j\beta\alpha}) = 0, \tag{5.3.33}$$

for all $\alpha \neq \beta$, and this too can only be satisfied by

$$B_{i\alpha\beta} = B_{i\beta\alpha}, \tag{5.3.34}$$

i.e. the matrix $\mathbf{B}$ is symmetric.

The two conditions (5.3.32) and (5.3.34) then force the result

$$\mathbf{B}_i = B(V_i)(\mathbf{V}_i\mathbf{V}_i - \tfrac{1}{3}V_i^2\,\mathbb{1}), \tag{5.3.35}$$

where $B(V_i)$ is a proportionality scalar.

We now turn our attention to the evaluation of the expansion coefficients η_i, ξ_i and ζ_i. We substitute ϕ_i, together with the above $\mathbf{A}_i$, $\mathbf{B}_i$ and $\mathbf{D}_{ij}$, into the constraints (5.3.19–21), viz.

$$f_i^{(0)}\Big\{A(V_i)\mathbf{V}_i \cdot \nabla \ln T + B(V_i)(\mathbf{V}_i\mathbf{V}_i - \tfrac{1}{3}V_i^2\,\mathbb{1}) : \nabla\mathbf{V}$$
$$+ \sum_j D_j(V_i)\mathbf{V}_i \cdot \mathbf{d}_j\Big\}\, d\mathbf{v}_i = -\int f_i^{(0)}[\eta_i m_i + \boldsymbol{\xi}_i \cdot m_i\mathbf{v}_i + \zeta_i \tfrac{1}{2}m_i V_i^2]\, d\mathbf{v}_i, \tag{5.3.36}$$

$$\sum_{i \neq I} m_i \int f_i^{(0)} \mathbf{v}_i \Big\{ A(V_i)\mathbf{V}_i \cdot \nabla \ln T + B(V_i)(\mathbf{V}_i\mathbf{V}_i - \tfrac{1}{3}V_i^2 \mathbb{1}) : \nabla\mathbf{V} + \sum_j D_j(V_i)\mathbf{V}_i \cdot \mathbf{d}_j \Big\} \, \mathrm{d}\mathbf{v}_i = - \sum_{i \neq I} m_i \int f_i^{(0)} \mathbf{v}_i \times [\eta_i m_i + \boldsymbol{\xi}_i \cdot m_i \mathbf{v}_i + \zeta_i \tfrac{1}{2} m_i V_i^2] \, \mathrm{d}\mathbf{v}_i \tag{5.3.37}$$

$$\sum_{i \neq I} \tfrac{1}{2} m_i \int f_i^{(0)} V_i^2 \Big\{ A(V_i)\mathbf{V}_i \cdot \nabla \ln T + B(V_i)(\mathbf{V}_i\mathbf{V}_i - \tfrac{1}{3}V_i^2 \mathbb{1}) : \nabla\mathbf{V} + \sum_j D_j(V_i)\mathbf{V}_i \cdot \mathbf{d}_j \Big\} \mathrm{d}\mathbf{v}_i = - \sum_{i \neq I} \tfrac{1}{2} m_i \int f_i^{(0)} V_i^2 \times [\eta_i m_i + \boldsymbol{\xi}_i \cdot m_i v_i + \zeta_i \tfrac{1}{2} m_i V_i^2] \, \mathrm{d}\mathbf{v}_i. \tag{5.3.38}$$

Equation (5.2.37) shows that the integrals involving $A(V_i)$ and $D_j(V_i)$ appearing in both equations (5.3.36) and (5.3.38) are identically zero. The off-diagonal elements in $\mathbf{V}_i\mathbf{V}_i - \frac{1}{3}V_i^2 \mathbb{1}$ (i.e. $V_{ix}V_{iy}$, for example - note that $\frac{1}{3}V_i^2 \mathbb{1}$ is diagonal only) yield zero contribution in equations (5.3.36) and (5.3.38) because of equation (5.2.38). The diagonal terms of $\mathbf{V}_i\mathbf{V}_i - \frac{1}{3}V_i^2 \mathbb{1}$ produce symmetrical integrands so that, for example,

$$\iiint g(V_i)V_{ix}^2 \, \mathrm{d}V_{ix} \, \mathrm{d}V_{iy} \, \mathrm{d}V_{iz} = \iiint g(V_i)V_{iy}^2 \, \mathrm{d}V_{ix} \, \mathrm{d}V_{iy} \, \mathrm{d}V_{iz} = \iiint g(V_i)V_{iz}^2 \, \mathrm{d}V_{ix} \, \mathrm{d}V_{iy} \, \mathrm{d}V_{iz} = \tfrac{1}{3} \int g(V_i)V_i^2 \, \mathrm{d}\mathbf{V}_i. \tag{5.3.39}$$

Thus, including both diagonal and off-diagonal results, we must have

$$\int g(V_i)(\mathbf{V}_i\mathbf{V}_i - \tfrac{1}{3}V_i^2 \mathbb{1}) \, \mathrm{d}\mathbf{V}_i = 0. \tag{5.3.40}$$

Consequently, the entire LHS of equations (5.3.36) and (5.3.38) vanish. Thus, using $\mathbf{v}_i = \mathbf{V}_i + \mathbf{V}$ and equation (5.2.37), we are led to the two equations

$$(\eta_i m_i + \boldsymbol{\xi}_i \cdot m_i \mathbf{V}) \int f_i^{(0)} \, \mathrm{d}\mathbf{v}_i + \tfrac{1}{2} m_i \zeta_i \int f_i^{(0)} V_i^2 \, \mathrm{d}\mathbf{v}_i = 0, \tag{5.3.41}$$

and

$$\sum_{i \neq I} \tfrac{1}{2} m_i (\eta_i m_i + \boldsymbol{\xi}_i \cdot m_i \mathbf{V}) \int f_i^{(0)} V_i^2 \, \mathrm{d}\mathbf{v}_i + \sum_{i \neq I} \tfrac{1}{2} m_i (\tfrac{1}{2} m_i \zeta_i) \times \int f_i^{(0)} V_i^4 \, \mathrm{d}\mathbf{v}_i = 0. \tag{5.3.42}$$

All the above integrals are positive definite and we may therefore eliminate $\eta_i m_i + \boldsymbol{\xi}_i \cdot m_i \mathbf{V}$, for example, from these two equations, viz.

$$\sum_{i \neq I} \tfrac{1}{2} m_i (\tfrac{1}{2} m_i \zeta_i) \left\{ \left(\int f_i^{(0)} \, d\mathbf{v}_i \right) \left(\int f_i^{(0)} V_i^4 \, d\mathbf{v}_i \right) - \left(\int f_i^{(0)} V_i^2 \, d\mathbf{v}_i \right)^2 \right\} = 0. \tag{5.3.43}$$

Direct calculation shows that the factor in the braces is non-zero. Thus, we must have

$$\zeta_i = 0, \tag{5.3.44}$$

thence

$$\eta_i m_i + \boldsymbol{\xi}_i \cdot m_i \mathbf{V} = 0, \tag{5.3.45}$$

for all *i*.

We now turn our attention to the remaining equation (5.3.37). Again, writing $\mathbf{v}_i = \mathbf{V}_i + \mathbf{V}$, we see, using equations (5.3.44) and (5.3.45), that

$$\begin{aligned} \text{RHS } (5.3.37) &= -\sum_{i \neq I} m_i^2 \boldsymbol{\xi}_i \cdot \int \mathbf{V}_i \mathbf{v}_i f_i^{(0)} \, d\mathbf{v}_i \\ &\neq 0. \end{aligned} \tag{5.3.46}$$

Consequently, equating this with the LHS (5.3.37), we find $\boldsymbol{\xi}_i$ to be linearly dependent upon $\nabla \ln T$, $\nabla \mathbf{V}$ (although the coefficient of this term does vanish in equation (5.3.37)) and $\mathbf{d}_j$ for all *j*. Noting that equations (5.3.44) and (5.3.45) stipulate that only the $\boldsymbol{\xi}_i \cdot m_i \mathbf{V}_i$ contribution remains in the summationally invariant component of ϕ_i given by equation (5.3.22), the above dependence of $\boldsymbol{\xi}_i$ on $\nabla \ln T$ and $\mathbf{d}_j$ suggests that this contribution can be absorbed into the $\nabla \ln T$ and $\mathbf{d}_j$ terms already present in ϕ_i. Thus, ϕ_i has the reduced form

$$\phi_i = \mathbf{A}_i \cdot \nabla \ln T + \mathbf{B}_i \cdot \nabla \mathbf{V} + \sum_j \mathbf{D}_{ij} \cdot \mathbf{d}_j, \tag{5.3.47}$$

where the expansion coefficients still obey equations (5.3.24), (5.3.25–27), (5.3.28) and (5.3.35). Further, since we have removed the summationally invariant terms from ϕ_i, the constraint (5.3.37) yields

$$\sum_{i \neq I} m_i \int f_i^{(0)} A(V_i) \mathbf{v}_i \mathbf{v}_i \, d\mathbf{v}_i = 0, \tag{5.3.48}$$

$$\sum_{i \neq I} m_i \int f_i^{(0)} B(V_i) \mathbf{v}_i (\mathbf{V}_i \mathbf{V}_i - \tfrac{1}{3} V_i^2 \mathbb{1}) \, d\mathbf{v}_i = 0, \tag{5.3.49}$$

$$\sum_{i \neq I} m_i \int f_i^{(0)} D_j(V_i) \mathbf{v}_i \mathbf{V}_i \, d\mathbf{v}_i = 0, \tag{5.3.50}$$

where we have used the linear independence of $\nabla \ln T$, $\nabla \mathbf{V}$ and $\mathbf{d}_j$ in obtaining three separate equations.

We proceed by considering the off-diagonal elements of the general integrand $\mathbf{V}_i\mathbf{V}_i g(V_i)\mathbf{V}_i$; we have, for example

$$\iiint g(V_i)V_{ix}V_{iy}(V_{ix}, V_{iy}, V_{iz})\,\mathrm{d}V_{ix}\,\mathrm{d}V_{iy}\,\mathrm{d}V_{iz} = 0, \tag{5.3.51}$$

since the x, y and z components are odd functions of V_{ix}, V_{iy} and V_{iz} respectively. Similarly, the diagonal elements yield

$$\iiint g(V_i)V_{ix}^2(V_{ix}, V_{iy}, V_{iz})\,\mathrm{d}V_{ix}\,\mathrm{d}V_{iy}\,\mathrm{d}V_{iz} = 0. \tag{5.3.52}$$

We therefore have

$$\int g(V_i)\mathbf{V}_i\mathbf{V}_i\mathbf{V}_i\,\mathrm{d}\mathbf{V}_i = 0. \tag{5.3.53}$$

Consequently, putting $\mathbf{v}_i = \mathbf{V}_i + \mathbf{V}$ and using equation (5.2.37), we see that the LHS (5.3.49) vanishes, and this implies that equation (5.3.49) offers no further constraints on the scalar $B(V_i)$. However, the two equations

$$\sum_{i \neq I} m_i \int f_i^{(0)} A(V_i)\mathbf{V}_i\mathbf{V}_i\,\mathrm{d}\mathbf{V}_i = 0, \tag{5.3.54}$$

and

$$\sum_{i \neq I} m_i \int f_i^{(0)} D_j(V_i)\mathbf{V}_i\mathbf{V}_i\,\mathrm{d}\mathbf{V}_i = 0, \tag{5.3.55}$$

are sufficient conditions to ensure uniqueness of $A(V_i)$ and $D_j(V_i)$ - corresponding uniqueness for $B(V_i)$ is not a problem.

We are not concerned here with the explicit evaluation of $A(V_i)$, $B(V_i)$ and $D_j(V_i)$ - they are effectively 'constants' dependent upon the gas mixture and the defining macroscopic parameters N_i, $\mathbf{V}$ and T. Rather, we are interested in the *structure* of the macroscopic conservation equations specifically to first order accuracy.

5.3.2 First order conservation equations

We require the quantities $\bar{\mathbf{V}}_i$, $\mathbf{P}$ and $\mathbf{q}$ which, in the zeroth order approximation were 0, $p\mathbb{1}$ and 0 respectively.

First, equation (5.1.4) enables the diffusion velocity $\bar{\mathbf{V}}_i^{(1)}$ to be written as

$$\bar{\mathbf{V}}_i^{(1)} = \frac{1}{N_i}\int \mathbf{V}_i f_i\,\mathrm{d}\mathbf{v}_i, \tag{5.3.56}$$

which, using $f_i = f_i^{(0)} + f_i^{(0)}\phi_i$, becomes

$$\bar{\mathbf{V}}_i^{(1)} = \frac{1}{N_i}\int \mathbf{V}_i f_i^{(0)}\phi_i\,\mathrm{d}\mathbf{v}_i \tag{5.3.57}$$

(recall that the zeroth order term vanishes since the integrand is odd). We thence have

$$\bar{\mathbf{V}}_i = \frac{1}{N_i}\int \mathbf{V}_i f_i^{(0)}\Big\{A(V_i)\mathbf{V}_i\cdot\nabla\ln T + B(V_i)(\mathbf{V}_i\mathbf{V}_i - \tfrac{1}{3}V_i^2\,\mathbb{1}) : \nabla\mathbf{V}$$

$$+\sum_j D_j(V_i)\mathbf{V}_i\cdot\mathbf{d}_j\Big\}\,\mathrm{d}\mathbf{v}_i$$

$$= \bar{A}_i\,\nabla\ln T + \sum_j \bar{D}_{ij}^{(1)}\,\mathbf{d}_j, \tag{5.3.58}$$

where the $B(V_i)$ term vanishes using equation (5.3.40). The coefficients $\bar{A}_i$ and $\bar{D}_{ij}^{(1)}$ can be determined once $A(V_i)$ and $D_j(V_i)$ are known. As mentioned previously, we are not concerned with such evaluations in this monograph.

The pressure tensor $\mathbf{P}$ has the form

$$\mathbf{P} = \sum_{i\neq I} m_i \int \mathbf{V}_i\mathbf{V}_i f_i\,\mathrm{d}\mathbf{v}_i, \tag{5.3.59}$$

so that

$$\mathbf{P}^{(1)} = p\,\mathbb{1} + \sum_{i\neq I} m_i \int \mathbf{V}_i\mathbf{V}_i f_i^{(0)}\phi_i\,\mathrm{d}\mathbf{v}_i$$

$$= p\,\mathbb{1} + \sum_{i\neq I} m_i \int \mathbf{V}_i\mathbf{V}_i f_i^{(0)}\Big\{A(V_i)\mathbf{V}_i\cdot\nabla\ln T$$

$$+ B(V_i)(\mathbf{V}_i\mathbf{V}_i - \tfrac{1}{3}V_i^2\,\mathbb{1}) : \nabla\mathbf{V} + \sum_j D_j(V_i)\mathbf{V}_i\cdot\mathbf{d}_j\Big\}\,\mathrm{d}\mathbf{v}_i. \tag{5.3.60}$$

Use of equation (5.3.53) then shows that

$$\mathbf{P}^{(1)} = p\,\mathbb{1} + \sum_{i\neq I} m_i \int f_i^{(0)}B(V_i)\mathbf{V}_i\mathbf{V}_i(\mathbf{V}_i\mathbf{V}_i - \tfrac{1}{3}V_i^2\,\mathbb{1}) : \nabla\mathbf{V}\,\mathrm{d}\mathbf{v}_i. \tag{5.3.61}$$

Thus, we see that the zeroth order term $p\,\mathbb{1}$ in the pressure tensor is now supplemented by another term proportional to $\nabla\mathbf{V}$. (Note that this supplementary term also depends on the macroscopic variables N_i, $\mathbf{V}$ and T through $f_i^{(0)}$ and $B(V_i)$.)

Another commonly used formulation of $\mathbf{P}^{(1)}$ is available. First, we note that any second rank tensor $\mathbf{w}$ may be written as the sum of its symmetric and anti-symmetric components, viz.

$$\mathbf{w} = \tfrac{1}{2}(\mathbf{w} + \mathbf{w}^{\mathrm{T}}) + \tfrac{1}{2}(\mathbf{w} - \mathbf{w}^{\mathrm{T}}), \tag{5.3.62}$$

where $\mathbf{w}^{\mathrm{T}}$ is the transpose of $\mathbf{w}$, i.e. the diagonal elements of $\mathbf{w} - \mathbf{w}^{\mathrm{T}}$ are zero.

Thus, recalling that the matrix $\mathbf{V}_i\mathbf{V}_i - \frac{1}{3}V_i^2\,\mathbb{1}$ is symmetric and has zero trace (i.e. the sum of its diagonal elements is zero), and noting equation (5.3.17) defining the double tensor product, we have

$$\begin{aligned}(\mathbf{V}_i\mathbf{V}_i - \tfrac{1}{3}V_i^2\mathbb{1}) : (\mathbf{w}-\mathbf{w}^{\mathrm{T}}) &= \sum_{\alpha\beta}(V_{i\alpha}V_{i\beta} - \tfrac{1}{3}V_i^2\delta_{\alpha\beta})(w_{\beta\alpha}-w_{\alpha\beta})\\ &= \sum_{\alpha\neq\beta} V_{i\alpha}V_{i\beta}(w_{\beta\alpha}-w_{\alpha\beta})\\ &= 0. \end{aligned} \tag{5.3.63}$$

We also have

$$\mathbf{w} : \mathbb{1} = \sum_{\alpha} w_{\alpha\alpha},$$

so that

$$\begin{aligned} V_i^2\mathbb{1} : \mathbf{w} &= V_i^2\mathbf{w} : \mathbb{1}\\ &= \left(\sum_\beta V_{i\beta\beta}^2\right)\left(\sum_\alpha w_{\alpha\alpha}\right)\\ &= (\mathbf{V}_i\mathbf{V}_i : \mathbb{1})(\mathbf{w} : \mathbb{1})\\ &= \mathbf{V}_i\mathbf{V}_i(\mathbf{w} : \mathbb{1}) : \mathbb{1}. \end{aligned} \tag{5.3.64}$$

This then allows the result

$$(\mathbf{V}_i\mathbf{V}_i - \tfrac{1}{3}V_i^2\mathbb{1}) : \mathbf{w} = \mathbf{V}_i\mathbf{V}_i : [\mathbf{w} - \tfrac{1}{3}(\mathbf{w} : \mathbb{1})\mathbb{1}], \tag{5.3.65}$$

which, using equations (5.3.62) and (5.3.63), becomes

$$(\mathbf{V}_i\mathbf{V}_i - \tfrac{1}{3}V_i^2\mathbb{1}) : \mathbf{w} = \mathbf{V}_i\mathbf{V}_i : \mathbf{S}, \tag{5.3.66}$$

where

$$\mathbf{S} = \tfrac{1}{2}(\mathbf{w}+\mathbf{w}^{\mathrm{T}}) - \tfrac{1}{3}(\mathbf{w} : \mathbb{1}). \tag{5.3.67}$$

Equation (5.3.61) can therefore be written in the form

$$\mathbf{P}^{(1)} = p\mathbb{1} + \sum_{i\neq I} m_i \int f_i^{(0)}B(V_i)\mathbf{V}_i\mathbf{V}_i\mathbf{V}_i\mathbf{V}_i : \mathbf{S}\,\mathrm{d}\mathbf{v}_i, \tag{5.3.68}$$

where

$$\mathbf{S} = \tfrac{1}{2}(\nabla\mathbf{V} + (\nabla\mathbf{V})^{\mathrm{T}}) - \tfrac{1}{3}(\nabla\mathbf{V} : \mathbb{1})\mathbb{1}, \tag{5.3.69}$$

i.e.

$$S_{\alpha\beta} = \tfrac{1}{2}\left(\frac{\partial V_\beta}{\partial x_\alpha} + \frac{\partial V_\alpha}{\partial x_\beta}\right) - \tfrac{1}{3}\nabla\cdot\mathbf{V}\delta_{\alpha\beta}.$$

The rate of shear tensor $\mathbf{S}$ is, of course, independent of $\mathbf{V}_i$ and thus the integral appearing in the above equation (5.3.68) may be explicitly evaluated once $B(V_i)$ is known [30]. Consequently, we write

$$\mathbf{P}^{(1)} = p\mathbb{1} - \mu_{\mathrm{V}}\mathbf{S}, \tag{5.3.70}$$

where μ_{V} is referred to as the coefficient of viscosity. Quasi-empirical expressions for μ_{V} are detailed in Appendix C.

We next evaluate $\mathbf{q}^{(1)}$. Equation (5.1.76), viz.

$$\mathbf{q} = \sum_{i \neq I} \tfrac{1}{2} m_i \int f_i V_i^2 \mathbf{V}_i \, \mathrm{d}\mathbf{v}_i, \tag{5.3.71}$$

yields

$$\mathbf{q}^{(1)} = \sum_{i \neq I} \tfrac{1}{2} m_i \int f_i^{(0)} V_i^2 \mathbf{V}_i \Big\{ A(V_i)\mathbf{V}_i \cdot \nabla \ln T + B(V_i)(\mathbf{V}_i\mathbf{V}_i - \tfrac{1}{3} V_i^2 \mathbb{1}) : \nabla \mathbf{V} + \sum_j D_j(V_i)\mathbf{V}_i \cdot \mathbf{d}_j \Big\} \, \mathrm{d}\mathbf{V}_i, \tag{5.3.72}$$

where, as shown previously, the zeroth order value of $\mathbf{q}$ vanishes. Reference to equations (5.2.37) and (5.3.53) shows that the integral involving $B(V_i)$ also vanishes. After some straightforward algebraic manipulation, and the application of equation (5.3.58), we find

$$\mathbf{q}^{(1)} = -k_c \nabla T + \sum_i \bar{D}_i^{(2)} \bar{\mathbf{V}}_i, \tag{5.3.73}$$

where k_c, the coefficient of heat conductivity, and $\bar{D}_i^{(2)}$, a thermal diffusion parameter, can be evaluated once $A(V_i)$ and $D_j(V_i)$ are known [30]. Quasi-empirical expressions for k_c are detailed in Appendix C.

We therefore have the first order macroscopic equations stipulating conservation of mass, linear momentum and energy:

$$\frac{\partial N_L}{\partial t} + \nabla \cdot [N_L(\mathbf{V} + \bar{\mathbf{V}}_L)] = N_U B_{UL} \bar{J}_{UL} + N_U A_{UL} + N_U C_{UL} - N_L B_{LU} \bar{J}_{LU} - N_L C_{LU}, \tag{5.3.74}$$

$$\frac{\partial N_U}{\partial t} + \nabla \cdot [N_U(\mathbf{V} + \bar{\mathbf{V}}_U)] = N_L B_{LU} \bar{J}_{LU} + N_L C_{LU} - N_U B_{UL} \bar{J}_{UL} - N_U A_{UL} - N_U C_L, \tag{5.3.75}$$

$$\frac{\partial N_e}{\partial t} + \nabla \cdot [N_e(\mathbf{V} + \bar{\mathbf{V}}_e)] = 0, \tag{5.3.76}$$

$$\rho \frac{\partial \mathbf{V}}{\partial t} + \rho(\mathbf{V} \cdot \nabla)\mathbf{V} + \nabla p - \nabla(\mu_V \mathbf{S}) = \sum_i N_i \mathbf{F}_i, \tag{5.3.77}$$

and

$$\rho \frac{\partial \epsilon}{\partial t} + \rho(\mathbf{V} \cdot \nabla)\epsilon + p \nabla \cdot \mathbf{V} - \mu_V \mathbf{S} : \nabla \mathbf{V} - \nabla \cdot \Big[k_c \nabla T - \sum_i \bar{D}_i^{(2)} \bar{\mathbf{V}}_i \Big] = \sum_{i \neq I} N_i \mathbf{F}_i \cdot \bar{\mathbf{V}}_i. \tag{5.3.78}$$

Note that the addition of equations (5.3.74-76), together with the use of equation (5.1.66), yields

$$\frac{\partial \rho}{\partial t} + \nabla \cdot (\rho \mathbf{V}) = 0, \tag{5.3.79}$$

as expected.

Higher ordered terms $f_i^{(2)}$, etc. may now be evaluated. Although not considered here, such an analysis shows [30] that the inclusion of these terms in the conservation equations generates second order derivatives in **V** and T as well as extra first order derivatives of degree 2 or more. These higher order terms only become important when the velocity and temperature gradients become exceedingly large (for example in the transition region between stellar chromospheres and coronae) and have little practical application in the discussions presented here.

Further, we have not considered the explicit evaluation of the transport coefficients μ, k_c, $\bar{D}_{ij}^{(1)}$ and $\bar{D}_i^{(2)}$ (nor, indeed, their radiative counterparts A_{UL}, B_{UL} and B_{LU}) since this lies outside the explicit scope of the interaction between radiation and matter. The main aspect to emerge from all the foregoing relates to the mathematical structure of the pressure tensor **P** and energy flux vector **q**. As we shall see, the dependence of these quantities on the *gradients* of the velocity and temperature has a significant influence on the mathematics, thence physics, of the transfer of radiation through a non-stationary non-uniform gas and thus, on the observed emergent radiation intensity within the spectral line.

5.3.3 Inclusion of inelastic terms

We complete this section by returning to the discussion of $f_i^{(1)}$ when the inelastic terms are not neglected. Equations (5.3.9) and (5.3.11) then become

$$\frac{\mathrm{d}N_i}{\mathrm{d}t} = -N_i \nabla \cdot \mathbf{V} + R_i, \tag{5.3.80}$$

and

$$\frac{\mathrm{d}T}{\mathrm{d}t} = -\tfrac{2}{3} T \nabla \cdot \mathbf{V} + \frac{2h\nu R_U}{2kN}, \tag{5.3.81}$$

where

$$R_L = N_U(B_{UL}\bar{J}_{UL} + A_{UL} + C_{UL}) - N_L(B_{LU}\bar{J}_{LU} + C_{LU}), \tag{5.3.82}$$

$$R_U = -R_L, \quad \text{and} \quad R_e = 0. \tag{5.3.83}$$

These extra terms on the RHS of equations (5.3.80) and (5.3.81) thence contribute the extra term

$$f_i^{(0)}\left[\frac{R_i}{N_i}+\frac{2h\nu R_{\rm U}}{3kNT}\left(\frac{m_iV_i^2}{2kT}-\tfrac{3}{2}\right)\right]$$

to the RHS (5.3.15). But this factor contains the summational invariants R_i (analogous to m_i) and $\frac{1}{2}m_iV_i^2$ and, as for the corresponding η_i and ζ_i terms appearing in equation (5.3.22), do not contribute to ϕ_i. Consequently, we obtain the same ϕ_i, thence $\bar{\mathbf{V}}_i^{(1)}$, $\mathbf{P}^{(1)}$ and $\mathbf{q}^{(1)}$ as determined above.

It is interesting to note that no radiation terms therefore appear in these first order quantities, and this implies that radiation pressure, in particular, is zero to first order in our analysis. It does appear in non-zero form, however, in the second order calculation (not presented here). Implicit radiation pressure terms also appear when one removes the assumption $\mathbf{v}_{\rm L}=\mathbf{v}_{\rm U}$ (see section 5.1.5), but this significantly complicates *all* the conservation equations.

5.4 The macroscopic velocity-dependent source function [22, 65]

The radiative transfer equation (1.2.8), (5.1.29), viz.

$$\frac{1}{c}\frac{\partial I}{\partial t}+(\mathbf{\Omega}\cdot\nabla)I=-\kappa(I-S), \tag{5.4.1}$$

was derived for the general time-dependent intensity $I(\mathbf{r},\Delta\nu,\mathbf{\Omega},t)$ where

$$\kappa=\frac{h\nu}{4\pi}\left[N_{\rm L}B_{\rm LU}\phi-N_{\rm U}B_{\rm UL}\psi\right], \tag{5.4.2}$$

and

$$S=\frac{N_{\rm U}A_{\rm UL}j}{N_{\rm L}B_{\rm LU}\phi-N_{\rm U}B_{\rm UL}\psi}. \tag{5.4.3}$$

The absorption and emission probabilities ϕ, ψ and j are general functions of $(\mathbf{r},\Delta\nu,\mathbf{\Omega},t)$ given by equations (5.1.23), (5.1.25) and (5.1.27), viz.

$$\phi=\frac{1}{N_{\rm L}}\int q(\gamma_{\rm L})f_{\rm L}\,d\mathbf{v}_{\rm L}, \tag{5.4.4}$$

and

$$j\equiv\psi=\frac{1}{N_{\rm U}}\int e(\gamma_{\rm U})f_{\rm U}\,d\mathbf{v}_{\rm U}, \tag{5.4.5}$$

where

$$\gamma_i=\Delta\nu-\frac{\nu}{c}\mathbf{v}_i\cdot\mathbf{\Omega}. \tag{5.4.6}$$

If we consider just the absorption profile for the moment, and put $\mathbf{v}_{\rm L}=\mathbf{V}_{\rm L}+\mathbf{V}$, we have

$$\phi=\frac{1}{N_{\rm L}}\int_{-\infty}^{\infty}q\left(\Delta\nu-\frac{\nu}{c}\mathbf{V}_{\rm L}\cdot\mathbf{\Omega}-\frac{\nu}{c}\mathbf{V}\cdot\mathbf{\Omega}\right)f_{\rm L}(\mathbf{V}_{\rm L})\,d\mathbf{V}_{\rm L}. \tag{5.4.7}$$

If we now put $\Delta\nu_V = \Delta\nu - \nu\mathbf{V}\cdot\mathbf{\Omega}/c$ and recall equation (5.1.2), i.e. $f_L = N_L F_L$, we see that equation (5.4.7) is identical to (1.4.5) with $\Delta\nu$ replaced by $\Delta\nu_V$. We therefore write

$$\phi(\Delta\nu_V) \equiv \phi(\Delta\nu, \mathbf{\Omega}) \equiv \phi\left(\Delta\nu - \frac{\nu}{c}\mathbf{V}\cdot\mathbf{\Omega}\right), \tag{5.4.8}$$

where we have explicitly stated the dependence of ϕ on the macroscopic velocity $\mathbf{V}$.

The absorption profile (1.4.14) for Doppler broadening using $f_L^{(0)}$ given by equation (5.2.28) then becomes

$$\phi\left(\Delta\nu - \frac{\nu}{c}\mathbf{V}\cdot\mathbf{\Omega}\right) = \frac{1}{\Delta\nu_D\sqrt{\pi}}\exp\left[-\left(\frac{\Delta\nu - \frac{\nu}{c}\mathbf{V}\cdot\mathbf{\Omega}}{\Delta\nu_D}\right)^2\right], \tag{5.4.9}$$

where

$$\Delta\nu_D = \frac{\nu}{c}\left(\frac{2kT}{m_L}\right)^{1/2}, \tag{5.4.10}$$

whereas, for Voigt broadening (1.4.18), we have

$$\phi\left(\Delta\nu - \frac{\nu}{c}\mathbf{V}\cdot\mathbf{\Omega}\right) = \frac{1}{\Delta\nu_D\sqrt{\pi}} H\left(a, \frac{\Delta\nu - \frac{\nu}{c}\mathbf{V}\cdot\mathbf{\Omega}}{\Delta\nu_D}\right), \tag{5.4.11}$$

where the Voigt function $H(a, \eta)$ is defined by equation (1.4.20).

However, it is not unreasonable to examine the higher order absorption probability obtained by using $f_L = f_L^{(0)} + f_L^{(0)}\phi_L$, not $f_L^{(0)}$ alone, in equation (5.4.7) where ϕ_L is given by equation (5.3.47). In this case, the absorption profile would include terms involving the velocity and temperature gradients along with the diffusion terms $\mathbf{d}_j$.

This then completes the specification of the absorption probability. The evaluation of the corresponding emission profiles is somewhat more difficult. As in chapter 1, we sidestep the issue here, however, by making the assumption of complete re-distribution, i.e.

$$\phi \equiv \psi \equiv j, \tag{5.4.12}$$

for all $\Delta\nu$ and $\mathbf{\Omega}$ (see section 1.5) so that the source function (5.4.3) has the far simpler form

$$S = \frac{\dfrac{N_U A_{UL}}{N_L B_{LU}}}{1 - \dfrac{N_U B_{UL}}{N_L B_{LU}}}. \tag{5.4.13}$$

We shall remove this assumption in chapter 7.

Clearly, we now need only find the density ratio N_U/N_L. The zeroth order mass conservation equations (5.2.42–44) are

$$\frac{\partial N_i}{\partial t} + \nabla \cdot (N_i \mathbf{V}) = R_i, \tag{5.4.14}$$

where

$$R_L = N_U(B_{UL}\bar{J}_{UL} + A_{UL} + C_{UL}) - N_L(B_{LU}\bar{J}_{LU} + C_{LU}), \tag{5.4.15}$$

$$R_U = -R_L, \quad \text{and} \quad R_e = 0, \tag{5.4.16}$$

and where $\bar{J}_{UL}$ and $\bar{J}_{LU}$ are now defined by

$$\bar{J}_{UL} = \frac{1}{4\pi}\int_{-\infty}^{\infty} d(\Delta\nu) \int_{4\pi} d\Omega\psi\left(\Delta\nu - \frac{\nu}{c}\mathbf{V}\cdot\mathbf{\Omega}\right) I(\mathbf{r}, \Delta\nu, \mathbf{\Omega}, t), \tag{5.4.17}$$

and

$$\bar{J}_{LU} = \frac{1}{4\pi}\int_{-\infty}^{\infty} d(\Delta\nu) \int_{4\pi} d\Omega\phi\left(\Delta\nu - \frac{\nu}{c}\mathbf{V}\cdot\mathbf{\Omega}\right) I(\mathbf{r}, \Delta\nu, \mathbf{\Omega}, t). \tag{5.4.18}$$

Our assumption of complete re-distribution (5.4.12) then enables us to write

$$\bar{J}_{UL} \equiv \bar{J}_{LU} \equiv \bar{J} \quad \text{say}, \tag{5.4.19}$$

so that equations (5.4.14–16) may then be combined to yield

$$\frac{N_U}{N_L} = \frac{B_{LU}\bar{J} + C_{LU} + R_L/N_L}{B_{UL}\bar{J} + C_{UL} + A_{UL}}. \tag{5.4.20}$$

Equation (5.4.13) then becomes

$$S = \frac{\bar{J} + \dfrac{2h\nu^3}{c^2}\cdot\dfrac{C_{LU}g_L}{A_{UL}g_U} + \dfrac{R_L}{N_L B_{LU}}}{1 + \dfrac{C_{UL}}{A_{UL}}\left(1 - \dfrac{C_{LU}g_L}{C_{UL}g_U}\right) - \dfrac{R_L g_L}{N_L A_{UL} g_U}}, \tag{5.4.21}$$

where we have used equations (1.6.5) and (1.6.6), viz.

$$\frac{A_{UL}}{B_{UL}} = \frac{2h\nu^3}{c^2}, \quad g_U B_{UL} = g_L B_{LU}. \tag{5.4.22}$$

If we now define ϵ' by

$$\epsilon' = \frac{C_{UL}}{A_{UL}}(1 - e^{-h\nu/kT}), \tag{5.4.23}$$

as in equation (1.6.11), and use equations (1.6.9) and (1.6.13), viz.

$$\frac{C_{LU}}{C_{UL}} = \frac{g_U}{g_L} e^{-h\nu/kT}, \quad B_\nu(T) = \frac{2h\nu^3/c^2}{e^{h\nu/kT} - 1}, \tag{5.4.24}$$

equation (5.4.21) then becomes

$$S = \frac{\bar{J} + \epsilon' B_\nu(T) + \dfrac{R_L}{N_L B_{LU}}}{1 + \epsilon' - \dfrac{R_L g_U}{N_L A_{UL} g_U}}, \tag{5.4.25}$$

which, after a slight re-arrangement, has the form

$$S = \frac{\bar{J} + \epsilon' B_\nu(T)}{1 + \epsilon'} + \frac{R_L g_L (S + \alpha)}{N_L A_{UL} g_U (1 + \epsilon')}, \tag{5.4.26}$$

where

$$\alpha = \frac{2h\nu^3}{c^2}. \tag{5.4.27}$$

Clearly, if $R_L \equiv 0$, and this corresponds to the stationary time-independent situation discussed in some detail in the first three chapters, then equation (5.4.25) reduces to the 2-level atom source function given by equation (1.6.12).

We now turn our attention to the evaluation of the ratio R_L/N_L. Equation (5.4.15) can be re-written as

$$\frac{R_L}{N_L} = \frac{1}{N_L}\frac{dN_L}{dt} + \nabla \cdot \mathbf{V}. \tag{5.4.28}$$

Noting that $N_L = N(N_L/N)$ where

$$N = N_U + N_L, \tag{5.4.29}$$

is the atom number density, then

$$\begin{aligned}\frac{dN_L}{dt} &= \frac{N_L}{N}\frac{dN}{dt} + N\frac{d}{dt}\left(\frac{N_L}{N}\right) \\ &= -N_L \nabla \cdot \mathbf{V} + N\frac{d}{dt}\left(\frac{N_L}{N}\right),\end{aligned} \tag{5.4.30}$$

where we have used the mass conservation equation for N, viz.

$$\frac{dN}{dt} + N\nabla \cdot \mathbf{V} = 0. \tag{5.4.31}$$

Equations (5.4.28) and (5.4.30) then yield

$$\begin{aligned}\frac{R_L}{N_L} = \frac{N}{N_L}\frac{d}{dt}\left(\frac{N_L}{N}\right) &= \left(1 + \frac{N_U}{N_L}\right)\frac{d}{dt}\left(\frac{1}{1 + \dfrac{N_U}{N_L}}\right) \\ &= \frac{-1}{1 + \dfrac{N_U}{N_L}}\frac{d}{dt}\left(\frac{N_U}{N_L}\right).\end{aligned} \tag{5.4.32}$$

The ratio N_U/N_L may be written in terms of S using equation (5.4.13), viz.

$$\frac{N_U}{N_L} = \frac{g_U S}{g_L(S+\alpha)}\,. \tag{5.4.33}$$

Consequently, equation (5.4.32) becomes

$$\frac{R_L}{N_L} = \frac{-g_U\alpha}{(S+\alpha)\,[(g_U+g_L)S+g_L\alpha]}\,\frac{dS}{dt}\,. \tag{5.4.34}$$

This last result, when substituted into equation (5.4.26), yields the required form for the velocity-dependent line source function, viz

$$S = \frac{\bar{J}+\epsilon' B_\nu(T)}{1+\epsilon'} - \frac{1}{A_{UL}(1+\epsilon')\left[1+\left(1+\dfrac{g_U}{g_L}\right)\dfrac{S}{\alpha}\right]}\,\frac{dS}{dt}\,, \tag{5.4.35}$$

or, using the more standard $\epsilon = \epsilon'/(1+\epsilon')$,

$$S = (1-\epsilon)\bar{J} + \epsilon B_\nu(T) - \frac{1-\epsilon}{A_{UL}\left[1+\left(1+\dfrac{g_U}{g_L}\right)\dfrac{S}{\alpha}\right]}\,\frac{dS}{dt}\,. \tag{5.4.36}$$

Again, if we consider the gas to be both time independent and stationary, the dS/dt term vanishes and we regain the source function studied in the previous chapters. Even if we allow the gas to be time independent but *non*-stationary, i.e. $\partial/\partial t$ terms disappear but not $\mathbf{V}\cdot\nabla$ terms (recall the definition (5.3.5) for the Lagrangian derivative), then there is still a non-zero contribution to S from the dS/dt factor which should not be ignored.

We already know that the first three terms on the RHS (5.4.36) correspond respectively to photon absorption followed by photon emission, photon destruction due to absorption followed by collisional de-excitation, and photon creation due to collisional excitation followed by spontaneous emission. It is of some benefit, therefore, to examine the physical basis of the fourth (dS/dt) term appearing in S. To do this, we take $S \ll \alpha$ so that equation (5.4.36) may be written in the form

$$\frac{1-\epsilon}{A_{UL}}\,\frac{dS}{dt} + S = (1-\epsilon)\bar{J} + \epsilon B_\nu(T). \tag{5.4.37}$$

For the purpose of this discussion, we put ϵ equal to a constant. This last equation can then be solved explicitly for S, viz.

$$S(t) = \frac{1}{\tau_l}\int_{-\infty}^{t} [(1-\epsilon)\bar{J}(t') + \epsilon B_\nu(t')]\exp\left[-\frac{t-t'}{\tau_l}\right]dt', \tag{5.4.38}$$

where $\tau_l = (1-\epsilon)/A_{UL}$.

This formal integration of equation (5.4.37) produces the value of the source function $S(t)$ in a particulr volume element as this element moves through the atmosphere. The atoms in this volume element will excite at some point in space, then, after a finite lifetime τ_l in the excited state (in which time the volume element moves to another location in the stellar atmosphere), de-excite under different physical conditions. The source function $S(t)$, which still incorporates the usual scattering ($\bar{J}$), destruction ($-\epsilon\bar{J}$) and creation ($\epsilon B_\nu(T)$) terms, will therefore involve contributions from previous points in time weighted by the factor $\exp\,[-(t-t')/\tau_l]$. When ϵ is small, the magnitude of this factor decreases exponentially over a time scale of $1/A_{\mathrm{UL}}$ which is simply the average lifetime of the excited state. Clearly, as A_{UL} increases (i.e. the time rate of spontaneous de-excitation increases), the lifetime of the excited state decreases and this, in turn, would suggest that the possibility of the atom de-exciting under different physical conditions from those under which it was excited must decrease. Indeed, in the limit as $A_{\mathrm{UL}} \to \infty$, the atom would have a zero lifetime in the excited state - consequently, the atom de-excites in exactly the same position (under exactly the same physical conditions) as it was excited. In this case

$$\frac{1}{\tau_l}\,\mathrm{e}^{-(t-t')/\tau_l} \to \delta(t-t'),$$

i.e.

$$S(t) \to (1-\epsilon)\bar{J} + \epsilon B_\nu(t),$$

and one regains the usual time-independent stationary source function. Of course, this limiting case could have been attained directly from equation (5.4.37) where $A_{\mathrm{UL}}^{-1}\,\mathrm{d}S/\mathrm{d}t \to 0$ as $A_{\mathrm{UL}} \to \infty$.

The factor $1-\epsilon$ appearing in τ_l takes into account the effect of collisional excitations and de-excitations on the lifetime of the excited state. As ϵ increases, for example, photon creation and destruction increases relative to photon scattering, and this therefore decreases the effective lifetime of the excited state. We re-write τ_l using equation (5.4.23), together with $\epsilon = \epsilon'/(1+\epsilon')$, i.e.

$$\tau_l = \frac{1}{A_{\mathrm{UL}} + C_{\mathrm{UL}}(1-\mathrm{e}^{-h\nu/kT})}\,. \tag{5.4.39}$$

Thus, as $A_{\mathrm{UL}} \to 0$, i.e. $\epsilon' \to \infty$ or $\epsilon \to 1$, then

$$\tau_l \to \tau_{\mathrm{c}} = \frac{1}{C_{\mathrm{UL}}(1-\mathrm{e}^{-h\nu/kT})}, \tag{5.4.40}$$

a finite value. Equation (5.4.38) then yields

$$S(t) \to \frac{1}{\tau_{\mathrm{c}}}\int_{-\infty}^{\infty} B_\nu(t')\mathrm{e}^{-(t-t')/\tau_{\mathrm{c}}}\,\mathrm{d}t'.$$

In the further limit as $C_{\mathrm{UL}} \to \infty$, $\tau_{\mathrm{c}}^{-1}\mathrm{e}^{-(t-t')/\tau_{\mathrm{c}}} \to \delta(t-t')$, so that $S(t) \to B_\nu(t)$, as expected from equilibrium considerations.

5.5 Numerical solution to the radiative transfer equation

We wish to solve

$$\frac{1}{c}\frac{\partial I}{\partial t}+(\boldsymbol{\Omega}\cdot\nabla)I=-\kappa(I-S), \tag{5.5.1}$$

where

$$\kappa=\frac{h\nu}{4\pi}\,(N_{\mathrm{L}}B_{\mathrm{LU}}-N_{\mathrm{U}}B_{\mathrm{UL}})\phi\left(\Delta\nu-\frac{\nu}{c}\mathbf{V}\cdot\boldsymbol{\Omega}\right), \tag{5.5.2}$$

and

$$S=\frac{1-\epsilon}{4\pi}\int_{-\infty}^{\infty}\mathrm{d}(\Delta\nu)\int_{4\pi}\mathrm{d}\Omega\phi\left(\Delta\nu-\frac{\nu}{c}\mathbf{V}\cdot\boldsymbol{\Omega}\right)I(\mathbf{r},\Delta\nu,\boldsymbol{\Omega},t)$$
$$+\,\epsilon B_{\nu}(T). \tag{5.5.3}$$

Note that we have again assumed complete re-distribution. Departures from complete re-distribution are examined in chapter 7.

For the moment, we consider 1-dimensional geometry so that $\mathbf{V}\cdot\boldsymbol{\Omega}\equiv V\mu$ and $\boldsymbol{\Omega}\cdot\nabla\equiv\mu\,\partial/\partial z$, where $\mu=\cos\theta$ (see figure 1.14). The application of macroscopic velocity field radiative transfer effects to multi-dimensional situations is discussed in section 5.7.

Further, we first consider the time-independent case for which all $\partial/\partial t$ and $\mathrm{d}/\mathrm{d}t$ terms are neglected. We include these terms in section 5.6.

Equation (5.5.1) then becomes

$$\mu\frac{\partial I}{\partial\tau}=\phi\left(\Delta\nu-\frac{\nu}{c}V\mu\right)(I-S), \tag{5.5.4}$$

where

$$S=\frac{1-\epsilon}{2}\int_{-\infty}^{\infty}\mathrm{d}(\Delta\nu)\int_{-1}^{1}\mathrm{d}\mu\phi\left(\Delta\nu-\frac{\nu}{c}V\mu\right)I(\tau,\Delta\nu,\mu)$$
$$+\,\epsilon B_{\nu}(T), \tag{5.5.5}$$

and

$$\tau(z)=-\frac{h\nu}{4\pi}\int_{0}^{z}[N_{\mathrm{L}}(z')B_{\mathrm{LU}}-N_{\mathrm{U}}(z')B_{\mathrm{UL}}]\,\mathrm{d}z', \tag{5.5.6}$$

is an optical depth variable independent of $\Delta\nu$ and μ (see section 1.9.1).

There exist several techniques for the numerical solution of equation (5.5.4). Here, we discuss only those which offer ease in operation, stability and accuracy.

5.5.1 Feautrier's technique

Feautrier's method of solution of the radiative transfer equation has been discussed in great detail in section 3.2 in its application to stationary media.

Its generalisation to non-stationary situations is rather straightforward when the time derivatives can be ignored [22, 24, 84]. We put

$$I^+(\tau, \Delta\nu, \mu) = I(\tau, \Delta\nu, \mu > 0), \tag{5.5.7}$$

and

$$I^-(\tau, \Delta\nu, \mu) = I(\tau, -\Delta\nu, \mu < 0), \tag{5.5.8}$$

for all $\Delta\nu$ across the entire spectral line profile. (Recall that in the stationary media calculations, symmetry in the spectral line profile about line centre enabled one to use only half the profile in performing the computations – this led to a significant saving in computer time and storage.)

The line source function (5.5.5) is both frequency and angle independent when complete re-distribution is assumed so that $S^+ = S^- = S(\tau)$. Consequently, equation (5.5.4) becomes

$$\mu \frac{\partial I^+}{\partial \tau} = \phi\left(\Delta\nu - \frac{\nu}{c} V\mu\right)(I^+ - S), \tag{5.5.9}$$

and

$$-\mu \frac{\partial I^-}{\partial \tau} = \phi\left(-\Delta\nu + \frac{\nu}{c} V\mu\right)(I^- - S), \ \forall\, \mu > 0. \tag{5.5.10}$$

The line absorption profile $\phi(x)$ is symmetric about line centre $x = 0$, i.e.

$$\phi\left(\Delta\nu - \frac{\nu}{c} V\mu\right) = \phi\left(-\Delta\nu + \frac{\nu}{c} V\mu\right). \tag{5.5.11}$$

Thus, writing

$$\Phi(\tau, \Delta\nu, \mu) = \tfrac{1}{2}(I^+ + I^-), \tag{5.5.12}$$

and

$$\Psi(\tau, \Delta\nu, \mu) = \tfrac{1}{2}(I^+ - I^-), \ \forall\, \mu > 0, \tag{5.5.13}$$

equations (5.5.9–10) become

$$\mu \frac{\partial \Psi}{\partial \tau} = \phi\left(\Delta\nu - \frac{\nu}{c} V\mu\right)(\Phi - S), \tag{5.5.14}$$

and

$$\phi \frac{\partial \Phi}{\partial \tau} = \phi\left(\Delta\nu - \frac{\nu}{c} V\mu\right)\Psi. \tag{5.5.15}$$

We can evaluate the source function S in terms of Φ by writing

$$S = \frac{1-\epsilon}{2} \int_{-\infty}^{\infty} \mathrm{d}(\Delta\nu) \int_0^1 \mathrm{d}\mu\phi\left(\Delta\nu - \frac{\nu}{c} V\mu\right) I^+$$

$$+ \frac{1-\epsilon}{2} \int_{\infty}^{-\infty} \mathrm{d}(-\Delta\nu) \int_{-1}^{0} \mathrm{d}\mu \left(-\Delta\nu - \frac{\nu}{c} V\mu\right) I^- + \epsilon B_\nu(T)$$

$$= \frac{1-\epsilon}{2}\int_{-\infty}^{\infty} d(\Delta\nu)\int_0^1 d\mu\phi\left(\Delta\nu - \frac{\nu}{c}V\mu\right)(I^+ + I^-) + \epsilon B_\nu(T)$$

$$= (1-\epsilon)\int_{-\infty}^{\infty} d(\Delta\nu)\int_0^1 d\mu\phi\left(\Delta\nu - \frac{\nu}{c}V\mu\right)\Phi(\tau, \Delta\nu, \mu) + \epsilon B_\nu(T). \tag{5.5.16}$$

Thus, substituting Ψ from equation (5.5.15) into (5.5.14) yields

$$\left[\frac{\mu}{\phi\left(\Delta\nu - \frac{\nu}{c}V\mu\right)}\frac{\partial}{\partial\tau}\right]^2 \Phi(\tau, \Delta\nu, \mu) = \Phi(\tau, \Delta\nu, \mu)$$
$$-(1-\epsilon)\int_{-\infty}^{\infty} d(\Delta\nu')\int_0^1 d\mu'\phi\left(\Delta\nu' - \frac{\nu}{c}V\mu'\right)\Phi(\tau, \Delta\nu', \mu')$$
$$-\epsilon B_\nu(T). \tag{5.5.17}$$

This second order integro-differential equation can then be solved numerically using finite difference formulae to approximate the derivatives and quadrature techniques to evaluate the integrals. The details have already been discussed explicitly in section 3.2. Again, the two required boundary conditions may be obtained from equation (5.5.15), viz.

$$\mu\frac{\partial\Phi(0, \Delta\nu, \mu)}{\partial\tau} = \phi\left(\Delta\nu - \frac{\nu}{c}V(0)\mu\right)[\Phi(0, \Delta\nu, \mu) - I^-(0, \Delta\nu, \mu)], \tag{5.5.18}$$

and

$$\mu\frac{\partial\Phi(\tau_\infty, \Delta\nu, \mu)}{\partial\tau} = \phi\left(\Delta\nu - \frac{\nu}{c}V(\tau_\infty)\mu\right)[I^+(\tau_\infty, \Delta\nu, \mu) - \Phi(\tau_\infty, \Delta\nu, \mu)], \tag{5.5.19}$$

where τ_∞ is the optical thickness of the gas (see equations (3.2.11) and (3.2.12) for the stationary situation). In particular, for a semi-infinite atmosphere, we have $I^+(\tau_\infty, \Delta\nu, \mu) = B_\nu(T)$.

A substantial advantage of the above method is the ease with which one can obtain the desired computer code simply by converting a program written for the 'stationary Feautrier technique'. Indeed, only a very few changes to the code are required. More importantly, stability is retained although this has only been shown using numerical experiments. The accuracy of the method depends upon the frequency grid $\{\Delta\nu_i\}$ chosen. Condition (3.2.90), viz. $\Delta\nu_i - \Delta\nu_{i-1} <$ min $(\Delta\nu_D)/2$ in the line core, is still required but this, together with the possibility of a very large frequency bandwidth due to very large velocity gradients (note that we must take $\{\Delta\nu_i\}$ such that any errors in the integral appearing in equation (5.5.17) satisfy

$$\left|\int_{-\infty}^{\infty} \phi\left(\Delta\nu - \frac{\nu}{c} V\mu\right) I(\tau, \Delta\nu, \mu)\, \mathrm{d}\mu - \sum_{i=1}^{N_\nu} w_i^{(1)} I(\tau, \Delta\nu_i, \mu)\right|$$
$$\leqslant \epsilon B_\nu(T),$$

could necessitate a very large number N_ν of frequency grid points. The inherent difficulties associated with computer time and storage requirements for large N_ν may be substantially removed, however, by use of the quadrature perturbation technique discussed in section 3.3 or the co-moving frame analysis presented in the following sub-section. Note, finally, that the Auer modification (section 3.2.9) can also be readily incorporated into the above method and this, too, increases the accuracy of the numerical solution significantly.

5.5.2 *The co-moving frame formulation [90]*

Equation (5.5.4) is an expression for $I(\tau, \Delta\nu, \mu)$ written in the rest frame of the observer. As we have already mentioned, large velocity gradients can lead to a large number N_ν of frequency quadrature points $\{\Delta\nu_i\}$ which, when used in equation (5.5.4), must also be measured in the rest frame of the observer. The subsequent difficulties generated by the need to invert large matrices using Feautrier's technique can be removed, in part, by use of the quadrature perturbation method (section 3.3) or, more fully, by re-writing the radiative transfer equation (5.5.4) in the local rest frame of the moving gas.

To do this, we define a new frequency variable ξ by

$$\xi = \nu - \frac{\nu}{c} V\mu. \tag{5.5.20}$$

One then proceeds by noting [114, 116] that I/ν^3, not I alone, is invariant under the transformation (5.5.20). This can be readily seen, for example, by recalling equation (5.4.33), viz.

$$\frac{N_\mathrm{U}}{N_\mathrm{L}} = \frac{g_\mathrm{U}/g_\mathrm{L}}{1 + \dfrac{2h\nu^3}{Sc^2}}. \tag{5.5.21}$$

Clearly, the number density ratio $N_\mathrm{U}/N_\mathrm{L}$ should be invariant under the above frequency transformation. This implies that S/ν^3 should be similarly invariant. Reference to any of the equations specifying the source function S shows that the radiation specific intensity I has the same dimensions as S and this, in turn, stipulates the invariance of I/ν^3. We therefore have

$$\frac{I_\mathrm{O}(\tau, \Delta\nu, \mu)}{\nu^3} = \frac{I_\mathrm{G}(\tau, \Delta\xi, \mu)}{\xi^3}, \tag{5.5.22}$$

where the subscripts O and G refer to the specific intensity I measured in the rest frame of the observer and gas respectively. Note that we have introduced the frequency variable $\Delta\xi$ (analogous to $\Delta\nu$) measured from line centre ν_0.

Consequently, partial differentiation yields

$$\frac{\partial I_O}{\partial\tau} = \frac{\nu^3}{\xi^3}\frac{\partial I_G}{\partial\tau} + \frac{3\nu^4}{\xi^4}\frac{\mu}{c}I_G\frac{dV}{d\tau} - \frac{\nu^3}{\xi^3}\frac{\mu\nu}{c}\frac{dV}{d\tau}\frac{\partial I_G}{\partial(\Delta\xi)}, \tag{5.5.23}$$

which, neglecting terms of order $(|V|/c)^2$, enables equation (5.5.4) to be written as

$$\alpha\mu\frac{\partial I_G}{\partial\tau} + \beta\frac{\partial I_G}{\partial(\Delta\xi)} + \gamma I_G = \phi(\Delta\xi)(I_G - S_G), \tag{5.5.24}$$

$$S_G = \frac{1-\epsilon}{2}\int_{-\infty}^{\infty} d(\Delta\xi)\phi(\Delta\xi)\int_{-1}^{1} d\mu\alpha I_G(\tau, \Delta\xi, \mu) + \epsilon B_\nu(T), \tag{5.5.25}$$

where

$$\left.\begin{aligned} &\alpha = 1 + \frac{3V\mu}{c}, \quad \beta = -\frac{\nu}{c}\alpha\mu^2\frac{dV}{d\tau},\\ &\text{and}\\ &\gamma = 3\left(1 + \frac{4V\mu}{c}\right)\frac{\mu^2}{c}\frac{dV}{d\tau}. \end{aligned}\right\} \tag{5.5.26}$$

If one treats I alone as the invariant, one finds $\alpha \equiv 1$ and $\gamma \equiv 0$, and this then yields the somewhat simpler equation normally considered in the literature.

To proceed further, we develop the perturbation series given by

$$I_G = I_G^{(0)} + \lambda I_G^{(1)} + \lambda^2 I_G^{(2)} + \cdots, \tag{5.5.27}$$

such that

$$S_G = S_G^{(0)} + \lambda S_G^{(1)} + \lambda^2 S_G^{(2)} + \cdots. \tag{5.5.28}$$

We write

$$\left.\begin{aligned} &\alpha = \alpha_0 + \lambda\alpha_1 \quad \alpha_0 \equiv 0,\\ &\beta = \beta_0 + \lambda\beta_1 \quad \beta_0 = -\frac{\nu\alpha_0\mu^2}{c}\frac{dV}{d\tau},\\ &\gamma = \gamma_0 + \lambda\gamma_1 \quad \gamma_0 \equiv 0, \end{aligned}\right\} \tag{5.5.29}$$

such that equation (5.5.24) reduces to the perturbation system

$$\mu\frac{\partial I_G^{(k)}}{\partial\tau} + \beta_0\frac{\partial I_G^{(k)}}{\partial(\Delta\xi)} = \phi(\Delta\xi)(I_G^{(k)} - S_G^{(k)}), \tag{5.5.30}$$

for all $k = 0, \infty$. The 'source functions' $S_G^{(k)}$ have the form

$$S_G^{(k)} = \frac{1-\epsilon}{2}\int_{-\infty}^{\infty} d(\Delta\xi')\phi(\Delta\xi')\int_{-1}^{1} d\mu'\alpha_0 I_G^{(k)}(\tau, \Delta\xi', \mu') + E_G^{(k)}(\tau, \Delta\xi, \mu), \tag{5.5.31}$$

where the inhomogeneous source terms are

$$E_{\mathrm{G}}^{(0)} = \epsilon B_\nu(T), \tag{5.5.32}$$

and, for $k = 1, \infty$,

$$E_{\mathrm{G}}^{(k)} = \frac{1}{\phi(\Delta\xi)}\left[\alpha_1\mu\frac{\partial I_{\mathrm{G}}^{(k-1)}}{\partial\tau} + \beta_1\frac{\partial I_{\mathrm{G}}^{(k-1)}}{\partial(\Delta\xi)} + \gamma_1 I_{\mathrm{G}}^{(k-1)}\right] + \frac{1-\epsilon}{2}\int_{-\infty}^{\infty}\mathrm{d}(\Delta\xi')\phi(\Delta\xi')\int_{-1}^{1}\mathrm{d}\mu'\alpha_1' I_{\mathrm{G}}^{(k-1)}(\tau, \Delta\xi', \mu'). \tag{5.5.33}$$

Clearly, the only difference between the zeroth order equation ($k = 0$) and the higher orders is in the source term $E_{\mathrm{G}}^{(k)}$ specified above. The integro-differential operator remains unaltered, and this leads to significant computational advantages (see section 3.3 detailing the quadrature perturbation technique and its applications).

We now require the zeroth order solution to equation (5.5.30); all higher order solutions thence follow immediately. Indeed, equation (5.5.30) with $k = 0$ is the basic equation one obtains by considering only I as the invariant under the transformation (5.5.20) and, unless $|V|/c$ is relatively large (~ 0.1), its solution contains the essential physics of interest.

(a) Feautrier's technique

We drop all unnecessary subscripts and write the equation (5.5.30) to be solved as

$$\mu\frac{\partial I}{\partial\tau} - \frac{\mu^2\nu}{c}\frac{\mathrm{d}V}{\mathrm{d}\tau}\frac{\partial I}{\partial(\Delta\xi)} = \phi(\Delta\xi)(I - S). \tag{5.5.34}$$

We define the $\pm$ components of I as

$$I^+ = I(\tau, \Delta\xi, \mu > 0), \quad I^- = I(\tau, \Delta\xi, \mu < 0), \tag{5.5.35}$$

so that equation (5.5.34) becomes

$$\pm\mu\frac{\partial I^\pm}{\partial\tau} - \frac{\mu^2\nu}{c}\frac{\mathrm{d}V}{\mathrm{d}\tau}\frac{\partial I^\pm}{\partial(\Delta\xi)} = \phi(\Delta\xi)(I^\pm - S), \forall\, \mu > 0, \tag{5.5.36}$$

where, again $S^+ = S^- = S$. Putting

$$\Phi = \tfrac{1}{2}(I^+ + I^-) \quad \text{and} \quad \Psi = \tfrac{1}{2}(I^+ - I^-), \forall\, \mu > 0, \tag{5.5.37}$$

equation (5.5.36) yields

$$\mu\frac{\partial\Psi}{\partial\tau} - \frac{\mu^2\nu}{c}\frac{\mathrm{d}V}{\mathrm{d}\tau}\frac{\partial\Phi}{\partial(\Delta\xi)} = \phi(\Delta\xi)(\Phi - S), \tag{5.5.38}$$

and

$$\mu\frac{\partial\Phi}{\partial\tau} - \frac{\mu^2\nu}{c}\frac{\mathrm{d}V}{\mathrm{d}\tau}\frac{\partial\Psi}{\partial(\Delta\xi)} = \phi(\Delta\xi)\Psi. \tag{5.5.39}$$

The Feautrier procedure outlined in the previous sub-section resulted in a de-coupling of Φ and Ψ so that only one equation (second order integro-differential) in Φ needed to be solved. Here, however, we see that the coupling between Φ and Ψ cannot be explicitly removed because of the existence of the non-zero velocity gradient $\mathrm{d}V/\mathrm{d}\tau$.

We therefore proceed by differencing equations (5.5.38) and (5.5.39) as they stand. We define $\{\tau_j\}$, $\{\Delta\xi_i\}$ and $\{\mu_k\}$ grids and write, for example, $\Phi_{ijk} = (\tau_j, \Delta\xi_i, \mu_k)$ so that, with $\Delta_j = \tau_{j+1} - \tau_j$ and $\Delta_i = \Delta\xi_{i+1} - \Delta\xi_i$, and $\nabla_j = \tau_j - \tau_{j-1}$ and $\nabla_i = \Delta\xi_i - \Delta\xi_{i-1}$, the above two equations become

$$\frac{\mu_k}{2}\left[\frac{\Psi_{i,j+1,k} - \Psi_{ijk}}{\Delta_j} + \frac{\Psi_{ijk} - \Psi_{i,j-1,k}}{\nabla_j}\right] - \frac{\mu_k^2 \nu_{ijk}}{2c}\left[\frac{\mathrm{d}V}{\mathrm{d}\tau}\right]_j \left[\frac{\Phi_{i+1,j,k} - \Phi_{ijk}}{\Delta_i} + \frac{\Phi_{ijk} - \Phi_{i-1,j,k}}{\nabla_i}\right] = \phi(\Delta\xi_i)(\Phi_{ijk} - S_j), \tag{5.5.40}$$

and

$$\frac{\mu_k}{2}\left[\frac{\Phi_{i,j+1,k} - \Phi_{ijk}}{\Delta_j} + \frac{\Phi_{ijk} - \Phi_{i,j-1,k}}{\nabla_j}\right] - \frac{\mu_k^2 \nu_{ijk}}{2c}\left[\frac{\mathrm{d}V}{\mathrm{d}\tau}\right]_j \left[\frac{\Psi_{i+1,j,k} - \Psi_{ijk}}{\Delta_i} + \frac{\Psi_{ijk} - \Psi_{i-1,j,k}}{\nabla_i}\right] = \phi(\Delta\xi_i)\Psi_{ijk}. \tag{5.5.41}$$

(Note that we write ν_{ijk} since, from equation (5.5.20), ν should be written in terms of ξ, μ and $V(\tau)$.)

If we now use quadrature as in equation (3.2.87) to evaluate the integral appearing in S_j, viz.

$$S_j = (1-\epsilon)\int_{-\infty}^{\infty} \mathrm{d}(\Delta\xi)\phi(\Delta\xi)\int_0^1 \Phi(\tau_j, \Delta\xi, \mu) + [\epsilon B_\nu(T)]_j$$

$$\approx (1-\epsilon)\sum_{i=1}^{N_\xi}\sum_{k=1}^{N_\mu} w_{ij}^{(1)} w_k^{(2)} \Phi(\tau_j, \Delta\xi_i, \mu_k) + [\epsilon B_\nu(T)]_j, \tag{5.5.42}$$

and define the vector $\boldsymbol{\Phi}_j$ using equation (3.2.40), viz.

$$\boldsymbol{\Phi}_j = \begin{pmatrix} \Phi(\tau_j, \Delta\xi_1, \mu_1) \\ \vdots \\ \Phi(\tau_j, \Delta\xi_{N_\xi}, \mu_1) \\ \vdots \\ \Phi(\tau_j, \Delta\xi_{N_\xi}, \mu_{N_\mu}) \end{pmatrix}, \tag{5.5.43}$$

with a similar expression for the $\boldsymbol{\Psi}_j$ vector, equations (5.5.40) and (5.5.41) may be written in the matrix form

$$\mathbf{A}_j\boldsymbol{\Psi}_{j-1} + \mathbf{B}_j\boldsymbol{\Psi}_j + \mathbf{C}_j\boldsymbol{\Psi}_{j+1} + \mathbf{D}_j^{(1)}\boldsymbol{\Phi}_j = L_j, \tag{5.5.44}$$

$$\mathbf{A}_j\Phi_{j-1} + \mathbf{B}_j\Phi_j + \mathbf{C}_j\Phi_{j+1} + \mathbf{D}_j^{(2)}\Psi_j = 0. \tag{5.5.45}$$

The $\mathbf{A}_j$, $\mathbf{B}_j$ and $\mathbf{C}_j$ matrices are diagonal with elements

$$(\mathbf{A}_j)_{ik} = \frac{\mu_k}{2\nabla_j\phi(\Delta\xi_i)}, \quad (\mathbf{C}_j)_{ik} = \frac{-\mu_k}{2\Delta_j\phi(\Delta\xi_i)}, \tag{5.5.46}$$

and

$$\mathbf{B}_j = -\mathbf{A}_j - \mathbf{C}_j. \tag{5.5.47}$$

Clearly, we must have

$$\mathbf{A}_1 \equiv \mathbf{0} \equiv \mathbf{C}_{N_\tau} \tag{5.5.48}$$

or, equivalently, $\nabla_{j=1} \to \infty$ and $\Delta_{j=N_\tau} \to \infty$. Similarly, we must have $\nabla_{i=1} \to \infty$ and $\Delta_{i=N_\tau} \to \infty$ so that, writing

$$e_{jik} = \frac{-v_{ijk}\mu_k^2}{2c\nabla_i\phi(\Delta\xi_i)}\left[\frac{\mathrm{d}V}{\mathrm{d}\tau}\right]_j, \quad f_{jik} = \frac{v_{ijk}\mu_k^2}{2c\Delta_i\phi(\Delta\xi_i)}\left[\frac{\mathrm{d}V}{\mathrm{d}\tau}\right]_j, \tag{5.5.49}$$

with

$$d_{jik} = 1 - e_{jik} - f_{jik}, \tag{5.5.50}$$

we have the tri-diagonal matrix

$$\mathbf{D}_j^{(2)} = \begin{pmatrix} d_{j11}, & f_{j11} & & & & \\ e_{j21}, & d_{j21}, & f_{j21} & & 0 & \\ & \cdot & & & & \\ & & \cdot & & & \\ & & \cdot & & & \\ 0 & & e_{jN_\xi 1}, & d_{jN_\xi 1}, & f_{jN_\xi 1} & \\ & & & \cdot & & \\ & & & & \cdot & \\ & & & & & \cdot \end{pmatrix}. \tag{5.5.51}$$

The matrix $\mathbf{D}_j^{(1)}$ is full because of the summation occurring in equation (5.5.42) for S_j. We thence have

$$\mathbf{D}_j^{(1)} = \mathbf{D}_j^{(2)} - (1 - \epsilon_j)\mathbf{W}_j, \tag{5.5.52}$$

where

$$\mathbf{W}_j = \begin{pmatrix} w_{1j}^{(1)}w_1^{(2)}, & w_{2j}^{(1)}w_1^{(2)}, & \ldots, & w_{N_\nu j}^{(1)}w_{N_\mu}^{(2)} \\ \vdots & & & \\ w_{1j}^{(1)}w_1^{(2)}, & w_{2j}^{(1)}w_k^{(2)} & & w_{N_\nu j}^{(1)}w_{N_\mu}^{(2)} \end{pmatrix} \tag{5.5.53}$$

- see equation (3.2.44). Finally, the vector $\mathbf{L}_j$ is given by

$$\mathbf{L}_j = [\epsilon B_\nu(T)]_j \mathbb{1}. \tag{5.5.54}$$

The two required boundary conditions may be obtained from equations (5.5.37) by writing

$$\Psi = \Phi - I^- \quad \text{and} \quad \Psi = I^+ - \Phi,$$

i.e.

$$\Psi(0, \Delta\xi, \mu) = \Phi(0, \Delta\xi, \mu) - I(0, \Delta\xi, \mu < 0), \tag{5.5.55}$$

and

$$\Psi(\tau_{N_\tau}, \Delta\xi, \mu) = I(\tau_{N_\tau}, \Delta\xi, \mu > 0) - \Phi(\tau_{N_\tau}, \Delta\xi, \mu), \tag{5.5.56}$$

where $I(0, \Delta\xi, \mu < 0)$ and $I(\tau_{N_\tau}, \Delta\xi, \mu > 0)$ are the given incident radiation intensities at $\tau = 0$ and $\tau = \tau_{N_\tau}$ respectively. These last two equations take the place of equation (5.5.45) for $j = 1$ and $j = N_\tau$.

We now solve the coupled matrix equations (5.5.44) and (5.5.45) by writing

$$\boldsymbol{\Phi}_j = \mathbf{U}_j^{(1)} + \mathbf{V}_j^{(1)}\boldsymbol{\Phi}_{j+1} + \mathbf{W}_j^{(1)}\boldsymbol{\Psi}_{j+1}, \tag{5.5.57}$$

$$\boldsymbol{\Psi}_j = \mathbf{U}_j^{(2)} + \mathbf{V}_j^{(2)}\boldsymbol{\Phi}_{j+1} + \mathbf{W}_j^{(2)}\boldsymbol{\Psi}_{j+1}. \tag{5.5.58}$$

Substitution of these two recursive expressions into equations (5.5.44) and (5.5.45) then yields

$$(\mathbf{D}_j^{(1)} + \mathbf{A}_j\mathbf{V}_{j-1}^{(2)})\boldsymbol{\Phi}_j + (\mathbf{B}_j + \mathbf{A}_j\mathbf{W}_{j-1}^{(2)})\boldsymbol{\Psi}_j = \mathbf{L}_j - \mathbf{A}_j\mathbf{U}_{j-1}^{(2)} - \mathbf{C}_j\boldsymbol{\Psi}_{j+1}, \tag{5.5.59}$$

and

$$(\mathbf{B}_j + \mathbf{A}_j\mathbf{V}_{j-1}^{(1)})\boldsymbol{\Phi}_j + (\mathbf{D}_j^{(2)} + \mathbf{A}_j\mathbf{W}_{j-1}^{(1)})\boldsymbol{\Psi}_j = -\mathbf{A}_j\mathbf{U}_{j-1}^{(1)} - \mathbf{C}_j\boldsymbol{\Phi}_{j+1}. \tag{5.5.60}$$

These two linear simultaneous equations can now be solved explicitly for $\boldsymbol{\Phi}_j$ and $\boldsymbol{\Psi}_j$. This then yields equations (5.5.57) and (5.5.58) with

$$\mathbf{U}_j^{(1)} = -\mathbf{X}_j[\mathbf{A}_j\mathbf{U}_{j-1}^{(1)} + \mathbf{Y}_j(\mathbf{L}_j - \mathbf{A}_j\mathbf{U}_{j-1}^{(2)})], \tag{5.5.61}$$

$$\mathbf{V}_j^{(1)} = -\mathbf{X}_j\mathbf{C}_j, \tag{5.5.62}$$

$$\mathbf{W}_j^{(1)} = \mathbf{X}_j\mathbf{Y}_j\mathbf{C}_j, \tag{5.5.63}$$

where

$$\mathbf{X}_j = [\mathbf{B}_j + \mathbf{A}_j\mathbf{V}_{j-1}^{(1)} - \mathbf{Y}_j(\mathbf{D}_j^{(1)} + \mathbf{A}_j\mathbf{V}_{j-1}^{(2)})]^{-1}, \tag{5.5.64}$$

$$\mathbf{Y}_j = (\mathbf{D}_j^{(2)} + \mathbf{A}_j\mathbf{W}_{j-1}^{(1)})(\mathbf{B}_j + \mathbf{A}_j\mathbf{W}_{j-1}^{(2)})^{-1}, \tag{5.5.65}$$

and

$$\mathbf{U}_j^{(2)} = (\mathbf{B}_j + \mathbf{A}_j\mathbf{W}_{j-1}^{(2)})^{-1}[\mathbf{L}_j - \mathbf{A}_j\mathbf{U}_{j-1}^{(2)} - (\mathbf{D}_j^{(1)} + \mathbf{A}_j\mathbf{V}_{j-1}^{(2)})\mathbf{U}_j^{(1)}], \tag{5.5.66}$$

$$\mathbf{V}_j^{(2)} = -(\mathbf{B}_j + \mathbf{A}_j\mathbf{W}_{j-1}^{(2)})^{-1}(\mathbf{D}_j^{(1)} + \mathbf{A}_j\mathbf{V}_{j-1}^{(2)})\mathbf{V}_j^{(1)}, \tag{5.5.67}$$

$$\mathbf{W}_j^{(2)} = -(\mathbf{B}_j + \mathbf{A}_j\mathbf{W}_{j-1}^{(2)})^{-1}[\mathbf{C}_j + (\mathbf{D}_j^{(1)} + \mathbf{A}_j\mathbf{V}_{j-1}^{(2)})\mathbf{W}_j^{(1)}]. \tag{5.5.68}$$

These recurrence relationships are started at $j = 1$ and continued through to $j = N_\tau$ in order to determine the $\mathbf{U}_j$, $\mathbf{V}_j$ and $\mathbf{W}_j$ quantities at all values of j. Once this has been completed, substitution of equations (5.5.57) and (5.5.58) for $j = N_\tau - 1$ into equations (5.5.44) and (5.5.56) yields $\boldsymbol{\Phi}_{N_\tau}$ and $\boldsymbol{\Psi}_{N_\tau}$. Knowledge of these latter two vectors enables one to determine $\boldsymbol{\Phi}_{N_\tau-1}$ and $\boldsymbol{\Psi}_{N_\tau-1}$, etc. by back-substitution into equations (5.5.57) and (5.5.58). The overall procedure described above is a simple modification of the process discussed in section 3.2.3. The reader is referred to that section for further computational details.

We now return our attention to the need to re-write the radiative transfer equation in the local rest frame of the moving gas. The elements in the $\mathbf{W}_j$ matrix generated in the observer's rest frame incorporate the factor $\phi(\Delta\nu - \nu V\mu/c)$ which varies from unity to a number $\ll\epsilon$. Consequently, the introduction of a non-zero velocity field can generate dominant terms in the far off-diagonal elements of the $\mathbf{V}_j$ matrix (here we would have i and k such that $\Delta\nu_i - \nu V_j \mu_k/c \approx 0$ leaving $\phi(\Delta\nu_i - \nu V_j \mu_k/c)$ a local maximum). Clearly, this off-diagonal dominance can lead to instabilities in the matrix inversion. In the co-moving frame representation, however, the corresponding factor occurs as $\phi(\Delta\xi)$ which produces dominant diagonal, or near-diagonal, elements (i such that $\Delta\xi_i \approx 0$). This does not assure stability, but it is certainly helpful in producing reasonably accurate matrix inverses.

This point can be seen more clearly if we consider the degenerate case in which the velocity field V is taken to be constant. Here, the gradient $\mathrm{d}V/\mathrm{d}\tau$ is zero so that the radiative transfer equation (5.5.34) reduces to that for the stationary problem. Physically, such a case simply represents the overall motion of the gas relative to the observer with no relative internal macroscopic motions. Consequently, as far as the gas is concerned, the interaction between the radiation and the particle matter takes place as if the gas was macroscopically completely stationary (this assumes, of course, zero incident radiation on the medium). Only once the radiation has escaped from the gas would it appear either red- or blue-shifted relative to the observer depending upon the sign of the velocity field. Mathematically, the dominant elements in the $\mathbf{V}_j$ matrix in the observer's rest frame representation would not be diagonal but, rather, those in off-diagonal locations corresponding to the frequency shift $\nu V\mu/c$. The co-moving frame representation, however, centres the frequency grid $\{\Delta\xi_i\}$ about this frequency shift and this, in turn, produces dominant diagonal elements. Thus, although one is solving precisely the same equation in both cases (no $\mathrm{d}V/\mathrm{d}\tau$ term in equation (5.5.34)), this difference in structure of the matrices to be inverted in the two representations can lead to significant differences in stability for large velocities.

Finally, we note the use of all the peripheral methods (improved boundary conditions, Auer's modification, Rybicki re-organisation, etc.) discussed in section 3.2 which can be incorporated into the co-moving representation. In particular, we recall the quadrature perturbation technique described in section 3.3. This, when coupled with the above co-moving frame formulation, increases stability further since the matrices requiring inversion are reduced in size. The savings in computer time and storage are correspondingly quite significant, particularly in view of the large number of frequency quadrature points N_ν one would otherwise need when the velocity gradient $\mathrm{d}V/\mathrm{d}\tau$ is large.

(b) Velocity perturbation technique [33]

An alternative to the method presented in the preceding sub-section may be considered where we again start from the co-moving frame representation of the radiative transfer equation (5.5.34), viz.

$$\mu \frac{\partial I}{\partial \tau} - \mu^2 \frac{\nu}{c} \frac{dV}{d\tau} \frac{\partial I}{\partial(\Delta\xi)} = \phi(\Delta\xi)(I - S). \tag{5.5.69}$$

We develop the perturbation series

$$I = I_0 + \lambda I_1 + \lambda^2 I_2 + \cdots, \tag{5.5.70}$$

so that

$$S = S_0 + \lambda S_1 + \lambda^2 S_2 + \cdots. \tag{5.5.71}$$

The above procedure should not be confused with that represented by equations (5.5.27) and (5.5.28) where we effectively expanded in orders of V/c. Recall that the quantity I appearing in equation (5.5.69) is identical to the $I_G^{(0)}$ appearing in equation (5.5.27). Indeed, the technique discussed here, and that given by series (5.5.27), may be easily combined (although we do not consider the pertinent coupling of the perturbations in this section since (a) it does not facilitate an understanding of the appropriate physics and mathematics and (b) it lends itself to expositional difficulties).

Thence, treating the $\partial I/\partial(\Delta\xi)$ term to be of order λ, substitution of the perturbation series (5.5.70) and (5.5.71) into equation (5.5.69), and equating coefficients of λ^n to zero, yields

$$\mu \frac{\partial I_0}{\partial \tau} = \phi(\Delta\xi)(I_0 - \bar{J}_0) - \phi(\Delta\xi)\epsilon B_\nu(T), \tag{5.5.72}$$

and, for all $k \geqslant 1$,

$$\mu \frac{\partial I_k}{\partial \tau} = \phi(\Delta\xi)(I_k - \bar{J}_k) + \mu^2 \frac{\nu}{c} \frac{dV}{d\tau} \frac{\partial I_{k-1}}{\partial(\Delta\xi)}, \tag{5.5.73}$$

where

$$\bar{J}_k = \frac{1-\epsilon}{2} \int_{-\infty}^{\infty} d(\Delta\xi)\phi(\Delta\xi) \int_{-1}^{1} d\mu I_k(\tau, \Delta\xi, \mu) \;\forall\; k \geqslant 0. \tag{5.5.74}$$

The first of the above equations is simply the ordinary equation of radiative transfer for stationary media. The other equations (5.5.73) for $k \geqslant 1$ then represent corrections to this 'stationary solution' due to the effect of the non-zero velocity field. Again, once the zeroth order solution I_0 has been obtained (presumably by the ordinary Feautrier technique - see section 3.2- note that only half the frequency bandwidth need be considered since here we are effectively solving a $V(\tau) \equiv 0$ problem), the higher order terms I_k ($k \geqslant 1$) may be readily evaluated since it is only the source terms $\epsilon B_\nu(T)$ for $k = 0$ and $-[\mu^2 \nu / c\phi(\Delta\xi)](dV/d\tau)[\partial I_{k-1}/\partial(\Delta\xi)]$ for $k \geqslant 1$ which differ from equation to equation - the Green's function solution is identical for all equations (5.5.72) and (5.5.73) - see section 3.3.

Several important points should be discussed. First, it would appear that large velocity gradients $dV/d\tau$ would perhaps violate the assumption that the

$\partial I/\partial(\Delta\xi)$ term is of order λ. However, the above specification of the perturbation series (5.5.70) has been explicitly written in terms of I whereas the actual computation of the solution involves the evaluation of the source function explicitly (via the $\Phi = (I^+ + I^-)/2$ term using the Feautrier technique). As we shall see, this source function is not affected by the velocity field to the same extent as is the specific radiation intensity basically because it is simply a non-local [34] 'average' of red and blue frequency radiation and an 'average' of $\mu > 0$ and $\mu < 0$ directional radiation (a decrease in the blue-wing specific intensity, for example, is usually associated with an increase in the red-wing radiation so that the difference in the sum of these two intensities is generally only slight). Indeed, if we write equation (5.5.69) in integral form, viz.

$$S = (1-\epsilon)\Lambda S + \epsilon B_\nu(T), \tag{5.5.75}$$

where the Λ operator now contains the non-zero velocity field $V(\tau)$, and replace Λ in this last equation by the approximate Λ^* operator where we put $V(\tau) \equiv 0$, i.e. $\Lambda^* \equiv \Lambda\,(V \equiv 0)$, so that

$$S = (1-\epsilon)\Lambda^* S + \epsilon B_\nu(T) + (1-\epsilon)(\Lambda - \Lambda^*)S, \tag{5.5.76}$$

then the above perturbation series (5.5.71) yields the recurrence relationships

$$S_l = (1-\epsilon)\Lambda^* S_l + \mathcal{E}_l, \tag{5.5.77}$$

where $\mathcal{E}_0 = \epsilon B_\nu(T)$ and $\mathcal{E}_l = (1-\epsilon)(\Lambda - \Lambda^*)S_{l-1}$ for all $l \geqslant 1$. Equation (5.5.77) is simply the integral form of the differential transfer equation (5.5.73); in particular, we find it more convenient to use the above $\mathcal{E}_l$ correction terms because of their non-local (ΛS_{l-1}) nature discussed above rather than those involving the more local $\mathrm{d}V/\mathrm{d}\tau$ quantity. Even so, convergence is not always attained.

Clearly, however, when $\mathrm{d}V/\mathrm{d}\tau$ is relatively small, or large over only a very small region of the gas (recall that the local solution of the radiative transfer equation is controlled *non-locally*), convergence difficulties do not arise. The use of the above velocity perturbation technique thence enables large savings in computer time and storage to be gained when solving time-dependent gas-dynamic problems. Here, we would wish to solve the radiative transfer equation at a large number of time steps (in order to examine the time evolution of a radiating shock in a plasma, for example). Consequently, if we solve the one stationary problem represented by equation (5.5.72), thence perturb these results at each time step using equation (5.5.73) where $N_\mathrm{L}, N_\mathrm{U}, N_\mathrm{e}$, $\mathbf{V}$ and T will change in each time interval, and note that the computer time required to solve equation (5.5.73) is significantly less than that required to solve equation (5.5.72), then the overall computer time needed for the entire time evolutionary gas-dynamic problem can be reduced by a factor approaching the number of time steps taken. It should be stressed here that the solution of the equation of radiative transfer at each point in time for such problems dominates the computations. The solution of the corresponding macroscopic conservation equations requires a relatively small computational effort, although the structuring of the computer code for these non-linear equations can be quite difficult. We shall examine time-dependent calculations in the following section.

Finally, it is obvious that any of the techniques discussed in chapter 3 may be used in solving equations (5.5.72) and (5.5.73). This enables solutions with non-zero velocity gradients to be checked by a variety of different methods. This is an important consideration since no exact closed form analytical solution (such as those presented in chapter 2) for the velocity-dependent equation of radiative transfer is available, even when density and temperature are assumed constant.

5.6 Time-dependent solutions

We have ignored all $\partial/\partial t$ and $\mathrm{d}/\mathrm{d}t$ terms in the preceding section. Here, we wish to include these time-dependent quantities in the equation of radiative transfer and, most importantly, determine the densities, velocities and temperatures appearing in this equation using the pertinent macroscopic conservation equations for mass, linear momentum and energy. We consider a reduced (but still self-consistent) problem consisting of the following 1-dimensional set of equations:

$$\frac{1}{c}\frac{\partial I}{\partial t}+\mu\frac{\partial I}{\partial x}=-\kappa_0\,\phi\left(\Delta\nu-\frac{\nu}{c}V\mu\right)(I-S), \tag{5.6.1}$$

where

$$\kappa_0=\frac{h\nu}{4\pi}(N_{\mathrm{L}}B_{\mathrm{LU}}-N_{\mathrm{U}}B_{\mathrm{UL}}), \tag{5.6.2}$$

$$S=\frac{1-\epsilon}{2}\int_{-\infty}^{\infty}\mathrm{d}(\Delta\nu)\int_{-1}^{1}\mathrm{d}\mu\,\phi\left(\Delta\nu-\frac{\nu}{c}V\mu\right)I(x,\Delta\nu,\mu)+\epsilon B_\nu(T)$$
$$-\frac{1-\epsilon}{A_{\mathrm{UL}}\left[1+\left(1+\dfrac{g_{\mathrm{U}}}{g_{\mathrm{L}}}\right)\dfrac{c^2S}{2h\nu^3}\right]}\frac{\mathrm{d}S}{\mathrm{d}t}, \tag{5.6.3}$$

$$\frac{\partial N_{\mathrm{L}}}{\partial t}+\frac{\partial}{\partial x}(N_{\mathrm{L}}V)=R_{\mathrm{L}}, \tag{5.6.4}$$

$$\frac{\partial N_{\mathrm{U}}}{\partial t}+\frac{\partial}{\partial x}(N_{\mathrm{U}}V)=-R_{\mathrm{L}}, \tag{5.6.5}$$

$$\frac{\partial N_{\mathrm{e}}}{\partial t}+\frac{\partial}{\partial x}(N_{\mathrm{e}}V)=0, \tag{5.6.6}$$

such that

$$\frac{\partial\rho}{\partial t}+\frac{\partial}{\partial x}(\rho V)=0, \tag{5.6.7}$$

where

$$R_{\mathrm{L}}=N_{\mathrm{U}}(B_{\mathrm{UL}}\bar{J}+A_{\mathrm{UL}}+C_{\mathrm{UL}})-N_{\mathrm{L}}(B_{\mathrm{LU}}\bar{J}+C_{\mathrm{LU}}), \tag{5.6.8}$$

$$\frac{\partial}{\partial t}(\rho V)+\frac{\partial}{\partial x}\left(\rho V^2+p-\frac{\mu_{\mathrm{V}}}{3}\frac{\partial V}{\partial x}\right)=0, \tag{5.6.9}$$

or

$$\rho\frac{\partial V}{\partial t}+\rho V\frac{\partial V}{\partial x}+\frac{\partial}{\partial x}\left(p-\frac{\mu_{\mathrm{V}}}{3}\frac{\partial V}{\partial x}\right)=0, \tag{5.6.10}$$

and

$$\frac{\partial}{\partial t}(\tfrac{1}{2}\rho V^2+\rho\epsilon)+\frac{\partial}{\partial x}\left[(\tfrac{1}{2}\rho V^2+\rho\epsilon+p)V-\frac{\mu_{\mathrm{V}}}{3}V\frac{\partial V}{\partial x}-k_{\mathrm{c}}\frac{\partial T}{\partial x}\right]=Q, \tag{5.6.11}$$

or

$$\rho\frac{\partial\epsilon}{\partial t}+\rho V\frac{\partial\epsilon}{\partial x}+p\frac{\partial V}{\partial x}-\frac{\mu_{\mathrm{V}}}{3}\left(\frac{\partial V}{\partial x}\right)^2-\frac{\partial}{\partial x}\left(k_{\mathrm{c}}\frac{\partial T}{\partial x}\right)=Q, \tag{5.6.12}$$

where

$$p=kNT, \tag{5.6.13}$$

$$\rho\epsilon=\tfrac{3}{2}kNT+N_{\mathrm{U}}h\nu, \tag{5.6.14}$$

and

$$Q=-4\pi\frac{\partial F_{\mathrm{RAD}}}{\partial x}-\frac{4\pi}{c}\frac{\partial\bar{I}}{\partial t}+h\nu(N_{\mathrm{L}}C_{\mathrm{LU}}-N_{\mathrm{U}}C_{\mathrm{UL}}). \tag{5.6.15}$$

Note that we have replaced the rate of shear tensor **S** appearing in the pressure tensor **P** (equation (5.3.70)) by $(\mu_{\mathrm{V}}/3)\partial V/\partial x$ - see equation (5.3.69). All external forces $\mathbf{F}_i$ have been put to zero.

There are a variety [96] of techniques developed for the solution of the above conservation equations. Here, we briefly consider just two methods simply to illustrate the types of problems peculiar to the computation of time-dependent solutions involving a non-zero radiation field. Cauchy characteristics [32, 45, 87], and their important specification of the domain of dependence of solutions, are not discussed here.

5.6.1 *A Eulerian finite difference approach*

The methods we present here require the macroscopic conservation equations (5.6.7), (5.6.9) and (5.6.11) to be written in the form

$$\frac{\partial \rho}{\partial t} + \frac{\partial M}{\partial x} = 0, \tag{5.6.16}$$

$$\frac{\partial M}{\partial t} + \frac{\partial Q_1}{\partial x} = 0, \tag{5.6.17}$$

$$\frac{\partial E}{\partial t} + \frac{\partial Q_2}{\partial x} = Q, \tag{5.6.18}$$

where

$$M = \rho V, \tag{5.6.19}$$

$$E = \tfrac{1}{2}\rho V^2 + \tfrac{3}{2}kNT + N_{\mathrm{U}}h\nu, \tag{5.6.20}$$

with

$$Q^{(1)} = kNT + \rho V^2 - \frac{\mu_{\mathrm{V}}}{3}\,\frac{\partial V}{\partial x}, \tag{5.6.21}$$

and

$$Q^{(2)} = kNTV + EV - \frac{\mu_{\mathrm{V}}}{3}\,V\,\frac{\partial V}{\partial x} - k_{\mathrm{c}}\,\frac{\partial T}{\partial x}. \tag{5.6.22}$$

We will solve the above three equations for the three unknowns ρ, M and E. In particular, we employ an iterative scheme where, at each time step, we assume an initial set of values for N_{U}, N_{L}, N_{e} and $\partial F_{\mathrm{RAD}}/\partial x$ (usually taken to be those values computed at the previous time step). These are then used to obtain Q and the internal energy term $N_{\mathrm{U}}h\nu$ in E. One then solves (as detailed below) the above three equations for ρ, M and E which, in turn, enables V (from equation (5.6.19)) and T (from equation (5.6.20)) to be evaluated. These velocity and temperature distributions may be substituted into the equation of radiative transfer (5.6.1) to obtain I, thence $\partial F_{\mathrm{RAD}}/\partial x$, whilst N_{U}, N_{L} and N_{e} may be obtained from equations (5.6.4–6). The entire process is then repeated until adequate convergence is attained.

Since our new unknowns to be determined are ρ, M and E, we must re-write $Q_1^{(1)}$ and $Q_2^{(2)}$ as

$$Q^{(1)}(\rho, M, E) = \tfrac{2}{3}\left[E - \frac{M^2}{2\rho} - N_{\mathrm{U}}h\nu\right] + \frac{M^2}{\rho} - \frac{\mu_{\mathrm{V}}}{3}(\rho, M, E)\,\frac{\partial}{\partial x}\left(\frac{M}{\rho}\right), \tag{5.6.23}$$

and

$$Q^{(2)}(\rho, M, E) = \tfrac{2}{3}\left[E - \frac{M^2}{2\rho} - N_{\mathrm{U}}h\nu\right]\frac{M}{\rho} - \frac{\mu_{\mathrm{V}}M}{3\rho}\,\frac{\partial}{\partial x}\left(\frac{M}{\rho}\right) + \frac{EM}{\rho} - \frac{2k_{\mathrm{c}}(\rho, M, E)}{3k}\,\frac{\partial}{\partial x}\left[\frac{1}{N}\left(E - \frac{M^2}{2\rho}\,N_{\mathrm{U}}h\nu\right)\right], \tag{5.6.24}$$

where

$$N = N_{\mathrm{L}} + N_{\mathrm{U}} + N_{\mathrm{e}} + N_{+}, \tag{5.6.25}$$

and

$$\rho = N_{\mathrm{L}} m_{\mathrm{L}} + N_{\mathrm{U}} m_{\mathrm{U}} + N_{\mathrm{e}} m_{\mathrm{e}} + N_{+} m_{+}. \tag{5.6.26}$$

We now examine two separate schemes for the numerical solution of equations (5.6.16-18).

(a) An implicit difference approach

We construct the space and time grids $\{x_j\}$ and $\{t_n\}$ so that, for any function $f(x, t)$, we write $f_{jn} = f(x_j, t_n)$. Time and space derivatives are approximated by

$$\left[\frac{\partial f}{\partial t}\right]_{jn} = \frac{1}{\Delta t}(f_{jn} - f_{j,n-1}), \tag{5.6.27}$$

and

$$\left[\frac{\partial f}{\partial x}\right]_{jn} = \frac{\Delta f_{jn}}{\Delta x} = \frac{1}{\Delta x}(f_{j+1,n} - f_{jn}). \tag{5.6.28}$$

Note that we take constant increments in Δx and Δt. It can be rigorously shown [87, 100] that stability of any numerical scheme to solve the conservation equations requires the Courant–Friedrichs–Lewy condition:

$$\frac{\Delta t}{\Delta x} \leqslant \frac{1}{c_{\mathrm{s}}}, \tag{5.6.29}$$

where c_{s} is the local speed of sound. This condition simply states that the time step Δt at time t must be less than the smallest time step characterising the various interaction processes inherent in the problem. Clearly, if Δt is greater than this characteristic time, an adequate representation of the essential physical time-dependent terms will not be attained, and this will obviously introduce errors into the computations.

We next write equations (5.6.16–18) in the form [19, 41]

$$\rho_{jn} = \rho_{j,n-1} - \frac{\Delta t}{\Delta x} \Delta M_{j-1,n}, \tag{5.6.30}$$

$$M_{jn} = M_{j,n-1} - \frac{\Delta t}{\Delta x} \Delta Q^{(1)}_{j-1,n}, \tag{5.6.31}$$

and

$$E_{jn} = E_{j,n-1} - \frac{\Delta t}{\Delta x} \Delta Q^{(2)}_{j-1,n} + \Delta t Q_{jn}. \tag{5.6.32}$$

These last three equations may be solved by an iterative procedure. First, at some nth point in time, one assumes a set of ρ_{jn}, M_{jn} and E_{jn} for all j in the $\{x_j\}$ grid, thereby enabling $\Delta M_{j-1,n}$, $\Delta Q^{(1)}_{j-1,n}$, $\Delta Q^{(2)}_{j-1,n}$ and ΔQ_{jn} are to be determined. Thus, since the $\rho_{j,n-1}$, $M_{j,n-1}$ and $E_{j,n-1}$ are known (they were evaluated at the preceding time step), a new set of ρ_{jn}, M_{jn} and E_{jn} may be computed. The process is repeated until convergence of ρ_{jn}, M_{jn} and E_{jn} is attained.

The details of the iterative procedure are quite important. One can obtain ρ_{jn} immediately from equation (5.6.30). This can then be substituted into equation (5.6.31) to obtain a new M_{jn}. Substitution back into equation (5.6.30) completes the first iteration loop. Once 'reasonably accurate' values of ρ_{jn} and M_{jn} are obtained, one can substitute into the third of the above three equations to compute E_{jn}. This forms the second iteration loop, the new E_{jn} being used to re-compute even more accurate ρ_{jn} and M_{jn}.

It is advantageous to employ a relaxation technique when solving equation (5.6.31). One determines a quantity $G_{jn}^{(k)}$ given by

$$G_{jn}^{(k)} = -(M_{jn}^{(k)} - M_{j,n-1}^{(k)}) - \frac{\Delta t}{\Delta x}\,\Delta Q_{j-1,n}^{(1)}, \tag{5.6.33}$$

where the superscript k refers to the kth iteration. One then writes

$$M_{jn}^{(k+1)} = M_{jn}^{(k)} + \delta G_{jn}^{(k)}. \tag{5.6.34}$$

Filler and Ludloff [41] derived bounds, depending upon μ_V and $\Delta t/\Delta x$, over which δ can vary for stability of the relaxation process for a somewhat simpler problem (zero radiation field, 1-species gas, constant viscosity) than that discussed above. They found that

$$\delta \leqslant \frac{1}{\frac{1}{2} + \left(\dfrac{2\mu_V}{\Delta x} + 4M\right)\dfrac{\Delta t}{3\rho\Delta x}}. \tag{5.6.35}$$

Test computations indicate that the above inequality on δ still yields stability of the relaxation (5.6.34) in all but the most extreme situations in which μ_V varies strongly with position.

Equation (5.6.32) can also present difficulties in its numerical solution for E_{jn}. Probably the most satisfactory approach is to first evaluate μ_V and k_c (detailed in Appendix C) as functions of E_{jn} using values obtained from the preceding iteration (in the second iteration loop) so that, noting the second order derivative of E arising from the $\partial[k_c(\partial T/\partial x)]\partial x$ term in equations (5.6.18) and (5.6.22), one can re-write equation (5.6.32) as

$$-a_{jn}E_{j-1,n} + b_{jn}E_{jn} - c_{jn}E_{j+1,n} = l_{jn}, \tag{5.6.36}$$

where the a_{jn}, b_{jn}, c_{jn} and l_{jn} are known at each iteration. One can solve this scalar equation in precisely the same manner as that presented in section 3.2.6(c) using a recursive scheme with back-substitution.

In practice, it has been found that two iterations in the first loop (coupled with the above relaxation process) to every one iteration in the second proves the most efficient method of solution. Approximately 10-20 iterations in all at each time step gives satisfactory results. More particularly, $\delta = 0.1$ gives adequate convergence of the relaxation process in most cases although, in some situations, the strong dependence of the viscosity on temperature (see section 5.7.2) necessitates the use of smaller δ.

It is, of course, necessary to couple the equation of radiative transfer into the above analysis. Since an iterative procedure is used to solve the time-dependent conservation equations, and the values of the radiation field entering those equations must therefore also be iterated upon, the $\partial I/\partial t$ and $\mathrm{d}S/\mathrm{d}t$ terms appearing in equations (5.6.1) and (5.6.3) may be evaluated using the results from the previous iteration. Thus, writing $I^{(k)}$ as the value of I at the kth iteration, we have

$$\mu \frac{\partial I^{(k+1)}}{\partial x} = -k_0 \phi\left(\Delta\nu - \frac{\nu V_\mu}{c}\right)\left[I^{(k+1)} - \frac{1-\epsilon}{2}\int_{-\infty}^{\infty} \mathrm{d}(\Delta\nu') \right.$$

$$\times \int_{-1}^{1} \mathrm{d}\mu' \phi\left(\Delta\nu' - \frac{\nu V_{\mu'}}{c}\right) I^{(k+1)}(x, \Delta\nu', \mu') - \epsilon B_\nu(T)$$

$$\left. - \frac{1-\epsilon}{A_{\mathrm{UL}}\left[1 + \left(1 + \frac{g_{\mathrm{U}}}{g_{\mathrm{L}}}\right)\frac{c^2 S^{(k)}}{2h\nu^3}\right]} \frac{\mathrm{d}S^{(k)}}{\mathrm{d}t}\right] - \frac{1}{c}\frac{\partial I^{(k)}}{\partial t}. \quad (5.6.37)$$

In this way, the source term $\epsilon B_\nu(T)$ is supplemented by the $\mathrm{d}S/\mathrm{d}t$ and $\partial I/\partial t$ terms so that one can use the Feautrier techniques as outlined in sections 5.5.1 and 5.5.2(a). Indeed, if one uses the velocity perturbation technique (section 5.5.2(b)), these two extra terms can be absorbed into the source term $\partial I_{k-1}/\partial(\Delta\xi)$ appearing in equation (5.5.73). As mentioned already, the use of the velocity perturbation technique can lead to enormous savings in computer time.

(b) An explicit difference approach

In writing equation (5.6.30), for example, the space derivative involves the quantities M_{jn} and $M_{j-1,n}$. Since these are expressed at the same time step as the ρ_{jn} on the LHS (5.6.30), there is an effective implicit relationship between ρ and M at t_n requiring iteration for its solution. We can, however, re-write equation (5.6.30) in explicit form by choosing to express the space derivative at the previous time step t_{n-1}. This is equivalent to choosing the forward difference formula

$$\left[\frac{\partial f}{\partial t}\right]_{jn} = \frac{1}{\Delta t}(f_{j,n+1} - f_{jn}), \quad (5.6.38)$$

rather than the backward difference (5.6.27). We then have

$$\rho_{j,n+1} = \rho_{jn} - \frac{\Delta t}{\Delta x} \Delta M_{j-1,n}, \tag{5.6.39}$$

$$M_{j,n+1} = M_{jn} - \frac{\Delta t}{\Delta x} \Delta Q^{(1)}_{j-1,n}, \tag{5.6.40}$$

and

$$E_{j,n+1} = E_{jn} - \frac{\Delta t}{\Delta x} \Delta Q^{(2)}_{j-1,n} + \Delta t Q_{jn}. \tag{5.6.41}$$

Here, all the terms on the RHS (5.6.39–41) have already been evaluated at the preceding time step. Consequently, these equations are explicit expressions for $\rho_{j,n+1}$, $M_{j,n+1}$ and $E_{j,n+1}$, and iteration is not necessary. However, the subsequent computational simplicity is somewhat nullified by certain restrictions placed on $\Delta t/\Delta x$ and Δx (over and above the Courant–Friedrichs–Lewy condition (5.6.29)). Filler and Ludloff [41] show that stability is attained for 'simple' gas-dynamic problems only if

$$\frac{\Delta t}{\Delta x} \leqslant \frac{1}{\dfrac{M}{\rho} + c_s + \dfrac{10\mu_V}{9\rho\Delta x}}, \tag{5.6.42}$$

and

$$\Delta x \leqslant \frac{10\mu_V}{27\rho c_s}, \tag{5.6.43}$$

where $c_s (\approx (kT/m_A)^{1/2})$ is the maximum speed of sound in the gas.

5.6.2 *A Lagrangian finite difference scheme [100]*

It is of some value to very briefly discuss a commonly used method for solving time-dependent gas-dynamic conservation equations in 1-dimensional geometry. Recalling the Lagrangian time derivative

$$\frac{d}{dt} = \frac{\partial}{\partial t} + V \frac{\partial}{\partial x}, \tag{5.6.44}$$

equations (5.6.7), (5.6.10) and (5.6.12) have the form

$$\frac{d\rho}{dt} = -\rho \frac{\partial V}{\partial x}, \tag{5.6.45}$$

$$\rho \frac{dV}{dt} = -\frac{\partial}{\partial x}\left(p - \frac{\mu_V}{3} \frac{\partial V}{\partial x}\right), \tag{5.6.46}$$

and

$$\rho \frac{d\epsilon}{dt} = -\left(p - \frac{\mu_V}{3} \frac{\partial V}{\partial x}\right) \frac{\partial V}{\partial x} + \frac{\partial}{\partial x}\left(k_c \frac{\partial T}{\partial x}\right) + Q. \tag{5.6.47}$$

Before actually describing the method in detail, we wish to discuss the thrust behind certain approximations which are usually made when the gas exhibits propagating shock waves. Generally, any non-attenuating collisional shock wave will develop to the stage where the shock front 'discontinuity' exists over a distance of only a few collisional mean free paths and, since these path lengths are necessarily very small in relation to the overall size of the gaseous ensemble, an accurate description of the density, temperature and velocity structure of the shock would require an enormously large number of spatial grid points $\{x_j\}$. In the absence of heat conductivity, the viscosity term $-\mu_V \partial V/\partial x$ acts (mathematically, physically and numerically) to smooth the shock in the vicinity of this discontinuity so that the smaller the value of μ_V the steeper the gradients in ρ, V and T. Indeed, in the limit as $\mu_V \to 0$, one has a rigorous mathematical discontinuity at the shock front. Numerically, the resultant non-uniqueness of ρ, V and T at this rigorous discontinuity will lead to instabilities in any finite differencing scheme used to solve the macroscopic conservation equations and this, in turn, will produce spurious fluctuations, for example, in ρ, V and T. Physically, the $\mu_V \partial V/\partial x$ term converts macroscopic translational (coherent) kinetic energy of the gaseous ensemble motion to microscopic kinetic energy of the individual particles thereby heating the gas. (In this way, the viscosity may be thought of as a frictional force. We shall discuss this in some detail in section 5.7.2.) Clearly, the limit as $\mu_V \to 0$ has no physical meaning.

The inherent numerical difficulties associated with realistic values of μ_V (almost always small) can be overcome, to some extent, by increasing μ_V to an appropriate value capable of artificially widening the shock front discontinuity from several collisional mean free paths to several spatial grid points. This can be done by replacing the real viscosity term $\mu_V \partial V/\partial x$ by an *artificial* counterpart such that

$$
\begin{aligned}
-\frac{\mu_V}{3}\frac{\partial V}{\partial x} &\to l^2 \rho \left(\frac{\partial V}{\partial x}\right)^2 \quad &\text{if} \quad \frac{\partial V}{\partial x} < 0, \\
&\to 0 \quad &\text{if} \quad \frac{\partial V}{\partial x} \geqslant 0,
\end{aligned}
\tag{5.6.48}
$$

where l (proportional to Δx) is a constant parameter to be suitably chosen to yield well-behaved numerical solutions. Clearly, in regions far removed from the shock front, the velocity gradient must be relatively small so that the above replacement (5.6.48) has little effect. Indeed, it can be shown that, for a 1-species (non-radiating) gas, the use of the above artificial viscosity produces the correct speed and energy dissipation of the shock.

We now proceed by defining the spatial coordinate X as the x location of a particular volume element of gas at time $t = 0$ for which the density is ρ_0.

This volume element will expand and contract as it moves through space depending upon the forces exerted on it so that, if we ignore any diffusion $\bar{V}_i$ of different ith species relative to the 'centre of mass' velocity V, the mass of this volume element will remain constant and we can write

$$\rho_0 \delta X = \rho \delta x, \tag{5.6.49}$$

where x now represents the position of this element at some later time t for which the density is ρ. This last result replaces equation (5.6.45) stipulating conservation of mass. It is further clear that the local velocity of this gaseous element is simply

$$V = \frac{\mathrm{d}x}{\mathrm{d}t}, \tag{5.6.50}$$

where $x = x(X, t)$, X and t being independent variables. Note that $x(X, 0) = X$.

If we now replace $(\mu_V/3)(\partial V/\partial x)$ by the artificial viscosity term (which we denote by q) appearing on the RHS (5.6.48), then the conservation of linear momentum equation (5.6.46) becomes

$$\frac{\mathrm{d}V}{\mathrm{d}t} = -\frac{1}{\rho_0} \frac{\partial}{\partial X}(p + q), \tag{5.6.51}$$

whilst the energy equation (5.6.47), using equation (5.6.45) to eliminate $\partial V/\partial x$, is just

$$\frac{\mathrm{d}\epsilon}{\mathrm{d}t} = -(p + q)\frac{\mathrm{d}}{\mathrm{d}t}\left(\frac{1}{\rho}\right) + \frac{1}{\rho_0}\frac{\partial}{\partial X}\left(\frac{k_c\rho}{\rho_0}\frac{\partial T}{\partial X}\right) + \frac{Q}{\rho}. \tag{5.6.52}$$

The above equations can then be differenced by constructing the space $\{X_j\}$ and time $\{t_n\}$ grids to yield the commonly used form

$$V_{j,n+1} = V_{jn} - \frac{\Delta t}{\rho_0 \Delta X}[p_{j+(1/2),n} - p_{j-(1/2),n} + q_{j+(1/2),n} - q_{j-(1/2),n}], \tag{5.6.53}$$

$$x_{j,n+1} = x_{jn} + \Delta t V_{j,n+1}, \tag{5.6.54}$$

$$\rho_{j+(1/2),n+1} = \frac{\rho_0 \Delta X}{x_{j+1,n+1} - x_{j,n+1}}, \tag{5.6.55}$$

$$\begin{aligned}\epsilon_{j+(1/2),n+1} = \epsilon_{j+(1/2),n} &- \left[\frac{p_{j+(1/2),n+1} + p_{j+(1/2),n}}{2} + q_{j+(1/2),n+1}\right]\left[\frac{1}{\rho_{j+(1/2),n+1}} - \frac{1}{\rho_{j+(1/2),n}}\right] \\ &+ \left(\frac{Q}{\rho}\right)_{j+(1/2),n} - \frac{1}{\rho_0^2(\Delta X)^2}\{(k_c\rho)_{j+(1/2),n}(T_{j+1,n} - T_{jn}) \\ &-(k_c\rho)_{j-(1/2),n}(T_{jn} - T_{j-1,n})\},\end{aligned} \tag{5.6.56}$$

where, for example,

$$q_{j+(1/2),n} = \frac{2a^2}{\dfrac{1}{\rho_{j+(1/2),n}} + \dfrac{1}{\rho_{j+(1/2),n-1}}} (V_{j+1,n} - V_{jn})^2, \tag{5.6.57}$$

if $V_{j+1,n} - V_{jn} < 0$ and, for $V_{j+1,n} - V_{jn} \geqslant 0$,

$$q_{j+(1/2),n} = 0, \tag{5.6.58}$$

where we have put $l = a\Delta x$ (a is a constant of order 2). The increment ΔX is simply $X_{j+1} - X_j$.

The above equations are basically explicit in structure although equation (5.6.56), when written in the form shown, must be solved iteratively with the equation of state

$$p_{j+(1/2),n+1} = \frac{k}{m_A} \rho_{j+(1/2),n+1} T_{j+(1/2),n+1}. \tag{5.6.59}$$

The ratio $\Delta t/\Delta X$ at time t is limited by the usual Courant condition

$$\frac{\Delta t}{\Delta X} \leqslant \frac{\rho_0}{(\rho c_s)_{\text{max at time } t}}, \tag{5.6.60}$$

defining the stability domain for smooth non-shocking flow. The introduction of artificial viscosity generates a further stability condition

$$\frac{\Delta t}{\Delta X} \lesssim \frac{\sqrt{5}\rho_0}{2\sqrt{3}a(\rho c_s)_{\text{max at time } t}}, \tag{5.6.61}$$

which, for practical values of a, is marginally more restrictive than inequality (5.6.60). The reader is referred to Richtmyer and Morton [100] for further details.

We complete this section by noting certain difficulties which arise using the above method when the gas radiates. As mentioned already, the speed of the shock front and the energy dissipated are exactly reproduced by the artificial viscosity method if the gas (non-radiating) consists of only one species. This is not true if there are at least two species with different mean free paths. For example, the analysis of the preceding section 5.5 showed that the transfer of radiation through the gas is affected by the velocity (and density and temperature) gradient. Consequently, an accurate description of this radiation energy transfer can only be obtained if the velocity gradient is similarly accurately determined. The method of artificial viscosity artificially widens the shock front 'discontinuity' by a factor of order 10^6 usually, i.e. the velocity gradient is reduced by 10^6. Thus, the transfer of radiation energy through the shock can be in considerable error and this, in turn, will lead to a completely erroneous picture of the energy structure of the gas being considered. If, however, the mean free path of photons (average distance between emission and absorption) is much greater than the width of even the artificially widened shock 'discontinuity', the velocity gradient will have only a small effect on the transfer of radiation through the medium [66]. This is generally true for continuum radiation (where most of the radiation energy is carried in stellar photospheres). But it is certainly not true for the majority of strong spectral lines formed in the outer regions of stellar atmospheres (viz. high photosphere, chromosphere, corona). In summary, therefore, there are two quite distinct points to be made here. First, care must be exercised when solving time-dependent gas-dynamic problems since the use of the artificially widened shock 'discontinuity' can lead to an incorrect energy structure of the radiating gas. Secondly, the introduction of a false velocity gradient can lead to significant errors in the computed shapes of the model emergent spectral line intensity profiles; the comparison of these incorrect model profiles with the data will clearly lead to an incorrectly diagnosed (ρ, V, T) distribution describing the radiating gas.

5.7 Velocity-dependent radiative transfer effects

There are several important transfer effects involving macroscopic motions within the radiating gas which, if not fully understood, can lead to a complete misinterpretation of spectral line data. We shall first ignore the time-dependent terms in the radiative transfer equation since these simply detail the manner in which the radiation field changes with respect to time (the effect of these time-dependent terms on the shape of the emergent line profile is minimal anyway) [22, 36, 77]. Our main concern here rests with the determination of the instantaneous state of the gas giving rise to particular emergent line profiles. This interpretation of the data, and the consequent specification of the densities, temperatures and macroscopic velocities describing the state of the gas giving rise to that data, might not necessarily be unique; this is a fundamental mathematical problem related both to the structured model of the interaction between radiation and matter, and to the inherent non-linearities of the equations describing that model – but at least some of the difficulties associated with an adequate interpretation of the data will be appreciated, along with some measure of reservation about the validity of any distribution of densities, temperatures and velocities which one might compute to explain those observations.

5.7.1 Emergent spectral line profiles

Here, we first consider an optically thin slab experiencing zero incident radiation. We have seen in section 1.10.2 that the emergent intensity from such a slab is an emission line profile symmetric about line centre as shown in figure 5.1(*a*) when there are zero macroscopic motions within the gas. We have also seen that if all the properties of the slab are symmetric about slab centre, then so is the source function as illustrated in figure 5.1(*b*).

It is instructive [56] to now consider the symmetric velocity field

$$V(z) = \frac{2V_0}{z_s}\left(\frac{z_s}{2} - z\right), \tag{5.7.1}$$

where z_s is the physical thickness of the slab. The above velocity field represents a constant velocity gradient with a velocity V_0 (toward the observer) at $z = 0$ and $-V_0$ at $z = z_s$. We find it easier to describe the effect of the velocity gradient on the normally ($\mu = 1$) emergent radiation intensity by considering a sub-slab in the region $z \approx 0$ shown as the shaded area in figure 5.2. The arrows indicate the relative magnitude and direction of the velocity. Clearly, the slab can be thought of as expanding away from slab centre.

This surface sub-slab is moving toward the observer with a speed approximately V_0 and therefore presents an absorption profile shifted to the blue wing in the observer's rest frame by a frequency equivalent of approximately $\nu V_0/c$, as shown in figure 5.3(*a*).

Consequently, this frequency-shifted absorption probability will now absorb more radiation in the blue wing of the spectral line than it would otherwise do if $V \equiv 0$ ($I(V \equiv 0)$ is shown as the dashed curve in figure 5.3(*b*)) and less in the red. The resultant emergent intensity is then shown as the solid curve in figure 5.3(*b*). Clearly, the net effect is to produce what appears to the observer to be

Fig.5.1. (*a*) The intensity emergent from an optically thin slab; (*b*) The corresponding source function.

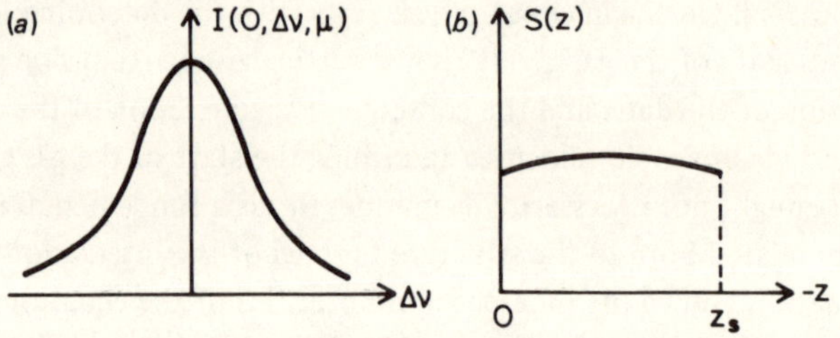

Fig.5.2. 1-dimensional geometry for an expanding slab. The arrows indicate the magnitude and direction of the macroscopic velocity of the radiating gas.

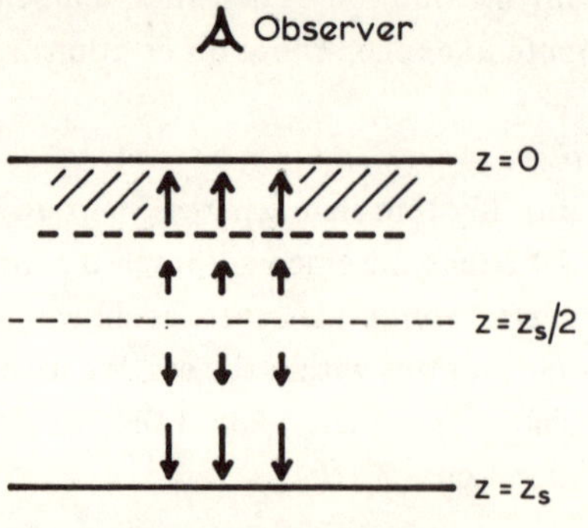

a *red*-shifted emission line profile. The observer could thence misinterpret the slab data by diagnosing slab motion away from the observer. Of course, one should discuss the influence on the radiation field of all the other sub-slabs within the slab; indeed, a full numerical computation is required. However, the essentials of the physics are contained in the above illustration, particularly in view of the need for the emergent radiation to pass through the surface sub-slab at $z \approx 0$ to reach the observer.

The above effect can yield more dramatic results when we consider an optically thick slab for which the spectral emergent line profile exhibits a central absorption feature when $V \equiv 0$, as shown by the dashed curve in figure 5.4 (this case has been discussed in section 1.10.2).

Here, we see that the blue-shifted (relative to the observer) absorption profile will reduce the blue emission which would otherwise ($V \equiv 0$) appear and enhance the red peak. Indeed, it is quite possible, depending upon the precise structure of the macroscopic velocity field within the slab, for the blue emission peak to

Fig.5.3. (*a*) The dotted curve represents the absorption profile for a stationary gas as a function of frequency measured in the rest frame of the observer. The full curve represents the corresponding absorption profile shifted to the blue for an ensemble of atoms moving toward the observer; (*b*) The emergent intensities for both stationary and expanding optically thin slabs are shown as dashed and full curves respectively. The absorption profile, shown as the dotted curve, has been included to illustrate the increased absorption of blue-wing photons by those surface regions of the slab expanding toward the observer, and the corresponding decreased absorption of red-wing photons.

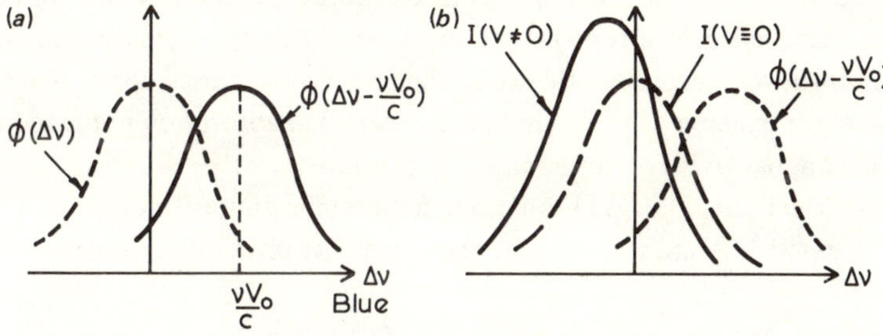

Fig.5.4. The emergent intensities for both stationary and expanding optically thick slabs are shown as dashed and full curves respectively. The dotted curve illustrates the blue shift of the absorption profile due to the motion of that part of the slab closest to the observer.

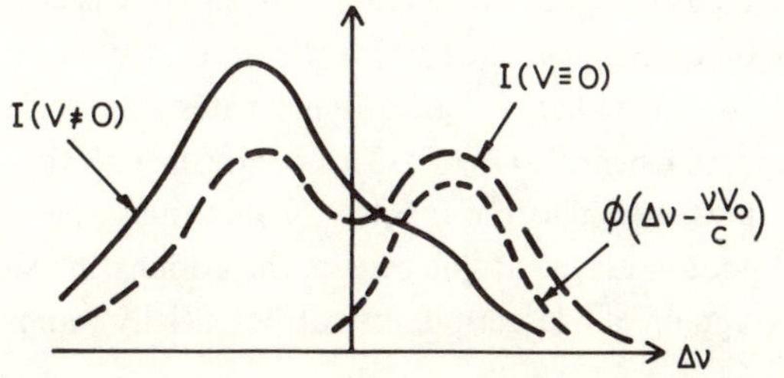

completely disappear (as shown in figure 5.4 - solid curve), thus yielding a red-shifted emergent spectral emission line profile. The emitting atmosphere in this case would appear both to be moving away from the observer (because of the resultant red shift) and to be optically thin (since there is no central absorption feature that one usually associates with optically thick slabs). The observer would be wrong on both counts. Of course, if the frequency dispersion of the observations is sufficiently high, the observer would be able to detect the asymmetry in the emergent line profile and this would at least indicate the need for the inclusion of a macroscopic velocity field in the diagnostic analysis. The above misinterpretation might then be avoided.

We have not discussed the effect of the velocity field on the source function. Clearly, applying the arguments detailed above for the surface sub-slab to internal sub-slabs, we see that the increase in the radiation field due to more trapping of photons in the blue wing of the line, for example, will be offset, to some degree, by the increased ability of photons to escape in the red wing. Indeed, for slab geometry, photon escape generally dominates photon trapping and the source function subsequently decreases to values below those in a velocity-independent, but otherwise identical, atmosphere. This can be seen by noting that photons emitted in the wings in a stationary finite slab escape more readily than line-centre photons. In fact, the radiation field, thence source function, in slab geometry is dominated by line-centre emission (following absorption or collisional excitation). Once a non-zero velocity field is introduced, thereby shifting the absorption profile maximum away from line centre, these line-centre photons will more readily escape from the slab and this, in turn, will decrease $\bar{J}$, thence S (the effect of increased absorption in the blue, for example, and the associated decreased absorption in the red has only a marginal effect on the source function).

Quite the reverse [70, 71] can occur for a semi-infinite atmosphere. If we take an expanding surface region of the somewhat physically idealistic form (see figure 5.5)

$$\left.\begin{aligned} V &= \frac{V_0}{z_s}(z_s - z), \quad & |z| < |z_s|, \\ &= 0, & |z| \geqslant |z_s|, \end{aligned}\right\} \tag{5.7.2}$$

with $|z_s| < \Theta$ (thermalisation path length - see section 1.8), then we again have increased absorption in the blue wing near the surface. However, for semi-infinite atmospheres, the blue-wing radiation basically emanates from regions deeper in the stellar atmosphere where $S \approx B_\nu(T)$. This contrasts with the relatively small number of line-centre photons ($\sim\sqrt{\epsilon}B_\nu(T)$) at the surface. Thus, although there is increased photon escape at line centre, the dominating factor is the increased blue-wing trapping of a larger number of potentially trappable photons. The fact that approximately half the blue-wing photons so absorbed will radiate back into the atmosphere (the other half will radiate toward the surface thereby probably escaping – they would escape from either surface in slab geometry) will suffice to more than counterbalance the increased escape of corresponding red-wing photons. This then produces an increase in the overall radiation field, thence source function, over that for the $V \equiv 0$, but otherwise identical, case. The results are illustrated in figure 5.5.

The effect of non-zero velocity fields on the spectral line intensity emergent from a semi-infinite atmosphere follows the same arguments as those presented for slab geometry. We again have increased blue-wing absorption and red-wing escape as shown in figure 5.6.

The increased source function near the surface of the semi-infinite atmosphere discussed above and illustrated in figure 5.5 can produce emission-type features in the very core of the emergent line profile. However, this is a very minor effect. The main result of the introduction of the velocity field (5.7.2) is the production of an asymmetric absorption line shifted to the observer's blue. It should be stressed here that one *cannot* associate a particular velocity with the observed shift of the minimum intensity in an asymmetric absorption line, i.e. one *cannot* deduce a velocity $V' = c\Delta\nu(I_{\min})\nu$ where $\Delta\nu(I_{\min})$ is the central absorption shift shown in figure 5.6. (Indeed, as we have seen, this type

Fig.5.5. The qualitative behaviour of the source function for stationary (dashed curve) and macroscopically moving (full curve) semi-infinite media.

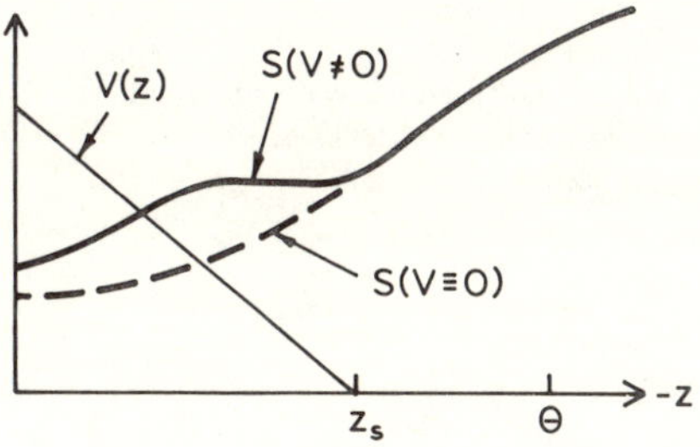

Fig.5.6. The emergent intensities for both stationary and macroscopically moving semi-infinite media are shown as dashed and full curves respectively.

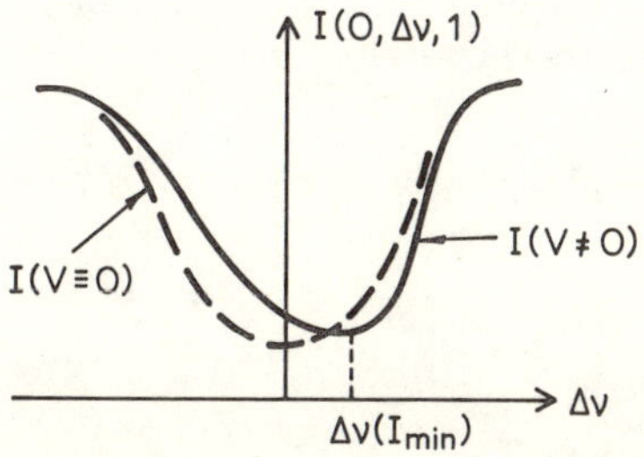

of deduction will lead to a velocity field having the wrong *sign* in slab configurations.) The asymmetry and apparent emergent absorption line shift are clearly a result of a velocity gradient and this necessarily implies the existence of at least two different velocity values in the gas. A geometry having only one macroscopic velocity value stipulates a zero velocity gradient and this produces an apparent shift but no asymmetry. Only then can one associate a frequency-shifted spectral line with a single particular velocity value.

This last point can be emphasised by examining the self-consistent solution of the time-dependent macroscopic conservation equations coupled to the equation of radiative transfer for a spectral emission line arising from a model optically thick slab. In particular, if we take a sequence of propagating shock waves passing through the slab due to a periodic pressure disturbance, for example, at the bottom edge of the slab, then the macroscopic velocity profile has the form shown in figure 5.7 where $t_2 > t_1$.

The corresponding density and temperature distributions have much the same structure as those for the velocity. It has been shown [19] that the resultant emergent spectral line profile almost always exhibits an enhanced red emission peak discussed above and shown as $I(t_1)$ in figure 5.8.

However, as the pulsating shock wave propagates through the slab (and we do not concern ourselves here with boundary effects such as reflection off the surface $z = 0$, for example), the emergent line profile can exhibit a blue peak enhancement shown as $I(t_2)$ in figure 5.8. Generally, if the blue peak enhancement does exist, it lasts for only a small fraction ($\sim\frac{1}{20}$) of the total period of

Fig.5.7. Velocity profiles for a propagating shock wave shown at two times where $t_2 > t_1$. The wave propagates in the positive x direction.

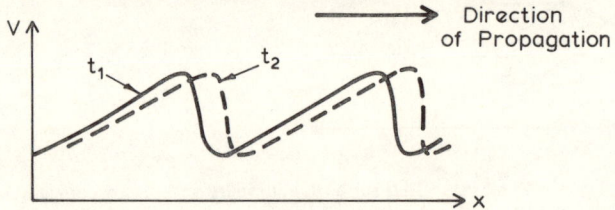

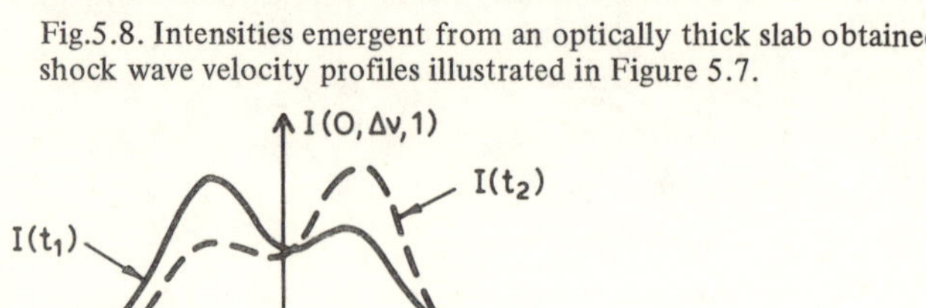

Fig.5.8. Intensities emergent from an optically thick slab obtained using the two shock wave velocity profiles illustrated in Figure 5.7.

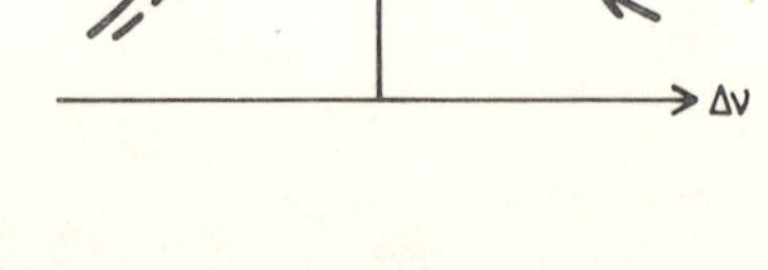

the pulsating shock before reverting to the more normal red peak enhanced emission profile. One could not possibly observe this type of time-dependent fluctuation in the emergent emission line peaks and thence diagnose the (ρ, V, T) structure of the gas giving rise to those fluctuations without computing the solutions to the coupled time-dependent conservation and radiative transfer equations. Indeed, it can be shown that the difference in the (ρ, V, T) distributions leading to the blue peak enhancement and those producing the enhanced red peak are barely discernible. The central emission peaks observed in the core of the Ca II H and K absorption lines (see figure 1.22 for $V \equiv 0$) exhibit a similar type of time-dependent fluctuation [127], and care must clearly be exercised in diagnosing the state of the gas giving rise to such observations.

5.7.2 *Atom and electron temperature differences*

All the foregoing discussion has assumed the temperature equality $T_L = T_U = T_e = T_+$. Although the temperatures of the excited and de-excited states and ions are approximately equal, the electron temperature can be quite different, particularly in the neighbourhood of a shock front 'discontinuity'. Here, we wish to examine these differences in some detail.

The non-relativistic masses of the excited and de-excited atoms are, of course, equal. Consequently, any gain or loss of kinetic energy by one of these two species will very quickly be shared by the other species through elastic collisions. This is an important point in the determination of the different temperatures since it is the temperature which effectively measures the microscopic kinetic energy of the gas; we shall therefore elaborate. The general equations stipulating conservation of linear momentum and energy of two particles of masses m_i and m_j with velocities $\mathbf{v}_{i1}$ and $\mathbf{v}_{j1}$ before an elastic encounter, and velocities $\mathbf{v}_{i2}$ and $\mathbf{v}_{j2}$ after, are

$$m_i\mathbf{v}_{i1} + m_j\mathbf{v}_{j1} = m_i\mathbf{v}_{i2} + m_j\mathbf{v}_{j2}, \tag{5.7.3}$$

and

$$\tfrac{1}{2}m_iv_{i1}^2 + \tfrac{1}{2}m_jv_{j1}^2 = \tfrac{1}{2}m_iv_{i2}^2 + \tfrac{1}{2}m_jv_{j2}^2. \tag{5.7.4}$$

These equations only apply, of course, if the resultant external force is zero in the case of equation (5.7.3) and the resultant force does zero work in the case of equation (5.7.4).

To illustrate the point of interest here, we take one of the particles (m_j, say) to be initially at rest. The above two equations, with $\mathbf{v}_{j1} = 0$, then yield

$$m_j(v_{i1}^2 - v_{i2}^2) = m_i\,(v_{i1}^2 - 2\mathbf{v}_{i1}\cdot\mathbf{v}_{i2} + v_{i2}^2). \tag{5.7.5}$$

Consequently, if $m_i = m_j$ (i.e. $m_U \equiv m_L$), then $\mathbf{v}_{i2} = 0$. If, in particular we consider only 1-dimensional motion, then $v_{i2} = 0$ so that the initial kinetic energy of the ith particle has been completely transferred to the jth particle.

This process then enables the de-excited and excited atoms to share their kinetic energies very efficiently.

If, on the other hand, we consider an atom–electron elastic encounter and take $m_i = m_A$ and $m_j = m_e$ so that $m_j \ll m_i$, then equation (5.7.5) states that $v_{i1} \approx v_{i2}$. Consequently, the energy of the atom is only slightly altered by the collision and it is therefore necessary for many of these atom–electron collisions to occur in general before the electron manages to extract a 'half-share' of the atomic kinetic energy. Of course, the reverse process also applies. The inelastic collisional de-excitation of the excited atom will increase the kinetic energy of the thermal field of free electrons. It requires many atom–electron elastic collisions ($\sim\sqrt{(m_A/m_e)}$) for this enhancement of the thermal field to be transmitted to the atoms.

Gas motions, such as shock waves, deposit energy far more efficiently into the heavier particles than into the electrons (via the viscosity terms - note that atoms are not as mobile as electrons so that μ_V(atoms) $>$ μ_V(electrons) - see appendix D). Thus, as the shock, for example, moves through the gas, the temperature (i.e. microscopic kinetic energy) of the atoms responds immediately (i.e. T_L and T_U increase) whilst the electron temperature must await the requisite number of elastic collisions with the atoms. Thus, in this transient configuration, the electrons and atoms have different temperatures. Consequently, we put $T_L = T_U = T$ (say), although in some cases even this is an unsatisfactory approximation, but stipulate $T_e \neq T$.

We thence have from equation (5.2.28)

$$f_L^{(0)} = N_L \left(\frac{m_A}{2\pi kT}\right)^{3/2} e^{-\frac{m_A V_L^2}{2kT}}, \tag{5.7.6}$$

$$f_U^{(0)} = N_U \left(\frac{m_A}{2\pi kT}\right)^{3/2} e^{-\frac{m_A V_U^2}{2kT}}, \tag{5.7.7}$$

and

$$f_e^{(0)} = N_e \left(\frac{m_e}{2\pi kT_e}\right)^{3/2} e^{-\frac{m_e V_e^2}{2kT_e}}. \tag{5.7.8}$$

These microscopic velocity distribution functions (or their first order equivalents) must be used in determining the absorption and emission probabilities ϕ, ψ and j, and the collisional excitation and de-excitation rates C_{LU} and C_{UL} - see equations (5.1.23), (5.1.25), (5.1.27) and (5.1.33). Thus, the introduction of these different species temperatures necessitates a re-examination of the equation of radiative transfer, viz.

$$\frac{1}{c}\frac{\partial I}{\partial t}+(\boldsymbol{\Omega}\cdot\nabla)I=-\frac{h\nu}{4\pi}\,[N_L B_{LU}\phi(T_L)I-N_U B_{UL}\psi(T_U)I$$
$$-N_U A_{UL}j(T_U)]. \tag{5.7.9}$$

The assumption of complete re-distribution now has the form

$$\phi(T_L)\equiv\psi(T_U)\equiv j(T_U), \tag{5.7.10}$$

which, coupled to the assumption $\phi(x)\equiv\psi(x)\equiv j(x)$, where x is a frequency variable, implies $T_L=T_U=T$ consistent with the above discussion. Consequently, noting the dependence of ϵ on C_{LU} and C_{UL} - see equations (1.6.19, 20), we must write the 2-level atom complete re-distribution source function as

$$S=[1-\epsilon(T_e)]\bar{J}(T)+\epsilon(T_e)B_\nu(T_e)+f(S,T_e)\frac{dS}{dt}, \tag{5.7.11}$$

where

$$\bar{J}(T)=\frac{1}{4\pi}\int_{-\infty}^{\infty}d(\Delta\nu)\int_{4\pi}d\Omega\phi(T)I. \tag{5.7.12}$$

Thus, if there does exist a significant difference between the temperature distributions of the atoms and electrons, this difference must be incorporated into the above equation of radiative transfer.

The macroscopic conservation equations are the same as those listed in section 5.3.2. However, since we have taken $T_e\neq T$, we require an extra expression for T_e. This may be simply obtained from the electron energy conservation equation (5.2.49), or its first order equivalent, with $i=e$ where $\rho_e\epsilon_e=\frac{3}{2}kN_eT_e$. Consequently, in a 1-dimensional model calculation, for example, the seven desired unknowns N_L, N_U, N_e, V, T, T_e and I may be determined from the six equations

$$\frac{\partial N_L}{\partial t}+\frac{\partial}{\partial x}[N_L(V+\bar{V}_L)]=R_L, \tag{5.7.13}$$

$$\frac{\partial N_U}{\partial t}+\frac{\partial}{\partial x}[N_U(V+\bar{V}_U)]=-R_L, \tag{5.7.14}$$

$$\frac{\partial N_e}{\partial t}+\frac{\partial}{\partial x}[N_e(V+\bar{V}_e)]=0, \tag{5.7.15}$$

$$\rho\frac{\partial V}{\partial t}+\rho V\frac{\partial V}{\partial x}+\frac{\partial}{\partial x}\left(p-\frac{\mu_V}{3}\frac{\partial V}{\partial x}\right)=0, \tag{5.7.16}$$

$$\rho\frac{\partial \epsilon}{\partial t}+\rho V\frac{\partial \epsilon}{\partial x}+\left(p-\frac{\mu_V}{3}\frac{\partial V}{\partial x}\right)\frac{\partial V}{\partial x}-\frac{\partial}{\partial x}\left(k_c\frac{\partial T}{\partial x}\right)+\sum_i\frac{\partial}{\partial x}(\bar{D}_i^{(2)}\bar{V}_i)=0, \tag{5.7.17}$$

and

$$\frac{\partial}{\partial t}(\tfrac{1}{2}\rho_e V_e^2 + \rho_e \epsilon_e) + \frac{\partial}{\partial x}[\tfrac{1}{2}\rho_e V^3 + \rho_e \epsilon_e V + p_e V] = \frac{T - T_e}{\tau_E}, \quad (5.7.18)$$

together with the radiative transfer equation

$$\frac{1}{c}\frac{\partial I}{\partial t} + \mu\frac{\partial I}{\partial x} = -\kappa_0 \phi(I - S). \quad (5.7.19)$$

Note that the first integral on the RHS (5.2.49) has been approximated in equation (5.7.18) by the relaxation term $(T - T_e)/\tau_E$ where τ_E represents the time taken for the atoms to complete the sharing of their kinetic energy with the free electrons via elastic collisions. Further, we have taken the electron viscosity to be zero.

Note also that we have not incorporated any continuum radiation into our model nor, indeed, have we explicitly included ions as a separate species (charge conservation would require this inclusion). The appropriate equations may be derived using the work already presented in sections 5.2 and 5.3. In this monograph, however, we have declined to detail any of the radiative transfer processes and effects relating to continuum radiation because of the added complexity and subsequent confusion this can precipitate – such derivations are therefore left to the reader.

Nevertheless, it is possible to discuss the physics of some of the results of the time-dependent radiating gas-dynamic problems when $T_e \neq T$ without recourse to such aspects of continuum radiation and its interaction with ionic particles.

We again consider a shock wave propagating through a radiating gas. The shock front dissipates energy via the $-(\mu_V/3)\partial V/\partial x$ term, this energy being transferred to the atoms. The resulting temperature T is shown as a solid curve in figure 5.9.

We include a temperature scale to indicate a typical magnitude [20] of the difference in temperature distribution of the atoms and electrons (shown as

Fig.5.9. Temperature profiles for a propagating shock wave. The full curve represents the atom and ion temperature distribution whereas the dashed curve illustrates the quite different electron temperature.

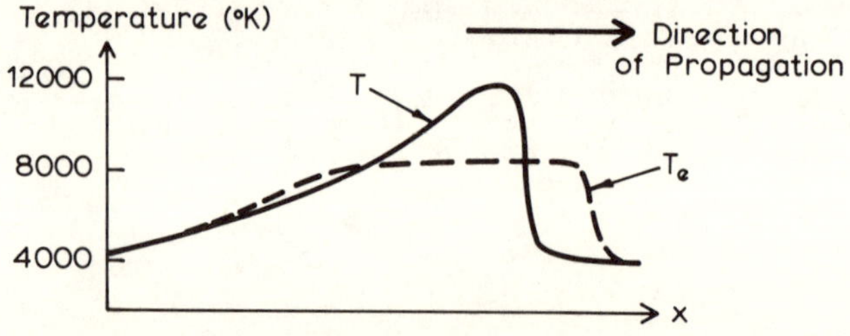

a dotted curve in figure 5.9). There are several points regarding T_e to be noticed here. First, $T_e > T$ in front of the shock. This may be explained by noting that the increase in T over that for the undisturbed gas (~4000 K in our model) increases the value of T_e (after approximately $\sqrt{(m_A/m_e)}$ number of elastic collisions) to about 8000 K. This T_e increase, in turn, increases $\epsilon(T_e)B_\nu(T_e)$ so that an increased number of photons are created by collisional excitation. These photons can then scatter through large distances relative to the mean free paths of elastic collisional events so that they can, in fact, travel ahead of the shock front (the photons actually pass through the shock front). The electron energy decrease resulting from these collisional excitations at the shock front partly accounts for the electron temperature plateau illustrated in figure 5.9 (as opposed to the peaking exhibited by T). However, once the photons have scattered through a thermalisation path length, they will be destroyed by collisional de-excitations, thus resulting in an increase in energy of the thermal fluid. This non-local transfer of thermal energy via photons explains the T_e increase before the shock. Precursor radiation 'preheats' the particles so that the shock does not pass through an undisturbed gas but, rather, through a gas already experiencing certain side effects of the shock and this will affect, for example, the subsequent speed of the shock. Of course, radiation will emit in any direction following a collisional excitation event so that $T_e > T$ behind the shock also. Finally, the flattening of the T_e distribution at the shock is due both to the thermal energy decrease resulting from collisional excitations followed by radiative de-excitation (already mentioned above), and to the effect of heat conductivity whereby any difference in T_e between two points is minimised, i.e. thermal energy flows from relatively 'hot' regions to relatively 'cool' via electron–electron collisions. Mathematically, this stipulation arises from the $k_c \partial T/\partial x$ term. Note that $k_c(\text{electrons}) \gg k_c(\text{atoms})$ since electrons, being smaller and lighter particles than atoms, are far more mobile (see appendix D).

Clearly, such large differences between T and T_e (~4000 K in our model) could be of some importance when included in the source function given by equation (5.7.11), particularly for transients involving violent mass motions. The diagnostic problem would again necessitate the solution of the appropriate equation of radiative transfer *coupled* to the corresponding macroscopic conservation equations.

5.8 Multi-dimensional radiative transfer

The analyses of chapters 1 through 4 concentrated on model atmospheres in only one dimension. The preceding analysis of the present chapter formulated the problem quite generally in three dimensions but, again, only considered the 1-dimensional configuration when discussing the qualitative behaviour of the solutions. However, both observational and theoretical arguments insist that the full multi-dimensional analysis is required for a better understanding of the structure and dynamics of stellar atmospheres. Indeed, we shall show that certain interpretations of stellar line spectra can yield completely erroneous models of the gas giving rise to those spectra if analysed using only 1-dimensional computations. This creates an awkward situation because of the extra care and difficulty associated with the solution of the multi-dimensional line transfer equation [62, 85]. We shall examine this in the following section. The numerical solution of the multi-dimensional gas-dynamic conservation equation is not straightforward either. However, the generalisation of the differencing of the 1-dimensional space derivative to those in the multi-dimensional gas-dynamic equations is conceptually (if not practically in terms of stability) simple and is not considered further here.

5.8.1 *The generalised Feautrier technique for stationary geometry [12]*

We first consider a numerical procedure for solving the time-independent (zero macroscopic velocity field) equation of radiative transfer

$$(\Omega \cdot \nabla) I = -\kappa (I - S), \tag{5.8.1}$$

where

$$S(\mathbf{r}) = \frac{1-\epsilon}{4\pi} \int_{4\pi} \mathrm{d}\Omega \int_{-\infty}^{\infty} \mathrm{d}(\Delta\nu)\phi(\Delta\nu) I(\mathbf{r}, \Delta\nu, \Omega) + \epsilon B_\nu(T), \tag{5.8.2}$$

in 2-dimensional form, viz.

$$\sin\theta \cos\phi \frac{\partial I}{\partial x} + \cos\theta \frac{\partial I}{\partial z} = -\kappa (I - S), \tag{5.8.3}$$

$$S(x, z) = \frac{1-\epsilon}{4\pi} \int_0^{2\pi} \mathrm{d}\phi \int_0^{\pi} \sin\theta \, \mathrm{d}\theta \times \int_{-\infty}^{\infty} \mathrm{d}(\Delta\nu)\phi(\Delta\nu) I(x, z, \Delta\nu, \theta, \phi) + \epsilon B_\nu(T). \tag{5.8.4}$$

The $\theta = \cos^{-1}\mu$ and ϕ are heliocentric and azimuthal angles, respectively, with $\theta = 0$ corresponding to normally emergent radiation – see figure 5.10. Equation (5.8.3) clearly reduces to the 1-dimensional equation of transfer already examined when only depth ($-z$) variations are considered in the model stellar atmosphere. Note that we have again assumed complete re-distribution for a model 2-level atom so that the source function $S(x, z)$ is independent of frequency and angle.

We now define

$$\left.\begin{aligned} I_+^+ &= I(x, z, \Delta\nu, \mu > 0, \cos\phi > 0), \\ I_+^- &= I(x, z, \Delta\nu, \mu < 0, \cos\phi > 0), \\ I_-^+ &= I(x, z, \Delta\nu, \mu > 0, \cos\phi < 0), \\ I_-^- &= I(x, z, \Delta\nu, \mu < 0, \cos\phi < 0), \end{aligned}\right\} \tag{5.8.5}$$

so that, with $S(x, z)$ independent of θ and ϕ, equation (5.8.3) becomes

$$\left.\begin{aligned} \frac{\mu}{\kappa}\frac{\partial I_+^+}{\partial z} + \frac{\sin\theta\cos\phi}{\kappa}\frac{\partial I_+^+}{\partial x} &= -I_+^+ + S, \\ -\frac{\mu}{\kappa}\frac{\partial I_+^-}{\partial z} + \frac{\sin\theta\cos\phi}{\kappa}\frac{\partial I_+^-}{\partial x} &= -I_+^- + S, \\ \frac{\mu}{\kappa}\frac{\partial I_-^+}{\partial z} - \frac{\sin\theta\cos\phi}{\kappa}\frac{\partial I_-^+}{\partial x} &= -I_-^+ + S, \\ -\frac{\mu}{\kappa}\frac{\partial I_-^-}{\partial z} - \frac{\sin\theta\cos\phi}{\kappa}\frac{\partial I_-^-}{\partial x} &= -I_-^- + S. \end{aligned}\right\} \tag{5.8.6}$$

We next define

$$\left.\begin{aligned} \Phi^{(1)} &= \tfrac{1}{2}(I_+^+ + I_-^-), \quad \Psi^{(1)} = \tfrac{1}{2}(I_+^+ - I_-^-), \\ \Phi^{(2)} &= \tfrac{1}{2}(I_-^+ + I_+^-), \quad \Psi^{(2)} = \tfrac{1}{2}(I_-^+ - I_+^-), \end{aligned}\right\} \tag{5.8.7}$$

so that equation (5.8.6) yields

$$\left.\begin{aligned} \left(\frac{\mu}{\kappa}\frac{\partial}{\partial z} + \frac{\sin\theta\cos\phi}{\kappa}\frac{\partial}{\partial x}\right)\Psi^{(1)} &= -\Phi^{(1)} + S, \\ \left(\frac{\mu}{\kappa}\frac{\partial}{\partial z} + \frac{\sin\theta\cos\phi}{\kappa}\frac{\partial}{\partial x}\right)\Phi^{(1)} &= -\Psi^{(1)}, \\ \left(\frac{\mu}{\kappa}\frac{\partial}{\partial z} - \frac{\sin\theta\cos\phi}{\kappa}\frac{\partial}{\partial x}\right)\Psi^{(2)} &= -\Phi^{(2)} + S, \\ \left(\frac{\mu}{\kappa}\frac{\partial}{\partial z} - \frac{\sin\theta\cos\phi}{\kappa}\frac{\partial}{\partial x}\right)\Phi^{(2)} &= -\Psi^{(2)}. \end{aligned}\right\} \tag{5.8.8}$$

We then have

$$\left.\begin{aligned} \left(\frac{\mu}{\kappa}\frac{\partial}{\partial z} + \frac{\sin\theta\cos\phi}{\kappa}\frac{\partial}{\partial x}\right)^2\Phi^{(1)} &= \Phi^{(1)} - S, \\ \left(\frac{\mu}{\kappa}\frac{\partial}{\partial z} - \frac{\sin\theta\cos\phi}{\kappa}\frac{\partial}{\partial x}\right)^2\Phi^{(2)} &= \Phi^{(2)} - S, \end{aligned}\right\} \tag{5.8.9}$$

similar in form to the 1-dimensional Feautrier equations studied in chapter 3.

The boundary equations may be obtained by noting, for example, that $\Psi^{(1)} = \Phi^{(1)} - I^-_- = I^+_+ - \Phi^{(1)}$, viz.

$$\left.\begin{aligned}\left(\frac{\mu}{\kappa}\frac{\partial}{\partial z} + \frac{\sin\theta\cos\phi}{\kappa}\frac{\partial}{\partial x}\right)\Phi^{(1)} &= -\Phi^{(1)} + I^-_-, \\ &= \Phi^{(1)} - I^+_+, \\ \left(\frac{\mu}{\kappa}\frac{\partial}{\partial z} - \frac{\sin\theta\cos\phi}{\kappa}\frac{\partial}{\partial x}\right)\Phi^{(2)} &= -\Phi^{(2)} + I^-_+, \\ &= \Phi^{(2)} - I^+_-.\end{aligned}\right\} \tag{5.8.10}$$

Typical boundary conditions would put I^-_- and I^-_+ equal to the appropriate incident radiation intensity at the surface with I^+_+ and I^+_- corresponding to the radiation field either at the other surface in the case of slab geometry or as $z \to -\infty$ for a semi-infinite model atmosphere.

The source function has the form

$$\begin{aligned}S(x,z) &= \frac{1}{4\pi}\int_0^\pi \mathrm{d}\phi \int_0^1 \mathrm{d}\mu \int_{-\infty}^{\infty} \mathrm{d}(\Delta\nu)\phi(\Delta\nu) \\ &\quad \times [I^+_+ + I^+_- + I^-_+ + I^-_-] + \epsilon B_\nu(T) \\ &= \frac{1}{2\pi}\int_0^\pi \mathrm{d}\phi \int_0^1 \mathrm{d}\mu \int_{-\infty}^{\infty} \mathrm{d}(\Delta\nu)\phi(\Delta\nu)[\Phi^{(1)} + \Phi^{(2)}] + \epsilon B_\nu(T).\end{aligned} \tag{5.8.11}$$

Clearly, then, equations (5.8.9) and (5.8.11) form a second order integro-differential set of equations to be solved for $\Phi^{(1)}$ and $\Phi^{(2)}$. The solution may be obtained numerically by differencing the spatial derivatives and replacing the integrals by suitable quadratures. In particular, we construct a grid of N_x points in the x direction ($r = 1, 2, \ldots, N_x$) and N in the z direction ($j = 1, 2, \ldots, N$) with N_μ, N_ϕ and N_ν grid points for μ, ϕ and $\Delta\nu$. Recall that we need only construct a frequency grid covering the half-range $\nu \in [0, \infty)$ for stationary

Fig.5.10. A right-handed orthogonal cartesian coordinate system with the direction of the radiation intensity represented by the solid arrow.

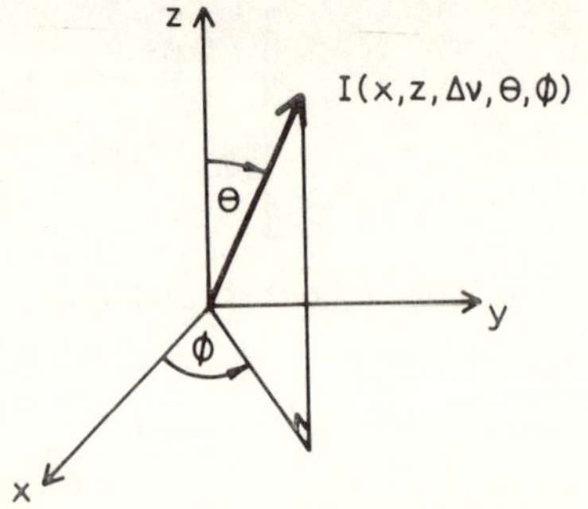

situations. Thus, we write

$$S(x,z) = (1-\epsilon)\sum_{il}^{N_\mu N_\phi}\sum_{k}^{N_\nu}\alpha_i\beta_k\gamma_l[\Phi^{(1)}(x,z,\Delta\nu_k,\mu_i,\phi_l) + \Phi^{(2)}(x,z,\Delta\nu_k,\mu_i,\phi_l)] + \epsilon B_\nu(T). \tag{5.8.12}$$

Note that the weights $\alpha_i\beta_k\gamma_l$ must be normalised, viz.

$$\alpha_i\beta_k\gamma_l = \frac{w_i^{(1)}w_k^{(2)}w_l^{(3)}}{\sum_{i'=1}^{N_\nu}\sum_{k'=1}^{N_\mu}\sum_{l'=1}^{N_\phi} w_{i'}^{(1)}w_{k'}^{(2)}w_{l'}^{(3)}} \tag{5.8.13}$$

- see equation (3.2.98) - where $w_i^{(1)}$, $w_k^{(2)}$ and $w_l^{(3)}$ are the individual weights for μ, $\Delta\nu$ and ϕ integrations. This has been discussed in some detail in section 3.2.5.

We next define the vector $\mathbf{\Phi}_j$ at each depth point z_j in the grid by

$$\mathbf{\Phi}_j = \begin{pmatrix} \Phi(x_1, z_j, \Delta\nu_1, \mu_1, \phi_1) \\ \vdots \\ \Phi(x_{N_x}, z_j, \Delta\nu_{N_\nu}, \mu_{N_\mu}, \phi_{N_\phi}) \end{pmatrix}, \tag{5.8.14}$$

so that the above two integro-differential equations can be written in the matrix form

$$\left.\begin{aligned} -\mathbf{A}_j^{(1)}\mathbf{\Phi}_{j-1}^{(1)} + \mathbf{B}_j^{(1)}\mathbf{\Phi}_j^{(1)} - \mathbf{C}_j^{(1)}\mathbf{\Phi}_{j+1}^{(1)} &= \mathbf{L}_j^{(1)} + \mathbf{D}_j^{(1)}\mathbf{\Phi}_j^{(2)}, \\ -\mathbf{A}_j^{(2)}\mathbf{\Phi}_{j-1}^{(2)} + \mathbf{B}_j^{(2)}\mathbf{\Phi}_j^{(2)} - \mathbf{C}_j^{(2)}\mathbf{\Phi}_{j+1}^{(2)} &= \mathbf{L}_j^{(2)} + \mathbf{D}_j^{(2)}\mathbf{\Phi}_j^{(1)}. \end{aligned}\right\} \tag{5.8.15}$$

Here, each matrix is of size $N_x N_{\Delta\nu} N_\mu N_\phi$. The $\mathbf{D}_j^{(k)}$ matrix is formed from the elements of $\alpha_i\beta_k\gamma_l$ in equation (5.8.12). In particular, if we write

$$\begin{aligned} \Delta_j^r &= \tfrac{1}{2}(\kappa_{j+1}^r + \kappa_j^r)(z_{j+1} - z_j), \\ \nabla_j^r &= \tfrac{1}{2}(\kappa_j^r + \kappa_{j-1}^r)(z_j - z_{j-1}), \\ \bar{\Delta}_j^r &= \tfrac{1}{2}(\kappa_j^{r+1} + \kappa_j^r)(x_{r+1} - x_r), \\ \bar{\nabla}_j^r &= \tfrac{1}{2}(\kappa_j^r + \kappa_j^{r-1})(x_r - x_{r-1}), \end{aligned}$$

where, for example, $\kappa_j^r \equiv \kappa(x_r, z_j, \Delta\nu)$, then

$$\left[\left(\frac{1}{\kappa}\frac{\partial}{\partial z}\right)^2 f\right]_{rj} = \frac{\dfrac{f_{j+1}^r - f_j^r}{\Delta_j^r} - \dfrac{f_j^r - f_{j-1}^r}{\nabla_j^r}}{\tfrac{1}{2}(\Delta_j^r + \nabla_j^r)},$$

$$\left[\left(\frac{1}{\kappa}\frac{\partial}{\partial x}\right)^2 f\right]_{rj} = \frac{\dfrac{f_j^{r+1} - f_j^r}{\bar{\Delta}_j^r} - \dfrac{f_j^r - f_j^{r-1}}{\bar{\nabla}_j^r}}{\tfrac{1}{2}(\bar{\Delta}_j^r + \bar{\nabla}_j^r)},$$

$$\left[\frac{1}{\kappa}\frac{\partial}{\partial z}\left(\frac{1}{\kappa}\frac{\partial}{\partial x}\right) f\right]_{rj} = \left\{\frac{1}{4}\left[\frac{f_{j+1}^{r+1} - f_{j+1}^r}{\bar{\Delta}_{j+1}^r} + \frac{f_{j+1}^r - f_{j+1}^{r-1}}{\bar{\nabla}_{j+1}^r} + \frac{f_j^{r+1} - f_j^r}{\bar{\Delta}_j^r}\right.\right.$$

(5.8.16)

(5.8.16 Contd.)

$$+\frac{f_j^r - f_j^{r-1}}{\bar{\nabla}_j^r}\Bigg] - \frac{1}{4}\Bigg[\frac{f_j^{r+1} - f_j^r}{\bar{\Delta}_j^r} + \frac{f_j^r - f_j^{r-1}}{\bar{\nabla}_j^r} + \frac{f_{j-1}^{r+1} - f_{j-1}^r}{\bar{\Delta}_{j-1}^r} + \frac{f_{j-1}^r - f_{j-1}^{r-1}}{\bar{\nabla}_{j-1}^r}\Bigg]\Bigg\} \frac{1}{\frac{1}{2}(\Delta_j^r + \nabla_j^r)},$$

with a similar (but not necessarily equal) expression for the other mixed spatial derivative. It is a straightforward but tedious task to then construct the $\mathbf{A}_j$, $\mathbf{B}_j$ and $\mathbf{C}_j$ matrices using the above differences, keeping in mind the internal structure or ordering of the defined $\mathbf{\Phi}_j$ vector.

The boundary equations can be put into the matrix form (5.8.15) but with

$$\mathbf{A}_1^{(k)} \equiv 0 \equiv \mathbf{C}_N^{(k)}, \quad k = 1, 2. \tag{5.8.17}$$

The coupled matrix equations (5.8.15) can then be solved using the recurrence relations

$$\left.\begin{aligned}\mathbf{\Phi}_j^{(1)} &= \mathbf{U}_j^{(1)} + \mathbf{V}_j^{(1)}\mathbf{\Phi}_{j+1}^{(1)} + \mathbf{W}_j^{(1)}\mathbf{\Phi}_{j+1}^{(2)},\\ \mathbf{\Phi}_j^{(2)} &= \mathbf{U}_j^{(2)} + \mathbf{V}_j^{(2)}\mathbf{\Phi}_{j+1}^{(2)} + \mathbf{W}_j^{(2)}\mathbf{\Phi}_{j+1}^{(1)}.\end{aligned}\right\} \tag{5.8.18}$$

Direct substitution of these last two expressions into equation (5.8.15) yields

$$\left.\begin{aligned}&-\mathbf{A}_j^{(1)}(\mathbf{U}_{j-1}^{(1)} + \mathbf{V}_{j-1}^{(1)}\mathbf{\Phi}_j^{(1)} + \mathbf{W}_{j-1}^{(1)}\mathbf{\Phi}_j^{(2)}) + \mathbf{B}_j^{(1)}\mathbf{\Phi}_j^{(1)} - \mathbf{C}_j^{(1)}\mathbf{\Phi}_{j+1}^{(1)}\\ &\quad = \mathbf{L}_j^{(1)} + \mathbf{D}_j^{(1)}\mathbf{\Phi}_j^{(2)},\\ &-\mathbf{A}_j^{(2)}(\mathbf{U}_{j-1}^{(2)} + \mathbf{V}_{j-1}^{(2)}\mathbf{\Phi}_j^{(2)} + \mathbf{W}_{j-1}^{(2)}\mathbf{\Phi}_j^{(1)}) + \mathbf{B}_j^{(2)}\mathbf{\Phi}_j^{(2)} - \mathbf{C}_j^{(2)}\mathbf{\Phi}_{j+1}^{(2)}\\ &\quad = \mathbf{L}_j^{(2)} + \mathbf{D}_j^{(2)}\mathbf{\Phi}_j^{(1)}.\end{aligned}\right\} \tag{5.8.19}$$

These equations can then be solved for $\mathbf{\Phi}_j^{(1)}$ and $\mathbf{\Phi}_j^{(2)}$. In particular, the second of equations (5.8.19) yields

$$\begin{aligned}\mathbf{\Phi}_j^{(2)} = [\mathbf{B}_j^{(2)} - \mathbf{A}_j^{(2)}\mathbf{V}_{j-1}^{(2)}]^{-1}[&\mathbf{L}_j^{(2)} + \mathbf{A}_j^{(2)}\mathbf{U}_{j-1}^{(2)}\\ &+ \mathbf{C}_j^{(2)}\mathbf{\Phi}_{j+1}^{(2)} + (\mathbf{D}_j^{(2)} + \mathbf{A}_j^{(2)}\mathbf{W}_{j-1}^{(2)})\mathbf{\Phi}_j^{(1)}],\end{aligned}$$

which, writing

$$\mathbf{P}_j^{(k)} = \mathbf{B}_j^{(k)} - \mathbf{A}_j^{(k)}\mathbf{V}_{j-1}^{(k)},$$

and

$$\mathbf{Q}_j^{(k)} = \mathbf{D}_j^{(k)} + \mathbf{A}_j^{(k)}\mathbf{W}_{j-1}^{(k)}, \tag{5.8.20}$$

becomes

$$\mathbf{\Phi}_j^{(2)} = [\mathbf{P}_j^{(2)}]^{-1}[\mathbf{L}_j^{(2)} + \mathbf{A}_j^{(2)}\mathbf{U}_{j-1}^{(2)} + \mathbf{C}_j^{(2)}\mathbf{\Phi}_{j+1}^{(2)} + \mathbf{Q}_j^{(2)}\mathbf{\Phi}_j^{(1)}].$$

Substitution of this last result into the first of equations (5.8.19) then yields

$$\begin{aligned}[\![\mathbf{P}_j^{(1)} - \mathbf{Q}_j^{(1)}[\mathbf{P}_j^{(2)}]^{-1}\mathbf{Q}_j^{(2)}]\!]\mathbf{\Phi}_j^{(1)} = \mathbf{L}_j^{(1)} &+ \mathbf{A}_j^{(1)}\mathbf{U}_{j-1}^{(1)}\\ + \mathbf{C}_j^{(1)}\mathbf{\Phi}_{j+1}^{(1)} + \mathbf{Q}_j^{(1)}[\mathbf{P}_j^{(2)}]^{-1}&(\mathbf{L}_j^{(2)} + \mathbf{A}_j^{(2)}\mathbf{U}_{j-1}^{(2)} + \mathbf{C}_j^{(2)}\mathbf{\Phi}_{j+1}^{(2)}),\end{aligned}$$

which is an expression involving just $\mathbf{\Phi}_j^{(1)}$, $\mathbf{\Phi}_{j+1}^{(1)}$ and $\mathbf{\Phi}_{j+1}^{(2)}$, and therefore is of the form of equation (5.8.18). It is not difficult to see, writing

$$\mathbf{R}_j^{(k)} = \{\mathbf{P}_j^{(k)} - \mathbf{Q}_j^{(k)}[\mathbf{P}_j^{(k')}]^{-1}\mathbf{Q}_j^{(k')}\}^{-1}, \tag{5.8.21}$$

where $k = 1, k' = 2$ or $k = 2, k' = 1$, that we then have

$$\mathbf{U}_j^{(k)} = \mathbf{R}_j^{(k)}[\mathbf{L}_j^{(k)} + \mathbf{A}_j^{(k)}\mathbf{U}_{j-1}^{(k)} + \mathbf{Q}_j^{(k)}[\mathbf{P}_j^{(k')}]^{-1}(\mathbf{L}_j^{(k')} + \mathbf{A}_j^{(k')}\mathbf{U}_{j-1}^{(k')})],$$
$$\mathbf{V}_j^{(k)} = \mathbf{R}_j^{(k)}\mathbf{C}_j^{(k)}, \tag{5.8.22}$$
$$\mathbf{W}_j^{(k)} = \mathbf{R}_j^{(k)}\mathbf{Q}_j^{(k)}[\mathbf{P}_j^{(k')}]^{-1}\mathbf{C}_j^{(k')}.$$

The source function is then obtained by direct substitution into equation (5.8.12). Clearly, the above solution is similar to that obtained in chapter 3 using the Feautrier technique in one dimension. It has only the extra term $\mathbf{W}_j^{(k)}$ due to the coupling of equations (5.8.9) by equation (5.8.12). It should be noted that a third spatial dimension requires the term $\sin\theta \sin\phi\, \partial I/\partial y$ to be added to the differential operator in equation (5.8.3). Continuing as in two dimensions, and by further separating the radiation field into positive and negative components of $\sin\phi$, we obtain four equations of the form (5.8.15). The corresponding solution is of the form given by equations (5.8.18), but with three coupling terms rather than the one obtained in two dimensions. However, these added recurrence relations are somewhat more algebraically complicated than those derived above - a simpler more general formulation of this full multi-dimensional solution will be considered in section 5.8.3 where non-zero macroscopic velocity fields will also be included.

Finally, we note that exact closed form analytical solutions of the multi-dimensional equation of radiative transfer, similar to those presented in chapter 2 for 1-dimensional geometry, have been obtained [18, 101] only for the case in which ϵ and $\phi(\Delta\nu)$, occurring in the integral term of equation (5.8.4), are independent of spatial coordinates. This is rather restrictive in providing bench-mark solutions against which numerical solutions of the type discussed above may be tested.

5.8.2 *Radiative channelling [12] in stationary geometry*

It has often been considered sufficient in the literature to assume that the spatially averaged emergent intensities obtained from a multi-dimensional analysis are identical to those computed from the corresponding spatially averaged 1-dimensional model atmosphere. To examine this more closely, we write the time-independent radiative transfer equation as

$$(\boldsymbol{\Omega} \cdot \nabla)I = -\kappa(x, y, z, \Delta\nu)[I(x, y, z, \Delta\nu, \theta, \phi) - S(x, y, z)], \tag{5.8.23}$$

with

$$S(x, y, z) = \frac{1-\epsilon(x, y, z)}{4\pi}\int_0^{2\pi} \mathrm{d}\phi \int_0^{\pi} \sin\theta\, \mathrm{d}\theta$$
$$\times \int_{-\infty}^{\infty} \mathrm{d}(\Delta\nu)\phi(\Delta\nu)\, I(x, y, z, \Delta\nu, \theta, \phi)$$
$$+ \epsilon(x, y, z)B_\nu(x, y, z), \tag{5.8.24}$$

where, using the Doppler profile, for example,

$$\phi(x, y, z, \Delta\nu) = \frac{1}{\sqrt{\pi}\Delta\nu_{\mathrm{D}}} \exp\left[-\left(\frac{\Delta\nu}{\Delta\nu_{\mathrm{D}}}\right)^2\right],$$

$$\Delta\nu_{\mathrm{D}} = \frac{\nu}{c}\left[\frac{2kT(x, y, z)}{m_{\mathrm{A}}}\right]^{1/2}. \tag{5.8.25}$$

The solution to the above equations will then yield the emergent intensity $I(x, y, 0, \Delta\nu, \theta, \phi)$. If we now consider the corresponding spatially averaged 1-dimensional equation with, for example,

$$\bar{T}(z) = \iint T(x, y, z)\, \mathrm{d}x\, \mathrm{d}y,$$

then it is clear from equation (5.8.25) that, writing $\phi(x, y, z, \Delta\nu) \equiv \phi(T(x, y, z), \Delta\nu)$,

$$\bar{\phi}(T(x, y, z), \Delta\nu) = \iint \phi(T(x, y, z), \Delta\nu)\, \mathrm{d}x\, \mathrm{d}y$$

$$\neq \phi(\bar{T}(z), \Delta\nu). \tag{5.8.26}$$

Indeed, if we consider a sinusoidally fluctuating temperature of the form

$$T(x, z) = \bar{T}(z)\left(1 + \beta\cos\left(\frac{2\pi x}{X}\right)\right),$$

then expression (5.8.26) is simply equivalent to

$$\frac{1}{X}\int_0^X \frac{\mathrm{e}^{-\frac{\alpha}{1+\beta\cos(2\pi x/X)}}}{\left[1 + \beta\cos\left(\frac{2\pi x}{X}\right)\right]^{1/2}}\, \mathrm{d}x \neq \mathrm{e}^{-\alpha}, \tag{5.8.27}$$

for all $\beta \neq 0$. Note that $\phi(\Delta\nu)$ appears *both* in $\kappa(x, z)$ and the integral $\bar{J}$. Consequently, the solution $I_1(z, \Delta\nu, \theta)$ obtained directly from the 1-dimensional transfer equation

$$\cos\theta\, \frac{\partial I_1}{\partial z} = -\bar{\kappa}(z)[I_1(z, \Delta\nu, \theta) - S_1(z)],$$

$$S_1(z) = \frac{1 - \bar{\epsilon}(z)}{2}\int_0^\pi \sin\theta\, \mathrm{d}\theta \int_{-\infty}^{\infty} \mathrm{d}(\Delta\nu)\, \phi(\bar{T}(z), \Delta\nu)$$

$$\times I_1(z, \Delta\nu, \theta) + \bar{\epsilon}(z) B_\nu(\bar{T}(z)),$$

is such that

$$\bar{I}(z, \Delta\nu, \theta) = \iiint I(x, y, z, \Delta\nu, \theta, \phi)\, \mathrm{d}x\, \mathrm{d}y\, \mathrm{d}\phi$$

$$\neq I_1(z, \Delta\nu, \theta)$$

and, in particular,

$$\bar{I}(0, \Delta\nu, \theta) \neq I_1(0, \Delta\nu, \theta). \tag{5.8.28}$$

The magnitude of the errors involved may be estimated by examining expression (5.8.27). However, it should be stressed that even if LHS (5.8.27) $\approx$ RHS (5.8.27) *any* departure from equality represents an effective source (or sink) of photons and, as we have discussed previously in relation to the normalisation of the angle and frequency weights, this can lead to a considerable change in the radiation field. For example, the $\epsilon B_\nu(T)$ term is generally very small when compared with the integral appearing in S (generally $\lesssim 10^{-4}$ for spectral lines of astrophysical interest) but it clearly imposes its influence on S throughout the entire stellar atmosphere ($\sqrt{\epsilon} B_\nu \lesssim S \lesssim B_\nu$) for semi-infinite geometry). Indeed, this emphasises the need for care when computing the weights appearing in equation (5.8.12). Consequently, the results obtained from a full multi-dimensional analysis can be significantly different from those computed using the corresponding spatially averaged 1-dimensional model atmosphere, and this can lead to considerable errors [13, 128] in the specification of the depth-dependent distribution of temperatures and densities used to describe the observed line spectra.

Keeping this in mind, we now turn to the examination of the channelling effect. We do this by considering two separate 2-dimensional atmospheres schematised in figure 5.11.

The shaded regions R^+ represent relatively opaque structures (i.e. $\kappa(R^+) > \kappa(R^-)$). Both model atmospheres are periodic in x with periodicities X_a and X_b, and depths Z_a and Z_b, so that $X_a \gg Z_a$ and $X_b \ll Z_b$. We take the atmosphere below the shaded regions to be homogeneous (i.e. 1-dimensional). We further take the line-centre photon mean free path $\lambda_0 \ll X$ in both cases.

If we now consider a photon created within a depth Z_a of the surface in one structure in case (a), we see that it has more chance of escaping from the atmosphere than scattering to an adjacent structure (only because $X_a \gg Z_a$).

Fig.5.11. (*a*) Shallow 2-dimensional structures in a semi-infinite medium; (*b*) Relatively deep 2-dimensional periodic structures.

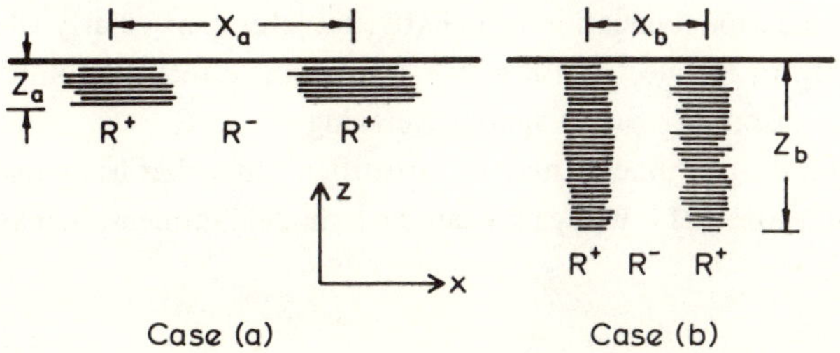

Consequently, when this photon leaves the atmosphere, and is thence observed, it will simply reflect the properties of the region from which it came. For example, if the more opaque regions are relatively hot (i.e. $T(R^+) > T(R^-)$), and this leads to a greater creation term $\epsilon B_\nu(T)$, then more photons would escape from R^+ than from R^- thus yielding the emergent intensity result $I(0, R^+, \Delta\nu, \Omega) > I(0, R^-, \Delta\nu, \Omega)$. The use of LTE arguments stipulates that a relatively bright radiation field *must* be associated with a relatively hot region, simply because the radiation field is represented by the Planck function in LTE and this, in turn, implies

$$B_\nu(T(R^+)) = I(0, R^+, \Delta\nu, \Omega) > I(0, R^-, \Delta\nu, \Omega) = B_\nu(T(R^-)),$$
$$\text{i.e. } T(R^+) > T(R^-).$$

In this case, LTE considerations would yield the correct interpretation $(T(R^+) > T(R^-))$ of the spectral line data.

If we now consider case (b) for which $Z_b \gg X_b$, then photons created deep $(\sim Z_b)$ within the atmosphere are capable of easily scattering to adjacent structures before escaping from the surface of the atmosphere. However, if $\kappa(R^+) > \kappa(R^-)$, such photons will 'bounce off' these R^+ structures and simply channel into the less opaque R^- regions. Indeed, a photon will find it easier to scatter, maybe once or twice, from the more opaque region into R^-, than to suffer all its scatters in R^+ before reaching the surface. This latter process would take many more scatters than the former, and is therefore less likely. We know, for a semi-infinite atmosphere, that $\sqrt{\epsilon} B_\nu(T) \lesssim S \lesssim B_\nu(T)$ so that the source term $\epsilon B_\nu(T)$ in the source function is dominated locally by the integral term $(1-\epsilon)\bar{J}$ for small ϵ ($\lesssim 10^{-4}$ for spectral lines of interest). Consequently, the radiation field, thence source function, is greater in R^- due to this channelling than in R^+ even if $T(R^+) > T(R^-)$ yields greater photon creation in R^+. Note that this increased photon creation in R^+ *competes* against the increased channelling into R^-. Consequently, we have $I(0, R^+, \Delta\nu, \Omega) < I(0, R^-, \Delta\nu, \Omega)$ so that relatively hot regions R^+ can, in fact, appear relatively dark to the observer. If the observer uses LTE arguments, however, it will be assumed that relatively dark corresponds to relatively cool and this, of course, will lead to a complete misinterpretation of the spectral line data in this case. Clearly, an adequate interpretation must therefore incorporate the solution to the full multi-dimensional equation of radiative transfer, particularly when the errors resulting from a neglect of the above channelling effect are compounded with those arising from premature spatial averaging.

The above physical arguments may be quantified somewhat using mathematical considerations [21]. We again examine a model 2-dimensional atmosphere for which

$$\sin\theta\cos\phi\frac{\partial I}{\partial x}+\cos\theta\frac{\partial I}{\partial z}=-\kappa(x,z,\Delta\nu)[I(x,z,\Delta\nu,\Omega)-S(x,z)]. \tag{5.8.29}$$

The characteristics for this partial differential equation are given by

$$\frac{\mathrm{d}x}{\sin\theta\cos\phi}=\frac{\mathrm{d}z}{\cos\theta}=\frac{-\mathrm{d}I}{\kappa(I-S)}. \tag{5.8.30}$$

We can then write the ordinary differential equation

$$\mu\frac{\mathrm{d}I}{\mathrm{d}z}=-\kappa(x,z,\Delta\nu)[I(x,z,\Delta\nu,\Omega)-S(x,z)], \tag{5.8.31}$$

but where now the differentiation $\mathrm{d}/\mathrm{d}z$ appearing in this last equation must take place along the path (characteristic) given by the first of equations (5.8.30), i.e. we must have

$$x=z\tan\theta\cos\phi+\text{constant}. \tag{5.8.32}$$

We then find

$$I(x,z,\Delta\nu,\Omega)=\int_{-\infty}^{z}\kappa(x',z',\Delta\nu)S(x',z')E(x',z',\Delta\nu,\Omega)\frac{\mathrm{d}z'}{\mu},$$

for $\mu>0$ and, for $\mu<0$,

$$I(x,z,\Delta\nu,\Omega)=\int_{0}^{z}\kappa(x',z',\Delta\nu)S(x',z')E(x',z',\Delta\nu,\Omega)\frac{\mathrm{d}z'}{\mu}, \tag{5.8.33}$$

where we have taken zero incident radiation flux at the surface of the atmosphere, and defined

$$E(x',z',\Delta\nu)=\exp\left[-\left|\frac{1}{\mu}\int_{z}^{z'}\kappa(x'',z'',\Delta\nu)\,\mathrm{d}z''\right|\right], \tag{5.8.34}$$

with

$$x'=x+(z'-z)\tan\theta\cos\phi,$$
$$x''=x+(z''-z)\tan\theta\cos\phi.$$

We may then write

$$S(x,z)=[1-\epsilon(x,z)]\Lambda S+\epsilon(x,z)B_{\nu}(x,z), \tag{5.8.35}$$

where the 2-dimensional Λ operator may be obtained from

$$\Lambda S\equiv\frac{1}{4\pi}\int_{-\infty}^{\infty}\phi(x,z,\Delta\nu)\,\mathrm{d}(\Delta\nu)\int_{0}^{2\pi}\mathrm{d}\phi\int_{-1}^{1}\mathrm{d}\mu I(x,z,\Delta\nu,\Omega),$$

i.e.

$$\Lambda \equiv \frac{1}{4\pi}\int_{-\infty}^{\infty} \phi(x, z, \Delta\nu)\,\mathrm{d}(\Delta\nu)\int_{0}^{2\pi}\mathrm{d}\phi \times \int_0^1 \frac{\mathrm{d}\mu}{\mu}\left\{\int_{-\infty}^{z} - \int_0^z\right\}\kappa(x', z', \Delta\nu)E(x', z', \Delta\nu, \Omega)\,\mathrm{d}z'$$

$$= \frac{1}{4\pi}\int_{-\infty}^{\infty} \phi(x, z, \Delta\nu)\,\mathrm{d}(\Delta\nu)\int_{0}^{2\pi}\mathrm{d}\phi \times \int_0^1 \frac{\mathrm{d}\mu}{\mu}\int_{-\infty}^{0}\kappa(x', z', \Delta\nu)E(x', z', \Delta\nu, \Omega)\,\mathrm{d}z'. \tag{5.8.36}$$

We now consider a perturbation solution of equation (5.8.35) about the corresponding spatially averaged 1-dimensional solution. We therefore write

$$S(x, z) = S_0(z) + \lambda S_1(x, z) + \cdots, \tag{5.8.37}$$

with, for example,

$$\kappa_0(x, z) = \bar{\kappa}_0(z) + \lambda[\kappa_0(x, z) - \bar{\kappa}_0(z)], \tag{5.8.38}$$

where we have put

$$\kappa(x, z, \Delta\nu) = \kappa_0(x, z)\phi(x, z, \Delta\nu),$$

and

$$\bar{\kappa}(z) = \frac{1}{X}\int_0^X \kappa(x, z)\,\mathrm{d}x. \tag{5.8.39}$$

The parameter λ has been included only to isolate terms of approximately equal magnitude. We thence have

$$S_0(z) = [1 - \bar{\epsilon}(z)]\Lambda_0 S_0 + \bar{\epsilon}(z)\bar{B}_\nu(z), \tag{5.8.40}$$

and

$$S_1(x, z) = [1 - \bar{\epsilon}(z)]\Lambda_0 S_1 + Q_1(x, z), \tag{5.8.41}$$

where

$$\Lambda_0 \equiv \tfrac{1}{2}\int_{-\infty}^{\infty} \bar{\phi}(z, \Delta\nu)\,\mathrm{d}(\Delta\nu)\int_0^1 \frac{\mathrm{d}\mu}{\mu} \times \int_{-\infty}^{0} \bar{\kappa}_0(z')\bar{\phi}(z', \Delta\nu)\bar{E}(z', \Delta\nu, \mu)\,\mathrm{d}z', \tag{5.8.42}$$

with

$$\bar{E}(z', \Delta\nu, \mu) = \exp\left[-\frac{1}{\mu}\left|\int_z^{z'} \bar{\kappa}_0(z'')\bar{\phi}(z'', \Delta\nu)\,\mathrm{d}z''\right|\right].$$

Clearly, the above $S_1(x, z)$ represents a 2-dimensional correction to the result $S_0(z)$ obtained from the corresponding 1-dimensional model atmosphere. Note that we do *not* expect $\int_0^X S_1(x, z)\,\mathrm{d}x = 0$. Further, the sign and magnitude of this correction will be controlled by the 'source term' $Q_1(x, z)$ given by

$$
\begin{aligned}
Q_1(x,z) = \frac{1-\bar{\epsilon}}{4\pi}\int_{-\infty}^{\infty} \mathrm{d}(\Delta\nu)\int_0^{2\pi}\mathrm{d}\phi\int_0^1\frac{\mathrm{d}\mu}{\mu}\int_{-\infty}^{0}\mathrm{d}z' S_0(z') \\
\times\{|\kappa_0(x',z') - \bar{\kappa}_0(z')|\,\bar{\phi}(z,\Delta\nu)\bar{\phi}(z',\Delta\nu)\bar{E}(z',\Delta\nu,\mu) \\
+\bar{\kappa}_0(z')[\phi(x,z,\Delta\nu) - \bar{\phi}(z,\Delta\nu)]\bar{\phi}(z',\Delta\nu)\bar{E}(z',\Delta\nu,\mu) \\
+\bar{\kappa}_0(z')\bar{\phi}(z,\Delta\nu)[\phi(x',z',\Delta\nu) - \bar{\phi}(z',\Delta\nu)]\bar{E}(z',\Delta\nu,\mu) \\
+\bar{\kappa}_0(z')\bar{\phi}(z,\Delta\nu)\bar{\phi}(z',\Delta\nu)\,[E(x',z',\Delta\nu,\Omega) - E(z',\Delta\nu,\mu)]\} \\
-[\epsilon(x,z) - \bar{\epsilon}(z)]\Lambda_0 S_0 + [\epsilon(x,z) - \bar{\epsilon}(z)]\bar{B}_\nu(z) \\
+\bar{\epsilon}(z)[B_\nu(x,z) - \bar{B}_\nu(z)]. \qquad (5.8.43)
\end{aligned}
$$

It should be stressed here that one should *not* use the above perturbation technique to actually compute solutions to a multi-dimensional radiative transfer equation - convergence is not always attained - we have included the series here simply to isolate, and thence illustrate, the various competing terms at work in multi-dimensional geometry.

Clearly, the sign of the above $Q_1(x, z)$ will determine if Q_1 acts as a source or sink of photons and this, in turn, depends upon the individual magnitudes of the seven competing difference terms appearing in equation (5.8.43). The first four terms are highly non-local in nature - they are, in fact, integrals over geometrical frequency and angle space - and thus care must be exercised in interpreting the actual structure of $Q_1(x, z)$.

The second, third and fourth terms represent channelling processes. For example, if at the x coordinates x^+ and x^- we have $T(x^+, z) > \bar{T}(z) > T(x^-, z)$, then an inspection of equation (5.8.25) shows that $\phi(x^+, z, \Delta\nu) < \bar{\phi}(z, \Delta\nu) <$ $\phi(x^-, z, \Delta\nu)$ at line centre but, in the wings, $\phi(x^+, z, \Delta\nu) > \bar{\phi}(z, \Delta\nu) >$ $\phi(x^-, z, \Delta\nu)$ - recall that $\phi(x, z, \Delta\nu)$ is normalised to unity. Consequently, the contribution to $Q_1(x, z)$ from this second difference term will be positive at x^+ for wing photons but negative for line-core radiation. Thus, in this case, line-core photons will channel into the cooler regions. The third term above has a similar behaviour. The fourth term may work in-phase or out-of-phase with the preceding two depending upon the variation of $\kappa_0(x, z)$ with density and temperature. For example, if we consider the situation in which $\kappa_0(x, z)$ increases with a corresponding increase in electron temperature, i.e. $\kappa_0(x^+, z) >$ $\bar{\kappa}_0(z) > \kappa_0(x^-, z)$, then we would expect $\kappa_0(x^+, z)\phi(x^+, z, \Delta\nu) > \bar{\kappa}_0(z)\bar{\phi}(z, \Delta\nu) >$ $\kappa_0(x^-, z)\phi(x^-, z, \Delta\nu)$ occurring in $E(x, z, \Delta\nu, \Omega)$ and $\bar{E}(z, \Delta\nu, \mu)$ in equation (5.8.43) in the wings of the line. (The situation is not quite as clear for line-centre photons.) Consequently, these wing photons would produce a negative component (i.e. sink) to $Q_1(x^+, z)$ in the fourth term, and this would act to compete against the channelling due to terms two and three. Again, it should be emphasised that the actual source term $Q_1(x, z)$ involves an integral over all dependent space, and thus we must integrate over various structures (line-of-sight characteristics) and over both line core and wing photons.

Next, consider the fifth term in $Q_1(x, z)$. It represents the relative destruction of photons in the various structures due to collisional de-excitation. One might expect that $\epsilon(x^+, z) > \bar{\epsilon}(z) > \epsilon(x^-, z)$ and this will tend to decrease the radiation field in the hotter structure more than in the cooler regions x^-. Since $B_\nu(T)$ is effectively the LTE source of photons, the sixth and seventh terms above specify the relative contributions to the radiation field from collisional excitation in the various structures. One would expect, in general, that these will be positive in x^+, thus basically increasing the radiation field in the hotter structure relative to that in the cooler structure.

Thus, we can see that there exist several quite different effects which can compete against or enhance one another in producing the line source function. Whether the effects work in-phase or out-of-phase depends substantially on the spectral lines in question. For example, the opacity in the resonance lines Ca II H and K will depend upon the number of atoms in the lower state of Ca II. Thus, if the temperature is sufficiently low, then a temperature increase will ionise more Ca I thereby increasing the opacity in these lines. However, a further temperature increase could result in Ca II ions being further ionised to Ca III thereby reducing the Ca II H and K opacity. In the first case, we would have $\kappa_0(x^+, z) > \bar{\kappa}_0(z) > \kappa_0(x^-, z)$ but, in the second, $\kappa_0(x^+, z) < \bar{\kappa}_0(z) < \kappa_0(x^-, z)$. This will clearly affect the first and fourth terms in $Q_1(x, z)$ above and significantly modify the channelling of the radiation field. Further, the precise form of $\phi(x, z, \Delta\nu)$ and the geometrical configuration of the atmosphere (we must integrate along line-of-sight characteristics) must also be important in detailing the relative magnitudes of the difference terms appearing in equation (5.8.43).

However, all the above discussion relates to the line source function. We must also consider the corresponding effect on the line emergent intensity. To do this, we write

$$I(x, 0, \Delta\nu, \Omega) = \int_{-\infty}^{0} \kappa(x', z', \Delta\nu) S(x', z') E(x', z', \Delta\nu, \Omega) \frac{dz'}{\mu} . \quad (5.8.44)$$

The previous perturbation series will then yield the 1-dimensional result

$$I_0(0, \Delta\nu, \mu) = \int_{-\infty}^{0} \kappa_0(z') \bar{\phi}(z', \Delta\nu) \bar{E}(z', \Delta\nu, \mu) S_0(z') \frac{dz'}{\mu} , \quad (5.8.45)$$

with the first order 'correction' of the form

$$I_1(x, 0, \Delta\nu, \Omega) = \int_{-\infty}^{0} \frac{dz'}{\mu}$$
$$\times \{[\kappa_0(x', z') - \bar{\kappa}_0(z')]\bar{\phi}(z', \Delta\nu)\bar{E}(z', \Delta\nu, \mu)S_0(z')$$
$$+ \bar{\kappa}_0(z')[\phi(x', z', \Delta\nu) - \bar{\phi}(z', \Delta\nu)]\bar{E}(z', \Delta\nu, \mu)S_0(z')$$
$$+ \bar{\kappa}_0(z')\bar{\phi}(z', \Delta\nu)[E(x', z', \Delta\nu, \Omega) - \bar{E}(z', \Delta\nu, \mu)]S_0(z')$$
$$+ \bar{\kappa}_0(z')\bar{\phi}(z', \Delta\nu)\bar{E}(z', \Delta\nu, \mu)[S_1(x', z') - S_0(z')]\}. \quad (5.8.46)$$

Thus we see that the source function 'correction' constitutes but one of four terms specifying the first order emergent intensity 'correction', the other terms being the same spatial channelling terms represented in the specification of $Q_1(x, z)$. Indeed, it is not inconceivable that the S_1 term in this last result could be quite insignificant when compared with the channelling terms and, as stated above, these may be in-phase or out-of-phase depending upon the radiative transition being considered and the geometrical configuration of the atmosphere [63].

It should be clear therefore that, by an appropriate choice of the geometry of the structures constituting the model atmosphere, one may realise a configuration for which the above competing effects produce relatively hot structures that appear to the observer to be relatively dark (and vice versa).

5.8.3 Multi-dimensional velocity fields [24]

We now include the effect of macroscopic multi-dimensional velocity fields on the transfer of spectral line radiation. We first develop a numerical method of solution consisting of a combination of the methods presented in sections 5.5.1 and 5.8.1 by starting with the time-independent complete re-distribution transfer equation

$$(\Omega \cdot \nabla)I = -\kappa_0(\mathbf{r})\phi(\mathbf{r}, \Delta\nu, \Omega)[I(\mathbf{r}, \Delta\nu, \Omega) - S(\mathbf{r})], \quad (5.8.47)$$

where

$$S(\mathbf{r}) = \frac{1-\epsilon}{4\pi}\int_{4\pi} d\Omega' \int_{-\infty}^{\infty} d(\Delta\nu')\phi(\mathbf{r}, \Delta\nu', \Omega')I(\mathbf{r}, \Delta\nu', \Omega') + \epsilon B_\nu(T), \quad (5.8.48)$$

where we have neglected the dS/dt term appearing in equation (5.4.41).

In particular, we note that

$$\phi(\mathbf{r}, \Delta\nu, \Omega) \equiv \phi\left(\Delta\nu - \frac{\nu_0}{c}\,\mathbf{V} \cdot \Omega\right), \quad (5.8.49)$$

where $\mathbf{V}$ is the macroscopic velocity field. Further, we assume the absorption probability $\phi(\gamma)$ to be symmetric about $\gamma = 0$ (this is satisfied by both Doppler and Voigt profiles) such that

$$\phi\left(-\Delta\nu + \frac{\nu}{c}\mathbf{V}\cdot\boldsymbol{\Omega}\right) = \phi\left(\Delta\nu - \frac{\nu_0}{c}\mathbf{V}\cdot\boldsymbol{\Omega}\right). \tag{5.8.50}$$

We now write

$$\left.\begin{aligned} I^+ &= I(\mathbf{r}, \Delta\nu, \boldsymbol{\Omega} > 0), \\ I^- &= I(\mathbf{r}, -\Delta\nu, \boldsymbol{\Omega} < 0), \end{aligned}\right\} \tag{5.8.51}$$

for all $\Delta\nu$ across the entire spectral line profile $\nu \in (-\infty, \infty)$. Equation (5.8.47) then becomes for all $\boldsymbol{\Omega} > 0$

$$\left.\begin{aligned} (\boldsymbol{\Omega}\cdot\nabla)I^+ &= -\kappa_0\phi\left(\Delta\nu - \frac{\nu_0}{c}\mathbf{V}\cdot\boldsymbol{\Omega}\right)(I^+ - S), \\ -(\boldsymbol{\Omega}\cdot\nabla)I^- &= -\kappa_0\phi\left(-\Delta\nu + \frac{\nu_0}{c}\mathbf{V}\cdot\boldsymbol{\Omega}\right)(I^- - S), \end{aligned}\right\} \tag{5.8.52}$$

where, since S is independent of $\Delta\nu$ and $\boldsymbol{\Omega}$, $S^+ \equiv S^- \equiv S$. Indeed, we have

$$\begin{aligned} S(\mathbf{r}) &= \frac{1-\epsilon}{4\pi}\int_{-\infty}^{\infty} \mathrm{d}(\Delta\nu')\left[\int_{\boldsymbol{\Omega}'>0} + \int_{\boldsymbol{\Omega}'<0}\right]\mathrm{d}\boldsymbol{\Omega}'\phi\left(\Delta\nu' - \frac{\nu_0}{c}\mathbf{V}\cdot\boldsymbol{\Omega}'\right) \\ &\quad\times I(\mathbf{r}, \Delta\nu', \boldsymbol{\Omega}') + \epsilon B_\nu(T) \\ &= \frac{1-\epsilon}{4\pi}\int_{-\infty}^{\infty}\mathrm{d}(\Delta\nu')\int_{2\pi}\mathrm{d}\boldsymbol{\Omega}'\phi\left(\Delta\nu' - \frac{\nu_0}{c}\mathbf{V}\cdot\boldsymbol{\Omega}'\right)I(\mathbf{r}, \Delta\nu', \boldsymbol{\Omega}') \\ &\quad+ \frac{1-\epsilon}{4\pi}\int_{\infty}^{-\infty}\mathrm{d}(-\Delta\nu')\int_{2\pi}\mathrm{d}\boldsymbol{\Omega}' \\ &\quad\times\phi\left(-\Delta\nu' + \frac{\nu_0}{c}\mathbf{V}\cdot\boldsymbol{\Omega}'\right)I(\mathbf{r}, -\Delta\nu', -\boldsymbol{\Omega}') + \epsilon B_\nu(T), \end{aligned}$$

which, using the symmetry relationship (5.8.50), becomes

$$\begin{aligned} S(\mathbf{r}) &= \frac{1-\epsilon}{4\pi}\int_{2\pi}\mathrm{d}\boldsymbol{\Omega}'\int_{-\infty}^{\infty}\mathrm{d}(\Delta\nu')\phi\left(\Delta\nu' - \frac{\nu_0}{c}\mathbf{V}\cdot\boldsymbol{\Omega}'\right)\Phi(\mathbf{r}, \Delta\nu', \boldsymbol{\Omega}') \\ &\quad+ \epsilon B_\nu(T), \end{aligned} \tag{5.8.53}$$

where we have defined

$$\Phi(\mathbf{r}, \Delta\nu, \boldsymbol{\Omega}) = \tfrac{1}{2}(I^+ + I^-) \quad \forall \quad \boldsymbol{\Omega} > 0. \tag{5.8.54}$$

If we further define $\Psi(\mathbf{r}, \Delta\nu, \boldsymbol{\Omega}) = (I^+ - I^-)/2$, then equations (5.8.52), together with relation (5.8.50), yield

$$\left.\begin{aligned}(\boldsymbol{\Omega}\cdot\nabla)\Psi &= -\kappa_0\phi\left(\Delta\nu-\frac{\nu_0}{c}\,\mathbf{V}\cdot\boldsymbol{\Omega}\right)(\Phi-S),\\ (\boldsymbol{\Omega}\cdot\nabla)\Phi &= -\kappa_0\phi\left(\Delta\nu-\frac{\nu_0}{c}\,\mathbf{V}\cdot\boldsymbol{\Omega}\right)\Psi,\end{aligned}\right\} \tag{5.8.55}$$

so that, eliminating Ψ, we have

$$\left[\frac{1}{\kappa_0\phi\left(\Delta\nu-\dfrac{\nu_0}{c}\,\mathbf{V}\cdot\boldsymbol{\Omega}\right)}\,\boldsymbol{\Omega}\cdot\nabla\right]^2\Phi = \Phi - S, \tag{5.8.56}$$

with the boundary equations

$$\left.\begin{aligned}\left[\frac{1}{\kappa_0\phi\left(\Delta\nu-\dfrac{\nu_0}{c}\,\mathbf{V}\cdot\boldsymbol{\Omega}\right)}\,\boldsymbol{\Omega}\cdot\nabla\right]\Phi &= I^- - \Phi,\\ &= \Phi - I^+.\end{aligned}\right\} \tag{5.8.57}$$

Equation (5.8.56) then constitutes the required second order integro-differential equation to be solved for the radiation term Φ. One proceeds in the numerical evaluation of the solution by suitably differencing the $(\boldsymbol{\Omega}/\kappa)\cdot\nabla$ term and replacing integrals over $\Delta\nu$ and $\boldsymbol{\Omega}$ by quadrature, as discussed previously. The resulting matrix equations then require the inverses of matrices of size $N_x N_y N_\nu N_\Omega$ to be calculated. The use of the quadrature perturbation techniques (AQPT and FQPT) discussed in chapter 3 and the velocity perturbation method (VPT) presented in section 5.5.2(b), can lead to significant savings in computer time and storage and, because of round-off error difficulties normally associated with the computation of large matrix inverses, increases the accuracy of the resulting solutions.

We now turn our attention to the qualitative effects of a multi-dimensional macroscopic velocity field on the transfer of spectral line radiation. We first notice that the non-linear appearance of $\mathbf{V}(\mathbf{r})$ in the absorption profile $\phi(\Delta\nu - \nu_0\mathbf{V}\cdot\boldsymbol{\Omega}/c)$ only acts to emphasise that the spatial average of the emergent intensity obtained from the multi-dimensional calculations is not equal to the emergent intensity obtained from the corresponding spatially averaged 1-dimensional model atmosphere; see equation (5.8.28). Thus, interpretations of spectral line data using only 1-dimensional computations can be quite misleading in detailing the average temperature, density and velocity structure of stellar atmospheres.

It is illustrative to next consider the two model 2-dimensional configurations shown in figure 5.12.

Case (*a*) represents vertical columns moving up and down whilst case (*b*) allows for convective-type motion. The length of the arrows shown in figure

5.12 specifies the relative strengths of the velocity field. Further, we model the gas such that the velocity increases (↑) or decreases (↓) with height (i.e. $\partial \mathbf{V}/\partial z \neq 0$). We first examine the physically idealistic situation for which the model atmosphere is assumed to have zero horizontal density and temperature variations (clearly, this configuration would not satisfy the gas-dynamic conservation equations in just the same manner that the model atmosphere schematised in figure 5.11 must, because of non-zero density gradients, exhibit mass motion).

Recall that in the preceding section the regions R^- of greater opacity scatter photons into regions R^- of lower opacity thus producing channelling of the radiation field into R^-. The velocity structure considered in case (*a*) above duplicates this process, to some extent, for line-centre photons since here these photons will experience a smaller probability of absorption due to the shift of the absorption profile away from line centre. Consequently, channelling of the radiation field into regions $R\uparrow$ and $R\downarrow$ of non-zero vertical velocity occurs. However, photons emitted in either the blue or red wings (depending on ↑ or ↓ in figure 5.12) of the line experience a greater opacity (as discussed in section 5.7) with the introduction of a velocity gradient than otherwise since the absorption profile is essentially shifted from line centre toward these wings. Thus, by the same argument as before, photons emitted in the wings are channelled into regions R_0 of zero or near-zero vertical velocity. This effect then competes against that involving line-centre photons mentioned above. However, this competition between the two effects is controlled by the magnitude of the frequency shift of the absorption profile and the optical thickness of the velocity structures. For example, if the structures have optical thickness of order unity or less measured in the wings of the line, these wing photons will merely scatter through the structures rather than 'bounce off' and this therefore decreases the channelling of the wing photons. In this case, line-centre photon channelling will dominate, i.e. $I(R\downarrow\uparrow) > I(R_0)$.

Fig.5.12. (*a*) A schematic view of a 2-dimensional oscillating atmosphere. The solid arrows indicate the direction and magnitude of the macroscopic velocity vector; (*b*) A model macroscopic velocity distribution for a quasi-convective configuration.

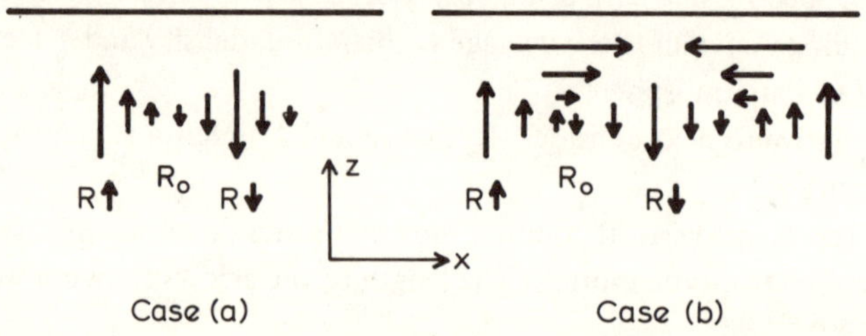

There is, however, a third effect which must be considered involving the radiation emitted deep down in the wings of the line being trapped near the surface in regions of non-zero vertical velocity (i.e. $R\uparrow\downarrow$) due to the effective shift of the line absorption profile into the wings, while escaping from the surface in regions R_0 where the vertical velocity is zero. This acts to increase the radiation field in these $R\uparrow\downarrow$ regions and is, therefore, in-phase with the channelling effect of the line-centre photons mentioned above (see figure 5.5).

It is interesting to consider the emergent spectral line intensity corresponding to an ordinary absorption line and to an absorption line exhibiting central emission peaks as illustrated in figure 1.25 for chromospheric-type configurations. For the first of the above-mentioned lines, the channelling of the line centre and red-wing photons into $R\uparrow$ (with the corresponding channelling of line centre and blue-wing photons into $R\downarrow$) will yield blue- (and red-) shifted minima in the asymmetric normally emergent line profile as illustrated in figure 5.13.

Thus, a blue-shifted absorption line, for example, would be correctly interpreted as motion toward the observer. Note, however, that the magnitude of this shift does *not* reflect the magnitude of the motion – the observed shift is a complicated folding of the velocity field, the absorption profile and the radiation field integrated over a considerable height range (perhaps of order of several hundred or thousand kilometres in stellar atmospheres) [14].

If we now consider the normally emergent intensity for the chromospheric-type spectral line shown as the dashed curve in figure 5.14 for a stationary atmosphere, then this increased line centre plus red-wing photon channelling into $R\uparrow$, for example, and a corresponding blue-wing channelling into $R\downarrow$ will tend to enhance the emission peak as shown.

Indeed, it is not difficult to compute emergent intensity line profiles for which the blue emission feature disappears entirely. Consequently, the extra emission of red-wing photons (with no corresponding blue-wing emission)

Fig.5.13. The asymmetric blue- and red-shifted emergent line intensities obtained using the monotonically decreasing source function $S^{(1)}$ of figure 1.20 but with a macroscopic velocity field included.

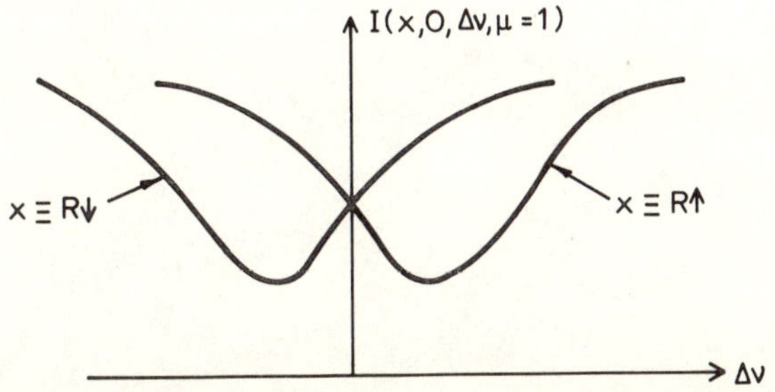

could, using LTE arguments, be erroneously interpreted as radiation being emitted from a region moving away from the observer (recall, however, that the velocity is, in fact, positive toward the observer in $R\uparrow$). Indeed, coupling the above channelling effects with those discussed in section 5.8.2, we can easily compute spectral line profiles in which relatively hot upward-moving regions, for example, can appear to the observer to be *red*-shifted and relatively *dark*. Clearly, a careful diagnostic analysis involving all the above-mentioned multi-dimensional radiative transfer effects is thence required.

Finally, a horizontal differential velocity, as shown in figure 5.12 for case (*b*), must decrease the horizontal line-centre opacity, but increases it in the wings, and this complicates the interpretation of the results even further. The regions of maximum horizontal velocity in case (*b*) correspond to regions R_0 of zero vertical velocity and it is these regions in case (*a*) that scatter line-centre photons into $R\uparrow\downarrow$. Although the 'vertical opacity' is not altered by these horizontal velocity fields, the 'horizontal opacity' for line-centre photons in R_0, for example, is decreased and therefore the channelling process mentioned above for these photons must be reduced (i.e. photons do not 'bounce off' the 'more opaque' structures as readily as before).

Fig.5.14. The symmetric emergent line intensity for a stationary chromospheric-type atmosphere is shown as a dashed curve. The full curve represents a typical asymmetric emergent intensity obtained when a macroscopic velocity field is included.

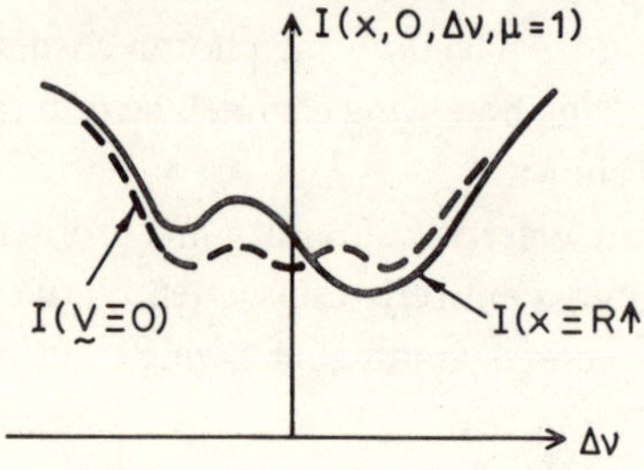

6

Quantum mechanical emission and absorption profiles

The equations of radiative transfer and statistical equilibrium were derived in chapter 1 by examining the five basic processes in which radiation and free electrons interact with atoms (viz. photon absorption, both stimulated and spontaneous photon emission, collisional excitation and collisional de-excitation). We effectively considered this derivation from a macroscopic point of view because the interactions involving the radiation field were characterised by the probabilities ϕ (absorption), ψ (stimulated emission) and j (spontaneous emission), and these only relate to the ensemble of atoms and photons rather than individual particles. The explicit functional dependencies of ϕ, ψ and j were not given – we overcame this omission by simply equating the three profiles (the assumption of complete re-distribution) so that they cancelled one another in the source function and thus (essentially) disappeared from the analysis. There was, of course, an advantage to be gained beyond that of the removal of the need for the actual specification of ϕ, ψ and j – it simplified the mathematics of the solution of the radiative transfer equation and, correspondingly, enabled an attempt at a clearer, more basic understanding of the physical fundamentals controlling the macroscopic transfer of radiation through, and escape from, a stellar atmosphere.

The thrust behind the present chapter is therefore two-fold. First, we wish to examine the above-mentioned five physical processes in more detail. We do this by studying the microscopic interaction of radiation with matter using quantum wave representations for both the radiation field and the particles. This then enables a more accurate equation of radiative transfer (chapter 8) to be derived, an equation exhibiting quantum mechanical interference effects which do not appear in the more classical derivation. Secondly, we wish to remove the assumption of complete re-distribution, and this can only be done if we have explicit expressions for the probabilities ϕ, ψ and j. Indeed, these quantities should appear quite naturally in the quantum mechanical transfer equation.

Further to both these aims, the above generalisation of the transfer equation should enable the ready inclusion of such important effects as magnetic field splitting of atomic energy levels (Zeeman effect) and polarisation of radiation during a scattering process. Consequently, a more solid foundation (relative to that presented in chapter 1) is constructed upon which even more-detailed physical processes, events and effects may be examined.

6.1 Preliminary discussion (first quantisation)

There exist many excellent books [11, 37, 76, 83, 106, 109] which detail the philosophy of modern quantum wave mechanics, and it is of little value therefore to attempt to repeat or collate their arguments here. Rather, we intend to merely summarise (and very briefly discuss) only those basic results of quantum wave mechanics which are pertinent to the derivations presented in the latter sections of this chapter and in chapter 8.

6.1.1 The Schrödinger wave equation

We begin by constructing a complex function $\psi(\mathbf{r}, t)$ for any particular particle dependent upon position coordinate $\mathbf{r}$ and time t. The origin of Schrödinger's wave equation for ψ, viz.

$$i\hbar \frac{\partial \psi}{\partial t} + \frac{\hbar^2}{2m} \nabla^2 \psi - V\psi = 0, \quad \hbar = \frac{h}{2\pi}, \tag{6.1.1}$$

where $V(\mathbf{r})$ is the potential energy of that particle moving in some force field, follows from the basic postulate that matter (particles) can be thought of as wave-like in nature. The function ψ is generally referred to as the wave function for that particle and is related to the amplitude of the wave that characterises that particle. Further, if the particle is to be thought of as a wave (or, more precisely, a wave packet), one can only attach a probability to its location at any time, and this emphasises our inherent lack of absolutely precise knowledge of both the position and time of any particle due to the interfering effect our observations must have on its subsequent motion. This, of course, is contained (in part) in the Heisenberg uncertainty principle.

Thus, if we denote by $P(\mathbf{r}, t)\,\mathrm{d}\mathbf{r}$ the probability of the particle being in the volume element $\mathrm{d}\mathbf{r}$ about $\mathbf{r}$ at time t, then we write

$$P(\mathbf{r}, t) = \psi^*(\mathbf{r}, t)\psi(\mathbf{r}, t) = |\psi|^2, \tag{6.1.2}$$

where the asterisk refers to the complex conjugate quantity. We can be more general by considering a system of particles for which the appropriate wave function $\psi(q_1, \ldots, q_N, t)$ depends upon the N generalised position coordinates q_r and time t. We then have

$$P(q_1, \ldots, q_N, t)\,\mathrm{d}q_1 \ldots \mathrm{d}q_N = |\psi|^2\,\mathrm{d}q_1 \ldots \mathrm{d}q_N,$$

as the probability of finding the system (or each part thereof) located within the region $\mathrm{d}q_1 \ldots \mathrm{d}q_N$ about $(q_1, \ldots, q_N)$ at time t.

The Hamiltonian $H(q_1, \ldots, q_N, p_1, \ldots, p_N, t)$, where the p_r $(r = 1, N)$ are the generalised momenta coordinates, offers a convenient means of studying systems of particles in classical mechanics. It similarly plays an important rôle in the structuring of wave mechanics.

First, the classical Hamiltonian is defined by

$$H = \sum_{r=1}^{N} \dot{q}_r p_r - L, \tag{6.1.3}$$

with

$$p_r = \frac{\partial L}{\partial \dot{q}_r}, \quad r = 1, N. \tag{6.1.4}$$

The Lagrangian $L(q_1, \ldots, q_N, p_1, \ldots, p_N, t)$ is defined by

$$L = T - V, \tag{6.1.5}$$

where T and V are the respective kinetic and potential energies of the system, and satisfies the Euler–Lagrange equation

$$\frac{\partial L}{\partial q_r} - \frac{\mathrm{d}}{\mathrm{d}t}\left(\frac{\partial L}{\partial \dot{q}_r}\right) = 0. \tag{6.1.6}$$

This last equation may be derived from Newton's equations of motion only for a system containing conservative forces. (An alternative meaning must be attached to V when the system is non-conservative.) If the transformation from the cartesian coordinate reference frame which realises the generalised system q_r is independent of time (i.e. the axes are fixed in space) and, again, we assume the system to be conservative (i.e. $\partial V/\partial \dot{q}_r = 0$), then one can readily show that the Hamiltonian H in classical mechanics is the total energy E of the system, i.e.

$$H = T + V. \tag{6.1.7}$$

Further, differentiating equation (6.1.3) with respect to time, noting that $H = H(q_1, \ldots, q_N, p_1, \ldots, p_N, t)$, and using equation (6.1.6), yields

$$\frac{\mathrm{d}H}{\mathrm{d}t} = -\frac{\partial L}{\partial t}, \tag{6.1.8}$$

so that H is a constant if L is explicitly independent of time. Thus, given the restrictions applicable in the derivation of the above equations, the Hamiltonian H has the well-defined physical interpretation of the total constant energy of the system. This interpretation, in fact, enables a prescription for the change-over from classical to quantum wave mechanics to be engineered. For example, the equation defining the energy E of a single particle in classical mechanics, i.e.

$$E = \sum_{j=1}^{3} \frac{p_j^2}{2m} + V, \tag{6.1.9}$$

where the p_j represent the linear momenta in a fixed cartesian reference frame, may be compared with equation (6.1.1) written as

$$\left(\mathrm{i}\hbar \frac{\partial}{\partial t}\right) \psi = \left(-\frac{\hbar^2}{2m} \nabla^2 + V\right) \psi. \tag{6.1.10}$$

We then have the correspondence rule which states that the classical equation (6.1.7), i.e.

$$H = E, \tag{6.1.11}$$

may be converted to the corresponding equation in quantum wave mechanics by associating E and H with the operators $\hat{E}$ and $\hat{H}$ acting on a wave function ψ, viz.

$$\hat{H}\psi = \hat{E}\psi, \tag{6.1.12}$$

where

$$\hat{E} \equiv \mathrm{i}\hbar \frac{\partial}{\partial t}, \tag{6.1.13}$$

$$\hat{H} \equiv \sum_{j=1}^{3} \frac{1}{2m} \hat{p}_j^2 + V, \tag{6.1.14}$$

and where, in cartesian coordinates,

$$\hat{p}_j \equiv \frac{\hbar}{\mathrm{i}} \frac{\partial}{\partial x_j}, \tag{6.1.15}$$

so that

$$\hat{H} \equiv -\frac{\hbar^2}{2m} \nabla^2 + V. \tag{6.1.16}$$

Note that the hat distinguishes operators from scalars. It should be mentioned that this correspondence rule given by equations (6.1.12–16) is not invariant under a change of coordinates (see, for example, Messiah [83], Vol. I, p. 69). The rule only applies in cartesian coordinates.

An equivalent alternative prescription for the change-over from classical to wave mechanics can be obtained using commutators. The expression

$$[\alpha, \beta] = \alpha\beta - \beta\alpha, \tag{6.1.17}$$

defines the commutator of the two quantities α and β. Clearly, if α and β are scalars, for example, the commutative law of scalar multiplication yields $[\alpha, \beta] = 0$, i.e. α and β commute. If, however, we consider the two operators $\hat{A}$ and $\hat{B}$, then, in general, we have

$$[\hat{A}, \hat{B}] = \hat{A}\hat{B} - \hat{B}\hat{A} \neq 0. \tag{6.1.18}$$

In particular, using equation (6.1.15), we have

$$\begin{aligned}[x_k, \hat{p}_j] &= x_k \frac{\hbar}{\mathrm{i}} \frac{\partial 1}{\partial x_j} - \frac{\hbar}{\mathrm{i}} \frac{\partial x_k}{\partial x_j} \\ &= \hbar \mathrm{i} \delta_{jk},\end{aligned} \tag{6.1.19}$$

whereas

$$[x_k, p_j] = 0. \tag{6.1.20}$$

Consequently, we may convert from classical to wave mechanics by simply replacing the linear momenta p_j by the operator $\hat{p}_j$ satisfying the non-zero commutator (6.1.19). This constitutes the basis of first quantisation whereby particles can be modelled to exist only in certain discrete states corresponding to quantised energy levels. As we shall see, second quantisation deals with the similar discrete structuring of the radiation field.

We complete this section by noting that the time-dependent Schrödinger wave equation is usually written as

$$i\hbar \frac{\partial \psi}{\partial t} = \hat{H}\psi, \tag{6.1.21}$$

whereas the time-independent equation

$$\nabla^2 \psi + \frac{2m}{\hbar^2}(E - V) = 0, \tag{6.1.22}$$

where E may also be thought of as an effective separation constant, has the form

$$\hat{H}\psi = E\psi. \tag{6.1.23}$$

Equations (6.1.21) and (6.1.23) apply equally for a single particle and a system of particles if the Hamiltonian is correctly constructed. In particular, it should again be noted that the notion of the Hamiltonian as the total energy of the system requires that system to be conservative. The motion of an electron in a magnetic field, for example, does not fulfil this requirement and, therefore, a modification to the Hamiltonian proves essential [78].

6.1.2 *Simple stationary states and orthogonality*

First, we briefly consider a simplified model of the hydrogen atom in which an electron moves about a nucleus of infinite mass, i.e. the influence of the electron on the nucleus is ignored. The time-independent Schrödinger wave equation (6.1.23) in spherical polar coordinates then has the form

$$\left[\frac{1}{r^2}\frac{\partial}{\partial r}\left(r^2\frac{\partial}{\partial r}\right) + \frac{1}{r^2 \sin\theta}\frac{\partial}{\partial\theta}\left(\sin\theta\frac{\partial}{\partial\theta}\right) + \frac{1}{r^2\sin^2\theta}\frac{\partial^2}{\partial\phi^2}\right]\psi + \frac{2m}{\hbar^2}(E - V)\psi = 0. \tag{6.1.24}$$

Almost all texts on quantum mechanics consider the solution to this last equation in some detail. It suffices here to state that, since E is constant and V is a function only of r (one typically assumes a coulombic potential here), one may use separation of variables by writing

$$\psi(r, \theta, \phi) = R(r)\Theta(\theta)\Phi(\phi). \tag{6.1.25}$$

One readily finds that the forms for Φ, Θ and R are given by

$$\Phi(\phi) = e^{im\phi}, \tag{6.1.26}$$

where m is an integer,

$$\Theta(\theta) = P_l^m(\cos\theta), \tag{6.1.27}$$

where P_l^m is the associated Legendre polynomial [125] ($|m| \leqslant l$, l integral), and

$$R(r) = R_n(r), \tag{6.1.28}$$

where $R_n(r)$ is a series in r which only converges if the total energy E is related to the integer n via the expression

$$E \propto \frac{1}{n^2}. \tag{6.1.29}$$

This last result is most important since it stipulates that the energy of the system is quantised, i.e. the electron bound to the nucleus of the atom can only exist in specific discrete energy levels. This, of course, was the basis of our discussion of energy spectra presented in chapter 1.

The required solution $\psi(r, \theta, \phi)$ of equation (6.1.24) depends quite explicitly on the three integers n, l and m and we may therefore write each individual component of the solution as $\psi_{nlm}(r, \theta, \phi)$. Similarly, we may emphasise that E has only discrete values by writing E_n. (Note that we should also consider electron spin here. However, this is not essential to our later requirements.) For the purposes of the present discussion we thence drop the lm subscripts and write equation (6.1.23) as

$$\hat{H}\psi_n = E_n\psi_n. \tag{6.1.30}$$

The complex conjugate of this last equation with n replaced by n', and treating $E_{n'}$ as complex, is just

$$(\hat{H}\psi_{n'})^* = E_{n'}^*\psi_{n'}^*. \tag{6.1.31}$$

Multiplying equation (6.1.30) by $\psi_{n'}^*$ and equation (6.1.31) by $\psi_{n'}$ subtracting and integrating over all space, then yields

$$\int \{\psi_{n'}^*(\hat{H}\psi_n) - (\hat{H}\psi_{n'})^*\psi_n\}\, d\tau = (E_n - E_{n'}^*)\int \psi_{n'}^*\psi_n\, d\tau, \tag{6.1.32}$$

where $d\tau$ represents the volume element of integration.

We proceed by directly substituting the Hamiltonian for this simplified stationary problem, viz.

$$\hat{H} = -\frac{\hbar^2}{2m}\nabla^2 + V, \tag{6.1.33}$$

into equation (6.1.32). The term in braces then becomes

$$\begin{aligned} -\frac{2m}{\hbar}\{-\} &= \psi_{n'}^*\nabla^2\psi_n - \psi_n\nabla^2\psi_{n'}^* \\ &= \nabla\cdot(\psi_{n'}^*\nabla\psi_n - \psi_n\nabla\psi_{n'}^*). \end{aligned} \tag{6.1.34}$$

Green's theorem [69] then shows that

$$-\frac{2m}{\hbar}\int_{\text{Vol}}\{-\}\,\mathrm{d}\tau = \int_{\text{Surf}}(\psi_{n'}^*\nabla\psi_n - \psi_n\nabla\psi_{n'}^*)\cdot\mathbf{n}\,\mathrm{d}S, \tag{6.1.35}$$

where Surf is the surface bounding the volume Vol of integration (**n** is the unit vector normal outward to that surface). If we take the volume element Vol to include all position space, the surface Surf obviously is at infinity where, because of the physical meaning of $|\psi|^2$, we must have vanishing ψ, ψ^* and their derivatives. Equation (6.1.32) then becomes

$$(E_n - E_{n'}^*)\int\psi_{n'}^*\psi_n\,\mathrm{d}\tau = 0. \tag{6.1.36}$$

The quantity $|\psi_n|^2\,\mathrm{d}\tau = \psi_n^*\psi_n\,\mathrm{d}\tau$ is the probability of finding the bound electron in the volume element $\mathrm{d}\tau$, the electron being in the energy level designated by E_n. Consequently, we must have the integral over all space

$$\int\psi_n^*\psi_n\,\mathrm{d}\tau = 1. \tag{6.1.37}$$

Thus, if $n = n'$, equation (6.1.36) then stipulates that $E_n = E_{n'}^*$ i.e. E_n is real. This is not surprising from a physical point of view.

If $n \neq n'$, i.e. $E_n \neq E_{n'}(=E_{n'}^*)$, then

$$\int\psi_{n'}^*\psi_n\,\mathrm{d}\tau = 0. \tag{6.1.38}$$

Equations (6.1.37) and (6.1.38) define the orthogonality of the wave functions ψ_n, i.e.

$$\int\psi_{n'}^*\psi_n\,\mathrm{d}\tau = \delta_{nn'}, \tag{6.1.39}$$

where $\delta_{nn'}$ is the usual Kronecker delta function. Functions satisfying equations of the type (6.1.39) are said to be normalised to unity, i.e. the ψ_n are orthogonal.

We have illustrated orthogonality of the wave functions ψ_n within the context of the model stationary (somewhat physically idealistic) hydrogen atom. As we shall see, however, the results obtained above are quite general.

6.1.3 *Elements of matrix mechanics*

It is particularly convenient to introduce the scalar product notation

$$\langle\alpha|\beta\rangle = \int\alpha^*\beta\,\mathrm{d}\tau, \tag{6.1.40}$$

where α and β are complex functions in Hilbert space, i.e. they are square integrable so that, for example, integrals of the form $\int|\alpha|^2\,\mathrm{d}\tau$ must exist. In this notation the LHS (6.1.39) is written as $\langle\psi_{n'}|\psi_n\rangle$ or, more generally, $\langle n'|n\rangle$ so that, from equation (6.1.39), we have the orthogonality relationship

$$\langle n'|n\rangle = \delta_{nn'}. \tag{6.1.41}$$

Further, it is not difficult to see that

$$(\langle n'|n\rangle)^* = \int \psi_{n'}\psi_n^* \, d\tau = \langle n|n'\rangle. \tag{6.1.42}$$

Equations (6.1.41) and (6.1.42) enable an analogy to be drawn between all the foregoing manipulations in Hilbert space (infinite dimensions) and those characterising vector and matrix algebra (finite dimensions).

For example, if we define the ket $|a\rangle$ by the column vector

$$|a\rangle = \begin{pmatrix} a_1 \\ \vdots \\ a_N \end{pmatrix} \tag{6.1.43}$$

in N dimensions, and its dual (bra) by the row vector

$$\langle a| = (a_1^*, \ldots, a_N^*), \tag{6.1.44}$$

then we have the scalar product

$$\langle a|b\rangle = \sum_{j=1}^{N} a_j^* b_j = s, \tag{6.1.45}$$

where s is a complex scalar. Further,

$$s^* = \sum_{j=1}^{N} a_j b_j^* = \sum_{j=1}^{N} b_j^* a_j = \langle b|a\rangle,$$

i.e.

$$(\langle a|b\rangle)^* = \langle b|a\rangle. \tag{6.1.46}$$

This last result should be compared with equation (6.1.42).

The scalar product notation is also used when the expressions involve operators. For example, we write

$$\langle\psi_{n'}|\hat{H}\psi_n\rangle \equiv \langle n'|\hat{H}|n\rangle = \int \psi_{n'}^* \hat{H}\psi_n \, d\tau. \tag{6.1.47}$$

Again, we may turn to matrix algebra for a simplification of the exposition of quantum wave mechanics if we allow $\hat{H}$ to be represented by the matrix

$$\hat{H} \equiv \begin{pmatrix} H_{11} & \cdots & H_{1N} \\ \vdots & & \\ H_{N1} & & H_{NN} \end{pmatrix}. \tag{6.1.48}$$

It is useful to define the adjoint ($^+$) of a row vector as the complex conjugate of the corresponding column vector, and vice versa, i.e.

$$(a_1, \ldots, a_N)^+ = \begin{pmatrix} a_1^* \\ \vdots \\ a_N^* \end{pmatrix}, \tag{6.1.49}$$

and

$$\begin{pmatrix} a_1 \\ \vdots \\ a_N \end{pmatrix}^+ = (a_1^*, \ldots, a_N^*), \tag{6.1.50}$$

so that

$$|a\rangle^+ = \langle a|, \quad \text{and} \quad \langle a|^+ = |a\rangle. \tag{6.1.51}$$

Correspondingly, we define the adjoint of a matrix A as the complex conjugate of its transpose, i.e.

$$\hat{A}^+ = \begin{pmatrix} A_{11}^* & \cdots & A_{N1}^* \\ \vdots & & \\ A_{1N}^* & & A_{NN}^* \end{pmatrix}. \tag{6.1.52}$$

Thus, if we write

$$|b\rangle = \hat{A}|a\rangle, \tag{6.1.53}$$

such that the elements b_i of $|b\rangle$ are

$$b_i = \sum_j A_{ij}a_j, \tag{6.1.54}$$

then

$$\langle b| = |b\rangle^+, \tag{6.1.55}$$

has elements b_i^* given by

$$b_i^* = \sum_j (A_{ij}a_j)^* = \sum_j a_j^* A_{ij}^*,$$

so that

$$(b_1^*, \ldots, b_N^*) = (a_1^*, \ldots, a_N^*)\begin{pmatrix} A_{11}^* & \cdots & A_{N1}^* \\ \vdots & & \\ A_{1N}^* & & A_{NN}^* \end{pmatrix}$$

$$= \langle a|\hat{A}^+. \tag{6.1.56}$$

We therefore have the important manipulative result

$$(\hat{A}|a\rangle)^+ = \langle a|\hat{A}^+. \tag{6.1.57}$$

We next construct the complex scalar Φ such that

$$\Phi = \langle a|\hat{A}|b\rangle. \tag{6.1.58}$$

If we write

$$|c\rangle = \hat{A}|b\rangle, \tag{6.1.59}$$

then

$$\Phi = \langle a|c\rangle, \tag{6.1.60}$$

so that

$$\Phi^* = \langle c|a\rangle, \tag{6.1.61}$$

from equation (6.1.46). But

$$|c\rangle^+ = \langle c| = (\hat{A}|b\rangle)^+ = \langle b|\hat{A}^+, \tag{6.1.62}$$

from equation (6.1.57). Thus,

$$\Phi^* = \langle b|\hat{A}^+|a\rangle = (\langle a|\hat{A}|b\rangle)^*. \tag{6.1.63}$$

A matrix is termed Hermitian if it is self-adjoint, i.e. if

$$\hat{A}^+ = \hat{A}, \tag{6.1.64}$$

and must therefore satisfy the important condition (see equation (6.1.63))

$$(\langle a|\hat{A}|b\rangle)^* = \langle b|\hat{A}|a\rangle. \tag{6.1.65}$$

Clearly, equation (6.1.64) stipulates that the diagonal elements of Hermitian matrices must be real.

The importance of this last result can be seen in several ways. First, the LHS (6.1.32) contains two integral terms, the second of which may be written as

$$\int (\hat{H}\psi_{n'})^*\psi_n \, \mathrm{d}\tau = \left(\int (\hat{H}\psi_{n'})\psi_n^* \, \mathrm{d}\tau\right)^* = \left(\int \psi_n^*\hat{H}\psi_{n'} \, \mathrm{d}\tau\right)^*$$

$$= (\langle n|\hat{H}|n'\rangle)^*. \tag{6.1.66}$$

Equation (6.1.32) then has the form

$$\langle n'|\hat{H}|n\rangle - (\langle n|\hat{H}|n'\rangle)^* = (E_n - E_{n'}^*)\langle n'|n\rangle. \tag{6.1.67}$$

Consequently, if the matrix $\hat{H}$ is Hermitian, i.e. satisfies equation (6.1.65), then we immediately find

$$(E_n - E_{n'}^*)\langle n'|n\rangle = 0, \tag{6.1.68}$$

with the subsequent results obtained before, viz. $E_n = E_n^*$ and $\langle n'|n\rangle = \delta_{nn'}$. Conversely, if E is to be a physically meaningful quantity, then it should be purely real, so that putting $n = n'$ in equation (6.1.67) yields

$$\langle n|\hat{H}|n\rangle = (\langle n|\hat{H}|n\rangle)^*, \tag{6.1.69}$$

i.e. $\hat{H}$ must have real diagonal elements and, indeed, must be Hermitian if the vectors $|n\rangle$ form an orthogonal set.

Alternatively, one could write the time-independent Schrödinger wave equation (6.1.23) in matrix form

$$\hat{H}|n\rangle = E_n|n\rangle. \tag{6.1.70}$$

Its adjoint (with n replaced by n') is then (see equation (6.1.57))

$$\langle n'|\hat{H}^+ = \langle n'|E_{n'}^*. \tag{6.1.71}$$

If we now multiply equation (6.1.70) on the left by $\langle n'|$ and equation (6.1.71) on the right by $|n\rangle$, and subtract, we find

$$\langle n'|\hat{H}|n\rangle - \langle n'|\hat{H}^{+}|n\rangle = (E_n - E_{n'}^{*})\langle n'|n\rangle, \qquad (6.1.72)$$

which, using equation (6.1.63), is identical to equation (6.1.67).

This is a more general result than that obtained previously. The ease with which the matrix formulation of quantum wave mechanics can deduce such general expressions is thereby emphasised.

Secondly, consider the time-dependent Schrödinger wave equation (6.1.21) in matrix form

$$i\hbar \frac{\partial}{\partial t}|n\rangle = \hat{H}|n\rangle. \qquad (6.1.73)$$

The adjoint of this last equation (with n replaced by n') is then

$$-i\hbar \frac{\partial}{\partial t}\langle n'| = \langle n'|\hat{H}^{+}. \qquad (6.1.74)$$

We may then write

$$\frac{\partial}{\partial t}(\langle n'|n\rangle) = \left(\frac{\partial}{\partial t}\langle n'|\right)|n\rangle + \langle n'|\frac{\partial}{\partial t}\langle n|$$

$$= -\frac{1}{i\hbar}[\langle n'|\hat{H}^{+}|n\rangle - \langle n'|\hat{H}|n\rangle]. \qquad (6.1.75)$$

Thus, if our interpretation of

$$\int |\psi_n|^2 \, d\tau = \langle n|n\rangle,$$

as a probability is to have any validity, the LHS (6.1.75) must be identically zero, and this again stipulates that $\hat{H}$ must be Hermitian.

6.1.4 *First quantisation*

We have used the solutions of the Schrödinger wave equation directly in section 6.1.3 to briefly illustrate the manner in which an electron bound to the nucleus of an atom exists only in discrete quantised energy levels. In the present section we wish to consider a more general force field (but, for ease in exposition, only in one dimension) to emphasise the effect of the commutator expression (6.1.19) on the development [126] of these quantised energy states. This latter approach is particularly useful in the presentation of second quantisation relating to the radiation field.

Consider a particle moving in the x direction under the influence of a general force field (in the direction x) having the potential $V(x)$. The classical Hamiltonian H has the form

$$H = \frac{p^2}{2m} + V(x), \qquad (6.1.76)$$

which, replacing V by $\bar{V}^2/2m$, becomes

$$H = \frac{1}{2m}(p^2 + \bar{V}^2)$$
$$= \frac{1}{2m}\{(p + \mathrm{i}\bar{V})(p - \mathrm{i}\bar{V}) - \mathrm{i}[\bar{V}, p]\}. \tag{6.1.77}$$

Clearly, the commutator term in this last equation vanishes since both $\bar{V}$ and p are scalars. However, if we now replace the scalar p by the operator $\hat{p}$ where $\hat{p}$ satisfies the commutator expression (6.1.19), then we see that this term must be retained, i.e. we have

$$\hat{H} = \frac{1}{2m}\{(\hat{P} + \mathrm{i}\bar{V})(\hat{p} - \mathrm{i}\bar{V}) - \mathrm{i}[\bar{V}, \hat{p}]\}. \tag{6.1.78}$$

We now expand $\bar{V}(x)$ in the Taylor series

$$\bar{V}(x) = \sum_{n=0}^{\infty} a_n x^n, \tag{6.1.79}$$

so that

$$[\bar{V}, \hat{p}] = \left[\sum_{n=0}^{\infty} a_n x^n, \hat{p}\right]$$
$$= \sum_{n=0}^{\infty} a_n [x^n, \hat{p}]. \tag{6.1.80}$$

Now

$$[x, \hat{p}] = \mathrm{i}\hbar, \tag{6.1.81}$$

so that

$$[x^2, \hat{p}] = x^2\hat{p} - \hat{p}x^2$$
$$= x(x\hat{p} - \hat{p}x) + (x\hat{p} - \hat{p}x)x$$
$$= 2x\mathrm{i}\hbar. \tag{6.1.82}$$

It is not difficult to repeat this process for $[x^3, \hat{p}]$, etc. to find the general result

$$[x^n, \hat{p}] = nx^{n-1}\mathrm{i}\hbar. \tag{6.1.83}$$

Equation (6.1.80) then yields

$$[\bar{V}, \hat{p}] = \mathrm{i}\hbar \sum_{n=0}^{\infty} na_n x^{n-1}$$
$$= \mathrm{i}\hbar \frac{\mathrm{d}\bar{V}}{\mathrm{d}x} \equiv \mathrm{i}\hbar \bar{V}'. \tag{6.1.84}$$

Thus, defining the operator $\hat{A}$ and its adjoint $\hat{A}^+$ by

$$\hat{A} = \hat{p} - \mathrm{i}\bar{V}, \quad \hat{A}^+ = \hat{p} + \mathrm{i}\bar{V}, \tag{6.1.85}$$

equation (6.1.78) becomes

$$\hat{H} = \frac{1}{2m}(\hat{A}^{+}\hat{A} + \hbar\bar{V}'). \tag{6.1.86}$$

The thrust behind the method leading to the quantisation of observable states lies in the construction of commutator relationships for $\hat{A}$ and $\hat{A}^{+}$. We readily find

$$\hat{A}^{+}\hat{A} = \hat{p}^2 + \bar{V}^2 + \mathrm{i}[\bar{V}, \hat{p}], \tag{6.1.87}$$

and

$$\hat{A}\hat{A}^{+} = \hat{p}^2 + \bar{V}^2 - \mathrm{i}[\bar{V}, \hat{p}], \tag{6.1.88}$$

so that

$$\begin{aligned}[\hat{A}, \hat{A}^{+}] = \hat{A}\hat{A}^{+} - \hat{A}^{+}\hat{A} &= -2\mathrm{i}[\bar{V}, \hat{p}], \\ &= 2\hbar\bar{V}',\end{aligned} \tag{6.1.89}$$

using equation (6.1.84). This last result may be put into a more convenient form by defining the new operator $\hat{B}$ such that

$$\hat{A} = \sqrt{(2\hbar\bar{V}')}\hat{B}, \tag{6.1.90}$$

yielding

$$[\hat{B}, \hat{B}^{+}] = \mathbb{1}, \tag{6.1.91}$$

where $\mathbb{1}$ is the identity matrix (unit operator) and, from equation (6.1.86),

$$\hat{H} = \frac{\hbar\bar{V}'}{m}(\hat{B}^{+}\hat{B} + \tfrac{1}{2}\mathbb{1}). \tag{6.1.92}$$

We now consider the time-independent Schrödinger wave equation

$$\hat{H}|n\rangle = E_n|n\rangle, \tag{6.1.93}$$

by defining the operator $\hat{R} = \hat{B}^{+}\hat{B}$ acting on the vector $|n\rangle$ such that

$$\hat{R}|n\rangle = \Theta_n|n\rangle, \tag{6.1.94}$$

where Θ_n is the eigenvalue of $|n\rangle$ corresponding to the operator $\hat{R}$. We proceed by noting that

$$\begin{aligned}\hat{R}\hat{B}|n\rangle &= \hat{B}^{+}\hat{B}\hat{B}|n\rangle \\ &= (\hat{B}\hat{B}^{+} - \mathbb{1})\hat{B}|n\rangle \\ &= \hat{B}\hat{R}|n\rangle - \hat{B}|n\rangle \\ &= (\Theta_n - 1)\hat{B}|n\rangle,\end{aligned} \tag{6.1.95}$$

whilst, similarly,

$$\hat{R}\hat{B}^{+}|n\rangle = (\Theta_n + 1)\hat{B}^{+}|n\rangle. \tag{6.1.96}$$

Reference to equation (6.1.95) reveals that $\hat{B}|n\rangle$ is an eigenvector of the operator $\hat{R}$ with eigenvalue $\Theta_n - 1$. We therefore represent it by

$$\hat{B}|n\rangle = c_n^{-}|\xi_1\rangle, \tag{6.1.97}$$

where $|\xi_1\rangle$ is an orthonormal vector satisfying

$$\langle\xi_1|\xi_1\rangle = 1, \tag{6.1.98}$$

and c_n^- is a complex proportionality scalar. Equation (6.1.95) then has the form

$$\hat{R}|\xi_1\rangle = (\Theta_n - 1)|\xi_1\rangle, \tag{6.1.99}$$

which is precisely the form of the defining equation (6.1.94).

We now wish to place restrictions on the values Θ_n may take. To do this, we write the adjoint of equation (6.1.97), viz.

$$\langle n|\hat{B}^+ = \langle\xi_1|c_n^{-*},$$

so that

$$\langle n|\hat{B}^+\hat{B}|n\rangle = \langle\xi_1|c_n^{-*}c_n^-|\xi_1\rangle = |c_n^-|^2\langle\xi_1|\xi_1\rangle$$
$$= |c_n^-|^2. \tag{6.1.100}$$

However,

$$\langle n|\hat{B}^+\hat{B}|n\rangle = \langle n|\hat{R}|n\rangle = \Theta_n, \tag{6.1.101}$$

so that

$$\Theta_n = |c_n^-|^2, \tag{6.1.102}$$

is both real and non-negative.

We now repeat the previous manipulations but on $|\xi_1\rangle$ rather than $|n\rangle$. We see that

$$\hat{R}\hat{B}|\xi_1\rangle = \hat{B}\hat{R}|\xi_1\rangle - \hat{B}|\xi_1\rangle$$
$$= (\Theta_n - 2)\hat{B}|\xi_1\rangle, \tag{6.1.103}$$

where we have used equation (6.1.99). Thus, we may construct a second eigenvector $|\xi_2\rangle$ such that

$$\hat{R}|\xi_2\rangle = (\Theta_n - 2)|\xi_2\rangle. \tag{6.1.104}$$

Repeated application of this process yields

$$\hat{R}|\xi_j\rangle = (\Theta_n - j)|\xi_j\rangle. \tag{6.1.105}$$

This last equation is structurally equivalent to equation (6.1.94) but with the eigenvalue $\Theta_n - j$. We have shown above, however, that all the eigenvalues are non-negative and thus, if the integer j in equation (6.1.105) is greater than Θ_n, then this condition would be violated. Note that the integer j is not restricted. Hence $\Theta_n - j$ must vanish for some value of j so that the repeated application of $\hat{R}$ on the vector $|n\rangle$ beyond the jth operation produces a null set. Consequently, Θ_n must be an integer. We thence put $\Theta_n = n$, so that equation (6.1.99) yields

$$\hat{R}|\xi_1\rangle = (n-1)|\xi_1\rangle. \tag{6.1.106}$$

However, equation (6.1.94) with n replaced by $n-1$ is

$$\hat{R}|n-1\rangle = (n-1)|n-1\rangle, \tag{6.1.107}$$

so that $|\xi_1\rangle \equiv |n-1\rangle$. Equations (6.1.97) and (6.1.102) then yield

$$\hat{B}|n\rangle = \sqrt{n}\,|n-1\rangle, \tag{6.1.108}$$

where we have ignored any complex component of c_n^- (it cancels out when we multiply by its complex conjugate).

Equation (6.1.96) becomes

$$\hat{R}\hat{B}^+|n\rangle = (n+1)\hat{B}^+|n\rangle, \tag{6.1.109}$$

so that, if we write

$$\hat{B}^+|n\rangle = c_n^+|\eta_1\rangle, \tag{6.1.110}$$

where $\langle\eta_1|\eta_1\rangle = 1$, we have

$$\hat{R}|\eta_1\rangle = (n+1)|\eta_1\rangle,$$

i.e.

$$|\eta_1\rangle \equiv |n+1\rangle. \tag{6.1.111}$$

It is therefore not difficult to obtain c_n^+:

$$\hat{B}^+|n\rangle = \sqrt{(n+1)}|n+1\rangle. \tag{6.1.112}$$

The purpose behind the foregoing discussion has now been realised. The eigenvalue $\Theta_n = n$ and equations (6.1.92-94) yield

$$\begin{aligned}\hat{H}|n\rangle &= \frac{\hbar\bar{V}'}{m}(\hat{R}+\tfrac{1}{2})|n\rangle \\ &= \frac{\hbar\bar{V}'}{m}(\Theta_n+\tfrac{1}{2})|n\rangle \\ &= \frac{\hbar\bar{V}'}{m}(n+\tfrac{1}{2})|n\rangle.\end{aligned} \tag{6.1.113}$$

However,

$$\hat{H}|n\rangle = E_n|n\rangle, \tag{6.1.114}$$

so that

$$E_n = \frac{\hbar\bar{V}'}{m}(n+\tfrac{1}{2}), \tag{6.1.115}$$

has only discrete values described by the integer *n*. This then completes the discussion of the quantisation of the particle system. Note that the above analysis may be generalised for multi-dimensional force fields.

6.2 Quantisation of the radiation field

We have seen that the energy of a particle system assumes only discrete values and that, in particular, the bound electron of an atom may only exist in certain prescribed quantised energy levels (relative to the atom). Consequently, the emission and absorption of radiation, which takes place via atom–photon interaction, can only occur for photons having the appropriate quantum of energy. If we are to proceed, therefore, in the examination of the quantum wave mechanical treatment of the transfer of radiation through a particle system, it is essential to develop a theory of radiation which incorporates the quantisation of the radiation energy.

Again, we refer the reader to the many excellent texts [49, 83, 103] on this subject. Here, we merely summarise those arguments which enable us to understand the derivations presented in the later sections.

6.2.1 *Classical equations for the electromagnetic field*

Maxwell's equations usually form the starting point for any discussion of the classical electromagnetic field. They may be written in the form

$$\nabla \times \mathbf{E} = -\frac{1}{c}\frac{\partial \mathbf{B}}{\partial t}, \tag{6.2.1}$$

$$\nabla \times \mathbf{B} = \frac{4\pi}{c}\,\mathbf{j} + \frac{1}{c}\frac{\partial \mathbf{E}}{\partial t}, \tag{6.2.2}$$

$$\nabla \cdot \mathbf{B} = 0, \tag{6.2.3}$$

and

$$\nabla \cdot \mathbf{E} = 4\pi\rho, \tag{6.2.4}$$

where the electric and magnetic fields $\mathbf{E}(\mathbf{r}, t)$ and $\mathbf{B}(\mathbf{r}, t)$ are functions of position $\mathbf{r}$ and time t. Conservation of charge stipulates that

$$\frac{\partial \rho}{\partial t} + \nabla \cdot \mathbf{j} = 0, \tag{6.2.5}$$

where ρ is the charge density and $\mathbf{j}$ the electric current density, i.e.

$$\rho = \sum_i q_i \delta(\mathbf{r} - \mathbf{r}_i), \tag{6.2.6}$$

and

$$\mathbf{j} = \sum_i q_i \mathbf{v}_i \delta(\mathbf{r} - \mathbf{r}_i), \tag{6.2.7}$$

where q_i is the charge on the ith particle and $\mathbf{v}_i$ is that particle's velocity.

One can derive the vector wave equations for $\mathbf{E}$ and $\mathbf{B}$ from equations (6.2.1) and (6.2.2), viz.

$$\nabla \times (\nabla \times \mathbf{E}) + \frac{1}{c^2}\frac{\partial \mathbf{E}}{\partial t} = -\frac{4\pi}{c^2}\frac{\partial}{\partial t}(\nabla \times \mathbf{j}), \tag{6.2.8}$$

and

$$\nabla \times (\nabla \times \mathbf{B}) + \frac{1}{c^2}\frac{\partial \mathbf{B}}{\partial t} = \frac{4\pi}{c}\nabla \times \mathbf{j}. \tag{6.2.9}$$

These last two equations may be reduced by noting that $\nabla \cdot (\nabla \times \mathbf{f}) = 0$, where $\mathbf{f}(\mathbf{r})$ is any differentiable function of $\mathbf{r}$. Equation (6.2.3) then enables a vector potential $\mathbf{A}$ to be defined so that

$$\mathbf{B} = \nabla \times \mathbf{A}. \tag{6.2.10}$$

Equation (6.2.1) then becomes

$$\nabla \times \left(\mathbf{E} + \frac{1}{c} \frac{\partial \mathbf{A}}{\partial t} \right) = 0, \tag{6.2.11}$$

which, noting that $\nabla \times (\nabla g) = 0$ where $g(\mathbf{r})$ is any differentiable function of $\mathbf{r}$, yields the result

$$\mathbf{E} = -\frac{1}{c} \frac{\partial \mathbf{A}}{\partial t} - \nabla \phi, \tag{6.2.12}$$

where ϕ is a scalar potential. Thus, once the three components of $\mathbf{A}$, and the scalar ϕ, are known, the six components of $\mathbf{E}$ and $\mathbf{B}$ may be immediately determined from equations (6.2.10) and (6.2.12). However, substitution of equation (6.2.12) into (6.2.4) shows that $\mathbf{A}$ and ϕ are not independent, i.e. we have

$$\frac{1}{c} \frac{\partial}{\partial t} (\nabla \cdot \mathbf{A}) + \nabla^2 \phi = -4\pi\rho. \tag{6.2.13}$$

Another equation in $\mathbf{A}$ and ϕ may be obtained by substituting equation (6.2.12) into (6.2.2), i.e.

$$\nabla \times (\nabla \times \mathbf{A}) = \frac{4\pi}{c} \mathbf{j} - \frac{1}{c^2} \frac{\partial^2 \mathbf{A}}{\partial t^2} - \frac{1}{c} \frac{\partial}{\partial t} (\nabla \phi). \tag{6.2.14}$$

This last result may be simplified by noting the general vector identity

$$\nabla \times (\nabla \times \mathbf{A}) = \nabla(\nabla \cdot \mathbf{A}) - (\nabla \cdot \nabla)\mathbf{A}. \tag{6.2.15}$$

Equation (6.2.14) then becomes

$$\nabla^2 \mathbf{A} - \nabla(\nabla \cdot \mathbf{A}) = \frac{1}{c^2} \frac{\partial^2 \mathbf{A}}{\partial t^2} - \frac{4\pi}{c} \mathbf{j} + \frac{1}{c} \frac{\partial}{\partial t} (\nabla \phi). \tag{6.2.16}$$

Clearly, there exists no unique value of $\mathbf{A}$ and ϕ which enable $\mathbf{E}$ and $\mathbf{B}$ to satisfy equations (6.2.3) and (6.2.11). For example, if we replace $\mathbf{A}$ by $\mathbf{A} + \nabla\eta$ where η is any differentiable scalar function, and noting that $\nabla \times \nabla\eta = 0$, then equation (6.2.3) is still satisfied. (In this case, however, we must similarly replace ϕ by $\phi - \dot{\eta}/c$ if $\mathbf{E}$ given by equation (6.2.12) is to remain unchanged.) Thus, we are free to place an extra constraint on the four quantities $\mathbf{A} = (A_x, A_y, A_z)$ and ϕ. Such constraints are called gauges – the occurrence of $\nabla \cdot \mathbf{A}$ in both equations (6.2.13) and (6.2.16) suggests that the coulomb gauge

$$\nabla \cdot \mathbf{A} = 0 \tag{6.2.17}$$

would be most convenient. Equation (6.2.13) then becomes

$$\nabla^2 \phi = -4\pi\rho, \tag{6.2.18}$$

which has the solution

$$\phi(\mathbf{r}, t) = \int \frac{\rho(\mathbf{r}', t)\, d\mathbf{r}'}{|\mathbf{r} - \mathbf{r}'|}. \tag{6.2.19}$$

Equation (6.2.16) yields

$$\nabla^2 \mathbf{A} = \frac{1}{c^2}\frac{\partial^2 \mathbf{A}}{\partial t^2} - \frac{4\pi}{c}\mathbf{j} + \frac{1}{c}\frac{\partial}{\partial t}(\nabla\phi), \qquad (6.2.20)$$

which, using the charge conservation equation (6.2.5) and equation (6.2.18), becomes the wave equation

$$\nabla^2 \mathbf{A} = \frac{1}{c^2}\frac{\partial^2 \mathbf{A}}{\partial t^2}. \qquad (6.2.21)$$

This last equation may be solved using a Fourier expansion technique. For example, if we consider equation (6.2.21) in cartesian coordinates **r**, and separate out the **r** and t dependence of **A**, we obtain a general solution for **A** consisting of the linear super-position of wave modes in the form

$$\mathbf{A}(\mathbf{r}, t) = \sum_{\gamma} a_\gamma(t)\mathbf{u}_\gamma(t), \qquad (6.2.22)$$

where

$$\nabla^2 \mathbf{u}_\gamma(\mathbf{r}) = -k_\gamma^2 \mathbf{u}_\gamma(\mathbf{r}), \qquad (6.2.23)$$

and

$$\frac{1}{c^2}\frac{\mathrm{d}^2 a_\gamma(t)}{\mathrm{d}t^2} = -k_\gamma^2 a_\gamma(t); \qquad (6.2.24)$$

k_γ^2 is the 'separation constant'. The $\mathbf{u}_\gamma(r)$ and $a_\gamma(t)$ then have the form

$$\mathbf{u}_\gamma(r) = \alpha_\gamma \mathbf{e}_\gamma \mathrm{e}^{\pm \mathrm{i}\mathbf{k}_\gamma\cdot\mathbf{r}}, \qquad (6.2.25)$$

and

$$a_\gamma(t) = a_\gamma(0)\mathrm{e}^{\pm \mathrm{i}\omega_\gamma t}, \qquad (6.2.26)$$

where

$$\omega_\gamma = ck_\gamma, \quad k_\gamma = |\mathbf{k}_\gamma|. \qquad (6.2.27)$$

The unit vector $\mathbf{e}_\gamma$ is yet to be determined. The wave vector $\mathbf{k}_\gamma$ denotes the direction of propagation of the γth mode of the wave.

If we take the electromagnetic field to be confined to a volume $\mathscr{V}$ where $x, y, z \in [0, l]$, i.e. $\mathscr{V} = l^3$, and we choose (for later convenience) the constant α_γ to be

$$\alpha_\gamma = \frac{1}{l^{3/2}}\left(\frac{2\pi\hbar c}{k_\gamma}\right)^{1/2}, \qquad (6.2.28)$$

we find the orthogonality relationships

$$\int_{\mathscr{V}} \mathbf{u}_\gamma^* \cdot \mathbf{u}_{\gamma'}\,\mathrm{d}^3\mathbf{r} = \frac{2\pi\hbar c}{k_\gamma}\,\delta_{\gamma\gamma'} \qquad (6.2.29)$$

Further, boundary conditions at the walls of the cavity constrain the possible values of $\mathbf{k}_\gamma$ to be

$$\mathbf{k}_\gamma = \frac{2\pi}{l}(n_x, n_y, n_z), \tag{6.2.30}$$

where the n_x, n_y and n_z are zero or integers. Clearly, therefore, the subscript γ is a composite term referring to the selection of integers n_x, n_y and n_z appearing in $\mathbf{k}_\gamma$, i.e. $\gamma = \gamma(n_x, n_y, n_z)$.

The unit vector $\mathbf{e}_\gamma$ must be so chosen that the gauge $\nabla \cdot \mathbf{A} = 0$ (equation (6.2.17)) is satisfied, i.e. we must have

$$\mathbf{e}_\gamma \cdot \mathbf{k}_\gamma = 0. \tag{6.2.31}$$

It is clear from equations (6.2.12), (6.2.22) and (6.2.25) that the electric vector $\mathbf{E}$ is in the direction of $\mathbf{e}_\gamma$ - this is defined to be the direction of polarisation - and thus, from equation (6.2.31), the coulomb gauge in cartesian coordinates restricts us to the study of plane waves perpendicular (i.e. transverse) to the electric field. It is common practice to define $\mathbf{e}_\gamma = (\mathrm{e}_{\gamma 1}, \mathrm{e}_{\gamma 2}, \mathrm{e}_{\gamma 3})$ such that $\mathbf{e}_{\gamma 3}$ is in the direction of $\mathbf{k}_\gamma$. Equation (6.2.31) then stipulates that $\mathrm{e}_{\gamma 3} \equiv 0$ with $\mathbf{e}_{\gamma 1}$ and $\mathbf{e}_{\gamma 2}$ perpendicular to one another, i.e. $\mathbf{e}_{\gamma 1} \cdot \mathbf{e}_{\gamma 2} = 0$, and perpendicular to $\mathbf{k}_\gamma$. Figure 6.1 shows the chosen configuration for $\mathbf{e}_{\gamma 1}$, $\mathbf{e}_{\gamma 2}$ and $\mathbf{k}_\gamma$.

We then have

$$\mathbf{A}(\mathbf{r}, t) = \sum_{\gamma\lambda} \alpha_\gamma \, a_{\gamma\lambda}(0) \mathbf{e}_{\gamma\lambda} \mathrm{e}^{\mathrm{i}(k_\gamma \cdot \mathbf{r} - \omega_\gamma t)}, \tag{6.2.32}$$

i.e.

$$\mathbf{u}_{\gamma\lambda} = \alpha_\gamma \mathbf{e}_{\gamma\lambda} \mathrm{e}^{\mathrm{i}\mathbf{k}_\gamma \cdot \mathbf{r}}, \tag{6.2.33}$$

where $\lambda = 1$ or 2, with the orthogonality property

$$\int_{\mathcal{V}} \mathbf{u}^*_{\gamma'\lambda'} \cdot \mathbf{u}_{\gamma\lambda} \, \mathrm{d}^3\mathbf{r} = \frac{2\pi\hbar c}{k} \delta_{\gamma\gamma'} \delta_{\lambda\lambda'}. \tag{6.2.34}$$

However, we are interested only in purely real $\mathbf{E}$, and thus we choose to re-write $\mathbf{A}$ in the form

$$\mathbf{A}(\mathbf{r}, t) = \sum_{\gamma\lambda} [a_{\gamma\lambda}(0) \mathbf{e}_{\gamma\lambda} \mathrm{e}^{\mathrm{i}(\mathbf{k}_\gamma \cdot \mathbf{r} - \omega_\gamma t)} + a^*_{\gamma\lambda}(0) \mathbf{e}_{\gamma\lambda} \mathrm{e}^{-\mathrm{i}(\mathbf{k}_\gamma \cdot \mathbf{r} - \omega_\gamma t)}], \tag{6.2.35}$$

Fig.6.1. The standard definition of the unit vectors.

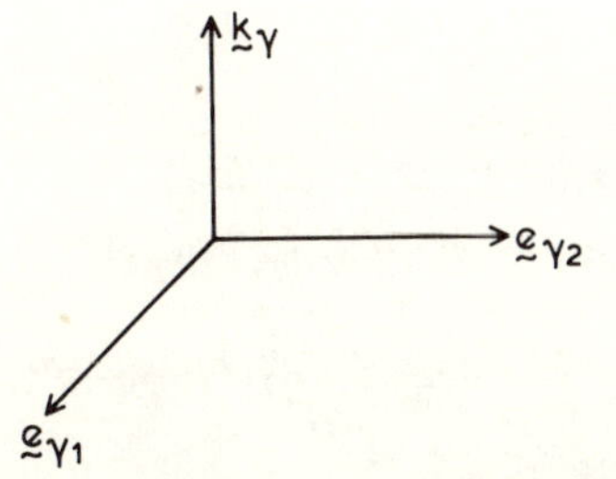

i.e.

$$\mathbf{A}(\mathbf{r}, t) = \sum_{\gamma\lambda} [a_{\gamma\lambda}(t)\mathbf{u}_{\gamma\lambda}(\mathbf{r}) + a^*_{\gamma\lambda}(t)\mathbf{u}^*_{\gamma\lambda}(\mathbf{r})]. \tag{6.2.36}$$

We may now use equations (6.2.10) and (6.2.12) to determine **B** and **E**, viz.

$$\begin{aligned}\mathbf{B} &= \sum_{\gamma\lambda} [a_{\gamma\lambda}\nabla \times \mathbf{u}_{\gamma\lambda} + a^*_{\gamma\lambda}\nabla \times \mathbf{u}^*_{\gamma\lambda}] \\ &= \mathrm{i}\sum_{\gamma\lambda} [a_{\gamma\lambda}\mathbf{k}_\gamma \times \mathbf{u}_{\gamma\lambda} - a^*_{\gamma\lambda}\mathbf{k}_\gamma \times \mathbf{u}^*_{\gamma\lambda}],\end{aligned} \tag{6.2.37}$$

and

$$\mathbf{E} = -\mathrm{i}\sum_{\gamma\lambda} \frac{\omega_\gamma}{c} (a_{\gamma\lambda}\mathbf{u}_{\gamma\lambda} - a^*_{\gamma\lambda}\mathbf{u}^*_{\gamma\lambda}) - \nabla\phi, \tag{6.2.38}$$

where ϕ is given by equation (6.2.19).

Equation (6.2.34) may be simplified further by noting from figure 6.1 that

$$\mathbf{k}_\gamma \times \mathbf{e}_{\gamma 1} = k_\gamma \mathbf{e}_{\gamma 2}, \quad \mathbf{k}_\gamma \times \mathbf{e}_{\gamma 2} = -k_\gamma \mathbf{e}_{\gamma 1}. \tag{6.2.39}$$

We then find

$$\mathbf{B} = \mathrm{i}\sum_{\gamma} k_\gamma [a_{\gamma 1}\mathbf{u}_{\gamma 2} - a_{\gamma 2}\mathbf{u}_{\gamma 1} - a^*_{\gamma 1}\mathbf{u}^*_{\gamma 2} + a^*_{\gamma 2}\mathbf{u}^*_{\gamma 1}]. \tag{6.2.40}$$

We are now in a position to determine the Hamiltonian of the field. It is convenient to first consider the free field Hamiltonian in which the charge density ρ, thence potential ϕ, is equated to zero. The Hamiltonian then becomes the energy of the field:

$$H = \frac{1}{8\pi}\int_{\mathscr{V}} (\mathbf{E}\cdot\mathbf{E} + \mathbf{B}\cdot\mathbf{B})\,\mathrm{d}^3\mathbf{r}. \tag{6.2.41}$$

The term in the above integrand may be readily evaluated so that, with the aid of the orthogonality relationships (6.2.34), we have

$$\begin{aligned}\int_{\mathscr{V}} \mathbf{E}\cdot\mathbf{E}\,\mathrm{d}^3\mathbf{r} &= -\int_{\mathscr{V}} \mathrm{d}^3\mathbf{r} \sum_{\gamma\lambda}\sum_{\gamma'\lambda'} \frac{\omega_\gamma\omega_{\gamma'}}{c^2} (a_{\gamma\lambda}\mathbf{u}_{\gamma\lambda} - a^*_{\gamma\lambda}\mathbf{u}_{\gamma\lambda}) \\ &\quad \times (a_{\gamma'\lambda'}\mathbf{u}_{\gamma'\lambda'} - a^*_{\gamma'\lambda'}\mathbf{u}_{\gamma'\lambda'}) \\ &= \sum_{\gamma\lambda} \left(\frac{\omega_\gamma}{c}\right)^2 \frac{2\pi\hbar c}{k_\gamma} [a_{\gamma\lambda}a^*_{\gamma\lambda} + a^*_{\gamma\lambda}a_{\gamma\lambda}],\end{aligned} \tag{6.2.42}$$

and

$$\begin{aligned}\int_{\mathscr{V}} \mathbf{B}\cdot\mathbf{B}\,\mathrm{d}^3\mathbf{r} &= -\int_{\mathscr{V}} \mathrm{d}^3\mathbf{r} \sum_{\gamma\gamma'} k_\gamma k_{\gamma'} \\ &\quad \times [a_{\gamma 1}\mathbf{u}_{\gamma 2} - a_{\gamma 2}\mathbf{u}_{\gamma 1} - a^*_{\gamma 1}\mathbf{u}^*_{\gamma 2} + a^*_{\gamma 2}\mathbf{u}^*_{\gamma 1}] \\ &\quad \times [a_{\gamma' 1}\mathbf{u}_{\gamma' 2} - a_{\gamma' 2}\mathbf{u}_{\gamma' 1} - a^*_{\gamma' 1}\mathbf{u}^*_{\gamma' 2} + a^*_{\gamma' 2}\mathbf{u}^*_{\gamma' 1}] \\ &= \sum_{\gamma} k_\gamma^2 \frac{2\pi\hbar c}{k_\gamma} [a_{\gamma 1}a^*_{\gamma 1} + a_{\gamma 2}a^*_{\gamma 2} + a^*_{\gamma 1}a_{\gamma 1} + a^*_{\gamma 2}a_{\gamma 2}] \\ &= \sum_{\gamma\lambda} 2\pi\hbar c k_\gamma [a_{\gamma\lambda}a^*_{\gamma\lambda} + a^*_{\gamma\lambda}a_{\gamma\lambda}].\end{aligned} \tag{6.2.43}$$

We then have the classical free field Hamiltonian in the form (recall that $\omega_\gamma = k_\gamma c$)

$$H = \tfrac{1}{2} \sum_{\gamma\lambda} \hbar\omega_\gamma [a_{\gamma\lambda} a^*_{\gamma\lambda} + a^*_{\gamma\lambda} a_{\gamma\lambda}]. \tag{6.2.44}$$

We now wish to quantise the radiation field using this last result.

6.2.2 *Second quantisation*

We saw in section 6.1.4 that the energy of a particle system can be quantised by constructing the appropriate commutation relationship between an operator and its adjoint (see equations (6.1.91) and (6.1.92)). We follow the same procedure here and replace the expansion coefficients $a_{\gamma\lambda}$ and $a^*_{\gamma\lambda}$ by the operators $\hat{a}_{\gamma\lambda}$ and its adjoint $\hat{a}^+_{\gamma\lambda}$. We further postulate the commutation relationship

$$[\hat{a}_{\gamma\lambda}, \hat{a}^+_{\gamma'\lambda'}] = \mathbb{1}\delta_{\gamma\gamma'}\delta_{\lambda\lambda'}, \tag{6.2.45}$$

this last step commonly being referred to as a second quantisation. However, it is not quite as *ad hoc* as it might appear. For example, the classical Hamiltonian (6.2.44) may be compared with that for the harmonic oscillator, viz.

$$H = \tfrac{1}{2}(p^2 + \omega^2 q^2) \tag{6.2.46}$$

(where, for convenience, we consider only one mode so that all subscripts may be dropped), if we make the transformation

$$a = \frac{1}{\sqrt{(2\hbar\omega)}} (p - \mathrm{i}\omega q), \quad a^* = \frac{1}{\sqrt{(2\hbar\omega)}} (p + \mathrm{i}\omega q). \tag{6.2.47}$$

Thus, if we now replace a and a^* by the operators $\hat{a}$ and $\hat{a}^+$ where p is replaced by $\hat{p}$, and use the prescription (6.1.19), viz.

$$[x_k, \hat{p}_j] = \hbar\mathrm{i}\mathbb{1}\delta_{jk}, \tag{6.2.48}$$

then

$$\begin{aligned}[\hat{a}, \hat{a}^+] &= \frac{1}{2\hbar\omega} [\hat{p} - \mathrm{i}\omega q, \hat{p} + \mathrm{i}\omega q] \\ &= -\frac{\mathrm{i}}{\hbar} [q, \hat{p}] \\ &= \mathbb{1}.\end{aligned}$$

Consequently, second quantisation is not an entirely separate postulate from first quantisation.

The quantum mechanical Hamiltonian operator $\hat{H}$ for the free field then becomes

$$\hat{H} = \tfrac{1}{2} \sum_{\gamma\lambda} \hbar\omega_\gamma [\hat{a}_{\gamma\lambda} \hat{a}^+_{\gamma\lambda} + \hat{a}^+_{\gamma\lambda} \hat{a}_{\gamma\lambda}], \tag{6.2.49}$$

which, using equation (6.2.45), is

$$\hat{H}=\sum_{\gamma\lambda}\hbar\omega_\gamma[\hat{a}^+_{\gamma\lambda}\hat{a}_{\gamma\lambda}+\tfrac{1}{2}\mathbb{1}]. \tag{6.2.50}$$

This last equation is of the same form as equation (6.1.92). We therefore proceed in a manner similar to that developed in section 6.1.4 by defining the operator $\hat{R}_{\gamma\lambda}$ by

$$\hat{R}_{\gamma\lambda}=\hat{a}^+_{\gamma\lambda}\hat{a}_{\gamma\lambda}, \tag{6.2.51}$$

so that Schrödinger's time-independent wave equation

$$\hat{H}|n\rangle=E_n|n\rangle, \tag{6.2.52}$$

has the form

$$\sum_{\gamma\lambda}\hbar\omega_\gamma(\hat{R}_{\gamma\lambda}+\tfrac{1}{2}\mathbb{1})|n\rangle=E_n|n\rangle. \tag{6.2.53}$$

The vector $|n\rangle$ in this last equation also satisfies

$$\hat{R}_{\gamma\lambda}|n\rangle=\Theta_{n\gamma\lambda}|n\rangle, \tag{6.2.54}$$

for all γ and λ so that the eigenvalues $\Theta_{n\gamma\lambda}$ of $\hat{R}_{\gamma\lambda}$, when appropriately summed, yield

$$E_n=\sum_{\gamma\lambda}\hbar\omega_\gamma(\Theta_{n\gamma\lambda}+\tfrac{1}{2}). \tag{6.2.55}$$

As stated previously, the existence of only discrete energy levels for an electron bound to an atom requires a corresponding existence of discrete energy levels for the radiation field. We therefore wish to show that the commutation relationship (6.2.45) which, in fact, led to equations (6.2.53) and (6.2.54), yields only integer values for $\Theta_{n\gamma\lambda}$.

We begin by defining the new vector $|\xi\rangle$ satisfying

$$c_n^-|\xi\rangle=\hat{a}|n\rangle, \tag{6.2.56}$$

where c_n^- is a complex constant and $\langle\xi|\eta\rangle=1$. We will drop all $\gamma\lambda$ subscripts for the moment. If we now operate on this last equation with $\hat{R}$, we obtain

$$\begin{aligned}c_n^-\hat{R}|\xi\rangle&=\hat{R}\hat{a}|n\rangle\\&=\hat{a}^+\hat{a}\hat{a}|n\rangle\\&=(\hat{a}\hat{a}^+-\mathbb{1})\hat{a}|n\rangle\\&=\hat{a}\hat{R}|n\rangle-\hat{a}|n\rangle\\&=(\Theta_n-1)\hat{a}|n\rangle\\&=(\Theta_n-1)c_n^-|\xi\rangle,\end{aligned}$$

i.e.

$$\hat{R}|\xi\rangle=(\Theta_n-1)|\xi\rangle. \tag{6.2.57}$$

Consequently, Θ_n-1 is also an eigenvalue of $\hat{R}$. The adjoint of equation (6.2.56) is

$$\langle\xi|c_n^{-*}=\langle n|\hat{a}^+, \tag{6.2.58}$$

so that

$$\langle \xi | c_n^{-*} c_n^- | \xi \rangle = \langle n | \hat{a}^+ \hat{a} | n \rangle,$$

which yields

$$\Theta_n = |c_n^-|^2 \geqslant 0. \tag{6.2.59}$$

This then shows that the eigenvalues of $\hat{R}$ are all real and non-negative.

If we next perform the manipulation

$$\begin{aligned} \hat{R}\hat{a} | \xi \rangle &= \hat{a}\hat{R} | \xi \rangle - \hat{a} | \xi \rangle \\ &= (\Theta_n - 2)\hat{a} | \xi \rangle, \end{aligned}$$

i.e.

$$\hat{R}(\hat{a})^2 | n \rangle = (\Theta_n - 2)(a)^2 | n \rangle, \tag{6.2.60}$$

we see that $\Theta_n - 2$ is also an eigenvalue of $\hat{R}$. We may continue this process;

$$\hat{R}(\hat{a})^m | n \rangle = (\Theta_n - m)(\hat{a})^m | n \rangle, \tag{6.2.61}$$

where, of course, m is an integer. Thus, successive application of the operator $\hat{a}$, which reduces the eigenvalue of $\hat{R}$ by an integer, will eventually produce a negative value of this eigenvalue when $m > \Theta_n$. However, we have shown that all eigenvalues of $\hat{R}$ must be real and non-negative, and this will only occur if Θ_n is itself an integer. For example, when $\Theta_n = m$, the application of the operator $\hat{a}$ reduces the eigenvalue to zero, thus producing a null vector from which no new state or vector (with negative eigenvalue) can be obtained, i.e.

$$\hat{R}(a)^{m+1} | n \rangle = 0.$$

We therefore write $\Theta_n = n$ where n is now an integer. Equation (6.2.57) then yields

$$\hat{R} | \xi \rangle = (n - 1) | \xi \rangle, \tag{6.2.62}$$

which, when compared with equation (6.2.54) with Θ_n (or n) replaced by $n - 1$, viz.

$$\hat{R} | n - 1 \rangle = (n - 1) | n - 1 \rangle, \tag{6.2.63}$$

yields

$$| \xi \rangle \equiv | n - 1 \rangle. \tag{6.2.64}$$

Equations (6.2.56) and (6.2.59) may then be combined to yield

$$\hat{a} | n \rangle = \sqrt{n} | n - 1 \rangle. \tag{6.2.65}$$

Thus, the result of the operator $\hat{a}$ on $|n\rangle$ is to produce a new lower state $|n - 1\rangle$. For this reason, it is called a lowering or destruction operator.

We next examine the effect of the $\hat{a}^+$ operator. We find

$$\begin{aligned} \hat{R}\hat{a}^+ | n \rangle &= \hat{a}^+ \hat{a} \hat{a}^+ | n \rangle \\ &= \hat{a}^+(\hat{a}^+\hat{a} + \mathbb{1}) | n \rangle \\ &= \hat{a}^+(\hat{R} + \mathbb{1}) | n \rangle \\ &= (n + 1)\hat{a}^+ | n \rangle. \end{aligned} \tag{6.2.66}$$

If we compare this last result with equation (6.2.54), but with n replaced by $n+1$, viz.

$$\hat{R}\,|n+1\rangle = (n+1)\,|n+1\rangle, \tag{6.2.67}$$

we see that

$$\hat{a}^{+}|n\rangle = c_n^{+}|n+1\rangle, \tag{6.2.68}$$

where c_n^{+} is a proportionality constant. If we now combine equation (6.2.68) with its adjoint, we have

$$|c_n^{+}|^2 = n+1, \tag{6.2.69}$$

which, apart from an indeterminable phase factor, yields

$$\hat{a}^{+}|n\rangle = \sqrt{(n+1)}\,|n+1\rangle. \tag{6.2.70}$$

Consequently, the operator $\hat{a}^{+}$ acting on the state $|n\rangle$ produces the new higher state $|n+1\rangle$ and, for this reason, is called a raising or creation operator.

Three quite general points are worth noting here. First, since $\Theta_n = n$ is an integer, we see from equation (6.2.55) that the energy of the free field consists of quanta $\hbar\omega_\gamma$, each of these quanta of energy being a photon of radiation. If there are $n_{\gamma\lambda}$ photons therefore of frequency ω_γ, then the total energy of those photons is $n_{\gamma\lambda}\hbar\omega_\gamma$. Indeed, we have

$$\begin{aligned}
\hat{H}\,|n\rangle &= \sum_{\omega\gamma} \hbar\omega_\gamma(\hat{a}^{+}_{\gamma\lambda}\hat{a}_{\gamma\lambda} + \tfrac{1}{2}\mathbb{1})\,|n\rangle \\
&= \sum_{\omega\gamma} \hbar\omega_\gamma(\hat{a}^{+}_{\gamma\lambda}\sqrt{n_{\gamma\lambda}}\,|n-1\rangle + \tfrac{1}{2}|n\rangle) \\
&= \sum_{\omega\gamma} \hbar\omega_\gamma(n_{\gamma\lambda}|n\rangle + \tfrac{1}{2}|n\rangle) \\
&= E_n|n\rangle,
\end{aligned}$$

so that

$$E_n = \sum_{\omega\gamma} \hbar\omega_\gamma(n_{\gamma\lambda} + \tfrac{1}{2}). \tag{6.2.71}$$

The second point relates to the physical interpretation of the $\hat{a}$ and $\hat{a}^{+}$ operators. As we have seen, the $\hat{a}$ operation reduces the $|n\rangle$ state to $|n-1\rangle$, i.e. one photon is removed from the photon radiation field. This corresponds to the absorption of a photon by an atom, the atom gaining the energy $\hbar\omega_\gamma$. The $\hat{a}^{+}$ process is the reverse – the $|n\rangle$ state changes to $|n+1\rangle$, i.e. one more photon is in the system. This then corresponds to emission of a photon with a subsequent loss in energy of the atom.

Thirdly, the eigenvalue for the destruction operator $\hat{a}$ is $\sqrt{n}$, but $\sqrt{(n+1)}$ for $\hat{a}^{+}$. As we shall see in a later section, this implies that the absorption process is proportional to the number n of $\hbar\omega_\gamma$ photons present which, of course, is proportional to the intensity of the radiation field. The emission term contains two components; one proportional to n as above and one proportional to unity.

The first corresponds to stimulated or induced emission (proportional to the specific radiation intensity), the second to spontaneous emission.

6.2.3 The general Hamiltonian

We have already obtained the Hamiltonian for the free radiation field and a free particle. In the present section, we wish to detail the composite Hamiltonian for a system of particles interacting with a radiation field. However, the non-conservative nature of a charged particle in an electromagnetic field introduces certain difficulties. For ease in exposition, therefore, we first consider a single charged particle. The force **F** on that particle is given by

$$\mathbf{F} = e\left(\mathbf{E} + \frac{1}{c}\,\mathbf{v} \times \mathbf{B}\right), \tag{6.2.72}$$

where e is the charge on the particle and **v** its velocity. If, as in section 6.2.1, we define the vector and scalar potentials **A** and ϕ by

$$\mathbf{B} = \nabla \times \mathbf{A}, \tag{6.2.73}$$

and

$$\mathbf{E} = -\frac{1}{c}\frac{\partial \mathbf{A}}{\partial t} - \nabla\phi, \tag{6.2.74}$$

we find

$$\frac{\mathbf{F}}{e} = -\frac{1}{c}\frac{\partial \mathbf{A}}{\partial t} - \nabla\phi + \frac{1}{c}\,\mathbf{v} \times (\nabla \times \mathbf{A}). \tag{6.2.75}$$

The vector identity [86]

$$\mathbf{v} \times (\nabla \times \mathbf{A}) = \mathbf{v} \cdot (\nabla\mathbf{A}) - (\mathbf{v} \cdot \nabla)\mathbf{A}, \tag{6.2.76}$$

together with the Lagrangian derivative

$$\frac{\mathrm{d}\mathbf{A}}{\mathrm{d}t} = \frac{\partial \mathbf{A}}{\partial t} + (\mathbf{v} \cdot \nabla)\mathbf{A},$$

enables **F** to be re-written in the form

$$\begin{aligned} \mathbf{F} &= e\left[-\frac{1}{c}\frac{\mathrm{d}\mathbf{A}}{\mathrm{d}t} - \nabla\phi + \frac{1}{c}\,\mathbf{v} \cdot (\nabla\mathbf{A})\right] \\ &= -\frac{e}{c}\frac{\mathrm{d}\mathbf{A}}{\mathrm{d}t} - e\nabla\left(\phi - \frac{1}{c}\,\mathbf{v} \cdot \mathbf{A}\right) \\ &= \left(\frac{\mathrm{d}}{\mathrm{d}t}\frac{\partial}{\partial \mathbf{v}} - \nabla\right)\left[e - \frac{e}{c}\,\mathbf{v} \cdot \mathbf{A}\right], \end{aligned} \tag{6.2.77}$$

where, consistent with the microscopic description of the gas detailed in section 5.1.1, we have taken **v** and **r** to be independent.

However, the resultant force **F** is just

$$\mathbf{F} = \frac{\mathrm{d}}{\mathrm{d}t}(m\mathbf{v}) = \frac{\mathrm{d}}{\mathrm{d}t}\frac{\partial}{\partial \mathbf{v}}\left(\tfrac{1}{2}mv^2\right), \tag{6.2.78}$$

so that equation (6.2.77) becomes

$$\left(\frac{\mathrm{d}}{\mathrm{d}t}\frac{\partial}{\partial \mathbf{v}} - \nabla\right)\left[\tfrac{1}{2}mv^2 - e\left(\phi - \frac{1}{c}\mathbf{v}\cdot\mathbf{A}\right)\right] = 0. \tag{6.2.79}$$

This last expression is of the Euler–Lagrange form

$$\left(\frac{\mathrm{d}}{\mathrm{d}t}\frac{\partial}{\partial \dot{q}_r} - \frac{\partial}{\partial q_r}\right)L_{\mathrm{p}} = 0, \tag{6.2.80}$$

where the Lagrangian L_{p} for the particle has the form

$$L_{\mathrm{p}} = \tfrac{1}{2}mv^2 - e\left(\phi - \frac{1}{c}\mathbf{v}\cdot\mathbf{A}\right). \tag{6.2.81}$$

The corresponding Hamiltonian H_{p} is given by

$$H_{\mathrm{p}} = \sum_r p_r\dot{q}_r - L_{\mathrm{p}}, \tag{6.2.82}$$

where the generalised momenta $p_r = \partial L/\partial \dot{q}_r$. In vector form, we then have

$$\mathbf{p} \equiv \frac{\partial L_{\mathrm{p}}}{\partial \mathbf{v}} = m\mathbf{v} + \frac{e}{c}\mathbf{A}, \tag{6.2.83}$$

and so

$$\begin{aligned} H_{\mathrm{p}} &= \left(m\mathbf{v} + \frac{e}{c}\mathbf{A}\right)\cdot\mathbf{v} - L_{\mathrm{p}} = \tfrac{1}{2}mv^2 + e\phi \\ &= \frac{1}{2m}\left(\mathbf{p} - \frac{e}{c}\mathbf{A}\right)^2 + e\phi. \end{aligned} \tag{6.2.84}$$

The Hamiltonian has been written in terms of $\mathbf{p}$ rather than $\mathbf{v}$ since it is the generalised momenta p_{r} which are conjugate to the generalised coordinates q_{r} and, as we saw in section 6.1.4, the prescription for conversion from classical to quantum mechanics involves the replacement of p_{r} by $\hat{p}_{\mathrm{r}} = -i\hbar\,\partial/\partial q_r$. Consequently, the quantum mechanical Hamiltonian operator $\hat{H}_{\mathrm{p}}$ for the particle is just

$$\hat{H}_{\mathrm{p}} = \frac{1}{2m}\left(\hat{p} - \frac{e}{c}\hat{A}\right)^2 + e\phi, \tag{6.2.85}$$

where

$$\hat{\mathbf{p}} \equiv -i\hbar\nabla \tag{6.2.86}$$

and, from equation (6.2.36),

$$\hat{\mathbf{A}} = \sum_{\gamma\lambda}\left[\hat{a}_{\gamma\lambda}\mathbf{u}_{\gamma\lambda} + \hat{a}^{+}_{\gamma\lambda}\mathbf{u}^{*}_{\gamma\lambda}\right], \tag{6.2.87}$$

where the operators $\hat{a}_{\gamma\lambda}$ and $\hat{a}^{+}_{\gamma\lambda}$ satisfy the commutation relationship (6.2.45).

This Hamiltonian must now be supplemented by the contribution $\hat{H}_{\mathrm{f}}$ arising from the energy of the free radiation field given by equation (6.2.50). Consequently, the complete Hamiltonian $\hat{H}$ for the single charged particle in an electromagnetic field is

$$\hat{H} = \hat{H}_{\mathrm{p}} + \hat{H}_{\mathrm{f}}$$
$$= \frac{1}{2m}\left(\hat{p} - \frac{e}{c}\hat{\mathbf{A}}\right)^2 + e\phi\mathbb{1} + \sum_{\gamma\lambda} \hbar\omega_\gamma [\hat{a}^+_{\gamma\lambda}\hat{a}_{\gamma\lambda} + \tfrac{1}{2}\mathbb{1}]. \quad (6.2.88)$$

The Hamiltonian for a system of particles may be similarly derived;

$$\hat{H} = \sum_i \left\{ \frac{1}{2m_i}\left(\hat{p}_i - \frac{e}{c}\hat{\mathbf{A}}_i\right)^2 + e\phi_i\mathbb{1} + \mathcal{G}_i \right.$$
$$\left. + \sum_{\gamma\lambda} \hbar\omega_\gamma [\hat{a}^+_{\gamma\lambda}\hat{a}_{\gamma\lambda} + \tfrac{1}{2}\mathbb{1}] \right\}, \quad (6.2.89)$$

where the term $\mathcal{G}_i$ represents the gravitational potential energy of the individual particles.

6.3 Perturbation methods [83]

As we have seen in the preceding section, the complete Hamiltonian operator for the radiation field (photons) and particles (electrons bound to atoms) is quite complicated, and thus the solution to the appropriate wave equation is almost impossible except for the simplest of physical situations. It is possible, however, to obtain approximate solutions to the more complicated physical configurations by expanding (perturbing) about known solutions to simpler, but related, problems. The purpose of this section therefore is to detail such methods for later use.

6.3.1 *Time-dependent perturbation theory*

Here we wish to solve the time-dependent Schrödinger wave equation

$$i\hbar \frac{\partial}{\partial t} |\psi(t)\rangle = \hat{H}(t)|\psi(t)\rangle, \quad (6.3.1)$$

where we take the Hamiltonian operator $\hat{H}(t)$ to consist of a time-independent part $\hat{H}_0$ and a time-dependent term $\hat{H}_1(t)$, viz.

$$\hat{H}(t) = \hat{H}_0 + \hat{H}_1(t). \quad (6.3.2)$$

We further assume that we know the time-dependent solution $|n(t)\rangle$ to the (zeroth order) equation

$$i\hbar \frac{\partial}{\partial t} |n(t)\rangle = \hat{H}_0|n(t)\rangle, \quad (6.3.3)$$

which, because of the time-independence of $\hat{H}_0$, is just

$$i\hbar \frac{\partial}{\partial t} |n(t)\rangle = E_n|n(t)\rangle. \quad (6.3.4)$$

We then have

$$|n(t)\rangle = |n(t_0)\rangle e^{-\frac{i}{\hbar}E_n(t-t_0)}, \tag{6.3.5}$$

where t_0 refers to some specified initial time.

The required unknown is then expanded in a series of these basis vectors $|n(t)\rangle$ such that

$$|\psi(t)\rangle = \sum_n a_n(t)|n(t)\rangle, \tag{6.3.6}$$

where, it should be emphasised, the coefficients $a_n(t)$ are time dependent. Further, we assume the orthogonality relationship

$$\langle n(t_0)|m(t_0)\rangle = \delta_{nm}, \tag{6.3.7}$$

although we note that it is always possible to construct a basis set to satisfy this last equation.

Substitution of equation (6.3.6) into (6.3.1) then yields

$$i\hbar \sum_n \dot{a}_n(t)|n(t)\rangle + i\hbar \sum_n a_n(t) \frac{\partial}{\partial t} |n(t)\rangle$$
$$= (\hat{H}_0 + \hat{H}_1(t)) \sum_n a_n(t)|n(t)\rangle,$$

which, using equation (6.3.3), becomes

$$i\hbar \sum_n \dot{a}_n(t)|n(t)\rangle = \hat{H}_1(t) \sum_n a_n(t)|n(t)\rangle \tag{6.3.8}$$

(note that $\hat{H}_0$ is time independent).

We now multiply equation (6.3.8) on the left by $\langle m(t)|$ to obtain

$$i\hbar\dot{a}_m(t) = \sum_n a_n(t)\langle m(t)|\hat{H}_1(t)|n(t)\rangle, \tag{6.3.9}$$

which, using equation (6.3.7) and writing

$$|n\rangle \equiv |n(t_0)\rangle \tag{6.3.10}$$

and

$$i\hbar(\omega_m - \omega_n) \equiv E_m - E_n, \tag{6.3.11}$$

becomes

$$i\hbar\dot{a}_m(t) = \sum_n a_n(t)\langle m|H_1(t)|n\rangle e^{i(\omega_m-\omega_n)(t-t_0)}. \tag{6.3.12}$$

We therefore have a set of ordinary first order differential equations to solve for the unknowns $a_m(t)$. Once these equations are solved, the required quantity $|\psi(t)\rangle$ follows immediately from equation (6.3.6). It should be stressed that at this point the analysis is exact. However, the solution of equations (6.3.12) is generally only available using an iterative or perturbation scheme. In particular, we assume that the time-dependent Hamiltonian $\hat{H}_1(t)$ represents only a weak interaction between the particles and the radiation field. This then enables us to replace $a_n(t)$ in the summation (6.3.12) by the corresponding initial value $a_n(t_0)$ so that, as a first approximation, we obtain

$$\dot{a}_m^{(1)}(t) = \frac{1}{i\hbar} \sum_n a_n(t_0) \langle m | \hat{H}_1(t) | n \rangle e^{i(\omega_m - \omega_n)(t - t_0)}. \tag{6.3.13}$$

If necessary, this process may be repeated so that, for the lth iteration, we have

$$a_n^{(l)}(t) = \frac{1}{i\hbar} \sum_n a_n^{(l-1)}(t) \langle m | \hat{H}_1(t) | n \rangle e^{i(\omega_m - \omega_n)(t - t_0)}. \tag{6.3.14}$$

Convergence of the above iterative scheme will be dominated by the influence of the operator $\hat{H}_1(t)$.

It is of some value to briefly discuss the physical significance of these time-dependent expansion coefficients $a_m(t)$. The physical interpretation of $|\psi|^2$ as a probability requires the normalisation

$$\langle \psi(t) | \psi(t) \rangle = 1. \tag{6.3.15}$$

Substitution of equations (6.3.6) and (6.3.7) then gives

$$\sum_n |a_n(t)|^2 = 1. \tag{6.3.16}$$

Thus, if we write

$$P_n(t) = |a_n(t)|^2, \tag{6.3.17}$$

so that

$$\sum_n P_n(t) = 1, \tag{6.3.18}$$

then we see that $P_n(t)$ may be similarly interpreted as a probability but, in this case, the probability of the system existing in the state defined by $|n\rangle$ at time t.

6.3.2 *Interaction representation*

We shall find it particularly convenient to re-formulate the time-dependent Schrödinger wave equation in a form which removes some of the time dependence. We begin by writing

$$i\hbar \frac{\partial}{\partial t} |\psi(t)\rangle = \hat{H} |\psi(t)\rangle, \tag{6.3.19}$$

where

$$\hat{H}(t) = \hat{H}_0 + \hat{H}_1(t). \tag{6.3.20}$$

Next, we define the wave function $|\psi^{\mathrm{I}}(t)\rangle$ in the interaction representation by

$$|\psi^{\mathrm{I}}(t)\rangle = e^{i\hat{H}_0 t/\hbar} |\psi(t)\rangle, \tag{6.3.21}$$

so that

$$\begin{aligned} i\hbar \frac{\partial}{\partial t} |\psi^{\mathrm{I}}(t)\rangle &= -e^{i\hat{H}_0 t/\hbar} \hat{H}_0 |\psi(t)\rangle + e^{i\hat{H}_0 t/\hbar} (\hat{H}_0 + \hat{H}_1) |\psi(t)\rangle \\ &= e^{i\hat{H}_0 t/\hbar} \hat{H}_1 e^{-i\hat{H}_0 t/\hbar} e^{i\hat{H}_0 t/\hbar} |\psi(t)\rangle \\ &= \hat{H}_1^{\mathrm{I}} |\psi^{\mathrm{I}}(t)\rangle, \end{aligned} \tag{6.3.22}$$

where we have defined the interaction Hamiltonian

$$\hat{H}_1^{\mathrm{I}}(t) = \mathrm{e}^{\mathrm{i}\hat{H}_0 t/\hbar}\hat{H}_1(t)\mathrm{e}^{-\mathrm{i}\hat{H}_0 t/\hbar}. \tag{6.3.23}$$

It is clear from equation (6.3.22) that $|\psi^{\mathrm{I}}(t)\rangle$ is independent of time when this interaction perturbation is zero. Thus, the interaction representation allows a perturbation about a stationary state.

We may now proceed as in section 6.3.1. We expand $|\psi^{\mathrm{I}}(t)\rangle$ in a set of basis vectors $|n^{\mathrm{I}}(t)\rangle$, viz.

$$|\psi^{\mathrm{I}}(t)\rangle = \sum_n a_n^{\mathrm{I}}(t)|n^{\mathrm{I}}(t)\rangle, \tag{6.3.24}$$

where

$$|n^{\mathrm{I}}(t)\rangle = \mathrm{e}^{\mathrm{i}\hat{H}_0 t/\hbar}|n(t)\rangle, \tag{6.3.25}$$

and $|n(t)\rangle$ satisfies equation (6.3.3), viz.

$$\mathrm{i}\hbar\frac{\partial}{\partial t}|n(t)\rangle = \hat{H}_0|n(t)\rangle. \tag{6.3.3}$$

It is not difficult to see that $|n^{\mathrm{I}}(t)\rangle$ is, in fact, time independent. For example, we have

$$\begin{aligned}\mathrm{i}\hbar\frac{\partial}{\partial t}|n^{\mathrm{I}}(t)\rangle &= -\mathrm{e}^{\mathrm{i}\hat{H}_0 t/\hbar}\hat{H}_0|n(t)\rangle + \mathrm{e}^{\mathrm{i}\hat{H}_0 t/\hbar}\mathrm{i}\hbar\frac{\partial}{\partial t}|n(t)\rangle\\ &= 0.\end{aligned} \tag{6.3.26}$$

Further, the $|n^{\mathrm{I}}\rangle$ are orthogonal (this is required if they are to form a basis set) since

$$\begin{aligned}\langle m^{\mathrm{I}}|n^{\mathrm{I}}\rangle &= \langle m|\mathrm{e}^{-\mathrm{i}\hat{H}_0 t/\hbar}\mathrm{e}^{\mathrm{i}\hat{H}_0 t/\hbar}|n\rangle\\ &= \langle m|n\rangle = \delta_{mn}.\end{aligned} \tag{6.3.27}$$

Equation (6.3.24) when substituted into (6.3.22) yields

$$\mathrm{i}\hbar\sum_n \dot{a}_n^{\mathrm{I}}(t)|n^{\mathrm{I}}\rangle = \sum_n a_n^{\mathrm{I}}(t)\hat{H}_1^{\mathrm{I}}|n^{\mathrm{I}}\rangle,$$

which, multiplying on the left by $\langle m^{\mathrm{I}}|$, and using equation (6.3.27), gives

$$\mathrm{i}\hbar\dot{a}_m^{\mathrm{I}}(t) = \sum_n a_n^{\mathrm{I}}(t)\langle m^{\mathrm{I}}|\hat{H}_1^{\mathrm{I}}|n^{\mathrm{I}}\rangle. \tag{6.3.28}$$

This last result may be compared with equation (6.3.12) for $a_m(t)$. Indeed, we see that

$$\begin{aligned}\langle m^{\mathrm{I}}|\hat{H}_1^{\mathrm{I}}|n^{\mathrm{I}}\rangle &= \langle m^{\mathrm{I}}|\mathrm{e}^{\mathrm{i}\hat{H}_0 t/\hbar}\hat{H}_1\mathrm{e}^{-\mathrm{i}\hat{H}_0 t/\hbar}|n^{\mathrm{I}}\rangle\\ &= \langle m(t)|\hat{H}_1|n(t)\rangle,\end{aligned} \tag{6.3.29}$$

so that equation (6.3.28) is identical to equation (6.3.9), i.e.

$$a_m^{\mathrm{I}}(t) \equiv a_m(t), \tag{6.3.30}$$

for all m. This last result also follows from our physical interpretation of $|\psi|^2$ as a probability distribution, i.e. we must have

$$\sum_m |a_m^{\mathrm{I}}(t)|^2 = 1. \tag{6.3.31}$$

Equations (6.3.4) and (6.3.25) yield

$$|n^{\mathrm{I}}\rangle = e^{i\hat{H}_0 t/\hbar}|n(t)\rangle$$
$$= e^{iE_n t/\hbar}|n(t)\rangle, \tag{6.3.32}$$

so that the interaction matrix elements $\langle m^{\mathrm{I}}|\hat{H}_1^{\mathrm{I}}|n^{\mathrm{I}}\rangle$ appearing in equation (6.3.28) may be written as

$$\langle m^{\mathrm{I}}|\hat{H}_1^{\mathrm{I}}|n^{\mathrm{I}}\rangle = e^{i(E_m - E_n)t/\hbar}\langle m^{\mathrm{I}}|\hat{H}_1|n^{\mathrm{I}}\rangle. \tag{6.3.33}$$

The results obtained above are formally equivalent to those presented in the preceding section. However, the interaction representation forces the time dependence of the wave function being considered to arise solely from the effect of the perturbing Hamiltonian - see equation (6.3.24) where $|n^{\mathrm{I}}\rangle$ are time independent. This then enables the time evolution due to $\hat{H}_1$ to be isolated from that arising from $\hat{H}_0$.

6.3.3 *Time-evolution operators*

The analysis presented in the preceding section can be generalised by writing

$$|\psi(t)\rangle = \hat{T}(t, t_0)|\psi(t_0)\rangle, \tag{6.3.34}$$

where $\hat{T}(t, t_0)$ is the time-evolution operator specifying the evolution of $|\psi\rangle$ from time t_0 to t.

If we substitute equation (6.3.34) into

$$i\hbar \frac{\partial}{\partial t}|\psi(t)\rangle = (\hat{H}_0 + \hat{H}_1(t))|\psi(t)\rangle, \tag{6.3.35}$$

we obtain

$$i\hbar \frac{\partial}{\partial t}\hat{T}(t, t_0)|\psi(t_0)\rangle = (\hat{H}_0 + \hat{H}_1(t))\hat{T}(t, t_0)|\psi(t_0)\rangle,$$

which shows that the operator $\hat{T}(t, t_0)$ satisfies the equation

$$i\hbar \frac{\partial}{\partial t}\hat{T}(t, t_0) = (\hat{H}_0 + \hat{H}_1(t))\hat{T}(t, t_0). \tag{6.3.36}$$

One then proceeds by defining the new operator $\hat{U}(t, t_0)$ given by

$$\hat{U}(t, t_0) = e^{i\hat{H}_0(t-t_0)/\hbar}\hat{T}(t, t_0). \tag{6.3.37}$$

We then have

$$\begin{aligned} i\hbar \frac{\partial}{\partial t}\hat{U}(t, t_0) &= -e^{i\hat{H}_0(t-t_0)/\hbar}\hat{H}_0\hat{T}(t, t_0) \\ &\quad + e^{i\hat{H}_0(t-t_0)/\hbar} i\hbar \frac{\partial}{\partial t}\hat{T}(t, t_0) \\ &= e^{i\hat{H}_0(t-t_0)/\hbar}\hat{H}_1(t)\hat{T}(t, t_0) \\ &= \hat{H}_1^{\mathrm{I}}(t)\hat{U}(t, t_0), \end{aligned} \tag{6.3.38}$$

where, analogous to equation (6.3.23), we have defined the interaction Hamiltonian

$$\hat{H}_1^{\mathrm{I}}(t) = \mathrm{e}^{\mathrm{i}\hat{H}_0(t-t_0)/\hbar}\hat{H}_1(t)\mathrm{e}^{-\mathrm{i}\hat{H}_0(t-t_0)/\hbar}. \tag{6.3.39}$$

Note that care must be exercised when manipulating the operators $\hat{H}_0, \hat{H}_1(t)$, $\hat{T}(t, t_0)$ and $\hat{U}(t, t_0)$ – these operators do not always commute.

Clearly, equation (6.3.38) for $\hat{U}(t, t_0)$ may be re-written as

$$\hat{U}(t, t_0) = \hat{U}(t_0, t_0) + \frac{1}{\mathrm{i}\hbar}\int_{t_0}^{t} \hat{H}_1^{\mathrm{I}}(t')\hat{U}(t', t_0)\,\mathrm{d}t'. \tag{6.3.40}$$

Equation (6.3.34) implies that

$$\hat{T}(t_0, t_0) = \mathbb{1}, \tag{6.3.41}$$

so that from equation (6.3.34), we have

$$\hat{U}(t_0, t_0) = \mathbb{1}. \tag{6.3.42}$$

We then write

$$\hat{U}(t, t_0) = \mathbb{1} + \frac{1}{\mathrm{i}\hbar}\int_{t_0}^{t} \hat{H}_1^{\mathrm{I}}(t')\hat{U}(t', t_0)\,\mathrm{d}t'. \tag{6.3.43}$$

This last equation may be solved by iteration such that

$$\hat{U}^{(l)}(t, t_0) = \mathbb{1} + \frac{1}{\mathrm{i}\hbar}\int_{t_0}^{t} \hat{H}_1^{\mathrm{I}}(t')\hat{U}^{(l-1)}(t', t_0)\,\mathrm{d}t', \tag{6.3.44}$$

where, again, the superscript l refers to the lth iteration. Thus, starting with $\hat{U}^{(0)}(t, t_0) = \mathbb{1}$ for all t, and this is the logical starting point in view of equation (6.3.43) if $\hat{H}_1$, thence $\hat{H}_1^{\mathrm{I}}$, is small compared with $\hat{H}_0$, i.e. $\hat{H}_1$ is indeed a perturbation, we have

$$\hat{U}^{(1)}(t, t_0) = \mathbb{1} + \frac{1}{\mathrm{i}\hbar}\int_{t_0}^{t} \hat{H}_1^{\mathrm{I}}(t')\,\mathrm{d}t', \tag{6.3.45}$$

$$\hat{U}^{(2)}(t, t_0) = \mathbb{1} + \frac{1}{\mathrm{i}\hbar}\int_{t_0}^{t} \hat{H}_1^{\mathrm{I}}(t')\,\mathrm{d}t' + \frac{1}{(\mathrm{i}\hbar)^2}\int_{t_0}^{t}\mathrm{d}t'\int_{t_0}^{t'}\mathrm{d}t''\hat{H}_1^{\mathrm{I}}(t')\hat{H}_1^{\mathrm{I}}(t''), \tag{6.3.46}$$

etc, where $t > t' > t'' > t_0$. We finally obtain

$$\hat{U}(t, t_0) = \hat{U}^{(\infty)}(t, t_0),$$

which is generally written as the Dyson exponential time operator

$$\hat{U}(t, t_0) = \hat{\mathscr{D}}\exp\left\{\frac{1}{\mathrm{i}\hbar}\int_{t_0}^{t} \hat{H}_1^{\mathrm{I}}(t')\,\mathrm{d}t'\right\}, \tag{6.3.47}$$

where the $\hat{\mathscr{D}}$ preceding the exponential term stipulates that the normal power series expansion of the exponential should be replaced by a series of repeated integrals, the first three terms of which appear on the RHS of equation (6.3.46).

Finally, the required evolution operator $\hat{T}(t, t_0)$ is given by

$$\hat{T}(t, t_0) = e^{-i\hat{H}_0(t-t_0)/\hbar} \hat{\mathscr{D}} \exp\left\{\frac{1}{i\hbar}\int_{t_0}^{t} \hat{H}_1^{\mathrm{I}}(t')\,dt'\right\}. \tag{6.3.48}$$

The adjoint of equation (6.3.34) is

$$\langle\psi(t)| = \langle\psi(t_0)|\hat{T}^{+}(t, t_0), \tag{6.3.49}$$

where, from equation (6.3.48), and using the Hermitian property of $\hat{H}_0$ and $\hat{H}_1$,

$$\begin{aligned}\hat{T}^{+}(t, t_0) &= \hat{U}^{+}(t, t_0)e^{i\hat{H}_0(t-t_0)/\hbar}\\ &= \hat{\mathscr{D}} \exp\left\{-\frac{1}{i\hbar}\int_{t_0}^{t} \hat{H}_1^{\mathrm{I}}(t')\,dt'\right\} e^{i\hat{H}_0(t-t_0)/\hbar}\\ &= \hat{U}(t_0, t)e^{i\hat{H}_0(t-t_0)/\hbar}.\end{aligned}$$

However,

$$\hat{T}(t, t_0)\hat{T}(t_0, t) = \mathbb{1},$$

or

$$e^{-i\hat{H}_0(t-t_0)/\hbar}\hat{U}(t, t_0)\hat{T}(t_0, t) = \mathbb{1},$$

so that with

$$\hat{U}(t_0, t)\hat{U}(t, t_0) = \mathbb{1},$$

we have

$$\begin{aligned}\hat{T}(t_0, t) &= \hat{U}(t_0, t)e^{i\hat{H}_0(t-t_0)/\hbar}\\ &= \hat{T}^{+}(t, t_0).\end{aligned} \tag{6.3.50}$$

This last result will be particularly important when examining the perturbing effects of collisions.

6.4 Spectral line profiles (no collisions)

The foregoing theory may now be used to obtain profiles of absorption, emission and scattering of radiation due to the interaction of that radiation with a single atom. We neglect the effects of collisions of the atom with other particles - these are considered in section 6.5.

As we have seen in the preceding section, the expansion coefficients $a_m(t)$ appearing in equation (6.3.6), for example, enable the probability $|a_m(t)|^2$ of the system being in the state $|m(t)\rangle$ at time t to be determined. However, it is important at this stage to define the eigenstates $|m(t)\rangle$ explicitly. Previously, we discussed first an atom with no radiation field, then a pure radiation field with no particles and, in each case, it was only necessary to use a single index to describe the state, i.e. we simply wrote $|m(t)\rangle$. Now our system includes both radiation and particles. We shall therefore first consider such a system to consist of a single atom in the state defined by the index m, and the radiation field to have n_r photons in the rth mode where r $(=1, \infty)$ is a composite subscript incorporating both frequency and polarisation. We then write the eigenstate of this entire system as $|m, n_1, n_2, \ldots, n_r, \ldots, t\rangle$.

Consequently, if the above system is altered by the atom absorbing a photon in the rth mode, so that it is excited to the $(m+1)$th level, then the new state is described by $|m+1, n_1, n_2, \ldots, n_r-1, \ldots, t\rangle$.

Similarly, if the atom emits a photon in the sth mode we have $|m-1, n_1, n_2, \ldots, n_s+1, \ldots, t\rangle$. As we proceed, we shall find it expedient to omit most of these index terms.

Reference to equation (6.3.12) shows that the matrix element $\langle m|\hat{H}_1(t)|n\rangle$ is required in the determination of the expansion coefficients $a_m(t)$. Equation (6.2.88) gives the appropriate form of the total Hamiltonian $\hat{H}$ of the system, viz.

$$\hat{H} = \frac{1}{2m}\left(\hat{\mathbf{p}} - \frac{e}{c}\hat{\mathbf{A}}\right)^2 + e\phi\mathbb{1} + \sum_{\gamma\lambda} \hbar\omega_\gamma\,[\hat{a}^+_{\gamma\lambda}\hat{a}_{\gamma\lambda} + \tfrac{1}{2}\mathbb{1}], \tag{6.4.1}$$

where $p = -i\hbar\nabla$ and

$$\hat{\mathbf{A}} = \sum_{\gamma\lambda}\,[\hat{a}_{\gamma\lambda}\mathbf{u}_{\gamma\lambda} + \hat{a}^+_{\gamma\lambda}\mathbf{u}^*_{\gamma\lambda}], \tag{6.4.2}$$

with

$$\mathbf{u}_{\gamma\lambda} = \alpha_\gamma \mathbf{e}_{\gamma\lambda} e^{i\mathbf{k}_\gamma\cdot\mathbf{r}}. \tag{6.4.3}$$

In particular, the $\hat{a}_{\gamma\lambda}$ and $\hat{a}^+_{\gamma\lambda}$ are destruction and creation operators respectively such that if $|n_{\gamma\lambda}\rangle$ represents a pure radiation eigenstate (i.e. no atoms present) then, from equations (6.2.65) and (6.2.70), we have

$$\hat{a}_{\gamma\lambda}|n_{\gamma\lambda}\rangle = \sqrt{n_{\gamma\lambda}}\,|n_{\gamma\lambda}-1\rangle, \tag{6.4.4}$$

and

$$\hat{a}^+_{\gamma\lambda}|n_{\gamma\lambda}\rangle = \sqrt{(n_{\gamma\lambda}+1)}\,|n_{\gamma\lambda}+1\rangle. \tag{6.4.5}$$

If we now refer to the notation introduced above, we have

$$\hat{a}\,|m, n_1, \ldots, n_r, \ldots, t\rangle = \sqrt{n_r}\,|m+1, n_1, \ldots, n_r-1, \ldots, t\rangle. \tag{6.4.6}$$

and

$$\hat{a}^+_s|m, n_1, \ldots, n_s, \ldots, t\rangle = \sqrt{(n_s+1)}\,|m-1, n_1, \ldots, n_s+1, \ldots, t\rangle. \tag{6.4.7}$$

According to the theory presented in section 6.3.1, the above total Hamiltonian $\hat{H}$ should be split into two components

$$\hat{H} = \hat{H}_0 + \hat{H}_1, \tag{6.4.8}$$

so that the perturbation term $\hat{H}_1$ refers simply to those quantities specifying the interaction of the radiation field with the matter, i.e. we have

$$\hat{H}_1 = -\frac{e}{2mc}(\hat{\mathbf{p}}\cdot\hat{\mathbf{A}} + \hat{\mathbf{A}}\cdot\hat{\mathbf{p}}) + \frac{e^2}{2mc^2}\hat{\mathbf{A}}\cdot\hat{\mathbf{A}}. \tag{6.4.9}$$

The $\hat{H}_0$ component obviously is then merely the remainder of the terms appearing in equation (6.4.1), i.e.

$$\hat{H}_0 = \frac{1}{2m}\hat{p}^2 + e\phi\mathbb{1} + \sum_{\gamma\lambda}\hbar\omega_\gamma[\hat{a}^+_{\gamma\lambda}\hat{a}_{\gamma\lambda} + \tfrac{1}{2}\mathbb{1}]. \tag{6.4.10}$$

It effectively defines the unperturbed basis set of vectors $|n(t)\rangle$ in which the perturbed wave function or eigenstate is expanded. Indeed, if we write

$$|m, n_r\rangle \equiv |m, n, \ldots, n_r, \ldots, t\rangle, \tag{6.4.11}$$

then

$$\hat{H}_0|m, n_r\rangle = E_m|m, n_r\rangle, \tag{6.4.12}$$

i.e.

$$E_m|m, n_r\rangle = \frac{1}{2m}\hat{p}^2|m, n_r\rangle + e\phi|m, n_r\rangle + \sum_{\gamma\lambda}\hbar\omega_\gamma(n_r + \tfrac{1}{2})|m, n_r\rangle,$$

where we have used both equations (6.4.6, 7). This then yields

$$\begin{aligned} E_m &= E_m^{(A)} + E_m^{(R)} \\ &= E_m^{(A)} + \sum_{\gamma\lambda}\hbar\omega_\gamma(n_r + \tfrac{1}{2}), \end{aligned} \tag{6.4.13}$$

where $E_m^{(A)}$ refers to the energy of the atomic system in state $|m, n_r\rangle$.

We now return to the evaluation of the interaction potential $\hat{H}_1$, viz.

$$\begin{aligned} \hat{H}_1 = &-\frac{e}{2mc}\sum_{\gamma\lambda}[\hat{a}_{\gamma\lambda}(\hat{p}\cdot\mathbf{u}_{\gamma\lambda} + \mathbf{u}_{\gamma\lambda}\cdot\hat{p}) + \hat{a}^+_{\gamma\lambda}(\hat{p}\cdot\mathbf{u}^*_{\gamma\lambda} + \mathbf{u}^*_{\gamma\lambda}\cdot\hat{p})] \\ &+ \frac{e^2}{2mc^2}\sum_{\gamma\lambda}\sum_{\gamma'\lambda'}[\hat{a}_{\gamma\lambda}\hat{a}_{\gamma'\lambda'}\mathbf{u}_{\gamma\lambda}\cdot\mathbf{u}_{\gamma'\lambda'} + \hat{a}_{\gamma\lambda}\hat{a}^+_{\gamma'\lambda'}\mathbf{u}_{\gamma\lambda}\cdot\mathbf{u}^*_{\gamma'\lambda'} \\ &+ \hat{a}^+_{\gamma\lambda}\hat{a}_{\gamma'\lambda'}\mathbf{u}^*_{\gamma\lambda}\cdot\mathbf{u}_{\gamma'\lambda'} + \hat{a}^+_{\gamma\lambda}\hat{a}^+_{\gamma'\lambda'}\mathbf{u}^*_{\gamma\lambda}\cdot\mathbf{u}^*_{\gamma'\lambda'}]. \end{aligned} \tag{6.4.14}$$

Since it is clear from equations (6.4.6, 7) that $\hat{a}_{\gamma\lambda}$ and $\hat{a}^+_{\gamma\lambda}$ are specifically destruction and creation operators corresponding to absorption and emission of photons respectively, the effect of the double operator terms on the RHS of equation (6.4.14) is quite straightforward. For example, writing

$$|n_r, n_s\rangle \equiv |m, n_1, \ldots, n_r, \ldots, n_s, \ldots, t\rangle, \tag{6.4.15}$$

$$\begin{aligned} \hat{a}_r\hat{a}^+_s|n_r, n_s\rangle &= \sqrt{(n_s+1)}\hat{a}_r|n_r, n_s+1\rangle \\ &= \sqrt{(n_s+1)}\sqrt{n_r}|n_r - 1, n_s + 1\rangle, \end{aligned} \tag{6.4.16}$$

so that an sth mode photon is emitted and an rth mode photon is absorbed. Such double operators therefore correspond to scattering processes involving two photons.

6.4.1 *Transition rate for photon emission [88]*

Consider a model atom as shown in figure 6.2 having just two levels in which the bound electron can exist.

The excited level is denoted by 2 and the lower by 1. We are interested in the probability of an electron initially in level 2 de-exciting to level 1 with the subsequent emission of a photon. This therefore restricts us to the evaluation of the rate at which this emission process takes place.

We define the state of the system initially as $|2, n_1, \ldots, n_{\gamma\lambda}, \ldots, 0\rangle$ and, finally, after the emission of a $(\gamma\lambda)$th mode photon, by $|1, n_1, \ldots, n_{\gamma\lambda}+1, \ldots, t\rangle$. For clarity in exposition, we write the initial state as $|2\rangle$ and the final state as $|1\rangle$, i.e. the photon states are implicit in this notation. However, it is important to realise that the de-excitation from level 2 to 1 gives rise to the emission of a $(\gamma\lambda)$th mode photon *only*, i.e. a photon in the $(\gamma\lambda)$th mode is associated *only* with the $|2\rangle \rightarrow |1\rangle$ transition.

We must therefore solve the system of equations (6.3.12), viz.

$$i\hbar\dot{a}_m(t) = \sum_{n=1}^{2} a_n(t)\langle m|\hat{H}_1|n\rangle e^{i(E_m - E_n)t/\hbar}. \tag{6.4.17}$$

for $m = 1, 2$. This, of course, will then yield the solution

$$|\psi(t)\rangle = \sum_{n=1}^{2} a_n(t)|n(t)\rangle, \tag{6.4.18}$$

to the time-dependent Schrödinger wave equation.

It is important to note now that the eigenstates $|1\rangle$ and $|2\rangle$ refer to the solutions of the unperturbed system in which atoms and photons are quite independent of one another. Thus, we may generally write

$$|m, n_r\rangle \equiv |m\rangle|n_r\rangle \equiv |n_r\rangle|m\rangle, \tag{6.4.19}$$

so that operators involving atom terms commute with those involving photons.

Next we notice that the emission process being studied in this section involves just one photon so that the scattering terms (6.4.16) are inapplicable here. The remaining terms in $\hat{H}_1$ are then

$$\hat{H}_1 = -\frac{e}{2mc}\sum_{\gamma\lambda}\left[\hat{a}_{\gamma\lambda}(\hat{\mathbf{p}}\cdot\mathbf{u}_{\gamma\lambda} + \mathbf{u}_{\gamma\lambda}\cdot\hat{\mathbf{p}}) + \hat{a}^{+}_{\gamma\lambda}(\hat{\mathbf{p}}\cdot\mathbf{u}^{*}_{\gamma\lambda} + \mathbf{u}^{*}_{\gamma\lambda}\cdot\hat{\mathbf{p}})\right], \tag{6.4.20}$$

Fig.6.2. The quantum representation of the upper and lower bound electron states.

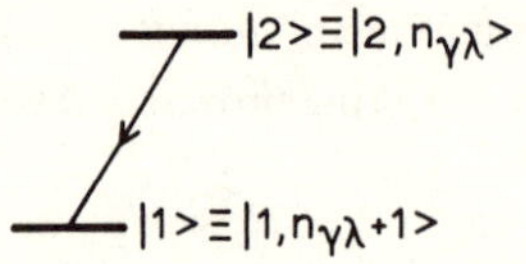

which, writing

$$\xi_{\gamma\lambda} = -\frac{e}{2mc}(\hat{p}\cdot\mathbf{u}_{\gamma\lambda} + \mathbf{u}_{\gamma\lambda}\cdot\hat{p}), \tag{6.4.21}$$

becomes

$$\hat{H}_1 = \sum_{\gamma\lambda}(\hat{a}_{\gamma\lambda}\xi_{\gamma\lambda} + \hat{a}^{+}_{\gamma\lambda}\xi^{*}_{\gamma\lambda}); \tag{6.4.22}$$

recall that $\hat{H}$ is Hermitian ($\hat{H}^+ = \hat{H}$) and, for example, $(\hat{p}\cdot\mathbf{u}_{\gamma\lambda})^* = (-i\hbar\nabla\cdot\mathbf{u}_{\gamma\lambda})^* = \hbar\mathbf{k}_\gamma\cdot\mathbf{u}^*_{\gamma\lambda}$.

Thus, the only non-zero matrix elements $\langle m|\hat{H}_1|n\rangle$ are

$$\begin{aligned}\langle 2|\hat{H}_1|1\rangle &\equiv \langle 2|\langle n_{\gamma\lambda}|\hat{a}_{\gamma\lambda}\xi_{\gamma\lambda}|n_{\gamma\lambda}+1\rangle|1\rangle\\ &= \langle 2|\xi_{\gamma\lambda}|1\rangle\langle n_{\gamma\lambda}|\hat{a}_{\gamma\lambda}|n_{\gamma\lambda}+1\rangle\\ &= Q_{21}\sqrt{(n_{\gamma\lambda}+1)}\langle n_{\gamma\lambda}|n_{\gamma\lambda}\rangle\\ &= Q_{21}\sqrt{(n_{\gamma\lambda}+1)},\end{aligned} \tag{6.4.23}$$

where

$$Q_{21} = \langle 2|\xi_{\gamma\lambda}|1\rangle, \tag{6.4.24}$$

and

$$\begin{aligned}\langle 1|\hat{H}_1|2\rangle &= \langle 1|\xi^*_{\gamma\lambda}|2\rangle\langle n_{\gamma\lambda}+1|\hat{a}_{\gamma\lambda}|n_{\gamma\lambda}\rangle\\ &= Q^*_{21}\sqrt{(n_{\gamma\lambda}+1)}.\end{aligned} \tag{6.4.25}$$

All other matrix elements are clearly zero. Equation (6.4.17) then yields

$$\begin{aligned}i\hbar\dot{a}_1(t) &= a_2(t)\langle 1|\hat{H}_1|2\rangle e^{i(E_1-E_2)t/\hbar}\\ &= Q^*_{21}\sqrt{(n_{\gamma\lambda}+1)}a_2(t)e^{i(\omega_\gamma-\omega_{21})t},\end{aligned} \tag{6.4.26}$$

where, using equation (6.4.13), we have written

$$\begin{aligned}E_1 - E_2 &= E_1^{(A)} - E_2^{(A)} + \sum_{\gamma\lambda}\omega_\gamma[(n_{\gamma\lambda}+1)+\tfrac{1}{2}] - \sum_{\gamma\lambda}\omega_\gamma[n_{\gamma\lambda}+\tfrac{1}{2}]\\ &= \hbar\omega_\gamma - \hbar\omega_{21},\end{aligned} \tag{6.4.27}$$

with

$$\omega_{21} = \frac{E^{(A)} - E^{(A)}}{\hbar} \tag{6.4.28}$$

representing the frequency difference between the two levels of the atom.

Since the system is in state $|2\rangle$ initially, we must have

$$|a_2(0)|^2 = 1, \tag{6.4.29}$$

which leads to the initial conditions

$$a_2(0) = 1, \quad a_1(0) = 0. \tag{6.4.30}$$

We next obtain approximate solutions of equation (6.4.26) by taking $a_2(t) = 1$ for all t. This is effectively the first step in the iterative scheme (6.3.14). Indeed, we shall remove this approximation in a later section. We then have

$$a_1(t) \approx \frac{Q_{21}^* \sqrt{(n_{\gamma\lambda}+1)}}{\hbar} \cdot \frac{1 - e^{i(\omega_\gamma - \omega_{21})t}}{\omega_\gamma - \omega_{21}}. \tag{6.4.31}$$

Thus, the probability of finding the system in the state $|1\rangle$ at time t is just

$$|a_1(t)|^2 \approx \frac{|Q_{21}|^2 (n_{\gamma\lambda}+1)}{\hbar^2} \frac{(1 - e^{i(\omega_\gamma - \omega_{21})t})(1 - e^{-i(\omega_\gamma - \omega_{21})t})}{(\omega_\gamma - \omega_{21})^2}$$

$$= \frac{2|Q_{21}|^2 (n_{\gamma\lambda}+1)}{\hbar^2} \cdot \frac{1 - \cos(\omega_\gamma - \omega_{21})t}{(\omega_\gamma - \omega_{21})^2}. \tag{6.4.32}$$

The above result clearly depends upon the frequency ω_γ of the photon being emitted. The overall rate of transition from level 2 to level 1, taking into account all photon emissions possible, must therefore incorporate an averaging over these ω_γ. Consequently, we proceed by defining this rate of transition $w(2 \to 1)$ from state $|2\rangle$ to state $|1\rangle$ as the suitably weighted probability per unit time of the system being in state $|1\rangle$, viz.

$$w(2 \to 1) = \frac{1}{4\pi t} \sum_{\gamma\lambda} \int d\Omega \rho(\omega_\gamma) |a_1(t)|^2,$$

$$\to \frac{1}{4\pi t} \sum_{\lambda} \int d\Omega \int_0^\infty \rho(E) |a_1(t)|^2 \, dE, \tag{6.4.33}$$

where $\rho(E)$ is the energy density of states with energy E. (The above manipulation (6.4.33) only has physical validity when applied to the broadened states needed to describe natural decay – this is discussed in section 6.4.3.) Note that we have also summed over solid angle Ω of photon emission and polarisation index λ. We then have

$$w(2 \to 1) \approx \frac{1}{2\pi t} \sum_{\lambda} \int d\Omega |Q_{21}|^2 (n_{\gamma\lambda}+1) f(t, E_{21}^{(A)}), \tag{6.4.34}$$

where, with $E_{21}^{(A)} = \hbar\omega_{21}$,

$$f(t, E_{21}^{(A)}) = \int_0^\infty \frac{1 - \cos\left(\dfrac{E - E_{21}^{(A)}}{\hbar} \cdot t\right)}{(E - E_{21}^{(A)})^2} \rho(E) \, dE.$$

If we change the variable of integration in this last equation from E to $x = (E - E_{21}^{(A)})t/\hbar$, noting that we can put $\rho(E) = 0$ for all $E < 0$, we have

$$f(t, E_{21}^{(A)}) = \frac{t}{\hbar} \int_{-\infty}^{\infty} \rho\left(E_{21}^{(A)} + \frac{\hbar x}{t}\right)(1 - \cos x) \frac{dx}{x^2}. \tag{6.4.35}$$

We now construct the contour $C = C_1 \cup C_2 \cup C_3$ in the complex z plane as shown in figure 6.3 and take the limits $R \to \infty$ and $\epsilon \to 0$. We clearly must have

$\rho(E) \to 0$ as $|E| \to \infty$. The integral over C_3 therefore tends to zero (note that $\exp(iz) = \exp(iRe^{i\theta}) = \exp(iR\cos\theta - R\sin\theta) \to 0$ as $R \to \infty$ for all $\theta \in [0, \pi]$). Cauchy's integral theorem then stipulates that

$$\int_{-\infty}^{\infty} (1-\cos x)\rho\left(E_{21}^{(A)} + \frac{x\hbar}{t}\right)\frac{dx}{x^2} + \mathcal{R}l \lim_{\epsilon\to 0} \times \int_{C_1} (1-e^{iz})\rho\left(E_{21}^{(A)} + \frac{z\hbar}{t}\right)\frac{dz}{z^2} = 0,$$

where, on $C_1, z = e^{i\theta}$. We then have

$$\int_{-\infty}^{\infty} (1-\cos x)\rho\left(E_{21}^{(A)} + \frac{x\hbar}{t}\right)\frac{dx}{x^2}$$
$$= -\mathcal{R}l \lim_{\epsilon\to 0} \int_{\pi}^{0} [-i\epsilon e^{i\theta} + O(\epsilon^2)]\rho\left(E_{21}^{(A)} + \frac{\epsilon e^{i\theta}\hbar}{t}\right)\frac{\epsilon e^{i\theta}\, i d\theta}{\epsilon^2 e^{2i\theta}}$$
$$= \pi\rho(E_{21}^{(A)}). \tag{6.4.36}$$

Equation (6.4.34) then yields

$$w(2\to 1) \approx \frac{1}{2\hbar}\sum_{\lambda}\int d\Omega\, |Q_{21}|^2 (n_{\gamma\lambda}+1)\rho(E_{21}^{(A)}), \tag{6.4.37}$$

which, as we should expect, is independent of time t. We further notice that the above approximate rate of transition $w(2\to 1)$ from level 2 to level 1 depends on the factor $n_{\gamma\lambda}+1$, the first part constituting the stimulated emission component ($n_{\gamma\lambda}$ is the number of ($\gamma\lambda$)th mode photons initially in the system) whilst the second part is simply the spontaneous emission component. Clearly, the $w(2\to 1)$ rate leads to the evaluation of the Einstein rate coefficients B_{21} and A_{21} for stimulated and spontaneous emission introduced in chapter 1.

The term $|Q_{21}|^2$ appearing in equation (6.4.37) involves the factors $\hat{\mathbf{p}}$ and $\mathbf{u}_{\gamma\lambda}$, and their complex conjugates, and may be evaluated to various orders of approximation – one has dipole radiation when only the first term in the

Fig.6.3. The contour in the upper-half complex z plane indented at the origin.

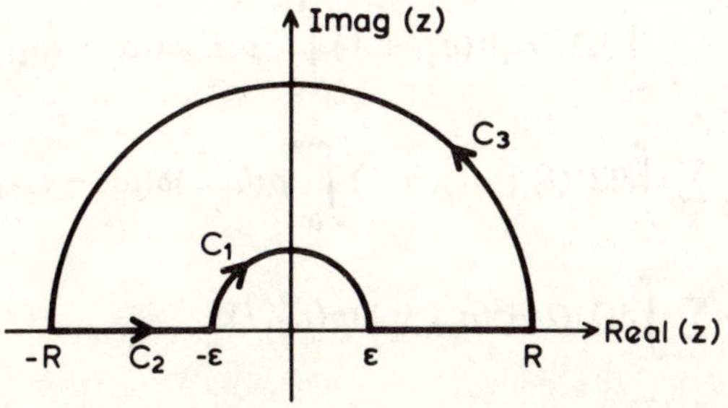

expansion of $u^*_{\gamma\lambda} = \alpha_\gamma \mathbf{e}_\gamma \mathrm{e}^{-\mathrm{i}\mathbf{k}_\gamma\cdot\mathbf{r}}$ is retained, multipole radiation otherwise. Our main concern revolves around the frequency and angle dependence of the emission, absorption and scattering processes, and thus the details of the assumptions and approximations leading to the evaluation of the $|Q_{21}|^2$ terms, which only result in effective constants anyway, will not be discussed here.

An alternative method for determining the transition rate $w(2\to 1)$ is often presented in the literature. Here, we re-define $w(2\to 1)$ by

$$w(2\to 1) = \frac{1}{4\pi}\sum_\lambda \int \mathrm{d}\Omega \int_0^\infty \rho(E) w_{\gamma\lambda}(2\to 1)\,\mathrm{d}E, \tag{6.4.38}$$

where

$$\begin{aligned} w_{\gamma\lambda}(2\to 1) &= \lim_{t\to\infty} \frac{\mathrm{d}}{\mathrm{d}t}|a_1(t)|^2 \\ &= \lim_{t\to\infty} (\dot{a}_1^*(t)a_1(t) + a_1^*(t)\dot{a}_1(t)). \end{aligned} \tag{6.4.39}$$

If we now substitute for $a_1(t)$ from equation (6.4.31), we have

$$w_{\gamma\lambda}(2\to 1) = \frac{|Q_{21}|^2(n_{\gamma\lambda}+1)}{\mathrm{i}\hbar^2} \times \lim_{t\to\infty}\left[\frac{\mathrm{e}^{\mathrm{i}(\omega_\gamma-\omega_{21})t}-1}{\omega_\gamma-\omega_{21}} + \frac{1-\mathrm{e}^{-\mathrm{i}(\omega_\gamma-\omega_{21})t}}{\omega_\gamma-\omega_{21}}\right]. \tag{6.4.40}$$

However, we have the basic expression [49]

$$\lim_{t\to\infty} \frac{1-\mathrm{e}^{\mathrm{i}xt}}{x} = \mathscr{P}\frac{1}{x} - \pi\mathrm{i}\delta(x) \tag{6.4.41}$$

(see appendix D) where $\mathscr{P}$ denotes the Cauchy principal value and $\delta(x)$ the Dirac delta function. The above integral then yields

$$\begin{aligned} w_{\gamma\lambda}(2\to 1) &= \frac{|Q_{21}|^2(n_{\gamma\lambda}+1)}{\mathrm{i}\hbar^2}\cdot 2\pi\mathrm{i}\delta(\omega_\gamma-\omega_{21}) \\ &= \frac{2\pi}{\hbar^2}|Q_{21}|^2(n_{\gamma\lambda}+1)\delta(\omega_\gamma-\omega_{21}). \end{aligned} \tag{6.4.42}$$

Thus,

$$\begin{aligned} w(2\to 1) &= \frac{1}{2\hbar^2}\sum_\lambda \int \mathrm{d}\Omega |Q_{21}|^2(n_{\gamma\lambda}+1)\int_0^\infty \rho(E)\delta(\omega_\gamma-\omega_{21})\,\mathrm{d}E \\ &= \frac{1}{2\hbar}\sum_\lambda \int \mathrm{d}\Omega |Q_{21}|^2(n_{\gamma\lambda}+1)\int_0^\infty \rho(\hbar\omega_\gamma)\delta(\omega_\gamma-\omega_{21})\,\mathrm{d}\omega_\gamma \\ &= \frac{1}{2\hbar}\sum_\lambda \int \mathrm{d}\Omega |Q_{21}|^2(n_{\gamma\lambda}+1)\rho(E_{21}^{(\mathrm{A})}), \end{aligned} \tag{6.4.43}$$

identical to equation (6.4.37).

6.4.2 *Transition rate for photon absorption*

Here we consider the same model 2-level atom studied in the preceding section, but with the bound electron initially in the de-excited state $|1\rangle$ as shown in figure 6.4. We then have as our initial conditions

$$a_1(0) = 1, \quad a_2(0) = 0. \tag{6.4.44}$$

In particular, as in section 6.4.1, we initially have $n_{\gamma\lambda}$ photons in the system. However, here, the final state will consist of just $n_{\gamma\lambda} - 1$ photons as opposed to $n_{\gamma\lambda} + 1$ photons for radiative emission.

Again, the set of equations (6.4.17) must be solved for $a_m(t)$ with the only non-zero matrix elements $\langle m|\hat{H}_1|n\rangle$ being given by equations (6.4.23) and (6.4.25). We then have

$$\begin{aligned} i\hbar\dot{a}_2(t) &= a_1(t)\langle 2|\hat{H}_1|1\rangle e^{i(E_2 - E_1)t/\hbar} \\ &= a_1(t)Q_{21}\sqrt{n_{\gamma\lambda}}\, e^{i(\omega_{21}-\omega_\gamma)t}, \end{aligned} \tag{6.4.45}$$

where

$$Q_{21} = -\frac{e}{2mc}\langle 2|(\hat{\mathbf{p}}\cdot\mathbf{u}_{\gamma\lambda} + \mathbf{u}_{\gamma\lambda}\cdot\hat{\mathbf{p}})|1\rangle.$$

We approximate $a_1(t)$ by unity for all times of interest to obtain

$$a_2(t) \approx \frac{Q_{21}\sqrt{n_{\gamma\lambda}}}{\hbar}\,\frac{1 - e^{i(\omega_{21}-\omega_\gamma)t}}{\omega_{21}-\omega_\gamma}, \tag{6.4.46}$$

so that

$$|a_2(t)|^2 \approx \frac{2|Q_{21}|^2 n_{\gamma\lambda}}{\hbar^2}\cdot\frac{1-\cos(\omega_{21}-\omega_\gamma)t}{(\omega_{21}-\omega_\gamma)^2}. \tag{6.4.47}$$

Following the preceding analysis, we then obtain the transition rate $w(1\to 2)$ from state $|1\rangle$ to state $|2\rangle$ in the form

$$w(1\to 2) \approx \frac{1}{2\hbar}\sum_\lambda \int d\Omega\, |Q_{21}|^2 n_{\gamma\lambda}\rho(E_{21}^{(A)}). \tag{6.4.48}$$

Here we see that the rate of photon absorption is proportional to the number $n_{\gamma\lambda}$ of photons in the $(\gamma\lambda)$th mode. This, of course, simply states that the quantity of radiation absorbed per unit time is proportional to the incident radiation field. Clearly, explicit evaluation of the above $w(1\to 2)$ term will lead to the Einstein rate coefficient B_{12} for radiative excitation.

Fig.6.4. The quantum representation of the upper and lower bound electron states.

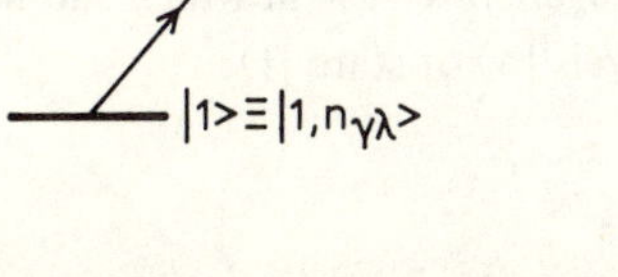

6.4.3 *Natural broadening – preliminary discussion [124]*

In the preceding sections, we obtained approximate solutions for the probability amplitude $|a_m(t)|^2$ of the final state by assuming the amplitude of the initial state to be unchanged for all times t of interest. If we consider an emission process, for example, this approximation breaks down for sufficiently large t since spontaneous emission by itself will eventually significantly deplete the number of atoms with bound electrons in the excited (initial) state, even in the absence of a radiation field which would act to stimulate such emission. Thus, the probability amplitude $|a_m(t)|^2$ for that initial state should be significantly different from unity. Indeed, if we do not allow reabsorption of the emitted photon, one should expect $|a_2(\infty)|^2 = 0$. Consequently, the work presented in the preceding two sections is really only applicable for t sufficiently large that $\omega_\gamma t \gg 1$, but also satisfying $t \ll 1/A_{21}$ the lifetime of the excited state.

In the present section, therefore, we wish to preliminarily discuss the removal of this approximation. A more detailed, and mathematically satisfying, account of the behaviour of the initial state probability amplitude is given in the following section.

We consider an emission process from the excited state $|2\rangle$ to the de-excited state $|1\rangle$ (see figure 6.2) and, using physical intuition, attempt a solution of the probability amplitude of the initial state by incorporating the exponential time decay

$$a_2(t) = \mathrm{e}^{-\eta t}, \tag{6.4.49}$$

where η is a complex quantity yet to be determined, but satisfying $\mathcal{R}l\ \eta > 0$ so that $|a_2(\infty)|^2 = 0$. Note that equation (6.4.49) satisfies the required initial condition for state $|2\rangle$, viz. $a_2(0) = 1$.

Further, we take into account the fact that the energy levels in which the bound electron exists are not infinitesimally sharp or discrete. This, of course, is stated in Heisenberg's uncertainty principle $\Delta E \cdot \Delta t \gtrsim h$ where ΔE and Δt refer to the accuracy of measurements of energy and time. Consequently, we allow the excited electron to exist initially in a specified sub-state of the broadened upper level $|2\rangle$ and, subsequently, to de-excite to any one of an infinite number of sub-states of the broadened lower (de-excited) state $|1\rangle$ as shown in figure 6.5

We can no longer, therefore, refer specifically to state $|2\rangle$ or state $|1\rangle$ and still retain an explicit description of the system. We thence introduce further indices α and β such that $|\beta\rangle$, for example, refers to the state of the system given the excited electron in the sub-state $|\beta, n_{\gamma\lambda}\rangle$ with $n_{\gamma\lambda}$ photons present. We are particularly interested in the non-degenerate case in which sub-levels $|\beta\rangle$ of state $|2\rangle$ do not overlap with sub-levels $|\alpha\rangle$ of state $|1\rangle$.

The basic equation (6.3.12) stipulating the time derivative of the probability amplitudes is repeated here:

$$i\hbar\dot{a}_m(t) = \sum_n a_n(t)\langle m|\hat{H}_1|n\rangle e^{i(E_m - E_n)t/\hbar}. \tag{6.4.50}$$

Thus, the appropriate equation for $a_\beta(t)$ is

$$i\hbar\dot{a}_\beta(t) = \sum_\alpha a_\alpha(t)\langle \beta|\hat{H}_1|\alpha\rangle e^{i(\omega_{\beta\alpha} - \omega_\gamma)t}, \tag{6.4.51}$$

where the excited state $|\beta\rangle$ can now de-excite to any of the $|\alpha\rangle$ states. However, the $|\alpha\rangle$ state can only arise from the $|\beta\rangle$ state explicitly in our model because of the initial conditions

$$a_\beta(0) = 1, \quad a_\alpha(0) = 0. \tag{6.4.52}$$

We therefore have

$$i\hbar\dot{a}_\alpha(t) = a_\beta(t)\langle \alpha|\hat{H}_1|\beta\rangle e^{i(\omega_\gamma - \omega_{\beta\alpha})t}. \tag{6.4.53}$$

From equations (6.4.23) and (6.4.25) we have

$$\langle \beta|\hat{H}_1|\alpha\rangle = Q_{\beta\alpha}\sqrt{(n_{\gamma\lambda} + 1)}, \tag{6.4.54}$$

$$\langle \alpha|\hat{H}_1|\beta\rangle = Q^*_{\beta\alpha}\sqrt{(n_{\gamma\lambda} + 1)} \tag{6.4.55}$$

- note that the matrix elements $\langle \beta|\hat{H}_1|\beta\rangle$ and $\langle \alpha|\hat{H}_1|\alpha\rangle$ are always zero for single photon processes.

The above equations then become

$$i\hbar\dot{a}_\beta(t) = \sum_\alpha a_\alpha(t)Q_{\beta\alpha}\sqrt{(n_{\gamma\lambda} + 1)}e^{i(\omega_{\beta\alpha} - \omega_\gamma)t}, \tag{6.4.56}$$

$$i\hbar\dot{a}_\alpha(t) = a_\beta(t)Q^*_{\beta\alpha}\sqrt{(n_{\gamma\lambda} + 1)}e^{i(\omega_\gamma - \omega_{\beta\alpha})t}. \tag{6.4.57}$$

If we now use the attempted solution (6.4.49), viz.

$$a_\beta(t) = e^{-\eta t}, \tag{6.4.58}$$

and the boundary conditions (6.4.52), we find

$$a_\alpha(t) = \frac{Q^*_{\beta\alpha}\sqrt{(n_{\gamma\lambda} + 1)}}{\hbar} \cdot \frac{1 - e^{i(\omega_\gamma - \omega_{\beta\alpha})t - \eta t}}{\omega_\gamma - \omega_{\beta\alpha} + i\eta}. \tag{6.4.59}$$

Fig.6.5. The quantum representation of the upper and lower bound electron states with both levels being naturally broadened.

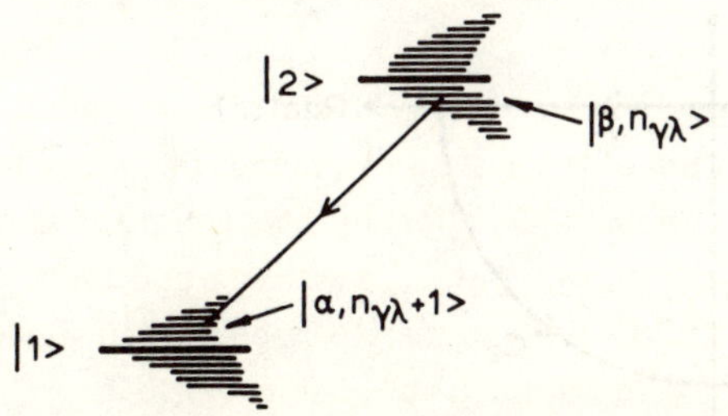

The decay constant η may then be obtained by substituting equations (6.4.48, 59) back into (6.4.56), i.e.

$$\eta = \frac{i}{\hbar^2} \sum_{\alpha} |Q_{\beta\alpha}|^2 (n_{\gamma\lambda} + 1) \frac{e^{i(\omega_{\beta\alpha} - \omega_\gamma)t + \eta t} - 1}{\omega_\gamma - \omega_{\beta\alpha} + i\eta}. \tag{6.4.60}$$

The form of η given by this last result is unsatisfactory since it implicitly appears to depend upon the time t, and this is obviously inconsistent with our attempted solution (6.4.58) for which η is constant. This difficulty may be 'partly' removed by replacing the summation over α in equation (6.4.60) by an integration of the form (see equation (6.4.33))

$$\eta = \frac{i}{4\pi\hbar} \sum_{\lambda} \int d\Omega \int_0^\infty |Q_{\beta\alpha}|^2 (n_{\gamma\lambda} + 1) \frac{e^{i(E_{\beta\alpha}^{(A)} - E)t/\hbar + \eta t} - 1}{E - E_{\beta\alpha}^{(A)} + i\hbar\eta} \cdot \rho(E)\, dE. \tag{6.4.61}$$

We next write the real and imaginary parts of η as

$$\eta = \frac{\Gamma}{2} + i\Delta\omega, \tag{6.4.62}$$

with $\Gamma > 0$ so that, from equation (6.4.58), we have

$$|a_\beta(t)|^2 = e^{-\Gamma t}, \tag{6.4.63}$$

which exhibits the required exponential time decay of the $|\beta\rangle$ state probability.

If we now define

$$f(E) = \sum_{\lambda} \int d\Omega |Q_{\beta\alpha}|^2 (n_{\gamma\lambda} + 1) \rho(E), \tag{6.4.64}$$

and change the variable of integration from E to $x = E - E_{\beta\alpha}^{(A)}$, equations (6.4.61) and (6.4.62) yield

Fig.6.6. The contour C in the lower-half complex z plane.

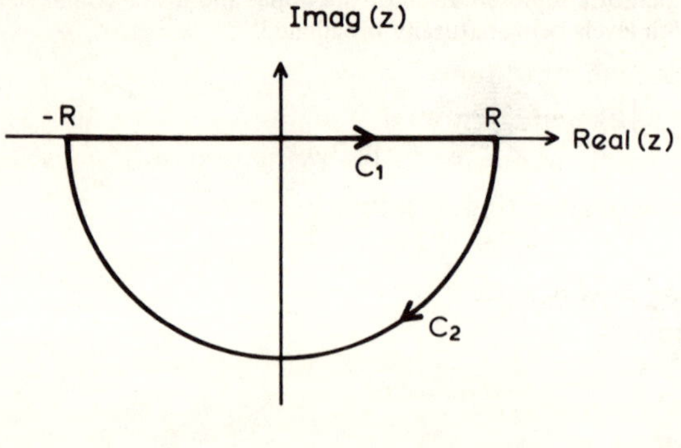

$$\frac{\Gamma}{2}+\mathrm{i}\Delta\omega=\frac{\mathrm{i}}{4\pi\hbar}\int_{-\infty}^{\infty}f(E_{\beta\alpha}^{(A)}+x)$$

$$\cdot\frac{\mathrm{e}^{\mathrm{i}(\hbar\Delta\omega-x)t/\hbar+\frac{\Gamma t}{2}}-1}{x-\hbar\Delta\omega+\mathrm{i}\hbar\Gamma/2}\cdot\mathrm{d}x, \qquad (6.4.65)$$

where the lower limit ($-\infty$) of integration has been obtained by effectively defining $\rho(E)$, thence $f(E)$, to be zero for all $E<0$.

Constructing the contour $C=C_1\cup C_2$ as shown in figure 6.6, letting $R\to\infty$, and using Cauchy's integral theorem, we find

$$\int_{-\infty}^{\infty}f(E_{\beta\alpha}^{(A)}+x)\cdot\frac{\mathrm{e}^{\mathrm{i}(\hbar\Delta\omega-x)t/\hbar+\frac{\Gamma t}{2}}-1}{x-\hbar\Delta\omega+\mathrm{i}\hbar\Gamma/2}\cdot\mathrm{d}x$$

$$+\lim_{R\to\infty}\int_{2\pi}^{\pi}f(E_{\beta\alpha}^{(A)}+R\mathrm{e}^{\mathrm{i}\theta})\cdot\frac{\mathrm{e}^{\mathrm{i}(\hbar\Delta\omega-R\mathrm{e}^{\mathrm{i}\theta})t/\hbar+\frac{\Gamma t}{2}}-1}{x-\hbar\Delta\omega+\mathrm{i}\hbar\Gamma/2}R\mathrm{e}^{\mathrm{i}\theta}\,\mathrm{i}\mathrm{d}\theta$$

$$=0. \qquad (6.4.66)$$

Note that the pole at $z=\hbar\Delta\omega-\mathrm{i}\Gamma/2$ is removable. Now $\mathrm{e}^{-\mathrm{i}R\mathrm{e}^{\mathrm{i}\theta}}=\mathrm{e}^{-\mathrm{i}R\cos\theta+R\sin\theta}\to 0$ as $R\to\infty$ for all $\theta\in[\pi, 2\pi]$. We next put

$$\lim_{R\to\infty}f(E_{\beta\alpha}^{(A)}+R\mathrm{e}^{\mathrm{i}\theta})\equiv f(E_{21}^{(A)}), \qquad (6.4.67)$$

i.e. we have merely replaced f by its average (resonance) value, so that

$$\lim_{R\to\infty}\sum_{C_2}=\pi\mathrm{i}f(E_{21}^{(A)}).$$

Equation (6.4.65) then becomes

$$\frac{\Gamma}{2}+\mathrm{i}\Delta\omega=\frac{1}{4\hbar}f(E_{21}^{(A)})$$

$$=\frac{1}{4\hbar}\sum_{\lambda}\int\mathrm{d}\Omega\,|Q_{21}|^2(n_{\gamma\lambda}+1)\rho(E_{21}^{(A)}), \qquad (6.4.68)$$

which, we note, is independent of time.

Further, reference to figure 6.5 shows that $\omega_{\beta\alpha}\approx\omega_{21}$ where, for example, state $|2\rangle$ refers to the 'centre of gravity' of the $|\beta\rangle$ state spread. Equation (6.4.68), together with equation (6.4.37), then becomes

$$\eta=\frac{\Gamma}{2}=\tfrac{1}{2}w(2\to 1). \qquad (6.4.69)$$

Note that in the above analysis, $\Delta\omega$ vanishes so that η is purely real. We shall return to this point in a moment.

Equation (6.4.59) then yields

$$|a_\alpha(\infty)|^2 \approx \frac{|Q_{\beta\alpha}|^2(n_{\gamma\lambda}+1)}{\hbar^2}\cdot\frac{1}{(\omega_\gamma-\omega_{\beta\alpha})^2+\left(\frac{\Gamma}{2}\right)^2}, \quad (6.4.70)$$

which is the usual Lorentzian probability distribution.

Since we have $|a_\beta(\infty)|^2 = 0$, we should expect

$$\sum_\alpha |a_\alpha(\infty)|^2 = 1. \quad (6.4.71)$$

We test this by integrating equation (6.4.70) over all the $|\alpha\rangle$ states to form the integral

$$\Phi = \frac{1}{4\pi}\sum_\lambda \int d\Omega \int_0^\infty |Q_{\beta\alpha}|^2(n_{\gamma\lambda}+1)\frac{\rho(E)\,dE}{(E-E_{\beta\alpha}^{(A)})^2+\left(\frac{\hbar\Gamma}{2}\right)^2}. \quad (6.4.72)$$

The last integral in equation (6.4.72) may be evaluated by changing the integration variable from E to $x = E - E_{\beta\alpha}^{(A)}$ and integrating around the contour $C = C_1 \cup C_2$ shown in figure 6.7. We then find

$$\lim_{R\to\infty}\int_C \frac{f(z)\,dz}{z^2+\left(\frac{\hbar\Gamma}{2}\right)^2} = \int_{-\infty}^{\infty}\frac{f(E_{\beta\alpha}^{(A)}+x)\,dx}{x^2+\left(\frac{\hbar\Gamma}{2}\right)^2} + \lim_{R\to\infty}$$

$$\times \int_0^\pi \frac{f(E_{\beta\alpha}^{(A)}+Re^{i\theta})\,Re^{i\theta}\,i d\theta}{R^2e^{2i\theta}+\left(\frac{\hbar\Gamma}{2}\right)^2}$$

$$= 2\pi i\cdot\frac{f(E_{\beta\alpha}^{(A)}+i\Gamma\hbar/2)}{i\hbar\Gamma},$$

i.e.

$$\Phi \approx \frac{1}{2\hbar\Gamma}\sum_\lambda \int d\Omega\,|Q_{21}|^2(n_{\gamma\lambda}+1)\rho(E_{21}^{(A)}), \quad (6.4.73)$$

Fig.6.7. The contour C in the upper-half complex z plane.

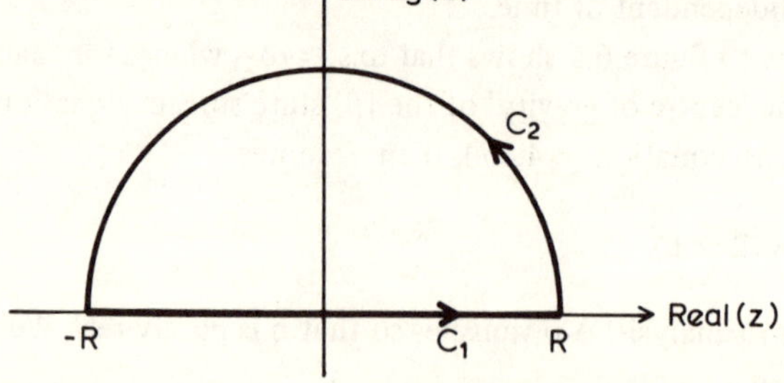

where, since we are considering only the non-degenerate case, we have taken $\hbar\Gamma \ll E_{21}^{(A)}$, i.e. $\Gamma \ll \omega_{21}$. We then have, from equations (6.4.37) and (6.4.69), $\Phi = 1$ as required.

The above analysis would appear to be quite valid since the integration replacement of the summation leading to equation (6.4.61) yields a time-independent η as required. However, a slightly different result, giving rise to a non-zero $\Delta\omega$, follows when we allow $t \to \infty$. For example, application of the formula (6.4.41) to equation (6.4.61), taking the limit as $t \to \infty$, yields

$$\eta = \frac{\mathrm{i}}{4\pi\hbar}\sum_{\lambda}\int \mathrm{d}\Omega \int_0^{\infty} |Q_{\beta\alpha}|^2(n_{\gamma\lambda}+1)\left\{-\mathscr{P}\frac{1}{E-E_{\beta\alpha}^{(A)}+\mathrm{i}\hbar\eta} - \pi\mathrm{i}\delta(E-E_{\beta\alpha}^{(A)}+\mathrm{i}\hbar\eta)\right\}\rho(E)\,\mathrm{d}E$$

$$= \frac{\Gamma}{2} + \mathrm{i}\Delta\omega. \tag{6.4.74}$$

Thus, separating out real and imaginary parts, and again recognising that $\eta \ll E_{\beta\alpha}^{(A)} \approx E_{21}^{(A)}$, we have

$$\frac{\Gamma}{2} \approx \frac{1}{4\hbar}\sum_{\lambda}\int \mathrm{d}\Omega\, |Q_{21}|^2(n_{\gamma\lambda}+1)\rho(E_{21}^{(A)}), \tag{6.4.75}$$

in agreement with equation (6.4.68), and

$$\Delta\omega \approx -\frac{1}{4\pi\hbar}\sum_{\lambda}\int \mathrm{d}\Omega\,\mathscr{P}\int_0^{\infty}\frac{|Q_{\beta\alpha}|^2(n_{\gamma\lambda}+1)\rho(E)\,\mathrm{d}E}{E-E_{\beta\alpha}^{(A)}+\mathrm{i}\hbar\eta}. \tag{6.4.76}$$

This last term constitutes an energy shift of the levels, but is usually small when compared to the energy difference $E_{21}^{(A)}$ between the levels for spectral lines of astrophysical interest. This may be more clearly seen if we now include $\Delta\omega$ via η in the equation (6.4.70) stipulating the Lorentzian profile:

$$|a_\alpha(\infty)|^2 \approx \frac{|Q_{\beta\alpha}|^2(n_{\gamma\lambda}+1)}{\hbar^2}\cdot\frac{1}{(\omega_\gamma-\omega_{\beta\alpha}-\Delta\omega)^2+\left(\frac{\Gamma}{2}\right)^2}. \tag{6.4.77}$$

The above analysis is a special case of the discussions presented in the following sections.

6.4.4 *Natural broadening – general method*

We wish to present a general method of solution for the probability amplitude $a_m(t)$ by which, not only emission and absorption processes, but scattering processes involving more than one photon, may be studied. We follow the general presentation of Heitler [49] (section 16).

We again wish to solve the system of equations

$$i\hbar\dot{a}_m(t) = \sum_n a_n(t)\langle m|\hat{H}_1|n\rangle e^{i(E_m-E_n)t/\hbar}, \tag{6.4.78}$$

where we stress that E_m, for example, refers to the total energy of the system (photons *and* particles) in state $|m\rangle$.

The initial conditions we state quite generally as

$$a_\beta(0) = 1, \quad a_n = 0 \quad (n \neq \beta), \tag{6.4.79}$$

i.e. the system is in the sub-state $|\beta\rangle$ initially. However, we wish to extend our domain of t from $[0, \infty)$ to $(-\infty, \infty)$ purely for mathematical considerations. A satisfactory solution of equation (6.4.78) for $t < 0$ is simply

$$a_m(t) = 0, \tag{6.4.80}$$

for all m. This, however, poses a difficulty at $t = 0$ since $a_\beta(t)$ jumps discontinuously from 0 to 1. Such a jump may be represented by the Heaviside unit step function, the derivative of which is the Dirac delta function. We then have

$$\dot{a}_\beta(t) = \delta(t) \quad \text{at } t = 0. \tag{6.4.81}$$

Equation (6.4.78) can then be re-written for all $t \in (-\infty, \infty)$ in the form

$$i\hbar\dot{a}_m(t) = \sum_n a_n(t)H_{mn}e^{i(E_m-E_n)t/\hbar}, \quad m \neq \beta, \tag{6.4.82}$$

and

$$i\hbar\dot{a}_\beta(t) = i\hbar\delta(t) + \sum_n a_n(t)H_{\beta n}e^{i(E_\beta-E_n)t/\hbar}, \tag{6.4.83}$$

where we have written the time-independent matrix elements $\langle m|\hat{H}_1|n\rangle$ as H_{mn}.

These last two equations can then be solved by constructing the transformation

$$a_n(t) = -\frac{1}{2\pi i}\int_{-\infty}^{\infty} G_n(E)e^{i(E_n-E)t/\hbar}\,dE. \tag{6.4.84}$$

The delta function $\delta(t)$ appearing in equation (6.4.83) may be put into a similar form by noting, using equation (6.4.41), that

$$\lim_{y\to\infty}\frac{1-e^{ixy}}{x} = -i\int_0^\infty e^{ixy}\,dy$$

$$= \mathscr{P}\frac{1}{x} - \pi i\delta(x), \tag{6.4.85}$$

so that

$$\int_{-\infty}^{\infty} e^{ixy}\,dy = 2\pi\delta(x). \tag{6.4.86}$$

Replacing y by $(E_\beta - E)/\hbar$, we find

$$i\hbar\delta(t) = -\frac{1}{2\pi i}\int_{-\infty}^{\infty} e^{i(E_\beta-E)t/\hbar}\,dE. \tag{6.4.87}$$

Substitution of equations (6.4.84) and (6.4.87) into (6.4.82) then yields

$$\frac{1}{2\pi i}\int_{-\infty}^{\infty}(E_m - E)G_m(E)e^{i(E_m - E)t/\hbar}\,dE$$

$$= -\frac{1}{2\pi i}\sum_n H_{mn}\int_{-\infty}^{\infty} G_n(E)e^{i(E_m - E)t/\hbar}\,dE, \tag{6.4.88}$$

for $m \neq \beta$, whilst equation (6.4.83) becomes

$$\frac{1}{2\pi i}\int_{-\infty}^{\infty}(E_\beta - E)G_\beta(E)e^{i(E_\beta - E)t/\hbar}\,dE$$

$$= -\frac{1}{2\pi i}\int_{-\infty}^{\infty} e^{i(E_\beta - E)t/\hbar}\,dE$$

$$-\frac{1}{2\pi i}\sum_n H_{\beta n}\int_{-\infty}^{\infty} G_n(E)e^{i(E_\beta - E)t/\hbar}\,dE. \tag{6.4.89}$$

We then write

$$(E - E_m)G_m(E) = \sum_n H_{mn}G_n(E), \quad m \neq \beta, \tag{6.4.90}$$

and

$$(E - E_\beta)G_\beta(E) = 1 + \sum_n H_{\beta n}G_n(E). \tag{6.4.91}$$

These last two equations are obviously sufficient to satisfy equations (6.4.88) and (6.4.89). However, since these quantities appear under an integral, it is not obvious that they are necessarily the required solutions. We shall return to this point later.

Clearly, the functions $G_m(E)$ are required, and these should be evaluated by dividing equation (6.4.90), for example, by $E - E_m$. However, we must allow for the possibility that $E - E_m = 0$ (see the singular eigenfunction technique presented in chapter 2). If we again return to equation (6.4.41) and define the new function $\zeta(x)$ such that

$$\zeta(x) = \mathscr{P}\frac{1}{x} - \pi i\delta(x), \tag{6.4.92}$$

then we have

$$x\zeta(x) = 1 - \pi i x\delta(x) = 1, \tag{6.4.93}$$

where, in this last equation, we have noted that $x\delta(x) = 0$ since $\delta(x) = 0$ for all $x \neq 0$.

Consequently, we see from equations (6.4.90) and (6.4.91), and replacing x by $E - E_m$ in equation (6.4.93), that the $G_m(E)$ function must simply be proportional to $\zeta(E - E_m)$. We then write the non-singular proportionality factor as $\bar{U}_m(E)$ such that

$$G_m(E) = \bar{U}_m(E)\zeta(E - E_m), \quad m \neq \beta. \tag{6.4.94}$$

Equation (6.4.90) then yields (note that $(E - E_m)\zeta(E - E_m) = 1$),

$$\bar{U}_m(E) = H_{m\beta}G_\beta(E) + \sum_{n \neq \beta} H_{mn}\bar{U}_n(E)\zeta(E - E_n). \tag{6.4.95}$$

It is convenient to define a further function $U_m(E)$ such that

$$\bar{U}_m(E) = U_m(E)G_\beta(E), \tag{6.4.96}$$

so that equation (6.4.95) becomes

$$U_m(E) = H_{m\beta} + \sum_{n \neq \beta} H_{mn}U_n(E)\zeta(E - E_n), \tag{6.4.97}$$

independent of $G_\beta(E)$. Equation (6.4.91) then yields

$$(E - E_\beta)G_\beta(E) = 1 + H_{\beta\beta}G_\beta(E) + \sum_{n \neq \beta} H_{\beta n}U_n(E)G_\beta(E)\zeta(E - E_n). \tag{6.4.98}$$

This immediately gives

$$G_\beta(E) = \frac{1}{E - E_\beta + i\hbar\eta_\beta(E)}, \tag{6.4.99}$$

where we have written

$$i\hbar\eta_\beta(E) = -H_{\beta\beta} - \sum_{n \neq \beta} H_{\beta n}U_n(E)\zeta(E - E_n). \tag{6.4.100}$$

Substitution of the above results into equation (6.4.84) gives

$$a_n(t) = -\frac{1}{2\pi i}\int_{-\infty}^{\infty} \frac{U_n(E)\zeta(E - E_n)e^{i(E_n - E)t/\hbar}}{E - E_\beta + i\hbar\eta_\beta(E)}\, dE, \tag{6.4.101}$$

for $n \neq \beta$, and

$$a_\beta(t) = -\frac{1}{2\pi i}\int_{-\infty}^{\infty} \frac{e^{i(E_n - E)t/\hbar}}{E - E_\beta + i\hbar\eta_\beta(E)}\, dE. \tag{6.4.102}$$

The initial condition $a_n(0) = 0$ for all $n \neq \beta$ allows equation (6.4.101) to be re-written as

$$a_n(t) - a_n(0) = -\frac{1}{2\pi i}\int_{-\infty}^{\infty} \frac{U_n(E)\zeta(E - E_n)}{E - E_\beta + i\hbar\eta_\beta(E)}$$

$$\times\, [e^{i(E_n - E)t/\hbar} - 1]\,\frac{E - E_n}{E - E_n}\cdot dE,$$

so that

$$a_n(t) = -\frac{1}{2\pi i}\int_{-\infty}^{\infty} \frac{U_n(E)[e^{i(E_n - E)t/\hbar} - 1]}{(E - E_\beta + i\hbar\eta_\beta(E))(E - E_n)}\, dE, \tag{6.4.103}$$

for $n \neq \beta$ (again note that $(E-E_n)\zeta(E-E_n)=1$). We shall be interested in times $t \to \infty$. From equations (6.4.85) and (6.4.92), we have

$$\lim_{t\to\infty} \frac{e^{i(E_n-E)t/\hbar}-1}{E_n-E} = -\zeta(E_n-E), \tag{6.4.104}$$

so that

$$a_n(\infty) = -\frac{1}{2\pi i}\int_{-\infty}^{\infty} \frac{U_n(E)\zeta(E_n-E)}{E-E_\beta+i\hbar\eta_\beta(E)}\,dE. \tag{6.4.105}$$

However, from equation (6.4.101), we have

$$a_n(0) = 0 = -\frac{1}{2\pi i}\int_{-\infty}^{\infty} \frac{U_n(E)\zeta(E-E_n)}{E-E_\beta+i\hbar\eta_\beta(E)}\,dE, \tag{6.4.106}$$

which, when added to equation (6.4.105) and noting from equation (6.4.92) that

$$\zeta(E_n-E)+\zeta(E-E_n) = -2\pi i\delta(E-E_n), \tag{6.4.107}$$

yields

$$\begin{aligned} a_n(\infty) &= \int_{-\infty}^{\infty} \frac{U_n(E)\zeta(E-E_n)}{E-E_\beta+i\hbar\eta_\beta(E)}\,dE \\ &= \frac{U_n(E_n)}{E_n-E_\beta+i\hbar\eta_\beta(E_n)}. \end{aligned} \tag{6.4.108}$$

We now wish to obtain expressions for the real and imaginary components of $\eta_\beta(E_n)$. First, we write

$$\eta_\beta(E_n) = \tfrac{1}{2}\Gamma_\beta(E_n) + i\Delta\omega_\beta(E_n), \tag{6.4.109}$$

so that equation (6.4.101) becomes

$$a_n(t) = -\frac{1}{2\pi i}\int_{-\infty}^{\infty} \frac{U_n(E)\zeta(E-E_n)e^{i(E_n-E)t/\hbar}}{E-E_\beta-\Delta E_\beta+i\hbar\Gamma_\beta(E_n)/2}\,dE, \tag{6.4.110}$$

and attempt to obtain Γ_β in terms of the transition rates discussed in sections 6.4.1 and 6.4.2.

Noting that $\zeta(x)$ also has the form [49] (see appendix D)

$$\zeta(x) = \lim_{\sigma\to\infty} \frac{1}{x+i\sigma}, \tag{6.4.111}$$

and recognising that $\hbar\Gamma_\beta(E_n) \ll E-E_\beta$ for most situations of physical interest thus allowing us to take $\Gamma_\beta(E_n)\to 0$ in equation (6.4.110), we find

$$\lim_{\Gamma_\beta\to 0} a_n(t) = -\frac{1}{2\pi i} \times \int_{-\infty}^{\infty} U_n(E)\zeta(E-E_n)\zeta(E-E_\beta-\Delta E_\beta)e^{i(E_n-E)t/\hbar}\,dE. \tag{6.4.112}$$

Now

$$\zeta(E-E_n)\zeta(E-E_\beta-\Delta E_\beta)$$
$$\equiv \lim_{\sigma_1\to 0}\frac{1}{E-E_n+\mathrm{i}\sigma_1}\cdot\lim_{\sigma_2\to 0}\frac{1}{E-E_\beta-\Delta E_\beta+\mathrm{i}\sigma_2}$$
$$=\lim_{\substack{\sigma_1\to 0\\ \sigma_2\to 0}}\frac{1}{E_\beta+\Delta E_\beta-E_n+\mathrm{i}(\sigma_1-\sigma_2)}$$
$$\times\left\{\frac{1}{E-E_\beta-\Delta E_\beta+\mathrm{i}\sigma_2}-\frac{1}{E-E_n+\mathrm{i}\sigma_1}\right\}$$
$$=\lim_{\substack{\sigma_1\to 0\\ \sigma_2\to 0}}\frac{1}{E_\beta+\Delta E_\beta-E_n+\mathrm{i}(\sigma_1-\sigma_2)}$$
$$\times[\zeta(E-E_\beta-\Delta E_\beta)-\zeta(E-E_n)]. \tag{6.4.113}$$

The remaining limit in this last equation yields $\zeta(E_\beta+\Delta E_\beta-E_n)$ if we take the limit $\sigma_1\to 0$ first, then $\sigma_2\to 0$, but gives $\zeta^*(E_\beta+\Delta E_\beta-E_n)$ if the reverse ordering is taken. Since σ_1 and σ_2 are quite independent, it does not matter which limit we choose first (the result should be the same regardless) and we therefore put

$$\zeta(E-E_n)\zeta(E-E_\beta-\Delta E_\beta)=\zeta(E_\beta+\Delta E_\beta-E_n)$$
$$\times[\zeta(E-E_\beta-\Delta E_\beta)-\zeta(E-E_n)]. \tag{6.4.114}$$

We will wish to take the limit as $t\to\infty$ in equation (6.4.112). The following result will prove useful:

$$\lim_{t\to\infty}\zeta(x)\mathrm{e}^{-\mathrm{i}xt}=x\zeta(x)\lim_{t\to\infty}\frac{1-\mathrm{e}^{-\mathrm{i}xt}}{-x}+\zeta(x)$$
$$=\zeta(-x)+\zeta(x)$$
$$=-2\pi\mathrm{i}\delta(x), \tag{6.4.115}$$

where we have noted that $x\zeta(x)=1$.

Hence, the behaviour of $a_n(t)$ for small $\Gamma_\beta(E_n)$ and large t is given by

$$a_n(t)\sim-\frac{1}{2\pi\mathrm{i}}\zeta(E_\beta+\Delta E_\beta-E_n)\int_{-\infty}^{\infty}U_n(E)$$
$$\times[\zeta(E-E_\beta-\Delta E_\beta)-\zeta(E-E_n)]\mathrm{e}^{\mathrm{i}(E_n-E)t/\hbar}\,\mathrm{d}E$$
$$\sim\zeta(E_\beta+\Delta E_\beta-E_n)\{U_n(E_\beta+\Delta E_\beta)\mathrm{e}^{\mathrm{i}(E_n-E_\beta-\Delta E_\beta)t/\hbar}-U_n(E_n)\}. \tag{6.4.116}$$

The transition probability $w(\beta\to n)$, discussed in sections 6.4.1 and 6.4.2, and defined by

$$w(\beta\to n)=\frac{1}{4\pi}\sum_\lambda\int\mathrm{d}\Omega\int_0^\infty\rho(E_n)w_{\gamma\lambda}(\beta\to n)\,\mathrm{d}E_n, \tag{6.4.117}$$

where

$$w_{\gamma\lambda}(\beta \to n) = \lim_{t\to\infty} \frac{\mathrm{d}}{\mathrm{d}t} (a_n^*(t)a_n(t)), \tag{6.4.118}$$

requires $\dot{a}_n(t)$ and $\dot{a}_n^*(t)$. From equation (6.4.116) we have for large t

$$\dot{a}_n(t) \sim -\frac{\mathrm{i}}{\hbar} U_n(E_\beta + \Delta E_\beta)\mathrm{e}^{\mathrm{i}(E_n - E_\beta - \Delta E_\beta)t/\hbar}, \tag{6.4.119}$$

so that

$$\begin{aligned} w_{\gamma\lambda}(\beta \to n) = \lim_{t\to\infty} \frac{\mathrm{i}}{\hbar} \{ & U_n^*(E_\beta + \Delta E_\beta)\zeta(E_\beta + \Delta E_\beta - E_n) \\ & \times [U_n(E_\beta + \Delta E_\beta) - U_n(E_n)\mathrm{e}^{-\mathrm{i}(E_n - E_\beta - \Delta E_\beta)t/\hbar}] \\ & - U_n(E_\beta + \Delta E_\beta)\zeta^*(E_\beta + \Delta E_\beta - E_n) \\ & \times [U_n^*(E_\beta + \Delta E_\beta) - U_n^*(E_n)\mathrm{e}^{\mathrm{i}(E_n - E_\beta - \Delta E_\beta)t/\hbar}]\}. \end{aligned} \tag{6.4.120}$$

This last expression may be simplified significantly by noting that

$$\begin{aligned} \lim_{t\to\infty} \zeta(x)\mathrm{e}^{\mathrm{i}xt} &= -x\zeta(x) \lim_{t\to\infty} \frac{1 - \mathrm{e}^{\mathrm{i}xt}}{x} + \zeta(x), \\ &= 0, \end{aligned} \tag{6.4.121}$$

and

$$\zeta^*(x) = \mathscr{P}\frac{1}{x} + \pi\mathrm{i}\delta(x) = -\zeta(-x). \tag{6.4.122}$$

We then have

$$\begin{aligned} w_{\gamma\lambda}(\beta \to b) &= \frac{\mathrm{i}}{\hbar} |U_n(E_\beta + \Delta E_\beta)|^2 [\zeta(E_\beta + \Delta E_\beta - E_n) \\ &\quad - \zeta^*(E_\beta + \Delta E_\beta - E_n)] \\ &= \frac{2\pi}{\hbar} |U_n(E_\beta + \Delta E_\beta)|^2 \delta(E_\beta + \Delta E_\beta - E_n). \end{aligned} \tag{6.4.123}$$

Consequently, $w(\beta \to n)$ given by equation (6.4.117) is simply

$$w(\beta \to n) = \frac{1}{2\hbar} \sum_\lambda \int \mathrm{d}\Omega\, |U_n(E_\beta + \Delta E_\beta)|^2 \rho(E_\beta + \Delta E_\beta). \tag{6.4.124}$$

This should be compared with equation (6.4.37) obtained using a somewhat simplified analysis.

Explicit expressions for $\Gamma_\beta(E)$ appearing in equation (6.4.109) may now be obtained. From equation (6.4.100) we find

$$\hbar\left(\frac{\Gamma_\beta(E)}{2} + \mathrm{i}\Delta\omega_\gamma(E)\right) = \mathrm{i}H_{\beta\beta} + \mathrm{i} \sum_{n\neq\beta} H_{\beta n}U_n(E)\zeta(E - E_n). \tag{6.4.125}$$

Now $H_{\beta\beta}$ is real since

$$H_{\beta\beta} = \langle\beta|\hat{H}_1|\beta\rangle = \langle\beta|\hat{H}_1^+|\beta\rangle = H_{\beta\beta}^* \tag{6.4.126}$$

- note that $\hat{H}_1$ is Hermitian.

We then have

$$\frac{\hbar\Gamma_\beta(E)}{2} = \tfrac{1}{2}\sum_{n\neq\beta}\{iH_{\beta n}U_n(E)\zeta(E-E_n) + [iH_{\beta n}U_n(E)\zeta(E-E_n)]^*\},$$

which, using equation (6.4.122), becomes

$$\hbar\Gamma_\beta(E) = i\sum_{n\neq\beta}[H_{\beta n}U_n(E)\zeta(E-E_n) + H_{\beta n}^*U_n^*(E)\zeta(E_n-E)]. \tag{6.4.127}$$

The terms appearing on the RHS of this last equation may be evaluated using equation (6.4.97) written in the forms

$$\begin{aligned} H_{n\beta} &= U_n(E) - \sum_{m\neq\beta} H_{nm}U_m(E)\zeta(E-E_m) \\ &= H_{\beta n}^*, \end{aligned} \tag{6.4.128}$$

and

$$\begin{aligned} H_{\beta n} &= H_{n\beta}^* \\ &= U_n^*(E) + \sum_{m\neq\beta} H_{mn}U_m^*(E)\zeta(E_m-E_n). \end{aligned} \tag{6.4.129}$$

We then have

$$\begin{aligned} \hbar\Gamma_\beta(E) = i\sum_{n\neq\beta}\Big\{&U_n^*(E)U_n(E)|\zeta(E-E_n)+\zeta(E_n-E)| \\ &+\sum_{m\neq\beta}\{H_{mn}U_n(E)U_m^*(E)\zeta(E-E_n)\zeta(E_m-E) \\ &-H_{nm}U_n^*(E)U_m(E)\zeta(E_n-E)\zeta(E-E_m)\}\Big\}. \end{aligned} \tag{6.4.130}$$

The summation term in the curly brackets is zero because of the symmetry of the m and n indices in the double summation. We then have

$$\Gamma_\beta(E) = \frac{2\pi}{\hbar}\sum_{n\neq\beta}|U_n(E)|^2\delta(E-E_n). \tag{6.4.131}$$

It is clear from equation (6.4.123) therefore that

$$\Gamma_\beta(E_\beta + \Delta E_\beta) = \sum_{n\neq\beta} w_{\gamma\lambda}(\beta\to n). \tag{6.4.132}$$

This then completes the general method of solution of the equations specifying $a_n(t)$.

6.4.5 *Natural broadening - 2-level atom*

In the present section we wish to illustrate the foregoing theory by applying it to the problem already studied in section 6.4.3 in which an atom initially in the excited state de-excites thereby emitting a photon. Again, we do not allow the atom to be re-excited after it has emitted the photon, i.e. we are effectively examining spontaneous emission only.

Recall that we denoted the excited state as $|2\rangle$ and de-excited state as $|1\rangle$ (see figure 6.2). Equation (6.4.108) then becomes

$$a_1(\infty) = \frac{U_1(E_1)}{E_1 - E_2 + i\hbar\eta_2(E_1)}, \tag{6.4.133}$$

where, from equation (6.4.97),

$$U_1(E_1) = H_{12} + \sum_{n\neq 2} H_{1n}U_n(E_1)\zeta(E_1 - E_n). \tag{6.4.134}$$

The summation term in this last equation makes no contribution to $U_1(E_1)$ since H_{12} is the only non-zero matrix element which allows a single photon transition to the $|1\rangle$ state given that this radiative de-excitation must arise from the $|2\rangle$ state, i.e. we have

$$U_1(E_1) = H_{12}. \tag{6.4.135}$$

The broadening parameter $\eta_2(E_1)$ given by equation (6.4.109) is simply

$$\eta_2(E_1) = \frac{\Gamma_2(E_1)}{2} + i\Delta\omega_2(E_1), \tag{6.4.136}$$

where, from equation (6.4.131),

$$\Gamma_2(E_1) = \frac{2\pi}{\hbar}|U_1(E_1)|^2 = \frac{2\pi}{\hbar}|H_{12}|^2. \tag{6.4.137}$$

The above equations then yield

$$|a_1(\infty)|^2 = \frac{|H_{12}|^2}{(E_1 - E_2 - \hbar\Delta\omega_2)^2 + (\hbar\Gamma_2/2)^2}. \tag{6.4.138}$$

The energy of the system in states $|1\rangle$ and $|2\rangle$ are

$$\left.\begin{aligned} E_1 &= E_1^{(A)} + \sum_{\gamma\lambda} \hbar\omega_\gamma|(n_{\gamma\lambda} + 1) + \tfrac{1}{2}|, \\ E_2 &= E_2^{(A)} + \sum_{\gamma\lambda} \hbar\omega_\gamma(n_{\gamma\lambda} + \tfrac{1}{2}), \end{aligned}\right\} \tag{6.4.139}$$

so that

$$E_1 - E_2 = \hbar\omega_\gamma - \hbar\omega_{21}, \tag{6.4.140}$$

where

$$\omega_{21} = (E_2^{(A)} - E_1^{(A)})/\hbar.$$

Equation (6.4.138) is then

$$|a_1(\infty)|^2 = \frac{|H_{12}|^2}{\hbar^2} \cdot \frac{1}{(\omega_\gamma - \omega_{21} + \Delta\omega_2)^2 + (\Gamma_2/2)^2}, \tag{6.4.141}$$

in agreement with equation (6.4.77), although the lifetime term Γ appearing in equation (6.4.77) is the average over angle, plus sum over λ, of Γ_2 (see equation (6.4.75)).

6.4.6 *Natural broadening profile*

In the preceding section, we evaluated the probability of photon emission given the atom initially in the excited state. We know, however, that the excited state is also broadened – this was not taken into account in the previous analysis since we explicitly fixed the initial state. The purpose of the present section, therefore, is to evaluate the probability of emission given the atom in any of the excited continuum of sub-levels of level 2. To do this, we first allow the absorption of a photon by the atom in the de-excited state $|1\rangle$ so that the probability of the atom being in these excited $|2\rangle$ states is then known. We then follow the subsequent emission of the photon due to de-excitation from that excited state. Figure 6.8 details the two processes.

The atom is in the $|1\rangle$ state initially, then absorbs a photon of energy $\hbar\omega_1$ thus being excited to the $|2\rangle$ state. The atom then de-excites to state $|3\rangle$, emitting a photon of energy $\hbar\omega_2$. We have introduced a third level, denoted by state $|3\rangle$, purely for purposes of exposition – state $|3\rangle$ refers to the final state, state $|1\rangle$ to the initial state – we shall equate level 3 with level 1 with the understanding, however, that the initially de-excited atom will, in general, have a slightly different energy from that of the atom in the final de-excited state following the above scattering process.

Equation (6.4.108) gives

$$a_2(\infty) = \frac{U_2(E_2)}{E_2 - E_1 + i\hbar\eta_1(E_2)}, \qquad (6.4.142)$$

and

$$a_3(\infty) = \frac{U_3(E_3)}{E_3 - E_1 + i\hbar\eta_1(E_3)}, \qquad (6.4.143)$$

where, from equation (6.4.97), we have

$$\begin{aligned} U_2(E) &= H_{21} + \sum_{n\neq 1} H_{2n}U_n(E)\zeta(E-E_n) \\ &= H_{21} + H_{23}U_3(E)\zeta(E-E_3), \end{aligned} \qquad (6.4.144)$$

Fig.6.8. The quantum representation of a model 3-level atom, each level being naturally broadened, is illustrated in which radiative excitation from state |1> to state |2> is followed by radiative de-excitation from state |2> to |3>.

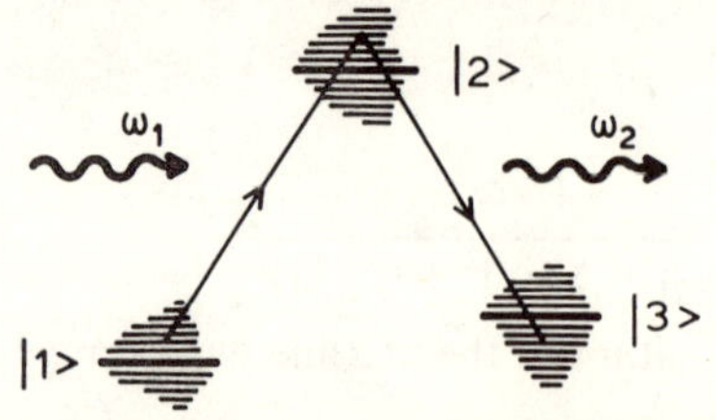

$$U_3(E) = H_{31} + \sum_{n \neq 1} H_{3n} U_n(E) \zeta(E - E_n)$$
$$= H_{32} U_2(E) \zeta(E - E_2). \qquad (6.4.145)$$

These last results follow since the matrix elements H_{22}, H_{31} and H_{33} are zero ($|1\rangle$ and $|3\rangle$ are states having the same atomic level) – we do not consider the scattering terms $\hat{\mathbf{A}} \cdot \hat{\mathbf{A}}$ given in equation (6.4.9) for $\hat{H}_1$ since they are of order $(e/c)^2$ compared with the terms of order e/c being considered here.

Substitution of equation (6.4.145) into (6.4.144) yields

$$U_2(E) = H_{21} + H_{23} H_{32} U_2(E) \zeta(E - E_2) \zeta(E - E_3). \qquad (6.4.146)$$

Noting, for example, that $(E - E_2)\zeta(E - E_2) = 1$, equation (6.4.146) becomes

$$(E - E_2) U_2(E) = (E - E_2) H_{21} + H_{23} H_{32} U_2(E) \zeta(E - E_3),$$

i.e.

$$U_2(E) = \frac{(E - E_2) H_{21}}{E - E_2 + i\hbar\eta_2(E)}, \qquad (6.4.147)$$

where we have defined

$$i\hbar\eta_2(E) = -H_{23} H_{32} \zeta(E - E_3). \qquad (6.4.148)$$

Equation (6.4.145) then yields

$$U_3(E) = \frac{H_{32} H_{21}}{E - E_2 + i\hbar\eta_2(E)}. \qquad (6.4.149)$$

Note that $U_2(E_2) = 0$ so that, from equation (6.4.142), $a_2(\infty) = 0$ as one would expect from the eventual decay due to spontaneous emission.

Equation (6.4.143), however, becomes

$$a_3(\infty) = \frac{H_{32} H_{21}}{[E_3 - E_2 + i\hbar\eta_2(E_3)][E_3 - E_1 + i\hbar\eta_1(E_3)]}. \qquad (6.4.150)$$

If we now put

$$\eta_k(E) = \frac{\Gamma_k(E)}{2} + i\Delta\omega_k(E), \qquad (6.4.151)$$

for $k = 1$ and 2, the probability amplitude for state $|3\rangle$ yields

$$|a_3(\infty)|^2 = \frac{|H_{32}|^2 |H_{21}|^2}{\left[(E_3 - E_2 - i\hbar\Delta\omega_2(E_3))^2 + \left(\dfrac{\hbar\Gamma_2(E_3)}{2}\right)^2\right]}$$
$$\times \frac{1}{\left[(E_3 - E_1 - i\hbar\Delta\omega_1(E_3))^2 + \left(\dfrac{\hbar\Gamma_1(E_3)}{2}\right)^2\right]}. \qquad (6.4.152)$$

We shall return to a more explicit analysis of $\eta_1(E_3)$ and $\eta_2(E_3)$ later in this section. First, we note that the initial energy E_1 of the system is

$$E_1 = E_1^{(A)} + E_0^{(R)}, \tag{6.4.153}$$

where $E_1^{(A)}$ is the energy of the atom in the de-excited $|1\rangle$ state and $E_0^{(R)}$ the energy of the initial radiation field. We then have

$$E_2 = E_2^{(A)} + E_0^{(R)} - \hbar\omega_1, \tag{6.4.154}$$

and

$$E_3 = E_3^{(A)} + E_0^{(R)} - \hbar\omega_1 + \hbar\omega_2. \tag{6.4.155}$$

Thus,

$$E_3 - E_2 = E_3^{(A)} - E_2^{(A)} + \hbar\omega_2 = \hbar(\omega_2 - \omega_{21}), \tag{6.4.156}$$

where we have put $E_3^{(A)} \equiv E_1^{(A)}$ and written $\omega_{21} = (E_2^{(A)} - E_1^{(A)})/\hbar$. Consequently, ignoring the relatively small terms $\Delta\omega_1(E_3)$ and $\Delta\omega_2(E_3)$, equation (6.4.152) becomes

$$|a_3(\infty)|^2 = \frac{|H_{32}|^2 |H_{21}|^2}{\hbar^4} p_1(\omega_1, \omega_2) p_2(\omega_2), \tag{6.4.157}$$

where

$$p_1(\omega_1, \omega_2) = \frac{1}{(\omega_2 - \omega_1)^2 + (\Gamma_1/2)^2}, \tag{6.4.158}$$

and

$$p_2(\omega_2) = \frac{1}{(\omega_2 - \omega_{21})^2 + (\Gamma_2/2)^2} \tag{6.4.159}$$

- it is implicitly understood that Γ_1 and Γ_2 are functions of E_3. Indeed, from equation (6.4.131), we have

$$\Gamma_1(E_3) = \frac{2\pi}{\hbar} |U_3(E_3)|^2,$$

which, using equation (6.4.149), is just

$$\Gamma_1 = \frac{2\pi}{\hbar} \cdot \frac{|H_{32}|^2 |H_{21}|^2}{(E_3 - E_2 - \hbar\Delta\omega_2(E_3))^2 + \left(\frac{\hbar\Gamma_2}{2}\right)^2}$$

$$= \frac{2\pi}{\hbar^3} \cdot \frac{|H_{32}|^2 |H_{21}|^2}{(\omega_2 - \omega_{21})^2 + (\Gamma_2/2)^2}, \tag{6.4.160}$$

where, again, we have neglected the level shift term $\Delta\omega_2(E_3)$.

We are primarily interested in the frequency dependence of $|a_3(\infty)|^2$ and, as can be seen from an examination of $p_1(\omega_1, \omega_2)$, this is hardly affected by the frequency dependence of Γ_1 above. We henceforth replace Γ_1 by its average over frequency, i.e. we put

$$\Gamma_1 = \frac{2\pi |H_{32}|^2 |H_{21}|^2}{\hbar^3} \int_0^\infty \frac{d(\hbar\omega_2)}{(\omega_2 - \omega_{21})^2 + (\Gamma_2/2)^2}. \tag{6.4.161}$$

The Γ_2 appearing in this last equation is given by (6.4.148), viz.

$$\frac{\Gamma_2}{2} = \text{Real}\, \eta_2(E_3) = \frac{\pi}{\hbar} |H_{32}|^2, \tag{6.4.162}$$

independent of ω_1 and ω_2.

We evaluate the above integral by defining the new variable $x_2 = \omega_2 - \omega_{21}$ which now measures the frequency difference from that attained for infinitesimally sharp levels 1 and 2. We then have

$$\int_0^\infty \frac{d\omega_2}{(\omega_2 - \omega_{21})^2 + (\Gamma_2/2)^2} = \int_{-\infty}^\infty \frac{dx_2}{x_2^2 + (\Gamma_2/2)^2} = \frac{2\pi}{\Gamma_2},$$

so that equation (6.4.161) yields

$$\Gamma_1 = \frac{(2\pi)^2}{\hbar^2} \cdot \frac{|H_{32}|^2 |H_{21}|^2}{\Gamma_2} = \frac{2\pi}{\hbar} |H_{21}|^2, \tag{6.4.163}$$

where we have used equation (6.4.162).

Returning to equation (6.4.157) we see that $|a_3(\infty)|^2$ gives the probability of the system being in state $|3\rangle$ after having absorbed a photon of frequency ω_1 and emitted a photon of frequency ω_2. Thus, to obtain the probability of emission we must now sum (integrate) over all the possible absorbed photons, i.e. we have the probability $P(2 \to 1)$ of emission in the form

$$P(2 \to 1) = \int_0^\infty |a_3(\infty)|^2 \, d(\hbar\omega_1) = \frac{|H_{32}|^2 |H_{21}|^2}{\hbar^4} p_2(\omega_2) \int_0^\infty p_1(\omega_1, \omega_2) \, d(\hbar\omega_1). \tag{6.4.164}$$

If we now replace the variable of integration ω_1 by $x_1 = \omega_1 - \omega_2$, and treat Γ_1 as a constant as above, we have

$$\int_0^\infty p_1(\omega_1, \omega_2) \, d\omega_1 = \int_0^\infty \frac{d\omega_1}{(\omega_2 - \omega_1)^2 + (\Gamma_1/2)^2} = \int_{-\infty}^\infty \frac{dx_1}{x_1^2 + (\Gamma_1/2)^2} = \frac{2\pi}{\Gamma_1}, \tag{6.4.165}$$

so that

$$P(2 \to 1) = \frac{2\pi |H_{32}|^2 |H_{21}|^2 p_2(\omega_2)}{\hbar^3 \Gamma_1}$$

$$= \frac{|H_{32}|^2}{\hbar^2} \cdot \frac{1}{(\omega_2 - \omega_{21})^2 + (\Gamma_2/2)^2}. \tag{6.4.166}$$

This is effectively the same result as that obtained in the preceding section.

We may similarly obtain the probability $P(1 \to 2)$ of photon absorption by summing over all the emitted photons. We then have

$$P(1 \to 2) = \int_0^\infty |a_3(\infty)|^2 \, d(\hbar\omega_3)$$

$$= \frac{|H_{32}|^2 |H_{21}|^2}{\hbar^3} \int_0^\infty p_1(\omega_1, \omega_2) p_2(\omega_2) \, d\omega_2$$

$$= \frac{|H_{32}|^2 |H_{21}|^2}{\hbar^3} I, \tag{6.4.167}$$

where the integral

$$I = \int_0^\infty \frac{d\omega_1}{[(\omega_2 - \omega_1)^2 + (\Gamma_1/2)^2][(\omega_2 - \omega_{21})^2 + (\Gamma_2/2)^2]}, \tag{6.4.168}$$

may be evaluated by replacing ω_2 by $x_2 = \omega_2 - \omega_{21}$ so that

$$I = \int_{-\infty}^\infty \frac{d\omega_2}{[(x_2 - \omega_1 + \omega_{21})^2 + (\Gamma_1/2)^2][x_2^2 + (\Gamma_2/2)^2]}$$

$$= \int_{-\infty}^\infty \frac{dx_2}{(x_2 - \bar{x}_1 + i\Gamma_1/2)(x_2 - \bar{x}_1 + i\Gamma_1/2)(x_2 + i\Gamma_2/2)(x_2 - i\Gamma_2/2)},$$

where we have put $\bar{x}_1 = \omega_1 - \omega_{21}$.

Integrating over the contour $C = C_1 \cup C_2$ shown in figure 6.9, noting that the integral over C_2 vanishes in the limit $R \to \infty$, and using Cauchy's integral theorem

Fig.6.9. The contour C in the upper-half complex z plane.

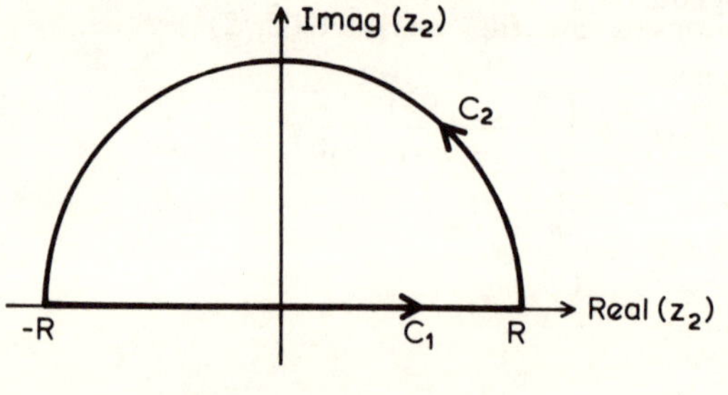

applied to the two poles $\bar{x}_1 + i\Gamma_1/2$ and $i\Gamma_2/2$ in the upper-half plane, we have

$$I = 2\pi i \left\{ \frac{1}{i\Gamma_1\left[\bar{x}_1 + \frac{i}{2}(\Gamma_1 + \Gamma_2)\right]\left[\bar{x}_1 + \frac{i}{2}(\Gamma_1 - \Gamma_2)\right]} + \frac{1}{\left[\frac{i}{2}(\Gamma_2 + \Gamma_1) - \bar{x}_1\right]\left[\frac{i}{2}(\Gamma_2 - \Gamma_1) - \bar{x}_1\right] i\Gamma_2} \right\}$$

$$= \frac{2}{\bar{x}_1 + \frac{i}{2}(\Gamma_1 - \Gamma_2)} \left\{ \frac{1}{\Gamma_1\left[\bar{x}_1 + \frac{i}{2}(\Gamma_1 + \Gamma_2)\right]} + \frac{1}{\Gamma_2\left[\bar{x}_1 - \frac{i}{2}(\Gamma_1 + \Gamma_2)\right]} \right\}$$

$$= \frac{2\pi}{\bar{x}_1 + \frac{i}{2}(\Gamma_1 - \Gamma_2)} \cdot \frac{(\Gamma_1 + \Gamma_2)\bar{x}_1 + \frac{i\Gamma_1}{2}(\Gamma_1 + \Gamma_2) - \frac{i\Gamma_2}{2}(\Gamma_1 + \Gamma_2)}{\Gamma_1\Gamma_2\left[\bar{x}_1^2 + \left(\frac{\Gamma_1 + \Gamma_2}{2}\right)^2\right]}$$

$$= \frac{2\pi(\Gamma_1 + \Gamma_2)}{\Gamma_1\Gamma_2} \frac{1}{\bar{x}_1^2 + \left(\frac{\Gamma_1 + \Gamma_2}{2}\right)^2},$$

i.e.

$$P(1 \to 2) = \frac{2\pi(\Gamma_1 + \Gamma_2)}{\Gamma_1\Gamma_2\hbar^3} \cdot \frac{|H_{32}|^2 |H_{21}|^2}{(\omega_1 - \omega_{21})^2 + \left(\frac{\Gamma_1 + \Gamma_2}{2}\right)^2}. \qquad (6.4.169)$$

Thus, the profile for absorption of a photon of frequency ω_1 is again a Lorentzian distribution, but the broadening parameter is now the sum $\Gamma_1 + \Gamma_2$.

6.4.7 *Scattering – no collisions*

We now consider the probability of the scattering of photons [72]. Specifically, we wish to evaluate the probability of a photon being absorbed and then re-emitted (at some other frequency). To do this, we consider a model 4-level atom as shown in figure 6.10. In particular, a photon of frequency ω_1 is first absorbed, then photons of frequency ω_2 and ω_3 are subsequently emitted.

Equation (6.4.108) yields

$$a_4(\infty) = \frac{U_4(E_4)}{E_4 - E_1 + i\hbar\eta_1(E_4)}, \qquad (6.4.170)$$

where, from equation (6.4.97),

$$U_4(E) = H_{41} + \sum_{n \neq 1} H_{4n}U_n(E)\zeta(E - E_n). \qquad (6.4.171)$$

If we allow transitions between the levels $1 \to 2$, $2 \to 3$ and $3 \to 4$ (as indicated by the solid arrowed lines in figure 6.10), then the only non-zero matrix elements are H_{12}, H_{21}, H_{23}, H_{32}, H_{34} and H_{43}. We then have

$$U_4(E) = H_{43}U_3(E)\zeta(E - E_3), \tag{6.4.172}$$

$$U_3(E) = H_{32}U_2(E)\zeta(E - E_2) + H_{34}U_4(E)\zeta(E - E_4), \tag{6.4.173}$$

and

$$U_2(E) = H_{21} + H_{23}U_3(E)\zeta(E - E_3). \tag{6.4.174}$$

Substitution of U_4 into U_3 yields

$$U_3(E) = H_{32}U_2(E)\zeta(E - E_2) + H_{34}H_{43}U_3(E)\zeta(E - E_3)\zeta(E - E_4),$$

so that

$$(E - E_3)U_3(E) = (E - E_3)H_{32}U_2(E)\zeta(E - E_2) + H_{34}H_{43}U_3(E)\zeta(E - E_4),$$

i.e.

$$U_3(E) = \frac{(E - E_3)H_{32}U_2(E)\zeta(E - E_2)}{E - E_3 + i\hbar\eta_3(E)}, \tag{6.4.175}$$

where we have defined

$$i\hbar\eta_3(E) = -|H_{34}|^2\zeta(E - E_4). \tag{6.4.176}$$

We then have for $U_2(E)$,

$$U_2(E) = H_{21} + \frac{H_{23}H_{32}U_2(E)\zeta(E - E_2)}{E - E_3 + i\hbar\eta_3(E)},$$

so that

$$(E - E_2)U_2(E) = (E - E_2)H_{21} + \frac{H_{23}H_{32}U_2(E)}{E - E_3 + i\hbar\eta_3(E)},$$

i.e.

$$U_2(E) = \frac{(E - E_2)H_{21}}{E - E_2 + i\hbar\eta_2(E)}, \tag{6.4.177}$$

where we have defined

$$i\hbar\eta_2(E) = \frac{-|H_{23}|^2}{E - E_3 + i\hbar\eta_3(E)}. \tag{6.4.178}$$

Fig.6.10. Scattering from state |1> to state |4> via states |2> thence |3>. All transitions are radiative.

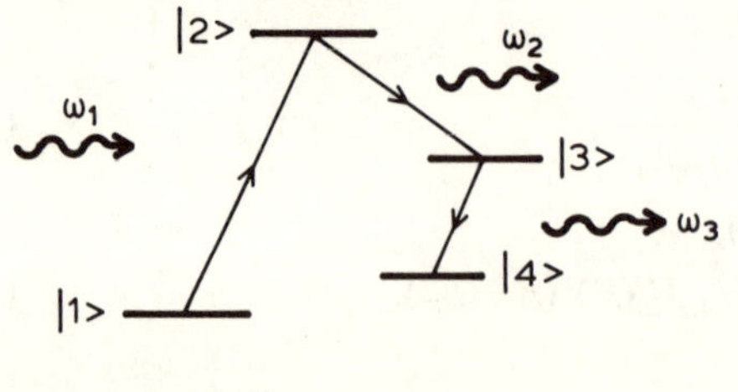

Substituting these results back into equation (6.4.172) then yields

$$U_4(E) = \frac{H_{43}(E-E_3)H_{32}U_2(E)\zeta(E-E_2)\zeta(E-E_3)}{E-E_3+i\hbar\eta_3(E)}$$

$$= \frac{H_{43}H_{32}H_{21}}{[E-E_3+i\hbar\eta_3(E)][E-E_2+i\hbar\eta_2(E)]}. \quad (6.4.179)$$

We now insert this last result into equation (6.4.170) so that

$$a_4(\infty) = \frac{H_{43}H_{32}H_{21}}{(E_4-E_1+i\hbar\eta_1)(E_4-E_2+i\hbar\eta_2)(E_4-E_3+i\hbar\eta_3)}. \quad (6.4.180)$$

If we again denote $E_1^{(A)}$, for example, as the energy of the atom in level 1 and $E_0^{(R)}$ the energy of the initial radiation field, then

$$\left.\begin{aligned} E_1 &= E_1^{(A)} + E_0^{(R)}, \\ E_2 &= E_2^{(A)} + E_0^{(R)} - \hbar\omega_1, \\ E_3 &= E_3^{(A)} + E_0^{(R)} - \hbar\omega_1 + \hbar\omega_2, \\ E_4 &= E_4^{(A)} + E_0^{(R)} - \hbar\omega_1 + \hbar\omega_2 + \hbar\omega_3. \end{aligned}\right\} \quad (6.4.181)$$

This then yields

$$E_4 - E_3 = \hbar\omega_3 - (E_3^{(A)} - E_4^{(A)}) = \hbar(\omega_3 - \omega_{34}) \quad (6.4.182)$$

and, similarly,

$$E_4 - E_2 = \hbar(\omega_2 - \omega_{23}) + \hbar(\omega_3 - \omega_{34}), \quad (6.4.183)$$

and

$$E_4 - E_1 = \hbar(\omega_2 + \omega_3 - \omega_1) + E_4^{(A)} - E_1^{(A)}$$

$$= \hbar(\omega_{21} - \omega_1) + \hbar(\omega_2 - \omega_{23}) + \hbar(\omega_3 - \omega_{34}). \quad (6.4.184)$$

Further, if we neglect the level shifts due to damping (i.e. the imaginary parts of η_1, η_2 and η_3 are put to zero), and write

$$\eta_k = \frac{\Gamma_k}{2}, \quad k = 1, 3, \quad (6.4.185)$$

equation (6.4.180) becomes

$$a_4(\infty) = \frac{H_{43}H_{32}H_{21}}{\left(E_4 - E_1 + \frac{i\hbar\Gamma_1}{2}\right)\left(E_4 - E_2 + \frac{i\hbar\Gamma_2}{2}\right)\left(E_4 - E_3 + \frac{i\hbar\Gamma_3}{2}\right)},$$

so that

$$|a_4(\infty)|^2 = \frac{|H_{43}|^2|H_{32}|^2|H_{21}|^2}{\hbar^6} p_1(\omega_1, \omega_2, \omega_3)p_2(\omega_2, \omega_3)p_3(\omega_3), \quad (6.4.186)$$

where

$$p_1(\omega_1, \omega_2, \omega_3) = \frac{1}{[(\omega_{21} - \omega_1) + (\omega_2 - \omega_{23}) + (\omega_3 - \omega_{34})]^2 + (\Gamma_1/2)^2}. \quad (6.4.187)$$

$$p_2(\omega_2, \omega_3) = \frac{1}{[(\omega_2 - \omega_{23}) + (\omega_3 - \omega_{34})]^2 + (\Gamma_2/2)^2}, \tag{6.4.188}$$

$$p_3(\omega_3) = \frac{1}{(\omega_3 - \omega_{34})^2 + (\Gamma_3/2)^2}. \tag{6.4.189}$$

The probability $p(\omega_1, \omega_2)$ of the scattering process whereby a photon of frequency ω_1 is absorbed *and* a photon of frequency ω_2 is emitted is then given by the 'sum'

$$p(\omega_1, \omega_2) = \int_0^\infty |a_4(\infty)|^3 \, d(\hbar\omega_3). \tag{6.4.190}$$

We next define $x_1 = \omega_1 - \omega_{21}$, $x_2 = \omega_2 - \omega_{23}$ and $x_3 = \omega_3 - \omega_{34}$ so that

$$p(\omega_1, \omega_2) = \frac{|H_{43}|^2 |H_{32}|^2 |H_{21}|^2}{\hbar^5} I, \tag{6.4.191}$$

where

$$I = \int_{-\infty}^{\infty} \frac{dx_3}{[(x_3 + x_2 - x_1)^2 + (\Gamma_1/2)^2][(x_3 + x_2)^2 + (\Gamma_2/2)^2][x_3^2 + (\Gamma_3/2)^2]}. \tag{6.4.192}$$

If we again ignore the frequency dependence of Γ_1, Γ_2 and Γ_3 as in the last section, this last integral may be evaluated by comparing it with the general result [50]

$$\frac{\delta_1\delta_2\delta_3}{\pi} \int_{-\infty}^{\infty} \frac{dy}{[(y-y_1)^2 + \delta_1^2][(y-y_2)^2 + \delta_2^2][(y-y_3)^2 + \delta_3^2]}$$
$$= \frac{4\delta_1\delta_2\delta_3(\delta_1 + \delta_2 + \delta_3)}{[(y_1 - y_2)^2 + (\delta_1 + \delta_2)^2][(y_2 - y_3)^2 + (\delta_2 + \delta_3)^2][(y_3 - y_1)^2 + (\delta_3 + \delta_1)^2]}$$
$$+ \frac{\delta_1\delta_2}{[(y_2 - y_3)^2 + (\delta_2 + \delta_3)^2][(y_3 - y_1)^2 + (\delta_3 + \delta_1)^2]}$$
$$+ \frac{\delta_2\delta_3}{[(y_3 - y_1)^2 + (\delta_3 + \delta_1)^2][(y_1 - y_2)^2 + (\delta_1 + \delta_2)^2]}$$
$$+ \frac{\delta_3\delta_1}{[(y_1 - y_2)^2 + (\delta_1 + \delta_2)^2][(y_2 - y_3)^2 + (\delta_2 + \delta_3)^2]}.$$

We therefore put $y \equiv x_3, y_1 \equiv x_1 - x_2, y_2 \equiv -x_2, y_3 \equiv 0$ and $\delta_k = \Gamma_k/2$ to obtain

$$\frac{\Gamma_1\Gamma_2\Gamma_3}{8\pi} I = \frac{\frac{1}{4}\Gamma_1\Gamma_2\Gamma_3(\Gamma_1 + \Gamma_2 + \Gamma_3)}{\left[x_1^2 + \left(\frac{\Gamma_1 + \Gamma_2}{2}\right)^2\right]\left[x_2^2 + \left(\frac{\Gamma_2 + \Gamma_3}{2}\right)^2\right]\left[(x_1 - x_2)^2 + \left(\frac{\Gamma_1 + \Gamma_3}{2}\right)^2\right]}$$

$$+\frac{\Gamma_1\Gamma_2/4}{\left[x_2^2+\left(\frac{\Gamma_2+\Gamma_3}{2}\right)^2\right]\left[(x_1-x_2)^2+\left(\frac{\Gamma_1+\Gamma_3}{2}\right)^2\right]}$$

$$+\frac{\Gamma_2\Gamma_3/4}{\left[(x_1-x_2)^2+\left(\frac{\Gamma_1+\Gamma_3}{2}\right)^2\right]\left[x_1^2+\left(\frac{\Gamma_1+\Gamma_2}{2}\right)^2\right]}$$

$$+\frac{\Gamma_1\Gamma_3/4}{\left[x_1^2+\left(\frac{\Gamma_1+\Gamma_2}{2}\right)^2\right]\left[x_2^2+\left(\frac{\Gamma_2+\Gamma_3}{2}\right)^2\right]}. \qquad (6.4.193)$$

We now turn to the evaluation of Γ_1, Γ_2 and Γ_3. Equation (6.4.131) immediately yields

$$\Gamma_1(E_4)=\frac{2\pi}{\hbar}\sum_{n\neq 1}|U_n(E_4)|^2\delta(E_4-E_n)$$

$$=\frac{2\pi}{\hbar}|U_4(E_4)|^2,$$

which, from equation (6.4.179), becomes

$$\Gamma_1=\frac{2\pi}{\hbar}\cdot\frac{|H_{43}|^2|H_{32}|^2|H_{21}|^2}{[(E_4-E_3)^2+(\hbar\Gamma_3/2)^2][(E_4-E_2)^2+(\hbar\Gamma_2/2)^2]}, \qquad (6.4.194)$$

where we have again neglected the level shifts corresponding to the imaginary parts of η_2 and η_3.

As in the preceding section, we have ignored the frequency dependence of Γ_1 (and Γ_2 and Γ_3). We therefore replace it by its average over ω_2 and ω_3 so that, using equations (6.4.182, 183) and changing the variable of integration in a manner similar to that leading to equation (6.4.193), we find

$$\Gamma_1=\frac{2\pi}{\hbar^5}|H_{43}|^2|H_{32}|^2|H_{21}|^2\int_0^\infty d(\hbar\omega_2)\int_0^\infty d(\hbar\omega_3)$$

$$\times\frac{1}{[(\omega_3-\omega_{34})^2+(\Gamma_3/2)][(\omega_2-\omega_{23}+\omega_3-\omega_{34})^2+(\Gamma_2/2)^2]}$$

$$=\frac{2\pi}{\hbar^3}|H_{43}|^2|H_{32}|^2|H_{21}|^2\int_{-\infty}^{\infty}\frac{dx_3}{x_3^2+(\Gamma_3/2)^2}$$

$$\times\int_{-\infty}^{\infty}\frac{dx_2}{(x_2+x_3)^2+(\Gamma_2/2)^2}$$

$$=\frac{(2\pi)^2}{\hbar^3\Gamma_2}|H_{43}|^2|H_{32}|^2|H_{21}|^2\int_{-\infty}^{\infty}\frac{dx_3}{x_3^2+(\Gamma_3/2)^2}$$

$$=\frac{(2\pi)^3}{\hbar^3\Gamma_2\Gamma_3}|H_{43}|^2|H_{32}|^2|H_{21}|^2. \qquad (6.4.195)$$

The damping parameter Γ_2 may be obtained from equation (6.4.178), i.e.

$$\begin{aligned}\Gamma_2 &= 2\ \text{Real}\ \eta_2(E_4)\\ &= \frac{2}{\hbar}\ \mathcal{R}l\ \frac{\mathrm{i}|H_{32}|^2}{E_4 - E_3 + \mathrm{i}\hbar\Gamma_3/2}\\ &= \frac{|H_{32}|^2\Gamma_3}{(E_4 - E_3)^2 + (\hbar\Gamma_3/2)^2}\\ &= \frac{|H_{32}|^2\Gamma_3}{\hbar^2}\cdot\frac{1}{(\omega_3 - \omega_{34})^2 + (\Gamma_3/2)^2}\,. \end{aligned} \tag{6.4.196}$$

If we again integrate over frequency, we find

$$\begin{aligned}\Gamma_2 &\rightarrow \frac{|H_{32}|^2\Gamma_3}{\hbar}\int_{-\infty}^{\infty}\frac{\mathrm{d}x_3}{x_3^2 + (\Gamma_3/2)^2}\\ &= \frac{2\pi|H_{32}|^2}{\hbar}\,. \end{aligned} \tag{6.4.197}$$

Finally, equation (6.4.176) yields

$$\Gamma_3 = 2\mathcal{R}l\ \eta_3(E_4) = \frac{2\pi}{\hbar}\ |H_{43}|^2\,. \tag{6.4.198}$$

Note that equation (6.4.195) may therefore be simplified, viz.

$$\Gamma_1 = \frac{2\pi}{\hbar}\ |H_{21}|^2\,. \tag{6.4.199}$$

To obtain the probability $p(\omega_1, \omega_2)$ of scattering, we now take level 3 to be the same as level 1, and level 4 the same as level 2. We then have $\Gamma_3 = \Gamma_1$ from equations (6.4.198) and (6.4.199). Equations (6.4.191) and (6.4.197-199) yield

$$p(\omega_1, \omega_2) = \frac{\Gamma_1\Gamma_2\Gamma_3}{\hbar^2(2\pi)^3}\ I,$$

which, using equation (6.4.193) with $x_2 = \omega_2 - \omega_{21}$ and $x_1 - x_2 = \omega_1 - \omega_2$, then gives [129]

$$\begin{aligned}&\hbar^2\pi^2 p(\omega_1, \omega_2)\\ &= \frac{\Gamma_1^2\Gamma_2(2\Gamma_1 + \Gamma_2)/4}{\left[(\omega_1 - \omega_{21})^2 + \left(\frac{\Gamma_1 + \Gamma_2}{2}\right)^2\right]\left[(\omega_2 - \omega_{21})^2 + \left(\frac{\Gamma_1 + \Gamma_2}{2}\right)^2\right] \times [(\omega_1 - \omega_2)^2 + \Gamma_1^2]}\\ &+ \frac{\Gamma_1\Gamma_2/4}{\left[(\omega_2 - \omega_{21})^2 + \left(\frac{\Gamma_1 + \Gamma_2}{2}\right)^2\right]\ [(\omega_1 - \omega_2)^2 + \Gamma_1^2]}\end{aligned}$$

$$+\frac{\Gamma_1\Gamma_2/4}{[(\omega_1-\omega_2)^2+\Gamma_1^2]\left[(\omega_1-\omega_{21})^2+\left(\frac{\Gamma_1+\Gamma_2}{2}\right)^2\right]}$$

$$+\frac{\Gamma_1^2/4}{\left[(\omega_1-\omega_{21})^2+\left(\frac{\Gamma_1+\Gamma_2}{2}\right)^2\right]\left[(\omega_2-\omega_{21})^2+\left(\frac{\Gamma_1+\Gamma_2}{2}\right)^2\right]}. \quad (6.4.200)$$

This then represents the probability of scattering whereby a photon of frequency ω_1 is first absorbed *and* then a photon of frequency ω_2 is emitted whilst the atom returns to the level (but not necessarily the same sub-state) it occupied initially.

6.5 Effect of perturbing collisions

We wish to formulate the scattering problem examined in the preceding section, but now with the effect of collisions included. We consider a 3-level atom as shown in figure 6.11 in which a photon of frequency ω_1 is absorbed, thus exciting the initial state $|1\rangle$ to the intermediate state $|2\rangle$. The atom subsequently de-excites to the final state $|3\rangle$ thereby emitting a photon of frequency ω_2.

We allow excitation and de-excitation processes to occur through collisions (inelastic), and also take into account the effect of elastic collisions which only alter the energy location of the bound electron within the level it occupies. These elastic encounters merely perturb the energy of the bound electron - they do not result in a direct transition to another level.

The analysis presented in the remainder of this chapter follows that of Omont, Smith and Cooper [92] who themselves drew upon the work of Fiutak and Van Kranendonk [43]. In particular, by way of illustrating the quantum mechanical exposition of collision theory as applied to atom-photon interactions, we only consider the case for which the assumption of impact broaden-

Fig.6.11. The quantum representation of scattering from state $|1>$ to $|3>$ via $|2>$. State $|2>$ is both naturally and collisionally broadened.

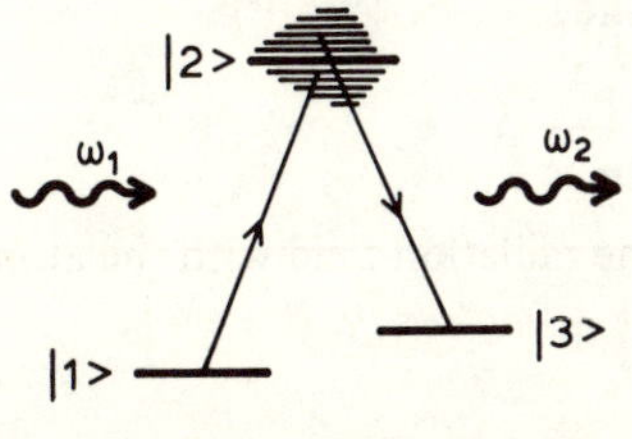

ing is applicable. Here one assumes either that the elastic collisions are well separated and therefore have a small time duration relative to the time between individual collisions, or that the collisions, if relatively long in duration, are sufficiently weak that the effect on the atom is relatively small (even if the collisions overlap in time). This is an extreme position, the other extreme being the quasi-static approximation whereby the perturbation is effectively constant over the lifetime of the individual states. In this latter case, the perturbers move relatively slowly (note that electrons are more mobile than ions) and this would be more appropriate the denser the medium under consideration. The impact approximation is correspondingly more applicable for regions higher in stellar atmospheres where smaller densities are encountered. The reader is referred to both Griem [46, 47] and Jefferies [59] for more-detailed discussions of these and other approximations used in studying collisional line broadening. Our aim here is to simply detail one of the results commonly used in the literature in order that a more general approach to the problem can be initiated if so desired.

6.5.1 The Schrödinger wave equation

We proceed by using a combination of the analyses presented in sections 6.3.1 and 6.3.3. We wish to solve the time-dependent Schrödinger wave equation

$$i\hbar \frac{\partial}{\partial t} |\Psi(t)\rangle = \hat{H} |\Psi(t)\rangle, \tag{6.5.1}$$

where $|\Psi(t)\rangle$ is the wave function for the entire system now including atoms, photons and perturbing particles (free electrons). We then have

$$\hat{H} = \hat{H}_{\mathrm{A}} + \hat{H}_{\mathrm{R}} + \hat{H}_{\mathrm{PA}} + \hat{H}_{\mathrm{RA}}, \tag{6.5.2}$$

where

$$\hat{H}_{\mathrm{A}} = \frac{1}{2m} \hat{\mathbf{p}} \cdot \hat{\mathbf{p}}, \tag{6.5.3}$$

is the Hamiltonian for the atoms excluding the radiation field and perturbing particles,

$$\hat{H}_{\mathrm{R}} = \sum_{\gamma\lambda} \hbar\omega_\gamma (\hat{a}^+_{\gamma\lambda}\hat{a}_{\gamma\lambda} + \tfrac{1}{2}\mathbb{1}), \tag{6.5.4}$$

is the Hamiltonian for the free radiation field,

$$\hat{H}_{\mathrm{RA}} = -\frac{e}{2mc}(\hat{\mathbf{p}} \cdot \hat{\mathbf{A}} + \hat{\mathbf{A}} \cdot \hat{\mathbf{p}}) + \frac{e^2}{2mc^2} \hat{\mathbf{A}} \cdot \hat{\mathbf{A}}, \tag{6.5.5}$$

is the Hamiltonian for the interaction of the radiation field with the atoms where

$$\hat{\mathbf{A}} = \sum_{\gamma\lambda} (\hat{a}_{\gamma\lambda}\mathbf{u}_{\gamma\lambda} + \hat{a}^+_{\gamma\lambda}\mathbf{u}^*_{\gamma\lambda}), \tag{6.5.6}$$

and finally $\hat{H}_{\mathrm{PA}}$, which we shall specify later, is the Hamiltonian for the perturbers and their interaction with the atoms. We assume no interaction between the free electrons and the radiation field although, clearly, this would be fundamental for some problems of significant physical interest.

We next expand $|\Psi(t)\rangle$ in a basis set of eigenstates $|\psi_n(t)\rangle$ satisfying

$$i\hbar \frac{\partial}{\partial t}|\psi_n(t)\rangle = (\hat{H}_A + \hat{H}_R + \hat{H}_{PA})|\psi_n(t)\rangle, \tag{6.5.7}$$

such that

$$|\Psi(t)\rangle = \sum_n a_n(t)|\psi_n(t)\rangle. \tag{6.5.8}$$

Substitution of equation (6.5.8) into (6.5.1) yields

$$i\hbar \sum_n \dot{a}_n(t)|\psi_n(t)\rangle + \sum_n a_n(t)(\hat{H}_A + \hat{H}_R + \hat{H}_{PA})|\psi_n(t)\rangle$$
$$= (\hat{H}_A + \hat{H}_R + \hat{H}_{PA} + \hat{H}_{RA}) \sum_n a_n(t)|\psi_n(t)\rangle,$$

so that

$$i\hbar \sum_n \dot{a}_n(t)|\psi_n(t)\rangle = \sum_n a_n(t)\hat{H}_{RA}|\psi_n(t)\rangle.$$

We now multiply this last result on the left by $\langle\psi_m(t)|$, noting that

$$\langle\psi_m(t)|\psi_n(t)\rangle = \delta_{mn}, \tag{6.5.9}$$

to obtain

$$i\hbar\dot{a}_m(t) = \sum_n a_n(t)\langle\psi_m(t)|\hat{H}_{RA}|\psi_n(t)\rangle. \tag{6.5.10}$$

We again ignore the two photon scattering term $\hat{\mathbf{A}} \cdot \hat{\mathbf{A}}$ in equation (6.5.5) so that

$$\hat{H}_{RA} = \sum_{\gamma\lambda} (\hat{a}_{\gamma\lambda}\xi_{\gamma\lambda} + \hat{a}^{+}_{\gamma\lambda}\xi^{*}_{\gamma\lambda}), \tag{6.5.11}$$

where

$$\xi_{\gamma\lambda} = -\frac{e}{2mc}(\hat{\mathbf{p}} \cdot \mathbf{u}_{\gamma\lambda} + \mathbf{u}_{\gamma\lambda} \cdot \hat{\mathbf{p}}). \tag{6.5.12}$$

However, if we are to proceed further, we must now solve equation (6.5.7) for the $|\psi_n(t)\rangle$. To do this we write

$$\hat{H}_0 = \hat{H}_A + \hat{H}_R, \tag{6.5.13}$$

so that

$$i\hbar \frac{\partial}{\partial t}|\psi_n(t)\rangle = (\hat{H}_0 + \hat{H}_{PA})|\psi_n(t)\rangle. \tag{6.5.14}$$

This last result is identical to equation (6.3.35) for which the solution was found to be (see section 6.3.3)

$$|\psi_n(t)\rangle = \hat{T}_0(t, t_0)|\psi_n(t_0)\rangle, \tag{6.5.15}$$

where

$$\hat{T}_0(t,t_0)=e^{-i\hat{H}_0(t-t_0)/\hbar}\hat{U}(t,t_0), \tag{6.5.16}$$

$$\hat{U}(t,t_0)=\mathscr{D}\exp\left\{\frac{1}{i\hbar}\int_{t_0}^{t}\hat{H}^{I}_{PA}(t')\,dt'\right\}, \tag{6.5.17}$$

with

$$\hat{H}^{I}_{PA}(t)=e^{i\hat{H}_0(t-t_0)/\hbar}\hat{H}_{PA}(t)e^{-i\hat{H}_0(t-t_0)/\hbar}. \tag{6.5.18}$$

We also found

$$\langle\psi_n(t)|=\langle\psi_n(t_0)|\hat{T}_0^{+}(t,t_0). \tag{6.5.19}$$

Noting that $\hat{H}_R$ commutes with $\hat{H}_A$ and $\hat{H}_{PA}$ (the radiation field involves operators independent of those for the atoms and perturbers), we have

$$\hat{H}^{I}_{PA}(t)=e^{i\hat{H}_A(t-t_0)/\hbar}\hat{H}_{PA}(t)e^{-i\hat{H}_A(t-t_0)/\hbar}, \tag{6.5.20}$$

with

$$\begin{aligned}\hat{T}_0(t,t_0)|\psi_n(t_0)\rangle&=e^{-i\hat{H}_A(t-t_0)/\hbar}\hat{U}(t,t_0)e^{-i\hat{H}_R(t-t_0)/\hbar}|\psi_n(t_0)\rangle\\&=e^{-iE_n^{(R)}(t-t_0)/\hbar}e^{-i\hat{H}_A(t-t_0)/\hbar}\hat{U}(t,t_0)|\psi_n(t_0)\rangle\\&=e^{-iE_n^{(R)}(t-t_0)/\hbar}\hat{T}(t,t_0)|\psi_n(t_0)\rangle,\end{aligned} \tag{6.5.21}$$

where

$$\hat{T}(t,t_0)=e^{-i\hat{H}_A(t-t_0)/\hbar}\hat{U}(t,t_0), \tag{6.5.22}$$

and

$$E_n^{(R)}=\hbar\omega_\gamma(n_{\gamma\lambda}+\tfrac{1}{2}), \tag{6.5.23}$$

is the energy of the radiation field in the $(\gamma\lambda)$th mode.

If we now write

$$\begin{aligned}|\psi_1(t)\rangle&=|1,n_1,\ldots,n_{\gamma\lambda},\ldots,t\rangle\\&\equiv|1(t)\rangle|n_{\gamma\lambda}\rangle,\end{aligned} \tag{6.5.24}$$

and

$$|\psi_2(t)\rangle\equiv|2(t)\rangle|n_{\gamma\lambda}-1\rangle, \tag{6.5.25}$$

where, for example, $|1(t)\rangle$ refers to the atom state and $|n_{\gamma\lambda}\rangle$ the photon state, and note that

$$E_1^{(R)}=\hbar\omega_\gamma(n_{\gamma\lambda}+\tfrac{1}{2}),\quad E_2^{(R)}=\hbar\omega_\gamma((n_{\gamma\lambda}-1)+\tfrac{1}{2}),$$

then (recall that we want to solve equation (6.5.10))

$$\begin{aligned}&\langle\psi_2(t)|\hat{H}_{RA}|\psi_1(t)\rangle\\&\quad=\langle\psi_2(t_0)|\hat{T}^{+}(t,t_0)e^{-i\omega_\gamma(t-t_0)}\\&\qquad\times\sum_{\gamma'\lambda'}(\hat{a}_{\gamma'\lambda'}\xi_{\gamma'\lambda'}+\hat{a}^{+}_{\gamma'\lambda'}\xi^{*}_{\gamma'\lambda'})\hat{T}(t,t_0)|\psi_1(t_0)\rangle\\&\quad=\sum_{\gamma'\lambda'}e^{-i\omega_\gamma(t-t_0)}[\langle 2(t)|\xi_{\gamma'\lambda'}|1(t)\rangle\langle n_{\gamma\lambda}-1|\hat{a}_{\gamma'\lambda'}|n_{\gamma\lambda}\rangle\\&\qquad+\langle 2(t)|\xi^{*}_{\gamma'\lambda'}|1(t)\rangle\langle n_{\gamma\lambda}-1|\hat{a}^{+}_{\gamma'\lambda'}|n_{\gamma\lambda}\rangle]\\&\quad=\sqrt{n_{\gamma\lambda}}\,e^{-i\omega_\gamma(t-t_0)}\langle 2(t)|\xi_{\gamma\lambda}|1(t)\rangle,\end{aligned} \tag{6.5.26}$$

where we have used equations (6.2.65) and (6.2.70). If we now consider the transition $|1\rangle \to |2\rangle$ arising explicitly from the absorption of a photon of frequency ω_1, then the only non-zero matrix element $\langle\psi_m(t)|\hat{H}_{\mathrm{RA}}|\psi_n(t)\rangle$ appearing in equation (6.5.10) is just

$$\langle\psi_2(t)|\hat{H}_{\mathrm{RA}}|\psi_1(t)\rangle = \surd n_1\, \mathrm{e}^{-\mathrm{i}\omega_1(t-t_0)}\langle 2(t)|\xi_1|1(t)\rangle. \tag{6.5.27}$$

Consequently, equation (6.5.10) yields

$$\dot{a}_2(t) = \frac{\surd n_1}{\mathrm{i}\hbar}\, \mathrm{e}^{-\mathrm{i}\omega_1(t-t_0)}\langle 2(t)|\xi_1|1(t)\rangle a_1(t). \tag{6.5.28}$$

Similarly, if we now consider the transition from the excited $|2\rangle$ state to the de-excited $|3\rangle$ state, with the subsequent emission of a photon of frequency ω_2, we find the only non-zero matrix element

$$\langle\psi_3(t)|\hat{H}_{\mathrm{RA}}|\psi_2(t)\rangle = \surd(n_2+1)\, \mathrm{e}^{\mathrm{i}\omega_2(t-t_0)}\langle 3(t)|\xi_2^*|2(t)\rangle, \tag{6.5.29}$$

- note that $E_2^{(\mathrm{R})} - E_1^{(\mathrm{R})} = -\hbar\omega_1$ whereas $E_3^{(\mathrm{R})} - E_2^{(\mathrm{R})} = \hbar\omega_2$, and $\hat{a}_1|n_1\rangle = \surd n_1|n_1-1\rangle$ whereas $\hat{a}_2|n_2\rangle = \surd(n_2+1)|n_2+1\rangle$.

Equation (6.5.10) for the $|2\rangle \to |3\rangle$ transition then yields

$$\dot{a}_3(t) = \frac{\surd(n_2+1)}{\mathrm{i}\hbar}\, \mathrm{e}^{\mathrm{i}\omega_2(t-t_0)}\langle 3(t)|\xi_2^*|2(t)\rangle a_2(t), \tag{6.5.30}$$

where we have not allowed direct transitions from $|1\rangle$ to $|3\rangle$.

Substitution for $a_2(t)$ from equation (6.5.28) written in the form

$$a_2(t) = \frac{\surd n_1}{\mathrm{i}\hbar}\int_{t_0}^{t} \mathrm{e}^{-\mathrm{i}\omega_1(t_1-t_0)}\langle 2(t_1)|\xi_1|1(t_1)\rangle a_1(t_1)\, \mathrm{d}t_1, \tag{6.5.31}$$

where we have used the initial condition $a_2(t_0) = 0$, then yields

$$\dot{a}_3(t) = \frac{\surd(n_1(n_2+1))}{(\mathrm{i}\hbar)^2}\, \mathrm{e}^{\mathrm{i}\omega_2(t-t_0')}\int_{t_0}^{t} \mathrm{e}^{-\mathrm{i}\omega_1(t_1-t_0)}$$
$$\times \langle 3(t)|\xi_2^*|2(t)\rangle\langle 2(t_1)|\xi_1|1(t_1)\rangle a_1(t_1)\, \mathrm{d}t_1,$$

so that

$$a_3(t) = \frac{\surd(n_1(n_2+1))}{(\mathrm{i}\hbar)^2}\int_{t_0'}^{t} \mathrm{d}t_2 \int_{t_0}^{t_2} \mathrm{d}t_1\, \mathrm{e}^{\mathrm{i}\omega_2(t_2-t_0')}\mathrm{e}^{-\mathrm{i}\omega_1(t_1-t_0)}$$
$$\times \langle 3(t_2)|\xi_2^*|2(t_2)\rangle\langle 2(t_1)|\xi_1|1(t_1)\rangle a_1(t_1) Y_{21}, \tag{6.5.32}$$

where we have used $a_3(t_0') = 0$. Without loss of generality we may take $t_0 = 0 = t_0'$. The requirement $t_2 \geqslant t_1$ in the above integral is accounted for by the introduction of the step function Y_{21} which is unity for $t_2 \geqslant t_1$ and zero otherwise.

Since we require the probability function $|a_3(t)|^2$ we note that

$$a_3^*(t) = \frac{\sqrt{(n_1(n_2+1))}}{(i\hbar)^2} \int_0^t \mathrm{d}t_4 \int_0^{t_4} \mathrm{d}t_3\, \mathrm{e}^{-\mathrm{i}\omega_2 t_4 + \mathrm{i}\omega_1 t_3}$$
$$\times \langle 1(t_3)|\xi_1^*|2(t_3)\rangle\langle 2(t_4)|\xi_2|3(t_4)\rangle a_1(t_3) Y_{43}, \qquad (6.5.33)$$

where we have changed the variables of integration from t_1 to t_3 and t_2 to t_4.

6.5.2 *The transition rate w(1 → 2 → 3)*

We now average this probability $|a_3(t)|^2$ to obtain an effective transition rate in much the same manner as that presented in section 6.4. We then write

$$w(1 \to 2 \to 2) = \lim_{t\to\infty} \frac{1}{t} \sum_i \rho_{ii} \{|a_3(t)|^2\}_{\mathrm{AV}}, \qquad (6.5.34)$$

where ρ_{ii} are the diagonal elements of the density matrix (population) of the initial state, and $\{-\}_{\mathrm{AV}}$ refers to the averaging over all the parameters describing the collisions.

Before substituting equations (6.5.32) and (6.5.33) into (6.5.34), we make certain simplifications. For example, we note that

$$\hat{T}^+(t_1, t_2) = \hat{U}^+(t_1, t_2)\mathrm{e}^{\mathrm{i}\hat{H}_{\mathrm{A}}(t_1 - t_2)/\hbar}$$
$$= \mathscr{D} \exp\left\{-\frac{1}{i\hbar}\int_{t_2}^{t_1} \hat{H}_{\mathrm{PA}}^{\mathrm{I}}(t')\,\mathrm{d}t'\right\} \mathrm{e}^{\mathrm{i}\hat{H}_{\mathrm{A}}(t_1 - t_2)/\hbar}, \qquad (6.5.35)$$

where we have used the Hermitian property of the Hamiltonian. We then have

$$\hat{T}^+(t_1, t_2) = \hat{T}(t_2, t_1), \qquad (6.5.36)$$

so that

$$\langle 3(t_2)|\xi_2^*|2(t_2)\rangle\langle 2(t_1)|\xi_1|1(t_1)\rangle$$
$$= \langle 3(t)|\hat{T}^+(t_2, t)\xi_2^*\hat{T}(t_2, t_1)|2(t_1)\rangle\langle 2(t_1)|\xi_1\hat{T}(t_1, 0)|1(0)\rangle$$
$$= \langle 3(t)|\hat{T}(t, t_2)\xi_2^*\hat{T}(t_2, t_1)|2(t_1)\rangle\langle 2(t_1)|\xi_1\hat{T}(t_1, 0)|1(0)\rangle. \qquad (6.5.37)$$

Further, we have the completeness relationship

$$\sum_\alpha |\alpha\rangle\langle\alpha| = \mathbb{1}, \qquad (6.5.38)$$

which may be easily proved by examining, for example, the *rs* element of the matrix on the LHS, viz.

$$\langle r|\sum_\alpha |\alpha\rangle\langle\alpha|s\rangle = \sum_\alpha \delta_{\alpha r}\delta_{\alpha s} = \delta_{rs}.$$

Substitution of the above results into equation (6.5.34) then yields

$$w(1 \to 2 \to 3) = \lim_{t\to\infty} \frac{n_1(n_2+1)}{t\hbar_4} \sum_i \rho_{ii} \left\{ \int_0^t \mathrm{d}t_4 \int_0^{t_4} \mathrm{d}t_3 \int_0^t \mathrm{d}t_2 \right.$$

$$\times \int_0^{t_2} \mathrm{d}t_1 \times \mathrm{e}^{\mathrm{i}(-\omega_1 t_1 + \omega_2 t_2 + \omega_1 t_3 - \omega_2 t_4)}$$

$$\left. \times \, \chi a_1(t_1) a_1^*(t_3) Y_{21} Y_{42} Y_{43} \right\}_{\mathrm{AV}}, \tag{6.5.39}$$

where

$$\chi = \langle 1(0) | \hat{T}(0, t_3) \xi_1^* | 2(t_3) \rangle \langle 2(t_3) | \hat{T}(t_3, t_4) \xi_2 T(t_4, t) | 3(t) \rangle$$
$$\times \langle 3(t) | \hat{T}(t, t_2) \xi_2^* T(t_2, t_1) | 2(t_1) \rangle \langle 2(t_1) | \xi_1 \hat{T}(t_1, 0) | 1(0) \rangle .$$

We could equally as well use Y_{24} in equation (6.5.39); this leads to the same result as that obtained using Y_{42}.

If we now recognise that the $\{-\}_{\mathrm{AV}}$ operation will include a summation whilst the atom is in the excited $|2\rangle$ and de-excited $|3\rangle$ states, then the completeness relationship (6.5.38) yields

$$\chi_{\mathrm{AV}} \to \{\langle 1(0) | \hat{T}(0, t_3) \xi_1^* \hat{T}(t_3, t_4) \xi_2 \hat{T}(t_4, t) \hat{T}(t, t_2)$$
$$\times \, \xi_2^* \hat{T}(t_2, t_1) \xi_1 \hat{T}(t_1, 0) | 1(0) \rangle\}_{\mathrm{AV}},$$

which, using

$$\hat{T}(t_4, t) \hat{T}(t, t_2) = \hat{T}(t_4, t_2), \tag{6.5.40}$$

becomes

$$\chi_{\mathrm{AV}} \to \{\langle 1(0) | \hat{T}(0, t_3) \xi_1^* \hat{T}(t_3, t_4) \xi_2 \hat{T}(t_4, t_2) \xi_2^* \hat{T}(t_2, t_1)$$
$$\times \, \xi_1 \hat{T}(t_1, 0) | 1(0) \rangle\}_{\mathrm{AV}}.$$

Equation (6.5.40) may be re-written, for example, as

$$\hat{T}(t_4, t_2) = \hat{T}(t_4, 0) \hat{T}(0, t_2) = \hat{T}(t_4, 0) \hat{T}^+(t_2, 0),$$

which, with obvious notation, we write as

$$\hat{T}(t_4, t_2) = \hat{T}_4 \hat{T}_2^+. \tag{6.5.41}$$

Note that

$$\hat{T}(t_4, t_4) \equiv \mathbb{1} = \hat{T}_4 \hat{T}_4^+ = \hat{T}_4^+ \hat{T}_4. \tag{6.5.42}$$

We then have

$$\chi_{\mathrm{AV}} \to \{\langle 1(0) | \hat{T}_3^+ \xi_1^* \hat{T}_3 \hat{T}_4^+ \xi_2 \hat{T}_4 \hat{T}_2^+ \xi_2^* \hat{T}_2 \hat{T}_1^+ \xi_1 \hat{T}_1 | 1(0) \rangle\}_{\mathrm{AV}},$$

which, again using the completeness relationship (6.5.38), becomes

$$\chi_{\mathrm{AV}} \to \left\{ \sum_{\substack{abcd \\ a'b'c'd'}} \sum_{jkl} \langle \mathrm{i} | \hat{T}_3^+ | d' \rangle \langle d' | \xi_1^* | d \rangle \langle d | \hat{T}_3 | l \rangle \langle l | \hat{T}_4^+ | c' \rangle \langle c' | \xi_2 | c \rangle \right.$$

$$\times \langle c | \hat{T}_4 | k \rangle \langle k | \hat{T}_2^+ | b' \rangle \langle b' | \xi_2^* | b \rangle \langle b | \hat{T}_2 | j \rangle \langle j | \hat{T}_1^+ | a' \rangle$$

$$\left. \times \langle a' | \xi_1 | a \rangle \langle a | \hat{T}_1 | \mathrm{i} \rangle \right\}_{\mathrm{AV}}, \tag{6.5.43}$$

where we have simply re-defined the initial state $|1(0)\rangle$ by $|\mathrm{i}\rangle$.

If we now define

$$\langle\langle ab|\hat{\Theta}_k|cd\rangle\rangle = \langle a|\hat{T}_k|c\rangle\langle d|\hat{T}_k^+|b\rangle, \tag{6.5.44}$$

for all $k = 1$ to 4, then

$$\chi_{AV} \to \sum_{\substack{abcd\\a'b'c'd'}} \{\langle d'|\xi_1^*|d\rangle\langle c'|\xi_2|c\rangle\langle b'|\xi_2^*|b\rangle\langle a'|\xi_1|a\rangle\Theta(a, \ldots, d')\}_{AV}, \tag{6.5.45}$$

where

$$\Theta(a, \ldots, d') = \sum_{jkl} \langle\langle dd'|\hat{\Theta}_3|li\rangle\rangle\langle\langle cc'|\hat{\Theta}_4|kl\rangle\rangle\langle\langle bb'|\hat{\Theta}_2|jk\rangle\rangle \times \langle\langle aa'|\hat{\Theta}_1|ij\rangle\rangle. \tag{6.5.46}$$

6.5.3 *Time ordering of $\Theta(a, \ldots, d')$*

The above product of the matrix elements appearing in $(a, \ldots, d')$ may be ordered in a variety of ways. However, as we shall see, certain bracketing together of these elements proves extremely fruitful in simplifying the analysis. In particular, the inequalities $t_2 > t_1$, $t_4 > t_2$ and $t_4 > t_3$ appearing in equation (6.5.39) allow three possible time sequences; $t_4 > t_3 > t_2 > t_1$, $t_4 > t_2 > t_3 > t_1$ and $t_4 > t_2 > t_1 > t_3$. In each case, the re-arrangement of the above time evolution product into non-overlapping operators enables their $\{-\}_{AV}$ averages to be obtained separately.

Sequence I *($t_4 > t_3 > t_2 > t_1$)*

We define

$$\begin{aligned}\bar{\Theta}_I &= \sum_i \rho_{ii}\Theta_I(a, \ldots, d')\\ &= \sum_{ijkl} \rho_{ii}\langle c|\hat{T}_4|k\rangle\langle l|\hat{T}_4^+|c'\rangle\langle d|\hat{T}_3|l\rangle\langle i|\hat{T}_3^+|d'\rangle\langle b|\hat{T}_2|j\rangle\langle k|\hat{T}_2^+|b'\rangle\\ &\quad \times \langle a|\hat{T}_1|i\rangle\langle j|\hat{T}_1^+|a'\rangle,\end{aligned}$$

which, using equation (6.5.42), becomes

$$\begin{aligned}\bar{\Theta}_I &= \sum_{ijkl} \rho_{ii}\langle c|\hat{T}_4\hat{T}_3^+\hat{T}_3|k\rangle\langle d|\hat{T}_3|l\rangle\langle l|\hat{T}_4^+|c'\rangle\langle k|\hat{T}_2^+|b'\rangle\langle i|\hat{T}_2^+\hat{T}_2\hat{T}_3^+|d'\rangle\\ &\quad \times \langle b|\hat{T}_2|j\rangle\langle j|\hat{T}_1^+|a'\rangle\langle a|\hat{T}_1|i\rangle\\ &= \sum_{ik} \rho_{ii}\langle c|\hat{T}_4\hat{T}_3^+\hat{T}_3|k\rangle\langle d|\hat{T}_3\hat{T}_4^+|c'\rangle\langle k|\hat{T}_2^+|b'\rangle\langle i|\hat{T}_2^+\hat{T}_2\hat{T}_3^+|d'\rangle\\ &\quad \times \langle b|\hat{T}_2\hat{T}_1^+|a'\rangle\langle a|\hat{T}_1|i\rangle\\ &= \sum_{ijkl} \rho_{ii}\langle c|\hat{T}_4\hat{T}_3^+|l\rangle\langle l|\hat{T}_3|k\rangle\langle d|\hat{T}_3\hat{T}_4^+|c'\rangle\langle k|\hat{T}_2^+|b'\rangle\langle i|\hat{T}_2^+|j\rangle\\ &\quad \times \langle j|\hat{T}_2\hat{T}_3^+|d'\rangle\langle b|\hat{T}_2\hat{T}_1^+|a'\rangle\langle a|\hat{T}_1|i\rangle\\ &= \sum_{ijkl} \rho_{ii}\langle c|\hat{T}_4\hat{T}_3^+|l\rangle\langle d|\hat{T}_3\hat{T}_4^+|c'\rangle\langle l|\hat{T}_3|k\rangle\langle k|\hat{T}_2^+|b'\rangle\langle j|\hat{T}_2\hat{T}_3^+|d'\rangle\\ &\quad \times \langle b|\hat{T}_2\hat{T}_1^+|a'\rangle\langle a|\hat{T}_1|i\rangle\langle i|\hat{T}_2^+|j\rangle\end{aligned}$$

$$= \sum_{ijl} \langle c|\hat{T}_4\hat{T}_3^+|l\rangle\langle d|\hat{T}_3\hat{T}_4^+|c'\rangle\langle l|\hat{T}_3\hat{T}_2^+|b'\rangle\langle j|\hat{T}_2\hat{T}_3^+|d'\rangle\langle b|\hat{T}_2\hat{T}_1^+|a'\rangle$$
$$\times \rho_{ii}\langle a|\hat{T}_1|i\rangle\langle i|\hat{T}_2^+|j\rangle. \qquad (6.5.47)$$

The term $\Sigma_i \rho_{ii}\langle a|\hat{T}_1|i\rangle\langle i|\hat{T}_2^+|j\rangle$ incorporates the time evolution of the (a_j)th element of ρ_{ii}. If we replace this evolution manifestation by the corresponding equilibrium representation assumed diagonal (note that we have already assumed the initial density operator to be diagonal), then the above term is simply $\rho_{aa}\langle a|\hat{T}_1\hat{T}_2^+|j\rangle$.

Further, from equations (6.5.40) and (6.5.42) we have, for example,

$$\hat{T}_1\hat{T}_2^+ = \hat{T}(t_1, t_2) = \hat{T}^+(t_2, t_1) = (\hat{T}_2\hat{T}_1^+)^+, \qquad (6.5.48)$$

so that defining $\hat{T}_{12} = \hat{T}_1\hat{T}_2$, and using definition (6.5.44), equation (6.5.47) becomes

$$\bar{\Theta}_{\text{I}}(a, \ldots, d') = \sum_{jl} \rho_{aa}\langle\langle cc'|\hat{\Theta}_{43}|ld\rangle\rangle\langle\langle ld'|\hat{\Theta}_{32}|b'j\rangle\rangle\langle\langle bj|\Theta_{21}|a'a\rangle\rangle. \qquad (6.5.49)$$

Sequence II $(t_4 > t_2 > t_3 > t_1)$

Equation (6.5.46) similarly yields

$$\bar{\Theta}_{\text{II}} = \sum_i \rho_{ii}\Theta_{\text{II}}(a, \ldots, d')$$
$$= \sum_{ik} \rho_{aa}\langle\langle cc'|\hat{\Theta}_{42}|b'k\rangle\rangle\langle\langle bk|\hat{\Theta}_{23}|id\rangle\rangle\langle\langle id'|\hat{\Theta}_{31}|a'a\rangle\rangle. \qquad (6.5.50)$$

Sequence III $(t_4 > t_2 > t_1 > t_3)$

The final required ordering of the matrix elements in equation (6.5.46) is just

$$\bar{\Theta}_{\text{III}} = \sum_i \rho_{ii}\Theta(a, \ldots, d')$$
$$= \sum_{ik} \rho_{aa}\langle\langle cc'|\hat{\Theta}_{42}|b'k\rangle\rangle\langle\langle bk|\hat{\Theta}_{21}|a'i\rangle\rangle\langle\langle ai|\hat{\Theta}_{13}|d'd\rangle\rangle. \qquad (6.5.51)$$

This then completes the specification of $\bar{\Theta}$ for the three time sequences I, II and III. Note that in each case there are no overlapping time terms. For example, $\bar{\Theta}_{\text{I}}$ contains only $\hat{\Theta}_{43}$, $\hat{\Theta}_{32}$ and $\hat{\Theta}_{21}$ corresponding to the time intervals $t_3 > t_4$, $t_2 > t_3$ and $t_1 > t_2$ with $t_4 > t_3 > t_2 > t_1$. This, of course, enables the averaging operator $\{-\}_{\text{AV}}$ to be taken separately for each component.

6.5.4 *Evaluation of the time-evolution operators*

To proceed further, we clearly need to evaluate terms of the form

$$\{\langle\langle aa'|\hat{\Theta}_{21}|bb'\rangle\rangle\}_{\text{AV}} = \{\langle a|\hat{T}_2\hat{T}_1^+|b\rangle\langle b'|\hat{T}_1\hat{T}_2^+|a'\rangle\}_{\text{AV}}. \qquad (6.5.52)$$

From equation (6.5.22) we have, for example,

$$\hat{T}_2 = \mathrm{e}^{-\mathrm{i}\hat{H}_{\text{A}}t_2/\hbar}\hat{U}_2, \qquad (6.5.53)$$

where

$$\hat{U}_2 = \hat{U}(t_2, 0). \tag{6.5.54}$$

Recognising that $\hat{H}_A$ is time independent and operates on the eigenstates $\langle a|$ and $|b\rangle$ such that

$$\langle a|e^{-i\hat{H}_A t/\hbar} = e^{-iE_a^{(A)}t/\hbar}\langle a|,$$

$$e^{i\hat{H}_A t/\hbar}|b\rangle = e^{iE_b^{(A)}t/\hbar}|b\rangle,$$

the first matrix element in equation (6.5.52) becomes

$$\langle a|\hat{T}_2\hat{T}_1^+|b\rangle = e^{iE_b^{(A)}t_1/\hbar - iE_a^{(A)}t_2/\hbar}\langle a|\hat{U}_2\hat{U}_1^+|b\rangle. \tag{6.5.55}$$

The $\{-\}_{AV}$ average of the reduced matrix element $\langle a|\hat{U}_2\hat{U}_1^+|b\rangle$ may be evaluated [46] by writing

$$\frac{d\hat{U}(t,0)}{dt} = \lim_{\Delta t\to 0}\frac{1}{\Delta t}[\hat{U}(t+\Delta t, 0) - \hat{U}(t,0)], \tag{6.5.56}$$

where, from equation (6.5.17),

$$\begin{aligned}\hat{U}(t+\Delta t, 0) &= \hat{\mathscr{D}}\exp\left\{\frac{1}{i\hbar}\int_0^{t+\Delta t}\hat{H}_{PA}^{I}(t')\,dt'\right\}\\ &= \hat{\mathscr{D}}\exp\left\{\frac{1}{i\hbar}\left(\int_t^{t+\Delta t} + \int_0^{t}\right)\hat{H}_{PA}(t')\,dt'\right\}\\ &= \hat{U}(t+\Delta t, t)\hat{U}(t,0).\end{aligned} \tag{6.5.57}$$

One should note, however, that this last step is non-trivial (see equation (6.3.46) for example).

Consequently, equation (6.5.56) yields

$$\frac{d\hat{U}_{AV}(t,0)}{dt} = \lim_{\Delta t\to 0}\frac{1}{\Delta t}\{[\hat{U}(t+\Delta t, t) - 1]\hat{U}(t,0)\}_{AV}. \tag{6.5.58}$$

We now make use of the impact approximation in which we assume the time duration of a collisional encounter is small relative to the time between collisions. In particular, we want the above Δt to be both (i) sufficiently large that the two factors constituting the product on the RHS (6.5.58) can be considered statistically independent and therefore averaged separately, and (ii) sufficiently small that the limit $\Delta t \to 0$ does indeed produce a valid time derivative. We then have

$$\frac{d\hat{U}_{AV}(t,0)}{dt} = \left\{\lim_{\Delta t\to 0}\frac{1}{\Delta t}[\hat{U}(t+\Delta t, t) - 1]\right\}_{AV}\hat{U}_{AV}(t,0). \tag{6.5.59}$$

We next note that, using equation (6.5.20),

$$\hat{U}(t+\Delta t, t) = \hat{\mathscr{D}}\exp\left\{\frac{1}{i\hbar}\int_t^{t+\Delta t} e^{iH_A t'/\hbar}\hat{H}_{PA}(t')e^{iH_A t'/\hbar}\,dt'\right\},$$

which, changing the variable of integration from t' to $t'' = t' - t$, becomes

$$\hat{U}(t+\Delta t, t) = \hat{\mathscr{D}} \exp\left\{\frac{1}{\mathrm{i}\hbar} \mathrm{e}^{\mathrm{i}\hat{H}_{\mathrm{A}}t/\hbar} \int_0^{\Delta t} \mathrm{e}^{\mathrm{i}\hat{H}_{\mathrm{A}}t''/\hbar} \hat{H}_{\mathrm{PA}}(t'') \mathrm{e}^{-\mathrm{i}\hat{H}_{\mathrm{A}}t''/\hbar}\, \mathrm{d}t'' \times \mathrm{e}^{-\mathrm{i}\hat{H}_{\mathrm{A}}t/\hbar}\right\}. \tag{6.5.60}$$

The general result

$$\mathrm{e}^{\mathrm{e}^{\hat{\alpha}}\hat{\beta}\mathrm{e}^{-\hat{\alpha}}} = \mathrm{e}^{\hat{\alpha}}\mathrm{e}^{\hat{\beta}}\mathrm{e}^{-\hat{\alpha}}, \tag{6.5.61}$$

obtained by directly expanding the LHS in terms of $(\mathrm{e}^{\hat{\alpha}}\hat{\beta}\mathrm{e}^{-\hat{\alpha}})^n$ and noting that $\mathrm{e}^{\hat{\alpha}}\mathrm{e}^{-\hat{\alpha}} = \mathbb{1}$, then enables equation (6.5.60) to be written as

$$\begin{aligned}\hat{U}(t+\Delta t, t) &= \mathrm{e}^{\mathrm{i}\hat{H}_{\mathrm{A}}t/\hbar} \hat{\mathscr{D}} \exp\left\{\frac{1}{\mathrm{i}\hbar}\int_0^{\Delta t} \hat{H}_{\mathrm{PA}}(t'')\,\mathrm{d}t''\right\} \mathrm{d}^{-\mathrm{i}\hat{H}_{\mathrm{A}}t/\hbar} \\ &= \mathrm{e}^{\mathrm{i}\hat{H}_{\mathrm{A}}t/\hbar} \hat{U}(\Delta t, 0) \mathrm{e}^{-\mathrm{i}\hat{H}_{\mathrm{A}}t/\hbar}.\end{aligned} \tag{6.5.62}$$

Equation (6.5.59) then becomes

$$\begin{aligned}\frac{\mathrm{d}\hat{U}_{\mathrm{AV}}(t,0)}{\mathrm{d}t} &= \left\{\mathrm{e}^{\mathrm{i}\hat{H}_{\mathrm{A}}t/\hbar} \lim_{\Delta t\to 0} \frac{1}{\Delta t}[\hat{U}(\Delta t, 0) - 1]\mathrm{e}^{-\mathrm{i}\hat{H}_{\mathrm{A}}t/\hbar}\right\}_{\mathrm{AV}} \hat{U}_{\mathrm{AV}}(t,0) \\ &= -\mathrm{e}^{\mathrm{i}\hat{H}_{\mathrm{A}}t/\hbar}\hat{\Phi}\mathrm{e}^{-\mathrm{i}\hat{H}_{\mathrm{A}}t/\hbar}\hat{U}_{\mathrm{AV}}(t,0),\end{aligned} \tag{6.5.63}$$

where we have defined the time-independent operator

$$\hat{\Phi} = \left\{\lim_{\Delta t\to 0}\frac{1}{\Delta t}[1 - \hat{U}(\Delta t, 0)]\right\}_{\mathrm{AV}}, \tag{6.5.64}$$

which is related to the scattering $\hat{S}$ matrix by $\hat{\Phi} = \mathbb{1} - \hat{S}$.

It is important to note that $\hat{H}_{\mathrm{A}}$ and $\hat{\Phi}$ do not generally commute. The solution to equation (6.5.63) is therefore

$$\hat{U}_{\mathrm{AV}}(t,0) = \mathrm{e}^{\mathrm{i}\hat{H}_{\mathrm{A}}t/\hbar}\mathrm{e}^{-(\mathrm{i}\hat{H}_{\mathrm{A}}/\hbar + \hat{\Phi})t}, \tag{6.5.65}$$

which may be easily verified by noting

$$\begin{aligned}\frac{\mathrm{d}}{\mathrm{d}t}[\mathrm{e}^{\mathrm{i}\hat{H}_{\mathrm{A}}t/\hbar}\mathrm{e}^{-(\mathrm{i}\hat{H}_{\mathrm{A}}/\hbar+\hat{\Phi})t}] &= \frac{\mathrm{i}}{\hbar}\hat{H}_{\mathrm{A}}\hat{U}_{\mathrm{AV}}(t,0) \\ &\quad -\mathrm{e}^{\mathrm{i}\hat{H}_{\mathrm{A}}t/\hbar}(\mathrm{i}\hat{H}_{\mathrm{A}}/\hbar + \hat{\Phi})\mathrm{e}^{-(\mathrm{i}\hat{H}_{\mathrm{A}}/\hbar+\hat{\Phi})t} \\ &= -\mathrm{e}^{\mathrm{i}\hat{H}_{\mathrm{A}}t/\hbar}\hat{\Phi}\mathrm{e}^{-(\mathrm{i}\hat{H}_{\mathrm{A}}/\hbar+\hat{\Phi})t} \\ &= -\mathrm{e}^{\mathrm{i}\hat{H}_{\mathrm{A}}t/\hbar}\hat{\Phi}\mathrm{e}^{-\mathrm{i}\hat{H}_{\mathrm{A}}t/\hbar}\hat{U}_{\mathrm{AV}}(t,0).\end{aligned}$$

Note further that the above solution satisfies the necessary condition $\hat{U}(0,0) = \mathbb{1}$.

Equation (6.5.53) yields

$$\hat{T}_{\mathrm{AV}}(t,0) = \mathrm{e}^{-(\mathrm{i}\hat{H}_{\mathrm{A}}/\hbar + \hat{\Phi})t}, \tag{6.5.66}$$

so that, with $\hat{T}_{\mathrm{AV}}\hat{T}^{+}_{\mathrm{AV}} \equiv \mathbb{1}$, i.e.

$$\mathrm{e}^{-(\mathrm{i}\hat{H}_{\mathrm{A}}/\hbar+\hat{\Phi})t}\mathrm{e}^{(\mathrm{i}\hat{H}_{\mathrm{A}}/\hbar-\hat{\Phi}^{+})t} = \mathbb{1},$$

where we have used $\hat{H}_A^+ = \hat{H}_A$, then

$$\hat{\Phi}^+ = -\hat{\Phi}, \tag{6.5.67}$$

i.e. $\hat{\Phi}$ is not Hermitian.

Consequently, the desired matrix elements become, for example,

$$\begin{aligned}\langle a|\hat{T}_2\hat{T}_1^+|b\rangle &= \langle a|e^{-(i\hat{H}_A/\hbar+\hat{\Phi})t_2}e^{(i\hat{H}_A/\hbar+\hat{\Phi})t_1}|b\rangle \\ &= \langle a|e^{-(i\hat{H}_A/\hbar+\hat{\Phi})(t_2-t_1)}|b\rangle.\end{aligned} \tag{6.5.68}$$

This last result clearly shows therefore that the matrix element $\langle a|T_2T_1^+|b\rangle$ has a time dependence involving simply $t_2 - t_1$. Similarly, we may show that

$$\langle b'|\hat{T}_1\hat{T}_2^+|a'\rangle = \langle b'|e^{(iH_A/\hbar+\Phi)(t_2-t_1)}|a'\rangle,$$

so that from equation (6.5.52) we have

$$\{\langle\langle aa'|\hat{\Theta}_{21}|bb'\rangle\rangle\}_{AV} = \langle\langle aa'|e^{-(i\hat{H}_A/\hbar+\hat{\Phi})(t_2-t_1)}|bb'\rangle\rangle, \tag{6.5.69}$$

which also has a time dependence involving only the difference $t_2 - t_1$.

In general, therefore, we may write

$$\hat{\Theta}_{mn} \equiv \hat{\Theta}(t_m - t_n) = e^{-(iH_A/\hbar+\hat{\Phi})(t_m-t_n)}. \tag{6.5.70}$$

The transition rate $w(1 \to 2 \to 3)$ given by equation (6.5.39) then has the form

$$\begin{aligned} w(1\to 2\to 3) = \lim_{t\to\infty} \frac{n_1(n_2+1)}{t\hbar^4} &\sum_{\substack{abcd\\a'b'c'd'}} \int_0^t dt_4 \int_0^{t_4} dt_3 \int_0^t dt_2 \int_0^{t_2} dt_1 \\ &\times Y_{21}Y_{42}Y_{43}e^{i(-\omega_1 t_1+\omega_2 t_2+\omega_1 t_3-\omega_2 t_4)}\langle d'|\xi_1^*|d\rangle\langle c'|\xi_2|c\rangle \\ &\times \langle b'|\xi_2^*|b\rangle\langle a'|\xi_1|a\rangle a_1(t_1)a_1^*(t_3) \\ &\times \sum_{\alpha=I}^{III} \{\bar{\Theta}_\alpha(a,\ldots,d')\}_{AV},\end{aligned} \tag{6.5.71}$$

where, from equations (6.5.49), (6.5.50) and (6.5.51), we have

$$\begin{aligned}\{\bar{\Theta}_I(a,\ldots,d')\}_{AV} &= \sum_{jl} \rho_{aa}\langle\langle cc'|\hat{\Theta}(t_4-t_3)|ld\rangle\rangle \\ &\quad\times \langle\langle ld'|\hat{\Theta}(t_3-t_2)|bj'\rangle\rangle\langle\langle bj|\hat{\Theta}(t_2-t_1)|a'a\rangle\rangle,\\ \{\bar{\Theta}_{II}(a,\ldots,d')\}_{AV} &= \sum_{ik} \rho_{aa}\langle\langle cc'|\hat{\Theta}(t_4-t_2)|b'k\rangle\rangle \\ &\quad\times \langle\langle bk|\hat{\Theta}(t_2-t_3)|id\rangle\rangle\langle\langle id'|\hat{\Theta}(t_3-t_1)|a'a\rangle\rangle,\\ \{\bar{\Theta}_{III}(a,\ldots,d')\}_{AV} &= \sum_{ik} \rho_{aa}\langle\langle cc'|\hat{\Theta}(t_4-t_2)|b'k\rangle\rangle \\ &\quad\times \langle\langle bk|\hat{\Theta}(t_2-t_1)|a'i\rangle\rangle\langle\langle ai|\hat{\Theta}(t_1-t_3)|d'd\rangle\rangle.\end{aligned} \tag{6.5.72}$$

Clearly, equation (6.5.71) may be re-written as

$$w(1\to 2\to 3) = \sum_{\alpha=I}^{III} w_\alpha(1\to 2\to 3). \tag{6.5.73}$$

To proceed further, we assume the width of the spectral line (related to the broadening of the energy levels of the atom) is small relative to the energy difference between these levels. This is equivalent to stating that the eigenstates of $\hat{H}_A$ are non-degenerate (i.e. no overlapping of energy levels). In this case, the off-diagonal terms of the above $\hat{\Theta}$ operators are dominated by the diagonal terms and one has

$$\langle\langle aa'|\hat{\Theta}(\tau)|bb'\rangle\rangle = \delta_{ab}\delta_{a'b'}e^{-(i\omega_{aa'}+\Phi_{aa'})\tau}. \qquad (6.5.74)$$

For purposes of illustration we write

$$\left.\begin{aligned} \Phi_{aa'} &= \Gamma_{aa'}^{(c)} + \tfrac{1}{2}(\Gamma_a + \Gamma_{a'}) + i\Delta\omega_{aa'}^{(c)}, \\ \Phi_{aa} &= \Gamma_a + \Gamma_a^{(c)} + i\Delta\omega_a^{(c)}, \end{aligned}\right\} \qquad (6.5.75)$$

where Γ_a and $\Gamma_{a'}$ have the same meaning as in the preceding sections, $\Gamma_{aa'}^{(c)}$ and $\Gamma_a^{(c)}$ refer to the broadening parameters due to collisions, and $\Delta\omega_{aa'}^{(c)}$ and $\Delta\omega_a^{(c)}$ refer to the corresponding level shifts.

Actually, the explicit evaluation of the above elements is beyond the scope of this presentation (see, for example, Baranger [10].) Nevertheless, we can see from equations (6.5.57) and (6.5.64) that if there are no perturbers (i.e. if $\hat{H}_{PA}$ does not occur in the total Hamiltonian), then $U(t, 0) = \mathbb{1}$ and this should produce a null operator $\hat{\Phi}$. However, equation (6.5.75) shows that even in the absence of collisions we still have non-zero $\Phi_{aa'}$. Thus, broadening due to spontaneous emission has been included phenomenologically as an effective perturbing encounter quite separate from collisional encounters. This, in turn, implies that such emissions are uncoupled from the collisions and must therefore be considered negligible in the above representation during any atom–particle perturbation. The above approximation can be removed by incorporating an analysis similar to that presented in section 6.4.3.

6.5.5 *The components of* $w(1 \to 2 \to 3)$

We are now in a position to evaluate the transition rates $w_\alpha(1 \to 2 \to 3)$ given by equation (6.5.71). We treat each of the sequences I, II and III separately.

Sequence I $(t_4 > t_3 > t_2 > t_1)$

We define the new variables of integration

$$\tau_3 = t_4 - t_3, \quad \tau_2 = t_3 - t_2, \quad \tau_1 = t_2 - t_1, \qquad (6.5.76)$$

so that

$$\begin{aligned} &-\omega_1 t_1 + \omega_2 t_2 - \omega_2 t_4 + \omega_1 t_3 \\ &\quad = \omega_1(t_2 - t_1) + \omega_1(t_3 - t_2) - \omega_2(t_3 - t_2) - \omega_2(t_4 - t_3) \\ &\quad = \omega_1\tau_1 + (\omega_1 - \omega_2)\tau_2 - \omega_2\tau_3. \end{aligned}$$

Since the quantities appearing in $\{\bar{\Theta}_I(a, \ldots, d')\}_{AV}$ involve only the time differences τ_3, τ_2 and τ_1, we have

$$w_{\mathrm{I}}(1 \to 2 \to 3) = \frac{n_1(n_2+1)}{\hbar^4} \sum_{\substack{abcd \\ a'b'c'd'}} \sum_{jl} \rho_{aa}$$

$$\times \langle d'|\xi_1^*|d\rangle\langle c'|\xi_2|c\rangle\langle b'|\xi_2^*|b\rangle\langle a'|\xi_1|a\rangle$$

$$\times \int_0^\infty \mathrm{d}\tau_3 \int_0^\infty \mathrm{d}\tau_2 \int_0^\infty \mathrm{d}\tau_1 \mathrm{e}^{\mathrm{i}[\omega_1\tau_1 + (\omega_1 - \omega_2)\tau_2 - \omega_2\tau_3]}$$

$$\times \langle\langle cc'|\hat{\Theta}(\tau_3)|ld\rangle\rangle\langle\langle ld'|\hat{\Theta}(\tau_2)|b'j\rangle\rangle\langle\langle bj|\hat{\Theta}(\tau_1)|a'a\rangle\rangle,$$

which, using equation (6.5.74), becomes

$$w_{\mathrm{I}}(1 \to 2 \to 3) = \frac{n_1(n_2+1)}{\hbar^4} \sum_{\substack{abcd \\ a'b'c'd'}} \sum_{jl} \rho_{aa}$$

$$\times \langle d'|\xi_1^*|d\rangle\langle c'|\xi_2|c\rangle\langle b'|\xi_2^*|b\rangle\langle a'|\xi_1|a\rangle$$

$$\times \int_0^\infty \mathrm{d}\tau_3 \int_0^\infty \mathrm{d}\tau_2 \int_0^\infty \mathrm{d}\tau_1 \mathrm{e}^{\mathrm{i}[\omega_1\tau_1 + (\omega_1 - \omega_2)\tau_2 - \omega_2\tau_3]}$$

$$\times \delta_{cl}\delta_{c'd}\mathrm{e}^{-(\mathrm{i}\omega_{cc'} + \Phi_{cc'})\tau_3}\delta_{lb'}\delta_{d'j}\mathrm{e}^{-(\mathrm{i}\omega_{ld'} + \Phi_{ld'})\tau_2}$$

$$\times \delta_{ba'}\delta_{ja}\mathrm{e}^{-(\mathrm{i}\omega_{bj} + \Phi_{bj})\tau_1}$$

$$= \frac{n_1(n_2+1)}{\hbar^4} \sum_{abcd} \rho_{aa}$$

$$\times \langle a|\xi_1^*|d\rangle\langle d|\xi_2|c\rangle\langle c|\xi_2^*|b\rangle\langle b|\xi_1|a\rangle$$

$$\times \int_0^\infty \mathrm{d}\tau_3 \int_0^\infty \mathrm{d}\tau_2 \int_0^\infty \mathrm{d}\tau_1 \exp\,[\mathrm{i}(\omega_1 - \omega_{ba} + \mathrm{i}\Phi_{ba})\tau_1$$

$$+ \mathrm{i}(\omega_1 - \omega_2 - \omega_{ca} + \mathrm{i}\Phi_{ca})\tau_2 + \mathrm{i}(-\omega_2 - \omega_{cd} + \mathrm{i}\Phi_{cd})\tau_3]$$

$$= -\frac{n_1(n_2+1)}{\hbar^4} \sum_{abcd} \rho_{aa}\mathrm{i}$$

$$\times \langle a|\xi_1^*|d\rangle\langle d|\xi_2|c\rangle\langle c|\xi_2^*|b\rangle\langle b|\xi_1|a\rangle$$

$$\times \frac{1}{[\omega_1 - \omega_{ba} + \mathrm{i}\Phi_{ba}][\omega_1 - \omega_2 - \omega_{ca} + \mathrm{i}\Phi_{ca}][-\omega_2 - \omega_{cd} + \mathrm{i}\Phi_{cd}]}, \tag{6.5.77}$$

where, of course, we take the damping factors $\mathcal{R}l\, \Phi > 0$.

We are only interested in the resonance terms here, i.e. those terms for which $\omega_1 \approx \omega_{ba}$, $\omega_1 - \omega_2 \approx \omega_{ca}$ and $\omega_2 \approx \omega_{dc}$. Thus, without loss of generality, we must have $b = 2$, $a = 1$, $d = 2$ and $c = 3$ where we note that level 3 is equivalent to level 1. We therefore find

$$w_{\mathrm{I}}(1 \to 2 \to 3) = -\frac{n_1(n_2+1)}{\hbar^4} \rho_{11}\mathrm{i}$$

$$\times \langle 1|\xi_1^*|2\rangle\langle 2|\xi_2|3\rangle\langle 3|\xi_2^*|2\rangle\langle 2|\xi_1|1\rangle$$

$$\times \frac{1}{[\omega_1-\omega_{21}+i\Phi_{21}][\omega_1-\omega_2+i\Phi_{31}][-\omega_2+\omega_{21}+i\Phi_{32}]}, \tag{6.5.78}$$

where we have put $\omega_{31}\equiv 0$ and $\omega_{32}=-\omega_{23}=-\omega_{21}$.

Sequence II $(t_4>t_2>t_3>t_1)$
Here we define

$$\tau_3=t_4-t_2, \quad \tau_2=t_2-t_3, \quad \tau_1=t_3-t_1, \tag{6.5.79}$$

so that

$$-\omega_1 t_1+\omega_2 t_2-\omega_2 t_4+\omega_1 t_3=\omega_1(t_3-t_1)-\omega_2(t_4-t_2),$$
$$=\omega_1\tau_1-\omega_2\tau_3.$$

We then have

$$w_{\mathrm{II}}(1\to 2\to 3)=\frac{n_1(n_2+1)}{\hbar^4}\sum_{\substack{abcd\\a'b'c'd'}}\sum_{ik}\rho_{aa}$$

$$\times \langle d'|\xi_1^*|d\rangle\langle c'|\xi_2|c\rangle\langle b'|\xi_2^*|b\rangle\langle a'|\xi_1|a\rangle$$

$$\times\int^{\infty}\mathrm{d}\tau_3\int^{\infty}\mathrm{d}\tau_2\int^{\infty}\mathrm{d}\tau_1\,e^{i(\omega_1\tau_1-\omega_2\tau_3)}$$

$$\times \langle\langle cc'|\hat{\Theta}(\tau_3)|b'k\rangle\rangle\langle\langle bk|\hat{\Theta}(\tau_2)|id\rangle\rangle\langle\langle id'|\hat{\Theta}(\tau_1)|a'a\rangle\rangle$$

$$=\frac{n_1(n_2+1)}{\hbar^4}\sum_{abcd}\rho_{aa}i$$

$$\times \langle a|\xi_1^*|d\rangle\langle d|\xi_2|c\rangle\langle c|\xi_2^*|b\rangle\langle b|\xi_1|a\rangle$$

$$\times\frac{1}{[\omega_1-\omega_{ba}+i\Phi_{ba}][-\omega_{bd}+i\Phi_{bd}][-\omega_2-\omega_{cd}+i\Phi_{cd}]}.$$

Again, we are interested only in the resonance terms for which $\omega_1\approx\omega_{ba}$, $\omega_2\approx\omega_{dc}$ and $\omega_{bd}\approx 0$. These last conditions are satisfied for $b=2, a=1$, $d=2$ and $c=3$. We therefore find

$$w_{\mathrm{II}}(1\to 2\to 3)=-\frac{n_1(n_2+1)}{\hbar^4}\rho_{11}$$

$$\times \langle a|\xi_1^*|2\rangle\langle 2|\xi_2|3\rangle\langle 3|\xi_2^*|2\rangle\langle 2|\xi_1|1\rangle$$

$$\times\frac{1}{(\omega_1-\omega_{21}+i\Phi_{21})\Phi_{22}(-\omega_2+\omega_{21}+i\Phi_{12})}. \tag{6.5.80}$$

Sequence III $(t_4>t_2>t_1>t_3)$
Here we define

$$\tau_3=t_4-t_2, \quad \tau_2=t_2-t_1, \quad \tau_1=t_1-t_2, \tag{6.5.81}$$

so that

$$-\omega_1 t_1 + \omega_2 t_2 - \omega_2 t_4 + \omega_1 t_3 = -\omega_1(t_1 - t_3) - \omega_2(t_4 - t_2)$$
$$= -\omega_1\tau_1 - \omega_2\tau_3.$$

Equations (6.5.71) and (6.5.72) then yield

$$w_{\mathrm{III}}(1 \to 2 \to 3) = \frac{n_1(n_2+1)}{\hbar^4} \sum_{\substack{abcd \\ a'b'c'd'}} \sum_{ik} \rho_{aa}$$
$$\times \langle d'|\xi_1^*|d\rangle\langle c'|\xi_2|c\rangle\langle b'|\xi_2^*|b\rangle\langle a'|\xi_1|a\rangle$$
$$\times \int_0^\infty \mathrm{d}\tau_3 \int_0^\infty \mathrm{d}\tau_2 \int_0^\infty \mathrm{d}\tau_1 \mathrm{e}^{-(\omega_1\tau_1+\omega_2\tau_3)}$$
$$\times \langle\langle cc'|\hat{\Theta}(\tau_3)|b'k\rangle\rangle\langle\langle bk|\hat{\Theta}(\tau_2)|a'i\rangle\rangle\langle\langle ai|\hat{\Theta}(\tau_1)|d'd\rangle\rangle$$
$$= -\frac{n_1(n_2+1)}{\hbar^4} \sum_{abcd} \rho_{aa}\mathrm{i}$$
$$\times \langle a|\xi_1^*|d\rangle\langle d|\xi_2|c\rangle\langle c|\xi_2^*|b\rangle\langle b|\xi_1|a\rangle$$
$$\times \frac{1}{(-\omega_1 - \omega_{ad} + \mathrm{i}\Phi_{ad})(-\omega_{bd} + \mathrm{i}\Phi_{bd})(-\omega_2 - \omega_{cd} + \mathrm{i}\Phi_{cd})}.$$

Again, the resonance terms may be obtained by requiring $\omega_1 \approx \omega_{da}$, $\omega_2 \approx \omega_{dc}$ and $\omega_{bd} \approx 0$, i.e. $d = 2$, $a = 1$, $c = 3$ and $b = 2$. We therefore obtain

$$w_{\mathrm{III}}(1 \to 2 \to 3) = -\frac{n_1(n_2+1)}{\hbar^4} \rho_{11}$$
$$\times \langle 1|\xi_1^*|2\rangle\langle 2|\xi_2|3\rangle\langle 3|\xi_2^*|2\rangle\langle 2|\xi_1|1\rangle$$
$$\times \frac{1}{(-\omega_1 + \omega_{21} + \mathrm{i}\Phi_{12})\Phi_{22}(-\omega_2 + \omega_{21} + \mathrm{i}\Phi_{12})}. \qquad (6.5.82)$$

We now consider two quite distinct cases. First, we examine the system in which no collisions are allowed - this produces a result identical to that obtained in the preceding section. We then study the scattering process where collisional encounters are allowed, but where the de-excited state is infinitesimally sharp.

6.5.6 *Zero collisional perturbations [92]*

Here we consider the system in which collisions are neglected. This therefore represents the scattering problem already studied in section 6.4.7.

Reference to equations (6.5.75) shows that if there are no collisions, and we neglect the level shifts $\Delta\omega$ as in the preceding section, then

$$\Phi_{aa'} = \tfrac{1}{2}(\Gamma_a + \Gamma_{a'}), \quad \Phi_{aa} = \Gamma_a. \qquad (6.5.83)$$

Equation (6.5.78) then yields

$$w_{\mathrm{I}}(1\to 2\to 1) = \mathcal{R}l \frac{\mathrm{i}K}{\left[\omega_1-\omega_{21}+\frac{\mathrm{i}}{2}(\Gamma_1+\Gamma_2)\right][\omega_1-\omega_2+\mathrm{i}\Gamma_1]\left[\omega_2-\omega_{21}-\frac{\mathrm{i}}{2}(\Gamma_1+\Gamma_2)\right]}, \tag{6.5.84}$$

where the real constant K is

$$K = \frac{n_1(n_2+1)}{\hbar^4}\,\rho_{11}|\langle 2|\xi_1|1\rangle|^2\,|\langle 2|\xi_2|1\rangle|^2. \tag{6.5.85}$$

Similarly, equations (6.5.80) and (6.5.82) yield

$$w_{\mathrm{II}}(1\to 2\to 1) = \mathcal{R}l \frac{K}{\left[\omega_1-\omega_{21}+\frac{\mathrm{i}}{2}(\Gamma_1+\Gamma_2)\right]\left[\omega_2-\omega_{21}-\frac{\mathrm{i}}{2}(\Gamma_1+\Gamma_2)\right]\Gamma_2}, \tag{6.5.86}$$

$$w_{\mathrm{III}}(1\to 2\to 1) = \mathcal{R}l \frac{-K}{\left[\omega_1-\omega_{21}-\frac{\mathrm{i}}{2}(\Gamma_1+\Gamma_2)\right]\left[\omega_2-\omega_{21}-\frac{\mathrm{i}}{2}(\Gamma_1+\Gamma_2)\right]\Gamma_2}. \tag{6.5.87}$$

For convenience in exposition of the requisite algebra, we define the terms

$$X_1 = (\omega_1-\omega_{21})^2+\left(\frac{\Gamma_1+\Gamma_2}{2}\right)^2,$$

$$X_2 = (\omega_2-\omega_{21})^2+\left(\frac{\Gamma_1+\Gamma_2}{2}\right)^2,$$

$$X_{12} = (\omega_1-\omega_2)^2+\Gamma_1^2.$$

Equation (6.5.84) then yields

$$\begin{aligned}
w_{\mathrm{I}}(1\to 2\to 1) &= \frac{K}{X_1X_2X_{12}}\,\mathcal{R}l\,\mathrm{i}\left[\omega_1-\omega_{21}-\frac{\mathrm{i}}{2}(\Gamma_1+\Gamma_2)\right] \\
&\quad\times(\omega_1-\omega_2-\mathrm{i}\Gamma_1)\left[\omega_2-\omega_{21}+\frac{\mathrm{i}}{2}(\Gamma_1+\Gamma_2)\right] \\
&= \frac{K}{X_1X_2X_{12}}\,\mathcal{R}l\,\mathrm{i}\left\{(\omega_1-\omega_{21})(\omega_2-\omega_{21})+\left(\frac{\Gamma_1+\Gamma_2}{2}\right)^2\right. \\
&\quad\left.+\frac{\mathrm{i}}{2}(\Gamma_1+\Gamma_2)(\omega_1-\omega_{21}-\omega_2+\omega_{21})\right\}(\omega_1-\omega_2-\mathrm{i}\Gamma_1) \\
&= \frac{K}{X_1X_2X_{12}}\left\{\Gamma_1\left[(\omega_1-\omega_{21})(\omega_2-\omega_{21})+\left(\frac{\Gamma_1+\Gamma_2}{2}\right)^2\right]\right. \\
&\quad\left.-\frac{\Gamma_1+\Gamma_2}{2}\cdot(\omega_1-\omega_2)^2\right\}.
\end{aligned} \tag{6.5.88}$$

This last expression may be simplified by noting that

$$(\omega_1-\omega_{21})(\omega_2-\omega_{21})=(\omega_1-\omega_{21})(\omega_1-\omega_{21}+\omega_2-\omega_1)$$
$$=(\omega_1-\omega_{21})^2+(\omega_1-\omega_{21})(\omega_2-\omega_1),$$

but this is also equal to

$$[\omega_2-\omega_{21}-(\omega_2-\omega_1)](\omega_2-\omega_{21})=(\omega_2-\omega_{21})^2$$
$$-(\omega_2-\omega_1)(\omega_2-\omega_{21}).$$

Thus, adding these last two equations, we have

$$(\omega_1-\omega_{21})(\omega_2-\omega_{21})=\tfrac{1}{2}[(\omega_1-\omega_{21})^2+(\omega_2-\omega_{21})^2$$
$$-(\omega_1-\omega_2)^2]$$
$$=\tfrac{1}{2}\left[X_1+X_2-X_{12}-2\left(\frac{\Gamma_1+\Gamma_2}{2}\right)^2+\Gamma_1^2\right]. \tag{6.5.89}$$

Equation (6.5.88) then yields

$$w_{\mathrm{I}}(1\to2\to1)=\frac{K}{X_1X_2X_{12}}\left\{\frac{\Gamma_1}{2}(X_1+X_2)-\frac{\Gamma_1X_{12}}{2}-\Gamma_1\left(\frac{\Gamma_1+\Gamma_2}{2}\right)^2\right.$$
$$\left.+\frac{\Gamma_1^3}{2}+\Gamma_1\left(\frac{\Gamma_1+\Gamma_2}{2}\right)^2-\frac{\Gamma_1+\Gamma_2}{2}\cdot X_{12}+\frac{\Gamma_1^2(\Gamma_1+\Gamma_2)}{2}\right\}$$
$$=\frac{K}{X_1X_2X_{12}}\left\{\frac{\Gamma_1}{2}(X_1+X_2)-\frac{2\Gamma_1+\Gamma_2}{2}\cdot X_{12}+\frac{\Gamma_1^2(2\Gamma_1+\Gamma_2)}{2}\right\}. \tag{6.5.90}$$

Further, equations (6.5.86) and (6.5.87) give

$$w_{\mathrm{II}}(1\to2\to1)+w_{\mathrm{III}}(1\to2\to1)$$
$$=\mathcal{R}l\,\frac{K}{\Gamma_2}\,\frac{1}{\omega_2-\omega_{21}-\frac{\mathrm{i}}{2}(\Gamma_1+\Gamma_2)}$$
$$\times\left\{\frac{-1}{\omega_1-\omega_{21}-\frac{\mathrm{i}}{2}(\Gamma_1+\Gamma_2)}+\frac{1}{\omega_1-\omega_{21}+\frac{\mathrm{i}}{2}(\Gamma_1+\Gamma_2)}\right\}$$
$$=\frac{K}{\Gamma_2X_1}\,\mathcal{R}l\,\frac{-\mathrm{i}(\Gamma_1+\Gamma_2)}{\omega_2-\omega_{21}-\frac{\mathrm{i}}{2}(\Gamma_1+\Gamma_2)}$$
$$=\frac{-K}{\Gamma_2X_1X_2}\,\mathcal{R}l\,\mathrm{i}(\Gamma_1+\Gamma_2)\left[\omega_2+\omega_{21}+\frac{\mathrm{i}}{2}(\Gamma_1+\Gamma_2)\right]$$
$$=\frac{K(\Gamma_1+\Gamma_2)^2}{2\Gamma_2X_1X_2}. \tag{6.5.91}$$

Thus, the total transition rate is

$$w(1 \to 2 \to 1) = \sum_{\alpha=\mathrm{I}}^{\mathrm{III}} w_\alpha(1 \to 2 \to 1)$$

$$= \frac{K\Gamma_1}{2X_2X_{12}} + \frac{K\Gamma_1}{2X_1X_{12}} - \frac{(2\Gamma_1 + \Gamma_2)K}{2X_1X_2} + \frac{K\Gamma_1^2(2\Gamma_1 + \Gamma_2)}{2X_1X_2X_{12}}$$

$$+ \frac{K(\Gamma_1 + \Gamma_2)^2}{2\Gamma_2 X_1 X_2}$$

$$= \frac{K}{2\Gamma_2}\left\{\frac{\Gamma_1^2\Gamma_2(2\Gamma_1 + \Gamma_2)}{X_1X_2X_{12}} + \frac{\Gamma_1\Gamma_2}{X_2X_{12}} + \frac{\Gamma_1\Gamma_2}{X_1X_{12}} + \frac{\Gamma_1^2}{X_1X_2}\right\}. \qquad (6.5.92)$$

This expression, apart from a constant normalisation factor, is identical to equation (6.4.200). This is an interesting result in view of the approximations inherent in equations (6.5.74) and (6.5.75).

6.5.7 *Collisions - zero width ground state [92]*

Here we wish to determine the transition rate for the scattering process in which collisions must be included. Before actually examining the simplified system in which the ground state has zero width, we consider the more general result by writing equations (6.5.78), (6.5.80) and (6.5.82) in the form

$$w_{\mathrm{I}}(1 \to 2 \to 1) = \mathcal{R}l \frac{\mathrm{i}K}{[\omega_1 - \omega_{21} - \Delta\omega_{21} + \mathrm{i}\Gamma_{21}][\omega_1 - \omega_2 - \Delta\omega_{11} + \mathrm{i}\Gamma_{11}][\omega_2 - \omega_{21} - \Delta\omega_{21} - \mathrm{i}\Gamma_{21}]}, \qquad (6.5.93)$$

$$w_{\mathrm{II}}(1 \to 2 \to 1) = \mathcal{R}l \frac{K}{\Gamma_{22}[\omega_1 - \omega_{21} - \Delta\omega_{21} + \mathrm{i}\Gamma_{21}] \times [\omega_2 - \omega_{21} - \Delta\omega_{21} - \mathrm{i}\Gamma_{21}]}, \qquad (6.5.94)$$

$$w_{\mathrm{III}}(1 \to 2 \to 1) = \mathcal{R}l \frac{-K}{\Gamma_{22}[\omega_1 - \omega_{21} - \Delta\omega_{21} - \mathrm{i}\Gamma_{21}] \times [\omega_2 - \omega_{21} - \Delta\omega_{21} - \mathrm{i}\Gamma_{21}]}, \qquad (6.5.95)$$

where we have put

$$\Phi_{aa'} = \Gamma_{aa'} + \mathrm{i}\Delta\omega_{aa'} \equiv \Gamma_{a'a} - \mathrm{i}\Delta\omega_{a'a}, \qquad (6.5.96)$$

for all a and a', and where K is defined by equation (6.5.85). If we now define

$$\Delta_1 = \omega_1 - \omega_{21}, \quad \Delta_2 = \omega_2 - \omega_{21}, \quad \Delta_{12} = \omega_1 - \omega_2,$$

and equate all level shifts $\Delta\omega_{21}$ and $\Delta\omega_{11}$ to zero so that, from equation (6.5.89), we have

$$\Delta_1\Delta_2 = \tfrac{1}{2}(\Delta_1^2 + \Delta_2^2 - \Delta_{12}^2)$$
$$= \tfrac{1}{2}(X_1 + X_2 - X_{12} - 2\Gamma_{21}^2 + \Gamma_{11}^2),$$

where we have put

$$X_1 = \Delta_1^2 + \Gamma_{21}^2, \quad X_2 = \Delta_2^2 + \Gamma_{21}^2, \quad X_{12} = \Delta_{12}^2 + \Gamma_{11}^2,$$

then from equation (6.5.93), we find

$$\begin{aligned}
w_{\mathrm{I}}(1 \to 2 \to 1) &= K\,\mathcal{R}l\,\frac{\mathrm{i}}{(\Delta_1 + \mathrm{i}\Gamma_{21})(\Delta_2 - \mathrm{i}\Gamma_{21})(\Delta_{12} + \mathrm{i}\Gamma_{11})} \\
&= \frac{K}{X_1 X_2 X_{12}}\,\mathcal{R}l\,\mathrm{i}(\Delta_1 - \mathrm{i}\Gamma_{21})(\Delta_2 + \mathrm{i}\Gamma_{21})(\Delta_{12} - \mathrm{i}\Gamma_{11}) \\
&= \frac{K}{X_1 X_2 X_{12}}\,\mathcal{R}l\,\mathrm{i}[\Delta_1\Delta_2 + \Gamma_{21}^2 + \mathrm{i}\Gamma_{21}(\Delta_1 - \Delta_2)](\Delta_{12} - \mathrm{i}\Gamma_{11}) \\
&= \frac{K}{X_1 X_2 X_{12}}[\Gamma_{11}(\Delta_1\Delta_2 + \Gamma_{21}^2) - \Gamma_{21}\Delta_{12}^2] \\
&= \frac{K}{X_1 X_2 X_{12}}\left[\frac{\Gamma_{11}}{2}(X_1 + X_2 - X_{12} - 2\Gamma_{21}^2 + \Gamma_{11}^2)\right. \\
&\quad \left. + \Gamma_{11}\Gamma_{21}^2 - \Gamma_{21}X_{12} + \Gamma_{21}\Gamma_{11}^2\right] \\
&= \frac{K}{2X_1 X_2 X_{12}}[\Gamma_{11}X_1 + \Gamma_{11}X_2 - (2\Gamma_{21} + \Gamma_{11})X_{12} \\
&\quad + \Gamma_{11}^2(2\Gamma_{21} + \Gamma_{11})].
\end{aligned} \tag{6.5.97}$$

Equations (6.5.94) and (6.5.95) may be combined to yield

$$\begin{aligned}
&w_{\mathrm{II}}(1 \to 2 \to 1) + w_{\mathrm{III}}(1 \to 2 \to 1) \\
&= \frac{K}{\Gamma_{22}}\,\mathcal{R}l\left[\frac{1}{(\Delta_1 + \mathrm{i}\Gamma_{21})(\Delta_2 - \mathrm{i}\Gamma_{21})} - \frac{1}{(\Delta_1 - \mathrm{i}\Gamma_{21})(\Delta_2 - \mathrm{i}\Gamma_{21})}\right] \\
&= \frac{K}{\Gamma_{22}}\,\mathcal{R}l\,\frac{1}{\Delta_2 - \mathrm{i}\Gamma_{21}}\left[\frac{1}{\Delta_1 + \mathrm{i}\Gamma_{21}} - \frac{1}{\Delta_1 - \mathrm{i}\Gamma_{21}}\right] \\
&= \frac{-K}{\Gamma_{22}X_1X_2}\,\mathcal{R}l\,(\Delta_2 + \mathrm{i}\Gamma_{21})2\mathrm{i}\Gamma_{21} \\
&= \frac{2K\Gamma_{21}^2}{\Gamma_{22}X_1X_2}.
\end{aligned} \tag{6.5.98}$$

We finally have therefore

$$\begin{aligned}
w(1 \to 2 \to 1) &= \sum_{\alpha=\mathrm{I}}^{\mathrm{III}} w_\alpha(1 \to 2 \to 1) \\
&= \frac{K}{2\Gamma_{22}}\left\{\frac{\Gamma_{11}\Gamma_{22}}{X_2X_{12}} + \frac{\Gamma_{11}\Gamma_{22}}{X_1X_{12}}\right. \\
&\quad \left. + \frac{4\Gamma_{21}^2 - \Gamma_{22}(2\Gamma_{21} + \Gamma_{11})}{X_1X_2} + \frac{\Gamma_{11}^2\Gamma_{22}(2\Gamma_{21} + \Gamma_{11})}{X_1X_2X_{12}}\right\}.
\end{aligned} \tag{6.5.99}$$

If we now put

$$q(\omega_k) = \frac{\Gamma_{21}}{\pi X_k} = \frac{\Gamma_{21}/\pi}{(\omega_k - \omega_{21})^2 + \Gamma_{21}^2}, \tag{6.5.100}$$

for $k = 1$ and 2, with

$$\int_{-\infty}^{\infty} q(\omega_k)\,\mathrm{d}\omega_k = 1, \tag{6.5.101}$$

i.e. $q(\omega_k)$ is the usual Lorentzian profile for photon absorption ($k = 1$) and emission ($k = 2$), then

$$w(1 \to 2 \to 1) = \frac{2K\pi^2}{\Gamma_2}\, q(\omega_1) p(\omega_1, \omega_2), \tag{6.5.102}$$

where we have defined

$$\begin{aligned} p(\omega_1, \omega_2) = {} & \frac{\Gamma_{11}\Gamma_2 q(\omega_2)}{4\pi\Gamma_{21}X_{12}q(\omega_1)} + \frac{\Gamma_{11}\Gamma_2}{4\pi\Gamma_{21}X_{12}} \\ & + \frac{4\Gamma_{21}^2 - \Gamma_{22}(2\Gamma_{21} + \Gamma_{11})}{4\Gamma_{21}^2}\, \frac{\Gamma_2}{\Gamma_{22}}\, q(\omega_2) \\ & + \frac{\Gamma_{11}^2\Gamma_2(2\Gamma_{21} + \Gamma_{11})}{4\Gamma_{21}^2 X_{12}}\, q(\omega_2). \end{aligned} \tag{6.5.103}$$

This last result expresses the atomic scattering function $p(\omega_1, \omega_2)$ to be discussed qualitatively in some detail in the following chapter. Having completed the general specification of the scattering process in which collisions are included, we now turn to the simplified situation in which the ground state has zero width, i.e. we let $\Gamma_1 \to 0$ and $\Gamma_{11} \to 0$. Recall from equations (6.4.92) and (6.4.111) that

$$\zeta(x) = \mathscr{P}\frac{1}{x} - \pi\mathrm{i}\delta(x) = \lim_{\sigma \to 0} \frac{1}{x + \mathrm{i}\sigma},$$

leading to

$$\begin{aligned} \lim_{\Gamma_{11}\to 0} \frac{\Gamma_{11}}{x_{12}} &= \lim_{\Gamma_{11}\to 0} \frac{\Gamma_{11}}{\Delta_{12}^2 + \Gamma_{11}^2} \\ &= \lim_{\Gamma_{11}\to 0} \frac{1}{2\mathrm{i}}\left(\frac{1}{\Delta_{12} - \mathrm{i}\Gamma_{11}} - \frac{1}{\Delta_{12} + \mathrm{i}\Gamma_{11}}\right) \\ &= \frac{1}{2\mathrm{i}}\,[\zeta^*(\Delta_{12}) - \zeta(\Delta_{12})] \\ &= \pi\delta(\Delta_{12}) = \pi\delta(\omega_1 - \omega_2). \end{aligned}$$

Equation (6.5.103) then becomes

$$p(\omega_1, \omega_2) = \frac{\Gamma_2 q(\omega_2)}{4\Gamma_{21} q(\omega_1)} \delta(\omega_1 - \omega_2) + \frac{\Gamma_2}{4\Gamma_{21}} \delta(\omega_1 - \omega_2) + \frac{4\Gamma_{21} - 2\Gamma_{22}}{4\Gamma_{21}} \cdot \Gamma_2 q(\omega_2)$$

$$= \frac{\Gamma_2}{2\Gamma_{21}} \delta(\omega_1 - \omega_2) + \frac{\Gamma_2}{2\Gamma_{21}\Gamma_{22}} (2\Gamma_{21} - \Gamma_{22}) q(\omega_2). \quad (6.5.104)$$

We can write

$$\Gamma_{21} = \tfrac{1}{2}(\Gamma_1 + \Gamma_2) + \tfrac{1}{2}(Q_{\rm I} + Q_{\rm E}),$$

and

$$\Gamma_{22} = Q_{\rm I} + \Gamma_2, \quad (6.5.105)$$

where $Q_{\rm I}$ and $Q_{\rm E}$ are the rates for inelastic and elastic collisions. The derivation of equations (6.5.105) depends upon the precise model one takes for the collisional encounters and is beyond the scope of the present discussion.

Substitution into the above $p(\omega_1, \omega_2)$ then yields

$$p(\omega_1, \omega_2) = \frac{\Gamma_2}{Q_{\rm I} + Q_{\rm E} + \Gamma_2} \cdot \delta(\omega_1 - \omega_2) + \frac{\Gamma_2}{\Gamma_2 + Q_{\rm I}} \times \frac{Q_{\rm E}}{Q_{\rm I} + Q_{\rm E} + \Gamma_2} q(\omega_2), \quad (6.5.106)$$

note that we have taken $\Gamma_1 \to 0$.

Clearly,

$$\int_{-\infty}^{\infty} p(\omega_1, \omega_2)\, d\omega_2 = \frac{\Gamma_2}{\Gamma_2 + Q_{\rm I}}, \quad (6.5.107)$$

reflecting the fact that not all de-excitations are radiative; recall that Γ_2 is the radiative rate.

The above expression (6.5.106) for $p(\omega_1, \omega_2)$ indicates that radiative de-excitations are coherent in the rest frame of the atom (recall that we have modelled the lower level to be infinitesimally sharp) whereas elastic events (those collisions which merely re-distribute the bound electron in the excited state) produce the normal Lorentzian emission line profile as expected.

Note that the ratio $Q_{\rm E}/(Q_{\rm I} + Q_{\rm E} + \Gamma_2)$ gives the probability of any event involving the excited state being an elastic collisional process whereas $\Gamma_2/(\Gamma_2 + Q_{\rm I})$ states the probability of a de-excitation event being radiative. This then forms a basis for a more thorough understanding of the general atomic scattering function given by equation (6.5.103) and will be used in the following chapter when discussing departures from complete re-distribution.

7

Frequency and angle re-distribution

In the analyses of chapters 1 through 5 we have made the assumption of complete re-distribution whereby the probabilities of stimulated and spontaneous emission due to the gaseous ensemble of atoms have been equated to that for absorption. The thrust behind this simplification was purely mathematical – the spectral line source function then became independent of angle and frequency, and this facilitated considerably the solution of the resulting equation of radiative transfer (it also aids us in a development of the understanding of the many transfer effects arising from the interaction of radiation with matter). The physical consequences of this assumption are also quite clear. The emitted photon has a frequency completely independent of the frequency at which it was absorbed so that radiation is completely re-distributed throughout the spectral line.

The quantum mechanical analyses of chapter 6 detail the manner in which radiation interacts with an individual atom. In particular, it was shown that there exists a correlation between the frequency of the photon being emitted and that of the absorbed photon. It is not unreasonable to imagine that this non-zero correlation in the rest frame of the atom can also lead to a non-zero correlation in the rest frame of the observer. The present chapter therefore aims to remove the assumption of complete re-distribution by developing various re-distribution functions incorporating the above non-zero correlation and this, in turn, will allow the emitted photon to have some memory of the frequency with which it was absorbed. Indeed, we must also consider the coupling between the directions of the absorbed and emitted photon. As we shall see, this departure from complete re-distribution in both angle and frequency introduces further radiative transfer effects which are extremely important for diagnostic analyses of spectral line data.

7.1 The emission profile

The equation of radiative transfer has the form (see equation (1.2.8–10))

$$\frac{1}{c}\frac{\partial I}{\partial t}+(\boldsymbol{\Omega}\cdot\nabla)I=-\kappa(I-S), \tag{7.1.1}$$

where

$$\kappa=\frac{h\nu}{4\pi}(N_{\mathrm{L}}B_{\mathrm{LU}}\phi-N_{\mathrm{U}}B_{\mathrm{UL}}\psi), \tag{7.1.2}$$

and

$$S=\frac{N_{\mathrm{U}}A_{\mathrm{UL}}j}{N_{\mathrm{L}}B_{\mathrm{LU}}\phi-N_{\mathrm{U}}B_{\mathrm{UL}}\psi} \tag{7.1.3}$$

In particular, ϕ is the probability of photon absorption, j and ψ the probabilities for spontaneous and stimulated emission.

As shown in section 1.4, ϕ can be written in the form

$$\phi=\int F_{\mathrm{L}}(\mathbf{v}_{\mathrm{L}})q(\gamma')\mathrm{d}\mathbf{v}_{\mathrm{L}}, \tag{7.1.4}$$

where $q(\gamma')$ is the probability per unit frequency interval that a photon of frequency γ' measured in the rest frame of the atom will be absorbed by that atom, and $F_{\mathrm{L}}(\mathbf{v}_{\mathrm{L}})$ is the velocity distribution function for the de-excited (absorbing) atoms. Note that

$$\int q(\gamma')\mathrm{d}\gamma'=1, \tag{7.1.5}$$

and

$$\int F_{\mathrm{L}}(\mathbf{v}_{\mathrm{L}})\mathrm{d}\mathbf{v}_{\mathrm{L}}=1, \tag{7.1.6}$$

with $F_{\mathrm{L}}\equiv f_{\mathrm{L}}/N_{\mathrm{L}}$ where f_{L} is the velocity distribution function used in chapter 5. If we transform from the rest frame of the atom to that of the observer, equation (7.1.4) is written as

$$\phi(\nu')=\int F_{\mathrm{L}}(\mathbf{v}_{\mathrm{L}})q\left(\nu'-\frac{\nu'}{c}\mathbf{v}_{\mathrm{L}}\cdot\boldsymbol{\Omega}'\right)\mathrm{d}\mathbf{v}_{\mathrm{L}} \tag{7.1.7}$$

or, writing $\Delta\nu'=\nu'-\nu_0$ where ν_0 is the 'centre of gravity' frequency of the transition in question,

$$\phi(\Delta\nu')=\int F_{\mathrm{L}}(\mathbf{v}_{\mathrm{L}})q\left(\Delta\nu'+\nu_0-\frac{\nu'}{c}\mathbf{v}_{\mathrm{L}}\cdot\boldsymbol{\Omega}'\right)\mathrm{d}\mathbf{v}_{\mathrm{L}},$$

where $\boldsymbol{\Omega}'$ is the direction of the photon being absorbed. We have already derived the Doppler and Voigt profiles using $q_{\mathrm{D}}(\gamma')=\delta(\gamma'-\nu_0)$ and $q_{\mathrm{V}}(\gamma')=(\delta/\pi)/[(\gamma'-\nu_0)^2+\delta^2]$, viz

$$\left.\begin{aligned}\phi_{\mathrm{D}}(\Delta\nu')&=\frac{1}{\Delta\nu_{\mathrm{D}}\sqrt{\pi}}\mathrm{e}^{-\left(\frac{\Delta\nu'}{\Delta\nu_{\mathrm{D}}}\right)^2},\quad \Delta\nu_{\mathrm{D}}=\frac{\nu_0}{c}\left[\frac{2kT}{m_{\mathrm{A}}}\right]^{1/2},\\ \phi_{\mathrm{V}}(\Delta\nu')&=\frac{1}{\Delta\nu_{\mathrm{D}}\sqrt{\pi}}H\left(\frac{\delta}{\Delta\nu_{\mathrm{D}}},\frac{\Delta\nu'}{\Delta\nu_{\mathrm{D}}}\right),\end{aligned}\right\} \tag{7.1.8}$$

where $H(a,\eta)$ is the usual Voigt function - see equations (1.4.14) and (1.4.18). These last two results were obtained using the Maxwellian velocity distribution (5.2.36)

$$F(v) = \frac{1}{\bar{V}^3 \pi^{3/2}} \, \mathrm{e}^{-\left(\frac{v}{\bar{V}}\right)^2}, \qquad \bar{V} = \left[\frac{2kT}{m_A}\right]^{1/2}. \tag{7.1.9}$$

It should be noted, however, that the use of this last equation can be invalidated if, in particular, large velocity gradients exist in the medium under consideration. This has been discussed in some detail in chapter 5.

We now turn our attention to the emission profiles. We have already seen (equation (6.4.37), for example) that quantum mechanical considerations stipulate $j \equiv \psi$. Consequently, we need only derive one of these probabilities and, since ψ involves the complicating features arising from the indirect stimulating influence of the external radiation field, we henceforth choose to examine [93] the probability of spontaneous emission in the remainder of this section.

If we are to detail the frequency of photon emission, then we must have some information regarding the position of the excited electron in the upper level before it actually de-excites. Further, we recognise that this position can arise either from the previous absorption of a photon or by the collisional excitation of the atom in the de-excited state.

We first examine the former possibility by considering the composite process whereby an atom is excited by absorbing a photon *and* then de-excites by spontaneous emission. Given that this absorption followed by a spontaneous emission process takes place, we denote by $p(\gamma', \gamma)$ the probability per unit frequency interval that if the frequency of the absorbed photon is γ' (measured in the rest frame of the atom) then the frequency of the emitted photon is γ. The atomic re-distribution function $p(\gamma', \gamma)$ has been considered in some detail in chapter 6 (see, for example, equation (6.5.103)). Correspondingly, ignoring any coupling (polarisation) between the angle and frequency of individual photons, we denote the angle probability of the above process by $g(\Omega', \Omega)/4\pi$. We stress that we are studying the event whereby a photon is absorbed and then spontaneously re-emitted so that, summing over all possible absorbing and emitting parameters, we have

$$\left.\begin{aligned} \int p(\gamma', \gamma)\mathrm{d}\gamma' = 1 = \int p(\gamma', \gamma)\mathrm{d}\gamma, \\ \frac{1}{4\pi}\int g(\Omega', \Omega)\mathrm{d}\Omega' = 1 = \frac{1}{4\pi}\int g(\Omega', \Omega)\mathrm{d}\Omega. \end{aligned}\right\} \tag{7.1.10}$$

Note that the $p(\gamma', \gamma)$ defined above excludes the possibility of collisional de-excitation of the atom and therefore differs slightly from the $p(\omega_1, \omega_2)$

considered in the previous chapter. We therefore put $Q_I = 0$ so that equation (6.5.107) reduces to the above equation (7.1.10). Inelastic collisional de-excitation is specifically taken into account in the ϵ term appearing in the spectral line source function.

If we now take the motion of the atom into account as in section 1.4, then we have (see equation 1.4.1), for example)

$$\gamma' = \nu' - \frac{\nu'}{c}\mathbf{v}_L \cdot \mathbf{\Omega}', \quad \gamma = \nu - \frac{\nu}{c}\mathbf{v}_U \cdot \mathbf{\Omega}, \tag{7.1.11}$$

where ν' and ν now refer to frequencies measured in the rest frame of the gas. For purposes of exposition we take the gas to be macroscopically stationary so that the gas and observer's frame of reference are identical. This assumption may be easily removed at the end of the present section using the procedure outlined in section 5.4 (see equation (5.4.9) for example). Note that the atom will gain linear momentum of order $h\nu\mathbf{\Omega}/c$ during an absorption process so that one could not expect the velocity of the excited atom to be identical to that of the de-excited particle, i.e. $\mathbf{v}_U \neq \mathbf{v}_L$. Note, further, that we ignore relativistic effects so that angles are invariant under a change of reference frame.

We next define the single re-distribution function [53]

$$\begin{aligned} R_{LU}(\gamma';\gamma) &= \frac{1}{4\pi}g(\mathbf{\Omega}', \mathbf{\Omega})q(\gamma')p(\gamma', \gamma) \\ &= \frac{1}{4\pi}g(\mathbf{\Omega}', \mathbf{\Omega})q\left(\nu' - \frac{\nu'}{c}\mathbf{v}_L \cdot \mathbf{\Omega}'\right)p\left(\nu' - \frac{\nu'}{c}\mathbf{v}_L \cdot \mathbf{\Omega}', \nu - \frac{\nu}{c}\mathbf{v}_U \cdot \mathbf{\Omega}\right) \\ &= R_{LU}(\nu', \mathbf{\Omega}'; \nu, \mathbf{\Omega}; \mathbf{v}_L, \mathbf{v}_U), \end{aligned} \tag{7.1.12}$$

so that $R_{LU}(\nu', \mathbf{\Omega}'; \nu, \mathbf{\Omega}; \mathbf{v}_L, \mathbf{v}_U)\mathrm{d}\nu'\mathrm{d}\mathbf{\Omega}'\mathrm{d}\nu\mathrm{d}\mathbf{\Omega}/4\pi$ is the probability of a single atom absorbing a photon in the range $\mathrm{d}\nu'\mathrm{d}\mathbf{\Omega}'$ about $(\nu', \mathbf{\Omega}')$ *and* spontaneously emitting a photon in the range $\mathrm{d}\nu\mathrm{d}\mathbf{\Omega}$ about $(\nu, \mathbf{\Omega})$. The inclusion of $\mathbf{v}_L$ and $\mathbf{v}_U$ emphasises that this composite probability (7.1.12) only pertains to an individual atom with velocities $\mathbf{v}_L$ and $\mathbf{v}_U$ in the lower and upper states.

Recall, however, that $j(\nu, \mathbf{\Omega})$ is the probability of an ensemble of atoms in the excited state (due to both radiative and collisional excitations) spontaneously emitting a $(\nu, \mathbf{\Omega})$ photon. The above discussion refers only to that emission following a single previous absorption event. Consequently, the contribution to $j(\nu, \mathbf{\Omega})$, taking into account all possible absorptions by an atom having the velocity $\mathbf{v}_L$ may be obtained simply by summing over all ν' and $\mathbf{\Omega}'$. Thus, with $I(\mathbf{r}, \nu', \mathbf{\Omega}')$ the specific intensity of the radiation field capable of being absorbed, we have

$$N_L B_{LU} \iint F_L(\mathbf{v}_L)F_U(\mathbf{v}_U)R_{LU}(\nu', \mathbf{\Omega}'; \nu, \mathbf{\Omega}; \mathbf{v}_L, \mathbf{v}_U)I(\mathbf{r}, \nu', \mathbf{\Omega}')\mathrm{d}\nu'\mathrm{d}\mathbf{\Omega}', \tag{7.1.13}$$

as the total rate per unit volume (N_L is the number density of the de-excited state) of photon absorptions by atoms in the de-excited state with velocity $\mathbf{v}_L$ (hence the inclusion of the velocity probability function F_L) followed by the spontaneous emission of a $(\nu, \mathbf{\Omega})$ photon whilst the excited atom is moving with velocity $\mathbf{v}_U$ (hence the inclusion of F_U). Note that conservation of linear momentum during a photon–atom encounter stipulates that

$$m_A \mathbf{v}_L + \frac{h\nu'}{c}\mathbf{\Omega}' = m_A \mathbf{v}_U, \tag{7.1.14}$$

so that $\mathbf{v}_L$ and $\mathbf{v}_U$ are not independent.

We now consider the corresponding contribution to $j(\nu, \mathbf{\Omega})$ arising from a previous inelastic collisional excitation from the de-excited state. In a manner similar to that leading to equation (7.1.12) we write

$$R_e(\mathbf{v}_e; \nu, \mathbf{\Omega}; \mathbf{v}_L, \mathbf{v}_U) = \sigma_{eL} f_L(\mathbf{v}_L) f_e(\mathbf{v}_e) q_c(\gamma), \tag{7.1.15}$$

where $\sigma_{eL} f_L f_e$ represents the rate per unit volume of collisional excitations involving atoms of velocity $\mathbf{v}_L$ and free electons of velocity $\mathbf{v}_e$. The reader is referred to section 5.1.2 for further details. This excitation event will result in the bound electron existing in any of the continuum of sub-levels constituting the upper state dependent upon the actual energy deposited by the free electron. We assume these collisions to be completely random so that this position of the bound electron in the excited state gives rise to a spontaneous emission profile $q_c(\gamma)$ quite independent of the precise excitation details. Indeed, it is not unreasonable, therefore, to assume that the distribution of the bound electrons in these excited sub-levels mimics that following photon absorption and this, in turn, enables us to put $q_c(\gamma) \equiv q(\gamma)$. The reader is particularly referred to the third term of equation (6.5.103) involving $q(\omega_2)$. Further, this collisional randomness allows us to assume isotropy of the photon emission process in the rest frame of the atom. Consequently, no explicit angle dependence appears in equation (7.1.15).

We now sum over all electron velocities $\mathbf{v}_e$ so that, with $f_L = N_L F_L$, we have

$$N_L \int \sigma_{eL} f_e(\mathbf{v}_e) F_L(\mathbf{v}_L) F_U(\mathbf{v}_U) q\left(\nu - \frac{\nu}{c}\mathbf{v}_U \cdot \mathbf{\Omega}\right) d\mathbf{v}_e \tag{7.1.16}$$

as the total rate per unit volume of collisional excitations involving atoms in the de-excited state with velocity $\mathbf{v}_L$ followed by the spontaneous emission of a $(\nu, \mathbf{\Omega})$ photon whilst the excited atom has velocity $\mathbf{v}_U$. Again, $\mathbf{v}_L$ and $\mathbf{v}_U$ are not independent since conservation of linear momentum stipulates

$$m_A \mathbf{v}_L + m_e \mathbf{v}_e \text{ (before)} = m_A \mathbf{v}_U + m_e \mathbf{v}_e \text{ (after)}.$$

However, we usually take $\mathbf{v}_U$ to be independent of $\mathbf{v}_e$ so that, using equation (5.1.33) viz.

$$C_{LU} \equiv \int \sigma_{eL} f_e(\mathbf{v}_e) d\mathbf{v}_e, \tag{7.1.17}$$

the contribution (7.1.16) becomes

$$N_L C_{LU} F_L(\mathbf{v}_L) F_U(\mathbf{v}_U) q(\nu, \Omega; \mathbf{v}_U). \tag{7.1.18}$$

The two contributions (7.1.13) and (7.1.18) to $j(\nu, \Omega)$ must now be added together and summed over all the possible emitting atoms. Consequently, we can define an atomic emission probability $e(\nu, \Omega; \mathbf{v}_U)$, similar to $q(\nu', \Omega'; \mathbf{v}_L)$ appearing in equation (7.1.4) for $\phi(\nu', \Omega')$, such that

$$j(\nu, \Omega) = \int F_U(\mathbf{v}_U) e(\nu, \Omega; \mathbf{v}_U) \mathrm{d}\mathbf{v}_U. \tag{7.1.19}$$

Reference to contributions (7.1.13) and (7.1.18) then shows that

$$e(\nu, \Omega; \mathbf{v}_U) = \alpha \Big\{ N_L B_{LU} F_L(\mathbf{v}_L) \iint R_{LU}(\nu', \Omega'; \nu, \Omega; \mathbf{v}_L, \mathbf{v}_U) I(\mathbf{r}, \nu', \Omega') \times \mathrm{d}\nu' \mathrm{d}\Omega' + N_L C_{LU} F_L(\mathbf{v}_L) q(\nu, \Omega, \mathbf{v}_U) \Big\}, \tag{7.1.20}$$

where the proportionality factor α has been included to preserve the normalisation of $e(\nu, \Omega; \mathbf{v}_U)$. Indeed, we have

$$\frac{1}{4\pi} \iint j(\nu, \Omega) \mathrm{d}\nu \mathrm{d}\Omega = 1 = \int F_U(\mathbf{v}_U) \mathrm{d}\mathbf{v}_U, \tag{7.1.21}$$

which, noting the dependence of R_{LU} and q appearing in equation (7.1.20) on γ, thus enabling $e(\nu, \Omega; \mathbf{v}_U)$ to be equivalently written as $e(\gamma)$, yields

$$\frac{1}{4\pi} \iint e(\nu, \Omega; \mathbf{v}_U) \mathrm{d}\nu \mathrm{d}\Omega = 1 = \int e(\gamma) \mathrm{d}\gamma. \tag{7.1.22}$$

Substitution of equation (7.1.20) into this last result gives

$$4\pi = \alpha \Big\{ N_L B_{LU} F_L(\mathbf{v}_L) \iint I(\mathbf{r}, \nu', \Omega')\, \mathrm{d}\nu'\, \mathrm{d}\Omega' \times \iint R_{LU}(\nu', \Omega'; \nu, \Omega; \mathbf{v}_L, \mathbf{v}_U)\, \mathrm{d}\nu\, \mathrm{d}\Omega + N_L C_{LU} F_L(\mathbf{v}_L) \iint q(\nu, \Omega; \mathbf{v}_U)\, \mathrm{d}\nu\, \mathrm{d}\Omega \Big\}. \tag{7.1.23}$$

Equation (7.1.12), together with equations (7.1.5) and (7.1.10), and an appropriate change of integration variable, shows that

$$\iint R_{LU}(\nu', \Omega'; \nu, \Omega; \mathbf{v}_L, \mathbf{v}_U)\, \mathrm{d}\nu\, \mathrm{d}\Omega = \frac{1}{4\pi} \iint g(\Omega', \Omega) q(\gamma') p(\gamma', \gamma)\, \mathrm{d}\nu\, \mathrm{d}\Omega$$

$$= \frac{1}{4\pi} q(\gamma') \int g(\Omega', \Omega)\, d\Omega \int p(\gamma', \gamma)\, d\gamma,$$

$$= q(\gamma'), \tag{7.1.24}$$

and

$$\iint q(\gamma)\, d\nu\, d\Omega = \int d\Omega \int q(\gamma)\, d\gamma = 4\pi,$$

so that equation (7.1.23) yields

$$4\pi = \alpha N_L F_L(\mathbf{v}_L) \left\{ B_{LU} \iint q(\nu', \Omega'; \mathbf{v}_L) I(r, \nu', \Omega')\, d\nu'\, d\Omega' + 4\pi C_{LU} \right\}.$$

We then have the atomic emission profile

$$e(\nu, \Omega; \mathbf{v}_U) = \frac{B_{LU} \iint R_{LU}(\nu', \Omega'; \nu, \Omega; \mathbf{v}_L, \mathbf{v}_U) I(\mathbf{r}, \nu', \Omega')\, d\nu'\, d\Omega' + C_{LU} q(\nu, \Omega; \mathbf{v}_U}{\frac{B_{LU}}{4\pi} \iint q(\nu', \Omega'; \mathbf{v}_L) I(\mathbf{r}, \nu', \Omega')\, d\nu'\, d\Omega' + C_{LU}}. \tag{7.1.25}$$

This last result, when substituted into equation (7.1.19), then yields the functional form for the emission profiles $j(\nu, \Omega)$ and $\psi(\nu, \Omega)$. Clearly, if we are to proceed further, we require the velocity probability function $F_U(\mathbf{v}_U)$ and the re-distribution term R_{LU}. This latter quantity involves $q(\gamma')$ and $p(\gamma', \gamma)$ terms already considered in some detail in chapter 6. The $F_U(\mathbf{v}_U)$ function may be obtained from the gas-dynamic conservation equations discussed in chapter 5. In particular, the mass conservation equation (5.1.36), viz.

$$\frac{\partial f_U}{\partial t} + \mathbf{v}_U \cdot \nabla f_U + \mathbf{a}_U \cdot \nabla_{\mathbf{v}_U} f_U$$

$$= \sum_{j \neq I}^{(1)} \int \sigma_{Uj}(f'_U f'_j - f_U f_j)\, d\mathbf{v}_j$$

$$- \frac{B_{UL}}{4\pi} f_U \iint e(\nu, \Omega; \mathbf{v}_U) I(\mathbf{r}, \nu, \Omega)\, d\nu\, d\Omega - A_{UL} f_U - C_{UL} f_U$$

$$+ \frac{B_{LU}}{4\pi} f_L \iint q(\nu', \Omega'; \mathbf{v}_L) I(\mathbf{r}, \nu', \Omega')\, d\nu'\, d\Omega' + C_{UL} f_L \tag{7.1.26}$$

itself exhibits the radiation field $I(\mathbf{r}, \nu, \Omega)$ and this further increases the non-linear dependence of the emission probabilities $j(\nu, \Omega)$ and $\psi(\nu, \Omega)$ on the unknown $I(\mathbf{r}, \nu, \Omega)$ to be determined.

7.2 Source function simplifications

The analysis of the preceding section showed that the emission profiles are non-linear functionals of the radiation field. This, in turn, forces the spectral line source function to also exhibit non-linearities in $I(\mathbf{r}, \nu, \Omega)$. The solution of the corresponding equation of radiative transfer is far more difficult than that for complete re-distribution not only because of these non-linearities but because of the dependence of the source function on both frequency and angle as well as position $\mathbf{r}$. Consequently, it is of some benefit to examine possible simplifications to the $j(\nu, \Omega)$ and $\psi(\nu, \Omega)$ derived in section 7.1.

7.2.1 Complete re-distribution

We have already seen the mathematical advantages of assuming the equality of all the emission and absorption profiles. Here we examine more closely the postulates leading to this assumption. A comparison of equations (7.1.7) and (7.1.19) shows that a sufficient condition for $\phi \equiv j$ is

$$e(\nu, \Omega; \mathbf{v}_\mathrm{U}) \equiv q(\nu, \Omega; \mathbf{v}_\mathrm{U}), \tag{7.2.1}$$

and

$$F_\mathrm{U}(\mathbf{v}_\mathrm{U}) \equiv F_\mathrm{L}(\mathbf{v}_\mathrm{L}). \tag{7.2.2}$$

Clearly, if collisions dominate in equation (7.1.25) so that the term involving C_LU is much larger than the corresponding B_LU term, then equation (7.2.1) is retrieved. Similarly, equation (7.1.26) would yield

$$C_\mathrm{LU} f_\mathrm{U} - C_\mathrm{UL} f_\mathrm{L} = 0,$$

i.e.

$$N_\mathrm{L} C_\mathrm{LU} F_\mathrm{L} - N_\mathrm{U} C_\mathrm{UL} F_\mathrm{U} = 0. \tag{7.2.3}$$

However, if collisions dominate, then we would expect LTE conditions to apply, i.e. $N_\mathrm{L} \rightarrow N_\mathrm{L}^*$ and $N_\mathrm{U} \rightarrow N_\mathrm{U}^*$ where $N_\mathrm{U}^* C_\mathrm{UL} = N_\mathrm{L}^* C_\mathrm{LU}$. Thus, equation (7.2.3) reduces to $F_\mathrm{L} \equiv F_\mathrm{U}$ identical to equation (7.2.2). Consequently, complete re-distribution is attained if collisions dominate the radiative processes. This corresponds to case (i) discussed in section 1.5. It should be noted, however, that $F_\mathrm{L} \not\equiv F_\mathrm{U}$ in general (see, for example, equation (5.2.53)). Even a Maxwellian distribution E_L for the de-excited state does not necessarily imply a corresponding Maxwellian F_U for the excited atoms.

If instead we have $I(\mathbf{r}, \nu, \Omega) \equiv I(\mathbf{r})$ both isotropic and independent of frequency at all $\mathbf{r}$, then equation (7.1.25) stipulates that

$$e(\nu, \Omega; \mathbf{v}_\mathrm{U}) = \frac{B_\mathrm{LU} I(\mathbf{r}) \iint R_\mathrm{LU}(\nu', \Omega'; \nu, \Omega; \mathbf{v}_\mathrm{L}, \mathbf{v}_\mathrm{U})\, \mathrm{d}\nu'\, \mathrm{d}\Omega' + C_\mathrm{LU} q(\nu, \Omega; \mathbf{v}_\mathrm{U})}{\dfrac{B_\mathrm{LU}}{4\pi} I(\mathbf{r}) \iint q(\nu', \Omega'; \mathbf{v}_\mathrm{L})\, \mathrm{d}\nu'\, \mathrm{d}\Omega' + C_\mathrm{LU}},$$

which, using equations (7.1.5), (7.1.10) and (7.1.12) with an appropriate change of integration variable, becomes

$$e(\nu, \Omega; \mathbf{v}_U) = \frac{B_{LU} I(r) \int q(\gamma') p(\gamma', \gamma)\, d\gamma' + C_{LU} q(\gamma)}{B_{LU} I(\mathbf{r}) + C_{LU}}. \tag{7.2.4}$$

Thus, we see that $e(\gamma) \equiv q(\gamma)$ if and only if

$$\int q(\gamma') p(\gamma', \gamma)\, d\gamma' = q(\gamma). \tag{7.2.5}$$

Clearly, this will hold either for complete re-distribution in the rest frame of the atom, viz.

$$p(\gamma', \gamma) \equiv q(\gamma), \tag{7.2.6}$$

where the frequency of the emitted photon (measured in the rest frame of the atom) is completely independent of the frequency at which it was absorbed, or for coherence in the rest frame of the atom, viz.

$$p(\gamma', \gamma) \equiv \delta(\gamma - \gamma'), \tag{7.2.7}$$

where the frequencies of the emitted and absorbed photons are identical.

The above two limiting situations [93] correspond to cases (ii) and (iii) respectively discussed in section 1.5.

7.2.2 *A linear expression [54] for $S(\mathbf{r}, \nu, \Omega)$*

We have already seen in section 1.6 that the assumption of complete re-distribution enables the spectral line source function to be written as

$$S(\mathbf{r}) = (1 - \epsilon)\bar{J} + \epsilon B_\nu(T), \tag{7.2.8}$$

where

$$\bar{J}(\mathbf{r}) = \frac{1}{4\pi} \iint \phi(\nu) I(\mathbf{r}, \nu, \Omega)\, d\nu\, d\Omega, \tag{7.2.9}$$

is the frequency- and angle-averaged radiation field. The above linear form of $S(\mathbf{r})$ enabled a 'reasonably straightforward' solution to the corresponding equation of radiative transfer to be evaluated. It is therefore tempting to make various approximations regarding S given by equation (7.1.3) and $e(\nu, \Omega; \mathbf{v}_U)$ given by equation (7.1.25) in order to again achieve a linear dependence of S on $I(\mathbf{r}, \nu, \Omega)$.

One standard approach is to treat stimulated emissions as negative absorptions and write

$$\phi \equiv \psi \not\equiv j. \tag{7.2.10}$$

It is perhaps more satisfactory to ignore stimulated emissions altogether although this would not always be justified in the infra-red [112]. In either case we may write

$$S = \frac{N_U A_{UL} j}{N_L B_{LU}\phi - N_U B_{UL}\psi}$$

$$= \frac{N_U A_{UL}}{N_L B_{LU} - N_U B_{UL}} \cdot \frac{j}{\phi}. \tag{7.2.11}$$

The time-independent statistical equilibrium rate equation (1.3.13), which may be derived from equation (7.1.26), for example, is

$$N_L(B_{LU}\bar{J}_{LU} + C_{LU}) = N_U(B_{UL}\bar{J}_{UL} + A_{UL} + C_{UL}), \tag{7.2.12}$$

where

$$\bar{J}_{LU} = \bar{J},$$

$$\bar{J}_{UL} = \frac{1}{4\pi}\iint \psi(\nu, \Omega) I(\mathbf{r}, \nu, \Omega)\, d\nu\, d\Omega.$$

However, equation (7.2.10) stipulates $\bar{J}_{UL} = \bar{J}_{LU} = \bar{J}$ so that equations (7.2.11) and (7.2.12) become

$$S = \frac{(B_{LU}\bar{J} + C_{LU})\dfrac{j}{B_{LU}\phi}}{1 + \dfrac{C_{UL}}{A_{UL}}\left(1 - \dfrac{C_{LU}B_{UL}}{C_{UL}B_{LU}}\right)}. \tag{7.2.13}$$

We proceed by noting from equations (7.1.19) and (7.1.25) that, with $I' \equiv I(\mathbf{r}, \nu', \Omega')$,

$$j = \int F_U \frac{B_{LU}\iint R_{LU}(\gamma', \gamma) I'\, d\nu'\, d\Omega' + C_{LU}q(\gamma)}{\dfrac{B_{LU}}{4\pi}\iint q(\gamma') I'\, d\nu'\, d\Omega' + C_{LU}} \cdot d\mathbf{v}_U. \tag{7.2.14}$$

Clearly, if we replace $q(\gamma')$ by $\phi(\nu')$ in the denominator of this last equation, so that it may be taken outside the integral, then this denominator will exactly cancel the $B_{LU}\bar{J} + C_{LU}$ term appearing in the numerator of equation (7.1.13). This would then remove the non-linear dependence of $S(\mathbf{r}, \nu, \Omega)$ on $I(\mathbf{r}, \nu', \Omega')$ and could be accomplished by separately summing contributions (7.1.13) and (7.1.18) over $\mathbf{v}_U$, then adding to obtain $j(\nu, \Omega)$. This is essentially the reverse of the procedure used in section 7.1. We further assume $\mathbf{v}_U = \mathbf{v}_L$ and $F_U(\mathbf{v}_U) \equiv F_L(\mathbf{v}_L) \equiv F(\mathbf{v})$, say, so that $F_U F_L$ in contributions (7.1.13) and (7.1.18) may be simply replaced by $F(\mathbf{v})$. We then have

$$j(\nu, \Omega) = \beta\left\{N_L B_{LU}\int F(\mathbf{v})\, d\mathbf{v}\iint R_{LU}(\nu', \Omega'; \nu, \Omega; \mathbf{v}) I(\mathbf{r}, \nu', \Omega')\, d\nu'\, d\Omega' \right.$$

$$\left. + N_L C_{LU}\int F(\mathbf{v}) q(\nu, \Omega; \mathbf{v})\, d\mathbf{v}\right\}, \tag{7.2.15}$$

where β is a new normalisation factor enabling the first of equations (7.1.21) to be satisfied.

The above result can be written as

$$j(\nu, \Omega) = \beta\left\{\frac{N_L B_{LU}}{4\pi}\iint R(\nu', \Omega'; \nu, \Omega) I(\mathbf{r}, \nu', \Omega')\, d\nu'\, d\Omega' \right.$$

$$\left. + N_L C_{LU}\phi(\nu)\right\}, \tag{7.2.16}$$

where we have defined the re-distribution function

$$R(\nu', \Omega'; \nu, \Omega) = 4\pi \int F(\mathbf{v}) R_{LU}(\nu', \Omega; \nu, \Omega; \mathbf{v})\, d\mathbf{v}, \tag{7.2.17}$$

and used equation (7.1.7), viz.

$$\phi(\nu) = \int F(\mathbf{v}) q(\nu, \Omega; \mathbf{v})\, d\mathbf{v}. \tag{7.2.18}$$

It is interesting to note that

$$\frac{1}{4\pi}\iint R(\nu', \Omega'; \nu, \Omega)\, d\nu\, d\Omega$$

$$= \int F(\mathbf{v})\, d\mathbf{v} \iint \frac{1}{4\pi} g(\Omega', \Omega) q(\gamma') p(\gamma', \gamma)\, d\nu\, d\Omega$$

$$= \int F(\mathbf{v}) g(\gamma')\, d\mathbf{v} \int \frac{g(\Omega', \Omega)}{4\pi}\, d\Omega \int p(\gamma', \gamma)\, d\gamma$$

$$= \int F(\mathbf{v}) q(\gamma')\, d\mathbf{v}$$

$$= \phi(\nu'). \tag{7.2.19}$$

Consequently, the normalisation condition (7.1.21) yields

$$1 = \beta\left[\frac{N_L B_{LU}}{4\pi}\iint \phi(\nu') I(\mathbf{r}, \nu', \Omega')\, d\nu'\ d\Omega' + N_L C_{LU}\right]$$

$$= \beta N_L (B_{LU}\bar{J} + C_{LU}), \tag{7.2.20}$$

so that

$$j(\nu, \Omega) = \frac{\dfrac{B_{LU}}{4\pi}\displaystyle\iint R(\nu', \Omega'; \nu, \Omega) I(\mathbf{r}, \nu', \Omega')\, d\nu'\, d\Omega' + C_{LU}\phi(\nu)}{B_{LU}\bar{J} + C_{LU}}. \tag{7.2.21}$$

Equation (7.2.13) then becomes

$$S = \frac{\dfrac{1}{4\pi\phi(\nu)}\displaystyle\iint R(\nu',\Omega';\nu,\Omega;\mathbf{v})I(\mathbf{r},\nu',\Omega')\,\mathrm{d}\nu'\,\mathrm{d}\Omega' + \dfrac{C_{\mathrm{LU}}}{B_{\mathrm{LU}}}}{1+\dfrac{C_{\mathrm{UL}}}{A_{\mathrm{UL}}}\left(1-\dfrac{C_{\mathrm{LU}}B_{\mathrm{UL}}}{C_{\mathrm{UL}}B_{\mathrm{LU}}}\right)}. \qquad (7.2.22)$$

Thus, using the results of section 1.6, viz.

$$g_{\mathrm{U}}B_{\mathrm{UL}} = g_{\mathrm{L}}B_{\mathrm{LU}}, \qquad \frac{A_{\mathrm{UL}}}{B_{\mathrm{UL}}} = \frac{2h\nu^3}{c^2},$$

$$\frac{C_{\mathrm{LU}}}{C_{\mathrm{UL}}} = \frac{g_{\mathrm{U}}}{g_{\mathrm{L}}}\,\mathrm{e}^{-h\nu/kT}, \qquad \epsilon' = \frac{C_{\mathrm{UL}}}{A_{\mathrm{UL}}}(1-\mathrm{e}^{-h\nu/kT}),$$

$$B_\nu(T) = \frac{2h\nu^3/c^2}{\mathrm{e}^{h\nu/kT}-1}, \qquad \epsilon = \frac{\epsilon'}{1+\epsilon'},$$

equation (7.2.22) readily yields

$$S(\mathbf{r},\nu,\Omega) = \frac{1-\epsilon}{4\pi\phi(\nu)}\iint R(\nu',\Omega';\nu,\Omega)I(\mathbf{r},\nu',\Omega')\,\mathrm{d}\nu'\,\mathrm{d}\Omega' + \epsilon B_\nu(T). \qquad (7.2.23)$$

This clearly has the same structural linear form as the complete re-distribution source function (7.2.8), but now the angle and frequency dependence of $S(\mathbf{r},\nu,\Omega)$ arise from the re-distribution function $R(\nu',\Omega';\nu,\Omega)/\phi(\nu)$. Again, $\epsilon B_\nu(T)$ represents the source of spontaneously emitted photons due to creation by collisional excitation. The integral term represents the corresponding source of photons due to a previous photon absorption, but only $1-\epsilon$ of these radiative excitations will be followed by a photon emission; the fraction ϵ accounts for photon destruction due to collisional de-excitation. Clearly, the factor $R(\nu',\Omega';\nu,\Omega)/\phi(\nu)$ simply allocates the parameters (ν,Ω) of the emitted photon given that it was a (ν',Ω') photon that was absorbed. Indeed, it is interesting to note that the replacement of $R(\nu',\Omega';\nu,\Omega)$ by the uncorrelated product $\phi(\nu')\phi(\nu)$ in equation (7.2.23) reproduces the complete re-distribution source function (7.2.8).

Finally, we note that the above result (7.2.21) for $j(\nu,\Omega)$ may be directly obtained from equation (7.2.14) by performing the mathematically invalid replacement

$$\int\frac{\alpha(x)}{\beta(x)}\,\mathrm{d}x \rightarrow \frac{\int\alpha(x)\,\mathrm{d}x}{\int\beta(x)\,\mathrm{d}x}$$

and, as before, putting $\mathbf{v}_{\mathrm{U}} = \mathbf{v}_{\mathrm{L}} = \mathbf{v}$ and $F_{\mathrm{U}} \equiv F_{\mathrm{L}} \equiv F$, where again we use equation (7.2.17) to define the re-distribution function $R(\nu',\Omega';\nu,\Omega)$.

7.2.3 *Uncoupling of $\mathbf{v}_L$ and $\mathbf{v}_U$ [97]*

A linear functional form similar to equation (7.2.23) for the source function can be obtained if we allow $\mathbf{v}_L$ and $\mathbf{v}_U$ to be completely uncoupled. In this case, we must integrate over both $\mathbf{v}_U$ *and* $\mathbf{v}_L$, not just $\mathbf{v}_U$, in equations (7.1.19) and (7.1.20). We then have, with obvious notation,

$$e(\gamma) = \alpha\left\{N_L B_{LU}\int F_L\,\mathrm{d}\mathbf{v}_L \iint R_{LU}(\gamma',\gamma)I'\,\mathrm{d}\nu'\,\mathrm{d}\Omega' + N_L C_{LU}\int F_L(\mathbf{v}_L)q(\gamma)\,\mathrm{d}\mathbf{v}_L\right\}. \tag{7.2.24}$$

We next note that $q(\gamma)=q(\nu,\mathbf{\Omega};\mathbf{v}_U)$ is independent of $\mathbf{v}_L$ so that the second term on the RHS (7.2.24) is just $N_L C_{LU} q(\gamma)$. We further note that

$$\begin{aligned}
&\frac{1}{4\pi}\iint \mathrm{d}\nu\,\mathrm{d}\Omega \int F_L R_{LU}(\gamma',\gamma)\,\mathrm{d}\mathbf{v}_L\\
&= \frac{1}{4\pi}\int F_L\,\mathrm{d}\mathbf{v}_L \iint \mathrm{d}\nu\,\mathrm{d}\Omega\,\frac{1}{4\pi} g(\Omega',\Omega)q(\gamma')p(\gamma',\gamma)\\
&= \frac{1}{4\pi}\int F_L q(\gamma')\,\mathrm{d}\mathbf{v}_L \int \frac{g(\Omega',\Omega)}{4\pi}\,\mathrm{d}\Omega \int p(\gamma',\gamma)\,\mathrm{d}\gamma\\
&= \frac{1}{4\pi}\int F_L q(\gamma')\,\mathrm{d}\mathbf{v}_L\\
&= \frac{1}{4\pi}\phi(\nu').
\end{aligned} \tag{7.2.25}$$

Consequently, substitution of equation (7.2.24) into the normalisation condition (7.1.22) yields

$$1 = \alpha\left\{\frac{N_L B_{LU}}{4\pi}\iint \phi(\nu')I(\mathbf{r},\nu',\mathbf{\Omega}')\,\mathrm{d}\nu'\,\mathrm{d}\Omega' + N_L C_{LU}\right\},$$

so that

$$\alpha = \frac{1}{N_L(B_{LU}\bar{J}+C_{LU})}, \tag{7.2.26}$$

independent of $\mathbf{v}_U$.

We therefore have the emission profile

$$\begin{aligned}
j(\nu,\Omega) &= \int F_U e(\gamma)\,\mathrm{d}\mathbf{v}_U\\
&= \frac{\int F_U\left\{B_{LU}\int F_L\,\mathrm{d}\mathbf{v}_L \iint R_{LU}(\gamma',\gamma)I'\,\mathrm{d}\nu'\,\mathrm{d}\Omega' + C_{LU}q(\gamma)\right\}\mathrm{d}\mathbf{v}_U}{B_{LU}\bar{J}+C_{LU}}\\
&= \frac{\frac{B_{LU}}{4\pi}\iint R(\nu',\Omega';\nu,\Omega)I(\mathbf{r},\nu',\mathbf{\Omega}')\,\mathrm{d}\nu'\,\mathrm{d}\Omega' + C_{LU}\phi(\nu)}{B_{LU}\bar{J}+C_{LU}},
\end{aligned} \tag{7.2.27}$$

where we have defined the new re-distribution function

$$R(\nu',\Omega';\nu,\Omega)=4\pi\int F_{\mathrm{U}}\,\mathrm{d}\mathbf{v}_{\mathrm{U}}\int F_{\mathrm{L}}\,\mathrm{d}\mathbf{v}_{\mathrm{L}}R_{\mathrm{LU}}(\gamma',\gamma). \qquad (7.2.28)$$

Equation (7.2.27) is identical in structure to equation (7.2.21) so that one obtains the same linear structural form (7.2.23) for $S(\mathbf{r},\nu,\Omega)$ as before.

However, if we uncouple $\mathbf{v}_{\mathrm{L}}$ and $\mathbf{v}_{\mathrm{U}}$ as above, and this could only be a result of a large number of elastic collisions between free electrons and the atom in the excited state, then one should expect complete re-distribution in the rest frame of the atom and complete isotropy of the radiation emitted, i.e.

$$p(\gamma',\gamma)=q(\gamma), \qquad g(\Omega',\Omega)=1.$$

Equation (7.2.28) then yields

$$\begin{aligned}R(\nu',\Omega';\nu,\Omega)&=\iint g(\Omega',\Omega)g(\gamma')p(\gamma',\gamma)F_{\mathrm{L}}F_{\mathrm{U}}\,\mathrm{d}\mathbf{v}_{\mathrm{L}}\,\mathrm{d}\mathbf{v}_{\mathrm{U}}\\&=\int F_{\mathrm{L}}q(\gamma')\,\mathrm{d}\mathbf{v}_{\mathrm{L}}\cdot\int F_{\mathrm{U}}q(\gamma)\,\mathrm{d}\mathbf{v}_{\mathrm{U}}\\&=\phi(\nu')\phi(\nu). \qquad (7.2.29)\end{aligned}$$

This last result, when substituted into equation (7.2.23), reduces the partial re-distribution source function to that for complete re-distribution (7.2.8) as expected.

7.3 Several limiting cases of $R(\nu',\Omega';\nu,\Omega)$

The linear form of the source function (7.2.23) is extremely useful in computing solutions of those radiative transfer equations which have reasonably physical validity. Consequently, it is of some benefit to further examine the re-distribution function $R(\nu',\Omega';\nu,\Omega)$ defined by equation (7.2.17) for several limiting cases [53]. These can then be used to detail the qualitative behaviour of the emergent radiation intensity for partial re-distribution so that, given the subsequent understanding derived from such discussions, we can thence consider more complicated, but more physically realistic, configurations.

In particular, we will first examine the three physically idealistic cases corresponding to (I) an atom with both lower and upper states infinitesimally sharp (this leads to complete coherence in the rest frame of the atom), (II) an atom with an infinitesimally sharp lower level but a naturally broadened excited state (here we have a Lorentzian profile for $q(\gamma)$), and (III) an atom as in (II) above but where elastic collisions perturb the bound electron in the excited state (we have complete non-coherence in the rest frame of the atom here since the excited electron will lose all memory of the excitation process if sufficient elastic collisions occur during the lifetime of the excited state). We consider each of the above in turn.

Finally, we examine the re-distribution functions obtained using the quantum mechanical results of section 6.5.

7.3.1 *The re-distribution function $R_{\mathrm{I}}(\nu', \Omega'; \nu, \Omega)$*

Here we examine the function $R(\nu', \Omega'; \nu, \Omega)$ appearing in the source function (7.2.23) when we take both lower and upper levels of the transition in question to be infinitesimally sharp. This implies that, in the rest frame of the atom, photons can be absorbed and emitted at only the one frequency and we can then write

$$q(\gamma') = \delta(\gamma' - \nu_0), \qquad p(\gamma', \gamma) = \delta(\gamma - \gamma'), \tag{7.3.1}$$

where ν_0 is the line-centre frequency.

Reference to equation (7.2.17) shows that

$$R(\nu', \Omega'; \nu, \Omega) = 4\pi \int F(\mathbf{v}) R_{\mathrm{LU}}(\nu', \Omega'; \nu, \Omega; \mathbf{v})\, \mathrm{d}\mathbf{v}, \tag{7.3.2}$$

where $F \equiv F_{\mathrm{U}} \equiv F_{\mathrm{L}}$ and $\mathbf{v} = \mathbf{v}_{\mathrm{U}} = \mathbf{v}_{\mathrm{L}}$. Further, we have

$$\begin{aligned} R_{\mathrm{LU}}(\gamma', \gamma) &= \frac{1}{4\pi} g(\Omega', \Omega) q(\gamma') p(\gamma', \gamma) \\ &= \frac{1}{4\pi} g(\Omega', \Omega) q\left(\nu' - \frac{\nu'}{c} \mathbf{v} \cdot \Omega'\right) \\ &\quad \times p\left(\nu' - \frac{\nu'}{c} \mathbf{v} \cdot \Omega', \nu - \frac{\nu}{c} \mathbf{v} \cdot \Omega\right). \end{aligned} \tag{7.3.3}$$

Finally, we assume $F(\mathbf{v})$ has the Maxwellian form

$$F(\mathbf{v}) = \frac{1}{\bar{V}^3 \pi^{3/2}} \mathrm{e}^{-\left(\frac{\mathbf{v}}{\bar{V}}\right)^2}, \qquad \bar{V} = \left[\frac{2kT}{m_{\mathrm{A}}}\right]^{1/2}. \tag{7.3.4}$$

We are now in a position to determine $R_{\mathrm{I}}(\nu', \Omega'; \nu, \Omega)$. We proceed [84] in the evaluation of the integral in equation (7.3.2) by defining the orthogonal system of unit vectors $(\Omega_1 \Omega_2, \Omega_3)$ and choose, for later convenience, Ω_1 and Ω_2 to lie in the plane of Ω and Ω' as shown in figure 7.1. In particular, we take $\Omega \cdot \Omega_1 = \Omega' \cdot \Omega_1 = \cos \theta/2$ where $\theta = \cos^{-1}(\Omega \cdot \Omega')$.

Thus, we may write Ω and Ω' in terms of Ω_1 and Ω_2 only, so that

$$\left.\begin{aligned} \Omega' &= \Omega_1 \cos\frac{\theta}{2} + \Omega_2 \sin\frac{\theta}{2} = a\Omega_1 + b\Omega_2, \\ \Omega &= \Omega_1 \cos\frac{\theta}{2} - \Omega_2 \sin\frac{\theta}{2} = a\Omega_1 - b\Omega_2. \end{aligned}\right\} \tag{7.3.5}$$

Further, we choose to define $\mathbf{v}$ as

$$\mathbf{v} = v_1 \Omega_1 + v_2 \Omega_2 + v_3 \Omega_3, \tag{7.3.6}$$

and write $\mathbf{u} = \mathbf{v}/\bar{V}$, i.e. $d\mathbf{u} = du_1\, du_2\, du_3 = dv_1\, dv_2\, dv_3/\bar{V}^3 = d\mathbf{v}/\bar{V}^3$, so that

$$\left.\begin{aligned} \mathbf{v}\cdot\boldsymbol{\Omega}' &= \bar{V}(au_1 + bu_2),\\ \mathbf{v}\cdot\boldsymbol{\Omega} &= \bar{V}(au_1 - bu_2). \end{aligned}\right\} \tag{7.3.7}$$

Thence, writing $\Delta\nu_D = \nu_0\bar{V}/c \approx \nu'\bar{V}/c \approx \nu\bar{V}/c$ (see equation (7.1.8)), the above equations yield

$$\begin{aligned} R_I(\nu', \Omega'; \nu, \Omega) = \frac{g(\Omega', \Omega)}{\pi^{3/2}} &\int_{-\infty}^{\infty} du_1 \int_{-\infty}^{\infty} du_2 \int_{-\infty}^{\infty} du_3\, e^{-u_1^2 - u_2^2 - u_3^2}\\ &\times q(\nu' - (au_1 + bu_2)\Delta\nu_D)p(\nu' - (au_1 + bu_2)\Delta\nu_D,\\ &\nu - (au_1 - bu_2)\Delta\nu_D). \end{aligned} \tag{7.3.8}$$

However, equations (7.3.1) become

$$\left.\begin{aligned} q(\gamma') &= q(\nu' - (au_1 + bu_2)\Delta\nu_D)\\ &= \delta(\nu' - (au_1 + bu_2)\Delta\nu_D - \nu_0),\\ p(\gamma', \gamma) &= p(\nu' - (au_1 + bu_2)\Delta\nu_D, \nu - (au_1 - bu_2)\Delta\nu_D)\\ &= \delta(\nu' - \nu - 2bu_2\Delta\nu_D). \end{aligned}\right\} \tag{7.3.9}$$

These last two equations when substituted into (7.3.8), and noting that

$$\int_{-\infty}^{\infty} e^{-u_3^2}\, du_3 = \sqrt{\pi},$$

yield

$$\begin{aligned} R_I(\nu', \Omega'; \nu, \Omega) = \frac{g(\Omega', \Omega)}{\pi} &\int_{-\infty}^{\infty} du_1 \int_{-\infty}^{\infty} du_2\, e^{-u_1^2 - u_2^2}\\ &\times \delta(\nu' - \nu_0 - (au_1 + bu_2)\Delta\nu_D)\\ &\times \delta(\nu' - \nu - 2bu_2\Delta\nu_D), \end{aligned} \tag{7.3.10}$$

Fig.7.1. The directions of the absorbed $\boldsymbol{\Omega}'$ and emitted $\boldsymbol{\Omega}$ photons are shown in relation to the chosen orthogonal system of unit vectors ($\boldsymbol{\Omega}_1$, $\boldsymbol{\Omega}_2$, $\boldsymbol{\Omega}_3$). The vectors $\boldsymbol{\Omega}$ and $\boldsymbol{\Omega}'$ lie in the plane of $\boldsymbol{\Omega}_1$ and $\boldsymbol{\Omega}_2$.

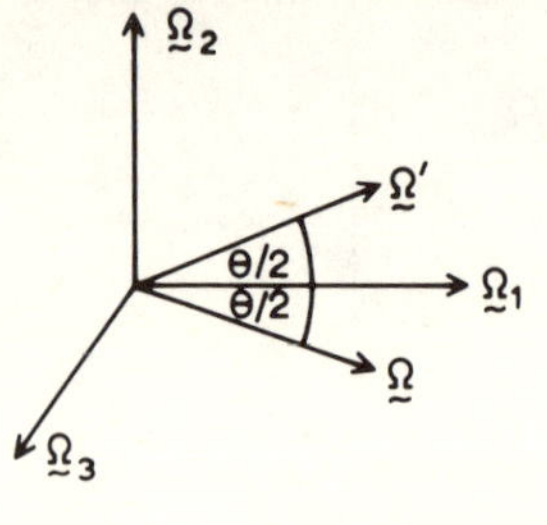

which, for $b \neq 0$, becomes

$$R_{\mathrm{I}}(\nu', \Omega'; \nu, \Omega) = \frac{g(\Omega', \Omega)}{2\pi b \Delta\nu_{\mathrm{D}}} \int_{-\infty}^{\infty} \mathrm{d}u_1 \int_{-\infty}^{\infty} \mathrm{d}u_2\, \mathrm{e}^{-u_1^2 - u_2^2}$$

$$\times\, \delta(\nu' - \nu_0 - (au_1 + bu_2)\Delta\nu_{\mathrm{D}})\delta\left(u_2 - \frac{\nu' - \nu}{2b\Delta\nu_{\mathrm{D}}}\right),$$

$$= \frac{g(\Omega', \Omega)}{2\pi b \Delta\nu_{\mathrm{D}}} \exp\left[-\left(\frac{\nu' - \nu}{2b\Delta\nu_{\mathrm{D}}}\right)^2\right] \int_{-\infty}^{\infty} \mathrm{d}u_1\, \mathrm{e}^{-u_1^2}$$

$$\times\, \delta\left(\frac{\nu' + \nu}{2} - \nu_0 - au_1\Delta\nu_{\mathrm{D}}\right)$$

$$= \frac{g(\Omega', \Omega)}{2\pi ab(\Delta\nu_{\mathrm{D}})^2} \exp\left[-\left(\frac{\nu' - \nu}{2b\Delta\nu_{\mathrm{D}}}\right)^2 - \left(\frac{\nu' + \nu - 2\nu_0}{2a\Delta\nu_{\mathrm{D}}}\right)^2\right]. \tag{7.3.11}$$

If $b = 0$, however, equation (7.3.10) yields

$$R_{\mathrm{I}}(\nu', \Omega'; \nu, \Omega) = \frac{g(\Omega', \Omega)}{\pi} \int_{-\infty}^{\infty} \mathrm{d}u_1 \int_{-\infty}^{\infty} \mathrm{d}u_2\, \mathrm{e}^{-u_1^2 - u_2^2}$$

$$\times\, \delta(\nu' - \nu_0 - au_1\Delta\nu_{\mathrm{D}})\delta(\nu' - \nu)$$

$$= \frac{g(\Omega', \Omega)}{a\Delta\nu_{\mathrm{D}}\sqrt{\pi}} \delta(\nu' - \nu) \exp\left[-\left(\frac{\nu' - \nu_0}{a\Delta\nu_{\mathrm{D}}}\right)^2\right]. \tag{7.3.12}$$

We now note that $2ab = \sin\theta$ and $a^2 - b^2 = \cos\theta$. Thus, putting $\Delta\nu = \nu - \nu_0$ and $\Delta\nu' = \nu' - \nu_0$, the above results have the form

$$R_{\mathrm{I}}(\nu', \Omega'; \nu, \Omega) = \frac{g(\Omega', \Omega)}{\pi(\Delta\nu_{\mathrm{D}})^2 \sin\theta} \exp\left[-\frac{a^2(\Delta\nu' - \Delta\nu)^2 + b^2(\Delta\nu' + \Delta\nu)^2}{\Delta\nu_{\mathrm{D}}^2 \sin^2\theta}\right]$$

$$= \frac{g(\Omega', \Omega)}{\pi(\Delta\nu_{\mathrm{D}})^2 \sin\theta}$$

$$\times \exp\left[-\left(\frac{\Delta\nu}{\Delta\nu_{\mathrm{D}}}\right)^2 - \left(\frac{\Delta\nu' - \Delta\nu\cos\theta}{\Delta\nu_{\mathrm{D}}\sin\theta}\right)^2\right], \tag{7.3.13}$$

for $\theta \neq 0, \pi$ whereas, for $\theta = 0$, for example, we have

$$R_{\mathrm{I}}(\nu', \Omega'; \nu, \Omega) = \frac{g(\Omega', \Omega)}{\Delta\nu_{\mathrm{D}}\sqrt{\pi}} \delta(\nu' - \nu)\, \mathrm{e}^{-(\Delta\nu/\Delta\nu_{\mathrm{D}})^2}. \tag{7.3.14}$$

It is clear from the above analysis that even coherence of the photon absorption - emission process in the rest frame of the atom does *not* imply corresponding coherence in the rest frame of the observer unless the photon is emitted in precisely the same direction as that occupied by the photon before it was absorbed, i.e. $\theta = 0$. Indeed, we see from equation (7.3.13) that if $\theta = \pi/2$, then

$$R_{\mathrm{I}}(\nu', \Omega'; \nu, \Omega) \rightarrow g(\Omega', \Omega) \frac{1}{\Delta\nu_{\mathrm{D}}\sqrt{\pi}} e^{-(\Delta\nu/\Delta\nu_{\mathrm{D}})^2}$$

$$\times \frac{1}{\Delta\nu_{\mathrm{D}}\sqrt{\pi}} e^{-(\Delta\nu'/\Delta\nu_{\mathrm{D}})^2}$$

$$= g(\Omega', \Omega)\phi(\Delta\nu)\phi(\Delta\nu'). \tag{7.3.15}$$

This last expression, apart from the factor $g(\Omega', \Omega)$, is of the form of $R(\nu', \Omega'; \nu, \Omega)$ which, when substituted into equation (7.2.23), yields the complete re-distribution source function. Consequently, $R_{\mathrm{I}}(\nu', \Omega'; \nu, \Omega)$ given by equations (7.3.13) and (7.3.14), may be thought of as a mixture of complete re-distribution and complete coherency dependent upon the angle θ of scatter. This can be more readily seen if we sketch $R_{\mathrm{I}}(\nu', \Omega'; \nu, \Omega)$ as a function of the emitting frequency $\Delta\nu$ for several values of both θ and the absorbing frequency $\Delta\nu'$. Figure 7.2, for example, illustrates the form $R_{\mathrm{I}}(\nu', \Omega'; \nu, \Omega)$ takes when the photon is absorbed at line centre. Note that the curve for $\theta = 0$ (equation (7.3.14)) lies along the ordinate and that $R_{\mathrm{I}}(\nu', \Omega'; \nu, \Omega)$ progressively changes from a delta function spike to the complete re-distribution result as θ increases to $\pi/2$.

It is clear from equation (7.3.13) that if we write

$$R(\Delta\nu', \Delta\nu, \theta) \equiv R(\nu', \Omega'; \nu, \Omega), \tag{7.3.16}$$

and assume the angular scattering function $g(\Omega', \Omega)$ depends only on the angle θ of scatter, and can therefore be written in the form $g(\Omega' \cdot \Omega) = g(\cos\theta)$, then

$$R_{\mathrm{I}}(\Delta\nu', -\Delta\nu, \pi - \theta) = R_{\mathrm{I}}(\Delta\nu', \Delta\nu, \theta), \tag{7.3.17}$$

as shown in figure 7.2 for $\theta = 0$ and π and $\theta = \pi/4$ and $3\pi/4$.

The $R_{\mathrm{I}}(\nu', \Omega'; \nu, \Omega)$ curves for photons absorbed in the wings ($\Delta\nu' \sim 3\Delta\nu_{\mathrm{D}}$) are sketched in figure 7.3. Again we have coherence at $\theta = 0$ shown as a spike

Fig.7.2. The re-distribution function R_{I} for several scattering angles when the photon is absorbed at line centre.

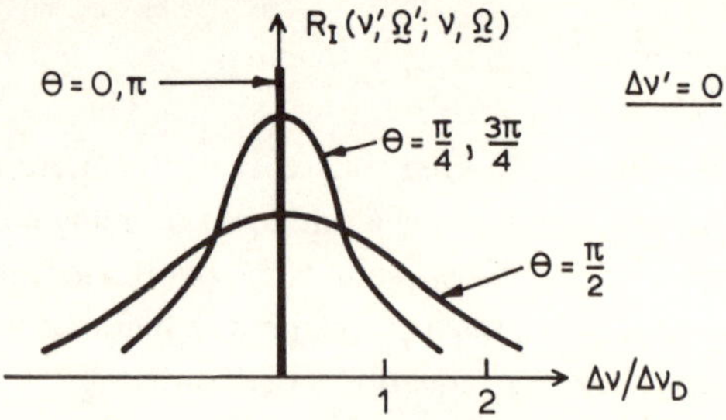

located at the frequencyof absorption $\Delta\nu' = 3\Delta\nu_D$, but complete re-distribution as $\theta \to \pi/2$.

In summary then, if the scattering angle $\theta \sim 0$, the photon will most probably be emitted at the frequency (in the rest frame of the observer) at which it was absorbed. If $\theta \sim \pi/2$, the photon will more likely be emitted at line centre (or at least in the line core $|\Delta\nu| < 3\Delta\nu_D$) independent of the frequency of absorption. However, a scattering angle $\theta \sim \pi$ produces a photon emission in the opposite wing to that in which the photon was absorbed.

7.3.2 *The re-distribution function $R_{II}(\nu', \Omega'; \nu, \Omega)$*

The second case of interest involves the scattering between an infinitesimally sharp lower level and a radiatively broadened upper state. Thus, for the absorption component we have

$$q(\gamma') = \frac{\delta/\pi}{(\gamma' - \nu_0)^2 + \delta^2}, \tag{7.3.18}$$

where δ is a measure of the width of the radiative broadening discussed in chapter 6 (see equation (6.4.141), for example). However, the atom in this case does not experience any perturbation whilst in the excited state. Since the lower level is perfectly sharp the photon is thence emitted at the same frequency (measured in the rest frame of the atom) at which it was absorbed so that

$$p(\gamma', \gamma) = \delta(\gamma' - \gamma), \tag{7.3.19}$$

as in case I above.

Following the procedure detailed in section 7.3.1, we substitute the above two expressions into equation (7.3.8) to find for $b \neq 0$

Fig.7.3. The re-distribution function R_I for several scattering angles when the photon is absorbed in the wings.

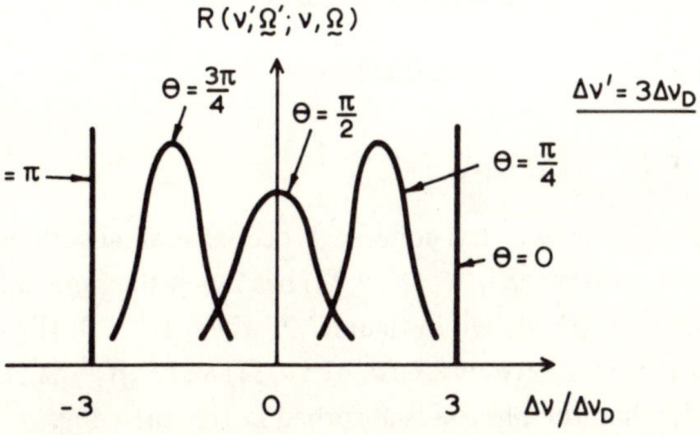

$$R_{\rm II}(\nu', \Omega'; \nu, \Omega) = \frac{g(\Omega', \Omega)}{2b\pi\Delta\nu_{\rm D}} \int_{-\infty}^{\infty} {\rm d}u_1 \int_{-\infty}^{\infty} {\rm d}u_2\, {\rm e}^{-u_1^2 - u_2^2}$$

$$\times \frac{\delta/\pi}{[\nu' - \nu_0 - (au_1 + bu_2)\Delta\nu_{\rm D}]^2 + \delta^2}\, \delta\left(u_2 - \frac{\nu' - \nu}{2b\Delta\nu_{\rm D}}\right)$$

$$= \frac{g(\Omega', \Omega)}{2b\pi\Delta\nu_{\rm D}} \exp\left[-\left(\frac{\nu' - \nu}{2b\Delta\nu_{\rm D}}\right)^2\right] \int_{-\infty}^{\infty} {\rm d}u_1\, {\rm e}^{-u_1^2}$$

$$\times \frac{\delta/\pi}{\left(\dfrac{\nu' + \nu}{2} - \nu_0 - au_1\Delta\nu_{\rm D}\right)^2 + \delta^2}$$

$$= \frac{g(\Omega', \Omega)}{2\pi ab(\Delta\nu_{\rm D})^2} \exp\left[-\left(\frac{\nu' - \nu}{2b\Delta\nu_{\rm D}}\right)^2\right] \int_{-\infty}^{\infty} {\rm d}u_1\, {\rm e}^{-u_1^2}$$

$$\times \frac{\delta/\pi a\Delta\nu_{\rm D}}{\left(\dfrac{\nu' + \nu - 2\nu_0}{2a\Delta\nu_{\rm D}} - u_1\right)^2 + \left(\dfrac{\delta}{a\Delta\nu_{\rm D}}\right)^2}$$

$$= \frac{g(\Omega', \Omega)}{2\pi ab(\Delta\nu_{\rm D})^2} \exp\left[-\left(\frac{\nu' - \nu}{2b\Delta\nu_{\rm D}}\right)^2\right]$$

$$\times H\left(\frac{\delta}{a\Delta\nu_{\rm D}}, \frac{\nu' + \nu - 2\nu_0}{2a\Delta\nu_{\rm D}}\right)$$

$$= \frac{g(\Omega', \Omega)}{\pi(\Delta\nu_{\rm D})^2 \sin\theta} \exp\left[-\left(\frac{\Delta\nu' - \Delta\nu}{2\Delta\nu_{\rm D} \sin\theta/2}\right)^2\right]$$

$$\times H\left[\frac{\delta}{\Delta\nu_{\rm D} \cos\theta/2}, \frac{\Delta\nu' + \Delta\nu}{2\Delta\nu_{\rm D} \cos\theta/2}\right], \tag{7.3.20}$$

where $H(\xi, \eta)$ is the Voigt function defined by

$$H(\xi, \eta) = \frac{\xi}{\pi} \int_{-\infty}^{\infty} \frac{{\rm e}^{-y^2}\, {\rm d}y}{(y - \eta)^2 + \xi^2}. \tag{7.3.21}$$

Again, it can be easily shown that coherency occurs at all absorbing frequencies $\Delta\nu'$ if $\theta = 0$. Further, $R_{\rm II}(\nu', \Omega'; \nu, \Omega)$ has much the same qualitative behaviour as $R_{\rm I}(\nu', \Omega'; \nu, \Omega)$, shown in figure 7.2, when $\Delta\nu' \sim 0$. However, there exist significant differences between $R_{\rm I}(\nu', \Omega'; \nu, \Omega)$ and $R_{\rm II}(\nu', \Omega'; \nu, \Omega)$, shown in figure 7.4, when the photon is absorbed in the line wings.

Note that although equation (7.3.17) in $R_{\mathrm{I}}(\Delta\nu', \Delta\nu, \theta)$ also holds for $R_{\mathrm{II}}(\Delta\nu', \Delta\nu, \theta)$ when $\theta = 0$ and π, it does not hold for $R_{\mathrm{II}}(\Delta\nu', \Delta\nu, \theta)$ for general θ. Here in figure 7.4 we see both the asymmetry exhibited by R_{II} for $\theta = \pi/2$ and the non-reflections of the curves for $\theta = \pi/4$ and $\theta = 3\pi/4$ about $\Delta\nu = 0$. Most important, however, we see that the probability of photon emission (given that the photon was absorbed in the wings $\Delta\nu' \sim 3\Delta\nu_{\mathrm{D}}$) is greatest at $\Delta\nu \sim 3\Delta\nu_{\mathrm{D}}$ for small θ but peaks at line centre $\Delta\nu = 0$ for $\theta = \pi/2$. Nevertheless, there still exists some measure of coherency reflected by the secondary peak in $R_{\mathrm{II}}(\nu', \boldsymbol{\Omega}'; \nu, \boldsymbol{\Omega})$ at $\Delta\nu \sim 3\Delta\nu_{\mathrm{D}}$ for $\theta = \pi/2$ in marked contrast to the results shown for $R_{\mathrm{I}}(\nu', \boldsymbol{\Omega}'; \nu, \boldsymbol{\Omega})$.

7.3.3 The re-distribution function $R_{\mathrm{III}}(\nu', \boldsymbol{\Omega}'; \nu, \boldsymbol{\Omega})$

The third example to be considered is similar to case II but includes elastic collisional perturbations of the atom during the life-time of the excited state. Consequently, if there are sufficient of these collisions, the excited atom loses all memory of the precise details of its excitation and it therefore de-excites emitting a photon completely de-coupled from the absorbed photon. Thus, we have

$$q(\gamma') = \frac{\delta/\pi}{(\gamma' - \nu_0)^2 + \delta^2}, \qquad p(\gamma', \gamma) = \frac{\delta/\pi}{(\gamma - \nu_0)^2 + \delta^2}. \tag{7.3.22}$$

See, for example, equation (6.5.102) and the second term of equation (6.5.106) with $Q_{\mathrm{I}} = 0$ and $Q_{\mathrm{E}} \gg \Gamma_2$.

We find it convenient to re-define the orthogonal set $(\boldsymbol{\Omega}_1, \boldsymbol{\Omega}_2, \boldsymbol{\Omega}_3)$ by $\boldsymbol{\Omega}_1 \equiv \boldsymbol{\Omega}'$ so that $\boldsymbol{\Omega}_1 \cdot \boldsymbol{\Omega} = \cos\theta$, $\boldsymbol{\Omega}_2 \cdot \boldsymbol{\Omega} = \sin\theta$ and $\boldsymbol{\Omega}_3 \cdot \boldsymbol{\Omega} = 0$. This then yields $\mathbf{v} \cdot \boldsymbol{\Omega}' = v_1$ with

$$\mathbf{v} \cdot \boldsymbol{\Omega} = v_1 \cos\theta + v_2 \sin\theta = av_1 + bv_2. \tag{7.3.23}$$

Fig.7.4. The re-distribution function R_{II} for several scattering angles when the photon is absorbed in the wings.

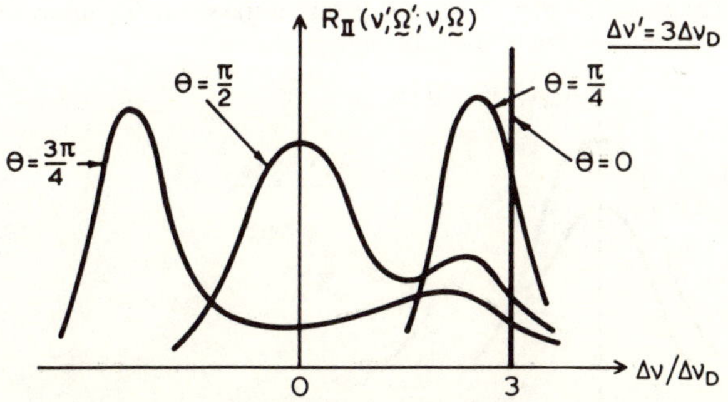

Consequently, equations (7.3.2-4) and (7.3.22) yield for non-zero b,

$$R_{\mathrm{III}}(\nu',\Omega';\nu,\Omega)=\frac{g(\Omega',\Omega)}{\pi^{3/2}}\int_{-\infty}^{\infty}\mathrm{d}u_1\int_{-\infty}^{\infty}\mathrm{d}u_2\int_{-\infty}^{\infty}\mathrm{d}u_3\,\mathrm{e}^{-u_1^2-u_2^2-u_3^2}$$

$$\times\frac{\delta/\pi}{(\nu'-\nu_0-u_1\Delta\nu_{\mathrm{D}})^2+\delta^2}\cdot\frac{\delta/\pi}{[(\nu-\nu_0-(au_1+bu_2)\Delta\nu_{\mathrm{D}}]^2+\delta^2}$$

$$=\frac{g(\Omega',\Omega)}{b\pi(\Delta\nu_{\mathrm{D}})^2}\int_{-\infty}^{\infty}\mathrm{d}u_1\,\mathrm{e}^{-u_1^2}\frac{\delta/\pi\Delta\nu_{\mathrm{D}}}{\left(\dfrac{\Delta\nu'}{\Delta\nu_{\mathrm{D}}}-u_1\right)^2+\left(\dfrac{\delta}{\Delta\nu_{\mathrm{D}}}\right)^2}$$

$$\times\int_{-\infty}^{\infty}\mathrm{d}u_2\,\mathrm{e}^{-u_2^2}\frac{\delta/\pi b\Delta\nu_{\mathrm{D}}}{\left(\dfrac{\Delta\nu}{b\Delta\nu_{\mathrm{D}}}-\dfrac{au_1}{b}-u_2\right)^2+\left(\dfrac{\delta}{b\Delta\nu_{\mathrm{D}}}\right)^2}$$

$$=\frac{g(\Omega',\Omega)\delta}{\pi^2(\Delta\nu_{\mathrm{D}})^3\sin\theta}\int_{-\infty}^{\infty}\frac{\mathrm{e}^{-u_1^2}\,\mathrm{d}u_1}{\left(u_1-\dfrac{\Delta\nu'}{\Delta\nu_{\mathrm{D}}}\right)^2+\left(\dfrac{\delta}{\Delta\nu_{\mathrm{D}}}\right)^2}$$

$$\times H\left[\frac{\delta}{\Delta\nu_{\mathrm{D}}\sin\theta},\frac{\Delta\nu}{\Delta\nu_{\mathrm{D}}\sin\theta}-\frac{u_1}{\tan\theta}\right].\tag{7.3.24}$$

This function behaves [99] more like complete re-distribution than does R_{I} or R_{II}. Even $\theta=0$ does not produce coherency in the rest frame of the observer. This can be seen in figure 7.5 where $R_{\mathrm{III}}(\nu',\Omega';\nu,\Omega)$ is sketched for $\Delta\nu'=3\Delta\nu_{\mathrm{D}}$. In particular, we have

$$R_{\mathrm{III}}(\Delta\nu',\Delta\nu,\pi/2)=\frac{g(0)}{\pi(\Delta\nu_{\mathrm{D}})^2}H\left(\frac{\delta}{\Delta\nu_{\mathrm{D}}},\frac{\Delta\nu'}{\Delta\nu_{\mathrm{D}}}\right)H\left(\frac{\delta}{\Delta\nu_{\mathrm{D}}},\frac{\Delta\nu}{\Delta\nu_{\mathrm{D}}}\right)$$
$$=g(0)\phi(\Delta\nu')\phi(\Delta\nu),\tag{7.3.25}$$

where we have used equation (1.4.18). Clearly, then, $\theta=\pi/2$ yields the com-

Fig.7.5. The re-distribution function R_{III} for several scattering angles when the photon is absorbed in the wings of the line.

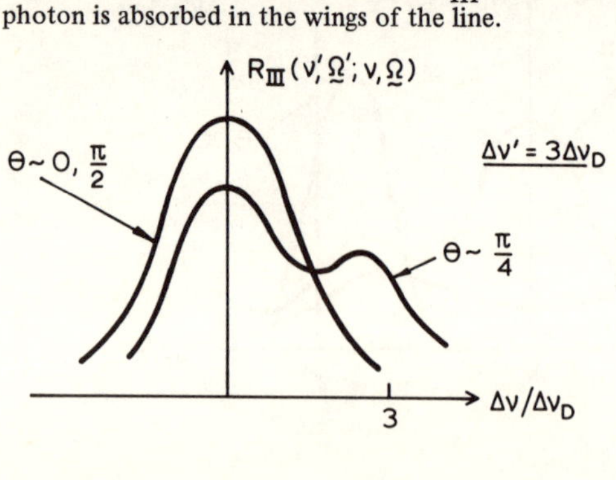

plete re-distribution result shown in figure 7.5 whereby photons are preferentially emitted at line centre independent of the frequency at which they were absorbed. Further, it is not difficult to see from equation (7.3.24) that

$$R_{\mathrm{III}}(\Delta\nu', -\Delta\nu, \pi-\theta) = R_{\mathrm{III}}(\Delta\nu', \Delta\nu, \theta), \tag{7.3.26}$$

as for $R_{\mathrm{I}}(\Delta\nu', \Delta\nu, \theta)$ but not $R_{\mathrm{II}}(\Delta\nu', \Delta\nu, \theta)$.

Finally, the qualitative behaviour of $R_{\mathrm{III}}(\nu', \Omega'; \nu, \Omega)$ for small θ may be seen by comparing

$$R_{\mathrm{III}}(\Delta\nu', \Delta\nu, 0) = \frac{g(1)}{\pi(\Delta\nu_{\mathrm{D}})^2}\int_{-\infty}^{\infty} \mathrm{e}^{-u_1^2}\,\mathrm{d}u_1 \frac{\delta/\pi\Delta\nu_{\mathrm{D}}}{\left(\dfrac{\Delta\nu'}{\Delta\nu_{\mathrm{D}}} - u_1\right)^2 + \left(\dfrac{\delta}{\Delta\nu_{\mathrm{D}}}\right)^2}$$

$$\times \int_{-\infty}^{\infty} \mathrm{e}^{-u_2^2}\,\mathrm{d}u_2 \frac{\delta/\pi\Delta\nu_{\mathrm{D}}}{\left(\dfrac{\Delta\nu}{\Delta\nu_{\mathrm{D}}} - u_1 - u_2\right)^2 + \left(\dfrac{\delta}{\Delta\nu_{\mathrm{D}}}\right)^2}, \tag{7.3.27}$$

with

$$R_{\mathrm{III}}(\Delta\nu', \Delta\nu, \pi/2) = \frac{g(0)}{\pi(\Delta\nu_{\mathrm{D}})^2}\int_{-\infty}^{\infty} \mathrm{e}^{-u_1^2}\,\mathrm{d}u_1 \frac{\delta/\pi\Delta\nu_{\mathrm{D}}}{\left(\dfrac{\Delta\nu'}{\Delta\nu_{\mathrm{D}}} - u_1\right)^2 + \left(\dfrac{\delta}{\Delta\nu_{\mathrm{D}}}\right)^2}$$

$$\times \int_{-\infty}^{\infty} \mathrm{e}^{-u_2^2}\,\mathrm{d}u_2 \frac{\delta/\pi\Delta\nu_{\mathrm{D}}}{\left(\dfrac{\Delta\nu}{\Delta\nu_{\mathrm{D}}} - u_2\right)^2 + \left(\dfrac{\delta}{\Delta\nu_{\mathrm{D}}}\right)^2}. \tag{7.3.28}$$

The only difference between these two expressions (the factors $g(0)$ and $g(1)$ can be considered as proportionality constants) is the appearance of u_1 in the second integral (over u_2) in equation (7.3.27). However, the $\mathrm{e}^{-u_1^2}$ term in both $R_{\mathrm{III}}(\Delta\nu', \Delta\nu, \mathrm{o})$ and $R_{\mathrm{III}}(\Delta\nu', \Delta\nu, \pi/2)$ above stipulates that the integral over u_1 is dominated by the integrand in the neighbourhood of $u_1 \sim 0$ so that

$$R_{\mathrm{III}}(\Delta\nu', \Delta\nu, 0) \approx R_{\mathrm{III}}(\Delta\nu', \Delta\nu, \pi/2)$$

as shown in figure 7.5.

7.3.4 *A simplified re-distribution function [61]*

Recall that equation (7.3.2) for $R(\nu', \Omega'; \nu, \Omega)$ was obtained in section 7.2.3 as a consequence of the desire to constrain the frequency- and angle-dependent source function $S(\mathbf{r}, \nu, \Omega)$ given by equation (7.2.23) to be a linear functional of the radiation field. Even so, we see from the preceding R_{I}, R_{II} and R_{III} that these re-distribution functions are quite difficult to manage numerically, particularly in view of their sometime coherent nature and the consequent need to construct a frequency *and* angle quadrature grid to adequately represent this coherency. Recall, further, that these R_{I}, R_{II} and R_{III} themselves represent only special limiting cases of the more physically realistic configurations given, for example, by equation (6.5.92). Indeed, we can use equation (6.4.200) stipulating $p(\gamma', \gamma)$, for example, to obtain an R_{IV} re-distribution function for the case in which both upper and lower levels are only radiatively broadened. Thus, some further form of simplification of $R(\nu', \Omega'; \nu, \Omega)$ is desirable.

To do this, we return to the quantum mechanical result of chapter 6 given by equation (6.5.106) for a radiatively *and* collisionally broadened upper state but an infinitesimally sharp lower level, viz.

$$w(1 \to 2 \to 1) \propto \frac{1}{\delta} q(\gamma') p(\gamma', \gamma), \tag{7.3.29}$$

where

$$p(\gamma', \gamma) = \frac{\delta}{Q_{\mathrm{I}} + Q_{\mathrm{E}} + \delta} \delta(\gamma' - \gamma) + \frac{\delta}{\delta + Q_{\mathrm{I}}} \frac{Q_{\mathrm{E}}}{Q_{\mathrm{I}} + Q_{\mathrm{E}} + \delta} q(\gamma).$$

The above formula has already been discussed in some detail at the end of chapter 6. For our purposes we put $Q_{\mathrm{I}} = 0$ and write $\beta = \delta/(Q_{\mathrm{E}} + \delta)$; note that the definition of $p(\gamma', \gamma)$ used here requires the atom to de-excite radiatively. We then have

$$p(\gamma', \gamma) = \beta\delta(\gamma' - \gamma) + (1 - \beta) q(\gamma). \tag{7.3.30}$$

Consequently, from equation (7.2.17),

$$\begin{aligned} R(\nu', \Omega'; \nu, \Omega) &= 4\pi \int F(\mathbf{v}) R_{\mathrm{LU}}(\nu', \Omega'; \nu, \Omega)\, \mathrm{d}\mathbf{v} \\ &= \beta R_{\mathrm{II}}(\nu', \Omega'; \nu, \Omega) + (1 - \beta) R_{\mathrm{III}}(\nu', \Omega'; \nu, \Omega). \end{aligned} \tag{7.3.31}$$

Thus a more physically realistic re-distribution function would incorporate a linear super-position of R_{II} and R_{III}, and this further increases the complexity of the numerical solution of the corresponding equation of radiative transfer.

Keeping in mind that the above $R(\nu', \Omega'; \nu, \Omega)$ are themselves a result of certain simplifications and therefore need not be considered as basic constituents of the problem, we thence replace R_{II} and R_{III} by numerically tractable angle-independent approximations. First, noting from section 7.3.3. that R_{III} has a qualitative behaviour similar to complete re-distribution for all scattering angles θ (see figure 7.5, for example, for $\theta = 0$ and $\pi/2$), we simply put

$$R_{\mathrm{III}}(\nu', \Omega'; \nu, \Omega) \to \bar{R}_{\mathrm{III}}(\Delta\nu'; \Delta\nu) = \phi(\Delta\nu')\phi(\Delta\nu). \tag{7.3.32}$$

Next, noting that R_{II} exhibits both a measure of coherency in the absorption wings even for $\theta \sim \pi/2$ (see figure 7.4) and a measure of non-coherency (recall that R_{II} has the same qualitative behaviour as R_{I} for $\Delta\nu' \sim 0$ - see figure 7.2), we write

$$\begin{aligned} R_{\mathrm{II}}(\nu', \Omega'; \nu, \Omega) \to \bar{R}_{\mathrm{II}}(\Delta\nu'; \Delta\nu) &= \alpha(\Delta\nu)\phi(\Delta\nu)\delta(\Delta\nu' - \Delta\nu) \\ &+ [1 - \alpha(\Delta\nu)]\phi(\Delta\nu')\phi(\Delta\nu), \end{aligned} \tag{7.3.33}$$

where we have clearly separated the coherent and complete re-distribution components. Since photons emitted in the wings under R_{II} were most probably absorbed in those wings, we have $\alpha(\Delta\nu \gtrsim 3\Delta\nu_D) \approx 1$. However, photon emission in the line core could have resulted from either absorption in the core followed by scattering through *any* angle θ or absorption in the wings followed by scattering through $\theta \sim \pi/2$. Clearly, $\alpha(\Delta\nu)$ details the fraction of total absorptions followed by this latter scattering event and should be small relative to the former process. It is therefore not unreasonable to have $\alpha(\Delta\nu)$ of the form illustrated in figure 7.6.

Note that $\alpha(\Delta\nu)$ is an even function since

$$R_{\text{II}}(-\Delta\nu', -\Delta\nu, \theta) = R_{\text{II}}(\Delta\nu', \Delta\nu, \theta), \tag{7.3.34}$$

from equation (7.3.20). Actually, the 'parameter' $\alpha(\Delta\nu)$ should [67] also depend upon $\Delta\nu'$ since $\bar{R}_{\text{II}}(\Delta\nu'; \Delta\nu)$ given by equation (7.3.33) does not satisfy the important normalisation condition (7.2.19), viz.

$$\frac{1}{4\pi}\iint R(\nu', \Omega'; \nu, \Omega)\; d\nu\, d\Omega = \phi(\nu'). \tag{7.3.35}$$

Indeed, the normalisation of $\phi(\nu')$ yields

$$\iiiint R(\nu', \Omega'; \nu, \Omega)\; d\nu\, \frac{d\Omega}{4\pi}\, d\nu'\, \frac{d\Omega'}{4\pi} = 1, \tag{7.3.36}$$

and this last result must be satisfied if the integral term in the spectral line source function is not to produce spurious sources and/or sinks of photons. However, the above approximation (7.3.33) already produces spurious photon channelling in both frequency and angle space (simply because it is an approximation) even if equation (7.3.35) was satisfied by $\bar{R}_{\text{II}}(\Delta\nu'; \Delta\nu)$ incorporating the appropriate modification of $\alpha(\Delta\nu)$. Since this effectively introduces spurious photon sources and sinks anyway, such a detailed concern for normalisation is not, in general, warranted.

Fig.7.6. The qualitative behaviour of the symmetric function $\alpha(\Delta\nu)$.

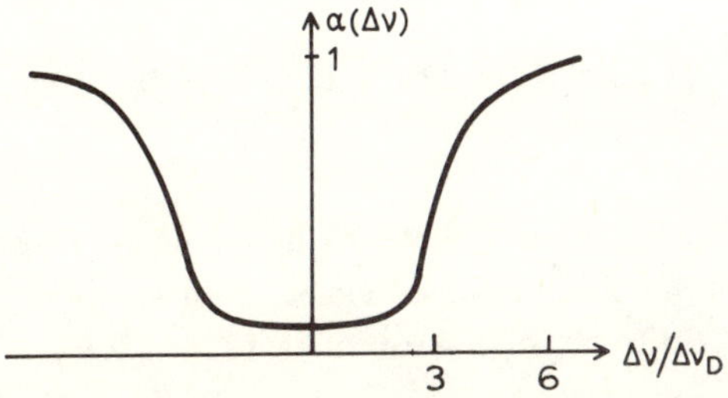

Thus, remaining with equations (7.3.32) and (7.3.33), equation (7.3.31) becomes

$$R(\nu', \Omega'; \nu, \Omega) = \Theta(\Delta\nu)\phi(\Delta\nu)\delta(\Delta\nu' - \Delta\nu) + [1 - \Theta(\Delta\nu)]\phi(\Delta\nu')\phi(\Delta\nu), \quad (7.3.37)$$

where $\Theta(\Delta\nu) = \beta\alpha(\Delta\nu)$.

The spectral line source function (7.2.23) then has the form

$$\begin{aligned} S(r, \Delta\nu) &= \frac{1-\epsilon}{4\pi\phi(\Delta\nu)} \iint \bar{R}(\Delta\nu', \Delta\nu) I(\mathbf{r}, \Delta\nu', \Omega')\, \mathrm{d}(\Delta\nu')\, \mathrm{d}\Omega + \epsilon B_\nu(T) \\ &= (1-\epsilon)\Theta(\Delta\nu) J(\mathbf{r}, \Delta\nu) + (1-\epsilon)[1 - \Theta(\Delta\nu)]\bar{J}(\mathbf{r}) + \epsilon B_\nu(T) \\ &= S_0(\mathbf{r}) + (1-\epsilon)\Theta(\Delta\nu)[J(\mathbf{r}, \Delta\nu) - \bar{J}(\mathbf{r})], \end{aligned} \quad (7.3.38)$$

where $J(\mathbf{r}, \Delta\nu)$ is the angle-averaged radiation field and $S_0(\mathbf{r})$ the complete re-distribution source function. Clearly, the solution of the equation of radiative transfer incorporating the above source function (7.3.38) does not entail the rather imposing numerical difficulties associated with the more general, though still linear, form (7.3.31).

7.4 Numerical methods of solution

The general non-linear form of the source function produces difficulties in the numerical solution of the equation of radiative transfer which are not apparent when using the linear form discussed in the preceding two sections. Consequently, we first develop the Feautrier technique for the simplified problem, and this then enables the generalisation to the more complicated equation to be more readily appreciated.

7.4.1 *The linear problem*

Here we wish to solve the time-independent transfer equation

$$(\Omega \cdot \nabla) I = -\kappa(I - S), \quad (7.4.1)$$

where

$$\kappa = \frac{h\nu_0}{4\pi}(N_{\mathrm{L}} B_{\mathrm{LU}} - N_{\mathrm{U}} B_{\mathrm{UL}})\phi(\nu), \quad (7.4.2)$$

and

$$S(\mathbf{r}, \nu, \Omega) = \frac{1-\epsilon}{4\pi\phi(\nu)} \iint R_i(\nu', \Omega'; \nu, \Omega) I(\mathbf{r}, \nu', \Omega')\, \mathrm{d}\nu'\, \mathrm{d}\Omega' + \epsilon B_\nu(T), \quad (7.4.3)$$

for $i = \mathrm{I}, \mathrm{II}, \mathrm{III}$ or any linear combination of these. We have noted in equation (7.4.2) that equation (7.4.3) was obtained in section 7.2.2. by assuming stimulated emissions to be equivalent to negative absorptions, i.e. $\phi \equiv \psi \not\equiv j$.

We wish to apply the Feautrier technique to the above equations [15]. We therefore divide the radiation field into its positive and negative components of $\boldsymbol{\Omega}$ so that with $\boldsymbol{\Omega} \equiv (\sin\theta \cos\phi, \sin\theta \sin\phi, \cos\theta)$ we have $\boldsymbol{\Omega} \gtrless 0 \equiv \cos\theta \gtrless 0$ *and* $\cos\phi \gtrless 0$. We then define the complete range $\boldsymbol{\Omega}_{\mathrm{C}}$ of $\boldsymbol{\Omega}$ and the two half-ranges $\boldsymbol{\Omega}_{\mathrm{H}}^{\pm}$ so that $\boldsymbol{\Omega}_{\mathrm{C}} \equiv \theta \in [0, \pi]$ *and* $\phi \in [0, 2\pi]$, $\boldsymbol{\Omega}_{\mathrm{H}}^{+} \equiv \theta \in [0, \pi/2]$ *and* $\phi \in [-\pi/2, \pi/2]$, $\boldsymbol{\Omega}_{\mathrm{H}}^{-} \equiv \theta \in [\pi/2, \pi]$ *and* $\phi \in [\pi/2, 3\pi/2]$.

Thus, we find

$$\pm\left(\boldsymbol{\Omega}\cdot\frac{1}{\kappa}\nabla\right)I(\mathbf{r}, \nu, \pm\boldsymbol{\Omega}) = -I(\mathbf{r}, \nu, \pm\boldsymbol{\Omega}) + S(\mathbf{r}, \nu, \pm\boldsymbol{\Omega}), \tag{7.4.4}$$

for all $\boldsymbol{\Omega} \in \boldsymbol{\Omega}_{\mathrm{H}}^{+}$ where we note that κ is independent of $\boldsymbol{\Omega}$ (we are not considering macroscopic velocity fields here).

We now define

$$\left.\begin{aligned} \Phi(\mathbf{r}, \nu, \boldsymbol{\Omega}) &= \tfrac{1}{2}[I(\mathbf{r}, \nu, \boldsymbol{\Omega}) + I(\mathbf{r}, \nu, -\boldsymbol{\Omega})], \\ \Psi(\mathbf{r}, \nu, \boldsymbol{\Omega}) &= \tfrac{1}{2}[I(\mathbf{r}, \nu, \boldsymbol{\Omega}) - I(\mathbf{r}, \nu, -\boldsymbol{\Omega})], \end{aligned}\right\} \tag{7.4.5}$$

for $\boldsymbol{\Omega} \in \boldsymbol{\Omega}_{\mathrm{H}}^{+}$ as usual, so that equation (7.4.4) yields

$$\left(\boldsymbol{\Omega}\cdot\frac{1}{\kappa}\nabla\right)\Phi = -\Psi + \tfrac{1}{2}[S(\mathbf{r}, \nu, \boldsymbol{\Omega}) - S(\mathbf{r}, \nu, -\boldsymbol{\Omega}], \tag{7.4.6}$$

$$\left(\boldsymbol{\Omega}\cdot\frac{1}{\kappa}\nabla\right)\Psi = -\Phi + \tfrac{1}{2}[S(\mathbf{r}, \nu, \boldsymbol{\Omega}) + S(\mathbf{r}, \nu, -\boldsymbol{\Omega})], \tag{7.4.7}$$

for all $\boldsymbol{\Omega} \in \boldsymbol{\Omega}_{\mathrm{H}}^{+}$. Further,

$$\begin{aligned} S(\mathbf{r}, \nu, \boldsymbol{\Omega}) \pm S(\mathbf{r}, \nu, -\boldsymbol{\Omega}) &= \frac{1-\epsilon}{4\pi\phi(\nu)}\iint_{\boldsymbol{\Omega}_{\mathrm{C}}} \mathrm{d}\nu'\,\mathrm{d}\boldsymbol{\Omega}' \\ &\times [R_i(\nu', \boldsymbol{\Omega}'; \nu, \boldsymbol{\Omega}) \pm R_i(\nu', \boldsymbol{\Omega}'; \nu, -\boldsymbol{\Omega})]I(\mathbf{r}, \nu', \boldsymbol{\Omega}') + [1 \pm 1]\epsilon B_\nu(T), \end{aligned} \tag{7.4.8}$$

for all $\boldsymbol{\Omega}' \in \boldsymbol{\Omega}_{\mathrm{C}}'$ and $\boldsymbol{\Omega} \in \boldsymbol{\Omega}_{\mathrm{H}}^{+}$.

The expressions on the RHS of this last result may be simplified by using certain general symmetry conditions satisfied by the re-distribution function. We know from equation (7.2.17) that

$$\begin{aligned} R(\nu', \boldsymbol{\Omega}'; \nu, \boldsymbol{\Omega}) &= 4\pi\int F(\upsilon)R_{\mathrm{LU}}(\nu', \boldsymbol{\Omega}'; \nu, \boldsymbol{\Omega}; \mathbf{v})\,\mathrm{d}\mathbf{v} \\ &= g(\boldsymbol{\Omega}', \boldsymbol{\Omega})\int F(\mathbf{v})q(\gamma')p(\gamma', \gamma)\,\mathrm{d}\mathbf{v} \\ &= g(\boldsymbol{\Omega}', \boldsymbol{\Omega})\int F(\mathbf{v})q\left(\nu' - \frac{\nu'}{c}\mathbf{v}\cdot\boldsymbol{\Omega}'\right) \\ &\quad\times p\left(\nu' - \frac{\nu'}{c}\mathbf{v}\cdot\boldsymbol{\Omega}', \nu - \frac{\nu}{c}\mathbf{v}\cdot\boldsymbol{\Omega}\right)\mathrm{d}\mathbf{v}. \end{aligned} \tag{7.4.9}$$

Thus, if we have the symmetry condition

$$F(-\mathbf{v}) = F(\mathbf{v}), \tag{7.4.10}$$

for all $\mathbf{v}$ (note that this is satisfied, for example, by the Maxwellian velocity distribution (7.3.4) used to obtain R_{I}, R_{II} and R_{III}), and assume the angular probability function $g(\boldsymbol{\Omega}', \boldsymbol{\Omega})$ depends solely on the angle of scatter $\cos^{-1}(\boldsymbol{\Omega}' \cdot \boldsymbol{\Omega})$ so that

$$g(-\boldsymbol{\Omega}', -\boldsymbol{\Omega}) = g(\boldsymbol{\Omega}', \boldsymbol{\Omega}), \tag{7.4.11}$$

for all $\boldsymbol{\Omega}' \in \boldsymbol{\Omega}'_{\mathrm{C}}$ and $\boldsymbol{\Omega} \in \boldsymbol{\Omega}_{\mathrm{C}}$, then equation (7.4.9) yields

$$\begin{aligned} R(\nu', -\boldsymbol{\Omega}'; \nu, -\boldsymbol{\Omega}) &= g(-\boldsymbol{\Omega}', -\boldsymbol{\Omega}) \int F(\mathbf{v}) q\left(\nu' + \frac{\nu'}{c} \mathbf{v} \cdot \boldsymbol{\Omega}'\right) \\ &\quad \times p\left(\nu' + \frac{\nu'}{\nu} \mathbf{v} \cdot \boldsymbol{\Omega}', \nu + \frac{\nu}{c} \mathbf{v} \cdot \boldsymbol{\Omega}\right) \mathrm{d}\mathbf{v} \\ &= g(\boldsymbol{\Omega}', \boldsymbol{\Omega}) \int F(-\mathbf{v}) q\left(\nu' - \frac{\nu'}{c} \mathbf{v} \cdot \boldsymbol{\Omega}'\right) \\ &\quad \times p\left(\nu' - \frac{\nu'}{c} \mathbf{v} \cdot \boldsymbol{\Omega}', \nu - \frac{\nu}{c} \mathbf{v} \cdot \boldsymbol{\Omega}\right) \mathrm{d}\mathbf{v} \\ &= R(\nu', \boldsymbol{\Omega}'; \nu, \boldsymbol{\Omega}), \end{aligned} \tag{7.4.12}$$

for all $\boldsymbol{\Omega}' \in \boldsymbol{\Omega}'_{\mathrm{C}}$ and $\boldsymbol{\Omega} \in \boldsymbol{\Omega}_{\mathrm{C}}$. Indeed, a similar analysis easily shows that

$$R(\nu', \pm\boldsymbol{\Omega}'; \nu, \mp\boldsymbol{\Omega}) = R(\nu', \mp\boldsymbol{\Omega}'; \nu, \pm\boldsymbol{\Omega}), \tag{7.4.13}$$

for all $\boldsymbol{\Omega}' \in \boldsymbol{\Omega}'_{\mathrm{C}}$ and $\boldsymbol{\Omega} \in \boldsymbol{\Omega}_{\mathrm{C}}$.

These results can then be used in equation (7.4.8) to yield

$$\begin{aligned} &S(\mathbf{r}, \nu, \boldsymbol{\Omega}) \pm S(\mathbf{r}, \nu, -\boldsymbol{\Omega}) \\ &= \frac{1-\epsilon}{4\pi\phi(\nu)} \int_{\nu'_{\mathrm{C}}} \mathrm{d}\nu' \left\{ \int_{\boldsymbol{\Omega}'_{\mathrm{H}}} \mathrm{d}\boldsymbol{\Omega}' \, [R_i(\nu', \boldsymbol{\Omega}'; \nu, \boldsymbol{\Omega}) \right. \\ &\quad \pm R_i(\nu', \boldsymbol{\Omega}'; \nu, -\boldsymbol{\Omega})] \, I(r, \nu', \boldsymbol{\Omega}') \\ &\quad \left. + \int_{\boldsymbol{\Omega}'_{\mathrm{H}}} \mathrm{d}\boldsymbol{\Omega}' [R_i(\nu', -\boldsymbol{\Omega}'; \nu, \boldsymbol{\Omega}) \pm R_i(\nu', -\boldsymbol{\Omega}'; \nu, -\boldsymbol{\Omega})] I(\mathbf{r}, \nu', -\boldsymbol{\Omega}') \right\} \\ &\quad + [1 \pm 1] \epsilon B_\nu(T) \\ &= \frac{1-\epsilon}{4\pi\phi(\nu)} \int_{\nu'_{\mathrm{C}}} \mathrm{d}\nu' \int_{\boldsymbol{\Omega}'_{\mathrm{H}}} \mathrm{d}\boldsymbol{\Omega}' \{ [R_i(\nu', \boldsymbol{\Omega}'; \nu, \boldsymbol{\Omega}) \\ &\quad \pm R_i(\nu', \boldsymbol{\Omega}'; \nu, -\boldsymbol{\Omega})] I(\mathbf{r}, \nu', \boldsymbol{\Omega}') \\ &\quad + [R_i(\nu', \boldsymbol{\Omega}'; \nu, -\boldsymbol{\Omega}) \pm R_i(\nu', \boldsymbol{\Omega}'; \nu, \boldsymbol{\Omega})] I(\mathbf{r}, \nu', -\boldsymbol{\Omega}') \} \\ &\quad + [1 \pm 1] \epsilon B_\nu(T), \end{aligned} \tag{7.4.14}$$

for all $\boldsymbol{\Omega} \in \boldsymbol{\Omega}_{\mathrm{H}}$ where the positive superscript on $\boldsymbol{\Omega}_{\mathrm{H}}$ and $\boldsymbol{\Omega}'_{\mathrm{H}}$ is now understood.

Consequently, we have

$$\begin{aligned}
&\tfrac{1}{2}[S(\mathbf{r}, \nu, \boldsymbol{\Omega}) + S(\mathbf{r}, \nu, -\boldsymbol{\Omega})] \\
&= \frac{1-\epsilon}{4\pi\phi(\nu)} \int_{\nu'_{\mathrm{C}}} \mathrm{d}\nu' \int_{\boldsymbol{\Omega}'_{\mathrm{H}}} \mathrm{d}\boldsymbol{\Omega}' [R_i(\nu', \boldsymbol{\Omega}'; \nu, \boldsymbol{\Omega}) + R_i(\nu', \boldsymbol{\Omega}'; \nu, -\boldsymbol{\Omega})] \\
&\quad \times \Phi(\mathbf{r}, \nu', \boldsymbol{\Omega}') + \epsilon B_\nu(T) \\
&= \frac{1-\epsilon}{4\pi} \int_{\nu'_{\mathrm{C}}} \mathrm{d}\nu' \int_{\boldsymbol{\Omega}'_{\mathrm{H}}} \mathrm{d}\boldsymbol{\Omega}' A_i^{+}(\nu', \boldsymbol{\Omega}'; \nu, \boldsymbol{\Omega}) \Phi(\mathbf{r}, \nu', \boldsymbol{\Omega}') + \epsilon B_\nu(T),
\end{aligned} \tag{7.4.15}$$

where we have defined

$$A_i^{\pm}(\nu', \boldsymbol{\Omega}'; \nu, \boldsymbol{\Omega}) = \frac{1}{\phi(\nu)} [R_i(\nu', \boldsymbol{\Omega}'; \nu, \boldsymbol{\Omega}) \pm R_i(\nu', \boldsymbol{\Omega}'; \nu, -\boldsymbol{\Omega})], \tag{7.4.16}$$

for all $\boldsymbol{\Omega}' \in \boldsymbol{\Omega}'_{\mathrm{H}}$ and $\boldsymbol{\Omega} \in \boldsymbol{\Omega}_{\mathrm{H}}$. Similarly, equation (7.4.14) yields

$$\begin{aligned}
&\tfrac{1}{2}[S(\mathbf{r}, \nu, \boldsymbol{\Omega}) - S(\mathbf{r}, \nu, -\boldsymbol{\Omega})] \\
&= \frac{1-\epsilon}{4\pi} \int_{\nu'_{\mathrm{C}}} \mathrm{d}\nu' \int_{\boldsymbol{\Omega}'_{\mathrm{H}}} \mathrm{d}\boldsymbol{\Omega}' A_i^{-}(\nu', \boldsymbol{\Omega}'; \nu, \boldsymbol{\Omega}) \Psi(\mathbf{r}, \nu', \boldsymbol{\Omega}'),
\end{aligned} \tag{7.4.17}$$

for all $\boldsymbol{\Omega} \in \boldsymbol{\Omega}'_{\mathrm{H}}$.

Equations (7.4.6) and (7.4.7) now become

$$\left(\boldsymbol{\Omega} \cdot \frac{1}{\kappa} \nabla\right) \Phi = -\Psi + \frac{1-\epsilon}{4\pi} \int_{\nu'_{\mathrm{C}}} \mathrm{d}\nu' \int_{\boldsymbol{\Omega}'_{\mathrm{H}}} \mathrm{d}\boldsymbol{\Omega}' A_i^{-}(\nu', \boldsymbol{\Omega}'; \nu, \boldsymbol{\Omega}) \Psi(\mathbf{r}, \nu', \boldsymbol{\Omega}'), \tag{7.4.18}$$

and

$$\begin{aligned}
\left(\boldsymbol{\Omega} \cdot \frac{1}{\kappa} \nabla\right) \Psi &= -\Phi + \frac{1-\epsilon}{4\pi} \int_{\nu'_{\mathrm{C}}} \mathrm{d}\nu' \int_{\boldsymbol{\Omega}'_{\mathrm{H}}} \mathrm{d}\boldsymbol{\Omega}' \\
&\quad \times A_i^{+}(\nu', \boldsymbol{\Omega}'; \nu, \boldsymbol{\Omega}) \Phi(\mathbf{r}, \nu', \boldsymbol{\Omega}') + \epsilon B_\nu(T),
\end{aligned} \tag{7.4.19}$$

for all $\boldsymbol{\Omega} \in \boldsymbol{\Omega}_{\mathrm{H}}$. These may then be put into the second order differential form

$$\begin{aligned}
&\left(\boldsymbol{\Omega} \cdot \frac{1}{\kappa} \nabla\right)\left(\boldsymbol{\Omega} \cdot \frac{1}{\kappa} \nabla\right) \Phi(\mathbf{r}, \nu, \boldsymbol{\Omega}) \\
&= \Phi(\mathbf{r}, \nu, \boldsymbol{\Omega}) - \frac{1-\epsilon}{4\pi} \int_{\nu'_{\mathrm{C}}} \mathrm{d}\nu' \int_{\boldsymbol{\Omega}'_{\mathrm{H}}} \mathrm{d}\boldsymbol{\Omega}' \\
&\quad \times A_i^{+}(\nu', \boldsymbol{\Omega}'; \nu, \boldsymbol{\Omega}) \Phi(\mathbf{r}, \nu', \boldsymbol{\Omega}') + Q(\mathbf{r}, \nu, \boldsymbol{\Omega}, \Psi),
\end{aligned} \tag{7.4.20}$$

with a similar equation for $\Psi(\mathbf{r}, \nu, \boldsymbol{\Omega})$, where the functional Q is defined by

$$\begin{aligned}
Q(\mathbf{r}, \nu, \boldsymbol{\Omega}, \Psi) &= -\epsilon B_\nu(T) + \left(\boldsymbol{\Omega} \cdot \frac{1}{\kappa} \nabla\right) \\
&\quad \times \left\{\frac{1-\epsilon}{4\pi} \int_{\nu'_{\mathrm{C}}} \mathrm{d}\nu' \int_{\boldsymbol{\Omega}'_{\mathrm{H}}} \mathrm{d}\boldsymbol{\Omega}' A_i^{-}(\nu', \boldsymbol{\Omega}'; \nu, \boldsymbol{\Omega}) \Psi(\mathbf{r}, \nu', \boldsymbol{\Omega}')\right\}.
\end{aligned} \tag{7.4.21}$$

The above second order differential equations are of the same basic form as the ordinary Feautrier equations discussed in chapter 3, and differ only in that they exhibit coupling terms in Φ and Ψ. However, this apparent difficulty can be easily overcome. The coupling terms are linear in Φ and Ψ and this, together with the finite difference approach, enables a system of linear matrix equations in Φ and Ψ to be specified. The resulting expressions are then readily uncoupled (see, for example, the uncoupling of $\Phi^{(1)}$ and $\Phi^{(2)}$ using the recursive relationship (5.8.18) in section 5.8.1), and thence the solution proceeds as in the ordinary Feautrier situation.

Further, it has been shown in section 3.3.3 that expressions of the form given by equation (7.4.20), for example, are stable under numerical reduction using the finite difference approach provided that

$$\frac{1}{\Phi(\mathbf{r},\nu,\Omega)}\left\{\frac{1-\epsilon}{4\pi}\int_{\nu'_{\mathrm{C}}}\mathrm{d}\nu'\int_{\Omega'_{\mathrm{H}}}\mathrm{d}\Omega' A_i^{+}(\nu',\Omega';\nu,\Omega)\Phi(\mathbf{r},\nu',\Omega')\right\}\lesssim 1, \tag{7.4.22}$$

at all $\mathbf{r}$. This will ensure that the diagonal elements of the matrices obtained are dominant, and therefore that the required matrix inversions are stable. The 'approximately equal to' sign in expression (7.4.22) allows for the possibility that the LHS is 'slightly' greater than unity for some values of ν and Ω. We know from equation (7.2.18) that

$$\phi(\nu')=\frac{1}{4\pi}\iint R(\nu',\Omega';\nu,\Omega)\,\mathrm{d}\nu\,\mathrm{d}\Omega. \tag{7.4.23}$$

Recognising that $R(\nu',\Omega';\nu,\Omega)$ is a function specifying the probability of a particular absorption process occurring followed by a particular radiative emission so that summing in this last equation (7.4.23) over all emitted photons yields the probability of absorption $\phi(\nu')$, then, correspondingly, summing over all the absorbed photons should yield the probability of emission $j(\nu,\Omega)$, viz.

$$j(\nu,\Omega)=\frac{1}{4\pi}\iint R(\nu',\Omega';\nu,\Omega)\,\mathrm{d}\nu'\,\mathrm{d}\Omega', \tag{7.4.24}$$

where $j(\nu,\Omega)<1$ since only the sum

$$\frac{1}{4\pi}\iint j(\nu,\Omega)\,\mathrm{d}\nu\,\mathrm{d}\Omega=1, \tag{7.4.25}$$

yields unity. Consequently, we have

$$\frac{1-\epsilon}{4\pi}\int_{\nu'_{\mathrm{C}}}\mathrm{d}\nu'\int_{\Omega'_{\mathrm{H}}}\mathrm{d}\Omega' A_i^{+}(\nu',\Omega';\nu,\Omega)$$
$$=\frac{1-\epsilon}{4\pi\phi(\nu)}\int_{\nu'_{\mathrm{C}}}\mathrm{d}\nu'\int_{\Omega'_{\mathrm{H}}}\mathrm{d}\Omega'\left[R_i(\nu',\Omega';\nu,\Omega)+R_i(\nu',\Omega';\nu,-\Omega)\right]$$

$$= \frac{1-\epsilon}{4\pi\phi(\nu)} \int_{\nu'_{\rm C}} {\rm d}\nu' \int_{\Omega'_{\rm H}} {\rm d}\Omega' [R_i(\nu', \Omega'; \nu, \Omega) + R_i(\nu', -\Omega'; \nu, \Omega)]$$

$$= \frac{1-\epsilon}{4\pi\phi(\nu)} \int_{\nu'_{\rm C}} {\rm d}\nu' \int_{\Omega'_{\rm H}} {\rm d}\Omega' R_i(\nu', \Omega'; \nu, \Omega)$$

$$= (1-\epsilon) \frac{j(\nu, \Omega)}{\phi(\nu)} . \tag{7.4.26}$$

Thus, taking $\Phi(\mathbf{r}, \nu', \Omega')$ outside the integral in expression (7.4.22) for the moment, we have LHS (7.4.22) $= (1-\epsilon)j/\phi$ which is not necessarily less than unity. Indeed, if we consider using photons for which $\phi(\nu) \ll 1$ but $j(\nu, \Omega) \lesssim 1$ (recall that photon coherence in the wings is a high probability) then even with $0 < \epsilon < 1$ we find $(1-\epsilon)j/\phi \gg 1$. However, stability problems should arise only in semi-infinite media (wing photons are relatively unimportant in detailing the radiation field in slab geometry) for which $\Phi(\text{wings}) \gg \Phi(\text{line core})$ in general and this will act to decrease the LHS (7.4.22). In fact, it is a measure of the power of the Feautrier technique that good stability is almost always attained even though condition (7.4.23) might be violated at particular values of ν and Ω. This, of course, simply reflects the fact that matrix inverses may be accurately computed even though some of the off-diagonal elements of the matrices are dominant.

It is numerically convenient to use the symmetry of the radiation field about line centre in the rest frame of the observer (in the absence of macroscopic velocity fields). This symmetry can be shown by writing equation (7.4.9) in the form

$$R(\Delta\nu', \Omega', \Delta\nu, \Omega) = g(\Omega', \Omega) \int F(\mathbf{v}) q \left(\Delta\nu' - \frac{\nu_0}{c} \mathbf{v} \cdot \Omega'\right)$$

$$\times p \left(\Delta\nu' - \frac{\nu_0}{c} \mathbf{v} \cdot \Omega', \Delta\nu - \frac{\nu_0}{c} \mathbf{v} \cdot \Omega\right) {\rm d}\mathbf{v}, \tag{7.4.27}$$

where we have put $\nu'/c \approx \nu_0/c \approx \nu/c$ (we are only interested in frequencies ν, for example, for which $\Delta\nu = \nu - \nu_0 \ll \nu$). We then have

$$R(-\Delta\nu', \Omega'; -\Delta\nu, \Omega) = g(\Omega', \Omega) \int F(\mathbf{v}) q \left(-\Delta\nu' - \frac{\nu_0}{c} \mathbf{v} \cdot \Omega'\right)$$

$$\times p \left(-\Delta\nu' - \frac{\nu_0}{c} \mathbf{v} \cdot \Omega', -\Delta\nu - \frac{\nu_0}{c} \mathbf{v} \cdot \Omega\right) {\rm d}\mathbf{v}$$

$$= g(\Omega', \Omega) \int F(-\mathbf{v}) q \left(-\Delta\nu' + \frac{\nu_0}{c} \mathbf{v} \cdot \Omega'\right)$$

$$\times p \left(-\Delta\nu' + \frac{\nu_0}{c} \mathbf{v} \cdot \Omega', -\Delta\nu + \frac{\nu_0}{c} \mathbf{v} \cdot \Omega\right) {\rm d}\mathbf{v}.$$

If we further assume symmetry about line centre in the rest frame of the atom for all photon-atom processes so that $q(-\Delta\gamma') = q(\Delta\gamma')$ and $p(-\Delta\gamma', -\Delta\gamma) = p(\Delta\gamma', \Delta\gamma)$, and note equation (7.4.10), then

$$R(-\Delta\nu', \Omega'; -\Delta\nu, \Omega) = R(\Delta\nu', \Omega'; \Delta\nu, \Omega). \tag{7.4.28}$$

This then implies that

$$\begin{aligned} S(\mathbf{r}, -\Delta\nu, \Omega) &= \frac{1-\epsilon}{4\pi} \iint R(\Delta\nu', \Omega'; -\Delta\nu, \Omega) I(\mathbf{r}, \Delta\nu', \Omega')\, \mathrm{d}(\Delta\nu')\, \mathrm{d}\Omega' \\ &\quad + \epsilon B_\nu(T) \\ &= \frac{1-\epsilon}{4\pi} \iint R(-\Delta\nu', \Omega'; -\Delta\nu, \Omega) I(\mathbf{r}, -\Delta\nu', \Omega') \\ &\quad \times \mathrm{d}(\Delta\nu')\, \mathrm{d}\Omega' + \epsilon B_\nu(T) \\ &= \frac{1-\epsilon}{4\pi} \iint R(\Delta\nu', \Omega'; \Delta\nu, \Omega) I(\mathbf{r}, -\Delta\nu', \Omega')\, \mathrm{d}(\Delta\nu')\, \mathrm{d}\Omega' \\ &\quad + \epsilon B_\nu(T). \end{aligned}$$

Thus, writing equation (7.4.1) in the form

$$(\Omega \cdot \nabla) I(\mathbf{r}, -\Delta\nu, \Omega) = -\kappa [I(\mathbf{r}, -\Delta\nu, \Omega) - S(\mathbf{r}, -\Delta\nu, \Omega)],$$

we immediately see that $I(\mathbf{r}, -\Delta\nu, \Omega)$ satisfies the same equation as $I(\mathbf{r}, -\Delta\nu, \Omega)$ so that

$$I(\mathbf{r}, -\Delta\nu, \Omega) = I(\mathbf{r}, \Delta\nu, \Omega). \tag{7.4.29}$$

Equations (7.4.5) then show that both $\Phi(\mathbf{r}, \Delta\nu, \Omega)$ and $\Psi(\mathbf{r}, \Delta\nu, \Omega)$ are also symmetric about line centre in the rest frame of the observer. Consequently, as in previous situations for which the macroscopic velocity field has been put equal to zero, we need only consider half the spectral line profile when using the Feautrier technique. As before, this produces significant savings in computer time and storage. It should be stressed, however, that great care should be exercised in constructing the angle and frequency quadrature grid points because of the coherency that can occur outside the line core for small scattering angles (see equation (7.3.18), for example).

Equation (7.4.20) may be further simplified when using the R_I and R_III redistribution functions if $g(\Omega', \Omega)$ is an even function of the scattering angle $\cos^{-1}(\Omega' \cdot \Omega)$. Complete coherence in the rest frame of the atom (this leads to R_I) implies $p(\gamma', \gamma) = \delta(\gamma' - \gamma)$ so that equation (7.4.9) becomes

$$R_\mathrm{I}(\Delta\nu', \Omega'; \Delta\nu, \Omega) = g(\Omega', \Omega) \int F(\mathbf{v}) q\left(\Delta\nu - \frac{\nu_0}{c} \mathbf{v} \cdot \Omega'\right) \mathrm{d}\mathbf{v}.$$

Thus, with

$$g(\Omega', -\Omega) = g(-\Omega', \Omega) = g(-\cos^{-1}(\Omega' \cdot \Omega) = g(\Omega', \Omega), \tag{7.4.30}$$

we have, using equations (7.4.13),

$$R_{\mathrm{I}}(-\Delta\nu', -\Omega'; \Delta\nu, \Omega) = R_{\mathrm{I}}(-\Delta\nu', \Omega'; \Delta\nu, -\Omega)$$

$$= g(\Omega', \Omega) \int F(\mathbf{v}) q\left(\Delta\nu - \frac{\nu_0}{c}\mathbf{v}\cdot\Omega'\right) \mathrm{d}\mathbf{v}$$

$$= R_{\mathrm{I}}(\Delta\nu', \Omega'; \Delta\nu, \Omega). \qquad (7.4.31)$$

The same result can be obtained when we assume complete re-distribution in the rest frame of the atom (this leads to R_{III}). Here, we have $p(\gamma', \gamma) = q(\gamma)$ so that equation (7.4.9) yields

$$R_{\mathrm{III}}(\Delta\nu', \Omega'; \Delta\nu, \Omega)$$

$$= g(\Omega', \Omega) \int F(\mathbf{v}) q\left(\Delta\nu' - \frac{\nu_0}{c}\mathbf{v}\cdot\Omega'\right) q\left(\Delta\nu - \frac{\nu_0}{c}\mathbf{v}\cdot\Omega\right) \mathrm{d}\mathbf{v}.$$

We again use equation (7.4.30) and assume symmetry of the scattering process about line centre in the rest frame of the atom, viz. $q(-\gamma') = q(\gamma')$, so that

$$R_{\mathrm{III}}(-\Delta\nu', -\Omega'; \Delta\nu, \Omega)$$

$$= g(\Omega', \Omega) \int F(\mathbf{v}) q\left(-\Delta\nu' + \frac{\nu_0}{c}\mathbf{v}\cdot\Omega'\right) q\left(\Delta\nu - \frac{\nu_0}{c}\mathbf{v}\cdot\Omega\right) \mathrm{d}\mathbf{v}$$

$$= g(\Omega', \Omega) \int F(\mathbf{v}) q\left(\Delta\nu' - \frac{\nu_0}{c}\mathbf{v}\cdot\Omega'\right) q\left(\Delta\nu - \frac{\nu_0}{c}\mathbf{v}\cdot\Omega\right) \mathrm{d}\mathbf{v}$$

$$= R_{\mathrm{III}}(\Delta\nu', \Omega'; \Delta\nu, \Omega), \qquad (7.4.32)$$

as in equation (7.4.31).

We now return to equation (7.4.18) and note, using equation (7.4.16), that

$$\int_{\nu'_{\mathrm{C}}} \mathrm{d}\nu' \int_{\Omega'_{\mathrm{H}}} \mathrm{d}\Omega' A_i^-(\nu', \Omega'; \nu, \Omega) \Psi(\mathbf{r}, \nu', \Omega')$$

$$\equiv \frac{1}{\phi(\nu)} \int_{-\infty}^{\infty} \mathrm{d}(\Delta\nu') \int_{\Omega'_{\mathrm{H}}} \mathrm{d}\Omega' [R_i(\Delta\nu', \Omega'; \Delta\nu, \Omega)$$
$$-R_i(\Delta\nu', \Omega'; \Delta\nu, -\Omega)] \Psi(\mathbf{r}, \Delta\nu', \Omega')$$

$$= \frac{1}{\phi(\nu)} \int_{-\infty}^{\infty} \mathrm{d}(\Delta\nu') \int_{\Omega'_{\mathrm{H}}} \mathrm{d}\Omega' [R_i(\Delta\nu', \Omega'; \Delta\nu, \Omega) \Psi(\mathbf{r}, \Delta\nu', \Omega')$$
$$-R_i(-\Delta\nu', \Omega'; \Delta\nu, -\Omega) \Psi(\mathbf{r}, -\Delta\nu', \Omega')]$$

$$= \frac{1}{\phi(\nu)} \int_{-\infty}^{\infty} \mathrm{d}(\Delta\nu') \int_{\Omega'_{\mathrm{H}}} \mathrm{d}\Omega' [R_i(\Delta\nu', \Omega'; \Delta\nu, \Omega) \Psi(\mathbf{r}, \Delta\nu', \Omega')$$
$$-R_i(-\Delta\nu', -\Omega'; \Delta\nu, \Omega) \Psi(\mathbf{r}, -\Delta\nu', \Omega')]$$

$$= 0, \qquad (7.4.33)$$

for $i = \textsc{i}$ and $\textsc{iii}$, where we have used equations (7.4.13), (7.4.31, 32) and the symmetry property of $\Psi(\mathbf{r}, \nu, \Omega)$ arising from equation (7.4.29).

Equation (7.4.20) then has the somewhat simpler form

$$\left(\Omega \cdot \frac{1}{\kappa}\nabla\right)\left(\Omega \cdot \frac{1}{\kappa}\nabla\right)\Phi(\mathbf{r}, \nu, \Omega)$$

$$= \Phi(\mathbf{r}, \nu, \Omega) - \frac{1-\epsilon}{2\pi}\int_{\nu'_{\mathrm{C}}} \mathrm{d}\nu' \int_{\Omega'_{\mathrm{H}}} \mathrm{d}\Omega'$$

$$\times R_i(\nu', \Omega'; \nu, \Omega)\Phi(\mathbf{r}, \nu', \Omega') - \epsilon B_\nu(T), \tag{7.4.34}$$

for $i = \textsc{i}$ and $\textsc{iii}$ where we have used an analysis similar to that leading to equation (7.4.33) to simplify the integral involving $A_i^+(\nu', \Omega'; \nu, \Omega)$ appearing in equation (7.4.21).

Thus, when using R_{I} and R_{III}, the determination of $\Phi(\mathbf{r}, \nu, \Omega)$ will completely determine the radiation field; $\Psi(\mathbf{r}, \nu, \Omega)$ need not be evaluated.

7.4.2 Inclusion of macroscopic velocities

All the analyses of the preceding sections have assumed the medium interacting with the radiation field to exhibit zero macroscopic velocity $\mathbf{V}$. We now include $\mathbf{V}$ in much the same manner as that presented in section 5.8.3. However, the differences are sufficiently important when one includes a general re-distribution function $R(\nu', \Omega'; \nu, \Omega)$, even for the simplified linear problem, that the development of the Feautrier technique for $\mathbf{V} \neq 0$ warrants attention here.

Clearly, $\mathbf{V}$ may be included in the analysis by replacing $\mathbf{v}$ by $\mathbf{v} + \mathbf{V}$. We then have

$$\phi(\Delta\nu) = \int F(\mathbf{v})q\left(\Delta\nu - \frac{\nu_0}{c}\mathbf{v}\cdot\Omega\right)\mathrm{d}\mathbf{v},$$

being replaced by

$$\int F(\mathbf{v})q\left(\Delta\nu - \frac{\nu_0}{c}(\mathbf{v}+\mathbf{V})\cdot\Omega\right)\mathrm{d}\mathbf{v}$$

– note that the velocity distribution $F(\mathbf{v})$ is a function of $\mathbf{v}$ measured relative to the moving gas (see equation (5.2.28), for example, and the beginning of section 5.4). We then have $\phi(\Delta\nu) \to \phi(\Delta\nu - \nu_0\mathbf{V}\cdot\Omega/c)$. Similarly we have

$$R(\Delta\nu', \Omega'; \Delta\nu, \Omega)$$

$$= g(\Omega', \Omega)\int F(\mathbf{v})q\left(\Delta\nu' - \frac{\nu_0}{c}\mathbf{v}\cdot\Omega'\right)$$

$$\times p\left(\Delta\nu' - \frac{\nu_0}{c}\mathbf{v}\cdot\Omega', \Delta\nu - \frac{\nu_0}{c}\mathbf{v}\cdot\Omega\right)\mathrm{d}\mathbf{v},$$

being replaced by

$$g(\Omega', \Omega) \int F(\mathbf{v}) q\left(\Delta\nu' - \frac{\nu_0}{c}(\mathbf{v}+\mathbf{V})\cdot\Omega'\right)$$
$$\times p\left(\Delta\nu' - \frac{\nu_0}{c}(\mathbf{v}+\mathbf{V})\cdot\Omega', \Delta\nu - \frac{\nu_0}{c}(\mathbf{v}+\mathbf{V})\cdot\Omega\right) \mathrm{d}\mathbf{v},$$

so that

$$R(\Delta\nu', \Omega'; \Delta\nu, \Omega) \to R(\Delta\nu' - \nu_0 \mathbf{V}\cdot\Omega'/c, \Omega'; \Delta\nu - \nu_0\mathbf{V}\cdot\Omega/c, \Omega).$$

We next define

$$I^+ = I(\mathbf{r}, \Delta\nu, \Omega) \quad \text{and} \quad I^- = I(\mathbf{r}, -\Delta\nu, -\Omega), \tag{7.4.35}$$

for all $\nu \in \nu_C$ and $\Omega \in \Omega_H$. The equation of radiative transfer then yields

$$\pm(\Omega\cdot\nabla)I = -\kappa_0\phi\left(\pm\Delta\nu \mp \frac{\nu_0}{c}\mathbf{V}\cdot\Omega\right)(I^\pm - S^\pm), \tag{7.4.36}$$

where $\kappa_0 = h\nu(N_L B_{LU} - N_U B_{UL})/4\pi$ and $S^\pm = S(\mathbf{r}, \pm\Delta\nu, \pm\Omega)$ for all $\Delta\nu \in (-\infty, \infty)$ and $\Omega \in \Omega_H$.

We assume symmetry of all radiative scattering processes about line centre in the rest frame of the atom so that

$$\phi\left(-\Delta\nu + \frac{\nu_0}{c}\mathbf{V}\cdot\Omega\right) = \phi\left(\Delta\nu - \frac{\nu_0}{c}\mathbf{V}\cdot\Omega\right). \tag{7.4.37}$$

Consequently, defining

$$\Phi(\mathbf{r}, \Delta\nu, \Omega) = \tfrac{1}{2}(I^+ + I^-), \tag{7.4.38}$$

and

$$\Psi(\mathbf{r}, \Delta\nu, \Omega) = \tfrac{1}{2}(I^+ - I^-), \tag{7.4.39}$$

for all $\Delta\nu \in (-\infty, \infty)$ and $\Omega \in \Omega_H$, equation (7.4.36) yields

$$\left[\Omega\cdot\frac{1}{\kappa_0\phi\left(\Delta\nu - \frac{\nu_0}{c}\mathbf{V}\cdot\Omega\right)}\nabla\right]\Psi = -\Phi + \frac{S^+ + S^-}{2}, \tag{7.4.40}$$

and

$$\left[\Omega\cdot\frac{1}{\kappa_0\phi\left(\Delta\nu - \frac{\nu_0}{c}\mathbf{V}\cdot\Omega\right)}\nabla\right]\Phi = -\Psi + \frac{S^+ - S^-}{2}. \tag{7.4.41}$$

We now examine the re-distribution function

$$R\left(-\Delta\nu' + \frac{\nu_0}{c}\mathbf{V}\cdot\Omega', -\Omega'; -\Delta\nu + \frac{\nu_0}{c}\mathbf{V}\cdot\Omega, -\Omega\right)$$
$$= g(\Omega', \Omega)\int F(\mathbf{v}) q\left(-\Delta\nu' + \frac{\nu_0}{c}(\mathbf{v}+\mathbf{V})\cdot\Omega'\right) \tag{7.4.42}$$

(7.4.42 Cont.)

$$\times p\left(-\Delta\nu' + \frac{\nu_0}{c}(\mathbf{v}+\mathbf{V})\cdot\boldsymbol{\Omega}', -\Delta\nu + \frac{\nu_0}{c}(\mathbf{v}+\mathbf{V})\cdot\boldsymbol{\Omega}\right)\mathrm{d}\mathbf{v}$$

$$= g(\boldsymbol{\Omega}', \boldsymbol{\Omega})\int F(\mathbf{v})q\left(\Delta\nu' - \frac{\nu_0}{c}(\mathbf{v}+\mathbf{V})\cdot\boldsymbol{\Omega}'\right)$$

$$\times p\left(\Delta\nu' - \frac{\nu_0}{c}(\mathbf{v}+\mathbf{V})\cdot\boldsymbol{\Omega}', \Delta\nu - \frac{\nu_0}{c}(\mathbf{v}+\mathbf{V})\cdot\boldsymbol{\Omega}\right)\mathrm{d}\mathbf{v}$$

$$= R\left(\Delta\nu' - \frac{\nu_0}{c}\mathbf{V}\cdot\boldsymbol{\Omega}', \boldsymbol{\Omega}'; \Delta\nu - \frac{\nu_0}{c}\mathbf{V}\cdot\boldsymbol{\Omega}, \boldsymbol{\Omega}\right), \qquad (7.4.42)$$

where we again assume scattering symmetry in the rest frame of the atom, i.e. $q(-\gamma')p(-\gamma', -\gamma) = q(\gamma')p(\gamma', \gamma)$.

If we now write $\xi = \Delta\nu - \nu_0\mathbf{V}\cdot\boldsymbol{\Omega}/c$, and ignore the usually unimportant $\mathrm{d}S/\mathrm{d}t$ term appearing in equation (5.4.36), then

$$\frac{S^+ + S^-}{2} = \frac{1-\epsilon}{8\pi\phi(\xi)}\int_{-\infty}^{\infty}\mathrm{d}(\Delta\nu')\int_{\Omega'_C}\mathrm{d}\boldsymbol{\Omega}'[R(\xi', \boldsymbol{\Omega}'; \xi, \boldsymbol{\Omega})$$
$$+ R(\xi', \boldsymbol{\Omega}'; -\xi, -\boldsymbol{\Omega})]I(\mathbf{r}, \Delta\nu', \boldsymbol{\Omega}') + \epsilon B_\nu(T)$$
$$= \frac{1-\epsilon}{8\pi\phi(\xi)}\int_{-\infty}^{\infty}\mathrm{d}(\Delta\nu')\int_{\Omega'_H}\mathrm{d}\boldsymbol{\Omega}'\Big\{\{R(\xi', \boldsymbol{\Omega}'; \xi, \boldsymbol{\Omega})$$
$$+ R(\xi', \boldsymbol{\Omega}'; -\xi, -\boldsymbol{\Omega})\}I(\mathbf{r}, \Delta\nu', \boldsymbol{\Omega}') + \{R(-\xi', -\boldsymbol{\Omega}'; \xi, \boldsymbol{\Omega})$$
$$+ R(-\xi', -\boldsymbol{\Omega}'; -\xi, -\boldsymbol{\Omega})\}I(\mathbf{r}, -\Delta\nu', -\boldsymbol{\Omega}')\Big\} + \epsilon B_\nu(T)$$
$$= \frac{1-\epsilon}{8\pi\phi(\xi)}\int_{-\infty}^{\infty}\mathrm{d}(\Delta\nu')\int_{\Omega'_H}\mathrm{d}\boldsymbol{\Omega}'\Big\{\{R(\xi', \boldsymbol{\Omega}'; \xi, \boldsymbol{\Omega})$$
$$+ R(\xi', \boldsymbol{\Omega}'; -\xi, -\boldsymbol{\Omega})\}I^+ + \{R(\xi', \boldsymbol{\Omega}'; -\xi, -\boldsymbol{\Omega})$$
$$+ R(\xi', \boldsymbol{\Omega}'; \xi, \boldsymbol{\Omega})\}I^-\Big\} + \epsilon B_\nu(T)$$
$$= \frac{1-\epsilon}{4\pi}\int_{-\infty}^{\infty}\mathrm{d}(\Delta\nu')\int_{\Omega'_H}\mathrm{d}\boldsymbol{\Omega}' B^+(\xi', \boldsymbol{\Omega}'; \xi, \boldsymbol{\Omega})\Phi(\mathbf{r}, \nu', \boldsymbol{\Omega}')$$
$$+ \epsilon B_\nu(T), \qquad (7.4.43)$$

where we define

$$B^\pm(\xi', \boldsymbol{\Omega}'; \xi, \boldsymbol{\Omega}) = \frac{1}{\phi(\xi)}[R(\xi', \boldsymbol{\Omega}'; \xi, \boldsymbol{\Omega}) \pm R(\xi', \boldsymbol{\Omega}'; -\xi, -\boldsymbol{\Omega})]. \qquad (7.4.44)$$

A similar analysis yields

$$\frac{S^+ - S^-}{2} = \frac{1-\epsilon}{4\pi}\int_{-\infty}^{\infty}\mathrm{d}(\Delta\nu')\int_{\Omega'_H}\mathrm{d}\boldsymbol{\Omega}' B^-(\xi', \boldsymbol{\Omega}'; \xi, \boldsymbol{\Omega})\Psi(\mathbf{r}, \nu', \boldsymbol{\Omega}'). \qquad (7.4.45)$$

Equations (7.4.43, 45) when substituted into (7.4.40, 41) yield expressions identical in structure to equations (7.4.18, 19). Thus, one may proceed to develop the second order Feautrier system of equations in exactly the same manner as that described in the preceding section. In fact, the simple re-definition (7.4.38, 39) of Φ and Ψ, along with the change $A^{\pm} \to B^{\pm}$ and $\Delta\nu \in (0, \infty) \to \Delta\nu \in (-\infty, \infty)$, converts a Feautrier radiative transfer computer program for $V \equiv 0$ to one for which $V \not\equiv 0$. Stability is hardly affected for sufficiently small velocity gradients.

However, all the above difficulties can be readily removed using the operator perturbation procedure outlined in the following section.

7.4.3 *The re-distribution perturbation technique [23]*

The large number of angle and frequency quadrature points required to adequately represent the coherency of wing photon scattering for small scattering angles suggests that the use of the angle and frequency quadrature perturbation techniques discussed in chapter 3 could be quite fruitful in reducing the size of the matrices requiring inversion. The introduction of these AQPT and FQPT into the re-distribution problem is straightforward and is not discussed here. In the present section we describe another perturbation technique, which we can couple to the AQPT and FQPT if need be, which enables a further reduction in the quantity of computing required to solve the pertinent equation of radiative transfer. Indeed, it is this technique which enables the general non-linear transfer equation [21, 93] (as opposed to the approximate linear form considered in the preceding section) to be solved. We first present the method applied to the linear equation, then shows its applicability to the more complicated non-linear problem.

We begin by defining the integral operator

$$\mathcal{L}_{\nu\Omega} \equiv \frac{1}{4\pi\phi(\nu)} \int_{\nu'_C} d\nu' \int_{\Omega'_C} d\Omega' R(\nu', \Omega'; \nu, \Omega), \tag{7.4.46}$$

so that the equation of radiative transfer has the form

$$(\Omega \cdot \nabla) I = -\kappa [I - (1-\epsilon)\mathcal{L}_{\nu\Omega} I - \epsilon B_\nu(T)]. \tag{7.4.47}$$

We thence define a simpler functional $\mathcal{L}^*_{\nu\Omega}$ where

$$\mathcal{L}^*_{\nu\Omega} = \frac{1}{4\pi\phi(\nu)} \int_{\nu'_C} d\nu' \int_{\Omega'_C} d\Omega' R^*(\nu', \Omega'; \nu, \Omega). \tag{7.4.48}$$

In this last equation we take the 'approximate' re-distribution function $R^*(\nu', \Omega'; \nu, \Omega)$ to be that specifying complete re-distribution, i.e.

$$R^*(\nu', \Omega'; \nu, \Omega) = \phi(\nu)\phi(\nu'). \tag{7.4.49}$$

Equation (7.4.48) then yields

$$\mathcal{L}^*_{\nu\Omega} \equiv \frac{1}{4\pi}\int_{\nu'_C} d\nu' \int_{\Omega'_C} d\Omega' \phi(\nu') \equiv \mathcal{L}^*, \tag{7.4.50}$$

where the result of the operator $\mathcal{L}^*$ produces a quantity independent of the emitted angle Ω and frequency ν.

We now construct a perturbation series in $I(\mathbf{r}, \nu, \Omega)$ of the form

$$I(\mathbf{r}, \nu, \Omega) = I^{(0)}(\mathbf{r}, \nu, \Omega) + \lambda I^{(1)}(\mathbf{r}, \nu, \Omega) + \cdots. \tag{7.4.51}$$

where λ specifies the order of the perturbation, and write the radiative transfer equation (7.4.47) in the form

$$(\Omega \cdot \nabla)I = -\kappa[I - (1-\epsilon)\mathcal{L}^*I - \epsilon B_\nu(T) - (1-\epsilon)(\mathcal{L}_{\nu\Omega} - \mathcal{L}^*)I]. \tag{7.4.52}$$

If we now consider the operator $(\mathcal{L}_{\nu\Omega} - \mathcal{L}^*)$ to be of order λ, then we find

$$(\Omega \cdot \nabla)I_0^{(0)} = -\kappa[I^{(0)} - (1-\epsilon)\mathcal{L}^*I^{(1)} - \epsilon B_\nu(T)], \tag{7.4.53}$$

and

$$(\Omega \cdot \nabla)I^{(k)} = -\kappa[I^{(k)} - (1-\epsilon)\mathcal{L}^*I^{(k)} - E^{(k)}(\mathbf{r}, \nu, \Omega)]. \tag{7.4.54}$$

for $k = 1, \infty$ where

$$E^{(k)}(\mathbf{r}, \nu, \Omega) = (1-\epsilon)(\mathcal{L}_{\nu\Omega} - \mathcal{L}^*)I^{(k-1)}. \tag{7.4.55}$$

We immediately notice that the linearity of $I(\mathbf{r}, \nu, \Omega)$ in the system of equations (7.4.53, 54) stipulates that the series, if it converges, converges to the correct solution. Indeed, one can discuss convergence of the above series in much the same manner as that used to examine the FQPT in chapter 3, although the situation is somewhat more complicated for the RPT. Numerical experiments indicate that convergence is *always* attained for problems of practical interest even when the $\mathcal{L}_{\nu\Omega}I^{(k)}$ term is *not* small relative to $\mathcal{L}^*I^{(k)}$.

Equation (7.4.53) is, of course, the standard equation of radiative transfer for which complete re-distribution applies, and may therefore be solved quite readily by a variety of numerical methods discussed in chapter 3. Equation (7.4.54) has precisely the same structure as equation (7.4.53) but a different 'error' term $E^{(k)}(\mathbf{r}, \nu, \Omega)$ involving the radiation field for the preceding term in the series. This inhomogeneous term is essentially an angle and frequency-dependent source or sink (depending on its sign) of photons at each point in the atmosphere and correspondingly yields an instant 'correction' to the preceding computed radiation field obtained using complete re-distribution.

The important point to note, however, is that one may use exactly the same process to solve equation (7.4.54) as that used for the solution of equation (7.4.53). In fact, since it is only the inhomogeneous term which differs for each of the k equations (7.4.54), one may use precisely the same numerical operations for both equations (7.4.53) and (7.4.54). For the Feautrier technique, for example, one would compute the required matrices and their corresponding inverses for equation (7.4.53), and then store these inverses for the solution of equation (7.4.54). The matrix inverses need only be computed once therefore and, since it is these which overwhelmingly constitute the bulk of the computing time required for the solution, the extra time required to solve equation (7.4.54) is minimal indeed.

One further advantage of the above RPT is its ability to use a smaller number of angle and frequency quadrature points than would otherwise be necessary if the full re-distribution function $R(\nu', \boldsymbol{\Omega}'; \nu, \boldsymbol{\Omega})$ were to be used as in the preceding section. This general $R(\nu', \boldsymbol{\Omega}'; \nu, \boldsymbol{\Omega})$ now only appears in the inhomogeneous source - sink term $E^{(k)}(\mathbf{r}, \nu, \boldsymbol{\Omega})$ and therefore offers fewer computational problems. Indeed, stability of the RPT is the same as that pertaining to complete re-distribution and, as we have seen, this poses no difficulties. This is a particularly important advantage when considering problems for which $\mathbf{V} \not\equiv 0$.

We complete this section by finally noting that the above perturbation method can be equivalently written as an iterative scheme using

$$I_l = \sum_{k=0}^{l} I^{(k)}. \tag{7.4.56}$$

Here the lth iterative value of $I(\mathbf{r}, \nu, \boldsymbol{\Omega})$ satisfies

$$(\boldsymbol{\Omega} \cdot \nabla) I_l = -\kappa [I_l - (1-\epsilon)\mathcal{L}^* I_l - \mathcal{E}_l], \tag{7.4.57}$$

where

$$\mathcal{E}_0 = \epsilon B_\nu(T),$$

and

$$\mathcal{E}_l = (1-\epsilon)(\mathcal{L}_{\nu\Omega} - \mathcal{L}^*) I_{l-1}, \quad \forall\, l \geqslant 1. \tag{7.4.58}$$

Clearly, the first iteration above yields the complete re-distribution solution. All the following iterations produce corrections due to departures from complete re-distribution. Thus, one can readily solve the non-linear problem, for which

$$S(\mathbf{r}, \nu, \boldsymbol{\Omega}) = \frac{N_U A_{UL}\, j(\nu, \boldsymbol{\Omega})}{N_L B_{LU} \phi(\nu) - N_U B_{UL} \psi(\nu, \boldsymbol{\Omega})} \tag{7.4.59}$$

and

$$j(\nu, \boldsymbol{\Omega}) = \int F_U(\mathbf{v}_U) \cdot \frac{B_{LU} \iint R_{LU}(\nu', \boldsymbol{\Omega}'; \nu, \boldsymbol{\Omega}; \mathbf{v}_L, \mathbf{v}_U) I(\mathbf{r}, \nu', \boldsymbol{\Omega}')\, d\nu'\, d\Omega' + C_{LU} q(\nu, \boldsymbol{\Omega}; \mathbf{v}_U)}{\dfrac{B_{LU}}{4\pi} \iint q(\nu', \boldsymbol{\Omega}'; \mathbf{v}_L) I(\mathbf{r}, \nu', \boldsymbol{\Omega}')\, d\nu'\, d\Omega' + C_{LU}} \cdot d\mathbf{v}_U, \tag{7.4.60}$$

with $\psi(\nu, \Omega) \equiv j(\nu, \Omega)$, using the above type of iterative scheme starting with the zeroth order complete re-distribution solution of the linear equation with subsequent iterations yielding corrections due to departures from both complete re-distribution *and* linearity. It is left to the reader to detail the expansion in λ of the above emission functional $j(\nu, \Omega)$ using the series (7.4.52).

7.5 Qualitative transfer effects due to partial re-distribution

We have already seen that the assumption of complete re-distribution is generally quite satisfactory in the core of the spectral line. Our main concern here is the influence departures from complete re-distribution have on the shape of the emergent intensity profiles in the wings of the line.

Complete re-distribution stipulated that photons were emitted quite independent of the frequencies at which they were absorbed. Thus, they were preferentially emitted at and near line centre. Photons absorbed in the wings of the line had little chance of being re-emitted in those wings. Departures from complete re-distribution, however, allow photons absorbed in the wings to be preferentially emitted there – note the exponential terms in equations (7.3.13) and (7.3.18), for example. This is a markedly different situation from complete re-distribution and leads to some rather interesting but relatively straightforward physical manifestations. Of course, this wing coherency depends upon the angle of scatter between the absorbed and emitted photon and this should be carefully taken into account when computing the solution of the pertinent equation of radiative transfer. Here, however, the frequency dependence of the overall scattering process is of prime importance in our qualitative discussion since the specific intensity can change by a factor $1/\sqrt{\epsilon}$ (typically $\epsilon \sim 10^{-4}$ or smaller for 'non-LTE spectral lines' of astrophysical interest) over the frequency range of the spectral line whereas the corresponding angular variation is no more than a factor of about two (except at the extreme 'surface', of course). Consequently, we shall restrict our discussion to the frequency dependence of the scattering process.

We denote by $S_0(\mathbf{r})$ the frequency- (and angle-) independent source function obtained assuming complete re-distribution. As we have already mentioned, departures from complete re-distribution now allow photons absorbed in the wings to be preferentially emitted there instead of the line core and this frequency channelling of the radiation field leads to an increase in photon escape since $\phi(\text{wings}) \ll \phi(\text{core})$. The source function is now frequency dependent so that this increased escape of wing photons yields $S(\text{wings}) < S_0$ at identical depths. This, in turn, leads to a decrease in the emergent radiation intensity observed in these wings.

The above effect is perhaps best illustrated by considering the chromospheric-type spectral line discussed in some detail in section 1.10.1 using complete re-distribution, and shown again in figure 7.7.

In particular, we have $|\Delta\nu_1| < |\Delta\nu_2| < |\Delta\nu_3| < |\Delta\nu_4|$, so that $\phi(\Delta\nu_1) > \phi(\Delta\nu_2) > \phi(\Delta\nu_3) > \phi(\Delta\nu_4)$, with $|z_1^*| < |z_2^*| < |z_3^*| < |z_4^*|$, where z_i^* corresponds to the physical depth at which the majority of the contribution to $I(0, \Delta\nu_i, 1)$ from $S_0(z)$ arises (see figure 1.10).

If we now allow departures from complete re-distribution [108, 122, 123] the source function must now be plotted for several values of frequency as shown in figure 7.8(*a*) where $S_i \equiv S(r, \Delta\nu_i, 1)$.

The net effect is a reduction of the wing emergent intensity shown as the solid curve in figure 7.8(*b*). Indeed, it is not difficult to see that it is quite possible to construct a model atmosphere which exhibits central emission peaks in the emergent line intensity when assuming complete re-distribution but no such peaks when the more physically realistic departures from complete re-distribution are incorporated into the analysis. This effect is also realised when the opacity term $h\nu N_L B_{LU}/4\pi$ is sufficiently small that the thermalisation depth Θ occurs below rather than above the temperature minimum. In this

Fig.7.7. (*a*) The source function as a function of depth in a chromospheric-type atmosphere assuming complete re-distribution; (*b*) The corresponding emergent intensity as a function of frequency from line centre.

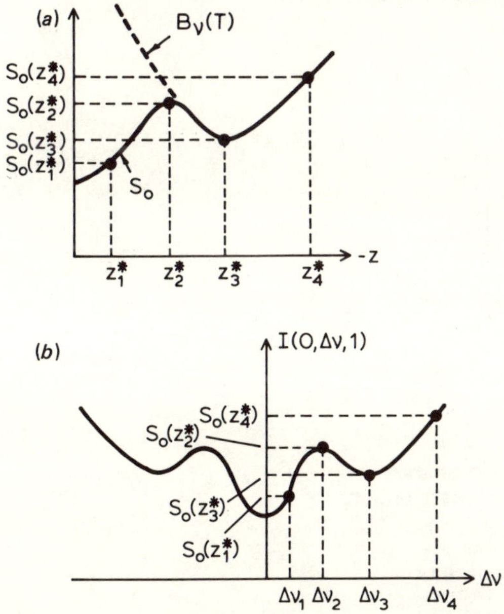

case, the rise in the source function above the temperature minimum shown in figure 7.7(*a*) would not occur (recall that $S_0 \approx B_\nu(T)$ for all depths $|z| > \Theta$) and neither would the central emission features. This has already been discussed in some detail in chapter 1.

An important side-effect now arises. The value of $I(0, \Delta\nu_3, 1)$ in figure 7.7(*b*) for complete re-distribution is predominantly determined by the value of the source function, thence Planck function, in the region z_3^* of the temperature minimum. Thus, as we observe at angles away from the normal, we still see central emission features in the emergent line profile with the secondary minima still controlled by the region at and near the temperature minimum.

Here, however, we must observe through a greater distance $|z_3^*|/\cos\theta_0$, as shown in figure 7.9, to actually 'see' this temperature minimum and we can only do this if we observe at frequencies $|\Delta\nu_3(\theta_0)|$, say, for which $\phi(\Delta\nu_3(\theta_0)) < \phi(\Delta\nu_3)$, i.e. $\Delta\nu_3(\theta_0) > \Delta\nu_3$. Under the assumption of complete re-distribution, therefore, the emergent spectral line intensity will have the centre-to-limb dependence illustrated in figure 7.10.

Fig.7.8. (*a*) The source function as a function of depth in a chromospheric-type atmosphere with departures from complete re-distribution for four different frequencies of emission; (*b*) The full curve represents the emergent intensity corresponding to the partial re-distribution source function shown in (*a*). The dashed curve enables a qualitative comparison between the emergent intensities for partial and complete re-distribution.

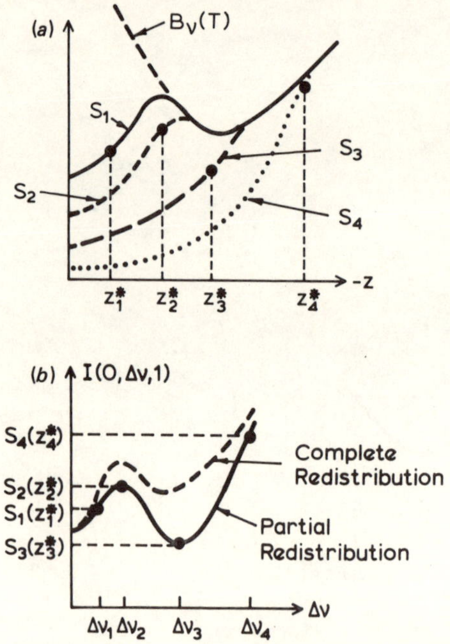

The interesting feature here is not the movement of the secondary minima from $|\Delta\nu_3|$ to $|\Delta\nu_3(\theta_0)|$ but the fact that $I(0, \Delta\nu_3, 1) \approx I(0, \Delta\nu_3(\theta_0), \cos\theta_0)$, i.e. the secondary minima do not exhibit limb-darkening.

We now turn our attention to the centre-to-limb behaviour of the emergent line profile for the same atmosphere, but with departures from complete re-distribution included. Again, we ignore the quantitatively important dependence of $S(\mathbf{r}, \Delta\nu, \mathbf{\Omega})$ on $\mathbf{\Omega}$. Thus, if we are to see to depths z_3^* at which the temperature minimum occurs, then we must again observe at frequencies $|\Delta\nu_3(\theta_0)|$ where $\phi(\Delta\nu_3(\theta_0)) \approx \phi(\Delta\nu_3)\cos\theta_0$. However, $S(r, \Delta\nu_3(\theta_0)) < S(r, \Delta\nu_3)$ for $|\Delta\nu_3(\theta_0)| > |\Delta\nu_3|$ because of the increased escape of wing photons due to increased wing coherency. We therefore have limb-darkening in the secondary minima of the emergent radiation intensity as illustrated in figure 7.11.

This effect had been observed for some time before the influence of departures from complete re-distribution was examined. Indeed, the secondary minima in $I(0, \Delta\nu, \mathbf{\Omega})$ had been used to model the region of the temperature minimum in stellar atmospheres using complete re-distribution. The above discussion therefore emphasises the care that must be exercised in such diagnostic analyses.

Fig.7.9. Radiation propagating at non-zero heliocentric angle emanates from regions higher in the atmosphere than does normally emergent radiation.

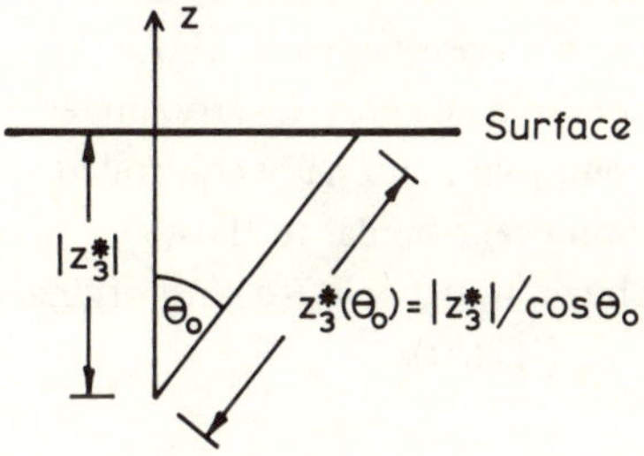

Fig.7.10. A qualitative comparison between the intensities at two different angles of emergence assuming complete re-distribution. Note that the minima located in the wings are approximately equal.

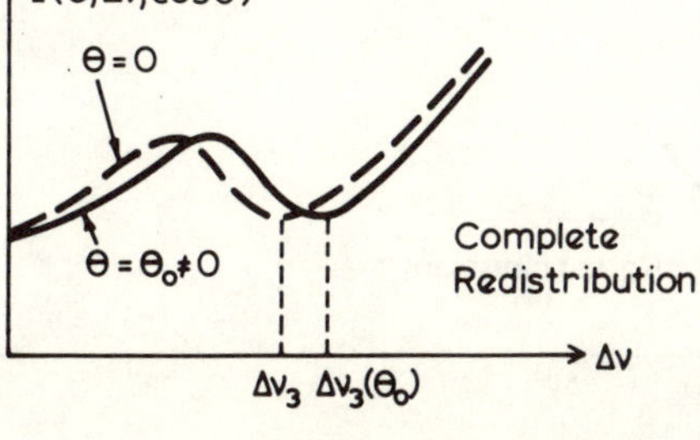

We stress this last point by considering the following example. The source function for a model 3-level atom has the form

$$S = \frac{\bar{\mathscr{J}} + \epsilon B_\nu(T) + \eta B^*}{1 + \epsilon + \eta},$$

where $\bar{\mathscr{J}}$ is a functional of the radiation field averaged over the frequencies and angles of absorption (ν', Ω'), and ηB^* and η involve the radiation field in the second transition. The reader is referred to chapter 4 for algebraic details. We have not discussed continuum radiation here, but if we consider the third level in this model 3-level atom to represent the state of the atom in which the outermost bound electron is no longer bound, then this third level effectively constitutes the continuum. A photo-ionisationally controlled line is one in which $|\eta B^*| \gg \epsilon B_\nu(T)$ such that the source function, thence emergent intensity, in that line is determined primarily by the continuum radiation. Specifically, photons in this case are created and destroyed more by processes involving the continuum (photo-ionisation and three-body re-combination) than by inelastic collisional events. A collisionally controlled line, on the other hand, has $\epsilon B_\nu(T) \gg |\eta B^*|$ so that collisional excitation and de-excitation processes are the dominant means of photon creation and destruction.

We now examine partial re-distribution effects for a photo-ionisationally controlled spectral line [121]. In particular, we usually have $|\eta B^*|$ simply decreasing with increasing height in the model atmosphere, since the continuum radiation field will be controlled more by the atmosphere at relatively large depths, whereas $\epsilon B_\nu(T)$ will perhaps increase with increasing height above the temperature minimum region for model stellar chromospheres. Clearly, under the assumption of complete re-distribution, a photo-ionisationally controlled spectral line will not exhibit any central emission peaks similar to those obtained for the collisionally controlled line shown in figure 7.7(*b*). An ordinary absorption line will appear (shown as I_0 in figure 7.12(*b*)).

Fig.7.11. A qualitative comparison between the intensities at two different angles of emergence assuming partial re-distribution. Note that the minima now exhibit limb-darkening.

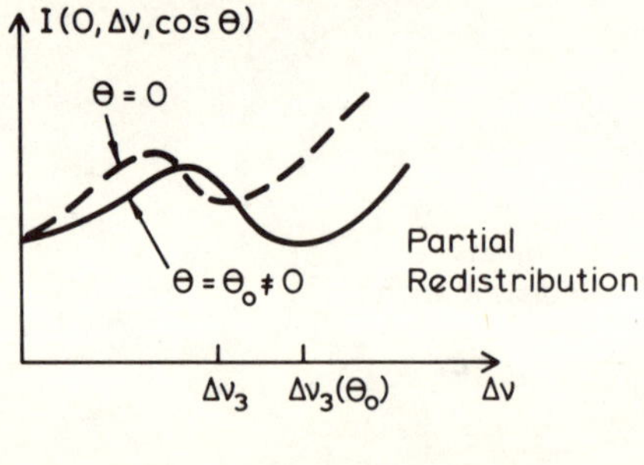

However, the frequency dependence of the line source function arising from the inclusion of departures from complete re-distribution can generate the somewhat different result illustrated in figure 7.12(*b*).

The emergent intensity will first increase away from line centre in the core of the line because of the increase of the core source function with increasing depth. However, the source function at any particular depth will decrease with increasing frequency $|\Delta\nu|$ outside this core due to the increased coherency in the line wings. Thus, the emergent intensity will correspondingly decrease thereby producing the secondary minima at $\Delta\nu_3$ as shown in figure 7.12(*b*). Consequently, one can also obtain central emission features even for atmospheres which do not exhibit a chromospheric-type temperature increase. This, of course, will depend upon the structure of the model atom and the relative magnitudes of the $\epsilon B_\nu(T)$, ηB^*, ϵ and η terms.

We finally consider an apparently anomalous effect which couples the above re-distribution effects with those pertaining to a non-zero macroscopic velocity field discussed in section 5.7.

Fig.7.12. (*a*) The partial re-distribution source function at four different frequencies of photon emission for a photo-ionisationally controlled spectral line; (*b*) A qualitative comparison between the emergent intensities for a photo-ionisationally controlled spectral line obtained using complete re-distribution and partial re-distribution.

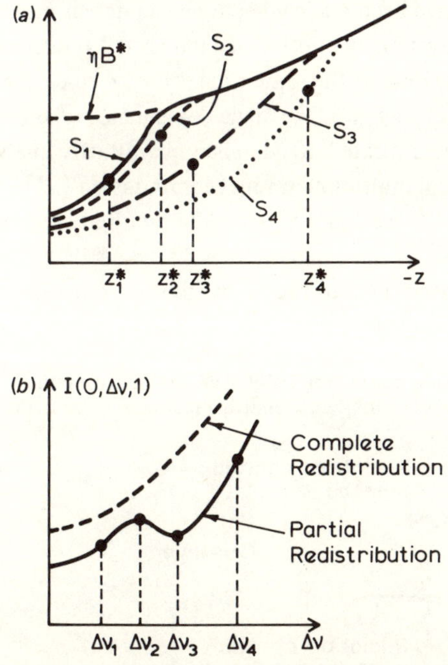

Consider a volume of gas (for example, a spicule) moving radially upwards. If it has an optical thickness of order unity, a photon should undergo only a few scatters at most before escaping from the structure. If one then considers a collisionally controlled line, photons that escape from that structure are those that most probably have been created by collisional excitation which, of course, yields photons predominantly emitted about line centre in the rest frame of the gas. If one views this structure at the limb as shown in figure 7.13 such that there is no velocity component of the structure in the observer's line of sight, one would observe an emission line centred about line centre.

If, however, the line is photo-ionisationally controlled, or the geometry is such that the photons one observes emerging from the structure are not a result of direct creation in the structure, but are a result of emission from the lower atmosphere followed by absorption thence emission in the structure (scattering through angles of order $\pi/2$), differences will occur. Since the structure is moving radially away from the lower atmosphere, it will absorb more in the blue wings (as discussed in some detail in section 5.7), and less in the red, than if the structure was stationary. Partial coherency (even for scattering through $\pi/2$ – see figure 7.4 for R_{II}) would then stipulate that these photons absorbed in the blue be preferentially emitted in the blue, leading to a blue-shifted emission profile at the limb as seen by the observer. If one were to assume complete re-distribution, one would incorrectly surmise, from such an observation, a non-zero velocity component of that structure along the observer's line of sight. Observations of this effect have been made where one can detect directly the relative shift of H_α (photo-ionisationally controlled) and Ca II I (collisionally controlled) at the same point at the limb at the same time. Clearly, since the photons of interest are scattering through $\pi/2$, the angle dependence of the re-distribution function must be incorporated in a detailed quantitative analysis of such observations. Indeed, the full multi-dimensional computation [21], including all those effects discussed in section 5.8, would be required here not only to ascertain the frequency shift of the emission lines discussed above, but to examine the effect the geometry coupled to the re-distribution processes have on the shape of the emergent intensity.

Fig.7.13. A sketch of a structure moving vertically, being irradiated by the underlying stationary atmosphere, and thence emitting radiation tangentially in the direction of the observer.

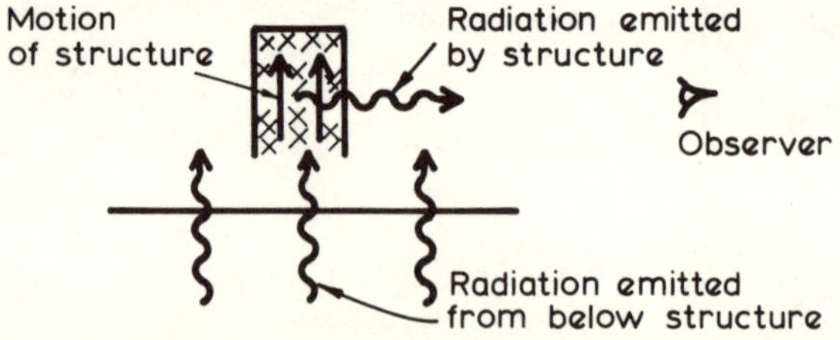

8

A quantum electrodynamical radiative transfer equation

We derived the equation of radiative transfer somewhat heuristically in chapters 1 and 5. We argued that one should expect the existence of a term representing the absorption of radiation with two more, detailing corresponding spontaneous and stimulated emission. We then obtained the appropriate probabilities of absorption and emission in chapter 7 using the quantum mechanical considerations of chapter 6. Indeed, we showed the natural appearance of spontaneous emission using second quantisation, and this lent some validity to the approach leading to the classical equation of radiative transfer. However, such a mixture of classical ideas regarding absorption and emission, and quantum wave mechanics to derive the particulars of these processes, is rather *ad hoc.* A more self-consistent approach is desirable not only because of the need to more accurately specify the functional dependence of the absorption and emission profiles, but as we shall see, to detail those higher order interference terms which do not appear in the classical derivation of the radiative transfer equation.

In the present chapter, we introduce the density matrix, its properties and its defining differential equation. This then enables the desired quantum electrodynamical equation of radiative transfer, together with the corresponding population rate equations, to be obtained.

8.1 The pure state density matrix

We begin by recalling the approach of chapter 6. The time-dependent Schrödinger equation in the wave function $|\psi(\mathbf{r}, t)\rangle$ is

$$i\hbar \frac{\partial}{\partial t} |\psi\rangle = \hat{H} |\psi\rangle, \tag{8.1.1}$$

where the Hamiltonian operator $\hat{H}$ consists of a time-independent (unperturbed) part $\hat{H}_0$ and the time-dependent perturbing component $\hat{H}_1$, viz.

$$\hat{H} = \hat{H}_0 + \hat{H}_1. \tag{8.1.2}$$

Thus, knowing the solution $|n\rangle$ to the reduced problem

$$i\hbar \frac{\partial}{\partial t}|n\rangle = \hat{H}_0|n\rangle = E_n|n\rangle, \tag{8.1.3}$$

where $\langle n|n\rangle = 1$ and E_n is the total energy of the system in the state $|n\rangle$, enables the required quantity $|\psi\rangle$ to be written as

$$|\psi\rangle = \sum_n a_n|n\rangle, \tag{8.1.4}$$

where the expansion coefficients $a_n(t)$ are time-dependent. Indeed, with

$$\langle\psi|\psi\rangle = \sum_n |a_n|^2 = 1, \tag{8.1.5}$$

we showed that $|a_n|^2$ may be simply interpreted as the probability of finding the system in state $|n\rangle$ at time t.

Section 6.4.1 detailed a general procedure which enables the $a_n(t)$ to be determined. An alternative approach using the density matrix [38, 115] is also available. We illustrate this method by considering the 2-level atom with states $|2\rangle$ and $|1\rangle$ representing the upper and lower states respectively, i.e.

$$|\psi\rangle = a_1|1\rangle + a_2|2\rangle. \tag{8.1.6}$$

We then define the pure state density matrix ρ as the outer product

$$\begin{aligned}\rho = |\psi\rangle\langle\psi| &= \begin{pmatrix} a_1 \\ a_2 \end{pmatrix} (a_1^* a_2^*) \\ &= \begin{pmatrix} |a_1|^2 & a_1 a_2^* \\ a_1^* a_2 & |a_2|^2 \end{pmatrix},\end{aligned} \tag{8.1.7}$$

so that the diagonal elements ρ_{11} and ρ_{22} can be *directly* related to the probabilities of the system being in state $|1\rangle$ and state $|2\rangle$.

An equation for ρ can then be obtained by noting

$$\begin{aligned} i\hbar \frac{\partial \rho}{\partial t} &= i\hbar\left(\frac{\partial}{\partial t}|\psi\rangle\right)\langle\psi| + i\hbar|\psi\rangle\frac{\partial}{\partial t}\langle\psi| \\ &= \hat{H}|\psi\rangle\langle\psi| - |\psi\rangle\langle\psi|\hat{H} \\ &= \hat{H}\rho - \rho\hat{H} \\ &= [\hat{H}, \rho], \end{aligned} \tag{8.1.8}$$

where the square parentheses indicate the commutator defined by equation (6.1.17). If we use equations (8.1.2) through (8.1.4), and define the mn matrix element of ρ by $\rho_{mn} = \langle m|\rho|n\rangle$, then we find

$$i\hbar \frac{\partial \rho_{mn}}{\partial t} = [\hat{H}_1, \rho]_{mn} + [\hat{H}_0, \rho]_{mn}, \tag{8.1.9}$$

where

$$
\begin{aligned}
[\hat{H}_0, \rho]_{mn} &= \langle m|\hat{H}_0|\psi\rangle\langle\psi|n\rangle - \langle m|\psi\rangle\langle\psi|\hat{H}_0|n\rangle \\
&= \langle m|\sum_{n'} a_{n'}E_{n'}|n'\rangle \sum_{n''} a_{n''}^* \langle n''|n\rangle \\
&\quad - \langle m|\sum_{n'} a_{n'}|n'\rangle \sum_{n''} a_{n''}^* E_{n''}\langle n''|n\rangle \\
&= (E_m - E_n)a_m a_n^* \\
&= (E_m - E_n)\rho_{mn}. \qquad (8.1.10)
\end{aligned}
$$

Clearly, then, the probabilities ρ_{nn} can be obtained from the simpler equation

$$
i\hbar \frac{\partial \rho_{nn}}{\partial t} = [\hat{H}_1, \rho]_{nn}. \qquad (8.1.11)
$$

Equation (8.1.8) can be further simplified using the interaction representation discussed in section 6.3.2. Here we define

$$
\rho^{\mathrm{I}} = \mathrm{e}^{\mathrm{i}\hat{H}_0 t/\hbar}\rho\,\mathrm{e}^{-\mathrm{i}\hat{H}_0 t/\hbar}, \qquad (8.1.12)
$$

so that

$$
\begin{aligned}
i\hbar \frac{\partial \rho^{\mathrm{I}}}{\partial t} &= -\mathrm{e}^{\mathrm{i}\hat{H}_0 t/\hbar}\hat{H}_0\rho\,\mathrm{e}^{-\mathrm{i}\hat{H}_0 t/\hbar} \\
&\quad + \mathrm{e}^{\mathrm{i}\hat{H}_0 t/\hbar}[(\hat{H}_0 + \hat{H}_1)\rho - \rho(\hat{H}_0 + \hat{H}_1)]\mathrm{e}^{-\mathrm{i}\hat{H}_0 t/\hbar} \\
&\quad + \mathrm{e}^{\mathrm{i}\hat{H}_0 t/\hbar}\rho\hat{H}_0\mathrm{e}^{-\mathrm{i}\hat{H}_0 t/\hbar} \\
&= \hat{H}_1^{\mathrm{I}}\rho^{\mathrm{I}} - \rho^{\mathrm{I}}\hat{H}_1^{\mathrm{I}} \\
&= [\hat{H}_1^{\mathrm{I}}, \rho^{\mathrm{I}}], \qquad (8.1.13)
\end{aligned}
$$

where

$$
\hat{H}_1^{\mathrm{I}} = \mathrm{e}^{\mathrm{i}\hat{H}_0 t/\hbar}\hat{H}_1\mathrm{e}^{-\mathrm{i}\hat{H}_0 t/\hbar}, \qquad (8.1.14)
$$

as in equation (6.3.23).

Equation (8.1.13) clearly has the exact solution

$$
\rho^{\mathrm{I}}(t) = \rho^{\mathrm{I}}(0) + \frac{1}{i\hbar}\int_0^t \mathrm{d}t'[\hat{H}_1^{\mathrm{I}}(t'), \rho^{\mathrm{I}}(t')]. \qquad (8.1.15)
$$

However, this last result is of little direct practical value for other than the most trivial problems. We usually prefer to solve equation (8.1.13) iteratively in much the same manner as that presented in section 6.3.1 by starting with

$$
\rho_0^{\mathrm{I}}(t) = \rho^{\mathrm{I}}(0) \quad \text{for all } t, \qquad (8.1.16)
$$

and using equation (8.1.15) written as

$$
\rho_l^{\mathrm{I}}(t) = \rho^{\mathrm{I}}(0) + \frac{1}{i\hbar}\int_0^t \mathrm{d}t'[\hat{H}^{\mathrm{I}}(t'), \rho_{l-1}^{\mathrm{I}}(t')], \qquad (8.1.17)
$$

where the l subscript refers to the lth iterative step. We then have

$$\rho_1^{\mathrm{I}}(t) = \rho^{\mathrm{I}}(0) + \frac{1}{i\hbar}\int_0^t \mathrm{d}t'[\hat{H}_1^{\mathrm{I}}(t'), \rho^{\mathrm{I}}(0)], \tag{8.1.18}$$

$$\rho_2^{\mathrm{I}}(t) = \rho^{\mathrm{I}}(0) + \frac{1}{i\hbar}\int_0^t \mathrm{d}t'[\hat{H}_2^{\mathrm{I}}(t'), \rho^{\mathrm{I}}(0)],$$

$$+ \frac{1}{(i\hbar)^2}\int_0^t \mathrm{d}t' \int_0^{t'} \mathrm{d}t''[\hat{H}_1^{\mathrm{I}}(t'), [\hat{H}_1^{\mathrm{I}}(t''), \rho^{\mathrm{I}}(0)]], \tag{8.1.19}$$

etc. We shall use the above series later in this chapter. For the moment, we complete this section by returning to equation (8.1.8) in order to illustrate the use of the density matrix in several rather simple examples [31, 91, 105].

8.1.1 Transition rate for photon emission

Here we consider the same 2-level atom as that examined in section 6.4.1 and shown in figure 8.1.

We again start with the atom in the excited state so that $|a_2(0)|^2 = 1 = \rho_{22}(0)$ with $\rho_{11}(0) = 0$ (see equations (6.4.29) and (6.4.30)). We then must solve equation (8.1.9), viz.

$$i\hbar \frac{\partial \rho_{mn}}{\partial t} = (E_m - E_n)\rho_{mn} + [\hat{H}_1, \rho]_{mn}, \tag{8.1.20}$$

where, from equation (6.4.22), we have

$$\hat{H}_1 = \sum_{\gamma\lambda} (\hat{a}_{\gamma\lambda}\xi_{\gamma\lambda} + \hat{a}_{\gamma\lambda}^{+}\xi_{\gamma\lambda}^{*}), \tag{8.1.21}$$

with

$$\xi_{\gamma\lambda} = -\frac{e}{2mc}(\hat{\mathbf{p}} \cdot \mathbf{u}_{\gamma\lambda} + \mathbf{u}_{\gamma\lambda} \cdot \hat{\mathbf{p}}). \tag{8.1.22}$$

In particular, we have

$$\left.\begin{aligned} \hat{a}_{\gamma\lambda}|1, n_{\gamma\lambda}+1\rangle &= \sqrt{(n_{\gamma\lambda}+1)}|2, n_{\gamma\lambda}\rangle, \\ \hat{a}_{\gamma\lambda}^{+}|2, n_{\gamma\lambda}\rangle &= \sqrt{(n_{\gamma\lambda}+1)}|1, n_{\gamma\lambda}+1\rangle. \end{aligned}\right\} \tag{8.1.23}$$

Fig.8.1. The quantum representation of the upper and lower bound electron states with photon emission.

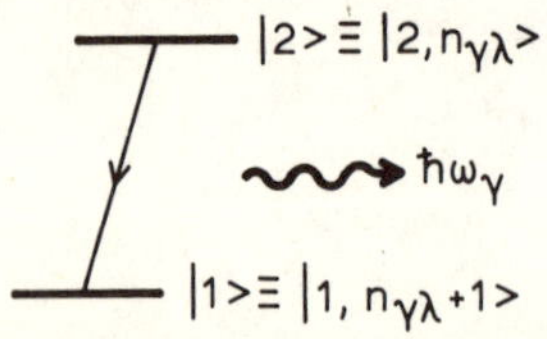

Thus, the only non-zero matrix elements $\langle m|\hat{H}_1|n\rangle$ of $\hat{H}_1$ are

$$\begin{aligned} H_{21} &= \langle 2|\hat{H}_1|1\rangle = Q_{21}\sqrt{(n_{\gamma\lambda}+1)}, \\ H_{12} &= \langle 1|\hat{H}_1|2\rangle = Q_{21}^*\sqrt{(n_{\gamma\lambda}+1)} \end{aligned} \qquad (8.2.24)$$

- see equations (6.4.23) and (6.4.25) - where $Q_{21} = \langle 2|\xi_{\gamma\lambda}|1\rangle$. Consequently, equation (8.1.20) may be written in the form

$$i\hbar \frac{\partial}{\partial t}\begin{pmatrix} \rho_{11} & \rho_{12} \\ \rho_{21} & \rho_{22} \end{pmatrix} = \begin{pmatrix} 0 & (E_1-E_2)\rho_{12} \\ (E_2-E_1)\rho_{21} & 0 \end{pmatrix} + \begin{pmatrix} 0 & H_{12} \\ H_{21} & 0 \end{pmatrix}\begin{pmatrix} \rho_{11} & \rho_{12} \\ \rho_{21} & \rho_{22} \end{pmatrix} - \begin{pmatrix} \rho_{11} & \rho_{12} \\ \rho_{21} & \rho_{22} \end{pmatrix}\begin{pmatrix} 0 & H_{12} \\ H_{21} & 0 \end{pmatrix}. \qquad (8.1.25)$$

In this last equation, we have E_1 and E_2 satisfying

$$\hat{H}_0|n\rangle = E_n|n\rangle = (E_n^{(A)} + E_n^{(R)})|n\rangle, \qquad (8.1.26)$$

where

$$E_n^{(R)} = \sum_{\gamma\lambda} \hbar\omega_\gamma(n_{\gamma\lambda} + \tfrac{1}{2}),$$

and

$$E_2^{(R)} - E_1^{(A)} = \hbar\omega_{21}. \qquad (8.1.27)$$

We therefore have

$$E_1 - E_2 = E_1^{(R)} - E_2^{(R)} + E_1^{(A)} - E_2^{(A)} = \hbar(\omega_\gamma - \omega_{21}), \qquad (8.1.28)$$

so that equation (8.1.25) yields the four equations

$$i\hbar \frac{\partial\rho_{11}}{\partial t} = H_{12}\rho_{21} - H_{21}\rho_{12}, \qquad (8.1.29)$$

$$i\hbar \frac{\partial\rho_{22}}{\partial t} = H_{21}\rho_{12} - H_{12}\rho_{21}, \qquad (8.1.30)$$

$$i\hbar \frac{\partial\rho_{12}}{\partial t} = \hbar(\omega_\gamma - \omega_{21})\rho_{12} + H_{12}\rho_{22} - H_{12}\rho_{11}, \qquad (8.1.31)$$

$$i\hbar \frac{\partial\rho_{21}}{\partial t} = -\hbar(\omega_\gamma - \omega_{21})\rho_{21} + H_{21}\rho_{11} - H_{21}\rho_{22}. \qquad (8.1.32)$$

We solve this system of equations by replacing ρ_{11} and ρ_{22} appearing on the RHS of equations (8.1.31) and (8.1.32) by their initial values (as in section 6.4.1 and suggested by equation (8.1.16)) so that

$$i\hbar\rho_{12} = \frac{H_{12}}{i(\omega_\gamma - \omega_{21})}\,[1 - e^{-i(\omega_\gamma - \omega_{21})t}], \qquad (8.1.33)$$

$$i\hbar\rho_{21} = \frac{H_{21}}{i(\omega_\gamma - \omega_{21})}\,[1 - e^{i(\omega_\gamma - \omega_{21})t}], \qquad (8.1.34)$$

where we have put $\rho_{12}(0) = a_1(0)a_2^*(0) = 0$ and $\rho_{21}(0) = a_1^*(0)a_2(0) = 0$. Note that $\rho_{21} = \rho_{12}^*$ as expected.

Equation (8.1.29) then becomes

$$i\hbar\frac{\partial\rho_{11}}{\partial t}=\frac{|H_{21}|^2}{i\hbar}\left[-\frac{e^{i(\omega_\gamma-\omega_{21})t}}{i(\omega_\gamma-\omega_{21})}+\frac{e^{-i(\omega_\gamma-\omega_{21})t}}{i(\omega_\gamma-\omega_{21})}\right],$$

i.e.

$$\rho_{11}(t)=\frac{|H_{21}|^2}{\hbar^2(\omega_\gamma-\omega_{21})^2}\left[2-e^{i(\omega_\gamma-\omega_{21})t}-e^{-i(\omega_\gamma-\omega_{21})t}\right]$$

$$\equiv\frac{2|Q_{21}|^2(n_{\gamma\lambda}+1)}{\hbar^2(\omega_\gamma-\omega_{21})^2}\left[1-\cos\left((\omega_\gamma-\omega_{21})t\right)\right], \tag{8.1.35}$$

identical to equation (6.4.32). One then follows the analysis detailed in section 6.4.1.

It is clear from all the foregoing that we are basically working in the more physically meaningful dependent variable $\rho_{nn}=|a_n|^2$ as opposed to $a_n(t)$ and, as we shall see, this can lead to a significant simplification of the problem involving the derivation of the equation of radiative transfer. The same procedure can, of course, be applied to the evaluation of the transition rate for photon absorption - see section 6.4.2.

8.1.2 A Laplace transform description of radiative decay

As a further example [94] of the density matrix approach we again consider an atom in the excited $|2\rangle$ state but allow a de-excitation to occur to any of the $|\alpha\rangle$ sub-states of the broadened $|1\rangle$ level as shown in figure 8.2.

We take the atom to be initially in the excited $|\beta\rangle$ sub-state of $|2\rangle$ so that $\rho_{\beta\beta}(0)=1$ with all other $\rho_{mn}(0)=0$. We further consider the situation in which the atom, once de-excited, does not re-excite, i.e. $\rho_{\beta\beta}(\infty)=0$ for all $|\beta\rangle$ in $|2\rangle$.

Thus, noting that the mnth element of the matrix product AB is

$$\{AB\}_{mn}=\sum_k A_{mk}B_{kn}, \tag{8.1.36}$$

equations (8.1.9) and (8.1.10) then yield

$$i\hbar\frac{\partial\rho_{\alpha\alpha}}{\partial t}=H_{\alpha\beta}\rho_{\beta\alpha}-\rho_{\alpha\beta}H_{\beta\alpha}, \tag{8.1.37}$$

Fig.8.2. The quantum representation of the upper and lower bound electron states with both levels being naturally broadened.

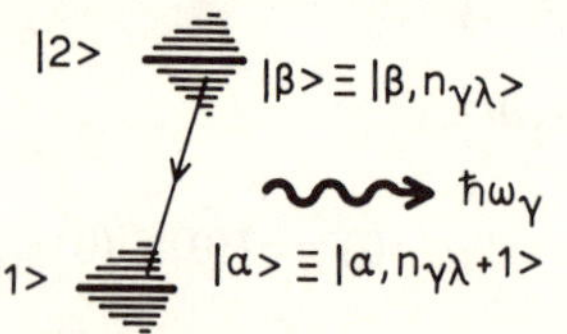

$$i\hbar\,\frac{\partial\rho_{\beta\beta}}{\partial t}=\sum_{\alpha}\,(H_{\beta\alpha}\rho_{\alpha\beta}-\rho_{\beta\alpha}H_{\alpha\beta}), \tag{8.1.38}$$

$$i\hbar\,\frac{\partial\rho_{\alpha\beta}}{\partial t}=\hbar(\omega_{\gamma}-\omega_{\beta\alpha})\rho_{\alpha\beta}+H_{\alpha\beta}\rho_{\beta\beta}-\sum_{\alpha'}\rho_{\alpha\alpha'}H_{\alpha'\beta}, \tag{8.1.39}$$

$$i\hbar\,\frac{\partial\rho_{\beta\alpha}}{\partial t}=-\hbar(\omega_{\gamma}-\omega_{\beta\alpha})\rho_{\beta\alpha}+\sum_{\alpha'}H_{\beta\alpha'}\rho_{\alpha'\alpha}-\rho_{\beta\beta}H_{\beta\alpha}, \tag{8.1.40}$$

where, as in the preceding section,

$$\left.\begin{aligned}H_{\alpha\beta}&=\langle\alpha|\hat{H}_1|\beta\rangle=Q^*_{\beta\alpha}\sqrt{(n_{\gamma\lambda}+1)},\\ H_{\beta\alpha}&=\langle\beta|\hat{H}_1|\alpha\rangle=Q_{\beta\alpha}\sqrt{(n_{\gamma\lambda}+1)}.\end{aligned}\right\} \tag{8.1.41}$$

Note that we do not include a summation over β in equation (8.1.37) corresponding to the summation over α appearing in equation (8.1.38) since the atom initially exists only in the $|\beta\rangle$ sub-state of $|2\rangle$. The summation over α must be included in the equation for $\rho_{\beta\beta}$ because the atom can de-excite to any of the $|\alpha\rangle$ sub-states of $|1\rangle$. Note further that the matrix elements $H_{\alpha\alpha'}=0$ for $|\alpha'\rangle$ in $|1\rangle$.

We proceed by again replacing $\rho_{\alpha\alpha'}(t)$ and $\rho_{\alpha'\alpha}(t)$ on the RHS of equations (8.1.39) and (8.1.40) by their corresponding initial values. Consequently, these two equations become

$$i\hbar\,\frac{\partial\rho_{\alpha\beta}}{\partial t}=\hbar(\omega_{\gamma}-\omega_{\beta\alpha})\rho_{\alpha\beta}+H_{\alpha\beta}\rho_{\beta\beta}, \tag{8.1.42}$$

$$i\hbar\,\frac{\partial\rho_{\beta\alpha}}{\partial t}=-\hbar(\omega_{\gamma}-\omega_{\beta\alpha})\rho_{\beta\alpha}-H_{\beta\alpha}\rho_{\beta\beta}. \tag{8.1.43}$$

If we now define the Laplace transform by

$$\mathcal{L}[\rho_{mn}(t)]\equiv\bar{\rho}_{mn}(s)=\int^{\infty}\mathrm{e}^{-st}\rho_{mn}(t)\,\mathrm{d}t, \tag{8.1.44}$$

with

$$\mathcal{L}\left[\frac{\partial\rho_{mn}}{\partial t}\right]=s\bar{\rho}_{mn}(s)-\rho_{mn}(0), \tag{8.1.45}$$

equations (8.1.37-38) and (8.1.42-43) then yield

$$i\hbar s\bar{\rho}_{\alpha\alpha}=H_{\alpha\beta}\bar{\rho}_{\beta\alpha}-H_{\beta\alpha}\bar{\rho}_{\alpha\beta}, \tag{8.1.46}$$

$$i\hbar s\bar{\rho}_{\beta\beta}=i\hbar+\sum_{\alpha}\,(H_{\beta\alpha}\bar{\rho}_{\alpha\beta}-H_{\alpha\beta}\bar{\rho}_{\beta\alpha}), \tag{8.1.47}$$

$$i\hbar s\bar{\rho}_{\alpha\beta}=\hbar(\omega_{\gamma}-\omega_{\beta\alpha})\bar{\rho}_{\alpha\beta}+H_{\alpha\beta}\bar{\rho}_{\beta\beta}, \tag{8.1.48}$$

$$i\hbar s\bar{\rho}_{\beta\alpha}=-\hbar(\omega_{\gamma}-\omega_{\beta\alpha})\bar{\rho}_{\beta\alpha}-H_{\beta\alpha}\bar{\rho}_{\beta\beta}, \tag{8.1.49}$$

where we have used $\rho_{\beta\beta}(0)=1$. These last two algebraic equations can easily be solved for $\bar{\rho}_{\alpha\beta}$ and $\bar{\rho}_{\beta\alpha}$ in terms of $\bar{\rho}_{\beta\beta}$, viz.

$$\bar{\rho}_{\alpha\beta} = \frac{H_{\alpha\beta}\bar{\rho}_{\beta\beta}}{i\hbar[s + i(\omega_\gamma - \omega_{\beta\alpha})]}, \quad \bar{\rho}_{\beta\alpha} = \frac{-H_{\beta\alpha}\bar{\rho}_{\beta\beta}}{i\hbar[s - i(\omega_\gamma - \omega_{\beta\alpha})]}. \tag{8.1.50}$$

These results can then be substituted into equation (8.1.47) to yield

$$i\hbar s\bar{\rho}_{\beta\beta} = i\hbar + \frac{\rho_{\beta\beta}}{i\hbar}\sum_\alpha |H_{\beta\alpha}|^2 \frac{2s}{s^2 + (\omega_\gamma - \omega_{\beta\alpha})^2},$$

i.e.

$$\bar{\rho}_{\beta\beta} = \frac{1}{s + \dfrac{2s}{\hbar^2}\displaystyle\sum_\alpha \frac{|H_{\beta\alpha}|^2}{s^2 + (\omega_\gamma - \omega_{\beta\alpha})^2}}. \tag{8.1.51}$$

The inverse Laplace transform is

$$\mathcal{L}^{-1}[\bar{\rho}_{mn}(t)] \equiv \rho_{mn}(t) = \frac{1}{2\pi i}\int_{\epsilon - i\infty}^{\epsilon + i\infty} \bar{\rho}_{mn}(s)e^{st}\,ds, \tag{8.1.52}$$

where $\epsilon > 0$ is to be considered sufficiently large that all poles of $\bar{\rho}_{mn}(s)$ are to the left of the line joining $\epsilon - i\infty$ to $\epsilon + i\infty$. Consequently, equation (8.1.52) yields

$$\rho_{\beta\beta}(t) = \frac{1}{2\pi i}\int_{\epsilon - i\infty}^{\epsilon + i\infty} \frac{e^{st}\,ds}{s + \dfrac{2s}{\hbar^2}\displaystyle\sum_\alpha \frac{|H_{\beta\alpha}|^2}{s^2 + (\omega_\gamma - \omega_{\beta\alpha})^2}}. \tag{8.1.53}$$

We now proceed in the same manner as that detailed in sections 6.4.1 and 6.4.3 by replacing the above summation over α by a suitably weighted integral over the energy distribution of the emitted photons averaged over emitted angle and summed over polarisation index λ. We then have

$$\frac{1}{\hbar^2}\sum_\alpha \frac{|H_{\beta\alpha}|^2}{s^2 + (\omega_\gamma - \omega_{\beta\alpha})^2} \to \frac{1}{4\pi}\sum_\lambda \int d\Omega \int_0^\infty \frac{\rho(E)|H_{\beta\alpha}|^2\,dE}{(E - E_{\beta\alpha}^{(A)})^2 + \hbar^2 s^2}. \tag{8.1.54}$$

The integral over E in this last expression may be evaluated by extending the range from $[0, \infty)$ to $(-\infty, \infty)$ taking $\rho(E) = 0$ for all $E < 0$, then constructing a contour incorporating the upper half complex plane in which there is a pole at $E = E_{\beta\alpha}^{(A)} + i\hbar s$. Cauchy's integral theorem then yields

$$\begin{aligned}
\frac{1}{\hbar^2}\sum_\alpha \frac{|H_{\beta\alpha}|^2}{s^2 + (\omega_\gamma - \omega_{\beta\alpha})^2} &\to \frac{1}{4\pi}\sum_\lambda \int d\Omega\,\rho(E_{\beta\alpha}^{(A)} + i\hbar s)|H_{\beta\alpha}|^2 \frac{2\pi i}{2i\hbar s} \\
&\cong \frac{1}{4hs}\sum_\lambda \int d\Omega\,\rho(E_{21}^{(A)})|H_{21}|^2 \\
&= \frac{\Gamma}{2s},
\end{aligned} \tag{8.1.55}$$

where we have replaced $\rho(E_{\beta\alpha}^{(A)} + i\hbar s)$ by its resonance value $\rho(E_{21}^{(A)})$ - we will discuss this approximation in a moment - and used equations (6.4.37) and (6.4.69) defining Γ, viz.

$$\Gamma = \frac{1}{2\hbar}\sum_{\lambda} \int d\Omega \, |H_{21}|^2 \rho(E_{21}^{(A)}). \tag{8.1.56}$$

We then have

$$\rho_{\beta\beta}(t) = \frac{1}{2\pi i}\int_{\epsilon - i\infty}^{\epsilon + i\infty} \frac{e^{st}\, ds}{s + \Gamma} = e^{-\Gamma t}, \tag{8.1.57}$$

where we have used the contour $C_1 \cup C_2$ illustrated in figure 8.3. Note that the contribution along the contour C_2 is zero since here $s = \epsilon + Re^{i\theta}$, $\theta \in [\pi/2, 3\pi/2]$ so that $e^{st} = e^{\epsilon t} \exp(Rt\cos\theta + iRt\sin\theta) \to 0$ as $R \to \infty$.

Clearly, in this last equation, we have evaluated the residue at the pole $s = -\Gamma$ where, for the non-degenerate case in which $|\beta\rangle$ sub-states of $|2\rangle$ do not overlap with any of the $|\alpha\rangle$ sub-states of $|1\rangle$, we have $\hbar\Gamma \ll E_{21}^{(A)}$. Thus, we have $|E_{\beta\alpha}^{(A)} + i\hbar s| = |E_{\beta\alpha}^{(A)} - i\hbar\Gamma| \approx E_{21}^{(A)}$ used above.

The result (8.1.57) is identical to that discussed in section 6.4.3 in which the excited state simply decays exponentially in time to the de-excited $|1\rangle$ state. It is interesting to evaluate the corresponding probability $\rho_{\alpha\alpha}$ of the individual de-excited $|\alpha\rangle$ sub-states. Equations (8.1.46) and (8.1.50) immediately yield

$$\bar{\rho}_{\alpha\alpha} = \frac{2|H_{\alpha\beta}|^2 \bar{\rho}_{\beta\beta}}{\hbar^2[s^2 + (\omega_\gamma - \omega_{\beta\alpha})^2]} = \frac{2|H_{\alpha\beta}|^2}{\hbar^2(s+\Gamma)(s^2+\xi^2)}, \tag{8.1.58}$$

where, for convenience, we have written $\xi = \omega_\gamma - \omega_{\beta\alpha}$. Consequently,

$$\bar{\rho}_{\alpha\alpha}(t) = \frac{|H_{\alpha\beta}|^2}{\pi i \hbar^2}\int_{\epsilon - i\infty}^{\epsilon + i\infty} \frac{e^{st}\, ds}{(s+\Gamma)(s^2+\xi^2)}. \tag{8.1.59}$$

Fig.8.3. The contour in the complex s plane used to determine the inverse Laplace transform.

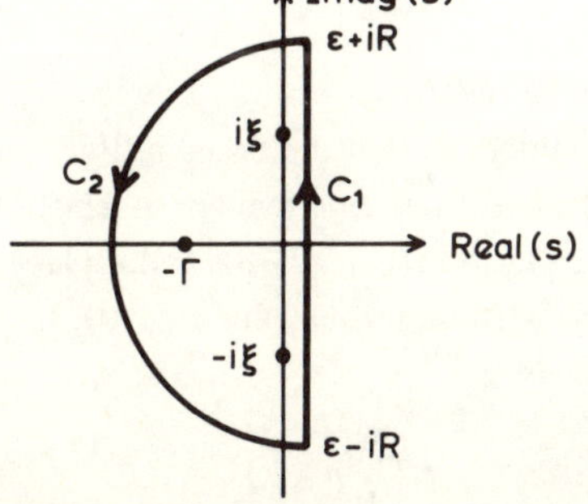

If we now consider the contour $C = C_1 \cup C_2$ as shown in figure 8.3, equation (8.1.59) becomes

$$\bar{\rho}_{\alpha\alpha}(t) = \frac{2|H_{\alpha\beta}|^2}{\hbar^2}\left\{\frac{e^{-\Gamma t}}{\xi^2+\Gamma^2} + \frac{e^{i\xi t}}{2i\xi(\Gamma+i\xi)} - \frac{e^{-i\xi t}}{2i\xi(\Gamma-i\xi)}\right\}. \quad (8.1.60)$$

Thus, recalling the limit

$$\lim_{t\to\infty} \frac{1-e^{ixt}}{x} = \mathscr{P}\frac{1}{x} - \pi i\delta(x) \quad (8.1.61)$$

(see equation (6.4.41), for example) and noting that $\rho_{\alpha\alpha}(0) = 0$, we have

$$\begin{aligned}\rho_{\alpha\alpha}(\infty) &= \lim_{t\to\infty}[\rho_{\alpha\alpha}(t) - \rho_{\alpha\alpha}(0)]\\ &= \frac{2|H_{\alpha\beta}|^2}{\hbar^2}\left\{-\frac{1}{\xi^2+\Gamma^2} - \lim_{t\to\infty}\left[-\frac{1}{2i(\Gamma+i\xi)}\cdot\frac{1-e^{i\xi t}}{\xi}\right.\right.\\ &\quad\left.\left.+\frac{1}{2i(\Gamma-i\xi)}\cdot\frac{1-e^{-i\xi t}}{\xi}\right]\right\}\\ &= \frac{2\pi|H_{\alpha\beta}|^2}{\hbar^2}\cdot\frac{\Gamma}{(\omega_\gamma-\omega_{\beta\alpha})^2+\Gamma^2}\cdot\delta(\omega_\gamma-\omega_{\beta\alpha}). \quad (8.1.62)\end{aligned}$$

Here we see that we do *not* obtain the Lorentzian profile but, instead, the coherent form $\delta(\omega_\gamma - \omega_{\beta\alpha})$. Of course, we have taken as an initial approximation $\rho_{\alpha\alpha}(t) = \rho_{\alpha\alpha}(0) = 0$ whereas, in deriving the Lorentzian given by equation (6.4.47), we made no such approximation. Indeed, the analysis of section 6.4.3 was effectively 'rigorous' regarding the functional form of $a_\alpha(t)$ once the stipulation $a_\beta(t) = e^{-\eta t}$ (see equation (6.4.49)) had been made. This therefore suggests a deficiency in the density matrix approach. However, more complicated problems, for which the simplified analyses of sections 6.4.1, 6.4.2 and 6.4.3 are not particularly appropriate, are more readily examined using the density matrix. Indeed, series solutions of the form (8.1.19) provide a systematic means of attacking such problems without recourse to separate individual techniques oriented toward individual situations. We shall be more interested in these series solutions in section 8.3 – however, we do examine the first and second order solutions of the above equations (8.1.37–40) in section 8.1.4.

8.1.3 An alternative approach to radiative decay

Here we consider the same problem as that discussed in the preceding sub-section, but detail a somewhat different mathematical approach to its resolution [107]. We again start with equations (8.1.37–40) and replace $\rho_{\alpha\alpha'}(t)$ on the RHS of these last two equations with its initial value $\rho_{\alpha\alpha'}(0)$. In particular, equation (8.1.42) can be written as

$$i\hbar\frac{\partial}{\partial t}[\rho_{\alpha\beta}e^{i(\omega_\gamma-\omega_{\beta\alpha})t}] = H_{\alpha\beta}e^{i(\omega_\gamma-\omega_{\beta\alpha})t}\rho_{\beta\beta}(t),$$

i.e.

$$\rho_{\alpha\beta}(t)=\frac{H_{\alpha\beta}}{i\hbar}\int_0^t e^{i(\omega_\gamma-\omega_{\beta\alpha})(t'-t)}\rho_{\beta\beta}(t')\,dt', \tag{8.1.63}$$

where we have used $\rho_{\alpha\beta}(0)=0$. Equation (8.1.38) is just

$$i\hbar\,\frac{\partial\rho_{\beta\beta}}{\partial t}=\sum_\alpha(H_{\beta\alpha}\rho_{\alpha\beta}-H_{\alpha\beta}\rho_{\beta\alpha}), \tag{8.1.64}$$

so that, using equation (8.1.63) and its complex conjugate, we have

$$\frac{\partial\rho_{\beta\beta}}{\partial t}=\frac{1}{(i\hbar)^2}\sum_\alpha|H_{\alpha\beta}|^2\int_0^t[e^{i(\omega_\gamma-\omega_{\beta\alpha})(t'-t)} + e^{-i(\omega_\gamma-\omega_{\beta\alpha})(t'-t)}]\rho_{\beta\beta}(t')\,dt'. \tag{8.1.65}$$

Again, we replace the summation over α by the suitably weighted average involving emitted photon angle Ω, polarisation index λ, and photon energy distribution. We thence have

$$\frac{\partial\rho_{\beta\beta}}{\partial t}=\frac{1}{4\pi(i\hbar)^2}\sum_\lambda\int d\Omega\int_0^\infty|H_{\alpha\beta}|^2\rho(E)\,dE \times\int_0^t[e^{i(E-\hbar\omega_{\beta\alpha})(t'-t)/\hbar}+e^{-i(E-\hbar\omega_{\beta\alpha})(t'-t)/\hbar}]\rho_{\beta\beta}(t')\,dt'. \tag{8.1.66}$$

We now approximate $\rho(E)$ by its resonance value $\rho(E_{21}^{(A)})$ and note from equation (6.4.86) that

$$\int_{-\infty}^{\infty}e^{\pm ix(t'-t)}\,dx=2\pi\delta(t'-t). \tag{8.1.67}$$

Consequently, extending the integral from $[0,\infty)$ to $(-\infty,\infty)$ in the usual manner, and with $\rho(E)=\rho(E_{21}^{(A)})$ now appearing outside the integral, we find

$$\begin{aligned}\frac{\partial\rho_{\beta\beta}}{\partial t}&=\frac{1}{4\pi(i\hbar)^2}\sum_\lambda\int d\Omega\,|H_{21}|^2\rho(E_{21}^{(A)})\int_0^t 2\pi\hbar\delta(t'-t)\rho_{\beta\beta}(t')\,dt'\\&=-\frac{1}{2\hbar}\sum_\lambda\int d\Omega\,|H_{21}|^2\rho(E_{21}^{(A)})\rho_{\beta\beta}(t)\\&=-\Gamma\rho_{\beta\beta}(t),\end{aligned} \tag{8.1.68}$$

so that

$$\rho_{\beta\beta}(t)=e^{-\Gamma t}, \tag{8.1.69}$$

as before. The approach detailed here and that given in section 8.1.2 are mathematically identical - recall that we also replaced $\rho(E_{\beta\alpha}^{(A)}+i\hbar s)$ by its resonance value $\rho(E_{21}^{(A)})$. Consequently, the subsequent $\rho_{\alpha\alpha}(t)$ will be the same as that determined by the second order result (8.1.60). Of course, higher order solutions may be readily obtained by substituting such $\rho_{\alpha\alpha}(t)$ and $\rho_{\beta\beta}(t)$, above, back into equations (8.1.39) and (8.1.40). This will yield third order expressions for $\rho_{\alpha\beta}(t)$ and its complex conjugate and these then enable the next order $\rho_{\alpha\alpha}(t)$ and $\rho_{\beta\beta}(t)$ to be determined from equations (8.1.37) and (8.1.38). The above ordering of the solutions for $\rho_{mn}(t)$ is discussed in the following section.

8.1.4 Ordering of solutions

The analyses of the two preceding sub-sections 8.1.2 and 8.1.3 relied rather heavily on the nature of the problem being considered there. Consequently, such 'exact' approaches to the solution of the Schrödinger wave equation in density matrix form are not readily applicable to more general problems. In the present section, we examine an iterative procedure (similar to that given by equation (8.1.17) in the interaction representation) where, in particular, the solutions for both $\rho_{\alpha\alpha}(t)$ *and* $\rho_{\beta\beta}(t)$, for example, are obtained accurate to the same order of approximation.

We demonstrate the ordering of these solutions by considering the same problem as that discussed in section 8.1.2, where we again write the equations to be solved;

$$i\hbar \frac{\partial \rho_{\alpha\alpha}}{\partial t} = H_{\alpha\beta}\rho_{\beta\alpha} - H_{\beta\alpha}\rho_{\alpha\beta}, \tag{8.1.70}$$

$$i\hbar \frac{\partial \rho_{\beta\beta}}{\partial t} = \sum_{\alpha} (H_{\beta\alpha}\rho_{\alpha\beta} - H_{\alpha\beta}\rho_{\beta\alpha}), \tag{8.1.71}$$

$$i\hbar \frac{\partial \rho_{\alpha\beta}}{\partial t} = \hbar\xi\rho_{\alpha\beta} + H_{\alpha\beta}\rho_{\beta\beta} - \sum_{\alpha'} \rho_{\alpha\alpha'}H_{\alpha'\beta}, \tag{8.1.72}$$

where, for convenience, we have put $\xi = \omega_\gamma - \omega_{\beta\alpha}$.

We thence put $\rho_{\alpha\alpha}(t) = \rho_{\alpha\alpha}(0) = 0$ and $\rho_{\beta\beta}(t) = \rho_{\beta\beta}(0) = 1$ for all t. [Note that we did *not* make this second stipulation regarding $\rho_{\beta\beta}(t)$ in sections 8.1.2 and 8.1.3.] Equation (8.1.72) then becomes

$$i\hbar \frac{\partial}{\partial t} (\rho_{\alpha\beta} e^{i\xi t}) = H_{\alpha\beta} e^{i\xi t},$$

i.e.

$$\rho_{\alpha\beta}(t) = -\frac{H_{\alpha\beta}}{\hbar\xi} (1 - e^{-i\xi t}). \tag{8.1.73}$$

This is a first order result for $\rho_{\alpha\beta}(t)$ – the corresponding zeroth order 'solution' is simply $\rho_{\alpha\beta}(t) = \rho_{\alpha\beta}(0) = 0$ for all t. The corresponding first order result for $\rho_{\beta\beta}(t)$ may be obtained by substituting equation (8.1.73) and its complex conjugate into equation (8.1.71). We then find

$$i\hbar\,\frac{\partial\rho_{\beta\beta}}{\partial t}=\sum_{\alpha}\frac{|H_{\alpha\beta}|^2}{\hbar\xi}\,(e^{-i\xi t}-e^{i\xi t})$$

$$\rightarrow\frac{1}{4\pi}\sum_{\lambda}\int d\Omega\int_{-\infty}^{\infty}|H_{\alpha\beta}|^2\,\frac{\rho(E)}{E-E_{\beta\alpha}^{(A)}}$$

$$\times\,[e^{-i(E-E_{\beta\alpha}^{(A)})t/\hbar}-e^{i(E-E_{\beta\alpha}^{(A)})t/\hbar}]\,dE, \qquad (8.1.74)$$

where, again, we put $\rho(E)=0$ for all $E<0$. This last integral may be evaluated by considering the two contours shown in figures 8.4(*a*) and 8.4(*b*).

Cauchy's integral theorem applied to the contour $C_1\cup C_2\cup C_3$ in figure 8.4(*a*) yields

$$\int_{C_1\cup C_2\cup C_3}\frac{|H_{\alpha\beta}|^2\rho(E)e^{i(E-E_{\beta\alpha}^{(A)})t/\hbar}}{E-E_{\beta\alpha}^{(A)}}\,dE=0. \qquad (8.1.75)$$

However, on C_2, $E-E_{\beta\alpha}^{(A)}=Re^{i\theta}$ for all $\theta\in[0,\pi]$ so that $\exp\,[i(E-E_{\beta\alpha}^{(A)})t/\hbar]=\exp\,[(iR\cos\theta-R\sin\theta)t/\hbar]\rightarrow 0$ as $R\rightarrow\infty$. Therefore, the integral over C_2 when $R\rightarrow\infty$ is zero in equation (8.1.75). On C_3 we have $E-E_{\beta\alpha}^{(A)}=\delta e^{i\theta}$ for all

Fig.8.4. (*a*) The contour in the upper-half complex E plane indented at $E_{\beta\alpha}^{(A)}$. (*b*) The corresponding contour in the lower-half plane.

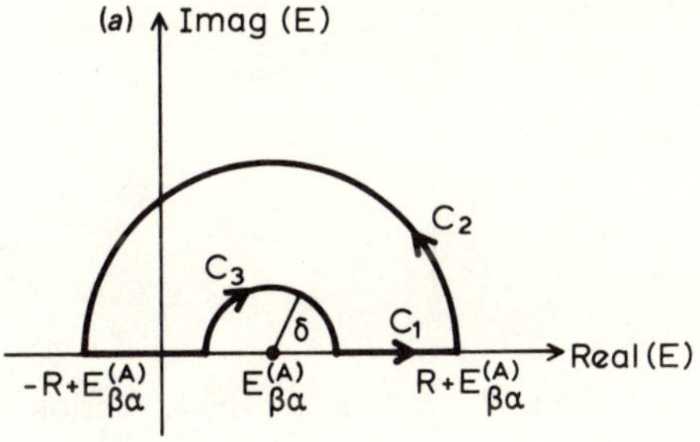

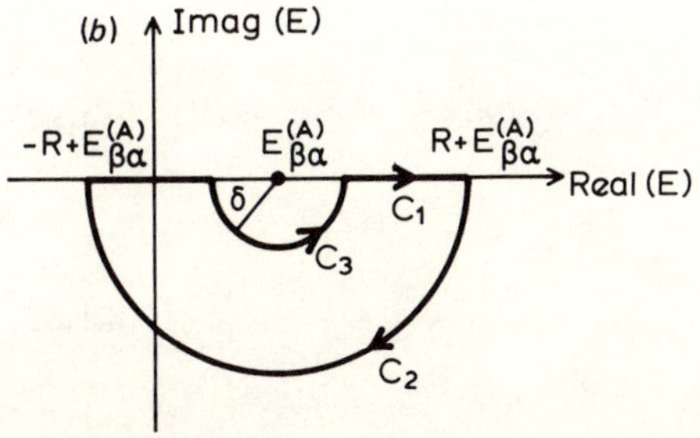

$\theta \in [\pi, 0]$ so that

$$\lim_{\delta \to 0} \sum_{C_3} \frac{|H_{\alpha\beta}|^2 \rho(E) e^{i(E - E_{\beta\alpha}^{(A)})t/\hbar}}{E - E_{\beta\alpha}^{(A)}} dE$$

$$= \lim_{\delta \to 0} \int_{\pi}^{0} \frac{|H_{\alpha\beta}|^2 \rho(E_{\beta\alpha}^{(A)} + \delta e^{i\theta}) e^{i\delta e^{i\theta} t/\hbar}}{\delta e^{i\theta}} i\delta e^{i\theta} \, d\theta$$

$$= -\pi i |H_{\alpha\beta}|^2 \rho(E_{\beta\alpha}^{(A)}). \tag{8.1.76}$$

Thus, taking the limit $R \to \infty$ and $\delta \to 0$ in equation (8.1.75) yields

$$\int_{-\infty}^{\infty} \frac{|H_{\alpha\beta}|^2 \rho(E) e^{i(E - E_{\beta\alpha}^{(A)})t/\hbar}}{E - E_{\beta\alpha}^{(A)}} dE = \pi i |H_{\alpha\beta}|^2 \rho(E_{\beta\alpha}^{(A)}). \tag{8.1.77}$$

Similarly, taking the contour $C_1 \cup C_2 \cup C_3$ illustrated in figure 8.4(*b*), we easily find

$$\int_{-\infty}^{\infty} \frac{|H_{\alpha\beta}|^2 \rho(E) e^{-i(E - E_{\beta\alpha}^{(A)})t/\hbar}}{E - E_{\beta\alpha}^{(A)}} dE = -\pi i |H_{\alpha\beta}|^2 \rho(E_{\beta\alpha}^{(A)}). \tag{8.1.78}$$

Consequently, replacing $\rho(E_{\beta\alpha}^{(A)})$ by its resonance value $\rho(E_{21}^{(A)})$, the above results yield

$$i\hbar \frac{\partial \rho_{\beta\beta}}{\partial t} = -\frac{i}{2} \sum_{\lambda} \int d\Omega |H_{21}|^2 \rho(E_{21}^{(A)}),$$

i.e.

$$\frac{\partial \rho_{\beta\beta}}{\partial t} = -\Gamma, \tag{8.1.79}$$

so that

$$\rho_{\beta\beta}(t) = 1 - \Gamma t. \tag{8.1.80}$$

This first order result for $\rho_{\beta\beta}(t)$ is simply the first two terms of the expansion $e^{-\Gamma t}$ obtained in sections 8.1.2 and 8.1.3. Clearly, this solution is invalid for $t > 1/\Gamma$. The corresponding first order result for $\rho_{\alpha\alpha}(t)$ may be similarly obtained by the substitution of $\rho_{\alpha\beta}(t)$ given by equation (8.1.73) into equation (8.1.70), i.e.

$$\rho_{\alpha\alpha}(t) = \frac{|H_{\alpha\beta}|^2}{\hbar^2(\omega_\gamma - \omega_{\beta\alpha})^2} [2 - e^{i(\omega_\gamma - \omega_{\beta\alpha})t} - e^{-i(\omega_\gamma - \omega_{\beta\alpha})t}], \tag{8.1.82}$$

so that

$$\rho_{\alpha\alpha}(\infty) = \frac{2|H_{\alpha\beta}|^2}{\hbar^2(\omega_\gamma - \omega_{\beta\alpha})^2}, \tag{8.1.82}$$

where we have used equation (8.1.61).

Further, it is not difficult to show that the second order result for $\rho_{\beta\beta}(t)$, obtained by first substituting the above first order $\rho_{\alpha\alpha}(t)$ and $\rho_{\beta\beta}(t)$ into equation (8.1.72) with the resulting second order $\rho_{\alpha\beta}(t)$ thence substituted into equation (8.1.71), contains the first three terms $1 - \Gamma t + \Gamma^2 t^2/2$ of $e^{-\Gamma t}$. Clearly, the above approach offers a systematic (but sometimes tedious) method for the evaluation of the required solution. We shall discuss this in more detail in section 8.3 where we obtain second order quantum electrodynamical equations of radiative transfer.

8.2 The mixed state density matrix [38, 115]

Section 8.1 discussed the fundamentals of the pure state density matrix for which the system was known to be in one of the eigenstates $|n\rangle$ of $|\psi\rangle$. There we did not know precisely which of these eigenstates of $|\psi\rangle$ the system might be in at any time t and we therefore developed a probability $|a_n(t)|^2 = \rho_{nn}(t)$ which overcame this uncertainty. Nevertheless, the corresponding $\rho(t)$ was deemed a pure state density matrix since it is only the quantum mechanical nature of the measurement process which produces the above uncertainty. Clearly then, this approach is quite satisfactory if we consider, for example, a single atom interacting with a radiation field. However, if there exists more than one interacting atom in the system, we not only have the above quantum mechanical uncertainty, but an uncertainty as to which atoms are being excited and de-excited at any particular point in time. For example, if we mix two groups of atoms where initially

$$\left.\begin{aligned} |\psi_1(0)\rangle &= \sum_n a_n^{(1)} |n\rangle, \\ |\psi_2(0)\rangle &= \sum_n a_n^{(2)} |n\rangle \end{aligned}\right\} \tag{8.2.1}$$

then the subsequent evolution of the system will clearly be described by a mixture of the eigenfunctions $|\psi_1(t)\rangle$ and $|\psi_2(t)\rangle$. Indeed, if we now introduce the quantity p_1 specifying the probability of the system being described solely by $|\psi_1(t)\rangle$ with a similar definition for p_2, then we define the corresponding *mixed state density matrix* by

$$\rho(t) = p_1 |\psi_1(t)\rangle\langle\psi_1(t)| + p_2 |\psi_2(t)\rangle\langle\psi_2(t)|, \tag{8.2.2}$$

where $p_1 + p_2 = 1$. If we have N_1 atoms of type 1 and N_2 atoms of type 2, then it is not unreasonable to suggest that $p_1 = N_1/(N_1 + N_2)$ for example.

Consequently, a distinction between the probabilities $|a_n(t)|^2$ and p_l must be drawn. The first simply corresponds to the uncertainties associated with the quantum mechanical measurement process, the second arises from the imprecise knowledge regarding the state $|\psi_1(t)\rangle$ or $|\psi_2(t)\rangle$ of the atoms interacting with the radiation field. Clearly, we can be more general by writing

$$\rho(t) = \sum_l p_l |\psi_l(t)\rangle\langle\psi_l(t)|. \tag{8.2.3}$$

The individual p_l must now be determined using standard statistical distribution theory. The pure state density matrix considered in the preceding section is simply a special case of equation (8.2.3) for which $p_l = 1$ for $l = m$, say, and zero otherwise.

Consequently, duplicating the analysis leading to equation (8.1.8), and noting the linearity of $|\psi_l(t)\rangle\langle\psi_l(t)|$ appearing in equation (8.2.3), we easily find the corresponding result for the mixed state density matrix, viz.

$$i\hbar\frac{\partial\rho}{\partial t} = [\hat{H}, \rho], \tag{8.2.4}$$

and this, in turn, gives

$$i\hbar\frac{\partial\rho^{\mathrm{I}}}{\partial t} = [\hat{H}_1^{\mathrm{I}}, \rho^{\mathrm{I}}] \tag{8.2.5}$$

(see equation (8.1.13)) where

$$\rho^{\mathrm{I}} = \mathrm{e}^{\mathrm{i}\hat{H}_0 t/\hbar}\rho\mathrm{e}^{-\mathrm{i}\hat{H}_0 t/\hbar}, \tag{8.2.6}$$

and $\hat{H} = \hat{H}_0 + \hat{H}_1(t)$. Equation (8.2.5) allows the exact solution

$$\rho^{\mathrm{I}}(t) = \rho^{\mathrm{I}}(0) + \frac{1}{i\hbar}\int_0^t \mathrm{d}t'[\hat{H}_1^{\mathrm{I}}(t'), \rho^{\mathrm{I}}(t')], \tag{8.2.7}$$

which, as in equation (8.1.9), produces the iterative series

$$\rho^{\mathrm{I}}(t) = \rho^{\mathrm{I}}(0) + \frac{1}{i\hbar}\int_0^t \mathrm{d}t'[\hat{H}_1^{\mathrm{I}}(t'), \rho^{\mathrm{I}}(0)]$$
$$+ \frac{1}{(i\hbar)^2}\int_0^t \mathrm{d}t'\int_0^{t'} \mathrm{d}t''[\hat{H}_1^{\mathrm{I}}(t'), [\hat{H}_1^{\mathrm{I}}(t''), \rho^{\mathrm{I}}(0)]] + \cdots. \tag{8.2.8}$$

Substitution of this last expression into equation (8.2.4) then yields

$$i\hbar\frac{\partial\rho^{\mathrm{I}}}{\partial t} = [\hat{H}_1^{\mathrm{I}}(t), \rho^{\mathrm{I}}(0)]$$
$$+ \frac{1}{i\hbar}\int_0^t \mathrm{d}t'[\hat{H}_1^{\mathrm{I}}(t), [\hat{H}_1^{\mathrm{I}}(t'), \rho^{\mathrm{I}}(0)]] + \cdots, \tag{8.2.9}$$

a result which will be of some value in the following sections. Note that the ρ satisfying these last equations is Hermitian (i.e. $\rho^+ = \rho$) as one would expect from equation (8.2.3) directly.

8.2.1 Expectation values

The thrust behind the above definition (8.2.3) is readily understood when we consider the expectation values of operators and observables. The expectation value of any observable quantity is that used in probability theory to describe the average of a number of measurements of that observable. Recognising that $\langle\psi|\psi\rangle$, given by

$$\langle\psi|\psi\rangle=\int\psi^*\psi\ \mathrm{d}\tau,$$

represents a probability, as discussed in section 6.1.1, for a pure state, then the expected value of an observable O is just

$$\langle O\rangle=\int\psi^*O\psi\ \mathrm{d}\tau.$$

The above Hilbert space notation can be carried over to matrix mechanics by considering the operator $\hat{O}$ such that its expectation $\langle\hat{O}\rangle$ in a pure state is

$$\langle\hat{O}\rangle=\langle\psi|\hat{O}|\psi\rangle, \tag{8.2.10}$$

whereas, for a mixed state described by the two state functions $|\psi_1(t)\rangle$ and $|\psi_2(t)\rangle$ for example,

$$\langle\hat{O}\rangle=p_1\langle\psi_1(t)|\hat{O}|\psi_1(t)\rangle+p_2\langle\psi_2(t)|\hat{O}|\psi_2(t)\rangle. \tag{8.2.11}$$

In general, we have

$$\langle\hat{O}\rangle=\sum_l p_l\langle\psi_l(t)|\hat{O}|\psi_l(t)\rangle, \tag{8.2.12}$$

so that, recalling the completeness relationship (6.5.38), viz.

$$\sum_\alpha|\alpha\rangle\langle\alpha|=\mathbb{1}, \tag{8.2.13}$$

we have

$$\begin{aligned}\langle\hat{O}\rangle&=\sum_{lm}p_l\langle\psi_l(t)|\hat{O}|m\rangle\langle m|\psi_l(t)\rangle\\&=\sum_{lm}p_l\langle m|\psi_l(t)\rangle\langle\psi_l(t)|\hat{O}|m\rangle\\&=\sum_m\langle m|\rho\hat{O}|m\rangle,\end{aligned} \tag{8.2.14}$$

where ρ is defined by equation (8.2.3). The trace of a matrix A is denoted by $\mathrm{tr}\,A$ and simply represents the sum of the diagonal elements of A. Clearly therefore, equation (8.2.14) is just

$$\langle\hat{O}\rangle=\mathrm{tr}\,(\rho\hat{O}). \tag{8.2.15}$$

The trace of ρ itself is easily evaluated:

$$\begin{aligned}\mathrm{tr}\,\rho&=\sum_m\langle m|\rho|m\rangle\\&=\sum_{ml}p_l\langle m|\psi_l(t)\rangle\langle\psi_l(t)|m\rangle\\&=\sum_{ml}p_l\langle\psi_l(t)|m\rangle\langle m|\psi_l(t)\rangle\\&=\sum_l p_l\langle\psi_l(t)|\psi_l(t)\rangle\\&=\sum_l p_l=1,\end{aligned} \tag{8.2.16}$$

where we recall that p_l is a probability.

It should be stressed here, however, that the wave vector $|\psi_I(t)\rangle$ used above will in general involve a specification of both the atom and radiation field states (and perturbers if included in the analysis). We should therefore write $|\psi_I(t)\rangle$ as $|i, n_1, \ldots, n_{\gamma\lambda}, \ldots, t\rangle$ where i specifically refers to the state of the atom (ith excited level) and $n_{\gamma\lambda}$ the state of the radiation field ($n_{\gamma\lambda}$ photons in the $(\gamma\lambda)$th mode). If, in particular, we consider just the one mode with $n_{\gamma\lambda} = n$, then $|\psi(t)\rangle \equiv |i, n, t\rangle \equiv |i, n\rangle = |i\rangle|n\rangle$. Thus, the outer product $|\psi\rangle\langle\psi|$ appearing in the definition (8.2.3) is a tensor involving all ith states of the atom and all n photon states. Consequently, we should examine the (in, jm) element $\rho_{in,jm} = \langle i, n|\rho|j, m\rangle$ of ρ rather than the superficial quantity ρ_{nm} discussed earlier. In this way, it is possible to use a reduced density matrix ρ_A say, which refers only to the atom states, or a ρ_R corresponding only to the radiation field.

For example, consider an observable $\hat{O}_R$ in n space which acts only on the variables describing the radiation field. Since ρ is a tensor in (i, n) space, equation (8.2.15) can only be used if we correspondingly put $\hat{O}_R$ into (i, n) space. This can be done by the outer product replacement $\hat{O}_R \to \hat{O}_R \otimes \mathbb{1}_A$ where $\mathbb{1}_A$ is the identity matrix in the atom space with elements $\langle i|\mathbb{1}_A|i'\rangle = \delta_{ii'}$. Note that we will have N_A coordinates to describe the atoms and N_R coordinates to describe the radiation field with $N_A \neq N_R$ in general. We then have

$$\begin{aligned}\langle\hat{O}_R\rangle &= \mathrm{tr}\,(\rho\hat{O}_R \otimes \mathbb{1}_A)\\ &= \sum_{in}\langle i, n|\rho\hat{O}_R \otimes \mathbb{1}_A|i, n\rangle\\ &= \sum_{ini'n'}\langle i, n|\rho|i', n'\rangle\langle i', n'|\hat{O}_R \otimes \mathbb{1}_A|i, n\rangle\\ &= \sum_{ini'n'}\rho_{in;i'n'}(\hat{O}_R)_{n'n}\delta_{ii'}\\ &= \sum_{inn'}\rho_{in;in'}(\hat{O}_R)_{n'n}.\end{aligned} \tag{8.2.17}$$

If we now define

$$\begin{aligned}(\rho_R)_{nn'} &= \sum_i \rho_{in;in'} = \sum_i \langle i, n|\rho|i, n'\rangle\\ &= (\mathrm{tr}_A\rho)_{nn'},\end{aligned} \tag{8.2.18}$$

then

$$\begin{aligned}\langle\hat{O}_R\rangle &= \sum_{nn'}(\rho_R)_{nn'}(\hat{O}_R)_{n'n}\\ &= \sum_{nn'}\langle n|\rho_R|n'\rangle\langle n'|\hat{O}_R|n\rangle\\ &= \sum_n\langle n|\rho_R\hat{O}_R|n\rangle\\ &= \mathrm{tr}_R(\rho_R\hat{O}_R),\end{aligned} \tag{8.2.19}$$

where tr_A and tr_R refer to the traces over the atom and radiation field coordinates respectively. Consequently, the expected value of $\hat{O}_R$ requires knowledge only of the reduced density matrix of the radiation field (although this, in turn, is obtainable directly only by an atom coordinate trace (8.2.18) over the entire mixed state density matrix ρ). Similarly, the expectation of an operator in atom space is easily found to be

$$\langle \hat{O}_A \rangle = \mathrm{tr}_A(\rho_A \hat{O}_A), \quad \rho_A = \mathrm{tr}_R \rho. \tag{8.2.20}$$

As we shall see in the following sections, the above definitions have great practical value in detailing both the equation of radiative transfer and the population rate equations for the energy levels of the atom.

Before proceeding, however, it is important to note the following points. The composite wave function $|\psi_{AR}\rangle$ for the entire atom plus radiation field system evolves according to the Schrödinger wave equation

$$i\hbar \frac{\partial}{\partial t} |\psi_{AR}\rangle = \hat{H} |\psi_{AR}\rangle, \tag{8.2.21}$$

where $\hat{H} = \hat{H}_A + \hat{H}_R + \hat{H}_{AR}$. If the atoms and radiation field are initially independent of one another, and there is no subsequent interaction between these two systems (i.e. $\hat{H}_{AR} \equiv 0$), then $|\psi_A\rangle$ and $|\psi_R\rangle$ evolve separately according to

$$i\hbar \frac{\partial |\psi_A\rangle}{\partial t} = \hat{H}_A |\psi_A\rangle, \quad i\hbar \frac{\partial |\psi_R\rangle}{\partial t} = \hat{H}_R |\psi_R\rangle. \tag{8.2.22}$$

Consequently, we can easily show that

$$i\hbar \frac{\partial}{\partial t} (|\psi_A\rangle |\psi_R\rangle) = (\hat{H}_A + \hat{H}_R) |\psi_A\rangle |\psi_R\rangle,$$

which, by comparison with equation (8.2.21) with $\hat{H}_{AR} \equiv 0$, yields

$$|\psi_{AR}\rangle \equiv |\psi_A\rangle |\psi_R\rangle. \tag{8.2.23}$$

(This last statement should not be confused with the expression $|i, n\rangle \equiv |i\rangle|n\rangle$ in which $|i\rangle$ and $|n\rangle$ are effectively basis vectors in the atom and radiation field coordinates respectively.) We then have

$$\begin{aligned} \rho = \rho_{AR} &= |\psi_A\rangle\langle\psi_A| \otimes |\psi_R\rangle\langle\psi_R| \\ &= \rho_A \otimes \rho_R, \end{aligned} \tag{8.2.24}$$

which, in turn, enables the expected value of the product $\hat{O}_A \hat{O}_R$ (where $\hat{O}_A$ acts only on atomic variables and $\hat{O}_R$ on radiation field variables) to be written as

$$\begin{aligned} \langle \hat{O}_A \otimes \hat{O}_R \rangle &= \sum_{in} \langle i, n | \rho \hat{O}_A \otimes \hat{A}_R | i, n \rangle \\ &= \sum_{ini'n'} \langle i, n | \rho | i', n' \rangle \langle i' | \hat{O}_A | i \rangle \langle n' | \hat{O}_R | n \rangle \\ &= \sum_{ini'n'} \langle i | \rho_A | i' \rangle \langle i' | \hat{O}_A | i \rangle \langle n | \rho_R | n' \rangle \langle n' | \hat{O}_R | n \rangle \end{aligned} \tag{8.2.25}$$

(8.2.25 cont.)
$$= \left(\sum_i \langle i|\rho_A\hat{O}_A|i\rangle\right)\left(\sum_n \langle n|\rho_R\hat{O}_R|n\rangle\right)$$
$$= \mathrm{tr}_A(\rho_A\hat{O}_A)\mathrm{tr}_R(\rho_R\hat{O}_R)$$
$$= \langle\hat{O}_A\rangle\langle\hat{O}_R\rangle.$$

Here we see that the expected value of $\hat{O}_A\hat{O}_R$ is simply the product of the individual expectations as one would expect from ordinary statistical theory for independent observables $\hat{O}_A$ and $\hat{O}_R$.

However, non-zero $\hat{H}_{AR}$ stipulates an evolution of the radiation field, for example, which is not independent of the atomic system. Mathematically, one cannot in general construct a Schrödinger wave equation of the form (8.2.21) for the product $|\psi_A\rangle|\psi_R\rangle$. Thus, $|\psi_{AR}\rangle \neq |\psi_A\rangle|\psi_R\rangle$, i.e.

$$\rho \equiv \rho_{AR} \neq \rho_A \otimes \rho_R. \tag{8.2.26}$$

This is an important qualification in the use of the mixed state density matrix.

Finally, noting that the trace operator is cyclical, viz.

$$\mathrm{tr}(ABC) = \mathrm{tr}(BCA) = \mathrm{tr}(CAB), \tag{8.2.27}$$

which is easily proved by writing, for example,

$$\begin{aligned}\mathrm{tr}(ABC) &= \sum_n \langle n|ABC|n\rangle \\ &= \sum_{nml} \langle n|A|m\rangle\langle m|B|l\rangle\langle l|C|n\rangle \\ &= \sum_{nml} \langle m|B|l\rangle\langle l|C|n\rangle\langle n|A|m\rangle \\ &= \sum_m \langle m|BCA|m\rangle \\ &= \mathrm{tr}(BCA),\end{aligned}$$

then we have

$$\begin{aligned}\langle\hat{O}\rangle &= \mathrm{tr}(\rho\hat{O}) \\ &= \mathrm{tr}(e^{i\hat{H}_0 t/\hbar}\rho e^{-i\hat{H}_0 t/\hbar}e^{i\hat{H}_0 t/\hbar}\hat{O}e^{-i\hat{H}_0 t/\hbar}) \\ &= \mathrm{tr}(\rho^I\hat{O}^I),\end{aligned} \tag{8.2.28}$$

where $\hat{O}^I$ is the operator $\hat{O}$ in the interaction representation. Clearly therefore, the expectation of an operator has the same structural form $\mathrm{tr}(\rho\hat{O})$ or $\mathrm{tr}(\rho^I\hat{O}^I)$ in either representation. In particular, we will be interested in the expectation of the number operator $\hat{a}^+_{\gamma\lambda}\hat{a}_{\gamma\lambda}$ discussed in section 6.2.3. We then have

$$\begin{aligned}n_{\gamma\lambda} &= \langle\hat{a}^+_{\gamma\lambda}\hat{a}_{\gamma\lambda}\rangle = \mathrm{tr}(\rho\hat{a}^+_{\gamma\lambda}\hat{a}_{\gamma\lambda}) \\ &= \mathrm{tr}(\rho^I e^{i\hat{H}_0 t/\hbar}\hat{a}^+_{\gamma\lambda}\hat{a}_{\gamma\lambda}e^{-i\hat{H}_0 t/\hbar}),\end{aligned} \tag{8.2.29}$$

where

$$\hat{H}_0 = \hat{H}_A + \sum_{\gamma'\lambda'} \hbar\omega_{\gamma'}(\hat{a}^+_{\gamma'\lambda'}\hat{a}_{\gamma'\lambda'} + \tfrac{1}{2}\mathbb{1}). \tag{8.2.30}$$

Clearly, $\hat{a}^+_{\gamma\lambda}\hat{a}_{\gamma\lambda}$ commutes with $\hat{H}_A$, $\hat{a}^+_{\gamma'\lambda'}\hat{a}_{\gamma'\lambda'}$ for all $\gamma \neq \gamma'$, $\lambda \neq \lambda'$ (note that, for example, $\hat{a}_{\gamma\lambda}$ commutes with $\hat{a}_{\gamma'\lambda'}$ even when $\gamma' \neq \gamma$ and $\lambda' \neq \lambda$) and with itself $\hat{a}^+_{\gamma\lambda}\hat{a}_{\gamma\lambda}$, so that

$$\langle n_{\gamma\lambda}\rangle = \mathrm{tr}(\rho\hat{a}^+_{\gamma\lambda}\hat{a}_{\gamma\lambda}) = \mathrm{tr}(\rho^{\mathrm{I}}\hat{a}^+_{\gamma\lambda}\hat{a}_{\gamma\lambda}). \tag{8.2.31}$$

Consequently, either ρ or ρ^{I} may be used to evaluate the expectation $\langle n_{\gamma\lambda}\rangle$.

8.2.2 A preliminary approach to the transfer equation

As an illustration of all the preceding theory of the mixed state density matrix, we next discuss a preliminary 'derivation' of the equation of radiative transfer using purely quantum electrodynamical considerations [94]. We take the absorbing - emitting atom to have a naturally broadened upper state $|2\rangle$ but a perfectly sharp lower state $|1\rangle$ as shown in figure 8.5. Note that we do not include elastic collisions whilst the atom is in the excited state. As in the preceding section the atomic states are denoted by $|i\rangle$ (or $|i'\rangle$) whereas the photon states are written as $|n\rangle \equiv |n_1, n_2, \ldots, n_\infty\rangle$ (or $|n'\rangle$) where, for example, there are n_1 photons in mode 1.

An equation of radiative transfer details the specific rate of change of photon intensity. Thus, the equation of transfer for the $(\gamma\lambda)$th mode must incorporate the rate of change of the expectation value $\langle n_{\gamma\lambda}\rangle$ of the number of photons in this mode. Equation (8.2.31) stipulates

$$\langle n_{\gamma\lambda}\rangle = \langle \hat{a}^+_{\gamma\lambda}\hat{a}_{\gamma\lambda}\rangle = \mathrm{tr}(\rho\hat{a}^+_{\gamma\lambda}\hat{a}_{\gamma\lambda}), \tag{8.2.32}$$

where $\hat{a}^+_{\gamma\lambda}$ and $\hat{a}_{\gamma\lambda}$ are the usual creation and annihilation operators satisfying

$$\hat{a}_{\gamma\lambda}|n_{\gamma\lambda}\rangle = \sqrt{n_{\gamma\lambda}}\,|n_{\gamma\lambda}-1\rangle, \quad \hat{a}^+_{\gamma\lambda}|n_{\gamma\lambda}\rangle = \sqrt{(n_{\gamma\lambda}+1)}\,|n_{\gamma\lambda}+1\rangle. \tag{8.2.33}$$

Consequently, taking the photon intensity in this mode to be proportional to the above expectation value, with J_0 the proportionality constant, we have

$$\begin{aligned} I &= J_0\langle n_{\gamma\lambda}\rangle = J_0\,\mathrm{tr}(\rho\hat{a}^+_{\gamma\lambda}\hat{a}_{\gamma\lambda}) \\ &= J_0 \sum_{ini'n'} \rho_{in;i'n'}(\hat{a}^+_{\gamma\lambda}\hat{a}_{\gamma\lambda})_{i'n';in} \end{aligned} \tag{8.2.34}$$

(see equation (8.2.17), for example). Further, we find from equations (8.2.33) that

$$(\hat{a}^+_{\gamma\lambda}\hat{a}_{\gamma\lambda})_{i'n';in} = \langle i'|\langle n'|\hat{a}^+_{\gamma\lambda}\hat{a}_{\gamma\lambda}|n\rangle|i\rangle = \delta_{ii'}\delta_{nn'}n_{\gamma\lambda}, \tag{8.2.35}$$

Fig.8.5. Radiative excitation and de-excitation where the upper state is naturally broadened.

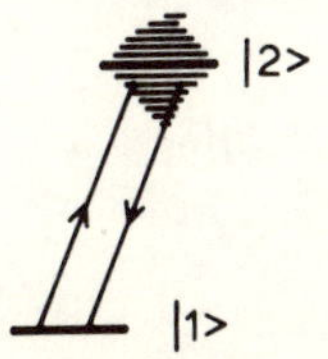

so that

$$I = J_0 \sum_{ini'n'} \rho_{in;i'n'} \delta_{ii'} \delta_{nn'} n_{\gamma\lambda}$$

$$= J_0 \sum_{in} \rho_{in;in} n_{\gamma\lambda}, \tag{8.2.36}$$

which finally yields

$$\frac{\partial I}{\partial t} = J_0 \sum_n n_{\gamma\lambda} \left[\frac{\partial \rho_{1n;1n}}{\partial t} + \sum_{i \in |2\rangle} \frac{\partial \rho_{in;in}}{\partial t} \right], \tag{8.2.37}$$

where the summation over i holds only for all sub-states $|i\rangle$ of the upper $|2\rangle$ level.

This last result then becomes a *quasi*-equation of radiative transfer. Clearly, if we are to proceed, we require the density matrix derivatives appearing on the RHS (8.2.37). These are obtained by first examining equation (8.2.4), viz.

$$i\hbar \frac{\partial \rho}{\partial t} = [\hat{H}, \rho] = [\hat{H}_0, \rho] + [\hat{H}_1, \rho], \tag{8.2.38}$$

where

$$\hat{H}_0 = \hat{H}_{\mathrm{A}} + \hat{H}_{\mathrm{R}}, \tag{8.2.39}$$

for which

$$\hat{H}_{\mathrm{A}} |i\rangle = E_i^{(\mathrm{A})} |i\rangle, \tag{8.2.40}$$

and

$$\hat{H}_{\mathrm{R}} |n\rangle = \sum_{\gamma\lambda} \hbar\omega_\gamma (\hat{a}^+_{\gamma\lambda} \hat{a}_{\gamma\lambda} + \tfrac{1}{2} \mathbb{1}) |n\rangle = \hbar\omega_\gamma (n_{\gamma\lambda} + \tfrac{1}{2}) |n\rangle, \tag{8.2.41}$$

with (see section 6.4.1)

$$\hat{H}_1 = \sum_{\gamma\lambda} (\hat{a}_{\gamma\lambda} \xi_{\gamma\lambda} + \hat{a}^+_{\gamma\lambda} \xi^*_{\gamma\lambda}), \tag{8.2.42}$$

$$\xi_{\gamma\lambda} = -\frac{e}{2mc} (\hat{\mathbf{p}} \cdot \mathbf{u}_{\gamma\lambda} + \mathbf{u}_{\gamma\lambda} \cdot \hat{\mathbf{p}}). \tag{8.2.43}$$

In particular, $\hat{a}_{\gamma\lambda}$ and $\hat{a}^+_{\gamma\lambda}$ depend only upon photon co-ordinates, and $\xi_{\gamma\lambda}$ and $\xi^*_{\gamma\lambda}$ only depend upon atom co-ordinates. Recognising that we can write the mixed state basis eigenfunction $|\psi\rangle \equiv |\psi_{in}\rangle$ as $|i\rangle|n\rangle$, the above results clearly yield

$$\hat{H}_0 |\psi_{in}\rangle = [E_i^{(\mathrm{A})} + \hbar\omega_\gamma (n_{\gamma\lambda} + \tfrac{1}{2})] |\psi_{in}\rangle,$$

so that

$$(\hat{H}_0)_{i'n';in} = \langle \psi_{i'n'} | \hat{H}_0 | \psi_{in} \rangle = [E_i^{(\mathrm{A})} + \hbar\omega_\gamma (n_{\gamma\lambda} + \tfrac{1}{2})] \delta_{ii'} \delta_{nn'}, \tag{8.2.44}$$

i.e. $\hat{H}_0$ is diagonal as expected. Consequently,

$$[\hat{H}_0, \rho]_{in;i'n'} = \sum_{i''n''} [(\hat{H}_0)_{in;i''n''}\rho_{i''n'';i'n'} - \rho_{in;i''n''}(\hat{H}_0)_{i''n'';i'n'}]$$
$$= (\hat{H}_0)_{in;in}\rho_{in;i'n'} - \rho_{in;i'n'}(\hat{H}_0)_{i'n';i'n'}$$
$$= [E_i^{(A)} - E_{i'}^{(A)} + \hbar\omega_\gamma(n_{\gamma\lambda} + \tfrac{1}{2})$$
$$-\hbar\omega_\gamma(n'_{\gamma\lambda} + \tfrac{1}{2})]\rho_{in;i'n'}$$
$$= -\hbar[(n'_{\gamma\lambda} - n_{\gamma\lambda})\omega_\gamma - \omega_{ii'}]\rho_{in;i'n'}, \tag{8.2.45}$$

where we have written

$$\omega_{ii'} = (E_i^{(A)} - E_{i'}^{(A)})/\hbar.$$

We now turn our attention to the interaction term $[\hat{H}_1, \rho]$ appearing in equation (8.2.38). Clearly,

$$(\hat{H}_1)_{in;i'n'} = \langle i|\langle n_{\gamma\lambda}|\hat{H}_1|n'_{\gamma\lambda}\rangle|i'\rangle$$
$$= \sum_{\gamma'\lambda'} [\langle n|\hat{a}_{\gamma'\lambda'}|n'\rangle\langle i|\xi_{\gamma'\lambda'}|i'\rangle + \langle n|\hat{a}^+_{\gamma'\lambda'}|n'\rangle\langle i|\xi^*_{\gamma'\lambda'}|i'\rangle]$$
$$= (\xi_{\gamma\lambda})_{ii'}\sqrt{n'}\,\langle n|n'-1\rangle + (\xi^*_{\gamma\lambda})_{ii'}\sqrt{(n'+1)}\,\langle n|n'+1\rangle$$
$$= (\xi_{\gamma\lambda})_{ii'}\sqrt{n'}\,\delta_{n,n'-1} + (\xi^*_{\gamma\lambda})_{ii'}\sqrt{(n'+1)}\delta_{n,n'+1}. \tag{8.2.46}$$

Thus, the non-zero elements of $\hat{H}_1$ have the form, for example,

$$\left.\begin{aligned}(\hat{H}_1)_{1n;2,n-1} = (\xi^*_{\gamma\lambda})_{12}\sqrt{n}, \quad (\hat{H}_1)_{1n;2,n+1} = (\xi_{\gamma\lambda})_{12}\sqrt{(n+1)},\\ (\hat{H}_1)_{2n;1,n-1} = (\xi^*_{\gamma\lambda})_{21}\sqrt{n}, \quad (\hat{H}_1)_{2n;1,n+1} = (\xi_{\gamma\lambda})_{21}\sqrt{(n+1)}.\end{aligned}\right\} \tag{8.2.47}$$

The above results may now be substituted into equation (8.2.38) to yield, for example,

$$i\hbar\frac{\partial\rho_{1n;1n}}{\partial t} = \sum_{i\in|2\rangle} [(\hat{H}_1)_{1n;i,n-1}\rho_{i,n-1;1n}$$
$$+ (\hat{H}_1)_{1n;i,n+1}\rho_{i,n+1;1n}$$
$$- \rho_{1n;i,n-1}(\hat{H}_1)_{i,n-1;1n}$$
$$- \rho_{1n;i,n+1}(\hat{H}_1)_{i,n+1;1n}]. \tag{8.2.48}$$

However, we are only interested in the situation whereby a photon absorption process corresponds to an atom excitation, and a photon emission corresponds to a de-excitation of the atom. These events are reflected in the terms $(\hat{H}_1)_{1n;i,n-1}$ and $(\hat{H}_1)_{i,n-1;1n}$ for which the system consisting of n photons in the radiation field with the atom in the de-excited state $|1\rangle$ changes to that with $n-1$ photons and the atom in the excited sub-state $|i\rangle$ of $|2\rangle$, or vice versa. The other two terms on the RHS (8.2.48) have little physical relevance to the present problem – they represent, for example, a system of $n+1$ photons with the atom in the excited state changing to a system in which there are only n photons with the atom now in the de-excited state! (Mathematically, the neglect of the two terms involving $(\hat{H}_1)_{1n;i,n+1}$ and $(\hat{H}_1)_{i,n+1;1n}$ is equivalent to the rotating wave approximation [88] for which subsequent time-dependent quantities of the form $e^{\pm(\omega_\gamma+\omega_{21})t}$ for $\rho_{i,n+1;1n}$ and $\rho_{1n;i,n+1}$, as opposed to $e^{\pm(\omega_\gamma-\omega_{21})t}$ for $\rho_{i,n-1;1n}$ and $\rho_{1n;i,n-1}$, are rapidly fluctuating and therefore provide unimportant contributions to the response of the system to the interaction Hamiltonian $\hat{H}_1$.) Equation (8.2.48) then becomes

$$i\hbar\frac{\partial\rho_{1n;1n}}{\partial t}=\sum_{i\in|2\rangle}[(\xi^*_{\gamma\lambda})_{1i}\sqrt{n}\,\rho_{i,n-1;1n}-(\xi_{\gamma\lambda})_{i1}\sqrt{n}\,\rho_{1n;i,n-1}]. \tag{8.2.49}$$

Similarly, we find

$$i\hbar\frac{\partial\rho_{in;in}}{\partial t}=(\xi_{\gamma\lambda})_{i1}\sqrt{(n+1)}\rho_{1,n+1;in}-(\xi^*_{\gamma\lambda})_{1i}\sqrt{(n+1)}\rho_{in;1,n+1}, \tag{8.2.50}$$

where no summation occurs here since the lower level $|1\rangle$ is not broadened.

It is clear from these last two equations that we can only proceed if the off-diagonal elements $\rho_{i,n-1;1n}$ and $\rho_{1n;i,n-1}$ are known. Equations (8.2.38) and (8.2.45) yield

$$\begin{aligned}i\hbar\frac{\partial\rho_{i,n-1;1n}}{\partial t}&=-\hbar(\omega_\gamma-\omega_{i1})\rho_{i,n-1;1n}\\&\quad+(\hat{H}_1)_{i,n-1;1n}\rho_{1n;1n}+(\hat{H}_1)_{i,n-1;1,n-2}\rho_{1,n-2;1n}\\&\quad-\sum_{i'\in|2\rangle}[\rho_{i,n-1;i',n+1}(\hat{H}_1)_{i',n+1;1n}\\&\quad+\rho_{i,n-1;i',n-1}(\hat{H}_1)_{i',n-1;1n}].\end{aligned} \tag{8.2.51}$$

As before, we retain only those terms of physical interest so that this last equation becomes

$$\begin{aligned}i\hbar\frac{\partial\rho_{i,n-1;1n}}{\partial t}&=-\hbar(\omega_\gamma-\omega_{i1})\rho_{i,n-1;1n}\\&\quad+(\xi_{\gamma\lambda})_{i1}\sqrt{n}\,\rho_{1n;1n}-\sum_{i'\in|2\rangle}(\xi_{\gamma\lambda})_{i'1}\sqrt{n}\,\rho_{i,n-1;i',n-1}.\end{aligned} \tag{8.2.52}$$

Similarly, we find

$$\begin{aligned}i\hbar\frac{\partial\rho_{1n;i,n-1}}{\partial t}&=\hbar(\omega_\gamma-\omega_{i1})\rho_{1n;i,n-1}\\&\quad-(\xi^*_{\gamma\lambda})_{1i}\sqrt{n}\,\rho_{1n;1n}+\sum_{i'\in|2\rangle}(\xi^*_{\gamma\lambda})_{1i'}\sqrt{n}\,\rho_{i',n-1;i,n-1}.\end{aligned} \tag{8.2.53}$$

Our purpose here in this sub-section is simply to illustrate the manner in which the various terms appearing in the equation of radiative transfer *may* be derived using the mixed state density matrix. We shall consider a more detailed analysis in section 8.3. For the present, however, we simplify the problem considerably by arbitrarily neglecting the secondary off-diagonal elements $\rho_{i',n-1;i,n-1}$ appearing on the RHS of equations (8.2.52) and (8.2.53), and replacing the remaining $\rho_{i,n-1;i,n-1}$ with their resonance values $\rho_{2,n-1;2,n-1}$. We then have

$$i\hbar\frac{\partial\rho_{i,n-1;1n}}{\partial t}=-\hbar(\omega_\gamma-\omega_{i1})\rho_{i,n-1;1n}+\sqrt{n}\,(\xi_{\gamma\lambda})_{i1}(\rho_{1n;1n}-\rho_{2,n-1;2,n-1}), \tag{8.2.54}$$

$$i\hbar\frac{\partial\rho_{1n;i,n-1}}{\partial t}=\hbar(\omega_\gamma-\omega_{i1})\rho_{1n;i,n-1}-\sqrt{n}\,(\xi^*_{\gamma\lambda})_{1i}(\rho_{1n;1n}-\rho_{2,n-1;2,n-1}). \tag{8.2.55}$$

Consequently, taking the Laplace transform, defined by equation (8.1.44), of equations (8.2.49), (8.2.50), (8.2.54) and (8.2.55) we have

$$i\hbar s\bar\rho_{1n;1n}=i\hbar\rho_{1n;1n}(0)+\sqrt{n}\sum_{i\in|2\rangle}[(\xi^*_{\gamma\lambda})_{1i}\bar\rho_{i,n-1;1n}-(\xi_{\gamma\lambda})_{i1}\bar\rho_{1n;i,n-1}], \tag{8.2.56}$$

$$i\hbar s\bar\rho_{in,in}=i\hbar\rho_{in;in}(0)+\sqrt{(n+1)}\,[(\xi_{\gamma\lambda})_{i1}\bar\rho_{1,n+1;in}-(\xi^*_{\gamma\lambda})_{1i}\bar\rho_{in;1,n+1}], \tag{8.2.57}$$

$$\hbar[\mathrm{i}s+(\omega_\gamma-\omega_{i1})]\bar\rho_{i,n-1;1n}=i\hbar\rho_{i,n-1;1n}(0)+\sqrt{n}\,(\xi_{\gamma\lambda})_{i1}[\bar\rho_{1n;1n}-\bar\rho_{2,n-1;2,n-1}] \tag{8.2.58}$$

$$\hbar[\mathrm{i}s-(\omega_\gamma-\omega_{i1})]\bar\rho_{1n;i,n-1}=i\hbar\rho_{1n;i,n-1}(0)-\sqrt{n}\,(\xi^*_{\gamma\lambda})_{1i}[\bar\rho_{1n;1n}-\bar\rho_{2,n-1;2,n-1}]. \tag{8.2.59}$$

These last two equations can be used to eliminate $\bar\rho_{i,n-1;1n}$ and $\bar\rho_{1n;i,n-1}$ from equation (8.2.56) so that

$$\begin{aligned}i\hbar s\bar\rho_{1n;1n}&=i\hbar\rho_{1n;1n}(0)\\&+\sqrt{n}\sum_{i\in|2\rangle}\left\{\frac{\mathrm{i}(\xi^*_{\gamma\lambda})_{1i}\rho_{i,n-1;1n}(0)}{\mathrm{i}s+(\omega_\gamma-\omega_{i1})}-\frac{\mathrm{i}(\xi_{\gamma\lambda})_{i1}\rho_{1n;i,n-1}(0)}{\mathrm{i}s-(\omega_\gamma-\omega_{i1})}\right\}_1\\&+n\sum_{i\in|2\rangle}\frac{|(\xi_{\gamma\lambda})_{1i}|^2}{\hbar}(\bar\rho_{1n;1n}-\bar\rho_{2,n-1;2,n-1})\\&\times\left[\frac{1}{\mathrm{i}s+(\omega_\gamma-\omega_{i1})}+\frac{1}{\mathrm{i}s-(\omega_\gamma-\omega_{i1})}\right].\end{aligned} \tag{8.2.60}$$

If, for convenience, we now take all off-diagonal elements of ρ to be initially zero (this is not unreasonable and, although unnecessary, it does simplify the algebra considerably, then $\{-\}_1=0$, and the above equation

becomes

$$i\hbar s\bar{\rho}_{1n;1n} = i\hbar\rho_{1n;1n}(0) - n\sum_{i\in|2\rangle}\frac{2is|(\xi_{\gamma\lambda})_{1i}|^2}{\hbar[s^2+(\omega_\gamma-\omega_{i1})^2]}(\bar{\rho}_{1n;1n}-\bar{\rho}_{2,n-1;2,n-1}). \tag{8.2.61}$$

We now replace the above summation over $i\in|2\rangle$ by the usual suitably weighted integral over E, Ω and λ to obtain

$$\begin{aligned} n\sum_{i\in|2\rangle}&\frac{2is|(\xi_{\gamma\lambda})_{1i}|^2}{\hbar[s^2+(\omega_\gamma-\omega_{i1})^2]}(\bar{\rho}_{1n;1n}-\bar{\rho}_{2,n-1;2,n-1}) \\ &\to\frac{n\hbar}{4\pi}\sum_\lambda\int d\Omega\int_{-\infty}^{\infty}dE\cdot\frac{\rho(E)2is|(\xi_{\gamma\lambda})_{1i}|^2}{[(E-E_{i1}^{(A)})^2+\hbar^2s^2]} \\ &\quad\times(\bar{\rho}_{1n;1n}-\bar{\rho}_{2,n-1;2,n-1}) \\ &=\frac{n\hbar}{4\pi}\sum_\lambda\int d\Omega\, 2\pi i\rho(E_{i1}^{(A)}+i\hbar s)2is|\xi_{\gamma\lambda}(E_{i1}^{(A)}+i\hbar s)|^2 \\ &\quad\times\frac{\bar{\rho}_{1n;1n}-\bar{\rho}_{2,n-1;2,n-1}}{2i\hbar s} \\ &\approx\frac{ni}{2}\sum_\lambda\int d\Omega\rho(E_{21}^{(A)})|(\xi_{\gamma\lambda})_{21}|^2(\bar{\rho}_{1n;1n}-\bar{\rho}_{2,n-1;2,n-1}) \\ &=\frac{i\hbar\Gamma_{n-1}}{2}(\bar{\rho}_{1n;1n}-\bar{\rho}_{2,n-1;2,n-1}), \end{aligned} \tag{8.2.62}$$

where, for convenience we define Γ_n by

$$\Gamma_n=\frac{1}{\hbar}\sum_\lambda\int d\Omega|(\xi_{\gamma\lambda})_{21}|^2(n_{\gamma\lambda}+1)\rho(E_{21}^{(A)}).$$

It should be mentioned here that the insertion of the $\rho(E)$ term into the above expression somewhat negates the thrust behind the use of the mixed state density matrix. Indeed, it effectively assumes the factorisation $\rho\equiv\rho_A\otimes\rho_R$ and, in the form shown, is not entirely consistent with the appearance of the $\bar{\rho}_{1n;1n}$ and $\bar{\rho}_{2,n-1;2,n-1}$ terms. However, this is a relatively minor point since here we are simply illustrating the use of the mixed state density matrix in deriving an equation of radiative transfer. A more rigorous treatment is presented in section 8.3.

We thence have

$$i\hbar s\bar{\rho}_{1n;1n}=i\hbar\rho_{1n;1n}(0)-\frac{i\hbar\Gamma_{n-1}}{2}(\bar{\rho}_{1n;1n}-\bar{\rho}_{2,n-1;2,n-1}). \tag{8.2.63}$$

If we now use equations (8.2.58) and (8.2.59) to similarly eliminate $\bar{\rho}_{i,n-1;1n}$ and $\bar{\rho}_{1n;i,n-1}$ from equation (8.2.57), and then sum both sides of the resulting equation over $i \in |2\rangle$ with the usual replacement

$$\begin{aligned}\sum_{i\in|2\rangle} \bar{\rho}_{i,n-1;i,n-1} &\to \frac{1}{4\pi}\sum_{\lambda}\int \mathrm{d}\Omega \int_{-\infty}^{\infty} \rho(E)\bar{\rho}_{i,n-1;i,n-1}(E)\,\mathrm{d}E \\ &\approx \frac{\bar{\rho}_{2,n-1;2,n-1}}{4\pi}\sum_{\lambda}\int \mathrm{d}\Omega \int_{-\infty}^{\infty} \rho(E)\,\mathrm{d}E, \\ &= \bar{\rho}_{2,n-1;2,n-1},\end{aligned} \tag{8.2.64}$$

then we easily find

$$\begin{aligned} i\hbar s\bar{\rho}_{2,n-1;2,n-1} &= i\hbar\rho_{2,n-1;2,n-1}(0) \\ &+ \frac{i\hbar\Gamma_{n-1}}{2}(\bar{\rho}_{1n;1n} - \bar{\rho}_{2,n-1;2,n-1}). \end{aligned} \tag{8.2.65}$$

Subtracting equation (8.2.65) from (8.2.59) yields

$$\begin{aligned}(s+\Gamma_{n-1})(\bar{\rho}_{1n;1n} - \bar{\rho}_{2,n-1;2,n-1}) &= \rho_{1n;1n}(0) - \rho_{2,n-1;2,n-1}(0) \\ &= \mathrm{A}\ (\text{say}),\end{aligned}$$

so that

$$\bar{\rho}_{1n;1n} - \bar{\rho}_{2,n-1;2,n-1} = \frac{A}{s+\Gamma_{n-1}},$$

i.e.

$$\rho_{1n;1n}(t) - \rho_{2,n-1;2,n-1}(t) = A\mathrm{e}^{-\Gamma_{n-1}t}. \tag{8.2.66}$$

Equation (8.2.54) then becomes

$$i\hbar\frac{\partial\rho_{i,n-1;1n}}{\partial t} + \hbar(\omega_\gamma - \omega_{i1})\rho_{i,n-1;1n} = \sqrt{n}\,(\xi_{\gamma\lambda})_{i1}A\mathrm{e}^{-\Gamma_{n-1}t},$$

so that

$$\begin{aligned}\rho_{i,n-1;1n}(t) &= \frac{-\sqrt{n}\,(\xi_{\gamma\lambda})_{i1}A\mathrm{e}^{-\Gamma_{n-1}t}}{i\hbar[\mathrm{i}(\omega_\gamma - \omega_{i1}) + \Gamma_{n-1}]} + \mathrm{B}\mathrm{e}^{i(\omega_\gamma - \omega_{i1})t} \\ &= \frac{-\sqrt{n}\,(\xi_{\gamma\lambda})_{i1}}{i\hbar[\mathrm{i}(\omega_\gamma - \omega_{i1}) + \Gamma_{n-1}]}(\rho_{1n;1n} - \rho_{2,n-1;2,n-1}) \\ &\quad + \mathrm{B}\mathrm{e}^{\mathrm{i}(\omega_\gamma - \omega_{i1})t},\end{aligned} \tag{8.2.67}$$

where B is an arbitrary constant. Similarly, equation (8.2.55) yields

$$\begin{aligned}\rho_{1n;i,n-1}(t) &= \frac{-\sqrt{n}(\xi^*_{\gamma\lambda})_{i1}}{i\hbar[\mathrm{i}(\omega_\gamma - \omega_{i1}) - \Gamma_{n-1}]}(\rho_{1n;1n} - \rho_{2,n-1;2,n-1}) \\ &\quad + B^*\mathrm{e}^{-\mathrm{i}(\omega_\gamma - \omega_{i1})t},\end{aligned} \tag{8.2.68}$$

as expected. The B and B^* constants may be obtained using the initial conditions, but we shall ignore these terms in all that follows.

Substitution of the above results into equations (8.2.49) and (8.2.50) then yields

$$\frac{\partial\rho_{1n;1n}}{\partial t}=\frac{-n}{(\mathrm{i}\hbar)^2}\sum_{i\in|2\rangle}|(\xi_{\gamma\lambda})_{i1}|^2\left[\frac{1}{\mathrm{i}(\omega_\gamma-\omega_{i1})+\Gamma_{n-1}}-\frac{1}{\mathrm{i}(\omega_\gamma-\omega_{i1})-\Gamma_{n-1}}\right](\rho_{1n;1n}-\rho_{2,n-1;2,n-1})$$
$$\approx-\frac{n}{\hbar^2}|(\xi_{\gamma\lambda})_{21}|^2\frac{2\Gamma_{n-1}}{(\omega_\gamma-\omega_{21})^2+\Gamma_{n-1}^2}(\rho_{1n;1n}-\rho_{2,n-1;2,n-1}) \tag{8.2.69}$$

and, similarly,

$$\frac{\partial\rho_{i,n-1;i,n-1}}{\partial t}=\frac{n}{\hbar^2}|(\xi_{\gamma\lambda})_{i1}|^2\frac{2\Gamma_{n-1}}{(\omega_\gamma-\omega_{21})^2+\Gamma_{n-1}^2}\times(\rho_{1n;1n}-\rho_{2,n-1;2,n-1}). \tag{8.2.70}$$

The quasi-transfer equation (8.2.37) now has the form

$$\frac{\partial I}{\partial t}=J_0\sum_n n\left[\frac{\partial\rho_{1n;1n}}{\partial t}+\sum_{i\in|2\rangle}\frac{\partial\rho_{in;in}}{\partial t}\right]$$
$$\approx-\frac{J_0}{h^2}\sum_n n\left\{n|(\xi_{\gamma\lambda})_{21}|^2\frac{2\Gamma_{n-1}}{(\omega_\gamma-\omega_{21})^2+\Gamma_{n-1}^2}\times(\rho_{1n;1n}-\rho_{2,n-1;2,n-1})-(n+1)\sum_{i\in|2\rangle}|(\xi_{\gamma\lambda})_{i1}|^2\frac{2\Gamma_n}{(\omega_\gamma-\omega_{i1})^2+\Gamma_n^2}\times(\rho_{1,n+1;1,n+1}-\rho_{2n;2n})\right\}$$
$$\approx-\frac{J_0}{\hbar^2}|(\xi_{\gamma\lambda})_{21}|^2\sum_n n\left\{\frac{2n\Gamma_{n-1}}{(\omega_\gamma-\omega_{21})^2+\Gamma_{n-1}^2}\times(\rho_{1n,1n}-\rho_{2,n-1;2,n-1})-\frac{2(n+1)\Gamma_n}{(\omega_\gamma-\omega_{21})^2+\Gamma_n^2}(\rho_{1,n+1;1,n+1}-\rho_{2n;2n})\right\},$$

so that, replacing n by $n+1$ in the first summation, we have

$$\frac{\partial I}{\partial t}=-\frac{J_0}{\hbar^2}|(\xi_{\gamma\lambda})_{21}|^2\sum_n\frac{2\Gamma_n}{(\omega_\gamma-\omega_{21})^2+\Gamma_n^2}\times\{(n+1)[(n+1)(\rho_{1,n+1;1,n+1}-\rho_{2n;2n})-n(\rho_{1,n+1;1,n+1}-\rho_{2n;2n})]\}$$

$$= -\frac{J_0}{\hbar^2}|(\xi_{\gamma\lambda})_{21}|^2 \sum_n \frac{2\Gamma_n}{(\omega_\gamma - \omega_{21})^2 + \Gamma_n^2}(n+1)$$
$$\times (\rho_{1,n+1;1,n+1} - \rho_{2n;2n})$$
$$= -\frac{2\pi J_0}{\hbar^2}|(\xi_{\gamma\lambda})_{21}|^2 \Big\{ \sum_n n\rho_{1n;1n} q(\Gamma_{n-1})$$
$$- \sum_n n\rho_{2n;2n} q(\Gamma_n) - \sum_n \rho_{2n;2n} q(\Gamma_n) \Big\}, \tag{8.2.71}$$

where we have defined the Lorentzian profile

$$q(\Gamma_n) = \frac{\Gamma_n/\pi}{(\omega_\gamma - \omega_{21})^2 + \Gamma_n^2}. \tag{8.2.72}$$

Equation (8.2.71) is then the desired equation of radiative transfer. The first term in the braces corresponds to photon absorption, the second to stimulated photon emission and the third to spontaneous photon emission. This can be more readily seen if we take $q(\Gamma_{n-1}) = q(\Gamma_n) = q(\Gamma)$ independent of n and then assume the factorisation of the density matrix ρ into the individual reduced atom and radiation field density, matrices ρ_A and ρ_R, i.e.

$$\rho = \rho_A \otimes \rho_R, \tag{8.2.73}$$

as discussed in the previous section (see equation (8.2.24) for example). We then have

$$\rho_{in;in} = \langle i|\langle n|\rho|n\rangle|i\rangle$$
$$= \langle i|\rho_A|i\rangle\langle n|\rho_R|n\rangle$$
$$= \rho_{ii}\rho_{nn}, \tag{8.2.74}$$

where, for example, ρ_{ii} is simply the probability density of the atom in state $|i\rangle$. It is not difficult to see that even without the above factorisation (8.2.73), we must have

$$\rho_{22} = \langle 2|\rho_A|2\rangle = \langle 2|\mathrm{tr}_R\rho|2\rangle$$
$$= \sum_n \langle 2|\langle n|\rho|n\rangle|2\rangle$$
$$= \sum_n \rho_{2n;2n}, \tag{8.2.75}$$

so that, with $q(\Gamma_n) \to q(\Gamma)$, the third term in the braces in equation (8.2.71) is just $\rho_{22}q(\Gamma)$.

If we now return to the factorisation (8.2.73), equation (8.2.74) yields

$$\sum_n n\rho_{1n;1n} q(\Gamma_{n-1}) \to q(\Gamma) \sum_n n\rho_{11}\rho_{nn}.$$

However, from equation (8.2.36) we have

$$I = J_0 \sum_n n \sum_i \rho_{in;in} = J_0 \sum_n n\rho_{nn}, \tag{8.2.76}$$

where we have used $\sum_i \rho_{ii} = \sum_i \langle i | \rho_A | i \rangle = 1$ (recall equation (8.2.16)) so that

$$J_0 \sum_n n\rho_{1n;1n} q(\Gamma_{n-1}) \rightarrow \rho_{11} I q(\Gamma)$$

and, similarly,

$$J_0 \sum_n n\rho_{2n;2n} q(\Gamma_n) \rightarrow \rho_{22} I q(\Gamma).$$

The quasi-transfer equation then has the more 'standard' form

$$\frac{\partial I}{\partial t} = -\frac{2\pi J_0 |(\xi_{\gamma\lambda})_{21}|^2}{\hbar^2} \left\{\rho_{11} \frac{I}{J_0} q(\Gamma) - \rho_{22} \frac{I}{J_0} q(\Gamma) - \rho_{22} q(\Gamma)\right\}, \tag{8.2.77}$$

Clearly, the $\rho_{11} I q(\Gamma)$ term contains the factor ρ_{11} related to the number of absorbing atoms (i.e. atoms in the de-excited state $|1\rangle$) together with the intensity I of photons and the absorption profile $q(\Gamma)$. This has been discussed phenomenologically in some detail in chapters 1 and 5. Similarly, the stimulated emission term contains $\rho_{22} I q(\Gamma)$ involving the number of atoms in the excited $|2\rangle$ state. Finally, we have the spontaneous emission component $\rho_{22} q(\Gamma)$ which, as expected, does not depend upon the radiation field I.

An important consideration now arises. The above transfer equation (8.2.77) was obtained using a variety of simplifications, approximations and assumptions. Yet it has the same basic form (even including the Lorentzian complete redistribution result) as the equation discussed in earlier chapters. Thus we ask the question; what is the 'correct' form of the equation of radiative transfer when all these simplifications, approximations and assumptions are removed? We shall address this problem in the following section.

8.3 Quantum transfer equations

We have already briefly examined a quantum transfer equation in the preceding section where it was shown that the usual photon absorption, stimulated emission and spontaneous emission terms appeared quite naturally without recourse to the phenomenological description of radiation – matter interaction discussed in chapters 1 and 5. However, a variety of approximations and simplifications were made, thereby allowing the specific use of Laplace transform techniques; but these are of little value for more general problems. The purpose of the present section, therefore, is to outline a procedure [74] which enables the quantum equation of radiative transfer to be derived for general situations of interest and this, in turn, affords an examination of all those first and higher order interference terms which do not appear in the above-mentioned phenomenological description.

8.3.1 Specification of the radiation intensity

We have already shown in section 6.2.1 that the electric and magnetic fields **E** and **B** satisfying Maxwell's equations (6.2.1–4) may be obtained from the Coulomb gauge using

$$\mathbf{E} = -\frac{1}{c}\frac{\partial \mathbf{A}}{\partial t}, \quad \mathbf{B} = \nabla \times \mathbf{A}, \tag{8.3.1}$$

$$\mathbf{A}(\mathbf{r}, t) = \sum_{\gamma\lambda} \alpha_\gamma [a_{\gamma\lambda}\mathbf{e}_{\gamma\lambda}\mathrm{e}^{\mathrm{i}(\mathbf{k}_\gamma\cdot\mathbf{r}-\omega_\gamma t)} + a^*_{\gamma\lambda}\mathbf{e}_{\gamma\lambda}\mathrm{e}^{-\mathrm{i}(\mathbf{k}_\gamma\cdot\mathbf{r}-\omega_\gamma t)}], \tag{8.3.2}$$

where $\alpha_\gamma = [2\pi\hbar c/k_\gamma l^3]^{1/2}$ with $k_\gamma = |\mathbf{k}_\gamma|$. The unit vectors $\mathbf{e}_{\gamma\lambda}$ are defined by equations (6.2.31), (6.2.39) and figure 6.1 with, in particular $\mathbf{e}_{\gamma\lambda}\cdot\mathbf{e}_{\gamma\lambda'} = \delta_{\lambda\lambda'}$.

Second quantisation (discussed in section 6.2.2) then enables the above classical expressions for **E** and **B** to be replaced by the operators

$$\hat{\mathbf{E}}(\mathbf{r}, t) = \sum_{\gamma\lambda} \frac{\mathrm{i}\omega_\gamma\alpha_\gamma}{c} [\hat{a}_{\gamma\lambda}\mathbf{e}_{\gamma\lambda}\mathrm{e}^{\mathrm{i}(\mathbf{k}_\gamma\cdot\mathbf{r}-\omega_\gamma t)} - \hat{a}^+_{\gamma\lambda}\mathbf{e}_{\gamma\lambda}\mathrm{e}^{-\mathrm{i}(\mathbf{k}_\gamma\cdot\mathbf{r}-\omega_\gamma t)}], \tag{8.3.3}$$

$$\hat{\mathbf{B}}(\mathbf{r}, t) = \sum_{\gamma\lambda} \mathrm{i}k_\gamma\alpha_\gamma(-1)^{\lambda+1}[\hat{a}_{\gamma\lambda}\mathbf{e}_{\gamma\lambda}\mathrm{e}^{\mathrm{i}(\mathbf{k}_\gamma\cdot\mathbf{r}-\omega_\gamma t)} - \hat{a}^+_{\gamma\lambda}\mathbf{e}_{\gamma\lambda}\mathrm{e}^{-\mathrm{i}(\mathbf{k}_\gamma\cdot\mathbf{r}-\omega_\gamma t)}] \tag{8.3.4}$$

– see equations (6.2.38) and (6.2.40) with equation (6.2.33) – where $\hat{a}_{\gamma\lambda}$ and $\hat{a}^+_{\gamma\lambda}$ are the usual time-independent annihilation and creation operators satisfying the commutation relationship (6.2.45), viz.

$$[\hat{a}_{\gamma\lambda}, \hat{a}^+_{\gamma'\lambda'}] = \mathbb{1}\delta_{\gamma\gamma'}\delta_{\lambda\lambda'}. \tag{8.3.5}$$

The total intensity $\mathscr{I}$ of the classical electro-magnetic field is related to the total energy of that field:

$$\mathscr{I}(\mathbf{r}, t) \propto \int_{\mathscr{V}} (\mathbf{E}\cdot\mathbf{E} + \mathbf{B}\cdot\mathbf{B})\,\mathrm{d}^3\mathbf{r}, \tag{8.3.6}$$

which, because of the similarity in structure of the two terms $\mathbf{E}\cdot\mathbf{E}$ and $\mathbf{B}\cdot\mathbf{B}$, and the orthogonality of $\mathrm{e}^{\mathrm{i}\mathbf{k}_\gamma\cdot\mathbf{r}}$ and $\mathrm{e}^{-\mathrm{i}\mathbf{k}_\gamma\cdot\mathbf{r}}$, can simply be written as the one statistical ensemble average (see, for example, equations (6.2.41) through (6.2.43) with $k_\gamma = \omega_\gamma/c$)

$$\mathscr{I}(\mathbf{r}, t) \propto \langle \mathbf{E}^*(\mathbf{r}, t)\cdot\mathbf{E}(\mathbf{r}, t)\rangle. \tag{8.3.7}$$

If we now write

$$\begin{aligned}\hat{E}_\lambda(\mathbf{r}, \mathbf{k}_\gamma, t) &= \mathrm{i}\omega_\gamma\alpha_\gamma\hat{a}_{\gamma\lambda}\mathbf{e}_{\gamma\lambda}\mathrm{e}^{\mathrm{i}(\mathbf{k}_\gamma\cdot\mathbf{r}-\omega_\gamma t)},\\ \hat{E}^+_\lambda(\mathbf{r}, \mathbf{k}_\gamma, t) &= -\mathrm{i}\omega_\gamma\alpha_\gamma\hat{a}^+_{\gamma\lambda}\mathbf{e}_{\gamma\lambda}\mathrm{e}^{-\mathrm{i}(\mathbf{k}_\gamma\cdot\mathbf{r}-\omega_\gamma t)},\end{aligned} \tag{8.3.8}$$

so that

$$\hat{E}(\mathbf{r},t) = \sum_{\gamma\lambda} [\hat{E}_\lambda(\mathbf{r},\mathbf{k}_\gamma,t) + \hat{E}_\lambda^+(\mathbf{r},\mathbf{k}_\gamma,t)]$$
$$= \sum_{\gamma} [\hat{E}_1 + \hat{E}_2 + \hat{E}_1^+ + \hat{E}_2^+], \tag{8.3.9}$$

the quantisation of the classical result (8.3.7) is then

$$\mathscr{I}(\mathbf{r},t) \propto \langle \hat{E}^+(\mathbf{r},t)\hat{E}(\mathbf{r},t)\rangle$$
$$= \langle \sum_{\gamma\gamma'} \sum_{\lambda\lambda'} [\hat{E}_\lambda^+(\mathbf{r},\mathbf{k}_\gamma,t)\hat{E}_{\lambda'}(\mathbf{r},\mathbf{k}_{\gamma'},t) + \hat{E}_{\lambda'}(\mathbf{r},\mathbf{k}_{\gamma'},t)\hat{E}_\lambda^+(\mathbf{r},\mathbf{k}_\gamma,t)]\rangle,$$

i.e.

$$\mathscr{I}(\mathbf{r},t) \propto \langle \sum_{\gamma\gamma'} \sum_{\lambda\lambda'} \omega_\gamma \omega_{\gamma'} \alpha_\gamma \alpha_{\gamma'} \mathbf{e}_{\gamma\lambda} \cdot \mathbf{e}_{\gamma'\lambda'} \mathrm{e}^{-\mathrm{i}[(\mathbf{k}_\gamma - \mathbf{k}_{\gamma'})\cdot\mathbf{r} - (\omega_\gamma - \omega_{\gamma'})t]}$$
$$\times [\hat{a}_{\gamma\lambda}^+ \hat{a}_{\gamma'\lambda'} + \hat{a}_{\gamma'\lambda'} \hat{a}_{\gamma\lambda}^+]\rangle. \tag{8.3.10}$$

Recalling the commutation relationship (8.3.5), we then have

$$\mathscr{I}(\mathbf{r},t) \propto \langle \sum_{\gamma\gamma'} \sum_{\lambda\lambda'} \omega_\gamma \omega_{\gamma'} \alpha_\gamma \alpha_{\gamma'} \mathbf{e}_{\gamma\lambda} \cdot \mathbf{e}_{\gamma'\lambda'} \mathrm{e}^{-\mathrm{i}[(\mathbf{k}_\gamma - \mathbf{k}_{\gamma'})\cdot\mathbf{r} - (\omega_\gamma - \omega_{\gamma'})t]}$$
$$\times [2\hat{a}_{\gamma\lambda}^+ \hat{a}_{\gamma'\lambda'} + \delta_{\gamma\gamma'}\delta_{\lambda\lambda'}]\rangle,$$

which, noting the minor variation of the factors ω_γ and α_γ compared with the exponential term in this last result, becomes

$$\mathscr{I}(\mathbf{r},t) \propto \langle(\hat{\mathscr{I}}_{11} + \hat{\mathscr{I}}_{12} + \hat{\mathscr{I}}_{21} + \hat{\mathscr{I}}_{22})\rangle, \tag{8.3.11}$$

where we have defined

$$\hat{\mathscr{I}}_{\lambda\lambda'} \equiv \hat{\mathscr{I}}_{\lambda\lambda'}(\mathbf{r},t) \propto \sum_{\gamma\gamma'} \hat{a}_{\gamma\lambda}^+ \hat{a}_{\gamma'\lambda'} \mathbf{e}_{\gamma\lambda} \cdot \mathbf{e}_{\gamma'\lambda'} \mathrm{e}^{-\mathrm{i}[(\mathbf{k}_\gamma - \mathbf{k}_{\gamma'})\cdot\mathbf{r} - (\omega_\gamma - \omega_{\gamma'})t]}. \tag{8.3.12}$$

We shall be interested in the equation specifying the transfer of the specific radiation intensity $I(\mathbf{r},\mathbf{k}_\gamma,t)$ at a particular frequency $\nu = \omega_\gamma/2\pi$ in the direction $\Omega = \mathbf{k}_\gamma/k_\gamma$ (rather than the 'total' intensity $\mathscr{I}(\mathbf{r},t)$). Consequently, we now write equation (8.3.12) as

$$\hat{\mathscr{I}}_{\lambda\lambda'}(\mathbf{r},t) \propto \langle \sum_{\gamma} \hat{I}_{\lambda\lambda'}(\mathbf{r},\mathbf{k}_\gamma t)\rangle, \tag{8.3.13}$$

with

$$\hat{I}_{\lambda\lambda'}(\mathbf{r},\mathbf{k}_{\gamma'},t) = J_0 \sum_{\gamma'} \hat{a}_{\gamma\lambda}^+ \hat{a}_{\gamma'\lambda'} \mathbf{e}_{\gamma\lambda} \cdot \mathbf{e}_{\gamma'\lambda'} \mathrm{e}^{-\mathrm{i}[(\mathbf{k}_\gamma - \mathbf{k}_{\gamma'})\cdot\mathbf{r} - (\omega_\gamma - \omega_{\gamma'})t]}, \tag{8.3.14}$$

where J_0 is the proportionality constant.

Clearly, equations (8.3.11) and (8.3.13) will yield the four polarisation intensities I_{11}, I_{12}, I_{21} and I_{22} (which can be related to the well-known Stokes' vector [73, 113] where

$$I_{\lambda\lambda'}(\mathbf{r},\mathbf{k}_\gamma,t) = \langle \hat{I}_{\lambda\lambda'}(\mathbf{r},\mathbf{k}_\gamma,t)\rangle$$
$$= \mathrm{tr}[\rho(t)\hat{I}_{\lambda\lambda'}(\mathbf{r},\mathbf{k}_\gamma,t)], \tag{8.3.15}$$

or, using equation (8.2.31) and noting the occurrence of $\hat{a}^{+}_{\gamma\lambda}\hat{a}_{\gamma'\lambda'}$ in equation (8.3.14),

$$I_{\lambda\lambda'}(\mathbf{r}, \mathbf{k}_\gamma, t) = \mathrm{tr}[\rho^{\mathrm{I}}(t)\hat{I}_{\lambda\lambda'}(\mathbf{r}, \mathbf{k}_\gamma, t)]. \tag{8.3.16}$$

This is a generalisation of the result (8.2.34).

The LHS of the equation of radiative transfer for each of the above intensities is

$$\frac{\partial I_{\lambda\lambda'}(\mathbf{r}, \mathbf{k}_\gamma, t)}{\partial t} = \frac{\partial I_{\lambda\lambda'}}{\partial t} + c(\boldsymbol{\Omega}\cdot\nabla)I_{\lambda\lambda'},$$

which, neglecting relativistic effects, has the explicit form

$$\begin{aligned}\frac{\partial I_{\lambda\lambda'}}{\partial t} &= \lim_{\Delta t\to 0}\frac{1}{\Delta t}[I_{\lambda\lambda'}(\mathbf{r}+c\boldsymbol{\Omega}\Delta t, \mathbf{k}_\gamma, t+\Delta t) - I_{\lambda\lambda'}(\mathbf{r}, \mathbf{k}_\gamma, t)]\\ &= \lim_{\Delta t\to 0}\frac{1}{\Delta t}\,\mathrm{tr}[\rho^{\mathrm{I}}(t+\Delta t)\hat{I}_{\lambda\lambda'}(\mathbf{r}+c\boldsymbol{\Omega}\Delta t, \mathbf{k}_\gamma, t+\Delta t)\\ &\quad -\rho^{\mathrm{I}}(t)\hat{I}_{\lambda\lambda'}(\mathbf{r}, \mathbf{k}_\gamma, t)].\end{aligned} \tag{8.3.17}$$

However, equation (8.3.14) with $\mathbf{k}_\gamma = \omega_\gamma\boldsymbol{\Omega}/c$, i.e. $\mathbf{k}_\gamma\cdot\boldsymbol{\Omega} = \omega_\gamma/c$, yields

$$\hat{I}_{\lambda\lambda'}(\mathbf{r}+c\boldsymbol{\Omega}\Delta t, \mathbf{k}_\gamma, t+\Delta t) = \hat{I}_{\lambda\lambda'}(\mathbf{r}, \mathbf{k}_\gamma, t), \tag{8.3.18}$$

so that the equation of transfer now has the form (note that the trace operator is cyclic – see equation (8.2.27))

$$\begin{aligned}\frac{\partial I_{\lambda\lambda'}}{\partial t} + c(\boldsymbol{\Omega}\cdot\nabla)I_{\lambda\lambda'} &= \lim_{\Delta t\to 0}\frac{1}{\Delta t}\,\mathrm{tr}\,\{\hat{I}_{\lambda\lambda'}(\mathbf{r}, \mathbf{k}_\gamma, t)[\rho^{\mathrm{I}}(t+\Delta t) - \rho^{\mathrm{I}}(t)]\}\\ &= \mathrm{tr}\left[\hat{I}_{\lambda\lambda'}(\mathbf{r}, \mathbf{k}_\gamma, t)\frac{\partial\rho^{\mathrm{I}}}{\partial t}\right].\end{aligned} \tag{8.3.19}$$

We already know from equation (8.2.9) that $\partial\rho^{\mathrm{I}}/\partial t$ has the series form

$$\begin{aligned}\frac{\partial\rho^{\mathrm{I}}}{\partial t} &= \frac{1}{i\hbar}[\hat{H}^{\mathrm{I}}_1(t), \rho^{\mathrm{I}}(0)]\\ &\quad + \frac{1}{(i\hbar)^2}\int_0^t \mathrm{d}t'[\hat{H}^{\mathrm{I}}_1(t), [\hat{H}^{\mathrm{I}}_1(t'), \rho^{\mathrm{I}}(0)]] + \cdots.\end{aligned} \tag{8.3.20}$$

Hence, we have

$$\begin{aligned}\frac{\partial I_{\lambda\lambda'}}{\partial t} + c(\boldsymbol{\Omega}\cdot\nabla)I_{\lambda\lambda'} &= \frac{1}{i\hbar}\mathrm{tr}\,\{\hat{I}_{\lambda\lambda'}[\hat{H}^{\mathrm{I}}_1(t), \rho^{\mathrm{I}}(0)]\}\\ &\quad + \frac{1}{(i\hbar)^2}\int_0^t \mathrm{tr}\,\{\hat{I}_{\lambda\lambda'}[\hat{H}^{\mathrm{I}}_1(t), [\hat{H}^{\mathrm{I}}_1(t'), \rho^{\mathrm{I}}(0)]]\}\,\mathrm{d}t' + \cdots.\end{aligned} \tag{8.3.21}$$

If we again recall equation (8.2.27), viz. $\mathrm{tr}(ABC) = \mathrm{tr}(BCA) = \mathrm{tr}(CAB)$, then we can clearly write

$$\begin{aligned}\mathrm{tr}\,\{\hat{I}_{\lambda\lambda}[\hat{H}^{\mathrm{I}}_1(t), \rho^{\mathrm{I}}(0)]\} &= \mathrm{tr}\,\{\hat{I}_{\lambda\lambda'}\hat{H}^{\mathrm{I}}_1(t)\rho^{\mathrm{I}}(0) - \hat{I}_{\lambda\lambda'}\rho^{\mathrm{I}}(0)\hat{H}^{\mathrm{I}}_1(t)\}\\ &= \mathrm{tr}\,\{\hat{I}_{\lambda\lambda'}\hat{H}^{\mathrm{I}}_1(t)\rho^{\mathrm{I}}(0) - \hat{H}^{\mathrm{I}}_1(t)\hat{I}_{\lambda\lambda'}\rho^{\mathrm{I}}(0)\}\\ &= \mathrm{tr}\,\{[\hat{I}_{\lambda\lambda'}, \hat{H}^{\mathrm{I}}_1(t)]\,\rho^{\mathrm{I}}(0)\}.\end{aligned} \tag{8.3.22}$$

Similarly, we find

$$\begin{aligned}
&\mathrm{tr}\,\{\hat{I}_{\lambda\lambda'}[\hat{H}^{\mathrm{I}}(t),[\hat{H}^{\mathrm{I}}(t'),\rho^{\mathrm{I}}(0)]]\}\\
&\quad=\mathrm{tr}\,\{\hat{I}_{\lambda\lambda'}\hat{H}_1^{\mathrm{I}}(t)\hat{H}_1^{\mathrm{I}}(t')\rho^{\mathrm{I}}(0)-\hat{I}_{\lambda\lambda'}\hat{H}_1^{\mathrm{I}}(t)\rho^{\mathrm{I}}(0)\hat{H}_1^{\mathrm{I}}(t')\\
&\qquad-\hat{I}_{\lambda\lambda'}\hat{H}_1^{\mathrm{I}}(t')\rho^{\mathrm{I}}(0)\hat{H}_1^{\mathrm{I}}(t)+\hat{I}_{\lambda\lambda'}\rho^{\mathrm{I}}(0)\hat{H}_1^{\mathrm{I}}(t')\hat{H}_1^{\mathrm{I}}(t)\}\\
&\quad=\mathrm{tr}\,\{[\hat{I}_{\lambda\lambda'}\hat{H}_1^{\mathrm{I}}(t)\hat{H}_1^{\mathrm{I}}(t')-\hat{H}_1^{\mathrm{I}}(t)\hat{I}_{\lambda\lambda'}\hat{H}_1^{\mathrm{I}}(t')\\
&\qquad-\hat{H}_1^{\mathrm{I}}(t')\hat{I}_{\lambda\lambda'}\hat{H}_1^{\mathrm{I}}(t)+\hat{H}_1^{\mathrm{I}}(t')\hat{H}_1^{\mathrm{I}}(t)\hat{I}_{\lambda\lambda'}]\,\rho^{\mathrm{I}}(0)\}\\
&\quad=\mathrm{tr}\,\{[\{\hat{I}_{\lambda\lambda'}\hat{H}_1^{\mathrm{I}}(t)-\hat{H}_1^{\mathrm{I}}(t)\hat{I}_{\lambda\lambda'}\}\hat{H}_1^{\mathrm{I}}(t')\\
&\qquad-\hat{H}_1^{\mathrm{I}}(t')\{\hat{I}_{\lambda\lambda'}\hat{H}_1^{\mathrm{I}}(t)-\hat{H}_1^{\mathrm{I}}(t)\hat{I}_{\lambda\lambda'}\}]\,\rho^{\mathrm{I}}(0)\}\\
&\quad=\mathrm{tr}\,\{[[\hat{I}_{\lambda\lambda'},\hat{H}_1^{\mathrm{I}}(t)],\hat{H}_1^{\mathrm{I}}(t')]\,\rho^{\mathrm{I}}(0)\}.
\end{aligned}\tag{8.3.23}$$

Our transfer equation (8.3.21) then has the more accessible form for future generalisation of the initial mixed state density matrix $\rho^{\mathrm{I}}(0)$:

$$\begin{aligned}
\frac{\partial I_{\lambda\lambda'}}{\partial t}+c(\boldsymbol{\Omega}\cdot\nabla)I_{\lambda\lambda'}&=\frac{1}{i\hbar}\,\mathrm{tr}\,\{[\hat{I}_{\lambda\lambda},\hat{H}_1^{\mathrm{I}}(t)]\,\rho^{\mathrm{I}}(0)\}\\
&\quad+\frac{1}{(i\hbar)^2}\int_0^t\mathrm{tr}\,\{[[\hat{I}_{\lambda\lambda'},\hat{H}_1^{\mathrm{I}}(t)],\hat{H}_1^{\mathrm{I}}(t')]\,\rho^{\mathrm{I}}(0)\}\,\mathrm{d}t'+\cdots.
\end{aligned}\tag{8.3.24}$$

Consequently, the task at hand is the evaluation of the various commutation expressions appearing on the RHS of equation (8.3.24). This, together with the determination of the interaction Hamiltonian $\hat{H}_1^{\mathrm{I}}(t)$, is discussed in the following sections.

8.3.2 The interaction Hamiltonian for zero elastic collisions

We already have the Hamiltonian operator

$$\hat{H}_1=\sum_{\gamma\lambda}(\hat{a}_{\gamma\lambda}\xi_{\gamma\lambda}+\hat{a}^{+}_{\gamma\lambda}\xi^{*}_{\gamma\lambda}),\tag{8.3.25}$$

where

$$\xi_{\gamma\lambda}=-\frac{e}{2mc}(\hat{\mathbf{p}}\cdot\mathbf{u}_{\gamma\lambda}+\mathbf{u}_{\gamma\lambda}\cdot\hat{\mathbf{p}})$$

depends only on the atom coordinates. The above $\hat{H}_1$ in the interaction representation is

$$\hat{H}_1^{\mathrm{I}}(t)=\mathrm{e}^{\mathrm{i}\hat{H}_0t/\hbar}\hat{H}_1(t)\mathrm{e}^{-\mathrm{i}\hat{H}_0t/\hbar},\tag{8.3.26}$$

where $\hat{H}_0$, in the absence of perturbing collisions, is the linear superposition of the time-independent operators $\hat{H}_{\mathrm{A}}$ for atoms and $\hat{H}_{\mathrm{R}}$ for the radiation field, viz.

$$\hat{H}_0=\hat{H}_{\mathrm{A}}+\hat{H}_{\mathrm{R}},\tag{8.3.27}$$

where

$$\hat{H}_{\mathrm{A}}|i\rangle=E_i^{(\mathrm{A})}|i\rangle,\tag{8.3.28}$$

and

$$\hat{H}_{\mathrm{R}}|n\rangle = \sum_{\gamma\lambda} \hbar\omega_\gamma(\hat{a}^+_{\gamma\lambda}\hat{a}_{\gamma\lambda} + \tfrac{1}{2}\mathbb{1})|n\rangle$$

$$= \hbar\omega_\gamma(n_{\gamma\lambda} + \tfrac{1}{2})|n\rangle. \tag{8.3.29}$$

Since atom and radiation field operators commute, i.e. $[\hat{O}_{\mathrm{A}}, \hat{O}_{\mathrm{R}}] = 0$, then the above $\hat{H}^{\mathrm{I}}_1(t)$ becomes

$$\hat{H}^{\mathrm{I}}_1 = \mathrm{e}^{\mathrm{i}(\hat{H}_{\mathrm{A}}+\hat{H}_{\mathrm{R}})t/\hbar} \sum_{\gamma\lambda} (\hat{a}_{\gamma\lambda}\xi_{\gamma\lambda} + \hat{a}^+_{\gamma\lambda}\xi^*_{\gamma\lambda})\mathrm{e}^{-\mathrm{i}(\hat{H}_{\mathrm{A}}+\hat{H}_{\mathrm{R}})t/\hbar}$$

$$= \sum_{\gamma\lambda} [\mathrm{e}^{\mathrm{i}\hat{H}_{\mathrm{A}}t/\hbar}\xi_{\gamma\lambda}\mathrm{e}^{-\mathrm{i}\hat{H}_{\mathrm{A}}t/\hbar}\mathrm{e}^{\mathrm{i}\hat{H}_{\mathrm{R}}t/\hbar}\hat{a}_{\gamma\lambda}\mathrm{e}^{-\mathrm{i}\hat{H}_{\mathrm{R}}t/\hbar}$$

$$+ \mathrm{e}^{\mathrm{i}\hat{H}_{\mathrm{A}}t/\hbar}\xi^*_{\gamma\lambda}\mathrm{e}^{-\mathrm{i}\hat{H}_{\mathrm{A}}t/\hbar}\mathrm{e}^{\mathrm{i}\hat{H}_{\mathrm{R}}t/\hbar}\hat{a}^+_{\gamma\lambda}\mathrm{e}^{-\mathrm{i}\hat{H}_{\mathrm{R}}t/\hbar}],$$

which we write as

$$\hat{H}^{\mathrm{I}}_1 = \sum_{\gamma\lambda} [\xi^{\mathrm{I}}_{\gamma\lambda}\hat{a}^{\mathrm{I}}_{\gamma\lambda} + (\xi^*_{\gamma\lambda})^{\mathrm{I}}(\hat{a}^+_{\gamma\lambda})^{\mathrm{I}}]. \tag{8.3.30}$$

(Note that operators involving atom coordinates do not commute because of first quantisation (viz. $[\hat{\mathbf{p}}, \mathbf{r}] = \mathrm{i}\hbar$) and that operators involving the radiation field do not commute because of second quantisation (viz. $[\hat{a}_{\gamma\lambda}, \hat{a}^+_{\gamma\lambda}] = \mathbb{1}$.)

Each of the individual expressions in equation (8.3.30) may be evaluated separately. For example, using the completeness relationship (6.5.38), viz.

$$\sum_i |i\rangle\langle i| = \mathbb{1},$$

we have

$$\xi^{\mathrm{I}}_{\gamma\lambda} = \mathrm{e}^{\mathrm{i}\hat{H}_{\mathrm{A}}t/\hbar}\,\mathbb{1}\,\xi_{\gamma\lambda}\,\mathbb{1}\,\mathrm{e}^{-\mathrm{i}\hat{H}_{\mathrm{A}}t/\hbar}$$

$$= \sum_{ij} \mathrm{e}^{\mathrm{i}\hat{H}_{\mathrm{A}}t/\hbar}|i\rangle\langle i|\xi_{\gamma\lambda}|j\rangle\langle j|\mathrm{e}^{-\mathrm{i}\hat{H}_{\mathrm{A}}t/\hbar}$$

$$= \sum_{ij} \mathrm{e}^{\mathrm{i}E^{(\mathrm{A})}_i t/\hbar}|i\rangle(\xi_{\gamma\lambda})_{ij}\langle j|\mathrm{e}^{-\mathrm{i}E^{(\mathrm{A})}_j t/\hbar}$$

$$= \sum_{ij} \mathrm{e}^{\mathrm{i}\omega_{ij}t}(\xi_{\gamma\lambda})_{ij}|i\rangle\langle j|, \tag{8.3.31}$$

where

$$\omega_{ij} = (E^{(\mathrm{A})}_i - E^{(\mathrm{A})}_j)/\hbar.$$

Next we find

$$\hat{a}^{\mathrm{I}}_{\gamma\lambda} = \mathrm{e}^{\mathrm{i}\hat{H}_{\mathrm{R}}t/\hbar}\hat{a}_{\gamma\lambda}\mathrm{e}^{-\mathrm{i}\hat{H}_{\mathrm{R}}t/\hbar}$$

$$= \exp\left[\mathrm{i}t\sum_{\gamma'\lambda'} \omega_{\gamma'}\hat{a}^+_{\gamma'\lambda'}\hat{a}_{\gamma'\lambda'}\right]\hat{a}_{\gamma\lambda}\exp\left[-\mathrm{i}t\sum_{\gamma'\lambda'} \omega_{\gamma'}\hat{a}^+_{\gamma'\lambda'}\hat{a}_{\gamma'\lambda'}\right],$$

which, using the commutation relationship (8.3.5), becomes

$$\hat{a}^{\mathrm{I}}_{\gamma\lambda} = \mathrm{e}^{\mathrm{i}\omega_\gamma t\hat{a}^+_{\gamma\lambda}\hat{a}_{\gamma\lambda}}\hat{a}_{\gamma\lambda}\mathrm{e}^{-\mathrm{i}\omega_\gamma t\hat{a}^+_{\gamma\lambda}\hat{a}_{\gamma\lambda}}. \tag{8.3.32}$$

This last equation may be simplified using a variety of methods. Perhaps the most straightforward approach is to operate with $\hat{a}^{\mathrm{I}}_{\gamma\lambda}$ on the vector $|n\rangle$ such that

$$\hat{a}^{\mathrm{I}}_{\gamma\lambda}|n\rangle = \mathrm{e}^{\mathrm{i}\omega_\gamma t\hat{a}^+_{\gamma\lambda}\hat{a}_{\gamma\lambda}}\hat{a}_{\gamma\lambda}\mathrm{e}^{-\mathrm{i}\omega_\gamma t\hat{a}^+_{\gamma\lambda}\hat{a}_{\gamma\lambda}}|n\rangle,$$

and note that $\hat{a}^+_{\gamma\lambda}\hat{a}_{\gamma\lambda}|n\rangle = \sqrt{n}\,\hat{a}^+_{\gamma\lambda}|n-1\rangle = n|n\rangle$. Thus, recalling that $\hat{a}^+_{\gamma\lambda}\hat{a}_{\gamma\lambda}$ commutes with itself, we then have

$$\begin{aligned}\hat{a}^{\mathrm{I}}_{\gamma\lambda}|n\rangle &= \sum_{\alpha\beta}\frac{1}{\alpha!\beta!}(\mathrm{i}\omega_\gamma t\hat{a}^+_{\gamma\lambda}\hat{a}_{\gamma\lambda})^\alpha\hat{a}_{\gamma\lambda}(-\mathrm{i}\omega_\gamma t\hat{a}^+_{\gamma\lambda}\hat{a}_{\gamma\lambda})^\beta|n\rangle \\ &= \sum_{\alpha\beta}\frac{1}{\alpha!\beta!}(\mathrm{i}\omega_\gamma t\hat{a}^+_{\gamma\lambda}\hat{a}_{\gamma\lambda})^\alpha\hat{a}_{\gamma\lambda}(-\mathrm{i}\omega_\gamma tn)^\beta|n\rangle \\ &= \mathrm{e}^{-\mathrm{i}\omega_\gamma tn}\sum_\alpha\frac{1}{\alpha!}(\mathrm{i}\omega_\gamma t\hat{a}^+_{\gamma\lambda}\hat{a}_{\gamma\lambda})^\alpha\sqrt{n}\,|n-1\rangle \\ &= \mathrm{e}^{-\mathrm{i}\omega_\gamma tn}\sqrt{n}\sum_\alpha\frac{1}{\alpha!}(\mathrm{i}\omega_\gamma t(n-1))^\alpha|n-1\rangle \\ &= \mathrm{e}^{-\mathrm{i}\omega_\gamma tn}\sqrt{n}\,\mathrm{e}^{\mathrm{i}\omega_\gamma t(n-1)}|n-1\rangle \\ &= \mathrm{e}^{-\mathrm{i}\omega_\gamma t}\sqrt{n}\,|n-1\rangle,\end{aligned}$$

which, noting that $\hat{a}_{\gamma\lambda}|n\rangle = \sqrt{n}\,|n-1\rangle$ again, becomes

$$\hat{a}^{\mathrm{I}}_{\gamma\lambda}|n\rangle = \mathrm{e}^{-\mathrm{i}\omega_\gamma t}\hat{a}_{\gamma\lambda}|n\rangle,$$

so that

$$\langle m|\hat{a}^{\mathrm{I}}_{\gamma\lambda}|n\rangle = \mathrm{e}^{-\mathrm{i}\omega_\gamma t}\langle m|\hat{a}_{\gamma\lambda}|n\rangle.$$

This last result simply states that the (mn)th element $(\hat{a}^{\mathrm{I}}_{\gamma\lambda})_{mn}$ of $\hat{a}^{\mathrm{I}}_{\gamma\lambda}$ is equal to $\mathrm{e}^{-\mathrm{i}\omega_\gamma t}(\hat{a}_{\gamma\lambda})_{mn}$, and this implies

$$\hat{a}^{\mathrm{I}}_{\gamma\lambda} = \mathrm{e}^{-\mathrm{i}\omega_\gamma t}\hat{a}_{\gamma\lambda}. \tag{8.3.33}$$

It is not difficult to similarly prove the adjoint results

$$(\xi^*_{\gamma\lambda})^{\mathrm{I}} = \sum_{ij}\mathrm{e}^{-\mathrm{i}\omega_{ij}t}(\xi^*_{\gamma\lambda})_{ji}|j\rangle\langle i|, \tag{8.3.34}$$

and

$$(\hat{a}^+_{\gamma\lambda})^{\mathrm{I}} = \mathrm{e}^{\mathrm{i}\omega_\gamma t}\hat{a}^+_{\gamma\lambda}, \tag{8.3.35}$$

so that equation (8.3.30) has the form

$$\hat{H}^{\mathrm{I}}_1(t) = \hat{V}(t) + \hat{V}^+(t), \tag{8.3.36}$$

where

$$\begin{aligned}\hat{V}(t) &= \sum_{\gamma\lambda}\xi^{\mathrm{I}}_{\gamma\lambda}\hat{a}^{\mathrm{I}}_{\gamma\lambda} \\ &= \sum_{\gamma\lambda}\sum_{ij}(\xi_{\gamma\lambda})_{ij}\mathrm{e}^{\mathrm{i}(\omega_{ij}-\omega_\gamma)t}\hat{a}_{\gamma\lambda}|i\rangle\langle j|,\end{aligned} \tag{8.3.37}$$

and

$$\hat{V}^+(t) = \sum_{\gamma\lambda}\sum_{ij}(\xi^*_{\gamma\lambda})_{ji}\mathrm{e}^{-\mathrm{i}(\omega_{ij}-\omega_\gamma)t}\hat{a}^+_{\gamma\lambda}|j\rangle\langle i|. \tag{8.3.38}$$

This completes the specification of $\hat{H}_1$ in the interaction representation.

8.3.3 Commutator evaluations

Having now determined the interaction Hamiltonian operator $\hat{H}_1^{\mathrm{I}}(t)$ appearing on the RHS (8.3.24), we next examine the explicit expressions for the commutators occurring there. We start with the first term on the RHS (8.3.24) and, for ease in exposition, consider the special case

$$\hat{I}_{\lambda\lambda'} = J_0 \hat{a}^+_{\gamma\lambda}\hat{a}_{\gamma\lambda'}, \tag{8.3.39}$$

of equation (8.3.14) corresponding to the single frequency ω_γ. We then have from equation (8.3.36)

$$\begin{aligned}[\hat{I}_{\lambda\lambda'}, \hat{H}_1^{\mathrm{I}}(t)] &= J_0\,[\hat{a}^+_{\gamma\lambda}\hat{a}_{\gamma\lambda'}, \hat{V}(t) + \hat{V}^+(t)]\\ &= J_0\,[\hat{a}^+_{\gamma\lambda}\hat{a}_{\gamma\lambda'}, \hat{V}(t)] + J_0\,[\hat{a}^+_{\gamma\lambda}\hat{a}_{\gamma\lambda'}, \hat{V}^+(t)],\end{aligned}$$

where, for example, noting that atom and radiation field operators commute,

$$[\hat{a}_{\gamma\lambda}\hat{a}_{\gamma\lambda'}, \hat{V}(t)] = \sum_{\gamma''\lambda''}\sum_{ij} (\xi_{\gamma''\lambda''})_{ij}\mathrm{e}^{\mathrm{i}(\omega_{ij}-\omega_\gamma)t}|\,i\rangle\langle j\,|\,[\hat{a}^+_{\gamma\lambda}\hat{a}_{\gamma\lambda'}, \hat{a}_{\gamma''\lambda''}]. \tag{8.3.40}$$

However, annihilation and creation operators commute with themselves of course, so that

$$\begin{aligned}[\hat{a}^+_{\gamma\lambda}\hat{a}_{\gamma\lambda'}, \hat{a}_{\gamma''\lambda''}] &= \hat{a}^+_{\gamma\lambda}\hat{a}_{\gamma\lambda'}\hat{a}_{\gamma''\lambda''} - \hat{a}_{\gamma''\lambda''}\hat{a}^+_{\gamma\lambda}\hat{a}_{\gamma\lambda'}\\ &= (\hat{a}^+_{\gamma\lambda}\hat{a}_{\gamma''\lambda''} - \hat{a}_{\gamma''\lambda''}\hat{a}^+_{\gamma\lambda})\hat{a}_{\gamma\lambda'}\\ &= [\hat{a}^+_{\gamma\lambda'}\hat{a}_{\gamma''\lambda''}]\hat{a}_{\gamma\lambda'}\\ &= -\hat{a}_{\gamma\lambda'}\delta_{\gamma\gamma''}\delta_{\lambda\lambda''}.\end{aligned} \tag{8.3.41}$$

Equation (8.3.40) then yields

$$[\hat{a}^+_{\gamma\lambda}\hat{a}_{\gamma\lambda'}, \hat{V}(t)] = -\sum_{ij} (\xi_{\gamma\lambda})_{ij}\mathrm{e}^{\mathrm{i}(\omega_{ij}-\omega_\gamma)t}|\,i\rangle\langle j\,|\hat{a}_{\gamma\lambda'}. \tag{8.3.42}$$

Similarly, we find

$$[\hat{a}^+_{\gamma\lambda}\hat{a}_{\gamma\lambda'}, \hat{a}^+_{\gamma''\lambda''}] = \hat{a}^+_{\gamma\lambda}\,[\hat{a}_{\gamma\lambda'}, \hat{a}^+_{\gamma''\lambda''}] = \hat{a}^+_{\gamma\lambda}\delta_{\gamma\gamma''}\delta_{\lambda'\lambda''}, \tag{8.3.43}$$

so that

$$[\hat{a}^+_{\gamma\lambda}\hat{a}_{\gamma\lambda'}, \hat{V}^+(t)] = \sum_{ij} (\xi^*_{\gamma\lambda'})_{ji}\mathrm{e}^{-\mathrm{i}(\omega_{ij}-\omega_\gamma)t}|\,j\rangle\langle i\,|\hat{a}^+_{\gamma\lambda'}. \tag{8.3.44}$$

The second order commutator appearing on the RHS (8.3.24) may be similarly evaluated. We write

$$\begin{aligned}[[\hat{I}_{\lambda\lambda'}, \hat{H}^{\mathrm{I}}(t)], \hat{H}^{\mathrm{I}}(t')] &= [[\hat{I}_{\lambda\lambda'}, \hat{V}(t) + \hat{V}^+(t)], \hat{V}(t') + \hat{V}^+(t')]\\ &= \hat{\Theta}_1 + \hat{\Theta}_2 + \hat{\Theta}_3 + \hat{\Theta}_4,\end{aligned} \tag{8.3.45}$$

where

$$\left.\begin{aligned}\hat{\Theta}_1 &= [[\hat{I}_{\lambda\lambda'}, \hat{V}(t)], \hat{V}(t')],\\ \hat{\Theta}_2 &= [[\hat{I}_{\lambda\lambda'}, \hat{V}(t)], \hat{V}^+(t')],\\ \hat{\Theta}_3 &= [[\hat{I}_{\lambda\lambda'}, \hat{V}^+(t)], \hat{V}(t')],\\ \hat{\Theta}_4 &= [[\hat{I}_{\lambda\lambda'}, \hat{V}^+(t)], \hat{V}^+(t')].\end{aligned}\right\} \tag{8.3.46}$$

Equation (8.3.44) may be used to simplify each of these terms. For example, we have

$$\hat{\Theta}_1 = [[J_0 \hat{a}^+_{\gamma\lambda}\hat{a}_{\gamma\lambda'}, \hat{V}(t)], \hat{V}(t')]$$
$$= \Big[-J_0 \sum_{ij} (\xi_{\gamma\lambda})_{ij} e^{i(\omega_{ij}-\omega_\gamma)t} |i\rangle\langle j|\hat{a}_{\gamma\lambda'},$$
$$\sum_{\gamma''\lambda''}\sum_{j'i'} (\xi_{\gamma''\lambda''})_{i'j'} e^{i(\omega_{i'j'}-\omega_{\gamma''})t'} |i'\rangle\langle j'|\hat{a}_{\gamma''\lambda''}\Big]$$
$$= -J_0 \sum_{iji'j'}\sum_{\gamma''\lambda''} (\xi_{\gamma\lambda})_{ij}(\xi_{\gamma''\lambda''})_{i'j'} e^{i(\omega_{ij}-\omega_\gamma)t+i(\omega_{i'j'}-\omega_{\gamma''})t'}$$
$$\times [|i\rangle\langle j|\hat{a}_{\gamma\lambda'}, |i'\rangle\langle j'|\hat{a}_{\gamma''\lambda''}]. \tag{8.3.47}$$

The commutator in this last equation can be simplified somewhat by noting that $\hat{a}_{\gamma\lambda'}$ and $\hat{a}_{\gamma''\lambda''}$ commute, i.e.

$$[|i\rangle\langle j|\hat{a}_{\gamma\lambda'}, |i'\rangle\langle j'|\hat{a}_{\gamma''\lambda''}] = \hat{a}_{\gamma\lambda'}\hat{a}_{\gamma''\lambda''}[|i\rangle\langle j|i'\rangle\langle j'| - |i'\rangle\langle j'|i\rangle\langle j|]$$
$$= \hat{a}_{\gamma\lambda'}\hat{a}_{\gamma''\lambda''}[|i\rangle\langle j'|\delta_{i'j} - |i'\rangle\langle j|\delta_{ij'}].$$

We then have

$$\hat{\Theta}_1 = -J_0 \sum_{ij}\sum_{\gamma''\lambda''} (\xi_{\gamma\lambda})_{ij}\hat{a}_{\gamma\lambda'}\hat{a}_{\gamma''\lambda''}\Big\{\sum_{j'}(\xi_{\gamma''\lambda''})_{jj'}$$
$$\times e^{i(\omega_{ij}-\omega_\gamma)t+i(\omega_{jj'}-\omega_{\gamma''})t'}|i\rangle\langle j'|$$
$$-\sum_{i'}(\xi_{\gamma''\lambda''})_{i'i} e^{i(\omega_{ij}-\omega_\gamma)t+i(\omega_{i'i}-\omega_{\gamma''})t'}|i'\rangle\langle j|\Big\}, \tag{8.3.48}$$

with a similar expression for $\hat{\Theta}_4$ but with the creation operator product $\hat{a}^+_{\gamma\lambda}\hat{a}^+_{\gamma''\lambda''}$ replacing $\hat{a}_{\gamma\lambda'}\hat{a}_{\gamma''\lambda''}$. As we shall see, the operator trace of these two $\hat{\Theta}_1$ and $\hat{\Theta}_4$ terms yield zero contribution in the transfer equation (8.3.24) and we do not therefore detail $\hat{\Theta}_4$ beyond the above-mentioned replacement.

The remaining two terms $\hat{\Theta}_2$ and $\hat{\Theta}_3$ appearing in equation (8.3.45) yield non-zero contributions in the transfer equation. Again using equation (8.3.42), we have

$$\hat{\Theta}_2 = [[J_0 \hat{a}^+_{\gamma\lambda}\hat{a}^+_{\gamma\lambda}, \hat{V}(t)], \hat{V}^+(t')]$$
$$= \Big[-J_0 \sum_{ij} (\xi_{\gamma\lambda})_{ij} e^{i(\omega_{ij}-\omega_\gamma)t} |i\rangle\langle j|\hat{a}_{\gamma\lambda'},$$
$$\times \sum_{\gamma''\lambda''}\sum_{i'j'} (\xi^*_{\gamma''\lambda''})_{j'i'} e^{-i(\omega_{i'j'}-\omega_{\gamma''})t'} |j'\rangle\langle i'|\hat{a}^+_{\gamma''\lambda''}\Big]$$
$$= -J_0 \sum_{iji'j'}\sum_{\gamma''\lambda''} (\xi_{\gamma\lambda})_{ij}(\xi^*_{\gamma''\lambda''})_{j'i'} e^{i(\omega_{ij}-\omega_\gamma)t-i(\omega_{i'j'}-\omega_{\gamma''})t}$$
$$\times [|i\rangle\langle j|\hat{a}_{\gamma\lambda'}, |j'\rangle\langle i'|\hat{a}^+_{\gamma''\lambda''}]. \tag{8.3.49}$$

The commutator in this last equation may be expanded as follows:

$$[|i\rangle\langle j|\hat{a}_{\gamma\lambda'}, |j'\rangle\langle i'|\hat{a}^+_{\gamma''\lambda''}] = |i\rangle\langle j|j'\rangle\langle i'|\hat{a}_{\gamma\lambda'}\hat{a}^+_{\gamma''\lambda''}$$
$$-|j'\rangle\langle i'|i\rangle\langle j|\hat{a}^+_{\gamma''\lambda''}\hat{a}_{\gamma\lambda'},$$

which, because $\hat{a}\hat{a}^+$ indirectly contains the intensity operator $\hat{I}$ via $J_0\hat{a}\hat{a}^+ = \hat{I} + J_0$ and this, as we shall see, will eventually produce the desired intensity scalar I on the RHS of the transfer equation, we write as

$$\begin{aligned}|i\rangle\langle i'|\delta_{jj'}(\delta_{\gamma\gamma''}\delta_{\lambda'\lambda''}\mathbb{1} + \hat{a}^+_{\gamma''\lambda''}\hat{a}_{\gamma\lambda'}) - |j'\rangle\langle j|\delta_{ii'}\hat{a}^+_{\gamma''\lambda''}\hat{a}_{\gamma\lambda'}\\ = |i\rangle\langle i'|\delta_{jj'}\delta_{\gamma\gamma''}\delta_{\lambda'\lambda''}\mathbb{1} + (|i\rangle\langle i'|\delta_{jj'} - |j'\rangle\langle j|\delta_{ii'})\hat{a}^+_{\gamma''\lambda''}\hat{a}_{\gamma\lambda'}.\end{aligned} \tag{8.3.50}$$

Equation (8.3.49) then becomes

$$\begin{aligned}\hat{\Theta}_2 &= -J_0 \sum_{iji'j'}\sum_{\gamma''\lambda''}(\xi_{\gamma\lambda})_{ij}(\xi^*_{\gamma''\lambda''})_{j'i'}\mathrm{e}^{\mathrm{i}(\omega_{ij}-\omega_\gamma)t-\mathrm{i}(\omega_{i'j'}-\omega_{\gamma''})t'}\\ &\quad\times\{|i\rangle\langle i'|\delta_{jj'}\delta_{\gamma\gamma''}\delta_{\lambda'\lambda''}\mathbb{1} + (|i\rangle\langle i'|\delta_{jj'} - |j'\rangle\langle j|\delta_{ii'})\hat{a}^+_{\gamma''\lambda''}\hat{a}_{\gamma\lambda'}\}\\ &= -J_0\sum_{ij}\Bigg\{\sum_{i'}(\xi_{\gamma\lambda})_{ij}(\xi^*_{\gamma\lambda'})_{ji'}\mathrm{e}^{\mathrm{i}(\omega_{ij}-\omega_\gamma)t-\mathrm{i}(\omega_{i'j'}-\omega_\gamma)t'}|i\rangle\langle i'|\\ &\quad+\sum_{\gamma''\lambda''}\sum_{i'}(\xi_{\gamma\lambda})_{ij}(\xi^*_{\gamma''\lambda''})_{ji'}\mathrm{e}^{\mathrm{i}(\omega_{ij}-\omega_\gamma)t-\mathrm{i}(\omega_{i'j}-\omega_{\gamma''})t'}|i\rangle\\ &\quad\times\langle i'|\hat{a}^+_{\gamma''\lambda''}\hat{a}_{\gamma\lambda'}\\ &\quad-\sum_{\gamma''\lambda''}\sum_{j'}(\xi_{\gamma\lambda})_{ij}(\xi^*_{\gamma''\lambda''})_{j'i}\mathrm{e}^{\mathrm{i}(\omega_{ij}-\omega_\gamma)t-\mathrm{i}(\omega_{ij'}-\omega_{\gamma''})t'}|j'\rangle\\ &\quad\times\langle j|\hat{a}^+_{\gamma''\lambda''}\hat{a}_{\gamma\lambda'}\Bigg\},\end{aligned} \tag{8.3.51}$$

which, for later convenience, we write as

$$\hat{\Theta}_2 = \hat{\Theta}_{21} + \hat{\Theta}_{22} + \hat{\Theta}_{23}. \tag{8.3.52}$$

The remaining term in equation (8.3.45) may now be evaluated by noting that

$$\begin{aligned}[\hat{P},\hat{Q}]^+ &= (\hat{P}\hat{Q})^+ - (\hat{Q}\hat{P})^+\\ &= \hat{Q}^+\hat{P}^+ - \hat{P}^+\hat{Q}^+\\ &= [\hat{Q}^+,\hat{P}^+]\\ &= -[\hat{P}^+,\hat{Q}^+].\end{aligned} \tag{8.3.53}$$

We then have

$$\begin{aligned}\hat{\Theta}_3^+ &= [[\hat{I}_{\lambda\lambda'},\hat{V}^+(t)],\hat{V}(t')]^+\\ &= -[[\hat{I}_{\lambda\lambda'},\hat{V}^+(t)],\hat{V}^+(t')]\\ &= -[-[\hat{I}^+_{\lambda\lambda'},\hat{V}(t)],\hat{V}^+(t')]\\ &= [[\hat{I}^+_{\lambda\lambda'},\hat{V}(t)],\hat{V}^+(t')].\end{aligned} \tag{8.3.54}$$

However, the adjoint of the intensity operator $\hat{I}_{\lambda\lambda'}$ yields

$$\hat{I}^+_{\lambda\lambda'} = J_0(\hat{a}^+_{\gamma\lambda}\hat{a}_{\gamma\lambda'})^+ = J_0\hat{a}^+_{\gamma\lambda'}\hat{a}_{\gamma\lambda} = \hat{I}_{\lambda'\lambda}, \tag{8.3.55}$$

so that

$$\hat{\Theta}_3^+ = [[\hat{I}_{\lambda'\lambda},\hat{V}(t)],\hat{V}^+(t')], \tag{8.3.56}$$

which is identical to $\hat{\Theta}_2$ but for the reverse ordering of λ and λ'. Thus, writing

$\hat{\Theta}_2 \equiv \hat{\Theta}_2(\lambda, \lambda')$, we have $\hat{\Theta}_3^+ = \hat{\Theta}_2(\lambda', \lambda)$ so that

$$\hat{\Theta}_3 = \hat{\Theta}_2^+(\lambda', \lambda), \tag{8.3.57}$$

which, from equation (8.3.51) yields

$$\begin{aligned}\hat{\Theta}_3 = -J_0 \sum_{ij} \Bigg\{ & \sum_{i'} (\xi^*_{\gamma\lambda'})_{ji}(\xi_{\gamma\lambda})_{i'j} \mathrm{e}^{-\mathrm{i}(\omega_{ij}-\omega_\gamma)t+\mathrm{i}(\omega_{i'j}-\omega_\gamma)t'} |i'\rangle\langle i| \\ & + \sum_{\gamma''\lambda''} \sum_{i'} (\xi^*_{\gamma\lambda'})_{ji}(\xi_{\gamma''\lambda''})_{i'j} \mathrm{e}^{-\mathrm{i}(\omega_{ij}-\omega_\gamma)t+\mathrm{i}(\omega_{i'j}-\omega_{\gamma''})t'} |i'\rangle \\ & \times \langle i|\hat{a}^+_{\gamma\lambda}\hat{a}_{\gamma''\lambda''} \\ & - \sum_{\gamma''\lambda''} \sum_{j'} (\xi^*_{\gamma\lambda'})_{ji}(\xi_{\gamma''\lambda''})_{ij'} \mathrm{e}^{-\mathrm{i}(\omega_{ij}-\omega_\gamma)t+\mathrm{i}(\omega_{ij'}-\omega_{\gamma''})t'} |j\rangle \\ & \times \langle j'|\hat{a}^+_{\gamma\lambda}\hat{a}_{\gamma''\lambda''} \Bigg\},\end{aligned} \tag{8.3.58}$$

which we write as

$$\hat{\Theta}_3 = \hat{\Theta}_{31} + \hat{\Theta}_{32} + \hat{\Theta}_{33}. \tag{8.3.59}$$

The above $\hat{\Theta}_3$, together with $\hat{\Theta}_2$ from equation (8.3.51) and $\hat{\Theta}_1$ and $\hat{\Theta}_4$, determines the second order commutator (8.3.45). This, and the first order commutation expressions (8.3.42) and (8.3.44) then enable the RHS of the transfer equation (8.3.24) to be explicitly detailed.

8.3.4 Derivation of the transfer equation

The theory presented in all the preceding sections may now be used to simplify the RHS of the radiative transfer equation (8.3.24). Clearly, we require terms of the form

$$T_1 = \{[\hat{I}_{\lambda\lambda'}, \hat{H}_1^{\mathrm{I}}(t)]\rho^{\mathrm{I}}(0)\}, \tag{8.3.60}$$

$$T_2 = \{[[\hat{I}_{\lambda\lambda'}, \hat{H}_1^{\mathrm{I}}(t)], \hat{H}_1^{\mathrm{I}}(t')]\rho^{\mathrm{I}}(0)\}. \tag{8.3.61}$$

In particular, equations (8.3.36), (8.3.39), (8.3.42) and (8.3.44) yield

$$\begin{aligned} T_1 &= \mathrm{tr} \Bigg\{ -J_0 \sum_{ij} \{(\xi_{\gamma\lambda})_{ij} \mathrm{e}^{\mathrm{i}(\omega_{ij}-\omega_\gamma)t} |i\rangle\langle j|\hat{a}_{\gamma\lambda'}\rho^{\mathrm{I}}(0) \\ &\qquad - (\xi^*_{\gamma\lambda'})_{ji} \mathrm{e}^{-\mathrm{i}(\omega_{ij}-\omega_\gamma)t} |j\rangle\langle i|\hat{a}^+_{\gamma\lambda}\rho^{\mathrm{I}}(0) \Bigg\} \\ &= -J_0 \sum_{ij} \{(\xi_{\gamma\lambda})_{ij} \mathrm{e}^{\mathrm{i}(\omega_{ij}-\omega_\gamma)t}\, \mathrm{tr}\, \{|i\rangle\langle j|\hat{a}_{\gamma\lambda'}\rho^{\mathrm{I}}(0)\} \\ &\qquad - (\xi^*_{\gamma\lambda'})_{ji} \mathrm{e}^{-\mathrm{i}(\omega_{ij}-\omega_\gamma)t}\, \mathrm{tr}\, \{|j\rangle\langle i|\hat{a}^+_{\gamma\lambda}\rho^{\mathrm{I}}(0)\}. \end{aligned} \tag{8.3.62}$$

We now assume that the atoms and radiation field are initially statistically independent of one another (this was discussed in section 8.2.1) so that

$$\rho^{\mathrm{I}}(0) = \rho^{\mathrm{I}}_{\mathrm{A}}(0) \otimes \rho^{\mathrm{I}}_{\mathrm{R}}(0). \tag{8.3.63}$$

We then have, for example,

$$\operatorname{tr}\{|i\rangle\langle j|\hat{a}_{\gamma\lambda'}\rho^{\mathrm{I}}(0)\} = \sum_{kn} \langle k|\langle n|i\rangle\langle j|\hat{a}_{\gamma\lambda'}\rho^{\mathrm{I}}_{\mathrm{A}}(0)\otimes\rho^{\mathrm{I}}_{\mathrm{R}}(0)|n\rangle|k\rangle$$
$$= \sum_{kn} \langle k|i\rangle\langle j|\rho^{\mathrm{I}}_{\mathrm{A}}(0)|k\rangle\langle n|\hat{a}_{\gamma\lambda'}\rho^{\mathrm{I}}_{\mathrm{R}}(0)|n\rangle$$
$$= \sum_{kn} \delta_{ki}\langle j|\rho^{\mathrm{I}}_{\mathrm{A}}(0)|k\rangle\sqrt{(n+1)}\,\langle n+1|\rho^{\mathrm{I}}_{\mathrm{R}}(0)|n\rangle, \tag{8.3.64}$$

where we have clearly taken $|k\rangle$ and $|n\rangle$ to represent atom and photon states respectively, and noted that $\langle n|\hat{a} \equiv (\hat{a}^{+}|n\rangle)^{+} = \sqrt{(n+1)}\,\langle n+1|$.

If we next assume that $\rho^{\mathrm{I}}_{\mathrm{R}}(0)$ is purely diagonal, and this simply states that we know the initial non-interacting distribution of the radiation field, then clearly the off-diagonal elements $\langle n+1|\rho^{\mathrm{I}}_{\mathrm{R}}(0)|n\rangle$ are zero so that

$$\operatorname{tr}\{|i\rangle\langle j|\hat{a}_{\gamma\lambda'}\rho^{\mathrm{I}}(0)\} = 0. \tag{8.3.65}$$

Similarly, we may show that the other trace appearing in equation (8.3.62) is also zero, and this then yields

$$T_1 = 0. \tag{8.3.66}$$

We now turn our attention to the T_2 term (8.3.61). Equations (8.3.45, 46), (8.3.52) and (8.3.59) yield

$$T_2 = \operatorname{tr}\left[(\hat{\Theta}_1 + \hat{\Theta}_4)\rho^{\mathrm{I}}(0) + \sum_{l=1}^{3} (\hat{\Theta}_{2l} + \hat{\Theta}_{3l})\rho^{\mathrm{I}}(0)\right]. \tag{8.3.67}$$

First, we see from equation (8.3.48) that the $\operatorname{tr}(\hat{\Theta}_1\rho^{\mathrm{I}}(0))$ term contains factors of the form

$$\operatorname{tr}\{\hat{a}_{\gamma\lambda'}\hat{a}_{\gamma''\lambda''}|i\rangle\langle j'|\rho^{\mathrm{I}}(0)\} = \sum_{kn} \langle k|i\rangle\langle j'|\rho^{\mathrm{I}}_{\mathrm{A}}(0)|k\rangle$$
$$\times \langle n|\hat{a}_{\gamma\lambda'}\hat{a}_{\gamma''\lambda''}\rho^{\mathrm{I}}_{\mathrm{R}}(0)|n\rangle,$$

where we have used the factorisation (8.3.63). However, we should be more explicit by writing

$$\sum_{n} \langle n|\hat{a}_{\gamma\lambda'}\hat{a}_{\gamma''\lambda''}\rho^{\mathrm{I}}_{\mathrm{R}}(0)|n\rangle \equiv \sum_{n_r} \langle n_r|\hat{a}_{\gamma\lambda'}\hat{a}_{\gamma''\lambda''}\rho^{\mathrm{I}}_{\mathrm{R}}(0)|n_r\rangle$$
$$= \sum_{n_r} \sqrt{(n_r+1)}\,\langle n_r+1|\hat{a}_{\gamma''\lambda''}\rho^{\mathrm{I}}_{\mathrm{R}}(0)|n_r\rangle\,\delta_{r\gamma}$$
$$= \sum_{n_r} \sqrt{[(n_r+1)(n_r+2)]}\,\langle n_r+2|\rho^{\mathrm{I}}_{\mathrm{R}}(0)|n_r\rangle\,\delta_{r\gamma}\delta_{r\gamma''}$$
$$\equiv \sum_{n} \sqrt{[(n+1)(n+2)]}\,\langle n+2|\rho^{\mathrm{I}}_{\mathrm{R}}(0)|n\rangle\,\delta_{\gamma\gamma''}, \tag{8.3.68}$$

which is zero as we again take $\rho^{\mathrm{I}}_{\mathrm{R}}(0)$ to be purely diagonal. The same arguments may be applied to all other traces over the terms involving $\hat{\Theta}_1$ and $\hat{\Theta}_4$ so that

$$\operatorname{tr}\{(\hat{\Theta}_1 + \hat{\Theta}_4)\rho^{\mathrm{I}}(0)\} = 0. \tag{8.3.69}$$

Each of the remaining six terms in equation (8.3.67) can be examined separately. First, using equations (8.3.51) and (8.3.63), we find

$$\mathrm{tr}\,\{\hat{\Theta}_{21}\rho^{\mathrm{I}}(0)\} = -J_0 \sum_{iji'} (\xi_{\gamma\lambda})_{ij}(\xi^*_{\gamma\lambda'})_{ji'} \times \mathrm{e}^{\mathrm{i}(\omega_{ij}-\omega_\gamma)t-\mathrm{i}(\omega_{i'j}-\omega_\gamma)t'} \times \sum_{kn} \langle k|i\rangle\langle i'|\rho_{\mathrm{A}}^{\mathrm{I}}(0)|k\rangle\langle n|\rho_{\mathrm{R}}^{\mathrm{I}}(0)|n\rangle. \quad (8.3.70)$$

We already know from equation (8.2.16) that the trace of ρ is unity or, in the above reduced case,

$$\mathrm{tr_R}\,[\rho_{\mathrm{R}}^{\mathrm{I}}(0)] = \sum_n \langle n|\rho_{\mathrm{R}}^{\mathrm{I}}(0)|n\rangle = 1, \quad (8.3.71)$$

so that

$$\sum_{kn} \langle k|i\rangle\langle i'|\rho_{\mathrm{A}}^{\mathrm{I}}(0)|k\rangle\langle n|\rho^{\mathrm{I}}(0)|n\rangle = \sum_k \delta_{ki}\langle i'|\rho_{\mathrm{A}}^{\mathrm{I}}(0)|k\rangle = \langle i'|\rho_{\mathrm{A}}^{\mathrm{I}}(0)|i\rangle.$$

Thus, taking the initial reduced atom density matrix to be purely diagonal also, i.e.

$$\langle i'|\rho_{\mathrm{A}}^{\mathrm{I}}(0)|i\rangle = \rho_{ii}\delta_{ii'}, \quad (8.3.72)$$

where ρ_{ii} is the atom population probability for state $|i\rangle$ (note that $\Sigma_i\rho_{ii} = \mathrm{tr_A}\,[\rho_{\mathrm{A}}^{\mathrm{I}}(0)] = 1$), then

$$\mathrm{tr}\,\{\hat{\Theta}_{21}\rho^{\mathrm{I}}(0)\} = -J_0 \sum_{ij} (\xi_{\gamma\lambda})_{ij}(\xi^*_{\gamma\lambda})_{ji}\mathrm{e}^{\mathrm{i}(\omega_{ij}-\omega_\gamma)(t-t')}\rho_{ii}. \quad (8.3.73)$$

Similarly, using equations (8.3.58) and (8.3.63), we find

$$\begin{aligned}\mathrm{tr}\,\{\hat{\Theta}_{31}\rho^{\mathrm{I}}(0)\} &= -J_0 \sum_{iji'} (\xi^*_{\gamma\lambda'})_{ji}(\xi_{\gamma\lambda})_{i'j}\mathrm{e}^{-\mathrm{i}(\omega_{ij}-\omega_\gamma)t+\mathrm{i}(\omega_{i'j}-\omega_\gamma)t'} \times \sum_{kn} \langle k|i'\rangle\langle i|\rho_{\mathrm{A}}^{\mathrm{I}}(0)|k\rangle\langle n|\rho_{\mathrm{R}}^{\mathrm{I}}(0)|n\rangle \\ &= -J_0 \sum_{ij} (\xi^*_{\gamma\lambda'})_{ji}(\xi_{\gamma\lambda})_{ij}\mathrm{e}^{-\mathrm{i}(\omega_{ij}-\omega_\gamma)(t-t')}\rho_{ii}. \quad (8.3.74)\end{aligned}$$

We now require the trace over the remaining four $\hat{\Theta}$ terms appearing in equation (8.3.67). We first note that

$$\mathrm{tr}\,\{\hat{\Theta}_{22}\rho^{\mathrm{I}}(0)\} = -J_0 \sum_{iji'} \sum_{\gamma''\lambda''} (\xi_{\gamma\lambda})_{ij}(\xi^*_{\gamma''\lambda''})_{ji'} \times \mathrm{e}^{\mathrm{i}(\omega_{ij}-\omega_\gamma)t-\mathrm{i}(\omega_{i'j}-\omega_{\gamma''})t'} \times \sum_{kn} \langle k|i\rangle\langle i'|\rho_{\mathrm{A}}^{\mathrm{I}}(0)|k\rangle\langle n|\hat{a}^+_{\gamma''\lambda''}\hat{a}_{\gamma\lambda'}\rho_{\mathrm{R}}^{\mathrm{I}}(0)|n\rangle. \quad (8.3.75)$$

However, following the same argument which led to equation (8.3.68), we have

$$\sum_n \langle n|\hat{a}^+_{\gamma''\lambda''}\hat{a}_{\gamma\lambda'}\rho_{\mathrm{R}}^{\mathrm{I}}(0)|n\rangle = \sum_n n\,\langle n|\rho_{\mathrm{R}}^{\mathrm{I}}(0)|n\rangle\,\delta_{\gamma\gamma''}. \quad (8.3.76)$$

The $\delta_{\gamma\gamma''}$ factor appearing in this last result then stipulates that only the $\gamma'' = \gamma$ term should be retained in the summation over γ'' in equation (8.3.75). However, we do not proceed directly with equation (8.3.76) but, rather, recall the definition (8.3.39):

$$\hat{I}_{\lambda\lambda'} = J_0 \hat{a}^{+}_{\gamma\lambda}\hat{a}_{\gamma\lambda'}, \tag{8.3.39}$$

so that equation (8.3.75) becomes

$$\begin{aligned}\operatorname{tr}\{\hat{\Theta}_{22}\rho^{\mathrm{I}}_{\mathrm{R}}(0)\} = &-\sum_{iji'}\sum_{\lambda''}(\xi_{\gamma\lambda})_{ij}(\xi^{*}_{\gamma\lambda''})_{ji'}\\ &\times \mathrm{e}^{\mathrm{i}(\omega_{ij}-\omega_{\gamma})t-\mathrm{i}(\omega_{i'j}-\omega_{\gamma})t'}\\ &\times \sum_{kn}\langle k|i\rangle\langle i'|\rho^{\mathrm{I}}_{\mathrm{A}}(0)|k\rangle\langle n|\hat{I}_{\lambda''\lambda'}\rho^{\mathrm{I}}_{\mathrm{R}}(0)|n\rangle,\end{aligned}$$

which, using equations (8.2.19) and (8.3.16) in the form

$$\begin{aligned}I_{\lambda''\lambda'} = \langle\hat{I}_{\lambda''\lambda'}\rangle &= \operatorname{tr}_{\mathbf{R}}[\hat{I}_{\lambda''\lambda'}\rho^{\mathrm{I}}_{\mathrm{R}}(0)]\\ &= \sum_{n}\langle n|\hat{I}_{\lambda''\lambda'}\rho^{\mathrm{I}}_{\mathrm{R}}(0)|n\rangle,\end{aligned} \tag{8.3.77}$$

where the factorisation (8.3.36) is implicit, yields

$$\begin{aligned}\operatorname{tr}\{\hat{\Theta}_{22}\rho^{\mathrm{I}}(0)\} = &-\sum_{ij}\sum_{\lambda''}(\xi_{\gamma\lambda})_{ij}(\xi^{*}_{\gamma\lambda''})_{ji}\\ &\times \mathrm{e}^{\mathrm{i}(\omega_{ij}-\omega_{\gamma})(t-t')}\rho_{ii}I_{\lambda''\lambda'}.\end{aligned} \tag{8.3.78}$$

Similarly, we find

$$\begin{aligned}\operatorname{tr}\{\hat{\Theta}_{32}\rho^{\mathrm{I}}(0)\} = &-\sum_{ij}\sum_{\lambda''}(\xi^{*}_{\gamma\lambda'})_{ji}(\xi_{\gamma\lambda''})_{ij}\\ &\times \mathrm{e}^{-\mathrm{i}(\omega_{ij}-\omega_{\gamma})(t-t')}\rho_{ii}I_{\lambda\lambda''},\end{aligned} \tag{8.3.79}$$

$$\begin{aligned}\operatorname{tr}\{\hat{\Theta}_{23}\rho^{\mathrm{I}}(0)\} = &\sum_{ij}\sum_{\lambda''}(\xi_{\gamma\lambda})_{ij}(\xi^{*}_{\gamma\lambda''})_{ji}\\ &\times \mathrm{e}^{\mathrm{i}(\omega_{ij}-\omega_{\gamma})(t-t')}\rho_{jj}I_{\lambda''\lambda'},\end{aligned} \tag{8.3.80}$$

$$\begin{aligned}\operatorname{tr}\{\hat{\Theta}_{33}\rho^{\mathrm{I}}(0)\} = &\sum_{ij}\sum_{\lambda''}(\xi^{*}_{\gamma\lambda'})_{ji}(\xi_{\gamma\lambda''})_{ij}\\ &\times \mathrm{e}^{-\mathrm{i}(\omega_{ij}-\omega_{\gamma})(t-t')}\rho_{jj}I_{\lambda\lambda''}.\end{aligned} \tag{8.3.81}$$

All the above terms appearing in T_2 must now be substituted into the integral in equation (8.3.24). We first recall equation (6.4.85), viz.

$$\lim_{y\to\infty}\frac{1-\mathrm{e}^{\mathrm{i}xy}}{x} = -\mathrm{i}\int_0^\infty \mathrm{e}^{\mathrm{i}xy}\,\mathrm{d}y = \mathscr{P}\frac{1}{x} - \pi\mathrm{i}\delta(x). \tag{8.3.82}$$

Consequently, taking the limit $t\to\infty$ in equation (8.3.24) – note that we are interested in the transfer equation involving time scales large compared with the period $2\pi/\omega_\gamma$ of the fluctuation of the quantised electromagnetic field – we find the integral of the time-dependent term in equation (8.3.73), for example, yields

$$\lim_{t\to\infty}\int_0^t e^{i(\omega_{ij}-\omega_\gamma)(t-t')}\,dt' = -\frac{1}{i}\lim_{t\to\infty}\frac{1-e^{i(\omega_{ij}-\omega_\gamma)t}}{\omega_{ij}-\omega_\gamma}$$

$$= i\mathscr{P}\frac{1}{\omega_{ij}-\omega_\gamma} + \pi\delta(\omega_{ij}-\omega_\gamma). \tag{8.3.83}$$

Replacing i by $-i$ produces the other integral term of interest. We then have

$$\lim_{t\to\infty}\int_0^t \text{tr}\,\{(\hat{\Theta}_{21}+\hat{\Theta}_{31})\rho^I(0)\}\,dt'$$

$$= -J_0\sum_{ij}(\xi_{\gamma\lambda})_{ij}(\xi^*_{\gamma\lambda'})_{ji}\rho_{ii}\left\{\left\{i\mathscr{P}\frac{1}{\omega_{ij}-\omega_\gamma}+\pi\delta(\omega_{ij}-\omega_\gamma)\right\}\right.$$

$$\left.+\left\{-i\mathscr{P}\frac{1}{\omega_{ij}-\omega_\gamma}+\pi\delta(\omega_{ij}-\omega_\gamma)\right\}\right\}$$

$$= -2\pi J_0\sum_{ij}(\xi_{\gamma\lambda})_{ij}(\xi^*_{\gamma\lambda'})_{ji}\rho_{ii}\delta(\omega_{ij}-\omega_\gamma). \tag{8.3.84}$$

Next, the $\hat{\Theta}_{22}$ and $\hat{\Theta}_{32}$ terms yield

$$\lim_{t\to\infty}\int_0^t \text{tr}\,\{(\hat{\Theta}_{22}+\hat{\Theta}_{32})\rho^I(0)\}\,dt'$$

$$= -\sum_{ij}\sum_{\lambda''}\rho_{ii}\left\{(\xi_{\gamma\lambda})_{ij}(\xi^*_{\gamma\lambda''})_{ji}I_{\lambda''\lambda'}\left\{i\mathscr{P}\frac{1}{\omega_{ij}-\omega_\gamma}+\pi\delta(\omega_{ij}-\omega_\gamma)\right\}\right.$$

$$\left.+(\xi_{\gamma\lambda''})_{ij}(\xi^*_{\gamma\lambda'})_{ji}I_{\lambda\lambda''}\left\{-i\mathscr{P}\frac{1}{\omega_{ij}-\omega_\gamma}+\pi\delta(\omega_{ij}-\omega_\gamma)\right\}\right\}, \tag{8.3.85}$$

whilst

$$\lim_{t\to\infty}\int_0^t \text{tr}\,\{(\hat{\Theta}_{23}+\hat{\Theta}_{33})\rho^I(0)\}\,dt'$$

$$= \sum_{ij}\sum_{\lambda''}\rho_{jj}\left\{(\xi_{\gamma\lambda})_{ij}(\xi^*_{\gamma\lambda''})_{ji}I_{\lambda''\lambda'}\left\{i\mathscr{P}\frac{1}{\omega_{ij}-\omega_\gamma}+\pi\delta(\omega_{ij}-\omega_\gamma)\right\}\right.$$

$$\left.+(\xi_{\gamma\lambda''})_{ij}(\xi^*_{\gamma\lambda'})_{ji}I_{\lambda\lambda''}\left\{-i\mathscr{P}\frac{1}{\omega_{ij}-\omega_\gamma}+\pi\delta(\omega_{ij}-\omega_\gamma\right\}\right\}. \tag{8.3.86}$$

Clearly, the expressions in braces in these last two equations are identical. Consequently, recalling $T_1 = 0$ from equation (8.3.66) and using equation

(8.3.84), the radiative transfer equation (8.3.24) becomes

$$\frac{\partial I_{\lambda\lambda'}}{\partial t} + c(\boldsymbol{\Omega}\cdot\nabla)I_{\lambda\lambda'}$$
$$= \frac{2\pi J_0}{\hbar^2}\sum_{ij}(\xi_{\gamma\lambda})_{ij}(\xi^*_{\gamma\lambda'})_{ji}\rho_{ii}\delta(\omega_{ij}-\omega_\gamma)$$
$$+\frac{\pi}{\hbar^2}\sum_{ij}\sum_{\lambda'}[(\xi_{\gamma\lambda})_{ij}(\xi^*_{\gamma\lambda''})_{ji}I_{\lambda''\lambda'}$$
$$+(\xi_{\gamma\lambda''})_{ij}(\xi^*_{\gamma\lambda'})_{ji}I_{\lambda\lambda''}](\rho_{ii}-\rho_{jj})\delta(\omega_{ij}-\omega_\gamma)$$
$$+\frac{\pi}{\hbar^2}\sum_{ij}\sum_{\lambda''}[(\xi_{\gamma\lambda})_{ij}(\xi^*_{\gamma\lambda''})_{ji}I_{\lambda''\lambda'}$$
$$-(\xi_{\gamma\lambda''})_{ij}(\xi^*_{\gamma\lambda'})_{ji}I_{\lambda\lambda''}](\rho_{ii}-\rho_{jj})\mathscr{P}\frac{1}{\omega_{ij}-\omega_\gamma}\,. \tag{8.3.87}$$

The above equation specifies the transfer of radiation for the four polarised intensities I_{11}, I_{12}, I_{21} and I_{22}. The three terms in this equation can be more readily understood if we now consider the unpolarised situation for which we write

$$I_{11}=I=I_{22},\quad I_{12}=0=I_{21} \tag{8.3.88}$$

– note, for example, that $\mathbf{e}_{\gamma\lambda}\cdot\mathbf{e}_{\gamma\lambda'}=\delta_{\lambda\lambda'}$. It is not difficult to then show that

$$\frac{\partial I}{\partial t}+c(\boldsymbol{\Omega}\cdot\nabla)I=\frac{2\pi J_0}{\hbar^2}\sum_{ij}\eta_{ij}(\mathbf{k}_\gamma)\left\{\rho_{ii}+(\rho_{ii}-\rho_{jj})\frac{I}{J_0}\ \delta(\omega_{ij}-\omega_\gamma)\right\}, \tag{8.3.89}$$

where

$$\eta_{ij}(\mathbf{k}_\gamma)=\tfrac{1}{2}\sum_{\lambda}(\xi_{\gamma\lambda})_{ij}(\xi^*_{\gamma\lambda})_{ji}. \tag{8.3.90}$$

We now see that the first term in braces in equation (8.3.89) corresponds to spontaneous photon emission, the second to stimulated photon emission and the third to photon absorption. Note that the term $\delta(\omega_{ij}-\omega_\gamma)$ with $\omega_{ij}=(E_i^{(A)}-E_j^{(A)})/\hbar$ stipulates that $i>j$, i.e. the $|i\rangle$ state is excited relative to the state $|j\rangle$. Equation (8.3.89) can, of course, be directly compared with the phenomelogically derived classical transfer equation (1.2.7) of chapter 1, viz.

$$\frac{\partial I}{\partial t}+c(\boldsymbol{\Omega}\cdot\nabla)I=\frac{h\nu c}{4\pi}\,[N_{\mathrm{U}}A_{\mathrm{UL}}j+N_{\mathrm{U}}B_{\mathrm{UL}}I\psi-N_{\mathrm{L}}B_{\mathrm{LU}}I\phi]. \tag{8.3.91}$$

In particular, we see that $\eta_{ij}\rho_{ii}\to N_{\mathrm{U}}A_{\mathrm{UL}}$, $\eta_{ij}\rho_{ii}I/J_0\to N_{\mathrm{U}}B_{\mathrm{UL}}I$ and $\eta_{ij}\rho_{jj}I/J_0\to N_{\mathrm{L}}B_{\mathrm{LU}}I$ so that $J_0=A_{\mathrm{UL}}/B_{\mathrm{UL}}=2h\nu^3_{\mathrm{UL}}/c^2$.

Several points should be mentioned here. First, we have only retained the first two terms on the RHS (8.3.24). This is equivalent to retaining, for example, the first two terms $1-\Gamma t$ in the expansion of $e^{-\Gamma t}$ discussed in section 8.1.4 in relation to the ordering of solutions. The inclusion of higher order terms in equation (8.3.24) will eventually yield a Lorentzian-type profile similar to that obtained in the simplified case examined using Laplace transforms in section 8.2.2 although the definition of Γ will be somewhat different in the present case. Indeed, the above retention of the first two terms in equation (8.3.24) is effectively equivalent to replacing $\rho^{\mathrm{I}}(t')$ in the exact result (8.2.7) by $\rho^{\mathrm{I}}(0)$. It could be feasible to better approximate this $\rho^{\mathrm{I}}(t')$ by a term containing the atomic factor $e^{-\Gamma t}$ as discussed in section 6.4.3. It is clear from equation (8.3.83) that such a replacement will immediately yield the expected Lorentzian rather than the delta function - however, this point needs further careful consideration particularly in view of the need to re-define Γ using the mixed state density matrix specifically rather than the factorisation (8.3.63).

Finally, we recall that elastic and inelastic collisions have been ignored here. If they were to be included, the same analysis as that presented above may be used but one should incorporate another Hamiltonian operator $\hat{H}^{\mathrm{I}}_{\mathrm{PA}}$ corresponding to particle–atom interactions and this, as in section 6.5, necessitates the replacement of the basis vectors $|k\rangle$ and $|n\rangle$ and the exponential time factors $e^{\pm i(\omega_{ij}-\omega_\gamma)t}$ by the appropriate time evolution operators. Further, one needs to both average over all the possible parameters describing these elastic and inelastic collisions (see equation (6.5.68), for example) and, of course, ensemble average over the microscopic motions of the individual atoms even in the absence of these collisions. This, and the fact that we have assumed the factorisation (8.3.63), not to mention the explicit diagonal structure of the initial reduced density matrices $\rho^{\mathrm{I}}_{\mathrm{A}}(0)$ and $\rho^{\mathrm{I}}_{\mathrm{R}}(0)$, certainly casts suspicion on the veracity of the classical equation of radiative transfer. Nevertheless, we see that even if we remove these simplifications, the method outlined above still provides a general procedure for the derivation of the quantum electrodynamical radiative transfer equation, and this enables magnetic fields, for example, and laser-type interactions, along with variable refractive index effects, to be readily incorporated into the analysis.

8.3.5 Population rate equations [75]

The radiative transfer equation (8.3.89) incorporates the population densities ρ_{ii} and ρ_{jj} analogous to the terms N_{L} and N_{U} appearing in the corresponding classical result (8.3.91). The ratio of these densities may be evaluated in much the same manner as that used to determine the ratio $N_{\mathrm{U}}/N_{\mathrm{L}}$ discussed in section 1.3 where, it will be recalled, we argued that in equilibrium the rate of population of a given atomic level by both radiative and inelastic collisional events must balance the corresponding rate of de-population. In the present section, however, we bypass this phenomenological approach by initiating our analysis from the quantum dynamical description of the radiating gas using the density matrix. For example, we have from equation (8.3.72)

$$\begin{aligned}\rho_{kk} &= \langle k|\rho_A^I(0)|k\rangle \\ &= \langle k|\rho_A^I(0)|k\rangle \sum_n \langle n|\rho_R^I(0)|n\rangle \\ &= \sum_n \langle n|\langle k|\rho_A^I(0)\otimes\rho_R^I(0)|k\rangle|n\rangle \\ &= \mathrm{tr}_R\,\{\langle k|\rho^I(0)|k\rangle\}. \end{aligned} \tag{8.3.92}$$

Equation (8.3.20) then yields

$$\frac{\partial\rho_{kk}}{\partial t} = \frac{1}{i\hbar}\,\mathrm{tr}_R\,\{\langle k|[\hat{H}_1^I(t),\rho^I(0)]|k\rangle\} + \frac{1}{(i\hbar)^2}\int_0^t \mathrm{d}t'\,\mathrm{tr}_R\,\{\langle k|[\hat{H}_1^I(t),[\hat{H}_1^I(t'),\rho^I(0)]]|k\rangle\} + \cdots. \tag{8.3.93}$$

We shall examine each of the terms on the RHS of this last equation separately. First, recalling $\hat{H}_1^I(t)$ from equations (8.3.36) through (8.3.38), viz.

$$\hat{H}_1^I(t) = \hat{V}(t) + \hat{V}^+(t),$$

where

$$\hat{V}(t) = \sum_{\gamma\lambda}\sum_{ij}(\xi_{\gamma\lambda})_{ij}\,\mathrm{e}^{\mathrm{i}(\omega_{ij}-\omega_\gamma)t}\hat{a}_{\gamma\lambda}|i\rangle\langle j|,$$

$$\hat{V}^+(t) = \sum_{\gamma\lambda}\sum_{ij}(\xi^*_{\gamma\lambda})_{ji}\,\mathrm{e}^{-\mathrm{i}(\omega_{ij}-\omega_\gamma)t}\hat{a}^+_{\gamma\lambda}|j\rangle\langle i|,$$

then, assuming the factorisation (8.3.63), we have

$$\begin{aligned}&\mathrm{tr}_R\,\{\langle k||\hat{H}_1^I(t),\rho^I(0)||k\rangle\} \\ &= \sum_n \langle n|\{\langle k|\hat{V}(t)\rho_A^I(0)\otimes\rho_R^I(0)|k\rangle - \langle k|\rho_A^I(0)\otimes\rho_R^I(0)\hat{V}(t)|k\rangle \\ &\quad + \langle k|\hat{V}^+(t)\rho_A^I(0)\otimes\rho_R^I(0)|k\rangle - \langle k|\rho_A^I(0)\otimes\rho_R^I(0)\hat{V}(t)|k\rangle\}|n\rangle \\ &= \sum_{\gamma\lambda}\sum_{ij}(\xi_{\gamma\lambda})_{ij}\,\mathrm{e}^{\mathrm{i}(\omega_{ij}-\omega_\gamma)t}\{\langle n|\hat{a}_{\gamma\lambda}\rho_R^I(0)|n\rangle\langle k|i\rangle\langle j|\rho_A^I(0)|k\rangle \\ &\quad - \langle n|\rho_R^I(0)\hat{a}_{\gamma\lambda}|n\rangle\langle k|\rho_A^I(0)|i\rangle\langle j|k\rangle\} + \sum_{\gamma\lambda}\sum_{ij}(\xi^*_{\gamma\lambda})_{ji}\,\mathrm{e}^{-(\omega_{ij}-\omega_\gamma)t} \\ &\quad \times \{\langle n|\hat{a}^+_{\gamma\lambda}\rho_R^I(0)|n\rangle\langle k|j\rangle\langle i|\rho_A^I(0)|k\rangle \\ &\quad - \langle n|\rho_R^I(0)\hat{a}^+_{\gamma\lambda}|n\rangle\langle k|\rho_A^I(0)|j\rangle\langle i|k\rangle\}.\end{aligned}$$

Again, we note that terms like $\langle n|\rho_R^I(0)\hat{a}_{\gamma\lambda}|n\rangle = \sqrt{n}\,\langle n|\rho_R^I(0)|n-1\rangle$ are zero if we take ρ_R^I to be initially diagonal only. Consequently, it is not difficult to show that all terms in this last equation are zero, i.e.

$$\mathrm{tr}_R\,\{\langle k|[\hat{H}_1^I(t),\rho^I(0)]|k\rangle\} = 0. \tag{8.3.94}$$

We now turn our attention to the trace operation appearing in the second term on the RHS (8.3.93). It is clear that the expansion of the double commutator will yield terms, amongst others, like

$$\begin{aligned}&\mathrm{tr_R}\,\{\langle k|\hat{V}(t)\hat{V}(t')\rho^{\mathrm{I}}(0)|k\rangle\}\\&\quad=\sum_n\sum_{\gamma\lambda\gamma'\lambda'}\sum_{iji'j'}(\xi_{\gamma\lambda})_{ij}(\xi_{\gamma'\lambda'})_{i'j'}\,\mathrm{e}^{\mathrm{i}(\omega_{ij}-\omega_\gamma)t+\mathrm{i}(\omega_{i'j'}-\omega_{\gamma'})t'}\\&\quad\times\delta_{ki}\delta_{i'j}\langle j'|\rho^{\mathrm{I}}_{\mathrm{A}}(0)|k\rangle\langle n|\hat{a}_{\gamma\lambda}\hat{a}_{\gamma'\lambda'}\rho^{\mathrm{I}}_{\mathrm{R}}(0)|n\rangle.\end{aligned}$$

Again, we use $\langle n|\hat{a}=(\hat{a}^+|n\rangle)^+=\sqrt{(n+1)}\,\langle n+1|$ so that $\langle n|\hat{a}\hat{a}\rho^{\mathrm{I}}_{\mathrm{R}}(0)|n\rangle=\sqrt{[(n+1)(n+2)]}\,\langle n+2|\rho^{\mathrm{I}}_{\mathrm{R}}(0)|n\rangle=0$ for $\rho^{\mathrm{I}}_{\mathrm{R}}(0)$ purely diagonal. This procedure may be applied to all the $\hat{V}\hat{V}$ and $\hat{V}^+\hat{V}^+$ terms appearing in the expansion of the double commutator in equation (8.3.93). We therefore have the remaining non-zero terms:

$$\begin{aligned}&\mathrm{tr_R}\,\{\langle k|[\hat{H}^{\mathrm{I}}_1(t),[\hat{H}^{\mathrm{I}}_1(t'),\rho^{\mathrm{I}}(0)]]|k\rangle\}\\&\quad=\mathrm{tr_R}\,\{\langle k|\{[\hat{V}(t),[\hat{V}^+(t'),\rho^{\mathrm{I}}(0)]]+[\hat{V}^+(t),[\hat{V}(t'),\rho^{\mathrm{I}}(0)]]\}|k\rangle\}\\&\quad=\mathrm{tr_R}\,\{\langle k|\{[\hat{V}(t),[\hat{V}^+(t'),\rho^{\mathrm{I}}(0)]]+[\hat{V}(t),[\hat{V}^+(t'),\rho^{\mathrm{I}}(0)]]^+\}|k\rangle\},\end{aligned}$$

where we have used equation (8.3.53). We then have

$$\mathrm{tr_R}\,\{\langle k|[\hat{H}^{\mathrm{I}}_1(t),[\hat{H}^{\mathrm{I}}_1(t'),\rho^{\mathrm{I}}(0)]]|k\rangle\}=\mathrm{tr_R}\left\{\langle k|\sum_{l=1}^{4}(\hat{\Phi}_l+\hat{\Phi}_l^+)|k\rangle\right\},\tag{8.3.95}$$

where, in order,

$$\begin{aligned}\sum_{l=1}^{4}\hat{\Phi}_l=\hat{V}(t)\hat{V}^+(t')\rho^{\mathrm{I}}(0)-\hat{V}(t)\rho^{\mathrm{I}}(0)\hat{V}^+(t')-\hat{V}^+(t')\rho^{\mathrm{I}}(0)\hat{V}(t)\\+\rho^{\mathrm{I}}(0)\hat{V}^+(t')\hat{V}(t).\end{aligned}\tag{8.3.96}$$

The first term in the above series yields

$$\begin{aligned}&\mathrm{tr_R}\,\{\langle k|\hat{\Phi}_1|k\rangle\}=\mathrm{tr_R}\,\{\langle k|\hat{V}(t)\hat{V}^+(t')\rho^{\mathrm{I}}(0)|k\rangle\}\\&\quad=\sum_n\sum_{\gamma\lambda\gamma'\lambda'}\sum_{iji'j'}(\xi_{\gamma\lambda})_{ij}(\xi^*_{\gamma'\lambda'})_{j'i'}\,\mathrm{e}^{\mathrm{i}(\omega_{ij}-\omega_\gamma)t-\mathrm{i}(\omega_{i'j'}-\omega_{\gamma'})t'}\\&\quad\times\langle k|i\rangle\langle j|j'\rangle\langle i'|\rho^{\mathrm{I}}_{\mathrm{A}}(0)|k\rangle\langle n|\rho^{\mathrm{I}}_{\mathrm{R}}(0)\hat{a}_{\gamma\lambda}\hat{a}^+_{\gamma'\lambda'}|n\rangle.\end{aligned}\tag{8.3.97}$$

If we now apply the second quantisation postulate $[\hat{a}_{\gamma\lambda},\hat{a}^+_{\gamma'\lambda'}]=\mathbb{1}\delta_{\gamma\gamma'}\delta_{\lambda\lambda'}$ and recall that $\langle n|\rho^{\mathrm{I}}_{\mathrm{R}}(0)\hat{a}_{\gamma\lambda}\hat{a}^+_{\gamma'\lambda'}|n\rangle$ is zero for all $\gamma\neq\gamma'$ (see equation (8.3.76) for example), then

$$\begin{aligned}&\mathrm{tr_R}\,\{\langle k|\hat{\Phi}_1|k\rangle\}=\sum_n\sum_{\gamma\lambda\lambda'}\sum_{ji'}(\xi_{\gamma\lambda})_{kj}(\xi^*_{\gamma\lambda'})_{ji'}\\&\times\mathrm{e}^{\mathrm{i}(\omega_{kj}-\omega_\gamma)t-\mathrm{i}(\omega_{i'j}-\omega_\gamma)t'}\\&\times\langle i'|\rho^{\mathrm{I}}_{\mathrm{A}}(0)|k\rangle\langle n|\rho^{\mathrm{I}}_{\mathrm{R}}(0)\left(\delta_{\lambda\lambda'}+\frac{I_{\lambda'\lambda}}{J_0}\right)|n\rangle,\end{aligned}$$

where we have also used equation (8.3.39). If we again take $\rho_A^I(0)$ to be purely diagonal, and recall equation (8.3.77), then this last result yields

$$\operatorname{tr}_R\{\langle k|\hat{\Phi}_1|k\rangle\} = \sum_{\gamma\lambda\lambda'}\sum_j (\xi_{\gamma\lambda})_{kj}(\xi^*_{\gamma\lambda'})_{jk} \times e^{i(\omega_{kj}-\omega_\gamma)(t-t')}\rho_{kk}\left(\delta_{\lambda\lambda'}+\frac{I_{\lambda'\lambda}}{J_0}\right). \tag{8.3.98}$$

We now integrate this term over t' as in equation (8.3.93), and use equation (8.3.83), to find

$$\frac{1}{(i\hbar)^2}\int_0^\infty dt' \operatorname{tr}_R\{\langle k|\hat{\Phi}_1|k\rangle\} = -\frac{1}{\hbar^2}\sum_{\gamma\lambda\lambda'}\sum_j (\xi_{\gamma\lambda})_{kj}(\xi^*_{\gamma\lambda'})_{jk}\rho_{kk}\left(\delta_{\lambda\lambda'}+\frac{I_{\lambda'\lambda}}{J_0}\right) f(\omega_{kj}-\omega_\gamma), \tag{8.3.99}$$

where

$$f(\omega_{kj}-\omega_\gamma) = i\mathscr{P}\frac{1}{\omega_{kj}-\omega_\gamma}+\pi\delta(\omega_{kj}-\omega_\gamma). \tag{8.3.100}$$

All seven other terms in equation (8.3.95) may be similarly obtained. In particular, we have

$$\frac{1}{(i\hbar)^2}\int_0^\infty dt' \operatorname{tr}_R\{\langle k|\hat{\Phi}_2|k\rangle\} = \frac{1}{\hbar^2}\sum_{\gamma\lambda\lambda'}\sum_j (\xi_{\gamma\lambda})_{kj}(\xi^*_{\gamma\lambda'})_{jk}\rho_{jj}\frac{I_{\lambda'\lambda}}{J_0} f(\omega_{kj}-\omega_\gamma)$$

$$\frac{2}{(i\hbar)^2}\int_0^\infty dt' \operatorname{tr}_R\{\langle k|\hat{\Phi}_3|k\rangle\} = \frac{1}{\hbar^2}\sum_{\gamma\lambda\lambda'}\sum_i (\xi_{\gamma\lambda})_{ik}(\xi^*_{\gamma\lambda'})_{ki}\rho_{ii}\left(\delta_{\lambda\lambda'}+\frac{I_{\lambda'\lambda}}{J_0}\right) f(\omega_{ik}-\omega_\gamma),$$

$$\frac{1}{(i\hbar)^2}\int_0^\infty dt' \operatorname{tr}_R\{\langle k|\hat{\Phi}_4|k\rangle = -\frac{1}{\hbar^2}\sum_{\gamma\lambda\lambda'}\sum_i (\xi_{\gamma\lambda})_{ik}(\xi^*_{\gamma\lambda'})_{ki}\rho_{kk}\frac{I_{\lambda'\lambda}}{J_0} f(\omega_{ik}-\omega_\gamma).$$

Consequently, summing these last four results and their complex conjugates (see equation (8.3.95)), and using equation (8.3.94), the desired rate equation

becomes

$$\frac{\partial \rho_{kk}}{\partial t} = \frac{2\pi}{\hbar^2} \sum_{\gamma\lambda} \sum_j \{(\xi_{\gamma\lambda})_{jk}(\xi^*_{\gamma\lambda})_{kj}\rho_{jj}\delta(\omega_{jk}-\omega_\gamma)$$
$$-(\xi_{\gamma\lambda})_{kj}(\xi^*_{\gamma\lambda})_{jk}\rho_{kk}\delta(\omega_{kj}-\omega_\gamma)\}$$
$$+\frac{2\pi}{\hbar^2 J_0} \sum_{\gamma\lambda\lambda'} \sum_j \{(\xi_{\gamma\lambda})_{kj}(\xi^*_{\gamma\lambda'})_{jk}(\rho_{jj}-\rho_{kk})I_{\lambda'\lambda}\delta(\omega_{kj}-\omega_\gamma)$$
$$+(\xi_{\gamma\lambda})_{jk}(\xi^*_{\gamma\lambda'})_{kj}(\rho_{jj}-\rho_{kk})I_{\lambda'\lambda}\delta(\omega_{jk}-\omega_\gamma)\}. \tag{8.3.101}$$

Again, we can examine a simplified version of this last equation by putting $I_{12}=0=I_{21}$ with $I_{11}=I=I_{22}$. We then have

$$\frac{\partial \rho_{kk}}{\partial t} = \frac{4\pi}{\hbar^2} \sum_{\gamma} \sum_j \Big\{\eta_{jk}(\mathbf{k}_\gamma)\rho_{jj}\delta(\omega_{jk}-\omega_\gamma)-\eta_{kj}(\mathbf{k}_\gamma)\rho_{kk}\delta(\omega_{kj}-\omega_\gamma)$$
$$+\eta_{kj}(\mathbf{k}_\gamma)(\rho_{jj}-\rho_{kk})\frac{I}{J_0}\delta(\omega_{kj}-\omega_\gamma)$$
$$+\eta_{jk}(\mathbf{k}_\gamma)(\rho_{jj}-\rho_{kk})\frac{I}{J_0}\delta(\omega_{jk}-\omega_\gamma)\Big\}, \tag{8.3.102}$$

where $\eta(\mathbf{k}_\gamma)$ is defined by equation (8.3.90).

Each of the terms on the RHS of this last equation has a specific physical interpretation and may be directly compared with those terms found in the corresponding classical complete re-distribution statistical equilibrium rate equation (4.4.17) for the kth level of the atom, viz

$$\sum_{j>k} N_j A_{jk} - \sum_{j<k} N_k A_{kj}$$
$$+\sum_{j<k}[N_j B_{jk}\bar{J}_{kj}-N_k B_{kj}\bar{J}_{kj}]+\sum_{j>k}[N_j B_{jk}\bar{J}_{jk}-N_k B_{kj}\bar{J}_{jk}]=0, \tag{8.3.103}$$

where, because of our neglect of collisional terms in obtaining equation (8.3.102), we have put all C_{ij} to zero.

We then see that the $\eta_{jk}\rho_{jj}\delta(\omega_{jk}-\omega_\gamma)$ term represents the rate of population of the $|k\rangle$ state due to spontaneous de-excitations from states $|j\rangle$, i.e. $\eta_{jk}\rho_{jj}\delta(\omega_{jk}-\omega_\gamma)\to N_j A_{jk}$. Note that the $\delta(\omega_{jk}-\omega_\gamma)$ function, together with $\omega_{jk}=(E_j^{(A)}-E_k^{(A)})/\hbar$, stipulates that $j>k$. Correspondingly, the $-\eta_{kj}\rho_{kk}\delta(\omega_{kj}-\omega_\gamma)$ term represents the rate of de-population of the kth level due to spontaneous de-excitations from level k to levels $j<k$. The remaining four terms in equation (8.3.102) are, in order, population of level k due to absorptions from levels $j<k$, de-population due to stimulated emission from k to levels $j<k$, population due to stimulated emission from levels $j>k$, and de-population due to absorptions from k to $j>k$. Note that the summation over γ in equation (8.3.102) is effectively equivalent to the integrals over $\Delta\nu$ and Ω which produce the $\bar{J}$.

APPENDIX A

Half-range orthogonality relationships

Here we wish to detail various half-range orthogonality properties. The integrals we wish to consider may be readily evaluated using the material already discussed in section 2.7.

First, consider the integral

$$I_1 = \int_0^\gamma W(\xi) g(\eta, \xi) g(-\eta_0, \xi)\, d\xi, \tag{A.1}$$

where $\eta \in [0, \gamma]$. We immediately obtain

$$\begin{aligned} I_1 &= \frac{\eta_0 \lambda(\eta) W(\eta)}{(\eta_0 + \eta)\Psi(\eta)} + \eta\eta_0 P \int_0^\gamma \frac{W(\xi)\, d\xi}{(\eta - \xi)(\eta_0 + \xi)} \\ &= \frac{\eta_0 \lambda(\eta) W(\eta)}{(\eta_0 + \eta)\Psi(\eta)} + \frac{\eta\eta_0}{\eta + \eta_0} P \int_0^\gamma W(\xi) \left[\frac{1}{\eta - \xi} + \frac{1}{\eta_0 + \xi} \right] d\xi. \end{aligned} \tag{A.2}$$

Equation (2.7.15), viz.

$$W(z) = Y^+(z) - Y^-(z), \tag{A.3}$$

then yields

$$P \int_0^\gamma \frac{W(\xi)\, d\xi}{\eta_0 + \xi} = \int_0^\gamma \frac{Y^+(\xi) - Y^-(\xi)}{\xi + \eta_0}\, d\xi = \int_{L_1} \frac{Y(\xi)\, d\xi}{\xi + \eta_0} \tag{A.4}$$

where L_1 is the contour shown in figure A.1.

Let L_2 be a circle of radius R with the origin as centre. Then, if $L = L_1 \cup L_2$, Cauchy's integral theorem yields

$$\int_L \frac{Y(\xi)\, d\xi}{\xi + \eta_0} \equiv \lim_{R \to \infty} \int_{L_1 \cup L_2} \frac{Y(\xi)\, d\xi}{\xi + \eta_0} = 2\pi i \sum_j R_j, \tag{A.5}$$

where R_j is the residue of the jth pole of $Y(\xi)/(\xi + \eta_0)$ contained within L. Note that the contour L_1 is assumed to be sufficiently close to the cut from O to γ that no poles are contained within L_1. In section 2.7, we showed that $Y(z) \sim 1/z$

as $|z| \to \infty$ so that the integral over L_2 vanishes as $R \to \infty$. We then obtain

$$\int_{L_1} \frac{Y(\xi)\,d\xi}{\xi + \eta_0} = 2\pi i Y(-\eta_0), \tag{A.6}$$

where we have used the fact that $Y(z)$ is analytic for all $z \notin [0, \gamma]$.

This last equation, together with equation (2.7.14), viz.

$$\frac{1}{\pi i} P \int_0^{\gamma} \frac{W(\xi)\,d\xi}{\xi - z} = Y^+(z) + Y^-(z), \tag{A.7}$$

then enables I_1 to be written in the form

$$I_1 = \frac{\eta_0 \lambda(\eta)[Y^+(\eta) - Y^-(\eta)]}{(\eta + \eta_0)\Psi(\eta)} + \frac{2\pi i \eta \eta_0 Y(-\eta_0)}{\eta + \eta_0} - \frac{\pi i \eta \eta_0}{\eta + \eta_0}[Y^+(\eta) - Y^-(\eta)]. \tag{A.8}$$

Further,

$$\lambda(\eta) = \frac{\Lambda^+(\eta) + \Lambda^-(\eta)}{2}, \tag{A.9}$$

and

$$\pi i \eta = \frac{\Lambda^+(\eta) - \Lambda^-(\eta)}{2\Psi(\eta)} \tag{A.10}$$

– see equations (2.3.19) and (2.3.21).

Equation (A.8) then has the form

$$I_1 = \frac{\eta_0}{\eta + \eta_0}\left\{2\pi i \eta Y(-\eta_0) + \frac{1}{\Psi(\eta)}[\Lambda^-(\eta)Y^+(\eta) - \Lambda^+(\eta)Y^-(\eta)]\right\}. \tag{A.11}$$

Equation (2.7.16) may be written as

$$\Lambda^-(\eta)Y^+(\eta) - \Lambda^+(\eta)Y^-(\eta) = \eta\Psi(\eta), \tag{A.12}$$

Fig.A.1. The contours L_1 and L_2 in the complex plane with L_1 surrounding the cut from 0 to γ.

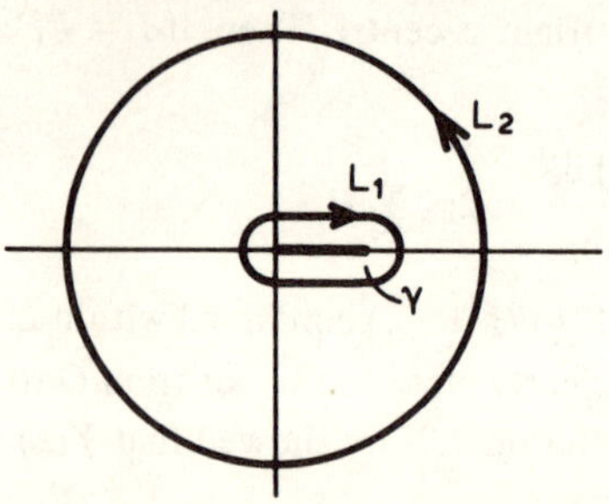

where, following the discussion presented in section 2.7, we have put the constant C_2 equal to unity. We then have

$$I_1 = \frac{\eta_0}{\eta + \eta_0}\,[2\pi i\eta Y(-\eta_0) + \eta], \tag{A.13}$$

which, using equation (2.7.32), viz.

$$Y(z) = \frac{1}{2\pi i}\left(\frac{z - \eta_0}{X(z)} - 1\right), \tag{A.14}$$

finally yields

$$I_1 = \frac{-2\eta\eta_0^2}{(\eta + \eta_0)X(-\eta_0)}. \tag{A.15}$$

This completes the derivation of equation (2.7.34).

The second integral of interest is

$$\begin{aligned} I_2 &= \int_0^\gamma W(\xi)g(\eta_0, \xi)g(-\eta_0, \xi)\,d\xi \\ &= \eta_0^2 \int_0^\gamma \frac{W(\xi)\,d\xi}{(\eta_0 + \xi)(\eta_0 - \xi)} \\ &= \frac{\eta_0}{2}\int_0^\gamma W(\xi)\left(\frac{1}{\eta_0 + \xi} + \frac{1}{\eta_0 - \xi}\right) d\xi \\ &= \frac{\eta_0}{2}\int_0^\gamma [Y^+(\xi) - Y^-(\xi)]\left(\frac{1}{\eta_0 + \xi} + \frac{1}{\eta_0 - \xi}\right) d\xi \\ &= \frac{\eta_0}{2}\int_{L_1} Y(\xi)\left(\frac{1}{\xi + \eta_0} - \frac{1}{\xi - \eta_0}\right) d\xi \\ &= \pi i\eta_0[Y(-\eta_0) - Y(\eta_0)]. \end{aligned} \tag{A.16}$$

Equation (A.14) then immediately yields

$$I_2 = \frac{-\eta_0^2}{X(-\eta_0)}. \tag{A.17}$$

We now consider the integral

$$I_3 = \int_0^\gamma W(\xi)g(\eta, \xi)g(-\eta', \xi)\,d\xi, \tag{A.18}$$

where $\eta, \eta' \in [0, \gamma]$. Since $\xi \in [0, \gamma]$ in the above integration, we must have

$$g(-\eta', \xi) = \frac{\eta'}{\eta' + \xi}, \tag{A.19}$$

i.e. there is no singular component. If we compare I_3 with I_1 but with η_0

replaced by η', we easily find that

$$I_3 = \frac{\eta'}{\eta + \eta'}[2\pi i \eta Y(-\eta') + \eta], \tag{A.20}$$

which, using equation (A.14), yields

$$I_3 = \frac{-\eta\eta'(\eta' + \eta_0)}{(\eta + \eta')X(-\eta')} . \tag{A.21}$$

Next put

$$I_4 = \int_0^{\gamma} W(\xi) g(\eta_0, \xi) g(-\eta', \xi)\, \mathrm{d}\xi, \tag{A.22}$$

where $\eta' \in [0, \gamma]$. Again, we have

$$g(-\eta', \xi) = \frac{\eta'}{\eta' + \xi}, \tag{A.23}$$

so that

$$\begin{aligned} I_4 &= \frac{\eta_0 \eta'}{\eta_0 + \eta'} \int_0^{\gamma} W(\xi) \left(\frac{1}{\eta_0 - \xi} + \frac{1}{\eta' + \xi} \right) \mathrm{d}\xi \\ &= \frac{2\pi i \eta_0 \eta'}{\eta_0 + \eta'} [Y(-\eta') - Y(\eta_0)] \\ &= \frac{-\eta_0 \eta'}{X(-\eta')} . \end{aligned} \tag{A.24}$$

A slightly more difficult integral is the following:

$$I_5 = \int_0^{\gamma} W(\xi) g^2(\eta_0, \xi)\, \mathrm{d}\xi = \eta_0^2 \int_0^{\gamma} \frac{W(\xi)\, \mathrm{d}\xi}{(\eta_0 - \xi)^2} . \tag{A.25}$$

We first define $I_5(\eta)$ such that

$$I_5(\eta) = \eta_0 \eta \int_0^{\gamma} \frac{W(\xi)\, \mathrm{d}\xi}{(\eta_0 - \xi)(\eta - \xi)}, \tag{A.26}$$

where, clearly

$$I_5 = I_5(\eta_0). \tag{A.27}$$

However, we know from equation (A.24) for example, with η' replaced by $-\eta$, that

$$I_5(\eta) = \frac{2\pi i \eta_0 \eta}{\eta_0 - \eta} [Y(\eta_0) - Y(\eta)] = \frac{-\eta_0 \eta}{X(\eta_0)},$$

i.e.

$$I_5 = \frac{-\eta_0^2}{X(\eta_0)}. \tag{A.28}$$

The final integral required is

$$I_6 = \int_0^{\gamma} W(\xi) g^2(\eta, \xi)\, \mathrm{d}\xi, \tag{A.29}$$

for $\eta \in [0, \gamma]$. To do this, we follow the procedure outlined in section 2.6 for the case of full-range orthogonality. We begin by considering the expansion

$$r(\xi) = a^{+}_{} g(\eta_0, \xi) + \int_0^\gamma A(\eta) g(\eta, \xi)\, d\eta, \tag{A.30}$$

so that, by multiplying by $W(\xi)g(\eta', \xi)$ with $\eta' \in [0, \gamma]$ and integrating over $\xi \in [0, \gamma]$, we have

$$\begin{aligned}\int_0^\gamma r(\xi)W(\xi)g(\eta', \xi)\, d\xi &= \int_0^\gamma W(\xi)g(\eta', \xi)\, d\xi \int_0^\gamma A(\eta)g(\eta, \xi)\, d\eta \\ &= \int_0^\gamma W(\xi)\left[P\left(\frac{\eta'}{\eta' - \xi}\right) + \frac{\lambda(\eta')}{\Psi(\eta')}\delta(\eta' - \xi)\right] \\ &\quad \times \int_0^\gamma A(\eta)\left[P\left(\frac{\eta}{\eta - \xi}\right) + \frac{\lambda(\eta)}{\Psi(\eta)}\delta(\eta - \xi)\right] d\eta\, d\xi \\ &= \eta' P \int_0^\gamma \frac{W(\xi)\, d\xi}{\eta' - \xi}\, P \int_0^\gamma \frac{\eta A(\eta)\, d\eta}{\eta - \xi} + A(\eta')\frac{\lambda^2(\eta')}{\Psi(\eta')} W(\eta') \\ &\quad + \int_0^\gamma \frac{A(\eta)}{\eta - \eta'}\left[\eta \frac{\lambda(\eta')W(\eta')}{\Psi(\eta')} - \eta' \frac{\lambda(\eta)W(\eta)}{\Psi(\eta)}\right] d\eta. \end{aligned} \tag{A.31}$$

The Poincaré–Bertrand formula has the form

$$\begin{aligned} P\int_0^\gamma \frac{dt}{t - t_0} \int_0^\gamma \frac{\phi(t, t_1)\, dt_1}{t_1 - t} &= -\pi^2 \phi(t_0, t_0) \\ &\quad + P\int_0^\gamma dt_1 P \int_0^\gamma \frac{\phi(t, t_1)\, dt_1}{(t - t_0)(t_1 - t)}, \end{aligned} \tag{A.32}$$

which, if we put $t \equiv \xi$, $t_1 \equiv \eta$ and

$$\phi(t, t_1) \equiv -W(\xi)\eta A(\eta), \tag{A.33}$$

with $t_0 \equiv \eta'$, yields

$$\begin{aligned} \text{RHS (A.31)} &= \pi^2 \eta'^2 W(\eta')A(\eta') + \frac{\lambda^2(\eta')W(\eta')A(\eta')}{\Psi^2(\eta')} \\ &\quad + P\int_0^\gamma \eta\eta' A(\eta)\, d\eta\, P\int_0^\gamma \frac{W(\xi)\, d\xi}{(\eta' - \xi)(\eta - \xi)} \\ &\quad + P\int_0^\gamma \frac{A(\eta)}{\eta - \eta'}\left[\eta \frac{\lambda(\eta')W(\eta')}{\Psi(\eta')} - \eta' \frac{\lambda(\eta)W(\eta)}{\Psi(\eta)}\right] d\eta \\ &= \left[\frac{\lambda^2(\eta')}{\Psi^2(\eta')} + \pi^2 \eta'^2\right] W(\eta')A(\eta') + T_3 + T_4, \end{aligned} \tag{A.34}$$

where T_3 and T_4 refer to the last two integrals appearing on the RHS of equation (A.34).

It is not difficult to see that

$$P\int_0^\gamma \frac{W(\xi)\,d\xi}{(\eta'-\xi)(\eta-\xi)} = \frac{1}{\eta-\eta'}P\int_0^\gamma W(\xi)\left(\frac{1}{\eta'-\xi}-\frac{1}{\eta-\xi}\right)d\xi, \quad (A.35)$$

which, using equation (A.7), yields

$$P\int_0^\gamma \frac{W(\xi)\,d\xi}{(\eta'-\xi)(\eta-\xi)} = \frac{\pi i}{\eta-\eta'}[Y^+(\eta)+Y^-(\eta)-Y^+(\eta')-Y^-(\eta')]. \quad (A.36)$$

We now turn to the terms appearing in T_4. Using equations (A.3), (A.9) and (A.12) we find

$$\begin{aligned}\lambda(\eta)W(\eta) &= \tfrac{1}{2}[\Lambda^+(\eta)+\Lambda^-(\eta)][Y^+(\eta)-Y^-(\eta)]\\ &= \tfrac{1}{2}[\Lambda^+(\eta)Y^+(\eta)-\Lambda^-(\eta)Y^-(\eta)+\eta\Psi(\eta)]\\ &= \tfrac{1}{2}[[\Lambda^+(\eta)-\Lambda^-(\eta)][Y^+(\eta)+Y^-(\eta)]+2\eta\Psi(\eta)]\\ &= \pi i\eta\Psi(\eta)[Y^+(\eta)+Y^-(\eta)]+\eta\Psi(\eta).\end{aligned} \quad (A.37)$$

The term in square parentheses in T_4 then has the form

$$\eta\frac{\lambda(\eta')W(\eta')}{\Psi(\eta')}-\eta'\frac{\lambda(\eta)W(\eta)}{\Psi(\eta)} = \pi i\eta\eta'[Y^+(\eta')+Y^-(\eta')-Y^+(\eta) - Y^-(\eta)]. \quad (A.38)$$

Comparison between T_3 and T_4 using equations (A.36) and (A.38) yields

$$T_3 = -T_4. \quad (A.39)$$

Equation (A.34) then gives

$$\int_0^\gamma W(\xi)g(\eta',\xi)\,d\xi\int_0^\gamma A(\eta)g(\eta,\xi)\,d\eta = \frac{N(\eta')W(\eta')A(\eta')}{\eta'\Psi(\eta')}, \quad (A.40)$$

where we have used the definition of $N(\eta)$ given by equation (2.6.23), viz.

$$N(\eta) = \left[\frac{\lambda^2(\eta)}{\Psi^2(\eta)}+\pi^2\eta^2\right]\eta\Psi(\eta). \quad (A.41)$$

Clearly, writing $A(\eta)=\delta(\eta-\eta'')$ in equation (A.40), we find

$$\int_0^\gamma W(\xi)g(\eta',\xi)g(\eta'',\xi)\,d\xi = \frac{N(\eta')W(\eta')}{\eta'\Psi(\eta')}\delta(\eta-\eta''), \quad (A.42)$$

which can also be written as

$$\int_0^\gamma W(\xi)g^2(\eta,\xi)\,d\xi = \frac{N(\eta)W(\eta)}{\eta\Psi(\eta)}. \quad (A.43)$$

The negative half-space orthogonality relations (2.7.60) through (2.7.65) may be readily derived in the same manner.

APPENDIX B

The $H(z)$ function

We wish to first determine a relationship between the half-range orthogonality weighting function $W(z)$ obtained in section 2.7 and the Chandrasekhar $H(z)$ function, then derive the non-linear integral equation which this latter quantity satisfies.

This may be done by first considering the function $f(z)$ given by

$$f(z) = -\frac{\Lambda(z)X(z)X(-z)}{\epsilon(z^2 - \eta_0^2)}, \tag{B.1}$$

where $X(z)$ is the positive half-range X function given by equation (2.7.18). Since $\Lambda(z)$ has two simple zeros at $\pm\eta_0$ and is analytic for all $z \notin [-\gamma, \gamma]$, and $X(z)$ is analytic for all $z \notin [0, \gamma]$ so that $X(-z)$ is analytic for all $z \notin [-\gamma, 0]$, then $f(z)$ is analytic for all $z \notin [-\gamma, \gamma]$. To understand the behaviour of $f(z)$ across the cut $[-\gamma, \gamma]$, we put

$$\frac{f^+(z)}{f^-(z)} = \frac{\Lambda^+(z)X^+(z)X^+(-z)}{\Lambda^-(z)X^-(z)X^-(-z)}. \tag{B.2}$$

If we consider only $z \geqslant 0$, then $X(-z)$ is analytic so that

$$\frac{f^+(z)}{f^-(z)} = \frac{\Lambda^+(z)X^+(z)}{\Lambda^-(z)X^-(z)} = 1, \tag{B.3}$$

where we have used equation (2.7.21). A similar argument holds for $z \leqslant 0$. We therefore have $f(z)$ also analytic across the cut so that $f(z)$ is analytic for all z in the complex plane.

Now $\Lambda(z) \to \epsilon$ whilst $X(z) \sim z$ and $X(-z) \sim z$ as $|z| \to \infty$. Thus, $f(z) \to 1$ as $|z| \to \infty$. Liouville's theorem then states that

$$f(z) \equiv 1, \tag{B.4}$$

i.e.

$$\frac{1}{X(z)X(-z)} = -\frac{\Lambda(z)}{\epsilon(z^2 - \eta_0^2)}. \tag{B.5}$$

Equation (2.7.33) stipulates that, for $z \in [0, \gamma]$,

$$W(z) = \frac{z-\eta_0}{2\pi i}\left(\frac{1}{X^+(z)} - \frac{1}{X^-(z)}\right)$$

$$= \frac{z-\eta_0}{2\pi i}\cdot\frac{\Lambda^+(z)-\Lambda^-(z)}{\Lambda^-(z)X^-(z)}$$

$$= (z-\eta_0)\frac{z\Psi(z)}{\Lambda^-(z)X^-(z)}, \tag{B.6}$$

where we have used equation (2.7.21) again, together with the result

$$\Lambda^\pm(z) = \lambda(z) \pm \pi i z\Psi(z). \tag{B.7}$$

However, equation (B.5) yields

$$\frac{1}{X^-(z)\Lambda^-(z)} = -\frac{X^-(-z)}{\epsilon(z^2-\eta_0^2)} = -\frac{X(-z)}{\epsilon(z^2-\eta_0^2)}, \tag{B.8}$$

where, since $X(-z)$ is analytic for all $z \notin [-\gamma, 0]$, we have $X^-(-z) \equiv X^+(-z) = X(-z)$. We therefore have, for $z \in [0, \gamma]$,

$$W(z) = -\frac{X(-z)z\Psi(z)}{\epsilon(z+\eta_0)} = \frac{Q(-z)z\Psi(z)}{\epsilon}, \tag{B.9}$$

where we have introduced the function $Q(z)$ defined by equation (2.7.23), viz.

$$Q(z) = \frac{X(z)}{z-\eta_0}. \tag{B.10}$$

We now define the $H(z)$ function such that

$$H(z) = \frac{Q(-z)}{\sqrt{\epsilon}} = -\frac{X(-z)}{\sqrt{\epsilon}(z+\eta_0)}, \tag{B.11}$$

i.e.

$$W(z) = \frac{z\Psi(z)H(z)}{\sqrt{\epsilon}}. \tag{B.12}$$

Note, from equation (B.5), that

$$\frac{1}{Q(z)Q(-z)} = \frac{\Lambda(z)}{\epsilon}, \tag{B.13}$$

whilst

$$\frac{1}{H(z)H(-z)} = \Lambda(z). \tag{B.14}$$

We now wish to derive an integral equation for $H(z)$.

If we take the contours L_1 and L_2 as shown in figure B.1 where L_2 is a circle centre the origin with radius R, then Cauchy's integral theorem gives

$$\lim_{\substack{R\to\infty \\ L_1\to \text{cut}}} \int_{L=L_1\cup L_2} \frac{dz'}{Q(z')(z'-z)} = 2\pi i \sum_j R_j, \tag{B.15}$$

where R_j is the residue of $1/Q(z')(z'-z)$ at the jth pole within L, i.e. we have

$$\lim_{L_1 \to \text{cut}} \frac{1}{2\pi i}\int_{L_1} \frac{dz'}{Q(z')(z'-z)} = \frac{1}{Q(z)} - 1, \tag{B.16}$$

where we have used the fact that $Q(z) \to 1$ as $|z| \to \infty$ to evaluate the integral over L_2. Re-arranging equation (B.16) yields

$$\frac{1}{Q(z)} = 1 + \frac{1}{2\pi i}\int_0^\gamma \left[\frac{1}{Q^+(z')} - \frac{1}{Q^-(z')}\right]\frac{dz'}{z'-z}. \tag{B.17}$$

Now we know that

$$\frac{1}{2\pi i}\left[\frac{1}{X^+(z)} - \frac{1}{X^-(z)}\right] = -\frac{X(-z)z\Psi(z)}{\epsilon(z^2-\eta_0^2)} \tag{B.18}$$

(see equations (B.6) and (B.8)), so that

$$\frac{1}{2\pi i}\left[\frac{1}{Q^+(z)} - \frac{1}{Q^-(z)}\right] = -\frac{X(-z)z\Psi(z)}{\epsilon(z+\eta_0)} = \frac{Q(-z)z\Psi(z)}{\epsilon}. \tag{B.19}$$

Equation (B.17) then becomes

$$\frac{1}{Q(z)} = 1 + \int_0^\gamma \frac{Q(-z')z'\Psi(z')\,dz'}{\epsilon(z'-z)}, \tag{B.20}$$

which, in terms of $H(z)$, has the form

$$\frac{1}{\sqrt{\epsilon}H(-z)} = 1 + \int_0^\gamma \frac{z'\Psi(z')H(z')\,dz'}{\sqrt{\epsilon}(z'-z)}. \tag{B.21}$$

If we now replace z by $-z$, we obtain

$$\frac{1}{H(z)} = \sqrt{\epsilon} + \int_0^\gamma \frac{z'\Psi(z')H(z')\,dz'}{z'+z}. \tag{B.22}$$

Fig.B.1. The contours L_1 and L_2 in the complex plane with L_1 surrounding the cut from 0 to γ.

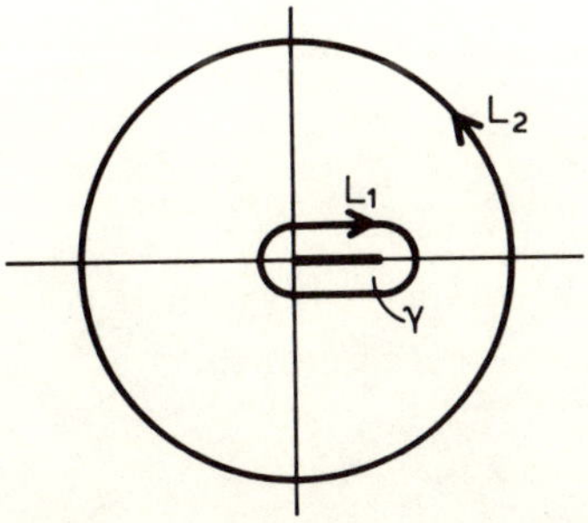

This equation may be solved numerically for $H(z)$ by simple iteration. The convergence of this iteration is extremely rapid.

A similar equation for $X(z)$ may be obtained by direct substitution using equation (B.11), viz.

$$1+\frac{\eta_0-z}{X(z)}=\int_0^{\gamma}\frac{z'\Psi(z')X(-z')\,\mathrm{d}z'}{\epsilon(z'+\eta_0)(z'-z)}. \tag{B.23}$$

APPENDIX C

Expressions for viscosity and heat conductivity

The coefficients of viscosity μ_V and heat conductivity k_c have been mentioned in section 5.3.2. Here we detail a set of quasi-empirical coefficients [57] which at least give some idea of the functional dependence of μ_V and k_c. (Subscripts A refer to atoms, I to ions and e to electrons.)

$$\mu_{VA} = \frac{5}{16Q_{AA}} (\pi m_I kT)^{1/2} \left[1 + \frac{N_I Q_{IA}}{N_A Q_{AA}}\right]^{-1},$$

$$\mu_{VI} = \frac{5N_I}{16Q_{IA}N_A} (\pi m_I kT)^{1/2} \left[1 + \frac{N_I Q_{II}}{N_A Q_{IA}}\right]^{-1},$$

$$\mu_{Ve} \equiv 0,$$

$$k_{cj} = \frac{15k\mu_{Vj}}{4m_j}, \quad j = A, I,$$

$$k_{ce} = \frac{75k}{64Q_{ee}(1+\sqrt{2})} \left(\frac{kT_e}{m_e}\right)^{1/2} \left[1 + \frac{\sqrt{2}N_A Q_{eA}}{(1+\sqrt{2})N_e Q_{ee}}\right]^{-1},$$

where we have put $T_A = T_I = T \neq T_e$, and

$$Q_{jj} = \frac{\pi e^4}{4(kT_j)^2} \ln\left(\frac{9k^3T_e^3}{4\pi N_e e^6}\right), \quad j = I, e,$$

$$Q_{eI} = Q_{ee},$$

$$Q_{AA} = 1.7 \times 10^{-14} T^{-1/4},$$

$$Q_{IA} = 1.4 \times 10^{-14},$$

$$Q_{eA} = (-0.35 + 0.775 \times 10^{-4} T_e)10^{-16}, \quad T_e > 10^4\text{K},$$

$$= (0.39 - 0.551 \times 10^{-4} T_e + 0.595\ 10^{-8} T_e^2)10^{-16}, \quad T_e < 10^4\text{K}.$$

All expressions are in cgs units.

APPENDIX D

Derivation of equation (6.4.41)

Here we wish to briefly discuss equation (6.4.41), involving a Cauchy principal value term and the Dirac delta function, and equation (6.4.111). We start by considering the integral of $e^{i\eta z}$ around the contour $C = C_1 \cup C_2$ shown in figure D.1. On C_2, $z = Re^{i\theta}$ so that $e^{i\eta z} = e^{\eta R(-\sin\theta + i\cos\theta)} \to 0$ as $R \to \infty$ $\forall\, \theta \in (0, \pi)$ *and* $\forall\, \eta > 0$. Thus, Cauchy's theorem yields

$$\lim_{R\to\infty} \int_C e^{i\eta z}\, dz = 0 = \lim_{R\to\infty} \int_{-R}^{R} e^{i\eta x}\, dx, \tag{D.1}$$

for all $\eta > 0$. But if $\eta = 0$, the integrals over each of the contours C_1 and C_2 will be divergent. The same argument can be applied for $\eta < 0$ but with C_2 now in the lower half-plane. We can then define the singular function

$$\delta(\eta) = \beta \lim_{R\to\infty} \int_{-R}^{R} e^{i\eta x}\, dx, \tag{D.2}$$

where β is a proportionality constant, so that

$$\delta(\eta) = \lim_{R\to\infty} \frac{\beta}{i\eta}\left(e^{i\eta R} - e^{-i\eta R}\right) = 2\beta \lim_{R\to\infty} \frac{\sin \eta R}{\eta}. \tag{D.3}$$

Fig.D.1. The contour C in the complex z plane.

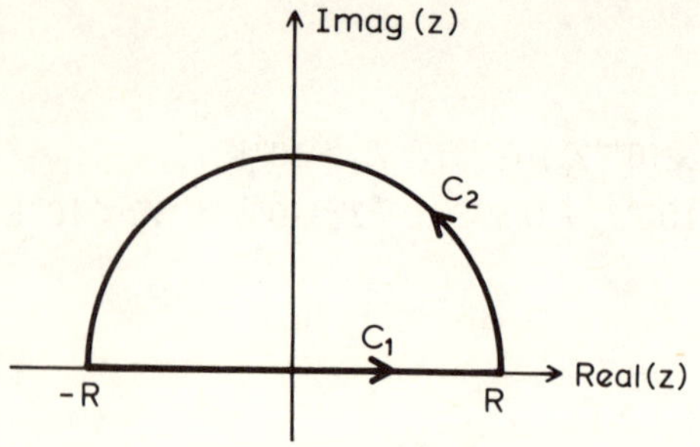

Consequently, for $f(\eta)$ any square integrable function, with $a<0$ and $b>0$,

$$\int_a^b f(\eta)\delta(\eta)\,\mathrm{d}\eta = 2\beta \lim_{R\to\infty} \int_a^b \frac{f(\eta)\sin\eta R}{\eta}\,\mathrm{d}\eta$$

$$= 2\beta \lim_{R\to\infty} \int_{aR}^{bR} f\left(\frac{\xi}{R}\right)\frac{\sin\xi}{\xi}\,\mathrm{d}\xi$$

$$= 2\beta f(0)\int_{-\infty}^{\infty}\frac{\sin\xi}{\xi}\,\mathrm{d}\xi$$

$$= 2\pi\beta f(0). \tag{D.4}$$

If we choose, as is customary, to write the RHS (D.4) as just $f(0)$, then $\beta = 1/2\pi$, i.e.

$$\delta(\eta) = \frac{1}{\pi}\lim_{R\to\infty}\frac{\sin\eta R}{\eta}. \tag{D.5}$$

Equations (6.4.41) and (6.4.92) of the text define

$$\zeta(\eta) = \lim_{R\to\infty}\frac{1-\mathrm{e}^{\mathrm{i}\eta R}}{\eta}$$

$$= \lim_{R\to\infty}\left[\frac{1-\cos\eta R}{\eta} - \frac{\mathrm{i}\sin\eta R}{\eta}\right]$$

$$= g(\eta) - \pi\mathrm{i}\delta(\eta), \tag{D.6}$$

where we have used equation (D.3) and defined

$$g(\eta) = \lim_{R\to\infty}\frac{1-\cos\eta R}{\eta}. \tag{D.7}$$

In particular, we find

$$\int_a^b f(\eta)g(\eta)\,\mathrm{d}\eta = \lim_{R\to\infty}\int_a^b \frac{1-\cos\eta R}{\eta} f(\eta)\,\mathrm{d}\eta$$

$$= \lim_{R\to\infty}\mathscr{P}\int_a^b \frac{1-\cos\eta R}{\eta} f(\eta)\,\mathrm{d}\eta$$

$$+ \lim_{\substack{\epsilon\to 0\\ R\to\infty}}\int_{-\epsilon}^{\epsilon}\frac{1-\cos\eta R}{\eta} f(\eta)\,\mathrm{d}\eta, \tag{D.8}$$

where we have used $\mathscr{P}$ to denote the Cauchy principal value (i.e. $\eta = 0$ has been removed from the region of integration). Clearly, in the limit as $R\to\infty$, the (infinitely) rapid oscillations of $\cos\eta R$, where $\eta \neq 0$, yield correspondingly zero contributions to the first term on the RHS (D.7). Consequently, expanding

$f(\eta)$ in a Taylor series about $\eta = 0$, we find

$$\int_a^b f(\eta)g(\eta)\,\mathrm{d}\eta = \mathscr{P}\int_a^b \frac{f(\eta)}{\eta}\,\mathrm{d}\eta + \lim_{\substack{\epsilon\to 0\\ R\to\infty}} \int_{-\epsilon}^{\epsilon} \frac{1}{\eta}\left[\frac{\eta^2R^2}{2!} - \frac{\eta^4R^4}{4!} + \cdots\right] \times [f(0) + \eta f'(0) + \cdots]\,\mathrm{d}\eta = \mathscr{P}\int_a^b \frac{f(\eta)}{\eta}\,\mathrm{d}\eta. \tag{D.9}$$

We then see that $g(\eta)$ may be written in the mnemonic form

$$g(\eta) \equiv \mathscr{P}\frac{1}{\eta}, \tag{D.10}$$

so that

$$\zeta(\eta) = \lim_{R\to\infty} \frac{1 - \mathrm{e}^{\mathrm{i}\eta R}}{\eta} = \mathscr{P}\frac{1}{\eta} - \pi\mathrm{i}\delta(\eta). \tag{D.11}$$

Note that the above approach is not markedly different from that leading to equation (2.3.33).

We now turn our attention to equation (6.4.111) by considering the contour $C = C_1 \cup C_2$ shown in figure D.2.

Since

$$\lim_{\epsilon\to 0} \int_{C_1} \frac{f(z)}{z}\,\mathrm{d}z = \mathscr{P}\int_a^b \frac{f(x)}{x}\,\mathrm{d}x, \tag{D.12}$$

and

$$\lim_{\epsilon\to 0} \int_{C_2} \frac{f(z)}{z}\,\mathrm{d}z = -\pi\mathrm{i}f(0) = -\pi\mathrm{i}\int_a^b f(x)\delta(x)\,\mathrm{d}x, \tag{D.13}$$

then

$$\int_a^b f(x)\zeta(x)\,\mathrm{d}x = \int_a^b f(x)\left[\mathscr{P}\frac{1}{x} - \pi\mathrm{i}\delta(x)\right]\mathrm{d}x = \lim_{\epsilon\to 0} \int_{C_1\cup C_2} \frac{f(z)}{z}\,\mathrm{d}z. \tag{D.14}$$

Fig.D.2. The contour C in the complex z plane indented at the origin.

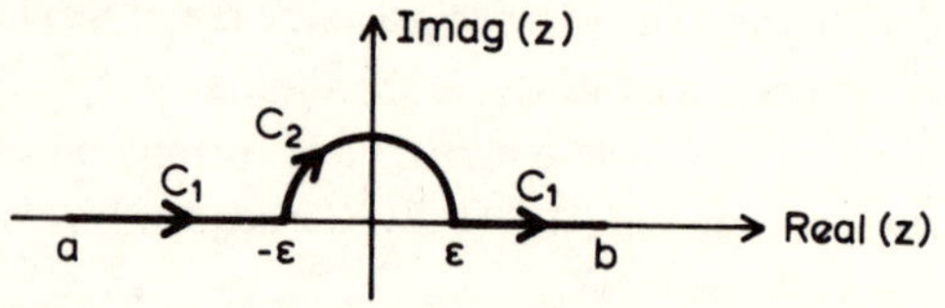

If we now consider the new contour $C' = C_1' \cup C_2'$ shown in figure D.3 where $C' \to C$ as $\sigma \to 0$, then

$$\int_a^b f(x)\zeta(x)\,\mathrm{d}x = \lim_{\substack{\epsilon\to 0\\ \sigma\to 0}} \int_{C_1'\cup C_2'} \frac{f(z)}{z}\,\mathrm{d}z$$

$$= \lim_{\substack{\epsilon\to 0\\ \sigma\to 0}} \mathscr{P}\int_a^b \frac{f(x+\mathrm{i}\sigma)}{x+\mathrm{i}\sigma}\,\mathrm{d}x$$

$$+ \lim_{\substack{\epsilon\to 0\\ \sigma\to 0}} \int_\pi^0 \frac{f(\mathrm{i}\sigma+\epsilon\mathrm{e}^{\mathrm{i}\theta})}{\mathrm{i}\sigma+\epsilon\mathrm{e}^{\mathrm{i}\theta}}\cdot \epsilon\mathrm{i}\mathrm{e}^{\mathrm{i}\theta}\,\mathrm{d}\theta$$

$$= \lim_{\substack{\epsilon\to 0\\ \sigma\to 0}} \mathscr{P}\int_a^b \frac{f(x)+\mathrm{i}\sigma f'(x)+\cdots}{x+\mathrm{i}\sigma}\cdot\mathrm{d}x$$

$$= \lim_{\sigma\to 0}\int_a^b \frac{f(x)}{x+\mathrm{i}\sigma}\,\mathrm{d}x. \tag{D.15}$$

Recall that $\mathscr{P}$ denotes the removal of the point $x = 0$ from the range of integration. Thus, if we choose to delete $\mathscr{P}$ from the integral specification (as in equation (D.15)), then we must correspondingly retain the $\mathrm{i}\sigma$ term in the denominator of the RHS (D.15). This, of course, does not apply to the $\mathrm{i}\sigma f'(x)$ etc terms appearing in the numerator. Equation (D.15) then suggests

$$\zeta(x) \equiv \lim_{\sigma\to 0}\frac{1}{x+\mathrm{i}\sigma}\,. \tag{D.16}$$

It should be stressed, however, that all the above is simply an outline of the properties of $\mathscr{P}$, $\delta(\eta)$ and $\zeta(\eta)$.

Fig.D.3. The contour C' in the complex z plane obtained by raising the contour C sketched in figure D.2.

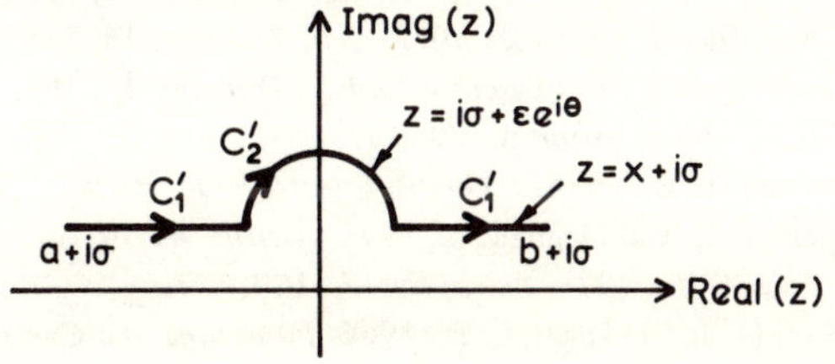

REFERENCES

[1] Abramowitz, M. and Stegun, I. A., 1965, *Handbook of Mathematical Functions*, Dover, N.Y.

[2] Athay, R. G. and Skumanich, A., 1967, *Ann. D'Astrophys.*, **30**, 669.

[3] Auer, L., 1967, *Astrophys. J. Letters*, **150**, L53.

[4] Auer, L., 1976, *J. Quant. Spectroscopy Radiat. Transfer*, **16**, 931.

[5] Auer, L. and Mihalas, D., 1969, *Astrophys. J.*, **158**, 641.

[6] Auer, L. H. and Mihalas, D., 1970, *Mon. Not. Roy. Astron. Soc.*, **149**, 65.

[7] Avrett, E. H., 1965, *Smithsonian Astrophys. Observ. Special Report*, **174**, 101.

[8] Avrett, E. H. and Kalkofen, W., 1968, *J. Quant. Spectroscopy Radiat. Transfer*, **8**, 219.

[9] Avrett, E. H. and Hummer, D. G., 1965, *Mon. Not. Roy. Astron. Soc.*, **130**, 295.

[10] Baranger, M., 1962, in *Atomic and Molecular Processes*, ed. D. R. Bates, Academic Press, N.Y.

[11] Bohm, D., 1951, *Quantum Theory*, Prentice-Hall, Englewood Cliffs, N.J.

[12] Cannon, C. J., 1970, *Astrophys, J.*, **161**, 255.

[13] Cannon, C. J., 1971, *Solar Phys.*, **16**, 314.

[14] Cannon, C. J., 1971, *Solar Phys.*, **21**, 82.

[15] Cannon, C. J., 1972, *Aust. J. Phys.*, **25**, 177.

[16] Cannon, C. J., 1973, *J. Quant. Spectroscopy Radiat. Transfer*, **13**, 627.

[17] Cannon, C. J., 1973, *Astrophys. J.*, **185**, 621.

[18] Cannon, C. J., 1973, *J. Quant. Spectroscopy Radiat. Transfer*, **13**, 1011.

[19] Cannon, C. J., 1974, *J. Quant. Spectroscopy Radiat. Transfer*, **14**, 745.

[20] Cannon, C. J., 1974, *J. Quant. Spectroscopy Radiat. Transfer*, **14**, 761.

[21] Cannon, C. J., 1976, *Astron. Astrophys.*, **52**, 337.

[22] Cannon, C. J. and Cram, L. E., 1974, *J. Quant. Spectroscopy Radiat. Transfer*, **14**, 93.

[23] Cannon, C. J., Lopert, P. B. and Magnan, C., 1975, *Astron. Astrophys.*, **42**, 347.

[24] Cannon, C. J. and Rees, D. E., 1971, *Astrophys. J.*, **169**, 157.

[25] Carrier, G. F., Krook, M. and Pearson, C. E., 1966, *Functions of a Complex Variable*, McGraw-Hill, N.Y.

[26] Case, K. M., 1960, *Ann. Phys.*, **9**, 1.

[27] Case, K. M. and Hazeltine, R. D., 1970, *J. Math. Phys.*, **11**, 1126.

[28] Case, K. M. and Zweifel, P. F., 1967, *Linear Transport Theory*, Addison-Wesley, Reading, Mass.

[29] Chandrasekhar, S., 1960, *Radiative Transfer*, Dover, N.Y.

[30] Chapman, S. and Cowling, T. G., 1939, *The Mathematical Theory of Non-Uniform Gases*, Cambridge University Press.

[31] Cohen-Tannoudji, C., 1977, in *Frontiers in Laser Spectroscopy*, Vol. 1, p. 1, North-Holland, Amsterdam.

[32] Courant, R. and Hilbert, D., 1962, *Methods of Mathematical Physics* Vol. II, Interscience, N.Y.

[33] Cram, L. E. and Lopert, P. B., 1975, *J. Quant. Spectroscopy Radiat. Transfer*, **16**, 347.

[34] Cram, L. E., 1977, *Astron. Astrophys.*, **56**, 401.

[35] Cuny, Y., 1967, *Ann. d'Astrophys.*, **30**, 143.

[36] Delache, P. and Froeschlé, C. F., 1972, *Astron. Astrophys.*, **16**, 348.

[37] Dicke, R. H. and Wittle, R. B., 1960, *Introduction to Quantum Mechanics*, Addison-Wesley, Reading, Mass.

[38] Fano, U., 1957, *Rev. Mod. Phys.*, **29**, 74.

[39] Feautrier, P., 1964, *Compt. Rend. Acad. Sci. Paris*, **258**, 3189.

[40] Ferziger, J. H. and Kaper, H. G., 1972, *Mathematical Theory of Transport Processes in Gases*, North-Holland, Amsterdam.

[41] Filler, L. and Ludloff, H. F., 1961, *Math. Comp.*, **15**, 261.

[42] Finn, G. D. and Jefferies, J. T., 1968, *J. Quant. Spectroscopy Radiat. Transfer*, **8**, 1675.

[43] Fiutak, J. and Van Kranendonk, J., 1962, *Canadian J. Phys.*, **40**, 1085.

[44] Fox, L., 1957, *The Numerical Solution of Two Point Boundary Problems in Ordinary Differential Equations*, Clarendon Press, Oxford.

[45] Garbedian, P. R., 1964, *Partial Differential Equations*, John Wiley, N.Y.

[46] Griem, H. R., 1964, *Plasma Spectroscopy*, McGraw-Hill, N.Y.

[47] Griem, H. R., 1974, *Spectral Line Broadening by Plasmas*, Academic Press, N.Y.

[48] Halliday, D. and Resnick, R., 1960, *Physics for Students of Science and Engineering*, John Wiley, N.Y.

[49] Heitler, W., 1954, *Quantum Theory of Radiation*, Clarendon Press, Oxford.

[50] Henyey, L. G., 1946, *Astrophys. J.*, **103**, 347.

[51] Hildebrand, F. B., 1976, *Advanced Calculus for Applications*, Prentice-Hall, Englewood Cliffs, N.J.

[52] Hirschfelder, J. O., Curtiss, C. F. and Bird, R. B., 1954, *The Molecular Theory of Gases and Liquids*, John Wiley, N.Y.

[53] Hummer, D. G., 1962, *Mon. Not. Roy. Astron. Soc.*, **125**, 21.

[54] Hummer, D. G., 1969, *Mon. Not. Roy. Astron. Soc.*, **145**, 95.

[55] Hummer, D. G. and Rybicki, G., 1967, *Methods Computational Phys.*, 7, 53.

[56] Hummer, D. G. and Rybicki, G., 1968, *Astrophys. J. Letters*, **153**, L107.

[57] Jaffrin, M. Y., 1965, *Phys. Fluids*, 8, 606.

[58] Jefferies, J. T., 1960, *Astrophys. J.*, **132**, 775.

[59] Jefferies, J. T., 1968, *Spectral Line Formation*, Blaisdel, Waltham, Mass.

[60] Jefferies, J. T. and Thomas, R. N., 1960, *Astrophys. J.*, **131**, 695.

[61] Jeffries, J. T. and White, O. R., 1960, *Astrophys. J.*, **132**, 726.

[62] Jones, H. P. and Skumanich, A., 1973, *Astrophys. J.*, **185**, 167.

[63] Jones, H. P. and Skumanich, A., 1980, *Astrophys. J. Suppl.*, **42**, 221.

[64] Kalkofen, W., 1974, *Astrophys. J.*, **188**, 105.

[65] Kalkofen, W. and Whitney, C. A., 1971, *J. Quant. Spectroscopy Radiat. Transfer*, **11**, 531.

[66] Klein, R. I., Stein, R. F. and Kalkofen, W., 1976, *Astrophys. J.*, **205**, 499.
[67] Kneer, F., 1975, *Astrophys. J.*, **200**, 367.
[68] Kourganoff, V., 1963, *Basic Methods in Transfer Problems*, Dover, N.Y.
[69] Kreyzig, E., 1979, *Advanced Engineering Mathematics*, John Wiley, N.Y.
[70] Kulander, J. L., 1967, *Astrophys. J.*, **147**, 1063.
[71] Kulander, J. L., 1968, *J. Quant. Spectroscopy Radiat. Transfer*, 8, 273.
[72] Lamb, F. K. and ter Haar, D., 1971, *Phys. Rept. C.*, **2**, 253.
[73] Landi Degl'Innocenti, E. and Landi Delg'Innocenti, M., 1972, *Solar Phys.*, **27**, 319.
[74] Landi Degl'Innocenti, E. and Landi Degl'Innocenti, M., 1975, *Nuovo Cimento*, **27B**, 134.
[75] Landi Degl'Innocenti, M., Landolfi, M. and Landi Degl'Innocenti, E., 1976, Nuovo Cimento, **35B**, 117.
[76] Landau, L. D. and Lifschitz, E. M., 1965, *Quantum Mechanics*, Addison-Wesley, Reading, Mass.
[77] Le Guet, F., 1972, *Astron. Astrophys.*, **16**, 356.
[78] Leech, J. W., 1963, *Classical Mechanics*, Methuen, London.
[79] Leighton, R. B., 1959, *Principles of Modern Physics*, McGraw-Hill, N.Y.
[80] Liboff, R. L., 1969, *Introduction to the Theory of Kinetic Equations*, John Wiley, N.Y.
[81] Linsky, J. L., 1968, *Smithsonian Astrophys. Observ. Special Report*, 274.
[82] McCormick, N. J. and Siewert, C. E., 1970, *Astrophys. J.*, **162**, 633.
[83] Messiah, A., 1966, *Quantum Mechanics*, North-Holland, Amsterdam.
[84] Mihalas, D., 1978, *Stellar Atmospheres*, Freeman, San Francisco.
[85] Mihalas, D., Auer, L. H. and Mihalas, R. B., 1978, *Astrophys. J.*, **220**, 1001.
[86] Milne, E. A., 1946, *Vectorial Mechanics*, Methuen, London.
[87] Mitchell, A. R., 1969, *Computational Methods in Partial Differential Equations*, John Wiley, N.Y.
[88] Mott, N. F. and Sneddon, I. N., 1963, *Wave Mechanics and its Applications*, Dover, N.Y.
[89] Muskhelishvili, N.I., 1953, *Singular Integral Equations*, P. Noordhoff, Groningen, The Netherlands.
[90] Noerdlinger, P. D. and Rybicki, G., 1974, *Astrophys. J.*, **193**, 651.
[91] Nussenzveig, H. M., 1973, *Introduction to Quantum Optics*, Gordon and Breach, London.
[92] Omont, A., Smith, E. W. and Cooper, J., 1972, *Astrophys. J.*, **175**, 185.
[93] Oxenius, J., 1965, *J. Quant. Spectroscopy Radiat. Transfer*, **5**, 771.
[94] Pantell, R. H. and Puthoff, H. E., 1969, *Fundamentals of Quantum Electronics*, John Wiley, N.Y.
[95] Pomraning, G. C., 1973, *The Equations of Radiation Hydrodynamics*, Pergamon Oxford.
[96] Potter, D. E., 1973, *Computational Physics*, John Wiley, London.
[97] Rees, D. E. and Reichel, A., 1968, *J. Quant. Spectroscopy Radiat. Transfer*, 8, 1795.
[98] Reichel, A., 1968, *J. Quant. Spectroscopy Radiat. Transfer*, **8**, 1601.
[99] Reichel, A. and Vardavas, I. M., 1975, *J. Quant. Spectroscopy Radiat. Transfer*, **15**, 929.

[100] Richtmyer, R. D. and Morton, K. W., 1967, *Difference Methods for Initial Value Problems*, Interscience, N.Y.

[101] Rybicki, G., 1965, Unpublished Ph.D. Thesis, Harvard University.

[102] Rybicki, G., 1971, *J. Quant. Spectroscopy Radiat. Transfer*, **11**, 589.

[103] Sakurai, J. J., 1967, *Advanced Quantum Mechanics*, Addison-Wesley, Reading, Mass.

[104] Sampson, D. H., 1965, *Radiative Contributions to Energy and Momentum Transport in a Gas*, Interscience, N.Y.

[105] Sargent, M. III, Scully, M. O. and Lamb, W. E. Jr., 1974, *Laser Physics*, Addison-Wesley, Reading, Mass.

[106] Schiff, L. T., 1968, *Quantum Mechanics*, McGraw-Hill, N.Y.

[107] Scully, M. O. and Lamb, W. E., Jr., 1967, *Phys. Rev.*, **159**, 208.

[108] Shine, R. A., Milkey, R. W. and Mihalas, D., 1975, *Astrophys.*, **199**, 724.

[109] Shore, B. W., and Menzel, D. H., 1968, *Principles of Atomic Spectra*, John Wiley, N.Y.

[110] Siewert, C. E. and Zweifel, P. F., 1966, *Ann. Phys.*, **36**, 61.

[111] Simon, R., 1963, *J. Quant. Spectroscopy Radiat. Transfer*, **3**, 1.

[112] Steinitz, R. and Shine, R. A., 1973, *Mon. Not. Roy. Astron. Soc.*, **162**, 197.

[113] Stenflo, J. O., 1971, in *Solar Magnetic Fields*, ed. R. Howard, IAU Symp., **43**, 101.

[114] Synge, J. L., 1957, *The Relativistic Gas*, North-Holland, Amsterdam.

[115] ter Haar, D., 1961, *Rept. Progr. Phys.*, **24**, 304.

[116] Thomas, L. H., 1930, *Quart. J. Math.*, **1**, 239.

[117] Thomas, R. N., 1957, *Astrophys. J.*, **125**, 260.

[118] Thomas, R. N., 1965, *Some Aspects of Non-Equilibrium Thermodynamics in the Presence of a Radiation Field*, Uni. Colorado Press.

[119] Thomas, R. N. and Athay, R. G., 1961, *Physics of the Solar Chromosphere*, Interscience, N.Y.

[120] Vardavas, I. M. and Cannon, C. J., 1974, *Aust. J. Phys.*, **27**, 157.

[121] Vardavas, I. M. and Cannon, C. J., 1976, *Astron. Astrophys.*, **53**, 107.

[122] Vardavas, I. M. and Cram, L. E., 1974, *Solar Phys.*, **38**, 367.

[123] Vernazza, J. E., 1972, Ph.D. Thesis, Harvard University.

[124] Weisskoff, V. and Wigner, E., 1930, *Zeits. f. Phys.*, **63**, 54.

[125] Whitaker, E. T. and Watson, M. A., 1915, *A Course of Modern Analysis*, Cambridge University Press.

[126] Weider, S., 1973, *The Foundation of Quantum Theory*, Academic Press, N.Y.

[127] Wilson, P. R. and Evans, C. D., 1971, *Solar Phys.*, **18**, 29.

[128] Wilson, P. R. and Williams, N. V., 1972, *Solar Phys.*, **26**, 30.

[129] Woolley, R.v.d.R. and Stibbs, D. W. N., 1953, *The Outer Layers of a Star*, Clarendon Press, Oxford.

[130] Zel'dovich, Y. B. and Raizer, Y. P., 1966, *Physics of Shock Waves and High Temperature Hydrodynamic Phenomena*, Academic Press, N.Y.